D0216273

Living Invertebrates

LIVING INVERTEBRATES

Vicki Pearse
John Pearse
Mildred Buchsbaum
Ralph Buchsbaum

DISCARD LCCC LIBRARY

BLACKWELL SCIENTIFIC PUBLICATIONS
Palo Alto, California

and

THE BOXWOOD PRESS
Pacific Grove, California

© 1987
by
The Boxwood Press
183 Ocean View Blvd.
Pacific Grove, CA 93950

All rights reserved.
No part of this publication
may be reproduced by any means
without the prior permission
of the copyright owner.

* * *

Library of Congress Cataloging in Publication Data:

Living Invertebrates.

Bibliography: p.
Includes index.
1. Invertebrates. I. Pearse, Vicki, 1942.
QL362.L58   1987   592   86-10790

ISBN: 0-86542-312-1

* * *

Distributors:

USA and CANADA

Blackwell Scientific Publications, Inc.
P.O. Box 50009
Palo Alto, CA 94303-9952

AUSTRALIA

Blackwell Scientific Publications (Australia) Pty Ltd
107 Barry St., Victoria 3053

UNITED KINGDOM

Blackwell Scientific Publications
Osney Mead, Oxford OX2 0EL

Editorial Offices:

667 Lytton Ave., Palo Alto, CA 94301
Osney Mead, Oxford, OX2 0EL, UK
8 John St., London WC1N 2ES, England
23 Ainslie Place, Edinburgh, EH3 6AJ, UK
52 Beacon St., Boston, MA 02108
107 Barry St., Carlton, Victoria 3053, Australia

Printed in U.S.A.

To

W.C. Allee
Libbie H. Hyman
Donald P. Abbott

who most influenced our thinking
about the invertebrates

and to

the animals themselves
who kept us thinking

PREFACE

L IVING INVERTEBRATES is an introduction to the invertebrates for students in colleges and universities and for educated people of any field who seek more familiarity with the diversity of animal life. This book has grown out of our experience with Ralph Buchsbaum's *Animals Without Backbones,* first published in 1938, which introduced generations of biologists as well as non-biologists to the major groups of invertebrates and which continues to be widely used (Third Edition, 1987). Many colleagues have told us that they would like a book for more advanced class levels with a more complete coverage of the groups of invertebrates, but with the same readable language and the minimum of technical detail and terminology appropriate to the new level. *Living Invertebrates* is our attempt at such a book.

Much of our effort in making this work comprehensive in coverage has gone into keeping it comprehensible. The amount of published material about invertebrates has reached such staggering levels that the selection of material is both extremely difficult and crucially important. We have continually resisted the temptation to incorporate every new fact, interesting to the specialist but more likely burdensome to the reader who is trying to obtain a fundamental understanding of invertebrate biology. We have selected observations and experiments that illustrate important methods, approaches, and principles. In particular we have tried to choose material that illuminates the functional morphology, physiology, ecology, and natural history of the members of the phyla, classes, orders, etc., in such a way as to bring out the inherent interest in each group.

Classifications immediately follow the treatment of all but some minor phyla. For each taxon we give a proper name, common name if any, brief nontechnical characterization, notes on habits and habitats, and examples of common or representative genera. For most invertebrate groups, several alternative taxonomic schemes and names compete in the current literature. We have sought to use those schemes and names that we believe best express the relationships among the animals.

IN THIS BOOK, as in *Animals Without Backbones,* we have substituted single vowels for some diphthongs (e.g., *ameba,* not *amoeba),* and we have chosen the regular English form of some commonly irregular plurals (e.g, *larvas,* not *larvae).* This usage conforms with long-established trends in standard English spelling and pronunciation among biologists (e.g., *ecology* from *aecology* or *oecology, estuary* from *aestuary, hemoglobin* from *haemoglobin, abdomens* from *abdomina, antennules* from *antennulae)* and with the regular formation of similar words in other sciences *(equator* from *aequator, sphere* from *sphaere, formulas* from *formulae, nebulas* from *nebulae)* and in other modern languages (German: *Larve, Larven;* French: *larve, larves;* Spanish: *larva, larvas;* etc.). Our adoption of regular English spellings is not intended only to introduce students to the plethora of biological terms more easily, though we think that it will have that effect. The more compelling reason for it is the dismaying frequency, in both spoken and written English, of grammatical errors specifically involving non-English plurals. Somehow the phrase "antennae-like structures" does not as immediately signal error in people's ears as does "antennas-like structures." Because such errors ("the larva feed," "the left nephridia develops," "each podia," "a pneumococci bacteria is," "cilia-bearing cells") appear increasingly—and not just from struggling students but from experienced scholars—we are convinced that the small effort of switching from *zoaeae* to *zoeas* is worthwhile. In almost all cases, both regular and irregular forms are, of course, already in dictionaries; we have not invented new words nor altered proper taxonomic names. Nor have we used regular English plural endings in every case (e.g., *cirri, nuclei, testes, mitochondria, septa, cilia).* To do so simply seemed too much all at once. We have consistently regularized all endings of the *-ae* type, reasoning hopefully that if these were to disappear from the language, so might "cecae," "septae," and "flagellae."

Our attention has been focused, however, on what seem to be more substantive issues of terminology, especially the precise application of the multiplicity of terms characteristic of invertebrate zoology. In bivalves, for example, the use of *inhalant* and *exhalant siphons* is unfortunate because *inhalant* and *exhalant* imply a flow of gas rather than of liquid and because *siphon* implies a siphoning action, which is in no way involved; also, *siphon* is often uncritically applied to *any* opening through which a bivalve directs its feeding currents, in addition to being used for unrelated structures in gastropods, in cephalopods (for two different structures), in arthropods (for several different structures), in echinoderms, and in tunicates. In this and similar tangles we can only try to find the best compromise, choosing terms that communicate as precisely and logically as is consistent with established usage; we try to omit terms that are highly specialized or contribute no information. In the example mentioned, we refer to *incurrent* and *excurrent openings* in describing a clam that has no siphons, and we define *siphon* separately in other bivalves, in gastropods, in echinoderms, and in tunicates. Similarly, in the chapter on sponges we refer to *incurrent pores* and *excurrent vents* instead of using *ostia* and *oscula,* terms (both derived from roots meaning "mouth") that give no information about the direction of water flow and are easily confused. In the ophiuroid section, we refer to *ambulacral ossicles* because the specialists' alternative *vertebras* seems a misnomer for structures in an *in*vertebrate.

SOME of the original drawings in *Animals Without Backbones,* by Elizabeth Buchsbaum Newhall, have been used here, but readers familiar with them will discover revisions in a great number. The many new drawings have been designed with the same style and conventions to make it easy to relate one to another, as well as to preserve the artistic coherence of the book. We are grateful to Mildred K. Waltrip for her thoughtful and patient attention to every detail and her superb execution of most of the new drawings in the book. In addition a number of new drawings have been executed by Jonathan Dimes.

The many photographs, nearly all of living animals, provide an approach to the shapes, textures, and aspects of invertebrates that cannot adequately be communicated through words or drawings. Some will be familiar but a large number have never before been published. We have striven to integrate them closely with the text.

The legends of the photographs and drawings, together with the fine-type notes, contain supplementary material—especially experimental, theoretical, speculative, or natural history information, specific examples, and terminology—intended to complement the text and to provide depth without interrupting or unduly lengthening the main story. This arrangement is designed to meet the needs of a variety of readers, including students at different levels or with different goals.

WE ARE GRATEFUL to the many colleagues and friends who read or discussed aspects of the text or illustrations, who supplied specimens to photograph, who gave us much needed encouragement, and who provided diverse kinds of support that were essential to the completion of this book: Donald P. Abbott, Alan N. Baker, Alan Baldridge, Richard L. Blanton, C.B. Calloway, R. Andrew Cameron, Richard D. Campbell, James T. Carlton, Helen E.S. Clark, C.C. Doncaster, Douglas J. Eernisse, Peter V. Fankboner, Kristian Fauchauld, Daphne Fautin, Laurel R. Fox, Lisbeth Francis, Michael T. Ghiselin, Arthur C. Giese, Keith Gillett, Martin F. Glaessner, Jefferson J. Gonor, Marcia Gowing, Karl G. Grell, Michael G. Hadfield, Norine D. Haven, Joel W. Hedgpeth, Robert R. Hessler, Galen H. Hilgard, Nicholas D. Holland, George O. Mackie, Meredith L. Jones, Bill Kennedy, Alan J. Kohn, Meredith Kusch, Herbert W. Levi, David R. Lindberg, Karl J. Marschall, Anthony Michaels, M. Patricia Morse, L. Michael Moser, A. Todd Newberry, William A. Newman, Claus Nielsen, Jan A. Nowell, Brent D. Opell, Jörg Ott, Christopher Reed, Henry M. Reiswig, Mary E. Rice, Jack Rudloe, Edward E. Ruppert, Mary Beth Saffo, Judy A. Sakanari, Christiane Schoepfer-Sterrer, Scott Smiley, Arthur C. Smith, Ralph I. Smith, Alan J. Southward, Eve C. Southward, Wolfgang Sterrer, Megami F. Strathmann, Richard R. Strathmann, Stephen A. Stricker, Neil R. Swanberg, Paul D. Taylor, Gotram Uhlig, Jean Vasserot, P.H. Wells, Adrian M. Wenner, Carroll B. Williams, Robert M. Woollacott, Russel L. Zimmer.

We thank those, including some of the above, who generously contributed many of the photographs that form such an important part of this book; too numerous to list here, the photographers are acknowledged in the legends accompanying the photographs. We are especially grateful for the large number of photographs by Edward S. Ross, Kjell B. Sandved, and Douglas P. Wilson.

We owe a special debt to John Staples of Blackwell Scientific Publications and his staff for their encouragement, support, and patience. William Wasley, Eleanor Coerr, Karen Morgan, Susan Hawthorne, Lucia Boyer, Sandy Pieper, and Ray Santella all provided help that freed us to work on the manuscript. And Estelle Campbell contributed importantly in her own special way, for which we will be grateful always.

Finally, we would like to thank in advance the readers of this book for calling to our attention any of the errors that must inevitably occur in a work of this scope.

Vicki Pearse
John Pearse
Mildred Buchsbaum
Ralph Buchsbaum

Santa Cruz and Pacific Grove,
  California

November, 1986

CONTENTS

Living Invertebrates

Five kingdoms of living things, distinguished by different levels of cellular organization and modes of nutrition, are here depicted graphically according to the estimated number of living, described species in each. They are far from equal. The 4 smaller kingdoms are dwarfed by the fifth, and largest, the kingdom of the animals. The further subdivision of the animals is still more unequal, with the 3% designated as vertebrates (animals with backbones) showing up as a tiny minority next to the 97% that are lumped together as invertebrates (animals without backbones). With invertebrates occupying the large white sector of this graph, and the rest of the living world (including the vertebrates) confined to the black, the *preponderance of invertebrates among living organisms* stands forth in stark contrast.

Of the 5 kingdoms, the two smallest also have the smallest members, most of them **unicellular** and microscopic. The least complex is the kingdom **MONERA,** which includes the archebacteria, eubacteria, and cyanobacteria (formerly called blue-green algas). Monerans differ radically from all the other living organisms in that they have no well-defined nucleus that holds the genetic material. Thus they are said to be **prokaryotic.** The kingdom **PROTISTA** (or Protoctista) consists of protozoans and unicellular algas. These have a membrane-bound nucleus that (between cell divisions) confines complex chromosomes. Such a cellular organization is said to be **eukaryotic.**

Organisms in these 2 kingdoms may be **autotrophs** ("self-nourishing"), synthesizing their own food from inorganic constituents with the aid of solar or chemically derived energy. Or they may be **heterotrophs** ("nourished by others"), obtaining their food from the bodies of other organisms. Or they may be both.

The members of the other 3 kingdoms are all eukaryotic like the protists, but they are **multicellular.** The three differ in mode of nutrition. In the kingdom **FUNGI** are the **saprobic heterotrophs**—the mushrooms, yeasts, molds, and others that take up nutrients from organic matter, either alive or decomposing. The kingdom **PLANTAE** is made up mostly of **photosynthetic autotrophs,** the familiar green plants. And the kingdom **ANIMALIA** consists primarily of **ingestive heterotrophs,** which live by eating other organisms.

Chapter 1

Seeking Perspective

NEARLY TWO MILLION SPECIES of living organisms have been described and named. The actual number may be closer to 10 to 20 million. Of these, only a small proportion are microscopic bacteria, protists, and fungi; most are large enough to see without a microscope. And only a modest proportion are photosynthetic protists and plants, though these are essential for storing solar energy in the organic molecules that nourish all but a few forms of life on Earth. Most species of living things are multicellular organisms in the **Kingdom Animalia.**

Within the animal kingdom, there are about 30 distinct groups, or **phyla.** Each phylum has a common underlying design, or *body plan,* with fundamental differences that distinguish it from all the others (though there are sometimes varying opinions as to what constitutes a fundamental difference). The features of each body plan must function together to allow individuals within the group to carry out the necessary life activities. All animals must obtain, digest, and assimilate food; circulate nutrients and gases within the body; dispose of metabolic wastes; coordinate activities with both nervous and endocrine systems; avoid predators and other hazards; grow; and reproduce.

The various kinds of animals carry out these activities in one way or another. But we humans are so accustomed to the particular plan and plumbing of our own bodies—in which even temporary failure of the brain, heart, lungs, liver, or kidneys can spell death—that we tend to equate these organs with the essential functions they perform and to think of them as universally necessary. Even when we let our imaginations "run wild" to invent extraterrestrial beings, most of them turn out with a body plan suspiciously like our own.

Yet many common invertebrates are "wilder" than anything we invent. Some of them carry out all the essential life activities without most of the organs that we regard as fundamental. Some have equipment for which we are still unable to find a function. And in between are those with the sorts of fascinating variations on recognizable parts that give our imaginations a welcome stretch. The invertebrates offer us a better perspective on what is truly indispensible to animal life.

Animal Diversity
Vertebrates and
Invertebrates
Classification
Body Plans:
materials,
habitats,
and heritage
Animal Origins

Animals are many-celled organisms that typically feed on other organisms. The name is derived from a Latin root meaning "a living being having breath or soul." An alternate name for the kingdom is **Metazoa,** referring to the idea that the multicellular organization of the animal body was derived from a unicellular organization, as found in protozoans.

1

Words like "body plan" and "design" need not mean that there is a designer or planner involved; the form and structure of an object, whether the meandering of a stream bed or the pattern of a butterfly's wing, reflect the properties of the materials involved, the interaction of physical forces, and the sequence of historical events.

Fully a third of all animal species are beetles. Another third is made up of flies, bees and ants, and butterflies and moths. Other insects, together with all the other species of animals, make up the remainder.

Our division of the animal world, into vertebrates and invertebrates, reflects to a large extent our own biased perspective. If this book had been written by and for individuals of an intelligent species of insects, for example, it might include vertebrates as just a subgroup among groups of animals that lack 6 pairs of legs, and the Animal Kingdom would be divided conveniently into insects and non-insects. Separate books would then deal with the different groups of insects, while a book like this, entitled *Living Non-hexapods,* would cover all the rest.

One phylum—our phylum, the **Chordata,** with about 43,000 species—includes the often fast-moving, large, and conspicuous mammals, birds, reptiles, amphibians, and fishes, along with some smaller, largely sedentary, filter-feeding animals. At some stage in their life history all share one distinctive feature: a cartilaginous supporting rod along the middle of the back, the *notochord,* for which the chordates are named. Nothing like it is found in any other phylum. The chordate body plan, which also includes pouches or slits (gill slits) on the sides of the pharynx, a ventral heart that pumps blood anteriorly, and a series of muscle blocks and associated nerves along the length of the body, can be viewed as a design especially suited for animals that swim and feed near the ocean floor, as do some of the apparently primitive chordates today. However, the chordate body plan is also readily modified for other types of life, not only for swimming in midwater as do many fishes, but for walking, hopping, or even flying over land as do reptiles, birds, and mammals. It is among the most adaptable of the animal designs.

Most chordates also have a row of bones that extend down the back and together form a backbone, or spine, which either augments or nearly replaces the notochord, and also encircles and protects the dorsal nerve cord. These bones are called *vertebras,* and those species that possess them are included in the subphylum **Vertebrata** and are commonly called *vertebras.* Only a relatively few kinds of chordates, considered in chapter 29, lack vertebras; these and all other animals are collectively called *invertebrates.* People commonly speak as if these two kinds of animals, vertebrates and invertebrates, are roughly comparable. But less than 3% of the species of animals are vertebrates (and these are mostly fishes), and they constitute only part of only one of some 30 animal phyla. This book is about the remaining 97%, the great spineless majority—the invertebrates.

To be called an invertebrate, an animal need not conform to any special shape or design, nor have any specified internal structure. It need not have any single positive attribute. It need only lack vertebras to be barred from the exclusive company of the vertebrates—exclusive because vertebrates include the one animal species that makes all the decisions about how to categorize animals and select names for the many groups.

BEFORE EXPLORING the different kinds of invertebrate animals, a word needs to be said about **biological classification,** a system for grouping individuals in an organized manner that reflects similarities and differences among them. The phylum is the highest category, or *taxon,* within the animal kingdom, and the taxons at successively lower levels contain animals that are increasingly similar and, presumably, more and more closely related. At the lowest level of the system is the **species** (plural, species), which, in concept, is a population of individuals that are able to interbreed and produce fertile offspring, and that are reproductively isolated.

In reality, individuals of relatively few species have been observed to interbreed, and of course no one has ever observed all the individuals of any species interbreeding (even if it were possible, who would want to do so?). Usually, however, all the individuals in a species are morphologically similar enough to each other that they can be recognized as an interbreeding unit with genetic exchange. In cases where different individuals in a species are markedly different, for example with dimorphic males and females or with adults and larvas, life-history details are needed to recognize the breadth of diversity within a species. Many larval forms were originally described in some detail as distinct "species" before being observed to transform into an adult of very different appearance and structure already assigned to another species. In other cases, closely similar individuals are recognized as belonging to different species only after detailed observations and genetic analyses have been made. And as biologists increase their ability to discriminate differences, more species are recognized.

Species of animals that are very similar to each other, reflecting a close relationship, are grouped together into taxonomic units called **genera** (singular, genus). For example, the American and European lobsters are different species of the same genus, *Homarus.* In the same way, genera that are similar to each other are grouped together into taxonomic units called **families.** Families can then be grouped together into **orders,** orders into **classes,** and classes into phyla. These categories are often further divided or grouped into sub- and super-units; for example, the Crustacea, to which lobsters belong, is one of three subphyla of the phylum Arthropoda.

The system of biological classification was developed by the Swedish naturalist Carl von Linné (or Carolus Linnaeus in Latin) in the 18th century and has been adopted universally by biologists. It recognizes and distinguishes natural groupings of organisms in an orderly hierarchical manner. The system took on an evolutionary rationale reflecting relationships through common ancestors in the 19th century, following Charles Darwin's insight into the mechanism of evolution through natural selection.

In the biological literature, the scientific name of a species is latinized and printed in italics; it includes both the generic name, with the first letter capitalized, and the species name (not capitalized). For example, the common domestic honey bee is *Apis mellifera.* Once the generic name has been introduced in a publication, the generic initial may be used for brevity, for example *A. mellifera,* if its use is unambiguous. The specific name is rarely used alone (because the same specific name may be attached to species of other genera). On the other hand, generic names are unique, and because different species in a genus are often very similar and difficult to distinguish, commonly only the generic name is used, even when referring only to one species, for example, *Apis*. (But similar forms need not imply similar biology, and generic reference used alone can be misleading.)

Classification of several edible invertebrates, all commonly known as "shellfish," illustrating how species can be grouped into genera, families, orders, classes, and phyla, depending on similarities—and inferred evolutionary relationships—among them.

Common name	American Lobster	Market Squid	Blue Mussel	Rock Scallop	European Oyster	Japanese Oyster	Virginia Oyster
Phylum	Arthropoda	Mollusca	Mollusca	Mollusca	Mollusca	Mollusca	Mollusca
Class	Malacostraca	Cephalopoda	Bivalvia	Bivalvia	Bivalvia	Bivalvia	Bivalvia
Order	Decapoda	Decapoda	Mytiloida	Pterioida	Pterioida	Pterioida	Pterioida
Family	Nephropidae	Loliginidae	Mytilidae	Pectinidae	Ostreidae	Ostreidae	Ostreidae
Genus	*Homarus*	*Loligo*	*Mytilus*	*Hinnites*	*Ostrea*	*Crassostrea*	*Crassostrea*
species	*americanus*	*opalescens*	*edulis*	*giganteus*	*edulis*	*gigas*	*virginica*

The names of genera are required by the rules of nomenclature to be unique. But such rules do not apply to other taxons. For example, in the table above, the order name Decapoda occurs in both arthropods and molluscs, and the species name *edulis* occurs in bivalves of different genera.

When all the members of a group are descendants of a common ancestor, the group is said to be *monophyletic*. When some members descend from one ancestor and others from other ancestors, the group is said to be *di-* or *polyphyletic*.

Biologists who study how organisms are classified generally agree on where most species should be placed within the classification system. However, sometimes there are strong differences among workers about where particular groupings belong within the system. These problems occur at all levels of classification. There can be disagreement over whether some group of individuals is a separate species or part of another species, as well as whether a group of species is so different from all other species that it should constitute a separate phylum rather than a subphylum or class. Thus, depending on viewpoint, biologists today usually group animal species in about 25 to 35 phyla. Such controversies do not mean that the system of classification is arbitrary, artificial, or simply human invention; rather, they reflect the real complexities resulting from past and continuing animal evolution.

All the species within each taxonomic unit are assumed to be descendants from a common ancestor, and all the descendants of that common ancestor are, in theory, grouped together, so that each taxonomic unit is *monophyletic*. The classification system within a phylum is an attempt to portray evolutionary relationships. Of course, recognizing these relationships has been, and continues to be, a major challenge in biology. Evolutionary relationships *between* phyla are just as challenging if not more so (see chapter 30).

THE DIVERSITY of body plans within the animal kingdom, and how life activities are carried on within each phylum, is the major focus of this book. The body plan of each animal allows

it to live and reproduce within particular environments. And
as it grows and develops, it may move from one environment
to another, sometimes radically changing its form and the
function of many components of its body. So its body plan is
not just a static property of the adult, but incorporates all of
the changes that take place during the life-history of the
individual. In addition, some features of the body plan alter in
form and function or even disappear during the evolutionary
history of groups within a phylum. Such intricate transformations are useful clues to historical relationships, and can also
serve as "natural experiments" against which we can test our
fundamental understanding of form and function. Sorting
them out is a challenge to match any in biology.

sperm
      ↘
        zygote ⟶ embryo ⟶ *birth
                           or* ⟶ juvenile ⟶ adult...... ⟶ gametes
                           hatch   *death*
      ↗
egg

sperm
      ↘
        zygote ⟶ embryo ⟶ *birth
                           or* ⟶ larva ⟶ *metamorphosis* ⟶ juvenile ⟶ adult...... ⟶ gametes
                           hatch   *death*
      ↗
egg

The life history of an animal begins with one cell, the *zygote,* formed by the union of an egg and a sperm. The new
individual develops as nuclei and cells divide and differentiate, forming an *embryo* either within the egg covering or
within the parent's body. Eventually the embryo hatches or is released from the parent. Often the new individual
appears to be merely a small version of the adult (above); it is then said to be a *juvenile* that simply continues to grow
and develop until it achieves sexual maturity and becomes an *adult,* usually in the same habitat. However, within the
life history of many animals, a *larva* is interposed between the embryo and juvenile (below). The larva commonly lives
in a very different habitat from that of the juvenile and adult, and its form and structure are markedly different.
Transition from the larva to juvenile requires a drastic change, termed *metamorphosis,* which goes hand in hand with a
profound ecological shift. Embryonic, larval (if present), juvenile, and adult features of an organism, and how they
develop from one form to the next, are all part of the organism's body plan.

In addition, many invertebrates undergo *asexual reproduction* by budding, fragmentation, or other means. Various
organisms replicate asexually at different stages of their life history—as embryos, larvas, juveniles, or adults. Such
replication greatly increases the variety of invertebrate life histories.

Among the many facets that are integrated into a body plan,
two are especially influential: (1) the materials of the tissues,
skeletons, and coverings of the animals, and (2) the habitat
within which the animals live and evolve. Interplay between
these two continually shapes and transforms the resulting
arrangement of the body plan; the animals maintain a
continuing "conversation" with their environment. In addition, past interactions endow a particular design with a genetic
heritage that molds its future evolution.

Many species of animals go through one or more metamorphoses during their life history, and radically change habitats, such as marine species with pelagic larvas and benthic adults (or vice versa) and insects with aquatic larvas and winged terrestrial adults. The changing body plan of these must be able to accommodate the different demands of each habitat.

The **materials** that make up an animal give it substance, and influence its structure and shape. Different body plans reflect the use of different materials that enable the animals to live in particular environments. The properties of materials are thus central to the designs of living bodies, as they are to any other construction. An insect covered with a hard, waxy, chitinous exoskeleton, for example, is able to survive in dry terrestrial areas unsuitable for soft naked worms. On the other hand, a body ensheathed in a hard exoskeleton can enlarge and grow only by shedding the skeleton from time to time—a complex and risky business. These skeletal materials thus have both inherent advantages and limitations.

Every animal evolved and continues to exist within the potentials and confines of a distinct **habitat,** and past and present habitats in large part shape the body plan. The body plans of the different phyla thus can be viewed as adaptations to major habitats. For example, the chordate combination of a streamlined muscular body with a stiffening notochord or backbone is suited especially well for controlled swimming and feeding in shallow marine or freshwater habitats. Further exploiting the possibilities of the chordate design, many land vertebrates (including such awkward swimmers as ourselves) have taken advantage of the backbone to provide support on land. In shifting to erect stance, however, humans may have pushed it to its limits: a painful slipped disk can be a powerful constraint.

A variety of habitats have influenced the body plans of different phyla. Many or most species of most phyla are **marine** and **benthic,** living on ocean bottoms. Many of these live within *soft sediments* and have a wormlike form that allows them to slip among the particles, or to actively burrow. Other animals cling to *hard surfaces* and have a variety of features that help them to hang on, such as the legs of crabs or the muscular foot of snails. Marine animals also can be **pelagic,** inhabiting *open water* with body designs that allow them to float and swim. Each of these habitats makes very different demands on animal structure. In addition, living on **land** without all the life-easing attributes of water, requires special design features. Indeed, of the approximately 30 phyla, only 2 have numerous species living in dry terrestrial habitats (arthropods and chordates). Other habitats that make special demands, and are characterized by species of a limited number of phyla, are **freshwaters** and the bodies of **host organisms** in which symbionts, especially parasites, display bizarrely modified body plans and life histories.

Habitats of adult animals. Note that most phyla are predominantly marine and benthic, some exclusively so. Only a few span the spectrum of different habitats.

Phyla / Subphyla	Marine		Freshwater		Terrestrial		Symbiotic	
	Benthic	Pelagic	Benthic	Pelagic	Moist	Xeric	Ecto	Endo
Porifera	+++		+				+	
Placozoa	+							
Orthonectida								+
Dicyemida								+
Cnidaria	+++	++	+	+			+	
Ctenophora	+	+						
Platyhelminthes	+++	+	+++		++		+	++++
Gnathostomulida	++							
Nemertea	++	+	+		+		+	
Nematoda	+++	+	+++	+	+++	+	+++	+++
Nematomorpha								++
Acanthocephala								++
Rotifera	+	+	++	++	+		+	+
Gastrotricha	++		++					
Kinorhyncha	++							
Loricifera	+							
Tardigrada	+		++		+			
Priapula	+							
Mollusca	+++++	+	+++		+++	+	+	+
Kamptozoa	+		+				+	
Pogonophora	++							
Sipuncula	++				+			
Echiura	++							
Annelida	++++	+	++		+++		++	
Onychophora					+			
Arthropoda								
Crustacea	++++	+++	+++	++	++		++	++
Chelicerata	++	+	++	++	++++	+++	++	+
Uniramia	+	+	+++	++	+++++	+++	++	++
Chaetognatha	+	+						
Phoronida	+							
Brachiopoda	++							
Bryozoa	+++		+					
Echinodermata	+++	+						
Hemichordata	+							
Chordata								
Urochordata	+++	+						
Cephalochordata	+							
Vertebrata	+++	+++	++	+++	+++	+++	+	+

"Pluses" indicate approximate abundance of living described species: $+ = 1\text{-}100$; $++ = 100\text{-}1{,}000$; $+++ = 10^3$ to 10^4; $++++ = 10^4$ to 10^5; $+++++ = 10^5$ or greater.

The essential background that shapes the possibilities and restrictions of body plans is the **genetic heritage,** which directs development. This may be viewed as a genetic program that "instructs" the organism in its use of materials and the form they are to take in growth and development under the conditions of a given habitat. The contents of the program determine both the capabilities and limitations of any particular group of organisms. Among clams and their relatives, for example, the bivalved shell provides the animals with protection and an enclosed space in which to filter and sort out food. It is such an integral feature that it is never abandoned, and the heavy shell may prevent clams from developing pelagic, swimming forms (only a few kinds—scallops—swim briefly, by clapping the two halves of the shell). Even among clams that bore into wood or rock, which then provides protection, a small shell is retained in modified form as a boring device.

We can usually recognize closely related species because they have similar forms. Presumably, they share most of their genetic information, and in a few instances, the genetic similarity has been quantitatively measured and confirmed. More distant relationships, seen at higher levels of classification, are assumed to share relatively less genetic information, and to reflect more distant common ancestors.

The information in an organism's genetic program largely reflects its evolution in past habitats. Within a species, the genetic information is altered by mutation and mixed from generation to generation through sexual reproduction. It changes constantly as the new combinations formed and expressed by individuals of the species are tested by natural selection. As species change through time, however, the great bulk of the genetic information remains largely unchanged. Animals living today contain most of the genetic information of their extinct ancestors, as inferred from molecular analyses of closely related species. The shared body plan of the members of each phylum reflects their genetic endowment, much of it shaped at the very origin of the phylum.

THE ORIGINS of the phyla have been subject to much speculation and little resolution for over a century, as is discussed in chapter 30. Most kinds of animals with recognizable fossil remains, including most of the major phyla (sponges, cnidarians, molluscs, annelids, arthropods, bryozoans, echinoderms, chordates), were present near the beginning of the animal fossil record, about 600 million years ago. Some phyla can be grouped together based on shared characters that presumably reflect a remote common ancestry, but there is no convincing scheme that derives all phyla of animals from a single common ancestor. Nevertheless, the multicellular bodies of animals, with many cell types differentiated into discrete tissues and organs, must originally have been derived from a single-cell, protistan organization. Modern protists, themselves extremely diverse in form and life-history (and probably in their own origins), thus present the best place to search for forms that might be like those ancestral to animals.

The following chapter considers protozoans both for their diversity of unicellular animal-like forms and for their diverse trends toward multicellular forms that can serve as models of animal ancestors.

Chapter 2

Protozoans

Flagellates
Amebas
Sporozoans
Ciliates

T HE MINUTE SIZE of protozoans is central to their biology. The great majority are microscopic and consist of a single cell. Yet the slow, gracefully flowing amebas, the whirling flagellates, and the fast, busy ciliates are individuals with complex form and behavior, not entirely comparable to any one cell of a multicellular plant or animal. The first observers of protozoans visualized them in terms of animal anatomy and interpreted the internal structures they saw as tiny hearts, stomachs, and intestines. Eventually it was realized that these protists are constructed on a level of organization wholly different from that of animals, but we are still only beginning to appreciate how different are the constraints under which they live.

One must imagine the activities and habits of protozoans on a quite unfamiliar scale. In this microworld, bacteria are sizeable chunks of food, and the important predators are themselves microscopic. Protozoans may swim at any level of a deep ocean or a shallow puddle, or the merest film of water between particles of sand or soil. Gravity is a minor force to contend with, compared to the electrostatic forces of particles in the water or the forces of surface tension at the air-water interface. Even the hydrodynamics of the water takes on a different aspect for a swimmer the size of a protozoan. At this scale of things, for example, the viscosity of the water overpowers the relatively small momentum of a protozoan that is zooming along and makes it stop very abruptly when it ceases active swimming; it does not glide slowly to a halt as would a fast-moving boat. The different way in which water currents flow around very small objects must be taken into account when trying to understand the mechanics of protozoan locomotion.

The typical viscous forces that act on a human swimming in water are at least 6 orders of magnitude smaller than the inertial forces. To appreciate the viscous forces acting on a protozoan, a human would have to swim in a fluid a million times more viscous than water. According to the calculations of biophysicists, thick molasses would come close.

9

Protozoans present an equal challenge to many other zoological viewpoints, for example, to common concepts about individuality, sex and reproduction, community structure, population genetics, and evolution through natural selection. These concepts have been shaped largely around organisms, with highly differentiated structures, that live as separate, genetically unique individuals for months or years, mature as males or females, mate, produce young, and die. Ideas and modes of analysis fitted to this limited context need to be adjusted for critical application to protozoans and may be significantly refined in the process. Many protozoans, for example, multiply indefinitely by asexual division. Are the many cells that result more comparable to related individuals in a lineage of animals—or to the cells within the body of a single metazoan individual? Other protozoans reproduce sexually, but without male and female genders; have the zygote as the only diploid stage; or display some other variant of a bewildering spectrum of life-histories. For many protozoans, however, even the most basic life-history information is unknown, and surprising and illuminating patterns no doubt still await discovery.

Such knowledge is of more than academic interest, for protozoans are of enormous ecological and practical importance. All aquatic ecosystems, both freshwater and marine, are strongly influenced by and dependent upon the activities of protozoans: as primary producers and consumers, as vital links in food chains, and as crucial agents in nutrient cycles. In addition, on a more direct medical and economic level, humans are intimately and sometimes devastatingly affected by the many parasitic protozoans that specifically afflict people as well as the wild and domestic animals that people eat. Protozoan diseases range from mildly irritating to fatal, and the principal one, malaria, is among the most important medical problems in the world today.

However, most of the protozoans found living together with other, bigger organisms do no apparent harm and are hence not properly called *parasites*. Many are *commensals* that simply use the outside or inside of a bigger organism as living space, and sometimes share a bit of its food, without causing any significant effect. Others are *mutualists,* that is, their activities actually benefit the bigger organism in some way and may even become essential to its survival. All 3 categories of relationship are kinds of *symbiosis* ("living together"; see also chapter 9). That such small organisms as protozoans would find many opportunities in the protected, nutrient-rich habitats provided by the bodies of bigger ones is not surprising, and a large percentage of protozoans are regularly symbiotic. On the other hand, many protozoans themselves harbor smaller protists or bacteria as symbionts.

PROTOZOANS defy easy definition. Not long ago a phylum Protozoa ("first animals") invariably headed the list of phyla of the animal kingdom. Hence, groups of protists with uniformly photosynthetic or absorptive nutrition tended to be excluded from the protozoans and left to the botanists or mycologists, respectively. The rest, though often characterized as having animal-like nutrition—that is, moving about and ingesting their food—are in fact extremely variable in this respect. Even within a single protozoan species, the individuals may at different times depend on photosynthesis, on ingesting solid food, or on taking up dissolved nutrients; indeed, some or all of these activities may be going on at the same time. Unable to classify these protists by nutritional mode, as animal- or plant-like, specialists in protozoan biology have historically divided their microscopic subjects of study into four major groupings, based mostly on their mode of locomotion: the **flagellates,** the amebas and other **sarcodines,** the relatively immobile **sporozoans,** and the **ciliates.** These groupings still provide a convenient basis for approaching this enormous and diverse assemblage, though (except for the ciliates) they are no longer considered to represent natural relationships and hence do not correspond to formally recognized natural taxa.

In the first place, the filamentous locomotory structures called **flagella** and **cilia,** which were originally used to distinguish flagellates and ciliates, are now recognized to have a common ultrastructure, revealed by the electron microscope. There is no clear distinction between them. Although the term *undulipodium* has been proposed to embrace both, it is so awkward that most biologists have rejected it, and the two original terms continue in nearly universal use. Though indistinguishable in fine structure, flagella and cilia typically are of a different length and distribution, and display a different sort of motion.

A flagellum is typically longer, occurs singly or in small numbers, and moves with a symmetrical undulating beat, either in one plane or helical. A cilium tends to be shorter, to occur in coordinated fields of hundreds or thousands, and to have an asymmetrical and stiffer beat. Many intermediate conditions occur, and there are instances in which the type of motion depends on whether the organism is moving forward or backward. Yet, *cilium* and *flagellum* continue in the biological literature as descriptive terms

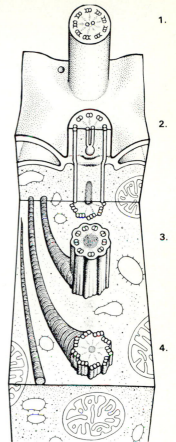

Ciliary structure as seen in 4 cilia cut at successively lower levels. **1.** Above cell surface. The cell membrane encloses a core of microtubules, 9 pairs surrounding 2 single ones. This core is common to virtually all motile cilia and flagella of protists, animals, and plants. A flagellum or cilium does not act like an inert whip, which flicks when the handle is jerked, but moves by means of interactions among the microtubules along its entire length. With few exceptions, deviations from the "9 + 2" microtubular pattern are found only in modified, *nonmotile* flagella or cilia. **2.** The 2 central microtubules end just above the cell surface; the 9 doublets continue inward and become 9 triplets. **3.** At the level of the cell surface, the outer shaft is partitioned by 1 or 2 shelves of dense matter parallel to the cell surface (one seen in center of section). **4.** Below the surface, the ciliary core continues as a basal body with 9 triplets of microtubules.

The fluid-filled pockets that flank the shaft in (**2**), and the striated fibers in (**3**) and (**4**), are features peculiar to ciliates. (After various sources)

useful for the majority of protists and animals. Bacterial flagella differ strikingly from those of eukaryotes in both structure and mode of operation; the term *bacterial flagellum* seems a sufficient distinction.

Meanwhile, other differences have been found to separate flagellates and ciliates, but the distinction between flagellates and amebas has blurred. As many life histories have become more completely known, it has been discovered that some flagellated forms lose the flagellum under certain conditions and become ameboid. They acquire a changing and irregular shape and move by means of the protoplasmic extensions called pseudopods, typical of amebas. Moreover, some predominantly ameboid organisms have flagellated stages in their life history.

Because of the unusual difficulty of determining relationships among protists, the classification of these organisms continues to be in rapid flux, so for most groups we have used informal designations that are more stable and more easily recognizable to nonspecialists. An abbreviated classification of protistan groups combined from several recent, comprehensive schemes is presented at the end of the chapter.

Because the main subjects of this book, the invertebrate animals, are metazoans, we treat the protozoans only briefly here and present but a sampling of the major or commonly seen types, and some of medical importance to humans. We will focus particularly on those forms that have evolved in the direction of large size and multicellular organization, in order to examine the ways in which protozoans have achieved this, compared to those of animals. The potential advantages of large size are presumably of the same sorts for members of both kingdoms: the ability to take larger prey and to escape smaller predators, the control of more resources and territory, and the creation of internal conditions that are to some extent independent of the external enviroment. No modern protozoan can be looked upon as an ancestor of animals—but the many sorts of multicellular protozoans may give us some idea of what the ancestors of animals may have been like.

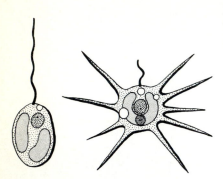

Ameboflagellate, *Chrysamoeba,* a photosynthetic chrysomonad that occurs in both an ovoid flagellate form and an ameboid form with long slender pseudopods and a short flagellum. The ameboid phase feeds and divides. (Modified after Lang)

FLAGELLATES

FLAGELLATES are protozoans that have one or more **flagella,** which beat with a planar or helical undulating motion. They propel the flagellate or, if the organism lives attached to a substrate, bring food-bearing water currents toward it.

Of the major protozoan groupings, it is only among the
flagellates that chloroplasts are found. Groups in which some
or all members have chloroplasts comprise the phytoflagellates (plant-flagellates). The remaining groups, by default,
together make up the zooflagellates (animal-flagellates). Both
are heterogeneous assemblages indeed. Many phytoflagellates
bear chloroplasts and are photosynthetic but have close
relatives that lack chloroplasts, or may themselves lose their
chloroplasts under certain conditions, or may ingest food in an
animal-like fashion while continuing to bear chloroplasts and
carry on photosynthesis. Among the best known and commonest phytoflagellates in freshwaters (with a few marine and
parasitic forms) are the **euglenids.** Euglenids with chloroplasts are sometimes so abundant as to produce a green scum
on ponds.

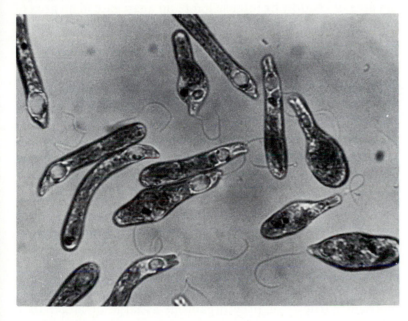

Euglenids, propelled by long flagella, display great flexibility and variety of
body shape.

The surface of a euglenid is covered by a cell membrane,
usually coated with a protective layer of mucilage. Just
beneath the cell membrane, a matrix of interconnected
proteinaceous strips constitutes a firm but flexible **pellicle**
with associated microtubules and fibrils. The pellicle gives the
euglenid a more or less definite shape when relaxed and
swimming. But if disturbed, as by a strong mechanical
stimulus, a euglenid may sharply contract the body, and
species with a flexible pellicle may take on varied shapes as
they wriggle along by characteristic contractions often referred
to as "euglenoid movement," and presumably mediated by

A succession of shapes seen in an
individual progressing by **euglenoid
movement.**

contractile fibrils. At the front end of the body, 1 or 2 long flagella emerge from a flask-shaped **flagellar pocket** (also called a reservoir, or anterior pocket). Into one side of the flagellar pocket, a large **contractile vacuole** discharges its contents at regular intervals (roughly every 20 to 30 seconds), ridding the cytoplasm of excess water.

On the other side of the flagellar pocket is an orange-red **eyespot** (or stigma), a mass of pigmented droplets that shields a sensitive photoreceptor on the flagellum. Such a light-sensing apparatus is absent from most colorless species and present in most green euglenids, which depend primarily on photosynthesis for their nutrition. It enables them to detect not only the intensity of light but also the direction from which it comes. Placed in a dish, green euglenids will move away from an area of shadow or intense light and aggregate in an area of moderate light. When exposed to light of suitable spectrum and intensity, green euglenids maintain themselves by photosynthesis and accumulate carbohydrates (specifically, paramylon) in conspicuous storage bodies on the chloroplasts and in the cytoplasm. However, they are also able to live by taking up dissolved nutrients through the surface and can survive in the dark if placed in a nutrient solution. In darkness, the organisms become colorless, but the chloroplasts are not destroyed by a brief dark period and will manufacture chlorophyll again when exposed to light. If the chloroplasts are destroyed (by prolonged darkness or by treatment with heat, antibiotics, or ultraviolet radiation), the euglenids become permanently colorless and heterotrophic, dependent on external sources of nutrients.

Some normally colorless euglenid species are able to ingest nutrient particles or prey organisms as large or larger than themselves through an anterior **cytostome** ("cell mouth") that can open widely. Food is enclosed and digested within membrane-bound **food vacuoles,** into which digestive enzymes are secreted.

The chloroplasts and photosynthetic pigments of euglenids closely resemble those of green algas and higher plants, although in most cellular features, euglenids are closer to some other flagellates. RNA-sequencing data also indicate an enormous genetic distance between euglenids and higher plants. This paradox may be explained by the hypothesis that euglenid chloroplasts were originally green algae or green algal chloroplasts, ingested by heterotrophic euglenids and retained as symbionts. This hypothesis also accounts for the extra membrane that is consistently found around the chloroplasts of euglenids. The extra membrane could be either the cell membrane of the algal cell or a host vacuolar membrane surrounding the symbiont as is usually seen in symbiotic associations. If this hypothesis is true, green euglenids arose from colorless ones. Of the existing lines of colorless

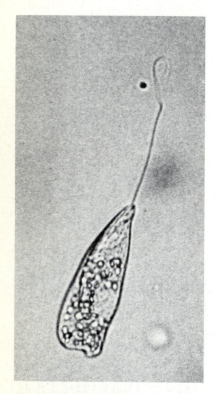

Heterotrophic euglenid, *Peranema trichophorum,* a freshwater species with a long, thick anterior flagellum held rigid and straight except for its tip, and a trailing flagellum pressed close against the body. This flexible euglenid can swim, glide on a substrate, and display vigorous euglenoid movement. There is no flagellar photoreceptor or eyespot, but the anterior flagellum acts as a sensory structure, being sensitive to both mechanical and chemical stimuli. The cytostome is a permanent opening, separate from the flagellar pocket. Two anterior rodlike structures, made of bundles of microtubules, can be protruded through the cytostome; attached to the surface of a prey organism, the rods appear to help pull it in, or they may pierce the prey, whose contents are then imbibed. Undigested remains are eliminated through a posterior anal pore. (R.B.)

euglenids, some may be primitively heterotrophic, while others probably arose later from intermediate green forms by loss of the symbiotic chloroplasts.

Contractile vacuoles are found in most freshwater protozoans, in which the higher ionic concentration of the cytoplasm (relative to the surrounding medium) results in a continual net uptake of water by osmosis, in addition to that taken in during feeding. In marine and parasitic protozoans, the ionic concentration of the cytoplasm is similar to that of the surrounding medium, there is little or no net uptake of water, and contractile vacuoles are generally absent.

Due to lack of clear evidence for a contractile mechanism in many "contractile" vacuoles, general terms such as "pulsating vacuole" or "water-expulsion vacuole" are often used.

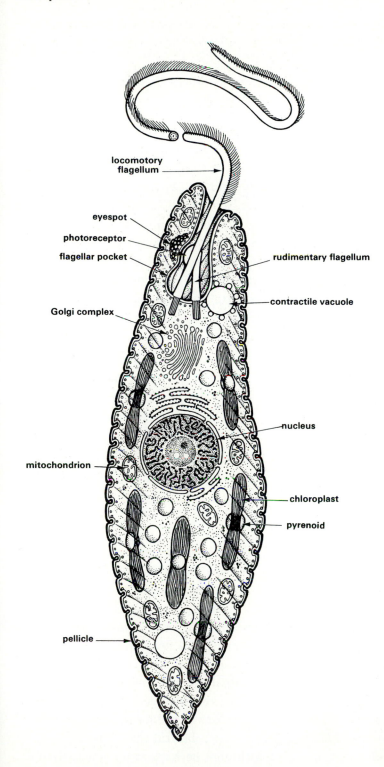

locomotory flagellum

eyespot

photoreceptor

flagellar pocket

rudimentary flagellum

contractile vacuole

Golgi complex

nucleus

mitochondrion

chloroplast

pyrenoid

pellicle

Euglenid, *Euglena gracilis* (diagrammatic, with some elements in section, others represented 3-dimensionally). A long locomotory *flagellum,* bearing a *photoreceptor* and a fringe of hair-like proteinaceous fibrils along one side, is accompanied by a second, different flagellum, sometimes short and rudimentary (as here), sometimes long and trailing as in other euglenids. Both sprout from the wall of the *flagellar pocket,* which is flanked by the *eyespot* on one side and the *contractile vacuole* on the other. The nucleus is surrounded by large disk-shaped *chloroplasts,* drawn in section to show their stacked membranes; each chloroplast bears a *pyrenoid,* in which carbohydrates are synthesized and stored. The grooves of the *pellicle* are seen on the surface of a euglenid as striations. Components of the pellicle, protective mucilaginous secretions, and various other cell products are made by the *Golgi complex.* The organism is about 50 μm long. (Based on G.F. Leedale, J.J. Wolken, D.E. Buetow, and others)

As mentioned above, euglenid nutrition is a mixture of photosynthetic autotrophy and of heterotrophy both by uptake of dissolved nutrients and by ingestion of food (though not all of these methods are practiced by any single species). Nutrition at least partly by heterotrophic means is probably common among many other green flagellates. Several species of freshwater photosynthetic flagellates studied (chrysomonads of the genus *Dinobryon*) ingest bacteria actively, obtaining more than half of their total nutrition in this way and proving to be more important consumers of bacteria (in the lake studied) than all the other larger and better recognized heterotrophs (ciliates, rotifers, and crustaceans) put together.

Euglenid flagellates multiply by repeated sequences of growth followed by **fission,** much as the cells of a growing multicellular animal alternately enlarge and divide. The nucleus of the flagellate divides and the two nuclei, each an identical genetic copy of the other, separate. Division of the cytoplasm proceeds down the length of the cell, and the two small flagellates go their own ways to grow and divide again. Such equal division is often referred to as *binary fission,* to distinguish it from the multiple fission seen in some protozoans, in which many asexual products form simultaneously.

Fission of a euglenid flagellate. Protozoans display a far wider spectrum of nuclear behavior during division than do animal cells. For example, division in euglenids differs from standard metazoan mitosis in that the nuclear membrane remains intact; the nucleolus remains discrete and divides; and there is no typical mitotic spindle. Nevertheless, an array of microtubules appears, and the filaments of chromatin replicate and separate to opposite poles of the nucleus. The 2 flagella may be distributed between the fission products and replicated afterwards; or they may be lost and regenerated as 4 in the course of division. Mitochondria and chloroplasts replicate independently, not necessarily in synchrony with the nucleus. (Combined from various sources)

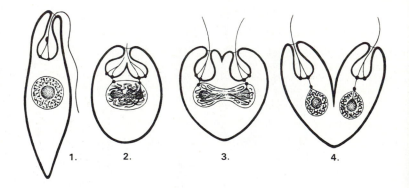

1.   2.   3.   4.

All the individuals in a lineage produced asexually by fission or some other means are said to be members of a **clone.** Although it is common to talk about "generations" of "parents and offspring" in clonal organisms, this confuses asexual processes with sexual reproduction. *No sexual reproduction* is known in euglenids (except for a few reports of uncertain significance). At no stage does the nucleus divide by meiosis to produce new genetic combinations, nor are gametes produced, nor does one individual exchange genetic material with another. However, as mutations introduce some genetic diversity into lineages of euglenids, natural selection can act and euglenids can evolve, although perhaps more slowly than do sexual organisms.

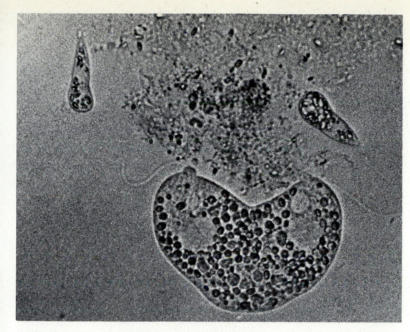

Euglena **dividing.** (R.B.)

Making many copies by continuing growth and division is the best defense of such small organisms as euglenids against predators and other environmental hazards. In addition, in some species, the coating of mucilage is elaborated into a protective **test** that surrounds the cell as an organic envelope that is sometimes hardened by mineral deposits and decorated with spiny projections. An opening in the test allows the flagellum to project and provides for the escape of new cells produced by division. Naked only briefly, each cell soon secretes a new test. Under unfavorable conditions such as drought or starvation, many species lose their flagella and secrete thick layers of mucilage that harden into a **cyst.** A euglenid may survive for many months within such a cyst, then emerge when it cracks open. Tests and cysts may also facilitate the spread of euglenids to new bodies of freshwater by wind or in mud carried on the feet of waterbirds and other animals.

Division sometimes takes place within cysts, but this is more characteristic of **palmella stages,** that lose the flagella, round up, coat themselves in a sheath of mucilage, and proceed to divide repeatedly. Sheets of such palmelloid cells may solidly cover a square meter or more of surface. When conditions ameliorate, the individual cells emerge from the mucilage, grow new flagella, and swim away. A palmelloid sheet is only a loose assemblage of cells, but some euglenids form more organized **multicellular colonies** in which the cells are arranged like fruits on a bush, connected by mucilaginous stalks. They retain short vestiges of the flagella, and these can be regrown to full length in about an hour. The flagellated cell then detaches and swims away to settle and found a new colony.

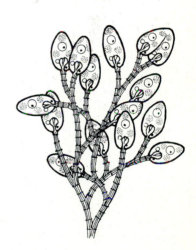

Colonial euglenid, *Colacium mucronatum,* forms large branching clusters up to 1 or 2 mm across, or may grow in bunches or sheets. The individual cells are only about 30 μm long. Several species of *Colacium* are found in freshwater, often growing on aquatic plants and animals. (Modified after G.F. Leedale)

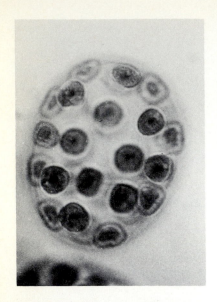

Volvocid colony, *Eudorina,* about 80 μm long, with several rows of cells embedded in a gelatinous matrix. The cells show little or no differentiation from each other in *Eudorina,* but in some species the anterior cells are slightly smaller and non-dividing and have larger eyespots than those at the posterior of the colony. Freshwater. Stained preparation. (R.B.)

Euglenid colonies show no evidence of coordinated behavior or differentiation among the cells. But in another group of mostly freshwater green flagellates, the **volvocids,** there is a spectrum of forms from dispersed clones of single cells, either green *(Dunaliella, Chlamydomonas, Carteria)* or colorless *(Polytoma, Polytomella),* to small colonies of cells with little or no differentiation *(Gonium, Eudorina, Pandorina),* to large spherical colonies *(Pleodorina, Volvox)* that are essentially **multicellular organisms.** A large individual of *Volvox* is a hollow spheroid 1 or 2 mm in diameter with up to 60,000 cells embedded in its mucilaginous wall. Each cell has a cup-shaped chloroplast, a red eyespot, 2 contractile vacuoles, and 2 flagella that project from the surface of the spheroid. Together, the flagella of all the cells propel the spheroid through the water, and it appears to respond in a coordinated fashion to stimuli such as light, swimming toward a well-lit area. Moreover, *differentiation* among the cells is evident. Those cells at one end of the spheroid have larger eyespots and presumably play a relatively more important role in directing the behavior. This end is always directed anteriorly as the spheroid is propelled through the water; thus, the organism shows a permanent *polarity.* Only a few of the cells at the rear end of the spheroid participate in the production of new individuals, either asexually or sexually. Unlike the solitary cells of euglenids, all of which have the potential for a sort of immortality through an indefinitely dividing lineage of descendants, most of the cells of a volvocid spheroid have a finite life span; only a few are privileged to replicate the genetic inheritance in the form of asexual offspring or to contribute to new genetic combinations in sexual progeny.

In the **asexual reproduction** of a new individual in *Volvox,* one of the rear cells enlarges, loses its flagella, and divides mitotically until a small ball of cells results. From a few to a dozen such replicate individuals are typically seen in the hollow interior of any large spheroid, which will eventually rupture and release them. Again, in **sexual reproduction,** certain of the rear cells differentiate into gametes; in different species, male and female gametes may develop in the same spheroid, in separate spheroids of the same clone, or only in separate clones. A cell forming a **macrogamete,** or egg, enlarges and becomes rounded and filled with nutrient reserves that are contributed in part by adjacent cells. A cell forming **microgametes,** or sperms, divides repeatedly and produces a sperm packet, which must be in contact with an egg-bearing

spheroid before the biflagellate sperms are released. The microgametes are not the highly specialized cells that animal sperms
are; they look much like any cell of the spheroid. In species
with separate sexes, male individuals release a substance that
induces gamete production in other individuals, both male and
female. A fertilized egg, or **zygote,** secretes around itself a
thick spiny covering, within which it can survive long periods
of unfavorable conditions such as drying or freezing. The
zygote divides by meiosis, and one of the resultant haploid cells
gives rise to a new spheroid.

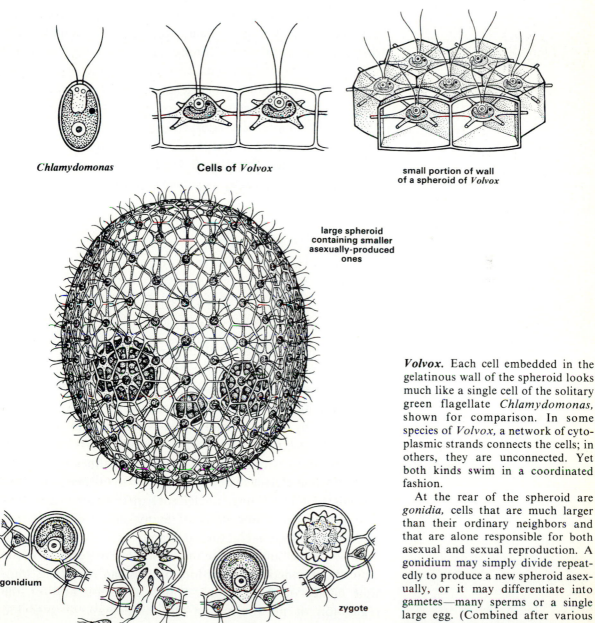

Chlamydomonas　　　　　　**Cells of** *Volvox*

small portion of wall
of a spheroid of *Volvox*

large spheroid
containing smaller
asexually-produced
ones

gonidium

sperms　　　　　　egg　　　　　　zygote

Volvox. Each cell embedded in the
gelatinous wall of the spheroid looks
much like a single cell of the solitary
green flagellate *Chlamydomonas,*
shown for comparison. In some
species of *Volvox,* a network of cytoplasmic strands connects the cells; in
others, they are unconnected. Yet
both kinds swim in a coordinated
fashion.

At the rear of the spheroid are
gonidia, cells that are much larger
than their ordinary neighbors and
that are alone responsible for both
asexual and sexual reproduction. A
gonidium may simply divide repeatedly to produce a new spheroid asexually, or it may differentiate into
gametes—many sperms or a single
large egg. (Combined after various
sources)

Each new individual spheroid of *Volvox* is produced from a single cell that divides repeatedly. The flagellated ends of the cells of the resultant ball are at first directed inward, so that the young spheroid must turn inside-out (through a small pore) in the course of attaining its mature organization. Because this inversion is strongly reminiscent of the development of the early stages of some sponges (see chapter 3), *Volvox* is the most commonly cited example of animal-like *development* in protozoans; but it is an unlikely model of animal ancestors in other respects, and equally complex developmental processes are found in a number of other protozoans.

Volvox demonstrates one way in which protozoans can become relatively large-bodied organisms. The dividing cells do not disperse but remain connected or closely associated as a spherical colony, which then becomes a multicellular individual by coordination and differentiation of its many cells. Another route to large size in protozoans is illustrated by a group of symbiotic flagellates, the **opalines,** named for their beautiful opalescence. Opalines live as commensals in the gut of frogs and fishes. The flagellates do not appear to harm the host and are nourished by dissolved materials in the host's gut.

Opalina **divides** and the products separate (arrows show direction of swimming). Opalines usually divide longitudinally, between the rows of cilia as shown here, although they sometimes divide across the rows, as ciliates do. More consistent characters that distinguish opalines from ciliates include differences in the basal structures associated with the cilia, a single type of nucleus (vs. two types in all ciliates), and the formation of gametes in sexual reproduction (vs. conjugation in ciliates).

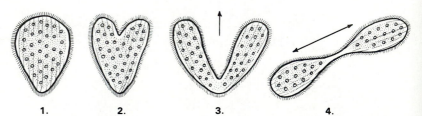

1.   2.   3.   4.

Extensive folding of the cell membrane increases the surface for nutrient uptake, as well as for exchange of respiratory gases, excretory products, and other materials. Opalines are giants among protozoans, some growing up to 2.8 mm long. They move by means of many rows of cilia covering the body, which is a *syncytium,* a membrane-bounded mass of continuous cytoplasm with **multiple nuclei.** Opalines have great numbers of nuclei (each about the size of an ordinary euglenid nucleus) or at least 2 huge nuclei, each up to 40 μm in diameter (the size of the whole body of many euglenids and other flagellates). Presumably, this large amount of genetic material does not reflect any increase in the number of genes or in the quantity of genetic information, but instead represents multiple copies of the genome, which are necessary for the regulation of metabolism in the large mass of syncytial cytoplasm.

The flagellates are a diverse assemblage of groups, a small sample of which is presented in the illustrations that follow.

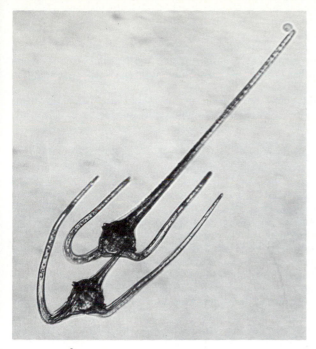

Dinoflagellates. *Ceratium,* common in freshwater and in seawater, depends on both photosynthesis and ingestion of food. The spines may discourage some predators and also retard sinking; they tend to be shorter in denser, colder waters and longer in less dense, warmer waters. These long-spined specimens were photographed in Bimini, Bahamas. (R.B.)

Most dinoflagellates occur as single cells, but in *Ceratium* and some others the dividing cells often remain together as loose, temporary chains; these 2 individuals represent part of such a chain. In *Dinothrix,* stable filaments of cells form branching, attached colonies.

Replication without division in the heterotrophic dinoflagellate *Polykrikos* results in compact, multinucleate, syncytial "colonies" of up to 16 units. *Polykrikos* and *Nematodinium* are also known for having elaborate capsules, each containing a coiled eversible thread that can be "fired" to the outside; these organelles resemble the nematocysts of cnidarians (chapter 5).

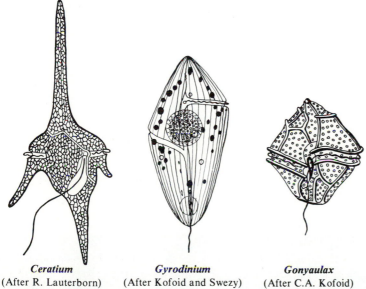

| *Ceratium* | *Gyrodinium* | *Gonyaulax* |
| (After R. Lauterborn) | (After Kofoid and Swezy) | (After C.A. Kofoid) |

Two flagella lie in grooves on the surface of a dinoflagellate, an encircling *transverse flagellum* and a trailing *longitudinal flagellum.* The cell is described as *armored* (or thecate) if the cell covering includes cellulose plates; these are not extracellular secretions, but lie between membranes and may be considered part of a pellicle. *Ceratium* and *Gonyaulax* are heavily armored. *Unarmored* (or naked) kinds such as *Gymnodinium* and *Noctiluca* (see next page) are covered by membranes alone. *Gyrodinium,* with a delicately striated covering, is intermediate.

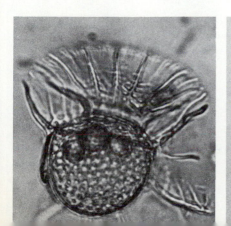
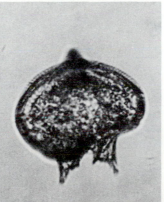

Left: **Dinophysid dinoflagellate** with elaborate flanges. *Ornithocercus,* heterotrophic. *Right:* **Peridinid dinoflagellate** with smooth outline except for small spines. *Peridinium* (or *Protoperidinium),* mostly heterotrophic. Both types are armored. (R.B.)

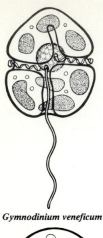

Gymnodinium veneficum

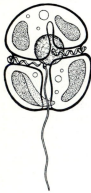

Gymnodinium vitiligo

Two species of *Gymnodinium,* both photosynthetic, distinguished by small morphological differences, but a major physiological one: *G. veneficum* releases into the water a toxin fatal to fishes and many invertebrates, whereas *G. vitiligo* is harmless. (After D. Ballantine)

Many kinds of dinoflagellates have similar "gymnodinioid" stages, including the **zooxanthellas,** photosynthetic dinoflagellates that are symbiotic in many other protists as well as in many sponges, corals, and other invertebrates.

Under favorable conditions, marine photosynthetic dinoflagellates multiply rapidly to form extremely dense populations, or blooms, in nearshore waters. Some of these are undetectable without microscopic examination of water samples. Others are conspicuous because they render the water red by day—**red tide**—or brilliantly luminescent by night, or because toxins released into the water cause massive, malodorous kills of fishes and/or invertebrates.

Some toxic dinoflagellate species do not cause such kills and even seem not to harm the aquatic animals that eat them; but the toxin that accumulates in the tissues of filter-feeding bivalve molluscs such as clams, oysters, and mussels (chapter 15) can be deadly to a human consumer. The agent in such **"paralytic shellfish poisoning"** is a powerful neurotoxin; the victim may die within a few hours of eating only a few toxic bivalves, and it is wise to respect the quarantines on collecting bivalves, usually in force only when blooms are likely, during the summer months. **"Ciguatera" poisoning,** which can also be fatal and has no predictable seasons, results from eating tropical fishes that acquire the toxin of benthic dinoflagellates. The dinoflagellates live on the surfaces of seaweeds, and the toxin is accumulated both by herbivorous fishes that browse on the seaweeds and by carnivorous fishes that prey on the herbivores.

Noctiluca is not a typical dinoflagellate in either its nuclear organization or the pattern of its life history, and some specialists suspect it should be placed in a separate group. The structure of the cell is also peculiar. There is only a single, longitudinal flagellum, and its deep groove is elaborated as a cytostome. A large mobile tentacle that extends from near the cytostome is used in feeding. (Combined after various sources)

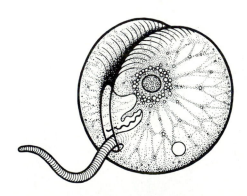

Field of *Noctiluca* can be observed during a bloom. When present in such large numbers, these big unarmored dinoflagellates present dazzling displays of luminescence at night. *Noctiluca* is completely heterotrophic, a voracious predator on various small organisms, including other dinoflagellates. The globular vacuolated cell sometimes reaches 2 mm across and divides by both binary and multiple fission. (D.P. Wilson)

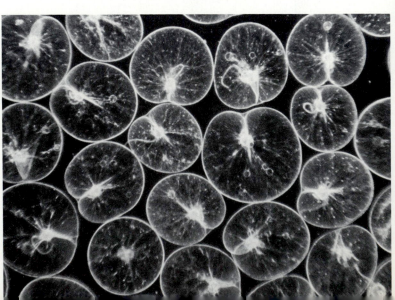

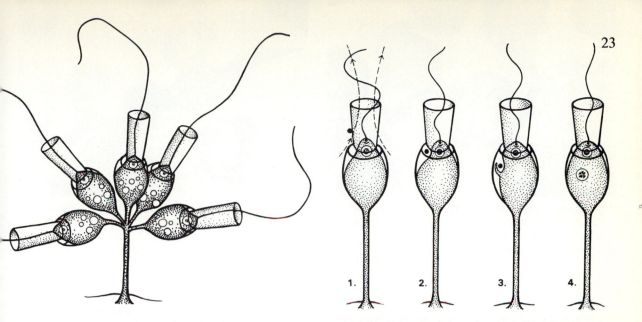

1. 2. 3. 4.

Collar flagellates (choanoflagellates) are a distinctive group without clear relationship to other flagellates. They are heterotrophic, feeding on bacteria and organic particles, and may be seen attached or free-swimming in fresh, brackish, and marine waters. Some are solitary *(Monosiga),* but in others the products of division remain together and form colonies of varied shapes: clusters *(Codosiga,* shown here), bushy branching forms *(Codono-cladium),* straight chains *(Desmarella),* and curved plates or spheres with the cells partially embedded in a gelatinous matrix, their distal ends projecting out in all directions *(Proterospongia, Sphaeroeca).* The life history of

A collar flagellate feeding. The collar that surrounds the single long flagellum is a device for filter-feeding. It is not a solid membrane, but a circlet of microvilli, long slender projections of cytoplasm, loosely bound together by organic material. The beating flagellum draws food-bearing water currents toward the base of the cell. Food particles are caught on the outside of the collar, while water passes through the collar between the microvilli and emerges from the open end. The food is moved down toward the cell body and taken into the cytoplasm. (Based on various sources)

Proterospongia choanojuncta includes both solitary and colonial phases. Collar flagellates bear a detailed resemblance to the collar cells of sponges (chapter 3) and both feed in the same way, suggesting a close evolutionary relationship. Some collar flagellates have a test, or lorica, of organic and/or siliceous composition, which may be compared to sponge skeletal materials. No other protist offers such convincing evidence of common ancestry with an animal group. (Modified after G. Lapage)

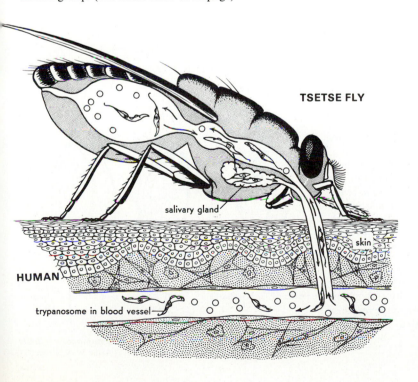

TSETSE FLY

salivary gland

skin

HUMAN

trypanosome in blood vessel

Trypanosomes include commensals and deadly parasites, in both animals and plants. In Africa, *Trypanosoma brucei* is transmitted to humans by tsetse flies *(Glossina)* and causes several forms of disease (trypanosomiasis), sometimes fatal, including African sleeping sickness. A biting fly injects the trypanosomes directly into the bloodstream of the vertebrate host or acquires trypanosomes from an already infected vertebrate. The flagellates multiply asexually in both hosts. Wild game animals act as a reservoir for the disease organism but are little affected. Humans and domestic stock, presumably because their contact with it is more recent, are devastated. In Central and South America, *T. cruzi* causes Chagas's disease, transmitted by blood-sucking reduviid bugs. Species of the related *Leishmania* also cause several kinds of disease (leishmaniasis). Together these diseases affect tens of millions of people around the world.

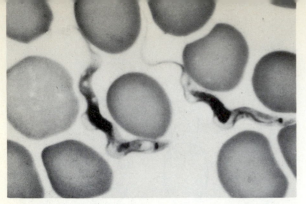

Trypanosomes in blood. This is the W. African form, *Trypanosoma brucei gambiense,* which causes African sleeping sickness. *T. brucei rhodesiense* of E. Africa causes a more virulent trypanosomiasis. Seen here among red blood cells, the flagellates are about 25 μm long. (Stained blood smear).

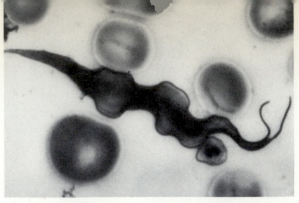

Dividing trypanosome, *Trypanosoma lewisi* in the blood of a rat. The flagellum is already replicated, while the nucleus still appears single. The flagellates multiply rapidly at first, but the host immune system is finally able to control them, and the rat is not seriously harmed. (Stained blood smear)

Giardia looks like a longitudinally split pear with spectacles. It is a diplomonad, so-called because of the doubling of organelles (2 nuclei, 2 sets of 4 flagella), which gives it a bilateral symmetry. All species are parasitic in vertebrates. The human parasite, *G. intestinalis* (shown here), lives in the upper small intestine. It attaches to the intestinal wall by an adhesive disk on its flattened surface and takes in nutrients by pinocytosis from the lumen of the host's intestine. It is estimated that about 10% of the people of the U.S. are infected. However, most infections cause no symptoms or only mild intestinal disturbance and diarrhea. Severe, chronic disease with impaired absorption of food seems to result only in particularly susceptible individuals, perhaps with inadequate immune responses, and such cases may be extremely serious, even fatal. Transmission from one host to another is by cysts, ingested in food or water that has been contaminated with fecal matter; hikers in the U.S. should be aware that this flagellate has been found even in remote mountain streams. The cysts are resistant to many disinfectants, but are killed by dessication or boiling. Nuclear division occurs within the cyst, giving rise to 4 nuclei, and cell division is completed after the flagellate emerges. (Combined from several sources)

Trichomonads are commensal or parasitic flagellates with a prominent axostyle (a supporting or locomotory organelle composed of microtubules) and parabasal body (rodlike Golgi apparatus). Shown here are the 3 species routinely found in a large percentage of the human population around the world. *Trichomonas tenax* lives in the mouth and is transmitted by kissing; its presence is positively correlated with the age of the host. *Pentatrichomonas hominis* is found in the large intestine. Both of these are generally regarded as harmless commensals. *Trichomonas vaginalis* is a parasite of the urogenital tract, found in both women and men (though men are usually without symptoms), and transmitted primarily by sexual intercourse. The severity of its effects depend both on the strain of the parasite and the response of the host, but it should not go untreated.

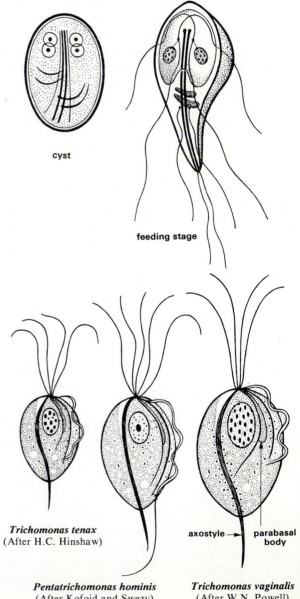

cyst

feeding stage

Trichomonas tenax
(After H.C. Hinshaw)

axostyle → | ← parabasal body

Pentatrichomonas hominis
(After Kofoid and Swezy)

Trichomonas vaginalis
(After W.N. Powell)

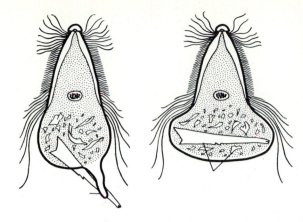

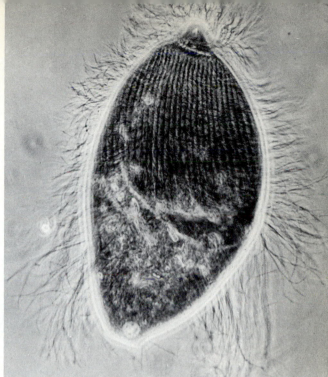

Trichonympha lives as a mutualistic symbiont in the hindgut of wood-eating insects (termites and wood roaches). The anterior end of these large, complex flagellates (sometimes over 300 μm long) is covered with an elaborate pellicle and hundreds of flagella, but the posterior end extends pseudopods and ingests bits of wood. The flagellate is able to produce cellulose-digesting enzymes, but the insect host cannot; both roaches and termites depend on carbohydrates released by their symbionts, though they probably digest some of the flagellates as well. As much as one-third of the weight of some termites *(Zootermopsis)* may consist of flagellates.

Symbiotic flagellate, *Trichonympha ampla,* is covered with flagella over most of its length (250 μm), and wood particles are visible in its posterior cytoplasm. It lives in the gut of the termite *Pterotermes occidentis* from the Sonoran desert of California, Arizona, and Sonora. (D.G. Chase)

Each time the insect molts, it loses the lining of the hindgut and all its symbionts. If it is unable to acquire new ones, it will starve to death, even though it continues to feed normally, for it cannot digest the wood. The flagellates are equally dependent on the mutualism and die within minutes outside the host. A young or newly-molted termite acquires symbionts by feeding directly from the anal opening of other termites in the colony. The flagellates in wood roaches are able to detect an insect hormone (ecdysone) secreted just before the molt and to respond by encysting; thus protected, the flagellates remain in the cast-off exoskeleton, which the roach eats and thereby restores the symbiosis. Presumably, newly-hatched wood roaches acquire symbionts by eating the molted exoskeletons of older ones.

AMEBAS AND OTHER SARCODINES

AN AMEBA, such as one might find among pond debris examined under a microscope, is an organism that well deserves its name, derived from a Greek word meaning "change." The single cell that constitutes its body is bounded only by a delicate *cell membrane,* and the ameba has no permanent shape, its outline changing constantly as it moves. Likewise, the nucleus and other organelles have no permanent position, but flow with the cytoplasm. The cytoplasm is differentiated into an inner, granular, more fluid **endoplasm** and an outer, clear, firmer **ectoplasm.**

Even this two-phase structure is dynamic, for ectoplasm (gel phase) and endoplasm (sol phase) are interconvertible. In a moving ameba, the ectoplasm in a given area of the periphery will suddenly become fluid, permitting the endoplasm to flow forward into a projecting lobe, or **pseudopod** ("false foot").

It has been suggested that amebas and other sarcodines evolved from flagellates by loss of the flagellated stage. Since this is often the sexual stage, such an origin might help to explain the absence of sexual reproduction in most groups of sarcodines.

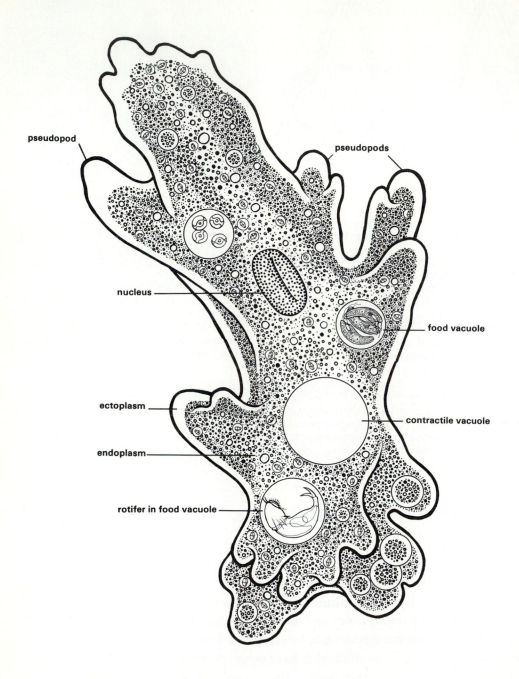

pseudopod

pseudopods

nucleus

food vacuole

ectoplasm

contractile vacuole

endoplasm

rotifer in food vacuole

Diagram of an **ameba** with details based mainly on a freshwater ameba, *Amoeba proteus.*

Near the tip of the pseudopod, the endoplasm flows to the sides and becomes firmer, so that the pseudopod extends as a tube of firm ectoplasm with endoplasm flowing in its lumen. An ameba extends pseudopods outward from any point on the temporary forward-moving region of the body, and endoplasm streams into each one successively, at the same time as it is being regenerated from ectoplasm near the rear of the body and also flowing back into the central mass from other, earlier pseudopods in the process of being withdrawn. In this way, the ameba moves slowly along (at about 5 μm per second, or 2 cm per hour) with frequent changes of direction. Such **ameboid movement** is seen also in ameboid cell types of multicellular animals.

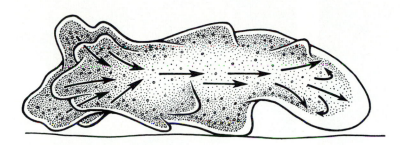

Ameboid movement. Though it appears to have nothing in common with muscular movement, ameboid movement probably depends on contractile components and mechanisms surprisingly similar to those in the muscle cells of animals. When fixed, sectioned, and examined by electron microscopy, the cytoplasm of an ameba is found to contain thick and thin microfilaments similar in appearance and dimensions to the thick (myosin) and thin (actin) microfilaments of striated muscle. Extracts of muscle myosin appear to cross-react with the thin filaments from amebas. And cytoplasm from amebas uses the nucleotide adenosine triphosphate (ATP) as an energy source for movement, as muscles do.

A major difference between muscles and ameboid pseudopods is that in muscle cells the systems of microfilaments are permanently differentiated and stable, whereas in amebas the rapid appearance and disappearance of the filaments may be the basis of the constant flux of sol/gel phase transformations between endoplasm and ectoplasm. Exactly how the microfilaments produce the cytoplasmic streaming of ameboid movement remains a subject of active research and discussion. And as the structure of pseudopods in the various sarcodines is quite diverse, it is unlikely that any one mechanism will account for them all.

Centrifugation of an ameba at high speeds results in stratification of the organelles. Those of a single type may be drawn out with a micropipet and the effect of their absence may be observed by comparing the amebas that lack them to controls that are centrifuged but left intact, a treatment from which amebas recover quickly. (Modified after Mast and Doyle)

Examining amebas with a microscope during centrifugation has provided estimates of the relative viscosity of different cytoplasmic regions. Cytoplasm isolated from cell membranes and nuclei by high-speed centrifugation has been used for electron microscope studies of contractile microfilaments, which seem to be less sensitive to destruction by chemical fixatives than when in the intact cell. Such isolated cytoplasm was also used to demonstrate the role of ATP as an energy source for cytoplasmic streaming; quiescent until the nucleotide is added, it then begins active streaming movements.

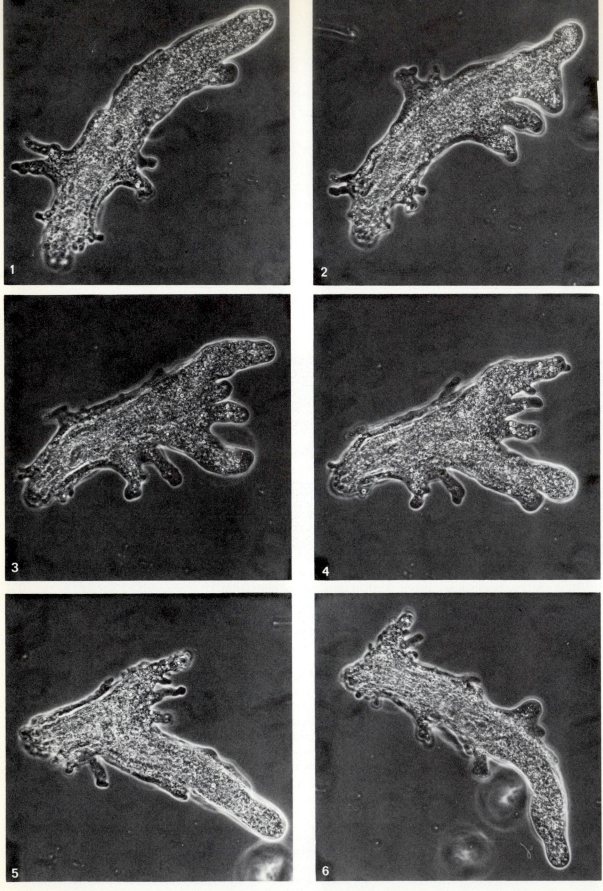

Ameboid movement illustrated by successive photos of an ameba, *Amoeba proteus*. (R.B.)

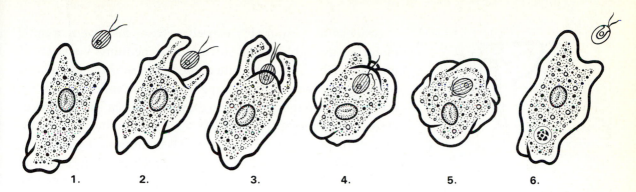

Capture and phagocytosis of prey. *Above,* **1.** to **5.** a pond ameba extends its pseudopods widely around an active prey, a fast-swimming flagellate; **6.** the ameba goes after another flagellate.

Below, **1.** to **3.** a soil ameba engulfs a slow-moving prey, another ameba; **4.** and **5.** the ameba eliminates undigested residue.

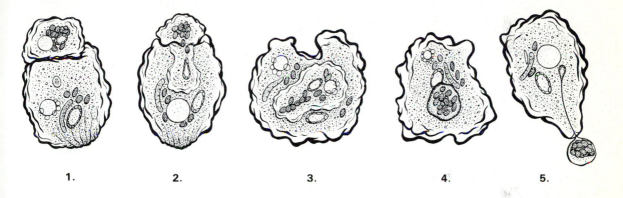

Pseudopods serve not only in locomotion, but also in *food capture.* An ameba can apparently detect vibrations in the water, for when it is stalking active prey, such as another moving protist or a minute animal, the pseudopods are extended widely around the prey and do not touch it until it is completely surrounded. When the ameba goes after a passive object, such as a nonmotile bacterial or algal cell, the pseudopods surround the food closely. Ingestion takes place by a process called **phagocytosis,** in which that portion of the cell membrane surrounding the food is pinched off to form a complete envelope, and the food is taken into the cytoplasm of the ameba in a drop of water bounded by cell membrane, a *food vacuole.* Digestive enzymes (already synthesized and packaged ready for use in small vesicles, the lysosomes) are released into the food vacuole and break the food down into small molecules that can pass directly into the cytoplasm to be used as sources of energy and materials for the activity and growth of the ameba.

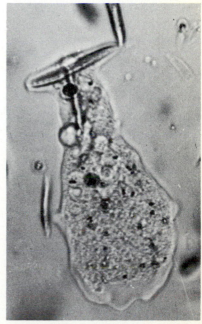

An **ameba ingesting** a diatom. (R.B.)

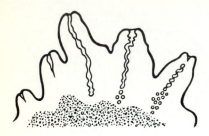

Pinocytosis. The cell membrane invaginates to form narrow canals which break up into small vesicles, much like small phagocytotic vacuoles but filled with liquid. The rate of vesicle formation is stimulated by protein in the surrounding medium. There is no absolute distinction between pinocytotic and phagocytotic vacuoles, because some liquid is inevitably taken in with food and solid particles are taken in during pinocytosis. Moreover, as digestion proceeds within food vacuoles, solubilized nutrients are incorporated from the vacuole into the cytoplasm by pinocytosis, small liquid-filled vesicles being pinched off from the vacuolar membrane. (Based on Mast and Doyle)

The contractile vacuole of an ameba occupies no fixed position. It is filled by fusion of many smaller vesicles and it empties when it fuses with the outer cell membrane. No contractile mechanism has been demonstrated, and it is likely that the vacuole collapses because of surface tension forces.

Amebas also nourish themselves from dissolved organic materials available in the surrounding medium. Vacuoles of nutrient-rich liquid are formed and taken into the cytoplasm by **pinocytosis** ("cell drinking"), much as solid objects are ingested by phagocytosis. Excess water taken in by phagocytosis or pinocytosis, and that which is produced in metabolism or passes in through the cell membrane by osmosis, accumulates in a *contractile vacuole,* which expels its contents to the outside at intervals. Indigestible particles are ejected by the ameba, and soluble nitrogenous wastes diffuse out through the cell membrane. Exchange of respiratory gases (oxygen and carbon dioxide) also occurs freely through the cell membrane. Transport of nutrients, dissolved gasses, and other materials within the cell occurs by diffusion and by cytoplasmic flow.

If food or water becomes scarce, or temperatures fall, many kinds of amebas are able to survive the adverse period by forming **cysts.** As long as conditions remain unfavorable, the ameba remains inactive within the secreted wall of the cyst, its metabolism reduced to minimal levels. When moisture or warm temperatures return, the ameba emerges and begins again to feed and grow.

Like euglenids, amebas grow and multiply by **fission.** The nucleus of the ameba divides by mitosis, nuclear division is followed by division of the cytoplasm, and the two new amebas separate to feed and grow. *No sexual reproduction* has been observed in amebas.

With its remarkably minimal level of permanent differentiation and structural organization, an ameba is able to move about, capture food, grow, replicate, and carry out all the physiological functions for which multicellular organisms have specialized equipment.

Binary fission of an ameba.

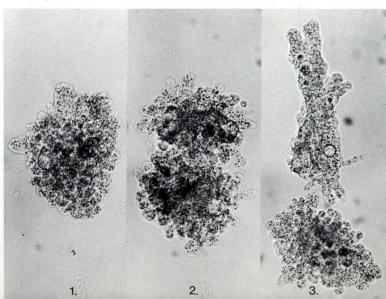

1.   2.   3.

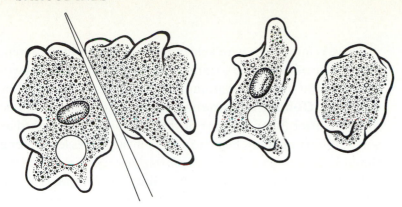

An ameba cut in two with a glass needle rapidly seals the cut surfaces. The portion with a nucleus behaves normally, feeds, grows, and divides. That without a nucleus may move and even feed for a short time, but it soon becomes inactive and is unable to digest food, to grow, or to divide.

TYPES OF ORGANIZATION other than that of an isolated cell with a single nucleus are often associated with increase in size, as seen in some flagellates, and are displayed by many kinds of amebas and other sarcodines. Some of these are quite large—larger than many multicellular animals. Giant amebas may measure 5 mm in length and appear little different from typical microscopic kinds except that they have multiple contractile vacuoles and *large numbers of nuclei* dispersed throughout the syncytial cytoplasm. A large individual of *Chaos carolinense* has about 1,000 nuclei which together occupy about 6% of the total cell volume. When such a giant ameba divides, all of its nuclei undergo mitosis simultaneously so that the number of nuclei is doubled. Then the cytoplasmic mass divides, into 2 to 5 parts, and each of the resulting amebas has approximately twice the number of nuclei per volume as did the original one. In time, growth restores the normal nuclear/cytoplasmic ratio.

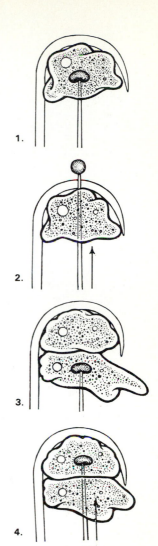

Transfer of a nucleus from a member of the same clone into an anucleate fragment restores normal capabilities. Transfer of a nucleus from another strain or species may be performed to examine the genetic compatibility of these different cellular components. Cell biologists are also able to work more selectively by transferring short chromosome fragments, bearing genes of known effect, from one organism into another. (Modified after Lorch and Danielli)

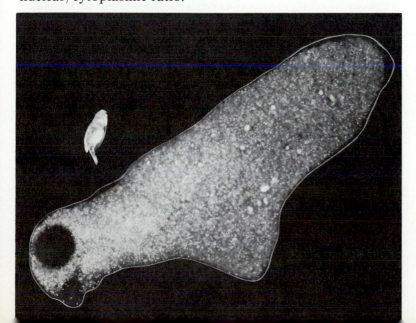

Giant multinucleate ameba, *Chaos carolinense,* dwarfs a many-celled rotifer (chapter 13). (P.S. Tice)

There is another way of achieving large size, quite different from the sorts discussed so far in which the products of growth and division remain continuously together as colonies or syncytia. **Collective amebas** (mycetozoans, or "slime molds"), after living for part of their life history dispersed as small solitary amebas, begin to collect in large masses. Among the collective amebas, there are *plasmodial types,* in which the individual amebas fuse into a syncytium, and *cellular* types, in which the individual cells remain discrete but closely associated as an integrated multicellular aggregate.

Dictyostelium discoideum is the best known of the collective amebas, because it has been widely used in laboratory culture as a model system for studying cellular differentiation. The single dispersed amebas (about 10 to 20 μm long) are found in the soil, on dung, or on decaying plants. They feed mainly on bacteria in these habitats, but can also take in dissolved organic material. When well fed, they multiply indefinitely by fission. In this phase, two amebas that come into contact appear to be mutually repelled. But when food becomes scarce, their behavior and physiology change. Contacts result in small groups sticking together, and these become centers of attraction toward which other amebas begin to migrate. The migrants are attracted by a chemical, called acrasin, produced by all the amebas but present in higher concentrations where a group of them has accumulated. Amebas continue to stream toward such centers of chemical attraction until a large aggregate has formed.

Acrasin may be considered a **pheromone.** A pheromone is a substance which one organism produces and secretes into the environment, which has a specific effect on the behavior or physiology of other members of the same species, and which thereby benefits the organism that originally produced it. Because the effect of acrasin is to bring together cells that will form a single, integrated multicellular organism, this substance might be considered a hormone instead.

The acrasin of *Dictyostelium discoideum* and a few other species has been identified as cyclic adenosine monophosphate, a small nucleotide. Curiously, cyclic AMP has been shown to be an essential agent in certain mammalian hormonal systems. However, other dictyostelids have several other, chemically different acrasins.

The aggregated amebas form a cohesive elongate mass, which commonly measures 1 to 2 mm long and may consist of up to 200,000 amebas, each still a discrete membrane-bounded cell. The aggregate creeps along for some distance like a tiny slug and is responsive to small differences in light and

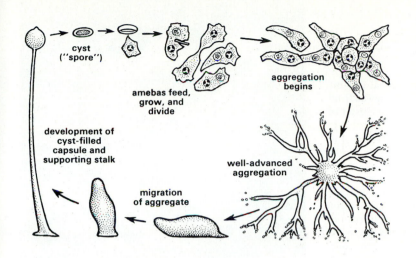

cyst ("spore")

amebas feed, grow, and divide

aggregation begins

development of cyst-filled capsule and supporting stalk

well-advanced aggregation

migration of aggregate

Asexual cycle of ***Dictyostelium discoideum.*** (Based on various sources)

The many small cysts dispersed from the sorocarp are referred to as *spores*, and two other kinds of cysts (not shown here) may also be formed during the life history. Aggregations of amebas in this species and in some related ones sometimes form large cysts *(macrocysts)* that are thought to represent a sexual stage. This may occur either among amebas of a single strain or only when strains of different mating types (comparable to separate sexes) are mixed. In some species (but not *D. discoideum*) each individual ameba in the feeding stage may enclose itself in a small cyst *(microcyst)*. Whether the amebas form spores, macrocysts, or microcysts can be influenced in the laboratory (and presumably in nature) by conditions of food, moisture, light, and temperature.

In *D. discoideum* over 75% of the cells in an aggregate end up as cysts, but in some other species the figure is closer to 10%. In either case, the process incurs considerable loss of life. However, still more amebas might die of starvation were the population to continue under the food-poor conditions that stimulate aggregation. And presumably the advantage gained by sorocarp formation (probably greater dispersal) more than offsets the sacrifice. Similar bodies are formed in response to food shortage by myxobacteria, by a variety of fungi, and by certain remarkable ciliates (see below).

temperature. Eventually it stops and begins to develop into a tall, cyst-forming structure (sorocarp) with a basal disk, a slender stalk, and a rounded top. The location of an individual ameba within the structure is determined by where it was in the migrating "slug," and this in turn determines its fate. Only the cells in the top develop into cysts that are dispersed and survive to hatch out again as single amebas. Those of the stalk and base die after the cysts are dispersed. However, as long as the amebas that join in an aggregation are members of a single clone, like the cells in a spheroid of *Volvox,* it makes no genetic difference which ones survive as cysts and which ones die after lending support as disk and stalk cells. The individual organism is probably best considered as a multicellular entity equivalent to the entire cyst-forming body, or even the entire clone. When the amebas in a single aggregation are not genetically identical, the situation becomes more complex.

Analysis of the genetics and evolutionary biology of an unconventional "dispersed organism" such as *Dictyostelium,* in which solitary unicellular units physically reconnect during one phase of the life history, may help clarify our thinking about clonal organisms in general. In all protozoans and in clonal fungi, plants, and animals, the units are unicellular or multicellular bodies that may remain physically separate but retain intimate evolutionary connections through their sharing of a common genome.

SOME OF THE DIVERSITY seen in ameboid protists with parasitic life histories, or with elaborate skeletal elements and well-differentiated cytoplasm, is illustrated in the following eight pages.

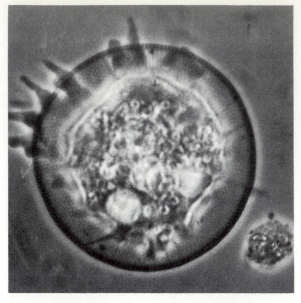

Shelled ameba, *Arcella,* has a hemispherical organic test. Visible in the photo *(right),* of an ameba viewed from above, is the textured surface of the test and a cluster of pseudopods, protruded through an opening in the lower surface. The ameba can rise up on its pseudopods and move, but is rather sedentary, remaining mostly in one place and using the pseudopods in feeding. Secretion of a gas bubble enables the ameba to float or, if turned over, to tip to one side and then right itself, after which the bubble disappears. When preparing to divide, the ameba secretes a new test with its opening apposed to the old one; one of the fission products inherits the old test, while the other occupies the new one. *Arcella* may have one, two, or many nuclei. (R.B.)

3 species of *Difflugia* are distinguished on the basis of shell shape, as well as by the form of the pseudopods as in naked amebas. These amebas build a shell of organic material covered with sand grains. *Difflugia* moves along by extending one or more pseudopods and attaching them to the substrate; the pseudopods contract, pulling the body and test after them, and are then released from their attachment. Testate amebas occur mainly in quiet freshwater habitats, sometimes in bogs or moist soil; a few are marine. (After J. Leidy)

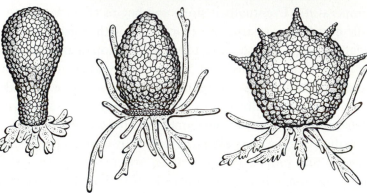

Cyst of *Acanthamoeba,* about 10 μm across, its thick irregular wall visible at the edge. Transmission electron micrograph (EM). *A. castellanii.* (B. Bowers)

Certain species of this and another genus of normally free-living amebas, commonly encountered in water and soil, have been found on rare occasions to infect humans and other animals. Several species of *Acanthamoeba* infect a variety of tissues in people already debilitated or immunologically deficient. *Naegleria fowleri* causes an acute form of meningoencephalitis in normal victims; people become infected while swimming, the amebas entering through the nasal mucosa and multiplying in the central nervous system. Though not a major public health problem, the diseases caused by these amebas are devastating to those who contract them and many have proved fatal. Medical scientists are trying to understand what causes the amebas to become pathogenic. Virulent strains of the amebas are resistant to 37°C, the temperature of the human body. Some evidence suggests that a combination of warm water and chlorine, as in swimming pools (or thermal and chlorine pollution in natural waterways), may favor the virulent strains, usually outcompeted by non-pathogenic amebas. Fatalities caused by pathogenic amebas in hot springs have been recorded.

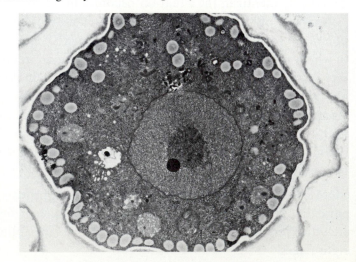

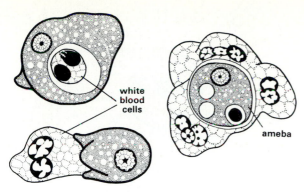

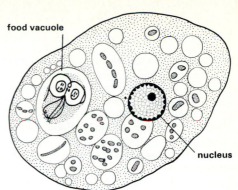

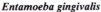

Entamoeba gingivalis

Entamoeba coli

Amebas symbiotic in humans are common and usually harmless commensals, a few species being widespread. The **mouth ameba,** *Entamoeba gingivalis,* is passed directly from person to person in kissing and in sharing food; it does not encyst. It feeds on loose cells and organic debris in the mouth. *Above left,* it is shown ingesting a leucocyte; *Above right,* several leucocytes cooperatively ingesting a mouth ameba. (Modified after H. Child).

The **intestinal ameba,** *Entamoeba coli,* lives in the lumen of the large intestine and is a voracious eater, but rarely ingests host cells. This one is packed with food vacuoles containing bacteria, organic particles, and a parasitic flagellate *(Giardia).* This ameba is transmitted as cysts that are ingested with food or water and can survive passage through the acidity of the stomach. (After F. Doflein).

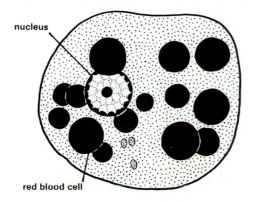

The **dysentery ameba,** *Entamoeba histolytica,* found in the lower part of the small intestine and in the large intestine, is a harmless commensal most of the time, but may cause serious or fatal disease under certain circumstances, invading the intestinal wall and ingesting red blood cells and other host tissues. It produces bleeding ulcers and may spread to the liver and other organs. It is transmitted by cysts ingested with fecal-contaminated food or water. It is distinguished from *E. coli* by the cytology of the cyst.

Entamoeba histolytica

Entamoeba histolytica. Left, Cyst; *E. coli* cyst. *Right,* Nuclei in mitosis.

Cysts of common human symbiotic amebas. (After Swezy)

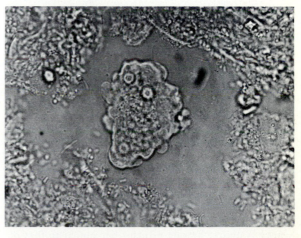

Living *Entamoeba histolytica.* The stools of infected persons contain both cysts and active feeding stages, but the latter quickly die outside the host's body. About 10% of the human population is estimated to harbor dysentery amebas, about 4% in the USA. (Army Medical Museum)

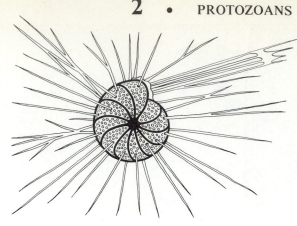

Foraminiferan has slender filamentous pseudopods that branch and anastomose, forming a network many times the diameter of the main body. Cytoplasmic flow in the pseudopods is bidirectional: mitochondria and other cytoplasmic inclusions flow simultaneously toward and away from the main cell body. Food (photosynthetic flagellates, diatoms, protozoans, tiny animals, particles of detritus) are caught by the pseudopods and moved into the central mass of cytoplasm. The pseudopods are also active in locomotion and in secreting the shell. The shell of this common type is made mostly of calcium carbonate with many chambers arranged in a spiral; pseudopods extend through the large opening and the tiny perforations (*foraminifera* = "pore-bearers"). In other types, the chambers are arranged differently and the shell may be made mostly of silica or of organic matter reinforced with sand grains. The largest living forams are 2 to 3 cm in diameter, but most are microscopic. Almost all are marine.

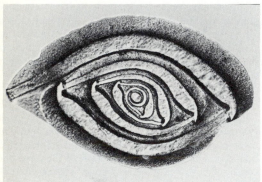

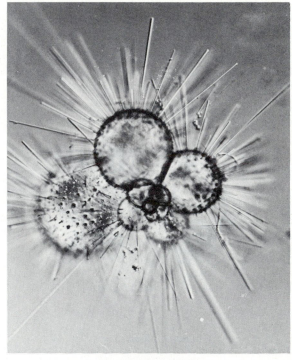

Chambers of increasing size are secreted as a foram grows; the organism continues to occupy all of the chambers. *Above:* The small, rounded original chamber is surrounded by larger, eye-shaped ones in the aptly named *Ophthalmium,* a benthic foraminiferan. (G. Uhlig) *Right:* Starting from a small rounded chamber in the center, a spiral series of larger, globular chambers is added; long spines project from the surface. (G. Uhlig). *Below right:* The largest, mostly recently formed chamber in this spiral test is conspicuous by its translucence. (R.B.)

The size of the original chamber and form of the shell are clues to the life-history stage in many forams which alternately undergo asexual replication and sexual reproduction. The *sexual stage* (gamont) is haploid and produces flagellated or ameboid gametes. The gametes (released into the water or brought together by union of two sexual individuals) fuse into zygotes. These develop into an *asexual stage* (agamont) that is diploid and often multinucleate; in some species, 2 kinds of nuclei differentiate, much as in ciliates (see below). The fully grown asexual stage may replicate, or the nuclei may undergo meiosis to produce haploid stages. This sort of life-history is unique among protists, but similar to that of many seaweeds and other plants.

Evidence of sexuality is lacking for many forams; others undergo a pseudosexual process (autogamy) in which pairs of gametes fuse within the shell of the same individual.

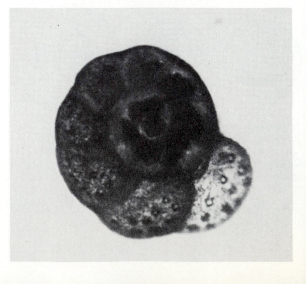

Treelike antarctic foraminiferan, its stem and branches up to nearly 2 cm high, grows rooted in soft bottoms. Buried in the sediment is a bulb containing the main cell body and single large nucleus, and from the base of the bulb a branched rootlike system extends downward. The whole is covered with an agglutinated test of sand grains and other foreign particles. Pseudopods extend from the upper branches into the water and capture suspended food organisms and organic particles, while the rooted portion takes up dissolved organic substances from the sediment. During a large part of the year when there is little suspended food in the water, these forams depend upon the uptake of dissolved nutrients, a mode of nutrition that probably supplements the diet of many forams. *Notodendrodes antarctikos.* McMurdo Sound, Antarctica. (T.E. DeLaca)

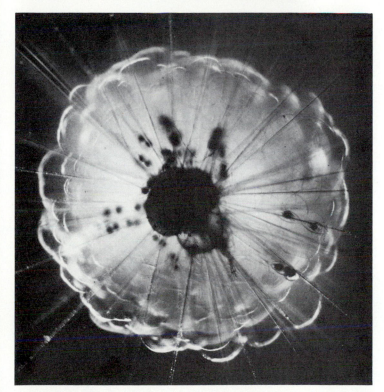

Planktonic foraminiferan, *Hastigerina pelagica.* The central test of globular chambers is surrounded by a flotation "bubble" (about 2 mm in diameter) of vacuolated cytoplasm. Spines, 7 mm long, support the pseudopods and aid in flotation. Fewer than 1% of foram species are planktonic, but these are often extremely abundant and important members of the plankton. Many planktonic forams, as well as some benthic ones, have photosynthetic symbionts (dinoflagellates, diatoms, chlorophytes, and others). The algal cells supply nutrients to the foram and may also be digested. Photosynthesis by the symbionts is correlated with increased calcification rates in the test, as found also in stony corals (chapter 6). On some coral reefs, forams contribute as much $CaCO_3$ to the reef as do corals and other animals and plants with calcareous skeletons. (A.W.H. Bé)

Fossil nummulitid forams, benthic types with calcareous tests that were sometimes 15 cm or more in diameter and 1.5 cm thick. They were so abundant in the early Tertiary that they formed thick deposits, now uplifted and exposed as great beds of limestone in Europe, Asia, and N. Africa. The Egyptian pyramids of Gizeh, near Cairo, are built of nummulitic limestone. Of over 40,000 described species of forams, about 90% are fossil; they are important aids to geologists in identifying and correlating rock layers. The calcareous tests of abundant planktonic forams such as *Globigerina* are today settling and accumulating over much of the ocean floor as thick deposits (called "globigerina ooze")—limestones of the distant future. (After Zittel)

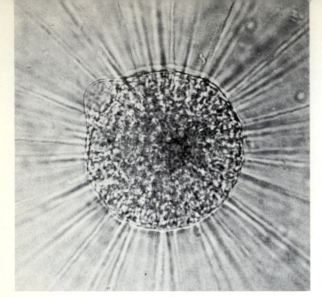

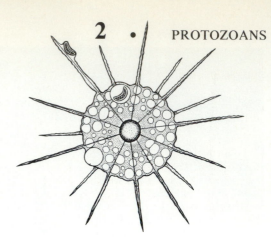

Heliozoan feeding. *Actinophrys sol.* Prey is caught on axopods, then enclosed in a food vacuole as the axopod is withdrawn to the central cytoplasm.

Heliozoan. The radiating pseudopods of these spectacular protozoans, reminiscent of the rays of the sun, give them their dramatic name, meaning "sun animals." The firm, straight pseudopods are of a type called *axopods,* supported by a central core (axoneme) of highly organized microtubules. Fluid-filled vacuoles, giving the cytoplasm a frothy appearance, make the heliozoan neutrally buoyant, so that it hangs suspended in the water near the bottom. It can also move slowly over a substrate with a rolling motion by shortening and lengthening its axopods one after another, a laborious process involving breakdown and reformation of the microtubules. Helio-zoans are all free-living, feeding on small organisms. In *A. sol* and some other heliozoans, several individuals sometimes cooperate to subdue and jointly digest a large prey within a common food vacuole, then separate again. Most heliozoans are freshwater, some are marine, and some live in moist soil. *Above,* the small heliozoan *Actinophrys sol* is common among aquatic plants in quiet freshwater ponds and lakes. Body, 50 μm in diameter. Lake Pymatuning, Pennsylvania. (R.B.)

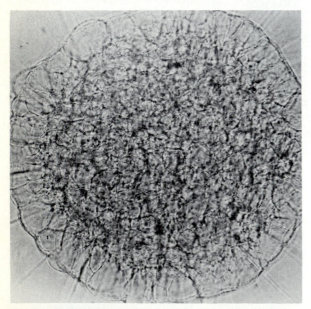

Multinucleate heliozoan, *Actinosphaerium,* reaches large sizes; the body of *A. eichhorni* can be 1,000 μm in diameter and has many nuclei in the periphery of the endoplasm; *A. arachnoideum* has fewer nuclei and is smaller. *Actinophrys sol* (above) is like most heliozoans in having a single nucleus, but lacks the zones of vacuolated ectoplasm and denser endoplasm in *Actinosphaerium* and typical of heliozoans. Lake Pymatuning, Pennsylvania. (R.B.)

Testate, stalked heliozoan. *Clathrulina,* with pseudopods extending through holes in the latticed organic test and with a hollow proteinaceous stalk. Repeated fission produces ame-boid or flagellated stages that squeeze out through the holes in the test. They wander off or settle nearby, attaching by a large pseudopod that secretes the stalk, then withdraws. Often the ame-boid stages attach to the parental test, and colonies of a dozen or so individuals may result. (Partly after J. Leidy)

The planktonic marine protozoans on this and the next 2 pages were named **radiolarians** from their spectacular arrays of radiating pseudopods and the radially symmetric skeletons in many. The Radiolaria is no longer formally recognized as a natural taxon by many specialists, but is split among the **acantharians, polycystines,** and **pheodarians**—distinctive groups that have much in common. All have a *central capsule* of cytoplasm, containing a nucleus or nuclei, and a region of *extracapsular cytoplasm*. These two main regions are usually separated by a *capsular membrane* (sometimes a substantial, organic wall) and are connected by cytoplasmic strands that penetrate the capsular boundary. The outer region is further organized into sub-zones and complex structures that aid in food capture and in regulating buoyancy. From the surface radiate *axopods,* supported by microtubular axonemes that originate within the central capsule. The elaborate siliceous skeletons of many polycystines are a major component of sea floor deposits and, like those of forams, serve geologists as distinctive markers in rock layers. Binary and multiple fission has been recorded in some forms; radiolarians may also release flagellated stages whose subsequent development is unknown. Radiolarians are the most intricately differentiated of the sarcodine protozoans.

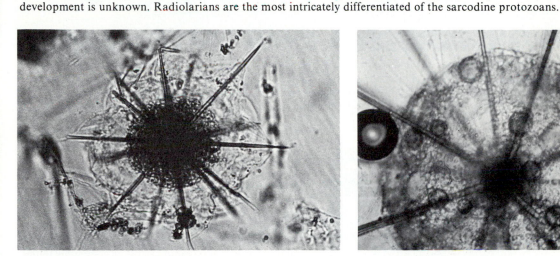

Acantharians differ from other radiolarians in having a skeleton of strontium sulfate typically consisting of 20 spines that radiate from the center of the cell in a geometrically regular arrangement. A thin capsular membrane present in most acantharians encloses many small nuclei in the central capsule. Contractile fibers allow an acantharian to expand or contract the extracapsular cytoplasm and control buoyancy. *Left,* acantharian from Bimini, Bahamas. (R.B.) *Right,* closeup of the central capsule of an acantharian containing photosynthetic symbionts, present in almost all species and probably an important source of nutrition. *Acanthometra.* (A. Michaels)

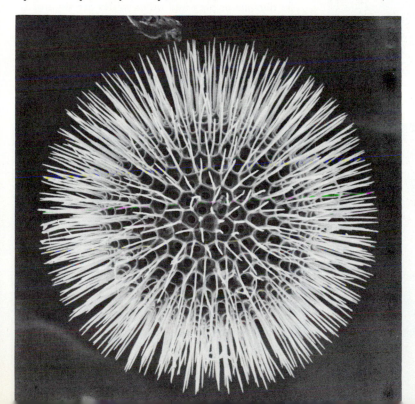

Spherical skeleton is typical of a spumellarid polycystine. The pseudopods radiate in all directions. In polycystines the central capsule is surrounded by an organic membrane, and the often elaborate skeleton is of opal (amorphous silica, hydrated SiO_2) plus small amounts of organic material. Different stages have a single, large, polyploid nucleus or many nuclei. Small flagellated forms are released, each carrying a large crystal of $SrSO_4$, which suggests a relationship to acantharians. Many polycystines have photosynthetic dinoflagellates as symbionts that live in the extracapsular cytoplasm. The symbionts release nutrients to the host and may be digested. In some forms, the symbionts are moved from daytime sites in the peripheral cytoplasm to a zone just outside the capsular membrane each night. About 200 μm in diameter. SEM (M. Gowing)

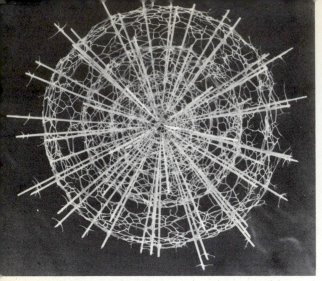
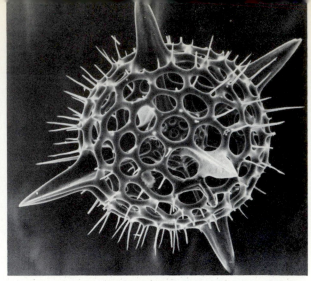

Concentric skeletal spheres, combined with radiating spines, are common patterns among spumellarid polycystines and range from complex delicate forms like the one *above left* (530 µm diam.) to more compact, robust kinds *above right* (180 µm diam.). The organisms that made these skeletons were captured in the North Pacific Central Gyre at 900 m and 1500 m, respectively. SEM (M. Gowing)

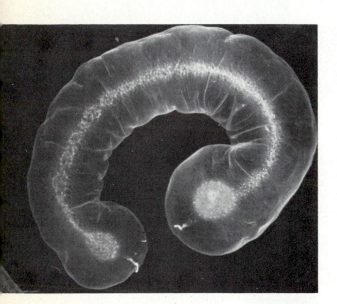

Radiolarian colonies seen here are spumellarid polycystines, with the hundreds to thousands of members appearing as bright bodies in a gelatinous matrix. A cytoplasmic network connects individual units. Pseudopods capture prey and provide surface for exchange of materials with the seawater. Photosynthesis by dinoflagellate symbionts is enhanced after prey capture; recycling nutrients between the radiolarian and its symbionts may be the clue to survival in the nutrient-poor waters where colonial radiolarians are commonly found. These are the largest living protozoans. **Sausage-shaped colony,** *Collozoum caudatum,* can be 2 m long, and a related species, *C. longiforme,* reaches 3 m. The central capsules are aligned in the axis of the colony, and cytoplasmic strands that run between them form a cablelike throughway for transport of shared nutrients. The remains of prey and other wastes are shunted to either end, where they are eliminated (stringy material seen emerging in photo). Equatorial Atlantic and Gulf Stream. (A.M. Brosius). **Pretzel-shaped colony,** *Solenosphaera collina,* is about 5 cm across. (Courtesy, N.R. Swanberg) **Spherical colony,** about 2 mm in diameter, was hand-collected in a jar by a SCUBA diver at about 50 m depth. Monterey Bay, California. (R.B.)

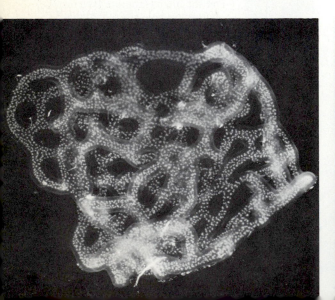
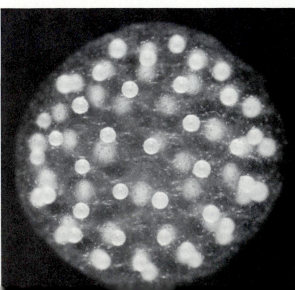

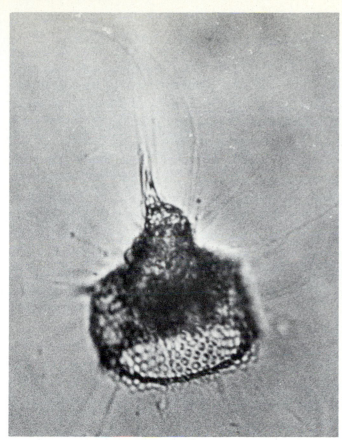
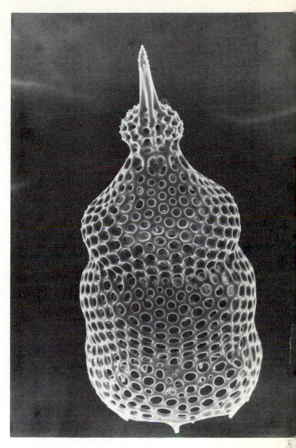

Radial symmetry of nassellarid polycystines contrasts with the spherical symmetry of most spumellarids. The axopods, instead of radiating equally in all directions, are clustered at one pole. Some species are further modified to biradial or bilateral symmetry. No colonial nassellarids are known. *Left,* live nassellarid. Bimini, Bahamas. (R.B.) *Right,* skeleton of a nassellarid, about 220 μm high. SEM (M. Gowing)

Pheodarians are named for the conspicuous deposit of dark-colored material (pheodium) that accumulates near the central capsule; of unknown function, it may be waste matter. In the photo of the preserved specimen *below left,* the pheodium is visible as a dark mass just below the central capsule. *Below right,* the skeleton of a pheodarian consists of silica combined with significant quantities of organic substances; SEM. Pheodarians generally live at considerable depths and lack photosynthetic symbionts. They have a single polyploid nucleus, and undergo both binary and multiple fission. Colonial pheodarians form complex aggregates joined by skeletal structures. (M. Gowing)

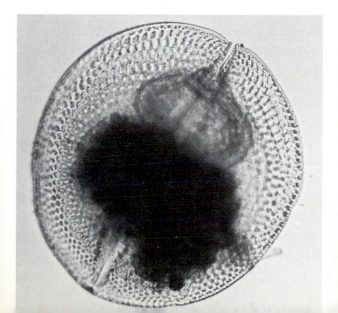
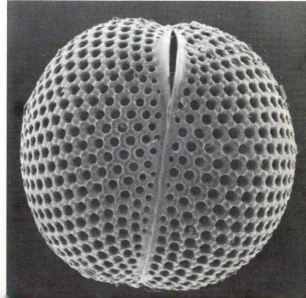

MYXOZOANS and APICOMPLEXANS

A MISCELLANY of parasitic groups were at one time assembled under the name **Sporozoa** on the basis that they form cysts, or "spores" (by analogy with fungi or plants), at certain stages of their life history—although not all of them meet even this criterion and in those that do, the "spores" are not directly comparable. Locomotion is likewise variable; there are some flagellated or ameboid stages, but sporozoans move mostly by body flexion, undulation of longitudinal ridges, or a sort of gliding. A more consistent feature found in sporozoans (but not unique to this group) is **multiple fission** (schizogony), in which a phase of cytoplasmic growth and nuclear division, resulting in a syncytium, is followed by cytoplasmic division into many discrete cells all at once. Multiple fission rapidly produces great numbers of individuals or gametes at various points in the life histories and presumably contributes to the abundance of these parasites. Protozoologists are still in the process of sorting out the sporozoans into a number of separate and, it is hoped, more natural groups. Examples of two of these, the myxozoans and apicomplexans, are described below.

OF ALL PROTOZOANS with multicellular stages, the **myxozoans** are perhaps the most highly differentiated, having several discrete cell types within a well-defined body, and protozoologists have more than once threatened to banish them to the animal kingdom. To retain myxozoans as protists, it is necessary to define animals not just in terms of being multicellular or having multiple differentiated cell types, but in terms of organizing those cell types into more than one kind of tissue and undergoing a certain sort of development.

The hosts of myxozoans are mostly freshwater fishes such as salmon and trout, many of them important to commercial or sport fisheries. In a common type of life history *(Myxobolus),* the principal growth stage, or **trophozoite,** is a syncytial ameboid form, which lives in various tissues of the host and may become quite massive, sometimes causing fatal disease, sometimes having little apparent effect. The syncytium undergoes multiple fission. Certain nuclei become surrounded by dense cytoplasm and discrete cell membranes. In this way, much of the syncytium breaks up into separate cells that give rise to cyst-forming stages. Still within the syncytium, these cells divide a fixed number of times, and the products form one or more cysts, called "spores," the stage responsible for dispersal and for infection of new hosts.

cellular envelope surrounds pair of developing spores

sporoplasm

polar capsule

wall

syncytial cytoplasm of trophozoite

Multicellular spores of a myxozoan, *Myxobolus.* Four different cell types differentiate as the spores develop within the syncytial cytoplasm of the trophozoite *(lower left).* Only the ameboid cell (sporoplasm) will emerge to propagate a new organism. The two nuclei of the ameboid cell fuse into one, but the evidence for sexual processes is still incomplete. Mature spores are freed *(upper right)* by the breakdown of host tissues and may remain alive for some time. The filaments of the polar capsules evert on contact with a new host. (Combined after various sources)

This part of the life history—the syncytial growth stage, and the way in which it produces dispersal stages— may be compared to that of orthonectid mesozoans (chapter 4). In orthonectids, however, sexual stages are involved, whereas no sexual processes have been definitively demonstrated in myxozoans.

The spores are small (typically about 10 to 15μm), but their cells are structurally complex and comprise 4 distinct, permanently differentiated cell types. A pair of cells form an envelope around the developing spores. In each spore another pair of cells produce and line the spore wall, which consists of two valves separated by an elaborate suture line. At one end of the spore are a pair of cells that secrete and hold the **polar capsules,** proteinaceous rounded or flask-shaped organelles each of which contains a spirally coiled, eversible hollow filament. The filament is everted upon contact with or ingestion by a fish, and apparently entangles or penetrates or adheres to host tissues, serving to anchor the myxozoan. Finally, there is a large ameboid cell (sporoplasm), the only part of the spore that will hatch out. The other cells of the spore, and probably the residual non-cellularized portion of the syncytium, will degenerate and die.

Anchored in the new host, the spore opens along the suture line between the valves, and the uninucleate ameboid cell emerges. This motile infective stage, or **sporozoite,** eventually gives rise to a multinucleate syncytium.

OTHER SPOROZOANS are no match for the myxozoans in terms of multicellular differentiation, but they display some wondrous life histories, typically including a sexual phase with highly dimorphic macrogametes and microgametes. The microgametes are not merely smaller but otherwise similar cells, as they are in *Volvox* and many other flagellates. In sporozoans the microgametes are the only flagellated stages in the life history, and they are highly differentiated sperms with from 1 to 3 flagella, a mitochondrion, and a nucleus. Although distinctive in structure and development, they are perhaps more like the sperms of animals than are those of any other protozoans.

Sporozoans are studied mostly because of the important diseases that they cause. The major group with some 4,000 described species (and thousands more not yet described) is the **apicomplexans,** named for the elaborate *apical complex,* a distinctive assemblage of microtubular arrays and filamentous or saclike organelles at one end of the cell. Most parts of the apical complex are present in the cells of at least some life history stages of all species. The function of all this apparatus, however, is simply unknown for the most part; possibly the various parts play a role in penetration of cells, in feeding, or in locomotion. Part of the apical complex has been shown to serve in attachment to the host in **gregarines,** which are

The size and morphology of the polar capsules and their manner of formation in polar cells are remarkably like the nematocysts of cnidarians (chapter 5). Some specialists insist the resemblance could not result from convergent evolution but must indicate a direct relationship, however difficult such might be to imagine. Encapsulated eversible filaments are found also in certain dinoflagellates and ciliates, but these do not appear quite so similar in their details to those of myxozoans.

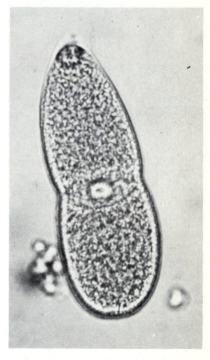

Gregarine. Several groups of gregarine parasites inhabit the gut, tissues, and body cavities of various invertebrates. The feeding stage (trophozoite) does not undergo schizogony and grows relatively large, to a length of 1 cm in some species. Two mature individuals (gamonts) pair and produce gametes by schizogony. As in all apicomplexans, the zygote divides by meiosis, and the rest of the life history is haploid. This gregarine was found in the digestive gland of an enteropneust (a hemichordate, chapter 28). The apical end, which was attached to a host cell, is at the top. Bermuda. (R.B.)

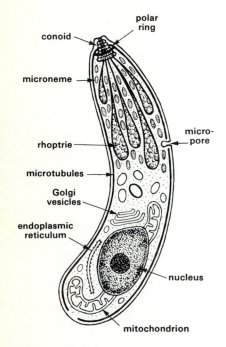

Coccidian, *Toxoplasma gondii,* with a typical *apical complex,* prominent in electron microscope sections but of uncertain function. The conoid, perhaps aided by secretions from the rhoptries, probably has a role in penetration of host cells. A series of microtubules beneath the pellicle extend backward from the polar ring nearly the length of the body and may serve in locomotion. The micropore is thought to be a site for ingestion of host cell contents. The whole body is minute, only about 6 μm long. (After various sources)

usually extracellular parasites of invertebrates; those specific to insect pests have been suggested as possible agents of biological control. The **coccidians** are typically intracellular parasites of vertebrates, sometimes with 2 different hosts in the life history of which one may be an invertebrate such as a biting fly that serves as a vector. The **piroplasms,** a small group closely related to coccidians, are also intracellular parasites of vertebrates, with ticks as vectors.

Many kinds of coccidians infect invertebrate and vertebrate animals, but few cause disease in humans. One of these is the coccidian *Toxoplasma gondii,* for which the final hosts are felines, including domestic housecats, but intermediate hosts are almost any bird or mammal, including humans. Infection rates of humans and their pet cats are high (20 to 80% in various human populations) but most cases are undetected and lead to permanent immunity. The disease, toxoplasmosis, is rare and seen most often in people with deficient immunity and in newborn children of women infected while pregnant. The fetus may suffer serious damage to the eyes and central nervous system; adults most often experience inflammation of the lymph nodes and a variety of other symptoms. Cats are infected by eating the raw flesh of an infected intermediate host. People also become infected by eating insufficiently cooked, infected meat; or by ingesting cysts from cat feces that have contaminated the hands or food; or by direct transmission from the mother across the placenta. Women should avoid handling cats and cat litter while pregnant, especially if they have not regularly been exposed to cats previously.

Pneumocystis, an organism that may be related to *Toxoplasma,* produces a form of pneumonia that is ordinarily rare but has become a major cause of death among people with acquired immune deficiency syndrome (AIDS). Similarly, a rare disease caused by the tick-borne piroplasm *Babesia* is seen mostly in people who have had the spleen removed. However, piroplasmosis caused by *Babesia* and *Theileria* is a widespread and serious problem in domestic cattle.

Hidden away among the many orders and dozens of families of these unfamiliar parasites is one large family of coccidians containing certain species whose members pose a threat to much of the human population. The potential victims are estimated at about a third to half of all people on Earth, living in a broad belt that extends around the world in tropical and subtropical zones and into some temperate regions. These protozoans are the agents of malaria, the **malarial parasites.** They are the cause of the single most important infectious disease of humankind, with roughly 150 million new cases per year, ending in 1.2 million deaths, in addition to further hundreds of millions of chronic or asymptomatic infections. The statistics, even higher in some times past, have been reduced only at the cost of vast amounts of money and effort that must continually be expended on treatment and on control, which is precarious at best. Malaria is currently on the rise.

Species of the large genus *Plasmodium* cause malaria not only in humans but in other primates, as well as in rodents, birds, and reptiles. The human malarial parasites *(P. vivax, P. malariae, P. ovale,* and the most deadly, *P. falciparum)* are transmitted only by certain mosquitoes (some 30 species of *Anopheles),* in which the sexual phase of the parasite's life history is completed. A person is infected when bitten by an infected mosquito, and the parasites invade host cells and feed on their contents. Development of a multinucleate syncytium (trophozoite, or plasmodium), followed by multiple fission (schizogony), occurs in repeated cycles, and enormous numbers of individual cells can be derived from each infective sporozoite. This tremendous multiplication within the vertebrate host provides a reasonable chance that the tiny sample of blood taken in the next random bite of a mosquito will contain representatives to complete the sexual phase of the life history. Differentiation of male and female gametes, fertilization, and development of the first syncytium take place in the mosquito host, followed once again by schizogony, which gets the new generation off to a good start.

Large-scale coordinated programs of mosquito control and treatment with drugs followed World War II and at first made spectacular reductions in rates of malarial infection. But ambitious plans for the eradication of the disease were abandoned when insecticide-resistant strains of mosquitos and drug-resistant strains of *Plasmodium* began to emerge. In recent decades, the job has been left to local governments, many of which have been unable to cope with the staggering combination of biological, technical, social, political, and economic challenges involved.

Hopes for control of the disease are now pinned on the development of a vaccine. These have been bolstered by progress in laboratory culture of *Plasmodium* and in genetic engineering techniques by which large quantities of specific antigenic proteins can be produced. However, the same problems that have always to some extent frustrated the development of natural immunity to these highly specialized parasites now face the developers of any vaccine. The parasite sequesters itself within cells, changes its antigenic properties at each stage of the life history, and occurs in many immunologically distinct strains. Even supposing that an effective vaccine can be developed, there remain all the problems of funding and delivering it in the poor and scattered communities of the huge malarial belt.

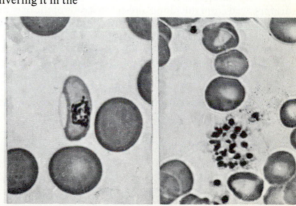

Malarial parasites. *Left,* female gametocyte of *Plasmodium falciparum* in a red blood cell; if ingested by a mosquito, the gametocyte will develop into a single egg. *Right,* merozoites of *P. vivax* being liberated by rupture of a red blood cell. Stained preparations. (U.S. Army Medical Museum)

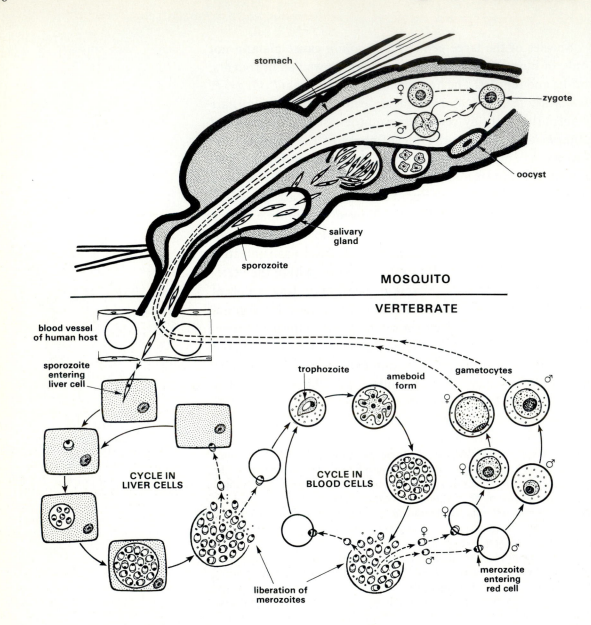

Life history of a malarial parasite, *Plasmodium vivax.* Fertilization takes place in the midgut of a female mosquito. The motile **zygote** (ookinete) burrows into the midgut wall and forms a cyst, the **oocyst,** which undergoes meiosis and multiple fission (schizogony) into large numbers of slender **sporozoites.** These are released into the body cavity of the mosquito and invade tissues, including the salivary glands. When the mosquito bites a human, saliva and sporozoites are injected into the blood stream of the vertebrate host. Within less than an hour, the sporozoites have disappeared from the blood and entered cells of the liver and other tissues. There they grow into syncytia and undergo schizogony. Great numbers of **merozoites** are liberated, some of which may enter new liver cells and repeat the cycle. Others initiate cycles of replication in red blood cells, with formation of a plasmodial **trophozoite.** These cycles of syncytial growth, schizogony, release into the plasma, and reinvasion of red blood cells are what produce the repeated, periodic attacks of chills and fever that characterize malaria. After one or more such cycles, some of the merozoites enter red blood cells and grow into male or female **gametocytes,** a stage that develops no further until ingested by a suitable mosquito taking a blood meal from a human host at this stage. Differentiation of the female gametocyte into an egg, and of the male gametocyte into 8 slender flagellated sperms, takes place in the midgut of the mosquito.

CILIATES

OF ALL THE PROTOZOANS, ciliates are most deceptively like active little animals. Watching the behavior of a ciliate busily exploring a microscopic landscape, it is difficult to keep in mind that the body is a single cell, unlike the bodies of those animals in the same habitat that are of much the same size and behave in much the same way but consist of many-celled organ-systems. Indeed, ciliates reach lengths of 4.5 mm, much larger than many small animals, and their cellular architecture is more complex than that of any animal cell.

Ciliates convenient to use as an example, because they have been much studied and display many typical features, are members of the freshwater genus *Paramecium*. In a drop of pond water under a microscope, parameciums may be seen rapidly swimming, each revolving on its long axis and advancing along a spiral path. Meeting some obstacle, a paramecium will back up, turn, and set off in a new direction. Or, encountering a bit of debris rich in the bacteria which are its main food, it will pause to take in a meal. A swimming paramecium is propelled by its **cilia,** thousands of them arranged in diagonal rows and uniformly covering the body.

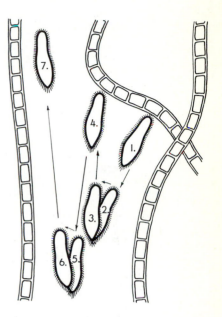

Avoidance reaction of a paramecium. On encountering an obstacle or a chemically noxious patch of water, the paramecium briefly reverses the beat of its cilia, which causes it to swim backward. Then it turns and proceeds forward again on a different course. It will repeat this behavior until it finds a free path. (Based on H.S. Jennings)

The direction of swimming depends upon the concentration of calcium ions in the cortical cytoplasm. Stimulation of the anterior cilia (as by an obstacle) causes the cell membrane to open channels through which calcium ions flow inward, resulting in ciliary reversal. Occasionally one discovers a mutant paramecium that cannot perform the avoidance reaction. Such *pawn* mutants that can move only forward (like chess pawns) have provided important clues to the mechanisms of the behavior. Their cilia are still able to reverse, for the injection of calcium ions into such mutant cells triggers normal backward swimming. Hence the defect in *pawn* mutants seems to lie in failure of the membrane to open its calcium channels. In a medium that lacks calcium, even normal parameciums cannot swim backward.

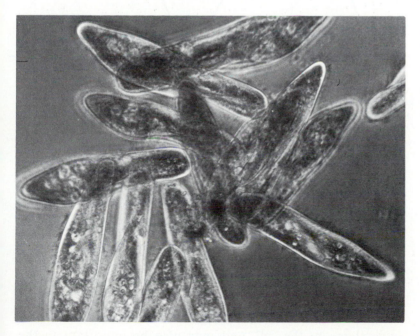

Parameciums move so rapidly that it is almost impossible for an observer to discern structural details without adding a viscous agent such as methyl cellulose to the water in order to slow the movement. However, because the viscous properties of the medium are so important to structures as small as cilia, the motion of the cilia is significantly modified by such a change in viscosity, as compared with that recorded by high-speed cinematography or stroboscopic analysis in a normal medium. (R.B.)

The pellicle and associated structures beneath the outer membrane together form the **cortex,** which is comparable to the ectoplasm of an ameba or the pellicle of a euglenid but is of an elaborate nature unique to ciliates. The outer region of a paramecium or other ciliate is more rigid than that of a euglenid but retains some flexibility. The most striking feature of the cortex is the exquisite geometry of its parts. The cilia are arrayed in precise rows and accompanied by a variety of other organelles with equally geometric placement. The exact spacing of the cilia (which must continually be regulated as the paramecium grows and divides) is crucial, because the elegant **coordination** of the thousands of beating cilia depends on mechanical, hydrodynamic interactions between them. Each cilium initiates a beat in response to that of its neighbor, so that *metachronal waves* of ciliary motion pass over the body surface. The speed and direction of the beat are controlled by changes in membrane potential, which in turn depend on the concentration and distribution of calcium and potassium ions in the underlying cytoplasm—a mechanism analogous to that in the nerve cells of animals.

Metachronal waves of ciliary beating in a row of cilia. These may be observed by focussing on one edge of the body of a paramecium or other ciliate, especially if the beat has been slowed by the addition of a viscous agent to the water.

A cilium has an asymmetrical beat, with an *effective stroke* in one direction, during which the relatively straight shaft moves swiftly in a large arc, and a *recovery stroke* in the other direction, as the cilium bends flexibly and swings close to the surface, returning more slowly to the starting position. Slowly is a relative term, for a cilium beats on the order of 20 times per second. Adjacent cilia beat just slightly out of phase with each other. (Based on M.A. Sleigh)

A meticulously regular network of microtubules and other fibrillar components runs between the ciliary bases, including overlapping fibers that parallel each row as well as tracts that cross-connect the rows. This elaborate *infraciliature* is of uncertain function. Although the network has the appearance of a coordinating system, the available experimental evidence is against such a role. It probably serves to anchor the cilia and give shape to the cell, and may also be important in the development of new cilia.

Distributed among the cilia, always in a precisely regular and species-specific fashion, are a variety of organelles, often including large mitochondria and bodies called **extrusomes,** which extrude mucus, toxic secretions, or other materials. In a paramecium, the extrusomes are *trichocysts,* each able to produce a long sticky proteinaceous thread when discharged. Parameciums fire their trichocysts under various conditions of stress, which the threads do nothing noticeable to relieve, and when attacked by predators, which are undeterred. It is thought that trichocysts may serve to anchor parameciums during feeding.

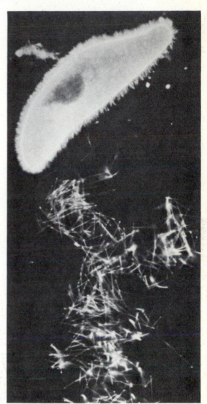

Trichocysts discharged by a paramecium after a drop of ink was added to the water. (P.S. Tice)

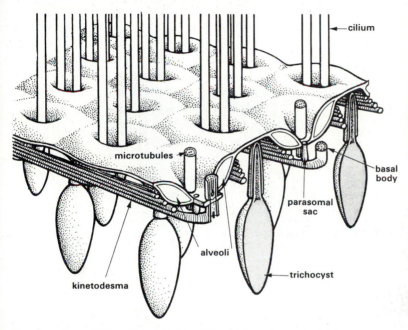

Cortex of a ciliate, *Paramecium.* The surface is marked by a regular hexagonal pattern of ridges and depressions, and from each depression project one or more *cilia,* the number depending on the species and the region of the body. Besides the outer cell membrane, two additional membranes envelope between them a system of fluid-filled pockets, the *alveoli.* Two of these pockets surround the base of each cilium and provide support. Each cilium sprouts from a *basal body* in the inner portion of the cortex, and a network of *microtubules* and other fibrous components runs among the basal bodies. Other organelles include *parasomal sacs,* which are sites of pinocytosis; *trichocysts* or other extrusomes; and often large mitochondria (but in *Paramecium* the mitochondria are not in the cortex).

Each region including a cilium with its basal body *(kinetosome),* alveoli, parasomal sac, *kinetodesma* (striated longitudinal fiber), and other associated structures is called a *kinetid.* A longitudinal row of kinetids is a *kinety,* and all the kineties taken together constitute the *kinetome* of the ciliate. (Modified after Ehret and Powers, and Jurand and Selman)

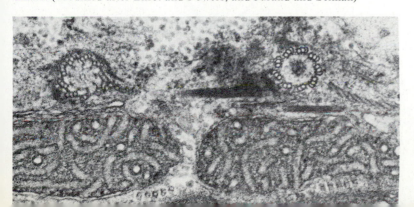

Section through cortex of a ciliate, *Tetrahymena pyriformis,* cuts across large mitochondria and reveals microtubular triplets in the basal bodies of cilia. Transmission electron micrograph (TEM). (J. Olmsted)

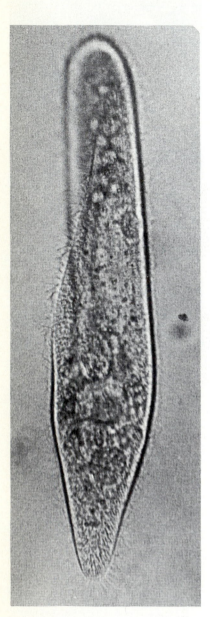

Paramecium caudatum has a slipper-shaped body and a "tail" of longer cilia at the rear end. (R.B.)

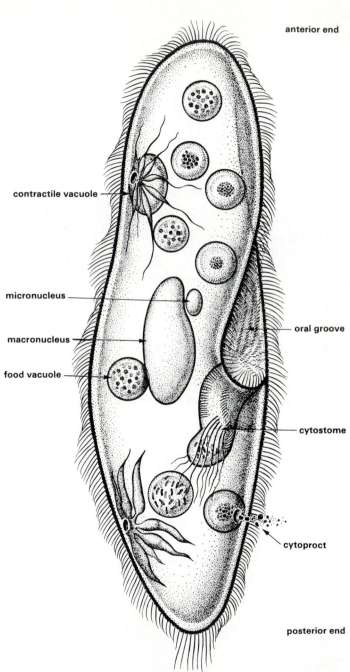

anterior end

contractile vacuole

micronucleus

macronucleus

food vacuole

oral groove

cytostome

cytoproct

posterior end

Diagram of a paramecium, *Paramecium caudatum.*

Feeding by a paramecium depends as much on cilia as does locomotion. The cell mouth, or cytostome, lies at the base of an elongate depression, the *oral groove,* on one side of the body (designated as "ventral"). The cilia of the oral groove create water currents that drive bacteria and other food particles deep into the funnel-shaped end of the groove, where specialized tracts of cilia swirl them around, concentrating them into a mass that is taken in through the mouth. This is not an opening but rather a portion of the surface specialized for phagocytosis. Just inside the mouth is a cytoplasmic region called the *cytopharynx,* reinforced with bundles of microtubules. Here the food mass is incorporated into a membrane-bounded *food vacuole* that is carried off by cytoplasmic streaming on a more or less definite course through the body, as digestion proceeds within it. Any undigested remains are eliminated through the *cytoproct* ("cell anus"), at a permanent site on the surface of the body. A well-fed paramecium undergoes regular cycles of growth and **fission.** In ciliates, the body divides *across* the long axis into anterior and posterior halves. Each half must then regenerate lost parts and reproportion itself before the normal structure and shape are again achieved.

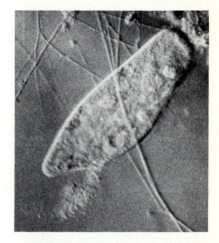

Solid wastes are ejected through the anal pore of a paramecium. (Photomicrograph from a motion picture. P.S. Tice)

Excess water is expelled by **contractile vacuoles,** like those of euglenids and amebas but (as with most ciliate structures) more elaborate. Whereas the contractile vacuoles of amebas (and perhaps those of euglenids) seem to be repeatedly formed anew by the coalescence of smaller vacuoles, contractile vacuoles in parameciums are part of a permanent system of organelles. Parameciums have 2 contractile vacuoles, one near each end of the body. The vacuole is filled from radiating canals, which appear to be filled in turn from a system of fine branching tubules that extend for some distance into the surrounding cytoplasm. Special mechanisms provide for contractility and prevent backflow. Each contractile vacuole discharges to the outside through a pore at a permanent, specialized site.

A diagnostic character of all ciliates is the presence of *2 different kinds of nuclei,* both of which usually divide at the time of cell fission. The one or more small nuclei, or **micronuclei,** are typically diploid (2n)—like those of animal cells (except gametes) but unlike those of protists such as sporozoans and many flagellates in which the only diploid stage is the zygote. At each cell fission, chromosomes in the micronucleus replicate, and it divides by mitosis. But it appears to have little or no role in the on-going metabolism of

Stephanopogon formerly was listed as a primitive and exceptional ciliate with only one kind of nucleus, from 2 to 16 in number. This benthic marine protozoan looks like a ciliate, but recent studies have shown that it lacks typical ciliate infraciliature and pellicle structure and that it divides like a flagellate. It has been removed from the ciliates and placed among the flagellates as a separate flagellate order or phylum, Pseudociliata.

the cell, and it lacks a nucleolus, which is the site of synthesis and storage of ribonucleic acid (RNA), the substance responsible for delivering metabolic instructions from nucleus to cytoplasm. Parameciums and other ciliates without a micronucleus may continue to swim and feed, and may give rise to a strain of individuals that undergo cycles of growth and fission in an apparently normal way.

One or more large nuclei, or **macronuclei,** contain multiple copies of the genome. The genetic material is not organized into typical chromosomes but dispersed in small bodies, and division is not by mitosis; following replication of the genetic material, the nucleus appears simply to pinch in two. In *Paramecium* the polyploidy of the macronucleus has been estimated at about 800n; in other ciliates the range is from 2n to nearly 13,000n. The amount roughly corresponds to the size of the cell. This and other evidence indicates that the macronucleus is responsible for regulating metabolism in the cytoplasm, and that its polyploidy is related to the large amount of cytoplasmic mass. The macronucleus has many nucleoli and actively carries on synthesis of RNA that has been shown to be active in the cytoplasm. If the macronucleus is lost or experimentally destroyed, the ciliate dies.

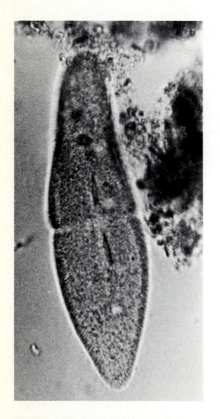

Fission. Both nuclei have divided, and the two new products of the macronucleus are visible as dark elongate structures, one on either side of the plane of fission, where the body has begun to constrict. (R.B.)

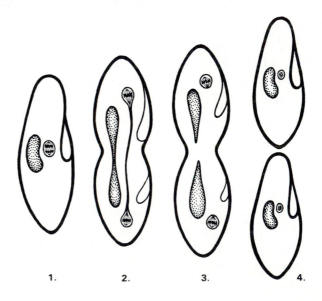

1.   2.   3.   4.

Fission of a paramecium is transverse and both kinds of nuclei divide, as in most other ciliates. Fission also involves regeneration of many cilia, the feeding apparatus, contractile vacuoles, and other structures.

The function of the micronucleus becomes evident only when two parameciums engage in the sexual process unique to ciliates, called **conjugation.** The two partners, or *conjugants,* attach by their ventral surfaces, and the micronucleus of each undergoes meiosis. After additional fissions and degenerations of nuclei, each conjugant is left with 2 haploid nuclei, one of which is kept and the other transferred to the partner. The nucleus that was retained now fuses with the newly received one, diploidy is restored, and the partners separate. Conjugation differs from other sexual processes in that there are no separate gametes to fuse into new cells, no zygotes. In a ciliate the new genetic combination in the *zygotic nucleus* finds itself in the already fully-differentiated body constructed under the direction of the former genome. The old genome exists no more, as the macronucleus disintegrates in the course of conjugation and must be reconstituted from a division product of the new micronucleus. Also, reproduction (increase in the number of individuals) is not an inherent part of the process, although conjugation is almost always followed promptly by one or more fissions.

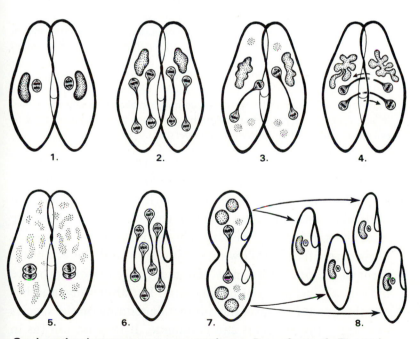

Conjugation between two parameciums, *P. caudatum.* **1.** The conjugants join. **2.** The macronuclei begin to break down; the micronuclei undergo meiosis. **3.** In each cell, 3 of the resulting haploid nuclei break down, and the 4th divides again. **4.** The conjugants exchange nuclei. **5.** The old and new haploid nuclei fuse into one. **6.** The conjugants separate and, in each, the new diploid nucleus divides 3 times. **7, 8.** Of the 8 nuclei, 3 break down, 4 become macronuclei, and the last remains a micronucleus. It divides twice more as the ex-conjugant fissions into 4 small individuals, co-founders of a new clone.

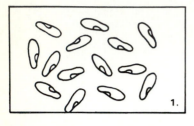

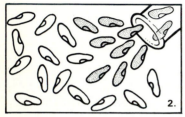

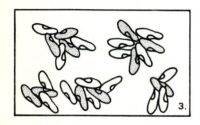

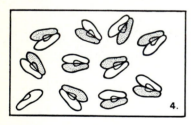

Mixing of different mating types.
To a laboratory culture of one strain
of *Paramecium* (**1**) is added another
of a complementary mating type (**2**).
Within minutes, the parameciums
have adhered into clumps (**3**). After
some hours all have segregated out
into conjugating pairs, each with both
types (**4**). This phase lasts for about 12
to 24 hours, depending on the species,
temperature, and other factors. In the
area of attachment, the conjugants
become precisely aligned, such that
the corresponding cortical structures
on each side come into intimate contact. The cilia are shed, and cytoplasmic fusion occurs between the
conjugants. (Based on a film by
R. Wichterman)

Conjugation also differs from sex in most organisms in that
individual parameciums are not differentiated as separate
males and females, nor are they hermaphrodites (male and
female combined). However, conjugants must be of *opposite
mating types,* and within a species there may be numerous
mating types, only some of which are compatible to conjugate.
The existence of mating types presumably increases the
probability of outbreeding. Groups (pairs or multiples) of
compatible mating types are termed *syngens,* and they constitute populations of interbreeding individuals that seem to fit
the standard species definition better than do the recognized
morphological species of parameciums.

In some other ciliates, such as sessile peritrichs and suctorians, conjugation regularly occurs between partners of
different sizes, and the smaller (microconjugant) is completely
incorporated into the larger (macroconjugant).

This sort of lopsided conjugation, although it occurs in a few free-
swimming ciliates, is most common and most easily understood in forms that
are sessile, that is, live permanently attached to some substrate. Sessile
ciliates such as vorticellids cannot seek out food or mates as free-ranging
kinds do, but create ciliary water-currents that bring food to the mouth,
much like many filter-feeding animals. An individual of *Vorticella* prepares
to conjugate by undergoing an unequal division that produces two
individuals of different size. The larger individual remains on the stalk; the
smaller one swims off. When the small conjugant locates a large one in
another clone, the bodies of the two fuse together, and fusion of their haploid
micronuclei follows.

In genetic terms, it does not really matter very much whether 1 or 2
conjugants remain at the end of conjugation. In both types of conjugation,
the original genomes of the partners recombine to become a single new one.
Where 2 partners separate, as in *Paramecium,* they do so as identical twins
(although cytoplasmic differences may persist).

Parameciums do not seem to have the same sort of
potential for immortality that amebas do. Along with the
specialization of the body, and perhaps of the nuclei in
particular, has come the development of a finite clonal life
span and a definite life history. The life history may be said to
begin immediately following conjugation and the establishment of a clone with a new genome. Early fissions produce
individuals incapable of conjugation. But after some number
of subsequent fissions—the number depending on the species
or strain—the products can conjugate. If they do not (as in
monoclonal laboratory cultures), but instead continue to
multiply by fission alone, the lineage begins to decline.
Senescence is marked by many signs, for example, fewer food
vacuoles, decreased rate of synthesis of many metabolic
substances, lowered capacity to repair damage, reduced ability
to conjugate (accompanied by lesser viability among the
offspring of conjugation), and eventually death.

The cytoplasm of juvenile individuals, incapable of conjugation, contains an extractable protein named "immaturin" not present in the sexually-competent cells that result from later fissions. However, if immaturin is experimentally injected into mature individuals, they no longer conjugate. How immaturin acts is unknown; perhaps it inhibits the manufacture of surface substances important in mating-type recognition or conjugal attachment.

Some ciliates appear not to undergo senescence or do so only after extremely long intervals. Rapid senescence may be a defect that persists because those ciliates that are saddled with it have evolved positive features that more than compensate. Or it may represent instead an adaptive outcome under the selective pressures of particular conditions that demand frequent sexual mixing.

In some respects, a single paramecium is comparable to an entire multicellular organism. It is many times larger than a typical animal cell and larger than many small animals. And it carries out all the behavioral and physiological activities necessary to individual survival. Moreover, the differentiation of the nuclei of parameciums and other ciliates into macro-nuclei and micronuclei corresponds, in animals, to the differentiation of the majority of *somatic cells* that make up the body and the few *germinal cells* that give rise to the gametes. In some animals in which this differentiation happens early and irreversibly, only the germinal cells retain the full genetic complement; the nuclei of somatic cells are found to be missing some portions, presumably those not relevant to their particular specialization (see nematodes, chapter 12). Similarly, in parameciums, the macronucleus has been shown to lack certain genetic sequences present in the micronucleus.

On the other hand, in terms of life-history and evolution, an entire clone of parameciums is more comparable to a multicellular animal. In both, development begins with the creation of a new genome and includes phases of *immaturity* marked by growth through active cell division, of *maturity* with sexual competence for the meiotic production of genetically different offspring, and finally of *senescence.*

BESIDES THESE TWO WAYS in which ciliates generally share common features with multicellular organisms, some ciliates develop in **colonial and aggregated forms.** In astomous ciliates (parasitic forms that lack a mouth) the products of fission do not separate but form chains of individuals. There are also a number of sessile colonial ciliates such as *Ophrydium,* a freshwater ciliate that contains photo-synthetic symbionts. Colonies of these ciliates form huge mounds up to 15 cm in diameter, containing tens of thousands of cells embedded in a gelatinous matrix and colored bright green by their algal symbionts.

Colonial ciliate, the marine vorticellid *Zoothamnium alternans,* grows
and branches in a closely regulated
manner to form an integrated multicellular structure. Conspicuous in this
colony near the branch bases are large
cells that will detach and swim off to
found new colonies. Bimini, Bahamas.
(R.B.)

Colonial vorticellids such as *Zoothamnium* grow permanently attached to the substrate by branching contractile
stalks. Through the stalks runs a strand of cytoplasm that
connects and coordinates the colony members, all of which
contract together at any disturbance. The colony grows in a
closely regulated manner, with a central stem from which
sprout side branches in a single plane. Individuals within the
colony are differentiated by size and function, and only those
cells in certain positions divide and conjugate. The founder of
the colony, which remains at the top of the central stem, is a
relatively large cell; it divides to initiate new branches and acts
as a macroconjugant. Smaller cells at the tips of branches
divide to produce new members along the branches and
eventually develop into microconjugants, which swim off to
mate with cells in other colonies. Large cells near the branch
bases are the only cells that can depart to found new colonies,
with or without conjugating. And finally, the many small
non-dividing cells along the branches seem to serve primarily
as "understudies" for the other types. If one of the dividing,
conjugating types is removed, an adjacent cell takes its place.
Evidently, there is a sort of hierarchy by which certain cells
inhibit the development and activities of others. Integration of
the colony seems to take place through the connecting
cytoplasmic strand, for if it is broken, the parts of the colony
on opposite sides of the break develop independently.

In addition to such colonial sorts that remain together after
fission, ciliates of the remarkable genus *Sorogena* aggregate
and form structures that look strikingly like the cyst-bearing
sorocarps of *Dictyostelium* and of other aggregating amebas.
Cysts of *Sorogena* have been found on terrestrial vegetation in
scattered parts of the world. Upon wetting, each cyst hatches
into a single ciliate, which preys voraciously on *Colpoda,*
another ciliate that encysts readily and is commonly found in
terrestrial situations. An individual of *Sorogena* may contain
as many as 5 large prey in various states of digestion. When
food is abundant, the ciliates grow and divide. When food runs
short, they aggregate on a substrate just beneath the water
surface and secrete a non-cellular stalk of protein and polysaccharide. The stalk material absorbs water, elongates, and
raises the clump of ciliates above the water surface. They
encyst, the cysts are eventually dispersed, and with luck some
reach more favorable sites.

1. Feeding stage of *Sorogena* with a rotund shape and conspicuous food vacuoles. (All photos on this page, R.L. Blanton)

2. Early aggregate. The ciliates normally aggregate during early morning hours. Aggregation requires a minimal density of cells and alternating periods of light and dark.

3. Hemispherical mound rises above the water level of the culture dish. Within the mound, up to several hundred aggregated cells mill about, coating themselves with a mucoid matrix that helps hold them together. They gradually move inward and upward.

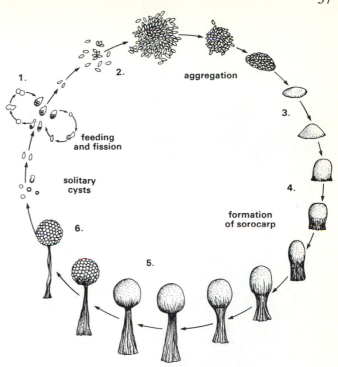

Asexual cycle of aggregating ciliate, *Sorogena stoianovitchae.* (From Blanton and Olive)

4. The cells form a rising column, continuing to secrete mucoid material that becomes a stalk, its outer portion hardening into a sheath. The mucoid matrix absorbs water and expands within the sheath, forcing the cells upward.

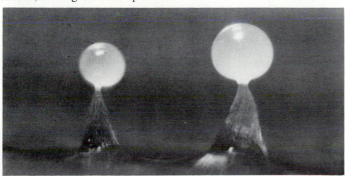

5. Developing stalk slowly dries in the air and contracts in folds around the enclosed matrix, further elevating the round mass of cells to a final height sometimes over 1 mm. Stalk development takes about 30 to 45 minutes.

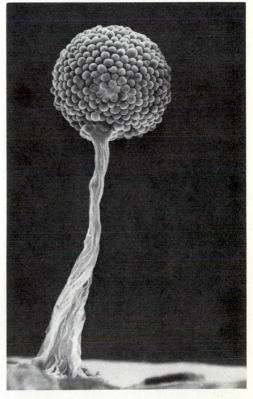

6. Mature sorocarp with furrowed, tapering stalk supports a rounded mass of encysted ciliates above the water surface, where they can be picked up and dispersed by air currents. Kept dry, they remain viable for 20 months or more. Scanning electron micrograph (SEM).

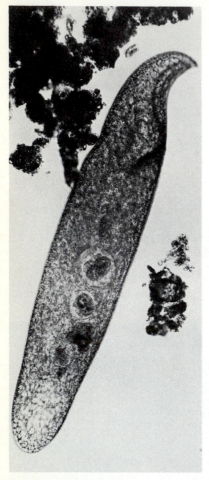

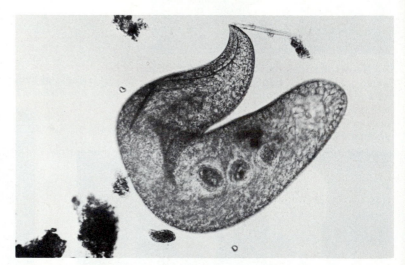

CILIATES are named for, and mostly readily recognized by, their many cilia, which may carpet the body uniformly or be bundled into tufts (cirri) or aligned in sheets (membranelles). Though most ciliates depend on their cilia for locomotion, sensory input, and food gathering, some types have few cilia or have cilia only at one stage of the life history. The defining character of ciliates is the presence of 2 kinds of nuclei in each cell.

The illustrations of ciliates that follow are arranged in sequence to correspond to the classification at the end of the chapter. Here, we have mentioned only a few of the most distinctive groups, because the characters separating many of the taxa are highly technical, and because ciliate classification is currently in such rapid flux that even the major ciliate subdivisions are not agreed upon by specialists.

Loxodes belongs to a group of distinctive ciliates (**karyorelictids**—"nuclear relicts") considered primitive because both the micronuclei and macronuclei are diploid and the latter never divide but are merely distributed between the two cells at each fission; the micronuclei undergo extra divisions and some of the products become new macronuclei. Found among sand grains, these ciliates are often large, flattened, and highly flexible and and contractile. Fragile and unable to form cysts, their means of dispersal is not obvious, yet many are cosmopolitan. Most karyorelictids are marine; *Loxodes* is exceptional in being a freshwater form. The mouth is located just behind the beaklike anterior region. (R.B.)

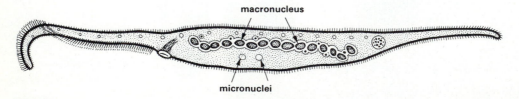

macronucleus

micronuclei

Dileptus has a long proboscis equipped with toxicysts, extrusomes that are both paralytic and proteolytic; discharged toxicysts penetrate, immobilize, and begin to cytolyze the prey. These large freshwater ciliates are omnivores, but feed mostly on other ciliates, flagellates, and amebas. They also attack small animals such as rotifers, and large numbers of the ciliates can together kill and devour bigger animals, even pond snails. The mouth at the base of the proboscis is highly distensible. *Dileptus monilatus* has a conspicuous beaded macronucleus and many micronuclei and contractile vacuoles. (Modified after Jones and Beers)

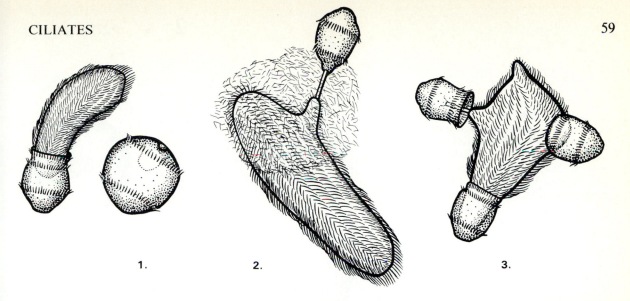

1.   2.   3.

A rapacious carnivore, *Didinium nasutum* attacks its preferred prey, *Paramecium,* with 3 kinds of extrusomes. This specialized predator "hunts" by random swimming, and when it happens to contact a paramecium, quickly attaches, discharges toxicysts, and proceeds to engulf it, even if smaller than its prey, as both mouth and body are highly distensible (**1**). Trichocysts discharged by the paramecium fail to protect it (**2**). Several individuals may feed on a large paramecium (**3**). The rate of prey capture depends on the relative densities of the predator and prey, and on temperature, which affects the rates of swimming (and hence, of contact), digestion, and regeneration of the feeding structures. At 20° C, if the density of prey is sufficiently high that no time is lost in hunting, *Didinium* may catch and eat a paramecium as often as every 2 hours, and fission after every third meal.

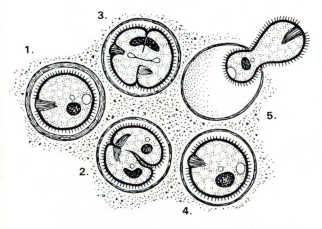

Cysts serve many ciliates as protection against adverse conditions, especially drought and lack of food. *Prorodon griseus,* found in small temporary pools, spends much of its life history encysted. When a rain has filled its pool, the ciliate swims freely and feeds. Between rains, its habitat dries completely. The ciliate assumes a spherical shape and rotates rapidly while secreting the cyst wall, the secretion being evenly spread by the beating cilia. Cysts are usually found in clumps on the bottom. The clump shown here illustrates: **1.** A thick-walled cyst that may survive long periods of drought. **2.** Two conjugants that paired off, sank to the bottom, and are completing conjugation within a cyst. **3.** Fission within a cyst. **4.** A thin-walled cyst from which the ciliate will escape after a brief period. **5.** Escape from a thin-walled cyst. (Modified after G.W. Tannreuther)

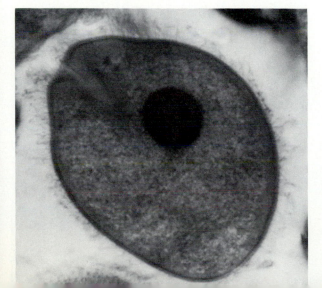

Balantidium coli, the only ciliate parasitic in humans, occurs also in some other mammals, especially other primates and pigs. It lives in the host's large intestine. In pigs *B. coli* is an extremely common and apparently harmless commensal. Cysts that pass out with the feces are the infective stage, and pig feces are the usual sources of human infections. In humans, this ciliate causes diarrhea and sometimes bloody dysentery, but is relatively rare. It is notably variable in size (30 to 300 μm long) and also in shape. Like many of its hosts, it is slender when deprived of starch, but grows rotund on a starchy diet, as in the plump individual seen here. Stained specimen. (Army Medical Museum)

Colpoda can form a thick-walled cyst capable of surviving prolonged drought. Dry cysts kept in the laboratory under vacuum have hatched out when wetted after 7 years and might well have survived longer. *Colpoda* often lives in freshwater or temporarily wetted terrestrial habitats, is readily and widely dispersed, and is a common contaminant in laboratory cultures of other protozoans.

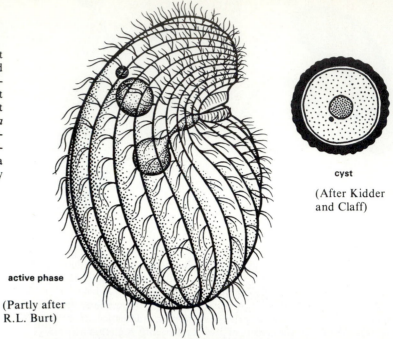

cyst

(After Kidder and Claff)

active phase

(Partly after R.L. Burt)

Suctorian lacks cilia in the mature stage but retains typical ciliate infraciliature and has a micronucleus and macronucleus. Attached by a stalk to some substrate (commonly an aquatic animal whose movements bring food-bearing water currents to the vicinity), it extends long knobbed feeding tentacles, with which it catches prey, chiefly other ciliates. The tentacles are supported by bundles of microtubules, somewhat like the axopods of heliozoans and radiolarians. Suctorians are solitary and colonial, freshwater and marine. *Acineta.* (R.B.)

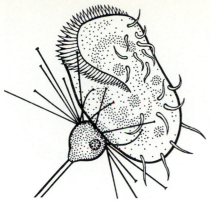

Suctorian with prey. This small suctorian, *Tokophrya lemnarum,* has captured a much larger ciliate *(Euplotes patella)* that blundered into the tentacles. Extrusomes (haptocysts) in the knobbed tips of the tentacles attach to the prey, paralyze it, and dissolve the prey pellicle. The tentacle knobs penetrate into the prey cytoplasm, and the tip of each knob invaginates to form a slim channel that extends down the center of the tentacle, surrounded by a cylinder of microtubules. Prey cytoplasm is drawn down the channel and incorporated into food vacuoles at the bottom. One imagines the predator sucking out the prey contents as if through so many drinking-straws, but the phagocytosis is probably mediated by the tentacular microtubules, not by suction. (Modified after A.E. Noble)

Suctorian budding. Suctorians undergo a sort of unequal fission, producing ciliated buds either from the external surface or in an internal pouch. In *Tokophrya lemnarum* (shown here) the bud escapes from an internal pouch through a narrow pore (**1**), swims off (**2**), settles on a substrate, and produces an attachment stalk (**3**). As the stalk elongates, the cilia are lost and tentacles appear (**4**), as the mature form develops (**5**). (Modified after A.E. Noble)

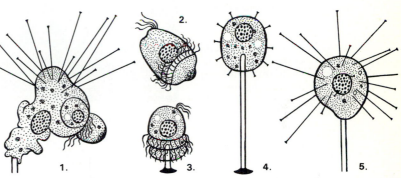

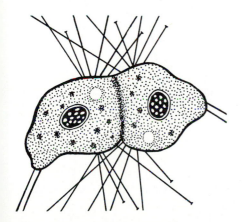

Suctorians conjugating. Two adjacent individuals lean toward each other, already prepared to conjugate and apparently guided by chemical signals. Initial contact is made by the tentacles, and eventually there forms a bridge, across which the cell membranes break down to allow exchange of nuclei. In most suctorians the conjugating individuals later separate, but in a few species one partner is pulled from its stalk and fuses completely with the other. *Tokophrya lemnarum.* (Modified after A.E. Noble)

Tetrahymena pyriformis, actually a complex of extremely similar species, is among the world's best known organisms— along with the bacterium *Escherichia coli,* the white rat, and a few other hardy species that survive well and behave cooperatively in the laboratory. *Tetrahymena* is readily reared in pure culture on a chemically defined medium and is a popular subject for both descriptive and experimental research in a great variety of fields. By applying temperature shock to cultures, the fission cycles of these ciliates can be synchronized, greatly facilitating study of the many synthetic and regenerative events necessary to replication. Conjugation is not essential to *Tetrahymena;* strains that lack micronuclei have been kept continuously in the laboratory for over 30 years. The last few decades have seen the publication of several thousand articles on the ultrastructure, biophysics, energetics, biochemistry, morphogenesis, physiology, genetics, systematics, evolution etc., of *Tetrahymena.* (Combined after various sources)

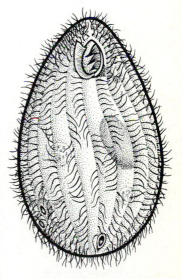

Vorticella is not colonial, but many separate stalked individuals often occur in dense clusters. When the ciliate is disturbed, a contractile strand quickly pulls the long stalk into a tight spiral, drawing the delicate bell-shaped body close to the substrate and, with luck, out of harm's way. The individual at the upper right of this photo is undergoing fission. One of the two products will remain attached to the stalk, while the other develops an extra whorl of cilia at the lower end of the bell and swims free. It may settle nearby, adding to the cluster; or, aided by water currents, travel some distance before attaching. Various other vorticellids are attached (with or without a stalk) or permanently free-swimming, solitary or colonial. Pennsylvania lake. (R.B.)

Contracted vorticellid, in addition to partly coiling the stalk, has protectively withdrawn and covered over its delicate feeding apparatus.

Free-swimming vorticellid with cilia at both ends of the body. (After Kepner and Pickens)

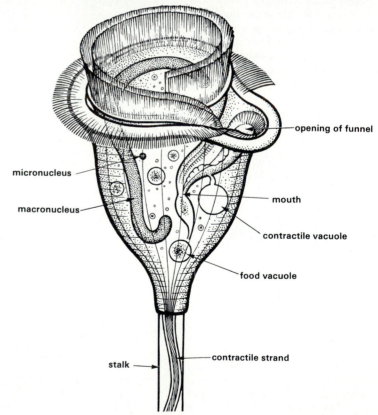

opening of funnel

micronucleus

macronucleus

mouth

contractile vacuole

food vacuole

stalk

contractile strand

Elaborate whorls of cilia around the mouth of *Vorticella* create a food-bearing water current, filter bacteria and other small food particles from the water, and convey food into the deep funnel that leads to the mouth. The contractile vacuole opens into this funnel, instead of directly to the outside as in other ciliates. Vorticellids are **peritrichs,** a large group of freshwater and marine ciliates that feed chiefly in this way. Many peritrichs are symbiotic in or on other protists or animals; such symbionts may be attached by a stalk to the host, or scurry over its surface by means of a band of cilia at the lower end of the body. (Combined after various sources)

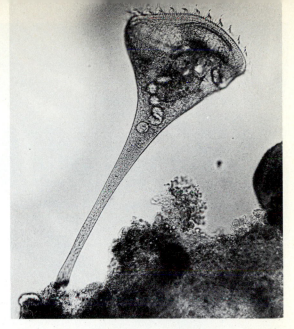

Stentor is a large trumpet-shaped ciliate found in freshwaters. The flexible body is highly contractile, and the ciliate can quickly change from its fully extended glory into a small rounded blob and back again. It can also detach and swim freely. When attached, it spends long periods extended and quietly feeding on bacteria, various protists, and even small animals such as rotifers (chapter 13). Surrounding the mouth area is a spiral of *ciliary membranelles,* flat triangular plates made of several rows of cilia that appear to be fused. The cilia are not really fused, but are held together by long slender projections of the ciliary membrane that interdigitate with those of adjacent cilia, and perhaps by an organic coating. Membranelles associated with the region surrounding the mouth, and other compound ciliary structures, are especially well-developed in **spirotrichs,** shown on this and the following 2 pages. The rest of the body of *Stentor* is uniformly covered with ordinary cilia, as is characteristic of the subgrouping of spirotrichs known as **heterotrichs.** (R.B.)

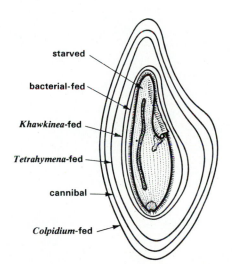

starved

bacterial-fed

Khawkinea-fed

Tetrahymena-fed

cannibal

Colpidium-fed

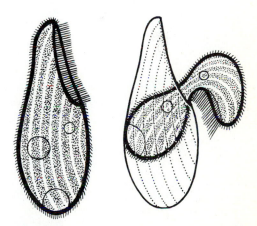

Blepharisma is primarily a bacterial-feeder, but if the laboratory cultures are provided with a suitable diet of small protists (such as the colorless euglenid *Khawkinea* or small individuals of *Tetrahymena),* the blepharismas become predaceous and grow to large sizes (represented by outlines in the figure *above).* Such giants can then eat relatively large prey (such as the ciliate *Colpidium)* and often turn into cannibals, feeding on smaller blepharismas. In nature, this dietary flexibility may sustain them through times when food runs low. They would cease to

Blepharisma emerging from a capsule of membranous protein and pigment granules, produced in response to certain substances (especially salts of certain alkaloids such as strychnine, morphine, cocaine, and novocaine). Because the shape and pattern of the body is so conspicuous in the firm capsule, observers at first concluded that the ciliate was shedding its pellicle, a remarkable act! Electron microscopy has demonstrated that the pellicle remains intact. (After J.E. Nadler)

divide while still growing slowly and thus become large enough to prey on various protists or on each other. When bacterial food is again available, blepharismas divide rapidly and return to normal size. These freshwater and marine heterotrichs have an unusual pink color. (Modified after A.C. Giese)

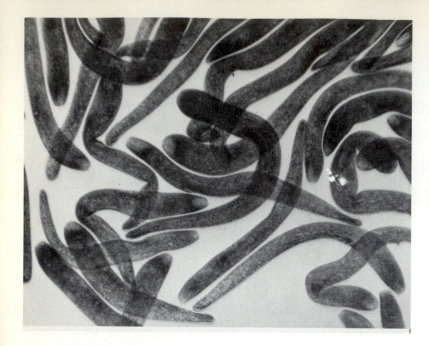

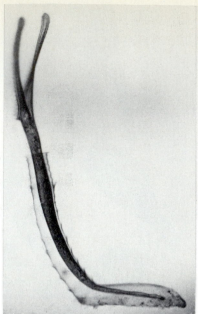

Spirostomum, a large freshwater heterotrich up to 4 mm long, is exceptionally flexible and contractile. (A.C. Lonert)

Folliculinid, *Metafolliculina andrewsi,* can withdraw into its protective lorica or (as here) extend 2 graceful winglike processes that bear the oral membranelles; the rest of the body is uniformly ciliated. In fission, the posterior half keeps the original lorica. The mouthless wormlike anterior half departs and migrates to a new site, often the surface of an aquatic plant or animal, where it secretes a new lorica and develops feeding structures. Folliculinids are almost all marine. (G. Uhlig)

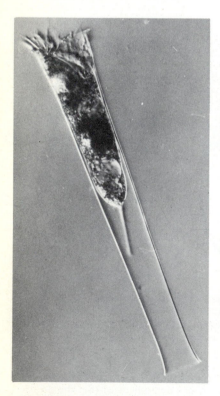

Tintinnid protrudes large locomotory and feeding membranelles from the opening of its vase-shaped lorica, about 350 μm long. The rest of the body is sparsely ciliated in these **oligotrichs** (*oligo* = "few," *trich* = "hairs, bristles, cilia"). Tintinnids are planktonic and primarily marine. Collected at 1,500 m. (G. Uhlig)

Lorica of a tintinnid. Length, about 265 μm. Collected at 1,000 m. Scanning electron micrograph. (M. Gowing)

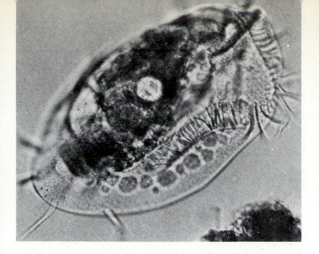
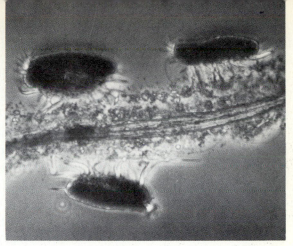

Hypotrichs are among the most complex, distinctive, and ubiquitous ciliates, and their movement and behavior are the most animal-like. They run about over both living and inert substrates on leglike tufts of cilia called *cirri*, attached to the lower surface of the flattened, oval or elongate body. Prominent membranelles and sensory cilia adorn the upper surface of the body, but there is little general body ciliation. As a group, they are best characterized as omnivores—gathering bacteria with ciliary currents, preying on various small protists, scavenging on plant or animal matter, or feeding on bottom detritus. *Above left,* a hypotrich showing conspicuous membranelles on the upper surface; from a Pennsylvania lake. *Above right,* marine hypotrichs from Bimini, Bahamas. (R.B.)

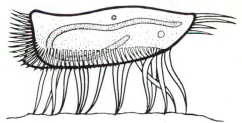

Stylonychia, a much-studied hypotrich, has a distinctive, low profile. Here it makes its way among debris from a freshwater lake. (R.B.)

Compound ciliary structures are not unique to hypotrichs, but this group specializes in them. They often give hypotrichs especially well-defined anterior and posterior ends, as in the one shown here with a series of large membranelles extending from the anterior end back along one side, and with a "tail" of sensory cilia at the rear end. The leglike cirri are especially long and stiff. *Euplotes neapolitanus.* (Modified after R. Wichterman)

Euplotes is a genus of common hypotrichs, predominantly marine and widely distributed, with heavy, readily observed cirri *(below center,* after C.V Taylor). A set of long, microtubular fibers run between the anterior and posterior cirri, and are visible in a silver-impregnated specimen *(below left,* from a rocky tidepool, Heligoland; G. Uhlig). When these fibers were cut in early microsurgical experiments, coordination between the widely separated cirri was reported to be lost *(below right,* after C.V. Taylor). This report led to the conclusion that the fibers were central to coordination, perhaps serving as a sort of simple nervous system. However, it now seems likely that the observed results of the early experiments were due to surgical trauma, as the experiment has been twice repeated, with the addition of cinematographic and electrical recording, and with the opposite result: cutting of the fibers did not destroy coordination. Coordination appears to be mediated by changes in membrane potential, as has been shown also in *Paramecium,* and the intercirral fibers of *Euplotes* have some other function, perhaps structural.

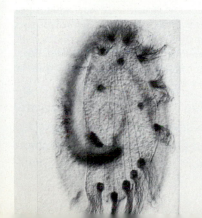
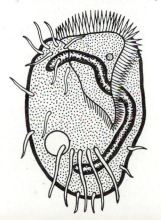

THE DIFFERENT WAYS in which protozoans produce large
bodies have been taken as models for the evolution of multicellularity in animals. The bodies encountered in the various
groups are of two main sorts: colonies of discrete cells and
syncytia with many nuclei in a continuous mass of cytoplasm.
Both sorts usually arise by *asexual fission* of, respectively, cells or
nuclei. But in some cases they result from *aggregation of
dispersed cells*. Each of the possible combinations seems to have
evolved independently many times.

Besides the aggregating forms already illustrated *(Dictyostelium,* an ameboid "cellular slime mold"; *Sorogena,* a ciliate),
many more are known to exist. The cellular slime molds are
considered by some specialists to include two independent groups
(dictyostelids and acrasids), not closely related. In addition, there
are the acellular slime molds (plasmodial or true slime molds, or
myxomycetes), in which the aggregating amebas do not remain as
discrete cells but form a syncytial mass. This seems to be the case
also in *Myxochrysis,* a chrysomonad ameboflagellate which has
chloroplasts but also ingests small prey. The life history of
Myxochrysis includes flagellated and ameboid stages, both with a
single nucleus; the amebas aggregate and form a multinucleate
syncytium. Normally solitary amebas may aggregate in small
clumps before encysting, suggesting how the behavior of collective amebas may have arisen; and several heliozoans sometimes
aggregate temporarily and together subdue a larger prey than any
one could handle alone. Aggregation has evidently evolved repeatedly among protozoans. Ironically, its significance in protistan development has been relatively little discussed, compared to
its popularity as a model for analyzing multicellular development.

Aggregation and integration of dispersed cells are essential features of
developmental events in many animals. For example, in the early development of salps (pelagic tunicates, chapter 29) the dividing cells disperse, and
clusters of them eventually spread into sheets that grow together to form the
definitive embryo. In vertebrate embryos, cells of the neural crest disperse,
migrate, and aggregate at particular sites, finally differentiating into
important components of the nervous, skeletal, and endocrine systems and
influencing the development of yet other structures. In planarians (flatworms, chapter 8), certain cells aggregate at the site of a wound, and together
proceed to repair the damage and replace lost structures, even if what is lost
should be the whole front end of the worm. Fully-developed, mature sponges
(chapter 3) and cnidarians (chapter 5) may be dissociated into separate cells,
which reaggregate into masses and regenerate the complex body form. Early
stages of sponges and cnidarians sometimes come together and fuse, with the
cells integrating themselves into a single body.

On the other hand, aggregating protozoans are not usually
considered as models for the evolution of multicellularity. This
is because cell division, not aggregation, is the dominant
pattern in the early development of multicellular animals.

Most animal embryos develop from a single cell, the zygote, that divides repeatedly, with the resulting cells remaining together and becoming differentiated into several types. This fact of development lends weight to the **colonial hypothesis** of the origin of animals—that multicellular animals arose from protistan ancestors that formed many-celled colonies in which the products of cell fissions remained attached. Proponents of this idea emphasize protists such as *Volvox,* which displays differentiation among the cells of its colonies. Related forms span a spectrum from single cells through increasingly large and complex colonies. The colonial habit of collar flagellates and their resemblance to the collar cells of sponges (chapter 3) forms another line of evidence. These and other colonial flagellates are the protists most often mentioned in connection with the colonial hypothesis. However, as we have seen, colonial growth is also a feature of ameboid protozoans (poly-cystine radiolarians), as well as of ciliates.

The other principal idea of the origins of multicellularity, the **syncytial hypothesis,** is that animals arose from a syncytial protist in which the many nuclei, each with its surrounding cytoplasm, became separated by cell membranes, and the cells remained together as an integrated organism. Although this idea is most commonly discussed in connection with ciliates (and arose at a time when *Opalina* and *Stephano-pogon* were still considered ciliates), living syncytial proto-zoans occur in many diverse groups. Examples include opalines and numerous other zooflagellates, as well as sporo-zoans and large ameboid forms such as the ameba *Chaos carolinense,* the foraminiferans, and some heliozoans. In myxozoans, nuclei of the syncytial phase become cellularized and develop into multicellular structures with several differen-tiated cell types. The idea that syncytial structure is a functional alternative to multicellularity in such large protis-tans does not preclude its evolution into multicellularity. Syncytial development is a feature of the early embryos of insects, and syncytial structures are found in parts of some adult multicellular animals as well.

Models of hypothetical ancestors are often too firmly attached to a particular living prototype and then debated unproductively on the basis of the specific characters of that organism. It does not matter that ciliates show particular symmetries, or that volvocid flagellates are photosynthetic and hence more apt as progenitors for plants than for animals. There is no reason to assume that the ancestors of animals were among any group of protists surviving today. However, modern

In certain large protozoans such as radiolarians the single nucleus is high-ly polyploid, and there may be little functional difference (or evolutionary distance) between this arrangement and others in which many small nuclei are present. In ciliates, for example, closely related forms have single or multiple micronuclei and macro-nuclei. Hence, although technically not syncytial, any polyploid proto-zoan might be considered as a candi-date for further evolution in the direc-tion of multicellularity.

protists do give us some concrete models for thinking about how animals may have evolved from protistan ancestors. And the evident readiness of protists of many kinds to produce large multinucleate or multicellular forms may help us to evaluate the possibility that this happened more than once.

CLASSIFICATION: Kingdom PROTISTA

IDEAS about how to classify protists are so diverse and are changing so rapidly that the choice of any particular scheme is inevitably somewhat arbitrary, especially for nonspecialists. The one listed below is a compromise among several published in recent biological literature. We have followed a current trend to list as separate phyla many groups that were formerly placed only at the class level or lower. This seems a useful approach in view of present uncertainties about protistan relationships, as well as an appropriate way to recognize the genuine diversity apparent within this kingdom of perhaps 120,000 species.

Some of the phyla are grouped under informal names in order to indicate similarities and possible relationships among them, and also to help integrate this classification with earlier ones. The attached notes are not formal definitions, but give only a few important characters of structure or habitat. Many variants of the names result from their having been originally assigned at different taxonomic levels, or as part of "algal" or "animal" series. Some small groups are omitted. (For further information on the classification and biology of protists, see sources given in the literature section at the end of this book.)

The red and brown seaweeds are sometimes classified as protists on the grounds that they are simply organized, that some show clear relationships to unicellular groups, and that the plant kingdom is a more natural (perhaps monophyletic) group without them. The simplest members of both groups are no more complex than some multicellular protists, but the majority of these seaweeds have levels of organization much more comparable to members of the multicellular kingdoms, and we consider these algal groups more appropriately placed with the plants.

The green algas pose a more difficult problem because there is a clear continuity from the many single-celled types through a series of increasingly complex colonial and multicellular forms to the green algas and green plants. The system of 5 kingdoms offers no simple solution to classifying this cohesive group. The boundaries between the protists and the multicellular kingdoms (animals, plants, and fungi) will probably always be subject to such border disputes.

NEARLY HALF of the protistan phyla consist of flagellates that are distinguished by structural and biochemical features, especially the form and number of flagella, nuclear organization (especially during division), composition of storage products and the cell covering, and (when present) photosynthetic pigments. Details of these should be sought in specialized sources.

FLAGELLATES

PHYTOFLAGELLATES, "plant" flagellates. Mostly photosynthetic, but some heterotrophic forms. Few flagella (commonly 2).

Phylum **Dinoflagellata** (Dinophyta, Pyrrhophyta, Mesokaryota), dinoflagellates. Large group with distinctive nuclear organization, in some ways unlike that of other eukaryotes. Some colonial forms. Mostly marine and planktonic, some freshwater. Photosynthetic and free-living: *Gymnodinium, Ceratium, Gonyaulax.* Photosynthetic and symbiotic: *Symbiodinium* (zooxanthellas in many invertebrates). Free-living and heterotrophic: *Polykrikos* (colonial), *Noctiluca.* Multinucleate, intracellular parasites in other protists and invertebrates: *Syndinium.*

Phylum **Cryptophyta** (Cryptomonadida). Mostly photosynthetic. Freshwater and marine.

Phylum **Euglenida** (Euglenophyta). Mostly freshwater. Mixture of many photosynthetic and heterotrophic forms. *Euglena, Peranema. Colacium* (colonial).

Chrysophyte group: photosynthetic golden-brown and yellow-green flagellates and related forms.

Phylum **Chrysophyta** (Chrysomonadida), mostly freshwater, plus marine silicoflagellates. Many colonial and multicellular forms. Solitary or colonial: *Dinobryon.* Ameboflagellates: *Myxochrysis* (aggregating), *Chrysamoeba.*

Phylum **Haptophyta** (Prymnesiophyta), coccolitho-phorids. Marine.

Phylum **Bacillariophyta** (Diatomea), diatoms. Large important group of primarily marine plankton. Some colonial forms.

Phylum **Xanthophyta**. Many multicellular and syncytial forms. Ameboid and flagellated stages. Mostly freshwater.

Phylum **Eustigmatophyta**. Mostly freshwater.

Chlorophyte group: green flagellates and varied multicellular forms.

Phylum **Chlorophyta** (Volvocida). Many colonial and multicellular forms. Mostly freshwater. *Chlamydomonas, Eudorina, Volvox.*

Phylum **Prasinophyta** (Prasinomonadida). Marine and freshwater. *Tetraselmis* (= *Platymonas),* endosymbiotic in acoel flatworms (chap. 9).

Phylum **Conjugatophyta** (Gamophyta). Large group of unicellular or filamentous forms, with no flagella at any stage. Freshwater. Desmids, *Spirogyra.*

Phylum **Charophyta**. Many large multicellular forms (stoneworts); usually considered plants. Mostly freshwater. *Chara, Nitella.*

Phylum **Glaucophyta**. No chloroplasts; contain symbiotic cyanobacteria. Freshwater. *Cyanophora.*

ZOOFLAGELLATES, "animal" flagellates. Heterotrophic. Few to many flagella.

Phylum **Choanoflagellata,** collar flagellates. With a single flagellum. Free-living. Feeding and structure similar to collar cells of sponges (chap. 3). Solitary: *Monosiga.* Colonial: *Codosiga, Proterospongia.*

Phylum **Kinetoplastida**. Flagellates having a DNA-rich body (kinetoplastid) associated with the single mitochondrion; 1 to 2 flagella. Some free-living, many commensal and parasitic forms. *Bodo* and relatives. Trypanosomes: *Trypanosoma, Leishmania* in invertebrates and vertebrates, *Phytomonas* in invertebrates and plants. May be related to euglenids.

Polymastigote group: some free-living, mostly symbiotic flagellates with several to many flagella.

Phylum **Metamonadida,** diverse assemblage of groups, including diplomonads: *Giardia.*

Phylum **Parabasalia**. With parabasal body. Some multinucleate forms. Trichomonads: *Trichomonas, Pentatrichomonas.* Hypermastigotes, all symbiotic in insects: *Trichonympha.*

Multinucleate, ciliated forms, both formerly classified with the ciliates, probably not related to each other.

Phylum **Opalinata,** opalines. Parasitic, mostly in amphibians. *Opalina.*

Phylum **Pseudociliata**. Free-living, marine, benthic. *Stephanopogon.*

SARCODINES, ameboid protozoans

RHIZOPODS. With broadly lobed or thin pseudopods at some stage of life history, used in feeding and locomotion.

Phylum **Karyoblastea**. Single genus of large amebas with peculiar cytology, having many nuclei but lacking most organelles and the process of division seen in other eukaryotes. Freshwater; free-living. *Pelomyxa.*

Phylum **Amoebozoa**. Common naked and testate amebas and ameboflagellates. Marine, freshwater, and soil; free-living and parasitic. *Amoeba, Chaos, Difflugia, Entamoeba, Naegleria, Acanthamoeba.*

Phylum **Mycetozoa**. Collective amebas, or "slime molds." Ameboid and flagellated stages called myxamebas and myxoflagellates. All free-living in soil or decaying vegetation.

Dictyostelids. Lack flagellated stages; aggregating amebas remain as discrete cells: *Dictyostelium, Polysphondylium.*

Myxogastrids, or myxomycetes. Have biflagellated stages; aggregating amebas form plasmodial (syncytial) masses: *Echinostelium, Lycogala, Trichia, Stemonitis, Physarum.*

Protostelids. Have flagellated stages; amebas do not aggregate but encyst singly on a simple acellular stalk; may be a primitive stem group: *Protostelium, Cavostelium.*

Phylum **Acrasia**. "Slime molds" in which amebas form cellular aggregates; sorocarps show little differentiation, compared to mycetozoans. Free-living in soil or decaying vegetation. *Acrasis, Guttulina.*

Phylum **Plasmodiophora.** Multinucleate plasmodial forms give rise to biflagellated stage, then to myxamebas; nuclear division (not aggregation) restores the plasmodial stage. Intracellular parasites of terrestrial plants. *Plasmodiophora* in cabbage, *Spongospora* in potato.

Phylum **Granuloreticulosa.** Foraminiferans and their relatives, a large group. Marine; free-living. *Globigerina, Hastigerina, Notodendrodes. Gromia* and related genera may be placed in separate group, Filosa.

ACTINOPODS. With stiff pseudopods (axopods) having a central core of microtubules, used primarily in feeding, little in locomotion. Polycystines and pheodarians (and often acantharians) are collectively called radiolarians; these groups have a central capsule containing the nucleus and usually walled off from the outer cytoplasm.

Phylum **Heliozoa.** Mostly freshwater, some marine, semiterrestrial. Mostly free-floating with pseudopods radiating in all directions. Some have organic or siliceous tests studded with hard organic or siliceous particles, in the form of scales or spicules. Some attached by inert or contractile cytoplasmic stalk. *Actinophrys, Actinosphaerium, Clathrulina.*

Phylum **Acantharia.** Marine, mostly planktonic. Many in shallow tropical waters with photosynthetic symbionts in central capsule. Skeleton of $SrSO_4$. *Acanthometra.*

Phylum **Polycystina.** Marine, planktonic. Many in shallow tropical waters with photosynthetic symbionts in extracapsular cytoplasm. Skeleton of silica.

Spumellarids: solitary, *Thecosphaera, Hexastylus;* colonial, *Collozoum, Solenosphaera.* Nassellarids are solitary: *Lamprocyclas, Theocorythium.*

Phylum **Phaeodaria.** Marine, planktonic. Skeleton of silica and organic matter. Mostly solitary. *Conchidium, Challengeron.*

Miscellaneous groups which (along with the mycetozoans) have been considered sarcodines or fungi.

Phylum **Labyrinthomorpha,** "slime-net molds," with spindle-shaped cells that move within an extracellular network, also flagellated stages.

Mastigomycetes, fungus-like organisms, saprobic or parasitic in freshwater and soil, with flagellated stages: Phyla **Chytridiomycota, Hyphochytridiomycota, Oomycota.**

SPOROZOANS, all parasitic

Phylum **Apicomplexa,** or Sporozoa. With distinctive apical complex of organelles. Gregarines. Coccidians: *Eimeria, Toxoplasma;* malarial parasites, *Plasmodium.* Piroplasms: *Babesia, Theileria.*

Phylum **Microspora.** Intracellular parasites of animals, especially arthropods; unicellular "spores." *Nosema.*

Phylum **Haplospora** (Ascetospora). Parasites of aquatic invertebrates; multicellular "spores." *Minchinia, Haplosporidium.*

Phylum **Myxozoa,** or Myxosporidia. Multicellular "spores" with polar capsules. *Myxobolus.*

CILIATES

Phylum **Ciliophora.** With cilia and 2 kinds of nuclei. Marine, freshwater, soil; free-living and parasitic.

Class **Kinetofragminophora** (formerly holotrichs), typically with uniform (holotrichous) body ciliation; mouth region with modified cilia, but no compound ciliary structures (4 subclasses).

Gymnostomes: *Loxodes* (karyorelictid). *Prorodon* (prostomatid). *Didinium, Dileptus* (haptorids).

Vestibuliferans: *Balantidium* (trichostome). *Colpoda, Sorogena* (colpodids).

Hypostomes.

Suctorians: *Tokophrya.*

Class **Oligohymenophora** (formerly holotrichs), typically with uniform body ciliation, but variable; mouth region with a few membranelles or other compound ciliary structures (2 subclasses).

Hymenostomes: *Paramecium, Tetrahymena.*

Peritrichs: *Vorticella, Zoothamnium, Ophrydium.*

Class **Polyhymenophora,** body ciliation variable; mouth region with many membranelles and other compound ciliary structures (1 subclass).

Spirotrichs: *Stentor, Blepharisma, Spirostomum, Metafolliculina* (heterotrichs). Tintinnids (oligotrichs). *Euplotes, Stylonychia* (hypotrichs).

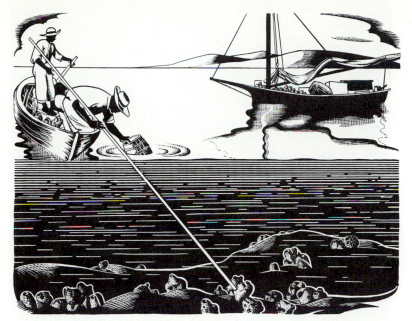

Sponges

*Cellular level of
construction*

SPONGES, or rather the skeletons of sponges, are mostly
familiar from the few species in which the cleaned and
dried meshwork of skeletal fibers is soft enough to be of use to
humans. Such sponges live only in warm shallow seas. The rest
of the more than 5,000 species of living sponges have skeletons
that are too hard or scratchy, too filled with gritty particles, or
too brittle and friable, to be commercially valuable. As a
group, sponges live in all seas, from pole to pole, and from
intertidal shores to abyssal depths. And from the earliest times
for which we have fossil-bearing rocks, they have been among
the most abundant creatures on sea bottoms.

Freshwater sponges comprise only about 150 species. They
are of no economic importance except when they occasionally
obstruct water conduits or reservoir drains, or when decaying
sponges impart a "swamp taste" to water supplies.

Sponges interest us for their many variations on a unique
porous structure, from which the phylum **PORIFERA,** the
"pore bearers," takes its name. Biochemists study poriferans
for the toxicity of particular species to humans and to other
animals, for their curious concentrations of certain chemicals
that occur naturally in seawater, and for the antimicrobial
properties of some species. And of course we appreciate
sponges for what their beautiful shapes and brilliant colors add
to the aquatic scenery visible at low tide, or from small boats,
or when snorkeling or scuba diving.

The use of fibrous sponge skeletons
goes back at least to the Bronze Age.
Paintings on ancient Greek vases
show sponges being used for bathing;
Cretan wall frescoes attest to their use
as paint rollers. From Greek writings
we learn that they were used for
scrubbing tables and floors, and for
padding helmets and leg armor. There
were therapeutic applications, such as
stopping blood flow. The ancient
Romans fashioned sponges into paint
brushes, tied them to wooden poles as
mops, and made them serve as sub-
stitutes for drinking cups. Today household "sponges" are usually synthetic,
but the superior qualities of natural
sponges are still prized by profes-
sional car or wall washers, and by
leather workers, potters, silver-smiths,
and lithographers. Sponges with
notable accumulations of bromine
and iodine compounds in the fibers of
the skeleton were charred and used, in
ancient times, for fumigation. In *Microciona prolifera* the iodine content can
be 0.3 percent of the total dry weight.

The only marine compound currently
used in cancer chemotherapy was synthesized in the laboratory based on a
nucleoside isolated from a Caribbean
sponge, *Cryptotethya crypta.*

Po-rif′-era

The shapes of sponges are distinctive in many species. These 3 marine species were growing suspended from the same piece of debris on a harbor wharf piling. The main mass is a simple, much-branched calcareous sponge, *Leucosolenia*. At the left protrude 3 large and 2 small vase-shaped individuals of *Sycon*, also calcareous, but with a more complex structure. At the top, surrounding the stalk to which all 3 species are attached, is the encrusting sponge *Halichondria panicea*, which has no distinctive shape. About 2× actual size. England. (D.P. Wilson)

Shape may vary with habitat, even within the same species. *Halichondria bowerbanki* has an encrusting growth form under wave-exposed boulders and rocky ledges, but takes on a slender fingerlike form, as shown here, when growing in quiet water. Plymouth, England. (D.P. Wilson)

Encrusting growth of *Halichondria panicea* growing in wave-exposed site. Wembury, England. (R.B.)

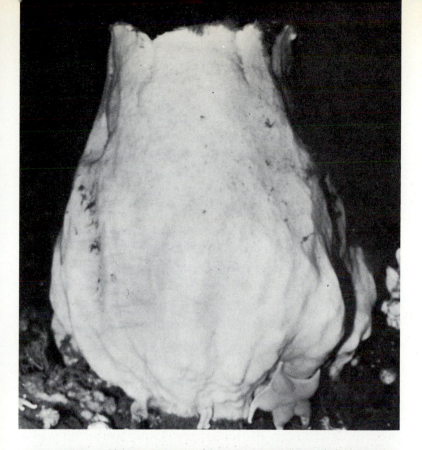

Giant sponge vase, *Scolymastra joubini,* is common at almost 50 m on the quiet bottom of McMurdo Sound, Antarctica. This specimen was about 2 m tall and, judging from the slow growth of such cold-water sponges, must be hundreds of years old. A sea star can be seen feeding on the sponge near its base. (P.K. Dayton)

Most massive sponge known is *Spheciospongia vesparia,* which looks like a large hassock, and is quite big enough for several people to sit on when it reaches a diameter of 2 m. Such a large sponge is a "living hotel." One specimen with a volume of 1.85 m³ was torn from its mooring at 3.5 m on the bottom of the Gulf of Mexico. Brought to shore and cut open, the sponge was examined for the abundance of animals that take shelter as juveniles, or live out their whole lives, in the protection of sponge cavities. Out of a total of more than 17,000 animal guests, of many phyla, about 16,000 were snapping shrimps of the genus *Synalpheus.* (From A.S. Pearse)

Tubular sponges, *Iophon laevistylus,* from New Zealand waters. (W. Doak)

A number of tropical and sub-tropical species are toxic. *Tedania toxicalis* kills fishes or invertebrates that share the same container. The "fire sponges" of the West Indies cause irritation of the hands merely on touching. And prolonged handling of *Tedania nigrescens* has caused both severe skin irritation and generalized symptoms.

A sponge that moves, *Tethya seychellensis,* is an exception. Young colonies up to 2-3 cm in diameter of this and another species of *Tethya* can detach from their sites of settlement and move slowly (10-15 mm/day) to new sites. The sponge protrudes long filamentous extensions of the body wall. Adhesive knobs at the tips attach to solid objects and the filaments contract, pulling the sponge along. Red Sea, Eilat, Israel. (L. Fishelson)

Sponges are sessile animals, living permanently affixed or partially embedded at the base. Although some have been observed to change position very slowly, when observed at intervals, they cannot be said to move about freely. Such animals cannot flee predators, but sponges often have bristly textures and strong chemicals that discourage most appetites. Nevertheless, they are fed on by certain fishes, sea stars, and molluscs, especially nudibranchs and snails.

Sponges live by pumping large volumes of water through their bodies and filtering out minute organisms and organic particles as food. They are not the only animals that make their living in this way, but they are especially interesting to biologists for the elegantly simple organization of their bodies.

It is traditional to look at sponges as examples of what primitive multicellular animals might have been like—living illustrations of a level of organization somewhere between that of protists and "true" animals. But there is no evidence that any other animals are even remotely related to sponges, and the modern descendants of this ancient group are in no way a primitive relic that evolution has by-passed. Sponges are an abundant and diverse group, uniquely specialized for their own way of life, which they continue to live as they have for hundreds of millions of years, side by side with more complex animals.

The behavioral repertoire of a sponge is hardly more varied than that of a rooted plant. And we can only sympathize with the early naturalists who were thoroughly confused about the place of such inert growths in the scheme of things. Aristotle concluded that their nature lay somewhere between plants and animals, but his successors could see no means by which sponges could feed without moving about, and they classed them as plants— or even as organic secretions deposited by the many kinds of small animals that find shelter in the channels and cavities of the larger sponges. By the 18th century a careful observer decided that sponges were animals, citing as evidence the water currents they produced and the movements that could be seen in the upstanding tissue that rimmed the one or more large openings through which water exited the sponge. But the controversy did not really end, giving sponges a clear title as animals, until the first half of the 19th century, when R.E. Grant of Edinburgh added a suspension of colored particles to the water and then saw, under the microscope, that these particles were taken into the animal, apparently through microscopic openings, and then came "vomiting forth, from a circular cavity, an impetuous torrent."

MULTICELLULARITY

PROTISTS are mostly undivided, microscopic organisms or loosely organized masses of similar units that result when the products of asexual multiplication remain connected as colonies. Only a few of the colonial forms show the beginnings of differentiation among the units of the colony.

On the other hand, most animals are subdivided into cells, and **differentiation among cells** is a hallmark of the animal kingdom. No one cell must carry on all the necessary tasks, and different cells become specialized for different functions. The possibilities opened by such division of labor among cells are seen in the great diversity of animal forms and the many animal life-styles that are not available to simpler organisms.

As cells become more specialized for certain functions, and less capable of surviving and multiplying independently, the need increases for **integration among cells,** so that their activities are coordinated throughout the many-celled organism. Communication among cells, by circulating chemicals such as hormones, by chemo-electrical transmission in a nervous system, or some combination of these, is a prominent character of most animals. In sponges, however, the mechanisms for integration are not conspicuous, and the levels of differentiation and of coordination among cells appear to be relatively less than in other animals, while the potential of individual cells is greater. Time-lapse films of sponge cells show them to be constantly changing in shape, in position, and in function.

Nevertheless, sponges demonstrate the beginnings of one important advantage of multicellularity—the production of an **internal environment** in which cells can create and maintain a contained volume of space with composition and properties that differ from those of the surrounding water. Through this medium, cells can share food and chemical communications.

In contrast to the cytoplasmic level of organization seen in most protists, and to the more complex levels of differentiation and integration seen in most animals, the many-celled sponges have been said to be constructed on a **cellular level of organization.**

The sponges are the only group of animals for which there are clear ancestral candidates among the protists—the collar flagellates. The most characteristic type of sponge cell, found in some form in all sponges, shows detailed similarities to a collar-flagellate. Some of these protozoans form large colonies of several thousand cells embedded in a gelatinous ball. And while it is a long time since there has been a serious suggestion that sponges were plants, it is still not easy to formally distinguish sponges from colonial protozoans. The common definitions are based on the manner of development and on the degree of specialization of cells.

The abundance of sponges suggests that we need to know much more than we do now about the part played by these filter-feeders in processing the the marine shore waters into which increasingly large and industrialized societies dump vast quantities of organic wastes and industrial chemicals. We do know that sponges ingest bacteria, and in the laboratory, at least, the common redbeard sponge, *Microciona prolifera,* has a voracious appetite for the bacteria of human feces.

Redbeard sponge, *Microciona prolifera,* has been long used in experimental studies of cell dissociation and reaggregation. Gulf of Mexico. (R.B.)

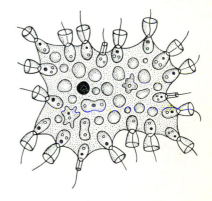

Proterospongia was described a century ago as a collar flagellate in which the gelatinous colony includes ameboid cells. (After W. Saville-Kent). Although recent observers of *Proterospongia* report no ameboid cells, colonial collar flagellates are still the most plausible intermediates in the evolution of sponges.

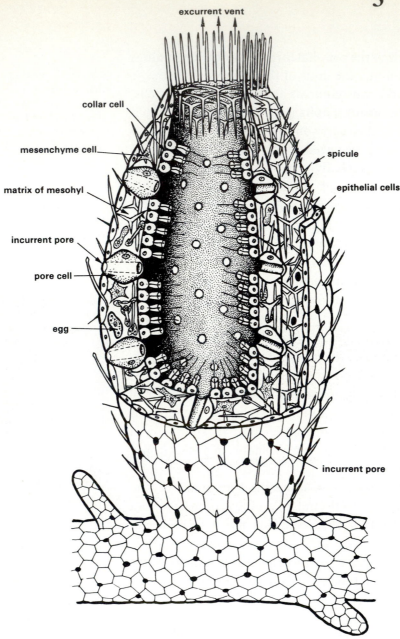

excurrent vent

collar cell

mesenchyme cell

matrix of mesohyl

incurrent pore

pore cell

egg

spicule

epithelial cells

incurrent pore

Diagram of a **simple sponge.** The upper part is cut away to show structure.

A simple sponge, *Leucosolenia,* natural size.

Individuality in sponges is harder to define than in most protozoans or other invertebrates. One concept defines most sponges as colonies, with each individual being that part served by one excurrent vent. Against this view is the evidence for interaction between such units. Water flow is often shunted from one set of channels to another; mesenchyme cells move about from old areas into new growth; and gametes are produced in all parts of a sponge at once. For these and other reasons it is argued that any continuous sponge mass is an individual, at least in terms of behavior or physiology. Another view, important in terms of population biology or evolution, is that an individual sponge consists of all the derivatives of a single fertilized egg, which share a genetic identity, even if they are physically separated through fragmentation or asexual reproduction. Under this view, a single sponge mass is *not* an individual if it arises (as some appear to do) by fusion of two or more sexually-produced offspring.

SPECIALIZATION OF CELLS

WE BEGIN HERE with the simplest type of sponge structure, which (inconveniently for us) few sponges display. Among those that do are the little sponges of the genus *Leucosolenia*, vase-shaped animals that are covered, inside and out, by **epithelial cells**. The outer epithelial cells are flattened, fit closely together, and are not very different from those in epithelia of many animals. But the epithelium of **collar cells** that lines the inside of the vase is unique to sponges.

The free surface of each collar cell bears a delicate ring of long microvilli, and these slender cytoplasmic extensions encircle a long flagellum. The spiral beat of the flagellum drives water past the cell in a continuous current moving through the collar and away from its tip, the slotted collar acting as a sieve. The spaces between the microvilli that form the collar are approximately 0.1 μm across, and are bridged with fine mucoid (glycoprotein) filaments between adjacent microvilli. Thus any organic particles or minute organisms larger than about 0.1 μm may be trapped and ingested by the collar cell, in the same way as by collar-flagellates.

A variety of invertebrate and vertebrate groups have cells that resemble collar cells in bearing a flagellum surrounded by a ring of microvilli. Many of these are sensory cells, and none has been shown to feed. Collar-flagellates and sponge collar cells not only feed in the same way, but are also alike in having a fringe of minute hairlike projections on the flagellum. These projections have not been found on the flagella of any metazoans other than sponges. Such detailed similarities in cellular structure and function support arguments for a close evolutionary relationship between collar-flagellates and sponges.

The combined action of the beat of the flagella of all the collar cells creates the water current which passes through the sponge. Although the flagella beat independently, not in synchrony with each other, a unidirectional flow results. The water enters the sponge by microscopic **incurrent pores** perforating the outside, is drawn through small channels that open among the bases of the collar cells, passes through the large internal cavity of the vase, and finally leaves by way of the large **excurrent vent** at the top of the sponge.

Measurements of water currents through sponges show that the flow rate is faster when a water current is flowing over a sponge than when the surrounding water is still. The faster the external water current, the faster is water flow through the sponge. The same can be seen in plastic models of sponges and is a purely physical effect. Presumably, real sponges take advantage of such passive flow induced by surrounding currents, and save some energy in this way.

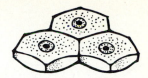

Epithelial cells of sponges lack a stable basement membrane, the thin collagenous layer that underlies the epithelia of other animals. The flattened epithelial cells that cover the outside and line the water channels of a sponge are termed **pinacocytes** *(pinaco* = "tablet," *cyte* = "cell").

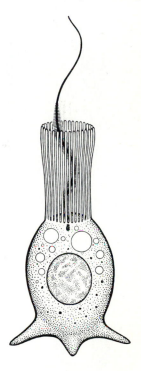

Collar cell of a sponge. The technical term, **choanocyte** *(choano* = "funnel," *cyte* = "cell"), wrongly suggests that water pours into the top of the funnel-shaped collar. Instead, water flows inward through the sides of the collar and out the top. Food particles are caught on the outside of the collar, moved downward, and ingested by the cell. (Based on R. Rasmont, on E.J. Fjerdingstad, and on B.A. Afzelius)

The large excurrent opening of a sponge is usually called the **oscule,** or osculum, but this is a particularly unfortunate technical term because it means "little mouth," and the sponges are unusual in the animal kingdom in that the largest external opening is *not* a mouth.

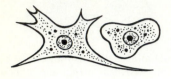

Mesenchyme cells.

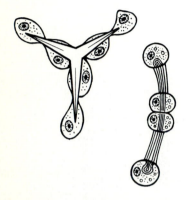

Mesenchyme cells secrete the skeleton. *Left,* **scleroblasts** secreting a spicule. (After W. Woodland). *Right,* **fibroblasts** secreting a spongin fiber. (After O. Tuzet)

Pore cell, or porocyte.

Contractile cells, or myocytes, surrounding a sponge opening. (Modified after A. Dendy)

The **mesohyl,** the part of a sponge between the outer epithelial cells and the collar cells, is filled with an organic, jellylike **extracellular matrix,** which supports the delicate cells. The polysaccharide ground substance of the matrix is laced with fine fibrils of a form of collagen which closely resembles that of the supporting basement membrane found underlying the epithelia of other animals. Abundant in the matrix are several types of **mesenchyme cells** that move about in ameboid fashion. Mesenchyme cells may ingest food particles, receive partly digested food from the collar cells, complete the digestion, carry the food from one place to another, and store food reserves. They also transport trapped inert particles and other solid waste materials to excurrent channels to be carried away by the outgoing current of water. Certain mesenchyme cells (archeocytes) are unspecialized and can develop into any of the more specialized cell types in the sponge. Such cells play a major role in repair of damage or in regeneration of a sponge.

Some of the mesenchyme cells secrete the **skeleton,** which helps to protect and support the soft cellular mass and enables sponges to grow to considerable size. In *Leucosolenia* and its relatives the skeleton consists of fine needlelike and multi-rayed **spicules** ("little spikes") of crystalline calcium carbonate. In other groups of sponges the mesenchyme cells produce a skeleton of siliceous spicules and/or **spongin,** tough elastic fibers of collagen. The proteins called collagens form the basis of connective tissues throughout the animal kingdom. A few sponges lack a skeleton and grow as thin amorphous encrustations, or are supported by an abundance of especially firm matrix material.

In *Leucosolenia* each pore through which water enters is a channel through the center of a **pore cell.** The pore cell is shaped like a short thick-walled tube and lies embedded in the mesohyl with its outer end opening among the epithelial cells and its inner end opening among the collar cells and into the central cavity of the sponge. The pore cells are contractile and can vary the size of pore openings or close them completely.

In some sponges there are special elongated **contractile cells,** which produce movement by becoming shorter (and thicker), thus drawing closer together adjacent structures. They are often arranged around the excurrent vents, and, in sponges that have no pore cells, around the pores; their contraction narrows the openings when irritating substances are present in the water or in response to mechanical stimuli.

STRUCTURAL TYPES

THE SIMPLEST, or **ascon type** of sponge structure is seen in *Leucosolenia,* one of the few genera that show asconoid construction. The body is a radially symmetrical vase or tube and the body wall is thin. The surface pores lead into the main central cavity, which is lined with collar cells. The number of collar cells in the central cavity is relatively low compared with the volume of water that must be moved out the single large excurrent opening at the top of the vase.

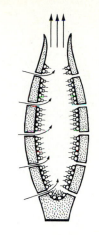

Ascon type.

The several grades of the **sycon type** of structure are also found only in relatively small calcareous sponges. The wall of the vase has been pushed out into radially arranged fingerlike projections, the **radial canals,** lined with collar cells. This greatly increases the surface (relative to the total mass) that is available for the location of the collar cells that propel the water currents. The main cavity of the sponge no longer has collar cells. A thickening of the body wall and the mesenchyme layer joins the outer ends of the radial projections, leaving the spaces between them as **incurrent canals,** into which the incurrent pores open. The incurrent canals are lined with epithelial cells, and they end blindly. Even thicker walls and mesenchyme, and increasingly branched incurrent canals, are seen in some calcareous sponges.

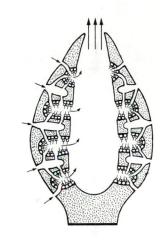

Sycon type.

The **leucon type** is the most complex type of sponge and carries to greater extremes the degree of branching of the incurrent channels and the enormous numbers of microscopic, rounded **flagellated chambers.** Some calcareous sponges and all the demosponges (these include a majority of sponges) are leuconoid. The compact leucon type has a higher pumping rate, brings in more food and oxygen, and expels the outgoing water with greater force, throwing the already filtered water farther away from the sponge's water intakes. Glass sponges, though without well-defined canals, are of a structure intermediate in complexity between typically syconoid and leuconoid types.

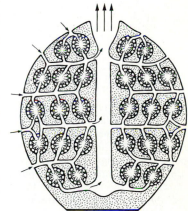

Leucon type.

The commonly observed **behavior** of sponges includes the pumping of water by the collar cells, the opening and closing of incurrent and excurrent openings, and the contraction of small portions of the surface or even the whole sponge. In many sponges, however, the response to a strong touch or even an actual cut does not appear to be transmitted beyond a short distance. All responses of sponges are slow, and in many species time-lapse photography must be used in order to observe any movement.

No specialized sensory or nerve cells have been identified in sponges, and evidence for a conduction system has been found in only one of the 4 sponge classes, the hexactinellids. Many attempts are being made to deduce the means of communication between cells from the observed behavior of sponges.

When **pumping rate** was monitored in 3 species of sponges in the West Indies they were found to follow different patterns. In *Mycale,* a red, tubelike sponge, the rate remains **constant day and night** in summer months. With the coming of lowered temperatures in winter, the rate decreases. During winter storms, when stirred-up particles clog the channels, the rate decreases even more. Such changes in sponge activity need not imply coordination of cells; each cell could respond independently to the same stimuli from the external environment.

A **diurnal rhythm** is shown by *Tethya crypta,* a black, globular sponge which relaxes the contractile membrane around the excurrent opening during daylight hours and gradually contracts it as sunset approaches. Any light sufficient to permit nighttime measurements of pumping rate inhibits contraction of the membrane. So, presumably, the pumping rhythm varies with the intensity of light. Again, no coordination of cells need be called on; light could act independently on each cell of the membrane.

Asynchronous cessations of pumping were recorded in populations of *Verongia gigantea,* a very large (gross volume up to 120 liters), yellow, cylindrical sponge. Individuals in a population cease pumping for brief periods (averaging 42 minutes) at intervals of about 19 hours. Since neighboring individuals of *Verongia* may at the same time be in either active or inactive phases of their cycle, it has not been possible to correlate the pumping rhythms with any changes in the external environment such as temperature, light, tides, lunar cycle, wave surge, or water currents. The only conclusion that can so far be drawn is that the periodic cessations are triggered internally and that there must be some communication, however slow, between cells in different parts of the sponge.

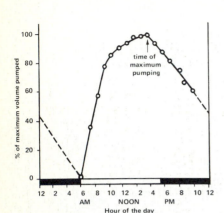

Pumping rate in *Tethya crypta.* Dotted lines represent hours of darkness when no measurements could be made. (After H.M. Reiswig)

In a 660 cm³ specimen of *Verongia* the change from normal pumping to full cessation took place in 3.8 to 6.4 minutes.

A conduction system has been demonstrated in a large hexactinellid sponge, *Rhabdocalyptus.* A mechanical disturbance or electrical shock applied at one point on a sponge causes the rate of water flow to decrease throughout the sponge. As the openings in this sponge are non-contractile, the decreased flow is probably caused by a cessation of flagellar beating. The rate of conduction is about 0.2 cm/second, much slower than most nervous conduction but not much slower than electrical conduction by epithelia in some animals.

No matter how massive the sponge, its **feeding** is restricted to food no larger than can be engulfed by a single cell: minute organic particles, bacteria, microscopic algal cells, small protozoans, and the gametes or early stages set free in the water by various plants and animals. Individual collar cells (as well as other epithelial and mesenchyme cells) feed independently. The food particles become enclosed within vacuoles into which digestive enzymes are secreted. As digestion proceeds, the contents of the food vacuoles are first acid and then alkaline. The method of ingestion and digestion is similar to that of protozoans; probably no sponge can handle the larger organisms ingested by some of the bigger or more voracious of the protozoans.

Sponges could also obtain a significant fraction of their nutrition by extracting such dissolved organic materials as amino acids from the great volumes of water that they pump through their bodies, and they have been shown to be capable of taking up such materials.

In addition to bringing a constant supply of food, the continuous current passing through a sponge furnishes oxygen to all the cells and carries away the carbon dioxide that results from cellular **respiration.** Consequently, not even very large sponges have any special mechanisms to aid in respiratory exchange, in the **elimination** of indigestible solid residues, or in the **excretion** of the nitrogenous by-products that result from protein metabolism. Such larger sponges do have more complex and efficient water-propelling channels.

Cells of freshwater sponges have contractile vacuoles, which presumably function like those of protozoans in regulating water content.

In **sexual reproduction,** sponges produce large, nutrient-rich **eggs** and small, tailed **sperms**—the kinds of gametes typical of metazoans and uncommon among protists. Sponge gametes usually arise from undifferentiated mesenchyme cells (archeocytes) or from collar cells that dedifferentiate, losing collar and flagellum. In the majority of sponges, both kinds of gametes arise in one individual, which is then known as a **hermaphrodite;** in some, they occur in separate male or female individuals that, except for their gametes, appear exactly alike. In almost all sponges, only the sperms are shed through the excurrent vent into the surrounding water, while the eggs are retained, and **fertilization** is internal, by sperms that enter with the water currents.

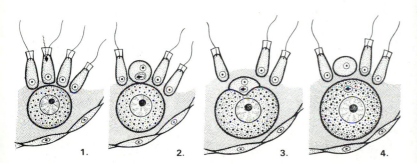

Fertilization is internal in almost all sponges. Sperms released into the surrounding water enter another sponge through the incurrent openings and canals, but cannot directly reach the egg cells, which are buried in the mesohyl. Instead, the sperms are brought to the eggs by collar cells. **1.** Sperm enters a collar cell. **2.** The sperm loses its flagellum and is enclosed in a vesicle. The collar cell loses its collar and flagellum, becomes rounded, and approaches an underlying egg. **3.** The transformed collar cell attaches itself to the egg. **4.** The sperm is transferred into the egg, and the collar cell departs. In some sponges, the collar cell is incorporated into the egg. (Based mostly on Tuzet and Paris)

Spawning of sperms, from *Verongia archeri,* a sponge 1.5 m long. Sperms issued from the excurrent opening, in a column 3 m high, for at least 10 minutes. Depth 49 m. Jamaica, W.I. (H.M. Reiswig)

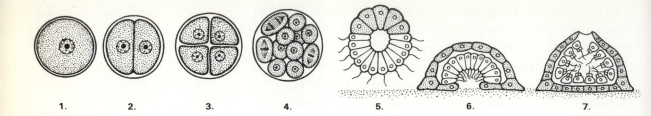

Early development and metamorphosis of a calcareous sponge. 1. Fertilized egg. **2.** Two-celled stage. **3.** Four-celled stage. **4.** Continued cell division produces a ball of cells. (After Tuzet and Paris) **5. Amphiblastula larva,** with hollow interior and half of surface flagellated, settles on anterior end. **6.** Flagellated cells are enclosed (through inversion or overgrowth) by the posterior cells. **7.** Young sponge with lining of collar cells; an excurrent opening has appeared at the free end. (Adapted from P.E. Fell)

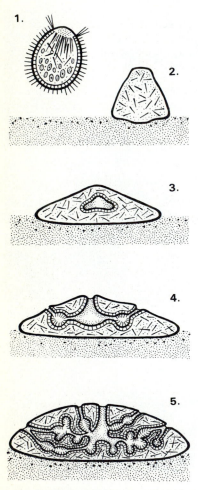

Metamorphosis of a demosponge. The free-swimming 2-layered **parenchymula larva** has a solid interior and is flagellated over most of its surface. It settles down on its anterior end and develops flagellated chambers and incurrent and excurrent openings. (After W. Marshall)

The fertilized egg cell, or **zygote,** develops within the mesohyl of the parent. There the developing **embryo** receives protection and, in some sponges, nourishment for continued growth. Finally, the embryonic sponge, now a ball of many cells, is released as a flagellated **larva.** The term "larva" is applied to any young free-living stage that is not just a miniature adult but will undergo dramatic changes in structure, **metamorphosis,** before attaining the adult form.

Sponge larvas range from simple sorts with a single layer of uniform cells surrounding a cavity, to solid forms with an outer cell layer surrounding a central mass of cells, spicules, and the beginnings of water canals. After swimming about for a short time, the sponge larva settles down, becomes firmly attached, and metamorphoses into a young sponge.

Sexual reproduction in sponges (as in other organisms) results in offspring with new genetic combinations, and perhaps with the potential to succeed in changing times, or in ways, or in places where their parents could not. But most of the energy of a sponge, beyond what is required for maintenance, goes into **growth.** Most sponges continue to grow indefinitely through budding and branching much as plants do. Fragments that break off, or special buds that become detached and are carried away, are able to attach and begin again to grow. This kind of growth, which results in the formation of new sponge masses that are genetically like the sponge from which they arose, is usually labelled **asexual reproduction.** Through growth by asexual reproduction, a single original zygote (a genetic individual) develops as a number of separate bodies that are distributed in time and space. This sort of dispersed individual is characteristic of sponges and occurs as one pattern of development in many other invertebrates.

All freshwater sponges and some marine ones produce units known as **gemmules.** These consist of a mass of food-filled mesenchyme cells, sometimes surrounded by a heavy protective coat strengthened with spicules. The gemmule survives drying and freezing and carries the sponge over the winter season or a dry period. Later, under favorable conditions, the sponge cells emerge through a pore in the gemmule coat, aggregate in a small mass, and grow into a sponge. Sometimes gemmules break away and are carried off by wind or water to establish a sponge in a new place; and in some marine sponges, gemmules may develop into asexual larvas that swim away and grow into new sponges. However, most field observations suggest that gemmules usually serve for survival of adverse conditions and only occasionally in dispersal of new units.

Some sponges are noteworthy for their **regeneration** of damaged or missing parts. More surprising is the ease with which they reorganize and grow as an integrated mass even after the cells are dissociated. When pressed through fine silk cloth, the sponge cells are separated and come through singly or in small groups. In a dish of seawater these cells creep about on the bottom in ameboid fashion. When they happen to come in contact, they stick together; and after some time, most of the cells are found to have united into small masses. Finally, these masses of aggregated cells reorganize into new sponges. If the dissociated cells of two species of sponges are mixed together in a dish of seawater, they sort themselves out to form separate masses, each composed of cells of only one species. Studies of reaggregation of dissociated sponge cells have shown that specific substances on the surfaces of the cells are important in cell recognition and adhesion. If those substances are removed by washing, and if their production is prevented by lowering the temperature of the seawater, the cells do not reaggregate.

Studies of cell-to-cell contacts in sponges may help us to understand such contacts in other animal groups and perhaps suggest means of coping with problems such as the spread of cancerous cells.

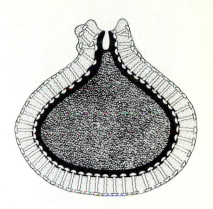

Section of a **gemmule** of a freshwater sponge. (After R. Evans)

The inner mass is a clump of about 500 cells. Throughout dormancy, even when it lasts for several years, these cells retain the ultrastructure of active cells. Within 24 hours of a rise in oxygen tension and water temperature (to 23°C) the membrane below the pore is digested by enzymes and the cells start escaping.

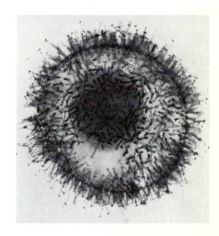

Gemmule of *Anheteromeyenia argyrosperma,* from a Wisconsin lake. Diam. 950μm. (Neidhoefer and Bautsch)

 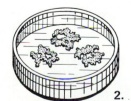 

1. Cell aggregations formed by reuniting dissociated cells. **2. Cell masses** formed by fusion of small cell aggregations. **3. Young sponges** formed by reorganization of cell masses. (Based on M.W. de Laubenfels)

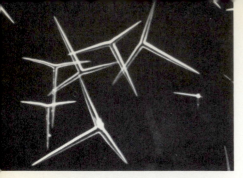

Calcareous spicules of *Leucilla*. As in most other calcareous sponges, the 3-rayed spicules predominate over 4-rayed and needlelike ones. (R.B.)

Calcareous sponges (class **Calcarea**) have a skeleton that consists of spicules of calcium carbonate. Though mostly of small size, rarely up to 15 cm, these sponges may spread out as irregular larger masses. They are exclusively marine and are found in all seas from the intertidal down to 200 m. A few calcareous sponges go down to at least 800 m. They do not flourish where salinity is diminished by influx of freshwater.

Above, **cluster of urns** of *Leucilla nuttingi,* 25 mm high. They hang upside down in shaded rocky crevices at extreme low-tide levels. *Below,* closeup of upper portion of *Leucilla* reveals pattern of 3-rayed spicules that lie with 2 of the rays parallel to the surface, giving it a smooth, shiny appearance. Monterey Bay, Calif. (R.B.)

Cluster of *Scypha (= Sycon).* Each urn is 1 cm high. The bristly surface is due to bundles of spicules that guard incurrent pores. Longer spicules form upstanding collar around excurrent opening. Monterey Bay, Calif. (R.B.)

Below, **purse sponge,** *Grantia compressa.* Generic name is from R.E. Grant, who cleared up the last uncertainties about sponges as animals and in 1836 gave them the phylum name Porifera. Species name is from the collapsed appearance of the creamy-white little sacs when the tide is out. *Grantia* drops body fragments that regenerate into new sponges. England. (D.P. Wilson)

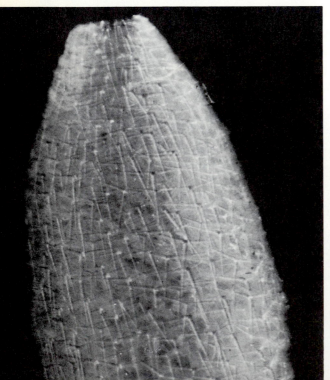

The **demosponges** (class **Demospongiae**) comprise 95% of sponge species, Most are found from the intertidal to the edge of the continental slope, but they continue down the slope in decreasing numbers and occur even at the greatest depths; all of the freshwater sponges are demosponges. A few genera have no skeleton at all. Some have siliceous spicules only, but they are never 6-rayed as in glass sponges. Some have spongin fibers only. A majority have a combination of siliceous spicules and spongin fibers.

Siliceous spicules are sharp and irritating. Those of freshwater sponges in eastern Europe and in the Amazon basin cause itchy skin infections in swimmers and fishermen, or when windblown from dried up ponds or streams, irritate the eyes, nose, or skin. These abrasive qualities have also been put to use. In the USSR dried freshwater sponges have been used to polish silver, brass, and copper. In the Amazon basin sponge spicules are added to clay to strengthen the finished pots.

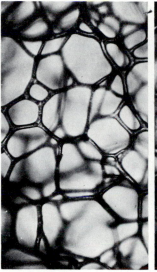

Spongin fiber network is elastic and absorptive. *Left*, a bit of dry sponge skeleton from *Hippospongia*, a commercial sponge. *Right*, water has been added and is taken up by the fibers as well as into the spaces between the fibers. (R.B.)

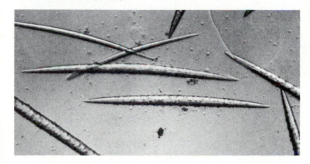

Siliceous spicules of demosponges are non-crystalline hydrated silica. *Above*, from *Lissodendoryx*, marine. *Below*, from *Spongilla*, freshwater; all the larger spicules are needlelike. (R.B.)

Freshwater sponge, *Ephydatia*. This specimen, encrusting a twig in standing shallow water in a Wisconsin lake is only 25 mm long, deeply lobed, and green in color from its green algal symbionts. The same species can be different in size, shape, and color under other conditions. Most freshwater sponges growing on the upper surfaces of objects are

green in color. Those growing on the undersides are brown, tan, gray, or fleshcolored. In quiet water they may be deeply lobed or covered with tufts of long fingerlike processes. In running water they grow as thin, flat encrusting mats, some of which reach 40 m^2. (Neidhoefer and Bautsch)

In one experiment with *Spongilla lacustris*, green specimens with symbionts grew 2 to 5 times faster in a sunny pond habitat than did similar nearby sponges that had lost their green symbionts after being shaded with black plastic. Both green and "white" *Spongilla* continued to feed on bacteria at the same rate.

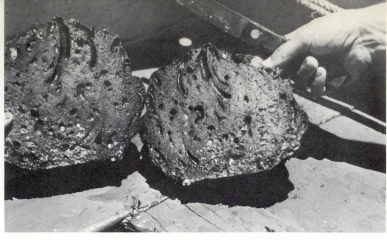

Living commercial sponge, *Hippospongia,* pulled up, by means of prongs on the end of a long pole, from its attachment on the sea bottom at Batabanó, Cuba. (R.B.)

The same sponge cut open. The cut surface looks like raw beef liver, and in this gross aspect the supporting framework of spongin fibers is not distinguishable from the mass of living cells. Several excurrent channels are visible. Batabanó, Cuba. (R.B.)

Yellow **bath sponges** are the product of cleaning, bleaching, and drying *Spongia officinalis.* In 1959, the year this photo was taken on a street in Athens, there were 105 sponge fishing boats, with about 500 divers, registered in Greece. Their catch weighed 100,000 kg. Greek sponge fishing is now in drastic decline. (M.B.)

Bahaman sponges were the largest portion of world production in 1938. In that year a fungal disease of sponges spread from the Bahamas to Cuba, Honduras, and Florida. This epidemic, overcollecting, and the advent of synthetic sponges reduced sponge trade to a small fraction of what it was. (W.C. Schroeder)

Sponge culture has been tried for restocking grounds depleted by over-collecting. A piece the size of that on the *left* was affixed to a tile, lowered to the bottom, off Florida, and harvested 3 years later, *right,* at an actual diameter of 15 cm. The epidemic put an end to sponge "farming" in the Bahamas, which yielded a harvest of 140,000 sponges from 1935 to 1939. (From *Bull. U.S. Bur. Fisheries*)

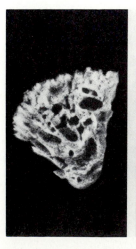

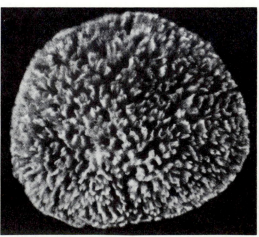

Preparing sponges for market at sponge exchange in Tarpon Springs, Florida. Cleaned sponges are pounded with a mallet (to break up shells or skeletons of invertebrates in sponge cavities) and trimmed to a regular shape with shears. (R.B.)

Dried **elephant's ear sponge,** *Spongia officinalis lamella,* valued for smooth texture and flatness when cut up into pieces. Used in pottery-making. Outer surface shows incurrent openings and inner wall of cup reveals many excurrent openings. N. Africa. (R.B.)

Sturdy **"wool"** sponge, *Hippospongia lachne,* from Florida waters makes up the bulk of world trade in sponges used for washing cars and cleaning walls. This sponge sells for more than $30. (R.B.)

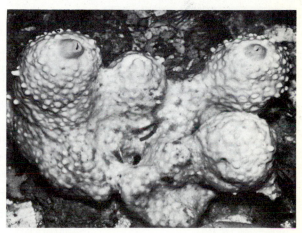

Boring sponge, *Cliona celata,* appears as bright yellow encrusting masses, as shown here, when the mature sponge has exhausted the calcareous shell or rock in which the young sponge grew. N.W. Florida. (R.B.)

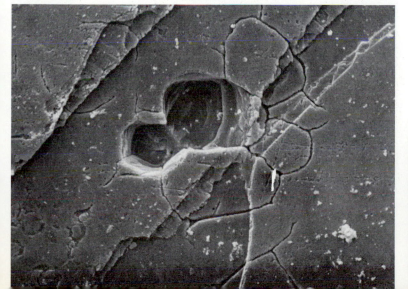

In the **boring process** of *Cliona lampa* chemically etched cracks are penetrated by pseudopods of mesenchyme cells that loosen chips, about 50μm in diameter. (SEM, K. Ruetzler and G. Rieger)

Abalone shell riddled with holes made by a boring sponge. (R.B.)

Sclerosponge, *Ceratoporella nicholsoni,* with tissues supported by a skeleton of siliceous spicules and organic fibers. (Both photos by T.F. Goreau, courtesy W.D. Hartman)

Underlying skeleton of *Ceratoporella* is of dense calcium carbonate—a skeletal feature that distinguishes sclerosponges and has earned them the alternate name, coralline sponges.

The **sclerosponges** (class **Sclerospongiae**) were long known as fossils, but the discovery of living specimens had to await the advent of scuba diving, which has made it possible to search for these marine sponges on steep rocky slopes or in dark underwater caves and crevices. They differ from all other sponges in that the supporting skeleton of siliceous spicules and organic fibers is underlain by a basal skeleton of dense calcium carbonate.

Closeup of glass sponge skeleton, *Euplectella.* (R.B.)

The **glass sponges** (class **Hexactinellida**), named from their siliceous, hexactine (6-rayed) spicules, are found in all seas, from pole to pole, but are most abundant on deep tropical bottoms. Dredged from such depths, hexactinellids come up in poor condition, but populations of *Rhabdocalyptus* in shallow waters off the west coast of Canada have provided material for

Glass sponge *(Staurocalyptus).* Long spicules protruding from the surface help prevent clogging of the small pores. Dried specimen. (R.B.)

detailed studies of structure and physiology. Glass sponges differ from all other sponges in that the bulk of the body is a **syncytial network,** a continuous mass of tissue strands containing many nuclei not separated by cell membranes. Even the fairly discrete mesenchyme cells of several kinds are often connected to one another and to the syncytial tissue by bridges of cytoplasm. The syncytium is supported by the skeleton and by a thin collagenous layer (mesolamella), which represents the thick mesohyl matrix present in other sponges. There is no outer epithelium, but only a thickened layer (dermal membrane) formed by the syncytium. The pores are simple holes, which are not surrounded by pore cells or muscle cells and are not capable of being closed. There are thimble-shaped flagellated chambers, but no epithelium of separate collar cells. Instead, "collar bodies," each consisting of a flagellum surrounded by a microvillar collar, arise from a syncytial network. As the pores are always open, water flow seems to be controlled entirely by stopping or starting of the flagellar beat, and this appears to be coordinated by signals conducted through the syncytium, the only conduction system that has so far been detected in any sponge.

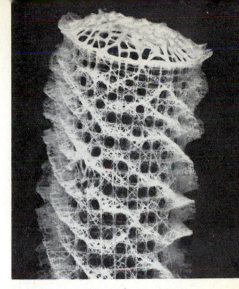

Euplectella. *Above,* upper end enlarged to show latticework of siliceous spicules and excurrent vent covered with a sieve plate. *Below,* whole skeleton. Long spicules at the base anchor the sponge in soft, deep Pacific bottoms, especially off Japan and the Philippines. Glass sponges on soft oozes may have anchoring spicules up to 3 m long. Others live attached directly to hard substrates. Cleaned and dried skeleton. (R.B.)

THE SPONGE BODY PLAN is unique. Water currents carrying food and oxygen enter through minute pores in the body surface, and wastes leave with outgoing water through one or more large vents. No other animals feed by means of collar cells or undergo a pattern of development in any way resembling that of sponges. Hence, it is thought that the sponges evolved from a group of protozoan ancestors different from the ones that gave rise to any of the other metazoans. And the phylum Porifera has sometimes been set aside in a separate subkingdom of animals, the PARAZOA.

The sponge body plan is important in a comparative sense, as an example of the cellular level of organization, with a high degree of independence among the various cells and relatively little intercellular coordination.

The sponges are interesting in their own right as ecologically significant members of marine and freshwater communities, and as members of an old and widespread phylum with an elegantly simple design as living filters.

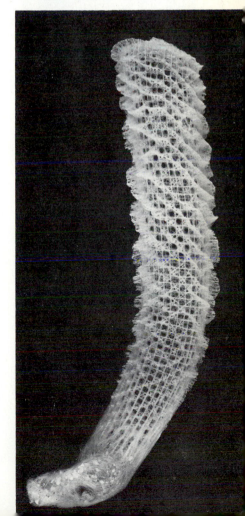

Calcareous spicules.

Left, **4-rayed spicules** from demosponges without spongin. *Right,* **demosponge combination** of siliceous spicules and spongin network.

Hexactine siliceous spicules.

CLASSIFICATION: Phylum PORIFERA

Subphylum CELLULARIA, with discrete cells.

Class CALCAREA

The **calcareous sponges.** Skeleton of discrete spicules of calcium carbonate. Includes all of the simpler (asconoid and syconoid) sponges as well as some complex (leuconoid) ones. Small, up to 15 cm high. Typically of drab colors. Single or clustered vase-shaped symmetrical forms. All marine, mostly in shallow water. *Grantia. Leucilla, Leucosolenia, Scypha* (= *Sycon*).

Class SCLEROSPONGIAE

The **coralline sponges.** Skeleton of siliceous spicules and organic fibers within the body of the sponge, overlying a basal skeleton of dense calcium carbonate. Leuconoid, with growth forms ranging from small thin crusts to massive domes a meter or more in diameter. All marine, usually found in deep water or in underwater caves and dark crevices. *Ceratoporella, Goreauiella, Merlia.*

Class DEMOSPONGIAE

The **demosponges.** When present, spicules are siliceous, but never 6-rayed, as in glass sponges. All of leuconoid structure. Demosponges include the majority of sponges and span the full range of sizes, shapes, and colors. Marine forms found from the intertidal to abyssal depths. Freshwater demosponges range from quiet ponds to running streams. Without a skeleton, *Halisarca.* Spicules, but no spongin fibers, *Tetilla.* Spongin fiber network only, *Hippospongia* and *Spongia* (the commercial sponges). Siliceous spicules combined with a spongin network, marine: *Axinella, Cliona* (boring sponge), *Halichondria, Haliclona, Microciona, Mycale, Reniera, Scolymastra, Spheciospongia, Tedania, Tethya, Verongia;* freshwater: *Ephydatia, Spongilla.*

Subphylum SYMPLASMA, largely syncytial.

Class HEXACTINELLIDA

The **glass sponges.** Skeleton of hexactine (6-rayed) siliceous spicules (or derivatives of these). Spicules discrete, and usually some also loosely bound or fused into a framework. No regular system of water channels; irregular spaces. Moderate size: mostly 10 to 30 cm high, up to 1 m. Radially symmetrical forms, attached to hard substrates or rooted at the base by a tuft of long flexible spicules. All marine, rarely in shallow water, mostly in moderately deep water, down to great ocean depths. *Euplectella, Staurocalyptus, Rhabdocalyptus.*

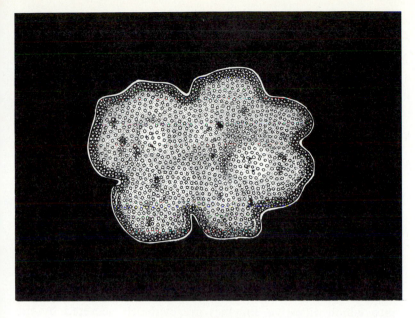

Chapter 4

Placozoans and Mesozoans

Barely metazoans

PLACOZOANS

DISCOVERED, described, and named about a hundred years ago (1883), *Trichoplax adhaerens* was seized upon as a rare bit of real live evidence in the then-current controversy about the nature of "the ancestral animal," which was otherwise fueled entirely by hypothetical models, protozoans, and early developmental stages of various animals. Soon, however, the excitement was dampened by an argument about whether the discovery was just a larval stage of a cnidarian (see chapter 5), and *Trichoplax* was subsequently more or less forgotten. Rediscovered in the early 1960s and cultured in the laboratory, it is now widely accepted as a simple metazoan, the only known member of the phylum **PLACOZOA.**

Trichoplax may be found gliding over the walls of seawater aquariums containing corals or other tropical marine organisms, the only kind of habitat in which it has so far been encountered. It looks like a giant, flattened ameba, measuring up to 2 or 3 mm across, and takes on variably lobed shapes as it moves this way and that in a strikingly ameboid manner. However, when examined more closely, it is found to consist of many tiny cells organized into two epithelia around a fluid middle layer of mesenchyme cells, like a filled pancake.

The **upper epithelium** is thin (1 μm) and loosely constructed of *cover cells,* rather flat cells, each with a central bulge containing the nucleus, as in the outer epithelium of sponges. Each cover cell bears a single flagellum. The only other cells in the upper epithelium are of a rounded type containing large lipid droplets, possibly degenerate cover cells.

Placozoan in the drawing that heads this chapter is from K.G. Grell, who obtained the animals from corals and other marine organisms transported from the Red Sea to his laboratory at Tübingen, West Germany, where he maintained the placozoans on cultures of flagellates.

placozoa = "plate-animals"

Placozoan moving along looks like a giant ameba. 2mm long. (R.B.)

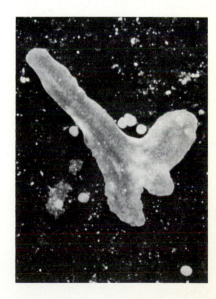

The **lower epithelium** is a thicker, denser layer of columnar *cylinder cells,* each with a single flagellum surrounded by microvilli, and non-flagellated *gland cells.* Between the epithelia, in a fluid-filled space, is a network of **mesenchyme cells** containing fibers that are probably contractile and contribute to the twisting and lifting movements of the flat body.

Differentiation of the two epithelia is evident, not only in the cellular makeup, but also in the *righting behavior* of a placozoan. If dislodged from the substrate, the animal promptly reorients itself as soon as it comes into contact with any surface so that the lower epithelium is against the substrate. Differentiation is also seen in the *feeding behavior.* An animal moving over an algal film will hump its body slightly so as to create beneath it a *temporary digestive pocket,* into which digestive enzymes are evidently secreted by the lower epithelium, for when the animal moves on again, a clear space is left where it rested. Sections of cells from feeding individuals, examined with the electron microscope for evidence of ingested particles, reveal no recognizable elements from the food material; digestion seems to be entirely extracellular. The externally digested nutrients are taken up by the lower epithelium. Thus, functionally, the lower epithelium corresponds to the inner or gut epithelium of other animals, while the upper epithelium corresponds to their outer or epidermal epithelium. The upper epithelium of placozoans probably also participates in nutrition by taking up dissolved organic materials from the seawater, as do epidermal surfaces of many other marine animals.

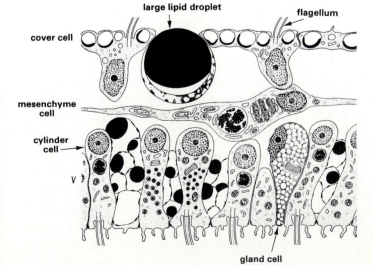

Section through a placozoan, showing the cells of the upper epithelium, lower epithelium, and mesenchyme. Of these, the mesenchyme cells are the most complex, each containing large mitochondria surrounded by vesicles, a vacuole filled with concretions, and symbiotic bacteria within the spaces of the endoplasmic reticulum. (From K.G. Grell)

PLACOZOANS

Placozoans have been seen to multiply asexually in three ways. One is simple **fission.** The animal elongates, constricts in the middle, and each half pulls away from the other until they are connected only by a thin strand of tissue. The strand finally ruptures and the two products, members of a clone, go their own ways. A second method is **fragmentation,** in which small fragments separate from the main body, or the whole body may change in shape from a disk to a donut to a horseshoe to a string of beads as it breaks up. The fragments become independent, feed, and grow. A third method is a kind of **budding** in which both upper and lower epithelia form a small balloon on the upper surface, as if pushed up by a finger from below. An ovoid bud soon pinches off. The upper epithelium completely covers the outside, the lower epithelium lines the hollow interior, and mesenchyme cells lie between the two layers. Thus in budding we see another correspondence of the upper and lower layers to outer and inner layers of other animals. The bud swims away, propelled by the flagella of the outer layer. Eventually, there develops an opening through which the inner layer evaginates, and the flattened, creeping form is restored.

Fission of a small placozoan, itself a product of fragmentation. (R.B.)

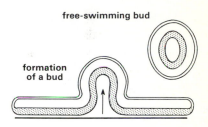

Diagrammatic representation of **budding** in placozoans. The bud incorporates all 3 layers: upper epithelium, mesenchyme, and lower epithelium. The innermost layer of the bud will evaginate to again become the lower epithelium of the creeping placozoan. (From K.G. Grell)

Fragmentation of placozoans. (R.B.)

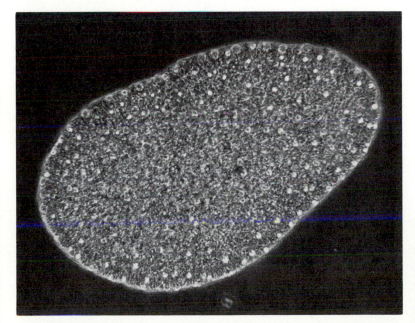

Resting placozoan seen in optical section (with the microscope lens focused close to, but below the upper surface) shows shiny lipid droplets in what are thought to be degenerating cover cells, mostly around the periphery of the animals. (R.B.)

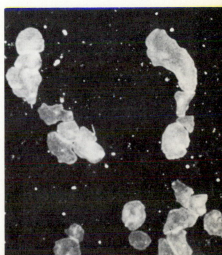

In the laboratory, it has been found that a culture of *Trichoplax* cannot go on indefinitely growing asexually, even though supplied with plenty of food. Eventually, the vessel becomes crowded, and the clone members begin to degenerate unless the culture is divided and put into fresh cultures. But when two different clones were mixed, a new development was seen, apparently a form of **sexual reproduction**. The animals still began to degenerate after a certain density was reached, but many of them contained a single huge cell (occasionally 2 or 3), many times the size of the ordinary cells, with a prominent nucleus and the cytoplasm filled with yolk granules— evidently an egg. Although no sperms were seen, the egg soon developed a raised membrane (resembling the "fertilization membrane" that many animal eggs form after sperm entry) and began to cleave into 2 cells, 4 cells, and so on. Cleavage of the egg continued within the parent or, if the egg was released by breakdown of the degenerating parent, in the surrounding seawater. In culture, cleaving eggs were seen with as many as 32 cells, but none developed beyond this stage, and the rest of the embryonic development (to a larva? or directly to an adult form?) remains unknown. Perhaps *Trichoplax* is a parasite and requires a host at this stage, like the mesozoans to be described in the next section. Or perhaps it is only that culture conditions were not quite right, and improvements in technique may permit us to view the full course of the life-history of a wonderfully simple metazoan.

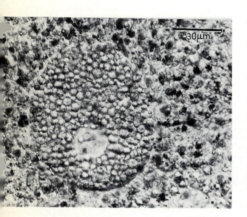

Egg of placozoan grows to about 120 μm in diameter. (K.G. Grell)

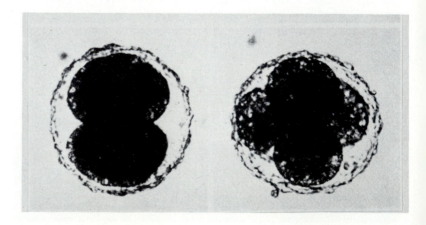

Cleavage of the egg into 2 and then 4 cells. (K.G. Grell)

MESOZOANS: Orthonectids and Dicyemids

THE NAME Mesozoa was originally intended to convey a level of organization somewhere between that of the simple unicellular protozoans and the complex multicellular metazoans. It was created for the second of the 2 small phyla described below, the dicyemids. Later, the orthonectids were added to the group, as well as a number of other small, simple organisms that didn't seem to belong anywhere else, including the placozoans. Eventually, most members of the medley were removed to other categories, while the orthonectids and dicyemids were left in the phylum Mesozoa. Recently, however, students of these animals have concluded that the 2 groups are not closely related and do not constitute a single phylum. They are placed together here as mesozoans in the original sense.

Orthonectids ("straight swimmers"). At the left are the free-living stages: a small male and a large female and, between them, a tiny bilobed larva. *Rhopalura granosa.* (After D. Atkins)

At the right is the parasitic stage, an ameboid syncytium, which contains several developing sexual forms within its cytoplasm.

THE ORTHONECTIDS swim free in the sea as adults, minute ciliated organisms less than a millimeter long. Only about 20 species comprise the phylum **ORTHONECTIDA,** and in most the sexes are separate. The bodies of the adults consist merely of an outer layer of ciliated *jacket cells* arranged in rings and surrounding *contractile cells,* which in turn enclose a mass of developing sperms or eggs. There is a constant number and arrangement of cells in each sex of each species, but the two sexes differ markedly, the male usually being much smaller than the female. The eggs are fertilized and develop within the body of the female. Two-layered ciliated larvas are released and enter a variety of invertebrate hosts: turbellarian flatworms, nemerteans, gastropod and bivalve molluscs, polychete annelids, brittle stars (echinoderms), and ascidians (chordates). Within the host, the orthonectid larva sheds its layer of ciliated cells, and the inner cells give rise to ameboid **syncytia** in the tissue spaces of the host. The syncytia grow and multiply by fragmentation, and the parasite may severely damage the host. After a time, a small number of discrete cells appear within the syncytium and these begin to divide, giving rise to adult forms, commonly of both sexes, which then escape from the host.

Dicyemids (dye-sye-eé-mids) attached to the wall of the excretory organ of their cephalopod host. The tiny free-swimming larvas will be released to the sea.

Dicyemids are found only in octopuses and cuttles that live on or near ocean bottoms, and are absent from squids except for one species with the unusual habit of swimming near the bottom. On temperate and polar sea bottoms adult benthic cephalopods generally are 100% infected, though without any apparent harm. The incidence is lower in the subtropics, and no cephalopod of the tropics or open ocean is known to harbor dicyemids.

THE DICYEMIDS are a little larger than orthonectids, up to 10 mm long, and more restricted in their range of hosts; they have been found only in the excretory organs of octopuses and some other cephalopod molluscs (chapter 15). The phylum **DICYEMIDA** includes about 65 species. The only known free-living stage is a ciliated larva, which is probably the stage that infects the host, a young cephalopod. The first stage that has been found in the host is a small wormlike form. The wormlike forms of dicyemids have an outer layer of ciliated *jacket cells*, specialized at one end for attachment to the wall of the host's excretory organ. The jacket cells enclose one or more long *axial cells*, and within each axial cell are small cells (axoblasts) that give rise to many small new wormlike individuals; these are released by rupture of the surrounding body wall. The wormlike forms grow by cell enlargement (the

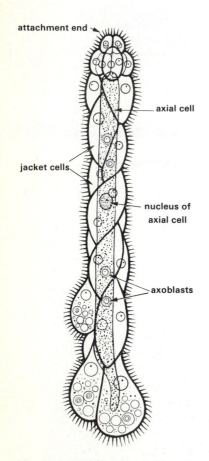

attachment end
axial cell
jacket cells
nucleus of axial cell
axoblasts

Young wormlike form of a dicyemid, *Dicyemennea abelis,* will grow by cell enlargement, becoming longer and slimmer, but the total number of cells will remain constant. A single long axial cell forms the core of the body. (After B.H. McConnaughey)

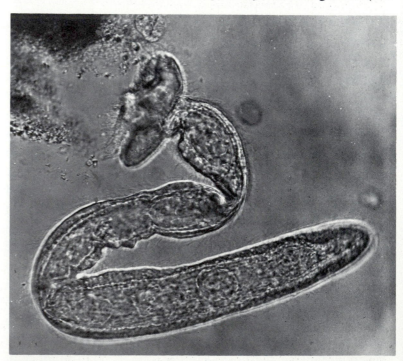

Dicyemid from a giant octopus of the U.S. Pacific Northwest. The attachment end (calotte) is fastened to a bit of tissue that has come loose from the wall of the host's excretory organ. Friday Harbor Laboratories, University of Washington. (R.B.)

final cell number is species-specific and is reached early in development) and continue to multiply asexually until the population of dicyemids in the excretory organ becomes quite dense. At this point, there arise hermaphroditic gamete-producing stages, which develop entirely within the axial cells of the wormlike forms and are never freed. Within the spherical axial cell of this stage, ameboid sperms develop, while the jacket cells turn into eggs. The sperms emerge from the axial cell and penetrate the eggs, and the resulting zygotes develop into ciliated larvas—all still within the axial cell of the final wormlike stage. The ciliated larvas break out through the body wall, and are released in the host's urine to the sea.

In the course of the life-history of a dicyemid, several different kinds of wormlike forms can be recognized. The earliest stage seen in a young newly infected cephalopod is the *stem nematogen,* in which the jacket cells enclose a row of 3 axial cells. The new wormlike forms to which the stem nematogen gives rise, called *nematogens,* are similar to the original form, but have only a single long axial cell. Asexual multiplication continues until the population of dicyemids becomes quite dense within the excretory organ. At this point a slightly different type of individual, the *rhombogen,* is produced, and within its axial cell the gamete-producing stages, or *infusorigens,* develop. Nematogens, rhombogens, and infusorigens are all asexually produced. The fertilized eggs develop into *infusoriform larvas.*

Although the larvas are usually said to be sexually produced, because they result from a fusion of sperm and egg, the system appears to permit self-fertilization only. Indeed, in one species, the sperm serves only to activate the egg (a process called pseudogamy). The egg becomes diploid by fusing with one of its polar bodies after meiosis, and it develops by parthenogenesis, that is, without fertilization. The products of self-fertilization and of diploid parthenogenesis are virtually identical in genetic terms, and neither process is sexual reproduction in the sense of recombining the genetic resources of two different individuals. The formation of eggs and sperms suggests that dicyemids may have evolved from ancestors that engaged in sexual reproduction with cross-fertilization.

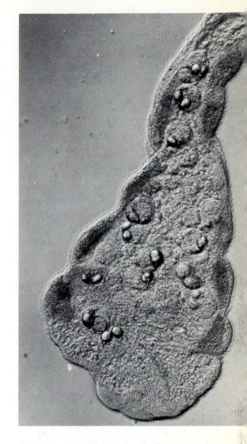

Developing larvas in the free end of a rhombogen stage of dicyemid. From a giant octopus, Friday Harbor, Washington. (R.B.)

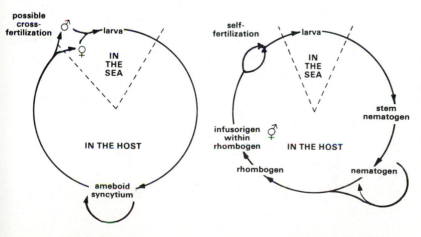

Orthonectid life cycle

Dicyemid life cycle

Life-histories of orthonectids and dicyemids. Most orthonectids have a sexual stage in which separate males and females are released, and cross-fertilization may occur. However, in those species that produce males and females from the same syncytium, or that produce hermaphroditic adults, self-fertilization is likely, unless there are specific barriers against it. Dicyemids, with their hermaphroditic infusorigens, routinely self-fertilize.

Orthonectids and dicyemids are similar in that both are endosymbiotic in other invertebrates. However, their host groups are different, and the orthonectids are demonstrably damaging parasites whereas the dicyemids appear to be harmless commensals. Constancy in the number and arrangement of cells is a feature of both groups, but is found also in a number of other animals of minute size. The life-histories of both groups include a free-swimming larva that infects new hosts, as well as multiple phases of asexual replication, but these are also common features of other parasites. The most striking and distinctive character of both mesozoan groups is the development of discrete cells that proceed to divide and form replicative individuals *intracellularly:* in the orthonectids, the cytoplasm of the syncytium harbors the developing sexual stages—in the dicyemids, the cytoplasm of large axial cells harbors developing nematogens, rhombogens, infusorigens (their own axial cells containing sperms), or larvas.

SIMPLE METAZOANS of several kinds stick out embarrassingly from the definitions that comfortably include most other animals. Like sponges, placozoans and mesozoans lack the nerve cells that are a convenient animal character. They have even fewer differentiated cell types than do sponges and in some respects a less complex organization. Some zoologists resolve this problem by shuffling all these groups into subkingdoms (Parazoa and Mesozoa) separate from the "true animals" (Eumetazoa). Others dismiss the placozoans as developmental stages of cnidarians (chapter 5) and the mesozoans as derivatives of flatworms (chapters 8 and 9), secondarily simplified for parasitism. Still other zoologists pounce on these simple metazoans as transitional animals, taking them (the placozoans especially) as models of what the first primitive animals might have been like or even as living descendants of the "ancestral animal" itself. The assumption that there *was* a single ancestor of all animals is difficult to defend, and in any case it seems highly improbable that any of its descendants survive unchanged today. Nevertheless, the existence of animals organized at the simple level that placozoans and mesozoans display helps us to refine our concepts of what defines an animal and of what forms may have been taken by the early animals that evolved from protists.

Chapter 5

Cnidarian body plan:

Medusas and Polyps

Radial symmetry
Two layers of cells
Tissue level of construction

MEDUSAS, more familiarly known as jellyfishes, are common off marine shores, where they may be seen and enjoyed for their delicate beauty and graceful pulsations, especially from the safety of a pier or a boat. Waders and swimmers have learned that medusas are generally to be avoided, because a few kinds can inflict painful or even fatal stings. Polyps are likely to be seen as the fleshy, colorful, flower-shaped sea anemones that carpet rocky tidepools and seldom or never move about. Around the world, snorkelers and scuba divers seek out tropical shallow sea bottoms for the incredibly beautiful pastel-hued reef corals, soft corals, sea whips, and sea fans—all of them permanently affixed. Despite differences in mobility and body form, medusas and polyps are basically similar in structure and both types occur in the life histories of many members of the phylum **CNIDARIA** (ny-dair´-i-a). Cnidarians are an almost exclusively marine group, and those few that have managed to invade fresh waters, like the hydras and freshwater medusas, are so small and translucent as to escape all but experienced eyes.

The name of the phylum is derived from the Latinized form of the Greek word *knide,* meaning "nettle," in recognition of the numerous microscopic stinging threads shot out by cnidarians in capturing prey or in defending themselves. A cnida, more often called a **nematocyst,** is a fluid-filled capsule enclosing a spirally coiled hollow thread, really a tube, that can be suddenly everted. Nematocysts are found in all cnidarians and are unique to this phylum.

Jellyfishes are related to fishes only insofar as both are animals. And so the term *medusa* is generally preferred despite its own origin from no more than a fancied resemblance of the waving tentacles to the snaky tresses of the mythical Gorgon Medusa.

To emphasize the fact that the main hollow in the body is a digestive cavity, medusas and polyps were long classed under the phylum name **COELENTERATA,** from *coel* ("hollow") and *enteron* ("gut"). This term is still used as an alternative name, but it has an unfortunate history. Originally, it also included the remotely related sponges as well as the gelatinous ctenophores.

nema = "thread"
cyst = "capsule"

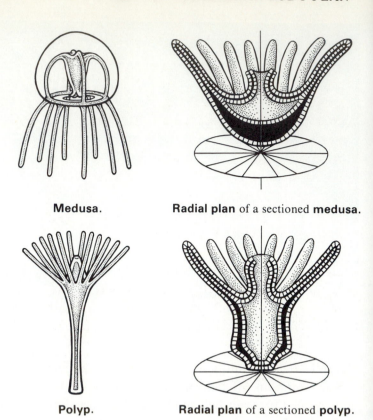

Medusa.

Radial plan of a sectioned **medusa.**

Polyp.

Radial plan of a sectioned **polyp.**

The most consistent external feature of cnidarians is the whorl (or whorls) of extensible tentacles with which they snare their prey. The tentacles range in number from 1, 2, or 4 in small medusas to hundreds in larger medusas and in anemones. The number of tentacles in any one species may be fixed or indefinite.

Unlike the sponges, in which the central hollow of the body does not serve for digestion, *the main hollow of the cnidarian body is a digestive cavity* which opens to the outside through a **mouth.** This important difference in structure allows cnidarians to ingest food of a size much larger than can be taken in by the independently feeding cells of protozoans or sponges.

A typical **medusa** looks like an inverted bowl of jelly, and it swims by gentle pulsations or merely drifts with the currents. The food-gathering tentacles hanging from the rim surround a centrally located and downwardly directed mouth. A typical **polyp** has a cylindrical body that is temporarily or permanently attached. The body column holds aloft an upwardly directed mouth surrounded by one or more rows of food-gathering tentacles radiating out in all directions. Polyps look deceptively flowerlike, but are as carnivorous as medusas.

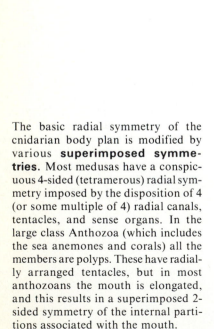

The basic radial symmetry of the cnidarian body plan is modified by various **superimposed symmetries.** Most medusas have a conspicuous 4-sided (tetramerous) radial symmetry imposed by the disposition of 4 (or some multiple of 4) radial canals, tentacles, and sense organs. In the large class Anthozoa (which includes the sea anemones and corals) all the members are polyps. These have radially arranged tentacles, but in most anthozoans the mouth is elongated, and this results in a superimposed 2-sided symmetry of the internal partitions associated with the mouth.

A basic **radial symmetry,** with *similar repeated parts radially arranged around a central body axis,* underlies the cnidarian body plan shared by both medusas and polyps. This central axis runs from the mouth at the **oral end** to the opposite or **aboral end.** The radiating pattern of repeated parts is a design that enables passively floating medusas or attached polyps to snare their prey or fend off enemies from whichever direction they may approach.

ORGANIZATION OF CELLS INTO TISSUES

One might suppose that in the course of evolution a great many different types of cells would arise; but when animals are examined microscopically, it is found that their cells may be classified into a few main types, usually about five: **epithelial, connective** (or **mesenchyme**), **muscular, nervous,** and **reproductive.** All but clearly recognizable nerve cells are present in sponges. Unquestioned nerve cells are also found in the cnidarians, as in nearly all other animals.

An association of cells of the same kind which work together to perform a common function is called a **tissue.** Thus, a mass of mesenchyme cells or some other type of connective cells is known as **connective tissue,** a bundle of muscle cells as **muscular tissue,** and a group of nerve cells as **nervous tissue.** Sponges were presented as animals organized on a cellular basis, but they have some beginnings of tissue formation. For instance, the flattened cells covering the exterior and lining some of the chambers are fitted closely together to form a covering. Such cells, clothing a free surface, are called epithelial cells, and the tissue they form is called an **epithelium.**

The *organization of cells into tissues* is an important innovation in structure, because various cell functions can be performed better by a group of cells of the same kind acting together than by separate cells. Scattered epithelial cells would be a poor protection for the surface of an animal, and epithelial cells almost always occur close together in a sheet as epithelial tissue. Similarly, a single muscle cell lacks the strength to produce much movement, while a bundle of muscle cells contracting together can lift a heavy weight. Because their cells act together in a much more coordinated fashion than do the cells of sponges, the cnidarians may be said to display the **tissue level of organization.**

Primitive cnidarian tissues may be *multifunctional.* The cells may be closely applied to each other, their free ends providing a covering layer comparable with a specialized epithelial tissue. The lower ends of the same cells may be elongated and contain contractile fibrils. These basal extensions, often two or more to a cell, are interdigitated with those of adjacent cells, forming a contractile layer functionally comparable with muscular tissue. Such muscularized epithelial cells may also transmit electrical impulses, produce adhesive substances, or engulf and digest food particles. Slender sensory cells, often with a hairlike protrusion, are more specialized, presumably for receiving information from

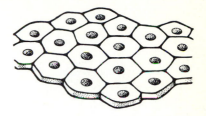

An **epithelium** is a group of cells covering a surface. Epithelial cells, especially those covering external surfaces, are often flattened like the tiles in a mosaic. Epithelial cells of the digestive lining or of glands are usually tall and columnar.

The more complex animals, the human animal among them, have many more different kinds of cells than a medusa or a polyp, but all are modifications of these same basic cell types. An epithelium covers the exterior of the human body, lines mouth and digestive tract, heart and blood vessels, and is folded in various places to form glands. Liver and thyroid cells are epithelial. Bone cells are connective tissue types. Because blood cells are produced in connective tissue, they are included in the connective tissue category.

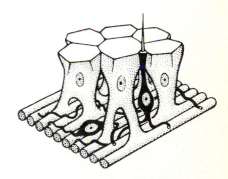

Group of **ectoderm cells** of a cnidarian showing *multifunctional* epithelial cells, with contractile fibrils in their elongate bases, and (in black) *specialized* sensory and nerve cells. (After Mackie and Passano)

the environment and transmitting it to the network of specialized nervous tissue that runs among the bases of the epithelial cells above the muscular layer. In more complex cnidarians, and in other more complex groups, the trend is toward specificity in the cells that work together as tissues.

A MEDUSA—GONIONEMUS

The marine genus *Gonionemus* belongs to a relatively primitive cnidarian family composed of shallow-water medusas. A medusa of *Gonionemus* clings to seaweeds by means of adhesive pads on the tentacles. The pads are located at angular bends near the tips of the slender tentacles, and this suggested the genus name, a combination of two Greek words meaning "angled threads."

The medusa consists of two well-developed layers of cells. The outer layer, or **ectoderm,** literally "outer skin," is a protective epithelium and contains several kinds of cells. The "inner skin," or **endoderm,** lines the internal cavity and is primarily a digestive epithelium, but also contains several kinds of cells. Between the two layers is a firm jellylike material, the **mesoglea** ("middle jelly"), which supports the thin cellular layers and helps to keep the animal's shape. The mesoglea of *Gonionemus* contains no cells, but it is laced with fine fibers that appear after the mesoglea is secreted, presumably by both ectoderm and endoderm.

A medusa of *Gonionemus* is shaped like a shallow bell, or umbrella, with a convex outer surface, the exumbrella, and a concave under surface, the subumbrella. From just above the rim, or margin, of the **bell** springs a fringe of long hollow **tentacles.** Extending inward from the margin is a thin muscular shelf, the **velum** ("veil"), which aids in the swimming movements. From the center of the concave surface, where one would look for the clapper of a bell, hangs a feeding tube, the **manubrium,** with the **mouth** at its tip. The upper end of the manubrium opens into an expanded **gastric cavity** (or stomach) from which lead four **radial canals** that arch through the mesoglea to the margin of the bell. There they join the circular or **ring canal** which runs around the margin and connects with the **tentacular canals** of the hollow tentacles. This continuous cavity through manubrium, gastric cavity, radial canals, ring canal, and tentacles is lined with digestive cells. It is called the **gastrovascular cavity** ("stomach-circulatory" cavity) because it serves the double function of digesting food and distributing it to all regions of the body.

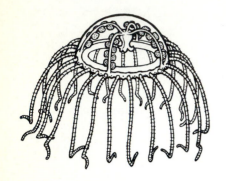

Gonionemus.

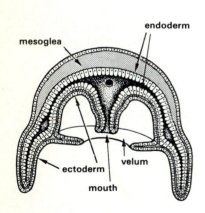

Diagrammatic section through a medusa, showing the two layers of cells and the mesoglea, a bulky mass in the bell and a thin sheet elsewhere.

The ectoderm may also be called the **epidermis** and the endoderm, the **gastrodermis.**

The gastrovascular cavity may also be called the **coelenteron,** or the **enteron,** or the **gastrocoel.**

The surfaces of the bell, and the external surfaces of the tentacles and manubrium, are covered with ectoderm. The velum consists of two opposing layers of ectoderm, held together by mesoglea. The gastrovascular cavity is lined with endoderm, and a single layer of endodermal cells lines the subumbrella of the bell, with only a thin sheet of mesoglea bonding it to the subumbrellar ectoderm. Except for the velum, then, a medusa of *Gonionemus* consists throughout of ectoderm and endoderm held together in different regions by a thin or by a bulky layer of mesoglea.

The mass of mesogleal substance in the bell provides bulk and tensile strength. Yet the water content of a medusa, averaging a variety of species, is about 96%. (This figure can be compared with a water content of about 65% for the human body.) The mesoglea of a medusa has a salt content of about 3% and an organic content of only about 1%, less than the proportion of gelatin needed to make a firm jelly. This small organic content enables a medusa composed of delicate cell layers to increase rapidly in size and bulk without the investment of time and energy that would be involved in producing an animal of equal dimensions by cell division. The main organic components of mesoglea are a matrix of acid mucopolysaccharide and fibers of collagenous proteinaceous substances, somehow structured so as to hold a surprisingly large amount of water. This watery mesoglea is rigid enough to serve as a firm bed against which muscles can pull, plastic enough to allow for the muscular contractions of the bell, and elastic enough to restore the bell's expanded shape between contractions. The mesoglea may also aid in buoyancy and may play a useful role in insulating certain nervous conduction pathways from adjacent ones. But the principal phylum-wide role of the mesoglea, beginning early in development and continuing throughout the life of the organism, is to provide a firm substrate for the delicate cells.

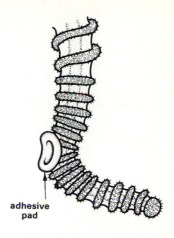

adhesive pad

Nematocyst batteries encircle the tentacles to their tips.

Because mesoglea consists mostly of water, it is probable that it presents no barrier to the passage or circulation of various molecules, whether they be of water, of food, or of nitrogenous compounds resulting from metabolic processes.

When certain medusas have been starved, they have decreased markedly in size, especially of the bell, which contains most of the mesoglea.

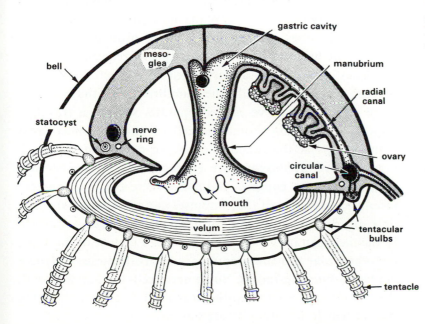

Diagram of a medusa of *Gonionemus* with 1/3 of the bell cut away. Statocysts are sensory structures that aid in maintaining balance. The tentacular bulbs are sites of nematocyst formation. Though the concentric lines representing the circular muscles of the velum appear to be on the surface, these muscles occur in the *inner* ectoderm of the velum.

The spindle-shaped contractile cells of sponges occur only locally and in small groupings, and there are none that deserve to be called *muscle fibers.*

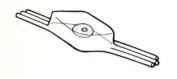

Epithelio-muscular cell with smooth muscle fibers.

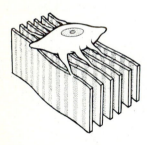

Epithelio-muscular cell with **striated** muscle fibers. (Both modified after L.A. Fraser)

The association of smooth muscle fibers with *slow movements* and of striated muscle fibers with *fast movements* of locomotion occurs in many other animal groups. Thus in vertebrates, the muscles of most internal organs consist of smooth fibers, and the powerful skeletal muscles of striated fibers.

Long **muscle fibers** that run in dense concentrations as bands or sheets of muscle make their first appearance in cnidarians. In *Gonionemus* the ectodermal epithelial cells, except those of the outer surface of the bell, differ from ordinary epithelial cells in that their bases are drawn out into long contractile muscle fibers. There is no separate muscular tissue, the muscle fibers occurring only in the bases of ectodermal epithelial cells. Because they cannot be strictly classed as purely epithelial or muscular, they are called **epithelio-muscular cells.** In the tentacles and manubrium, and beneath the radial canals, the **smooth muscle fibers** of the epithelio-muscular cells run longitudinally. When these lengthwise fibers contract, the tentacles and manubrium are shortened and withdrawn, then gradually extended when the muscle fibers relax again. In receiving prey from the tentacles, the manubrium can be bent and extended in any direction by unequal contraction of the muscle fibers on opposite sides. In the subumbrella and in the velum, the epithelio-muscular cells bear **striated muscle fibers** which run circularly and are responsible for the strong, rapid contractions of the bell in swimming. Primitive medusas are the simplest animals that show striation of muscle fibers.

Medusas possess all of the essentials of the kinds of **nervous mechanisms** present in any of the more complex many-celled animals. Impulses may be spread by the ordinary **epithelial cells** of the bell, or by **nerve cells,** specialized cells with long processes, which form a **nerve net** in the ectodermal epithelium of the subumbrella, the manubrium, and the tentacles. Each cell in the nerve net is separate, so that a traveling impulse, in passing from one cell to another, must cross definite gaps at the junctions between nerve endings. Such junctions, or **synapses,** are characteristic of the nervous systems of more complex animals, where the junctions are usually so constituted that impulses can pass across them in only one direction, making for definite specialized pathways. In the medusan nerve net, some synapses may be of this type, but others may transmit impulses in both directions, and there are *few definite pathways.* Some nerve cells connect also with the muscle processes of epithelio-muscular cells and with the cells in which nematocysts are lodged.

An electrical stimulus applied anywhere on the subumbrella will result in a swimming contraction. Normally, however, the swimming beat is initiated in the **nerve ring,** a concentration of nerve cells around the margin of the bell. Besides its role as a pacemaker for the swimming beat, it receives input from sensory cells and generally controls and coordinates the animal's behavior. Thus the nerve ring may be regarded as the beginnings of a **central nervous system.**

Conduction by both ectodermal and endodermal epithelia of the bell supplements and interacts with the faster activity of the specialized sensory and nervous elements. Conduction in both the nerve net and the epithelia is similar, in that an impulse picked up at any point will travel from one cell to another in all directions.

In many small medusas, epithelial conduction serves those parts of the subumbrella which lack a nerve net.

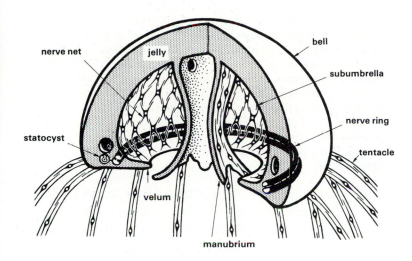

The **nervous elements** of a medusa of *Gonionemus* feature a **nerve ring** consisting of two portions. The larger, outer portion receives *sensory input* from the tentacles, statocysts, and epithelial sensory receptors, and coordinates the tentacles and the manubrium. The smaller, inner portion of the nerve ring is concerned mainly with *motor functions,* sending impulses that activate the swimming muscles of the bell and velum. Cells of the nerve ring connect with each other and with the ectodermal **nerve net,** which runs close to the muscle processes of the subumbrella, manubrium, and tentacles. Not shown in the diagram are the concentrations of nerve cells that run as **radial nerve tracts** beneath each of the 4 radial canals.

The nerve ring consists of nerve fibers that extend from cell bodies in the epidermis. Experimental removal of most of the nerve ring in a hydrozoan medusa does not stop the beating of the umbrella. However, complete removal of the nerve ring does bring an end to the contractions.

In summary, then, the excitable epithelia do not initiate behavior but only react to stimulation, mediating simple generalized responses. The specialized nervous elements initiate spontaneous behavior such as rhythmic pulsations; they integrate sensory information; they control local responses, and they provide pathways for certain complex responses.

The electrical events that accompany the **transmission of an impulse along nerve processes** appear to be similar to those in other animals, although the *rate* at which the impulse travels is much slower in cnidarians.

In some of the larger jellyfishes the rate of nervous conduction has been recorded at about 25 cm/sec. In humans it is close to 12,500 cm/sec, or about 500 times as fast.

The magnitude and speed of the electrical impulse that passes along a nerve are independent of the strength of the stimulus. If the stimulus is adequate to start a nerve impulse, the nerve cell responds completely. If the stimulus falls below a certain threshold of intensity, the nerve cell will not respond at all. This "all-or-none" response is a property of all nerve cells.

In more complex animals, transmission of an impulse across a synapse is typically chemical. A transmitter substance is released from tiny vesicles in one nerve ending and diffuses across the synapse to again initiate the electrical impulse in the other ending. In cnidarians, it is still uncertain whether all synaptic transmission is similarly chemical or whether the electrical event in the first ending directly generates an impulse in the next. In line with the multidirectional conduction in most cnidarian nerve nets, vesicles may be seen in *both* nerve endings at a synapse. However, in some cnidarians synapses have been found with vesicles on only one side, suggesting polarized, one-way synaptic transmission, as in more complex groups.

In at least some medusas, there is evidence that the nerve net and the several conducting epithelia represent functionally separate conducting systems. Still other cnidarians have two nerve nets. Such multiple, diffuse conducting systems appear to be the cnidarian "solution" to a "problem" that in most more complex animals has been solved by the building of specific pathways into their nervous systems. Thus interactions between manubrium and tentacles may occur while a medusa is actively swimming. Such a combination of activities would be impossible with only a single, diffuse conduction system. On the other hand, it has been shown that two systems may sometimes interact. Impulses carried by epithelial cells may stimulate conduction in nerve cells, or vice versa. Or, when the system controlling a feeding response is active, impulses stimulating swimming may be inhibited. Such coordinated **nervous inhibition,** which prevents the simultaneous occurrence of two actions which might oppose or interfere with each other, is a prominent feature of the nervous systems of more complex animals.

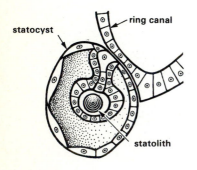

The **statocyst** of the medusa of a *Gonionemus* arises at the base of the velum and between tentacles. Suspended in the cavity of the statocyst ("balance sac") is a stalk that secretes and encloses a freely-moving ball, the **statolith** ("balance stone"). (Combined from L.J. Thomas and from H. Joseph)

Specialized **sensory structures** occur around the bell margin, embedded in the mesoglea between the bases of the tentacles. In *Gonionemus,* each of these is a **statocyst,** a small sac containing a calcareous concretion, the **statolith.** As the medusa swims about, the movements of the statolith within the sac initiate nervous impulses which modify the swimming movements. In addition, prominent swellings at the bases of the tentacles are abundantly supplied with sensory cells. These hollow **tentacular bulbs** communicate with the ring canal. They are lined with pigment but have not been shown to be light-receptive though the whole epithelium of the lower surface is generally sensitive to light. These swellings are filled with nematocysts, being sites of nematocyst formation.

Locomotion in a medusa is accomplished by slow rhythmic pulsations of the bell. In *Gonionemus,* the striated muscle fibers of the ectodermal cells of the bell and velum contract together, forcing water out from the concavity of the bell through the reduced opening in the contracted velum. This jet-propels the animal in the direction opposite to that in which the water is expelled. Between contractions the elasticity of the mesoglea restores the original shape of the bell. When a barrier or irritating stimulus is encountered, the bell and velum begin to contract more on one side than on the other, so that the medusa swims off in a new direction.

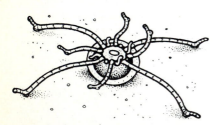

Young medusa of *Gonionemus* resting upside down on substrate with "knee-pads" adhering. If the statocysts are removed, a swimming medusa still appears to maintain its equilibrium, suggesting that there may be other sources of information, perhaps in the velum or in the pull of the pendant manubrium or tentacles. (After H.F. Perkins)

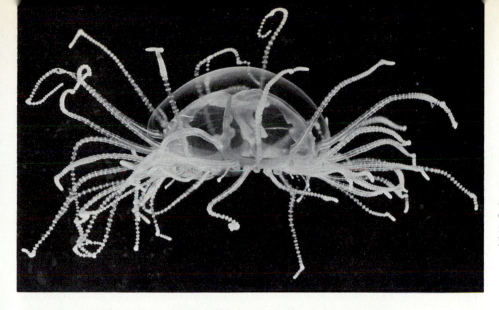

Gonionemus vertens. This medusa feeds on passing small organisms, such as fishes and crustaceans, that touch the outspread and nematocyst-ringed tentacles. Bell is 20 mm in diameter. England. (D.P. Wilson)

Feeding behavior in the genus *Gonionemus* is more active than in most medusas, which appear to swim or float about, tentacles widely extended, depending only on chance contact to capture prey. Except during bright sunlight hours, or in darkness, a medusa of *Gonionemus* may, for hours at a time, use a characteristic "fishing" technique. It swims upward, turns over on reaching the surface of the water, and then floats slowly downward with the bell inverted and the tentacles extended horizontally in a wide snare, from which passing worms, tiny shrimps, or small fishes seldom escape.

What makes this small fragile medusa an effective fisherman, well able to satisfy its carnivorous "appetite," is the special offensive equipment it carries. As the fishing medusa floats down through the water and a small animal brushes one of the tentacles in passing, the unlucky victim is suddenly riddled with a shower of poisonous, paralyzing hollow threads shot out from the **nematocysts,** with which the tentacles, especially, are heavily armed. Each nematocyst consists of a fluid-filled **capsule,** with an opening at one end capped by a **lid** (operculum) and containing a long, spirally coiled hollow thread or **tube.** Each nematocyst is produced within a cell which has become specialized as a **cnidocyte** (or nematocyte). Cnidocytes are continually produced in the tentacular bulbs from undifferentiated cells called interstitial cells. From the bulbs they migrate out along the tentacles to take their places in the raised batteries that coil around each tentacle. The cnidocyte has a **cnidocil,** a modified cilium, projecting from the cell surface. When the cnidocil is stimulated, the lid springs open and the coiled tube turns inside out. This has been compared to the way in which one everts the finger of a glove by blowing into the glove.

Gonionemus possesses only one kind of nematocyst, with large barbs on the bulbous base of the tube. The nematocyst is

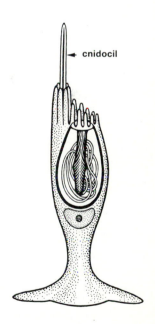

← cnidocil

Cnidocyte of *Gonionemus* with uneverted nematocyst and protruding cnidocil. (After J. Westfall)

Other cnidarians may have several kinds of nematocysts, and different types may occur in different parts of the life history. The distinctive characters of nematocysts are used in identification of species and in establishing relationships among cnidarians.

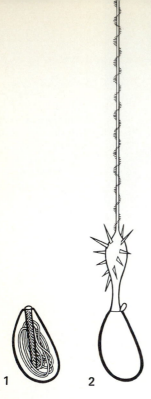

Nematocysts removed from the cells in which they occur. **1**. An undischarged nematocyst showing the coiled tube and the largest of the barbs compacted within. **2**. The discharged nematocyst with lid pushed aside, tube turned inside out, and barbs, large and small, protruding. (Based mostly on H. Joseph)

In some medusas, pores opening at the bases of the radial canals may also serve in elimination and excretion.

The medusas will spawn in the laboratory, even in early afternoon, if they are placed in darkness. The longer they are in the light, the more prompt and more precisely timed is their spawning after they are put in the dark. Thus, late in the afternoon, after a full day of sunlight, a group of medusas placed in darkness will all spawn within a few minutes of each other. This synchronous behavior makes it more likely that the eggs will be found and fertilized by sperms.

everted with explosive force, the barbed base emerging first, piercing the surface of the prey, and then enlarging the wound as the barbs turn away from each other. Through this wound the open-ended tube penetrates into the soft tissues within, injecting a toxin. After the toxin from a number of nematocysts has paralyzed a small animal, the tentacles are wrapped around the prey and contract, drawing it to the mouth.

Extracellular digestion begins in the gastric cavity. Gland cells in the endoderm secrete enzymes, chiefly of the protein- and fat-digesting types, which reduce the digestible parts of the prey to a thick suspension containing many small fragments. This material is then distributed throughout the gastrovascular cavity: to the radial canals, the ring canal, and the tentacular canals, both by muscular movements of the body and by the beating of flagella on the endoderm cells. Food particles are engulfed by the endoderm cells, and the process of **intracellular digestion** is completed within food vacuoles in these cells. For medusas have retained, in part, the protozoan method of food ingestion and digestion. Because preliminary digestion in a medusa takes place in the large digestive cavity, where enzymes are poured out by many cells acting together to disintegrate a food organism, a medusa can eat animals which are very large as compared with those that can be taken by a sponge. (In sponges, as in protozoans, the prey must be of a size that can immediately be engulfed by a single cell.)

Elimination of undigested residues, such as the hard coverings of small crustacean prey, is by way of the mouth, which serves both as entrance and exit for the gastrovascular cavity.

Respiratory exchange and the **excretion** of nitrogenous wastes (mostly in the form of ammonia) are presumably by diffusion. Because of the thinness of the body layers and the circulation in the flagellated gastrovascular cavity, most of the cells, inside and out, are exposed to circulating fluids.

Sexual reproduction in the medusas of *Gonionemus* occurs at certain times of the year, generally in summer. At this time, there appear hanging beneath the 4 radial canals the **gonads,** which are simply folded ribbons of ectoderm containing the sex cells, or **gametes.** Female and male gonads, **ovaries** and **testes,** occur on separate individuals, but look superficially alike. The gametes, **eggs** and **sperms,** develop from interstitial cells in the ectoderm. When ripe, they break through the surface of the gonads and are shed directly into the seawater. Shedding of the gametes, or **spawning,** normally occurs at dusk. Swimming actively through the water, many sperms may encounter each floating egg; but after one sperm enters the egg, effecting **fertilization,** all others are excluded. Each egg is enveloped in a jelly-coat. After fertilization, the egg-envelope immediately shrinks, and the egg sinks to the bottom. If not fertilized within a short time after it is released, the egg dies.

Development begins, as in sponges and in nearly all other many-celled animals, with division of the fertilized egg, or **zygote,** into 2 cells. These promptly divide again, forming 4 cells. Continued division results in a ball of cells, one cell-layer thick, called a **blastula.** The cells of the blastula are elongate in shape, one end tapering towards the center of the ball, the larger end exposed at the surface. Cilia develop on the surface of the blastula, and their beating causes it to rotate within the egg envelope. Finally, it breaks free and spins away. Now the cells begin to rearrange themselves, so that some withdraw completely to the interior of the ball, and the remaining ones spread out to cover the surface. The resulting 2-layered stage, or **gastrula,** becomes an oval, ciliated, actively swimming **planula,** the name given to any such free-swimming cnidarian larva.

After swimming about for up to several days, a planula of *Gonionemus* settles down and may creep on the bottom for awhile; it loses its cilia and finally fixes itself to one spot, attaching by the end that was foremost in swimming. Now the cells of the outer layer, or ectoderm, again begin to divide and

Zygote

2-cell stage

4-cell stage

Blastula

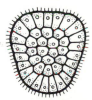

Young planula

Early development of *Gonionemus.* (Based partly on Bodo and Bouillon)

Mature medusa *(Gonionemus vertens),* seen in oral view, with mouth in center and frilled **gonads** on each of the 4 radial canals. 20 mm in diameter. England. (D.P. Wilson)

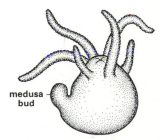

medusa bud

A **polyp** results after further development of the planula. The polyp is an immature, non-sexual stage in the life history of *Gonionemus.* It first asexually produces more polyps like itself and later, new medusas. (Based on H.Joseph)

1. One-day-old asexual bud of a polyp of *Gonionemus*. Ectoderm and endoderm cells multiply at bud site. They push out as a rounded bulge, carrying with them a layer of mesoglea. There is no cavity in the bud until later.

2. Five-day-old bud, approaching time of detachment, is held to parent by a strand of constricting ectoderm.

3. Detached bud, 15 minutes after stage 2. Usually bud creeps about for 2 to 4 days before settling down.

4. Young polyp with 2 tentacles and developing internal cavity. (Modified after H.F. Perkins)

flatten, increasing the outer surface. The inner mass of cells, or endoderm, arrange themselves in an organized layer around a developing central space which becomes the gastrovascular cavity. At the upper, unattached end, 2 pairs of tentacles push out; a mouth breaks through, and the young individual has become a **polyp,** the name applied to any tubular cnidarian which bears a whorl of tentacles around the mouth at the free (oral) end of the body and is attached at the other (aboral) end.

The polyp feeds and grows. Although it cannot swim about but must sit in one place, it catches small animals with its widely extended, nematocyst-armed tentacles, much as the adult medusa does. After some time, it begins **asexual reproduction** by a process called **budding.** On the side of the body both ectodermal and endodermal layers hump up, forming a projection which elongates, constricts close to the polyp body, and soon drops off. The **asexual bud** (frustule) looks very much like the sexually produced planula larva except that it is not ciliated. It creeps about and finally develops into a fixed polyp, which in turn produces more buds that pinch off, creep about, and develop into polyps. Such asexual multiplication is a form of growth in which the individual does not simply enlarge but instead produces a clone of similar bodies. Almost all cnidarians undergo **clonal growth** at some stage. Finally, each polyp of *Gonionemus* produces a bud that develops directly into a medusa, which completes the life history when it reproduces sexually.

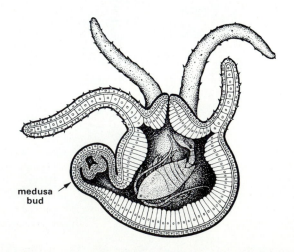

medusa bud

Diagrammatic section through a **polyp** of *Gonionemus* which has a **medusa bud** and has just ingested a copepod. Note that the tentacles of the polyp have a solid core of endoderm cells, in contrast to the hollow tentacles of the medusa. Both hollow and solid tentacles occur in various kinds of cnidarian polyps and medusas. (Mostly after H. Joseph)

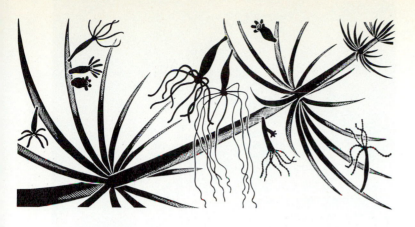

A SOLITARY POLYP—HYDRA

A hydra looks like a slimmer and more graceful version of the pear-shaped polyp of *Gonionemus*. But unlike their marine relatives, the hydras are among the very few cnidarians that live in fresh waters. Hydras are abundant in ponds, lakes, and streams, but are often overlooked because of their small size and habit of contracting down into little knobs when disturbed. Most inhabit quiet waters, and are readily collected by gathering aquatic plants for later examination. Those few species found in swift waters should be sought for on the undersides of loose stones. Hydras attach to aquatic plants or to stones by a sticky secretion from the expanded base. The body appears to the naked eye like a bit of translucent thread frayed out at its free end into several strands, the highly extensible tentacles. These may trail in running water, but otherwise radiate out in a wide snare, gathering passing prey with an impressive array of nematocysts.

The ability of these animals to replace lost or injured parts won for them the name "hydra." An early naturalist saw in this habit a resemblance to the mythical monster Hydra, which was finally slain by Hercules. Hydra had nine heads; and when Hercules cut one off, two grew in its place.

Hydras have 4 kinds of nematocysts, 2 of them concerned with entangling and paralyzing prey, 1 with defense, and another with adhesion of the tentacles in locomotion. Such a variety of nematocyst types, 3 more than the one kind seen in *Gonionemus,* suggests that the apparent simplicity of hydras is a result of secondary simplification from more complex cnidarian ancestors. Why then are hydras so consistently described in textbooks as examples of cnidarian structure, observed alive in the classroom, and favored for research studies of cnidarian ultrastructure, behavior, and regeneration? The answer seems to be that their very aberrancies make them convenient to work with. Their secondary simplification eases the task of detailed analysis. Their presence in fresh waters in most parts of the world makes hydras readily available. And animals that can be maintained in small dishes of infrequently changed water, at ordinary room temperatures, invite long-term experimental studies of growth, reproduction, and regeneration.

The various **regions of the body** consist throughout of 2 layers of cells, ectoderm and endoderm, supported and held together by a thin sheet of noncellular mesoglea in which the bases of the epithelial cells are firmly embedded. The presence of, or the proportion of, cell types differs from one body region

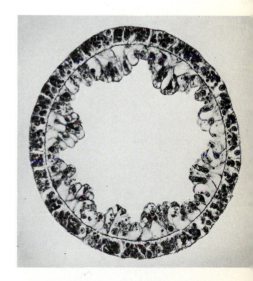

A hydra consists of **2 layers of cells.** Between the 2 layers is the thin mesoglea. Stained cross-section.
(A.C. Lonert)

Hydra with bud. This individual is of the species *Hydra littoralis,* common in running water. The body is about 12 mm long. When well extended, the tentacles are 1.5 times the length of the body. Other species vary in size and shape of body, number and length of tentacles (which may be shorter than, or 3 to 5 times longer than, the body), shape of nematocysts, color, habitat, etc. (P.S. Tice)

hypo = "below"
stome = "mouth"

In some species the stalk is demarcated by a change in diameter, while in others the transition is scarcely noticeable externally.

Some of the sensory cells protrude short, modified cilia, some have rounded tips, and still others end bluntly. The several types of sensory endings are unequally distributed in the various regions of the body and may transmit different kinds of information.

to the next. The round **mouth** is at the apex of a cone-shaped manubrium, which in hydras is called the **hypostome.** From its base arises the circlet of **tentacles,** usually 5 or 6 of them. The hypostome leads into an expanded and greatly expansible **gastric region,** the main part of the body column, where preliminary digestion occurs. Below this gastric ("stomach") region is a **stalk,** ending in an adhesive **basal disk.** At the junction of gastric region and stalk is the **budding zone,** where new polyps arise as asexual buds, grow rapidly, and then separate from the parent hydra.

The **ectoderm** directly faces the hazards, as well as the opportunities, of the external environment. It has long, slender sensory cells, especially numerous on the tentacles, around the mouth, and in the basal disk. These transmit information to a **nerve net** that runs throughout the ectoderm just above the layer of muscle processes. As in *Gonionemus,* there is no separate muscular tissue. Each **epithelio-muscular cell** of a hydra has basal muscle processes running parallel with each other and dovetailing with or overlapping those of adjacent cells. This functionally continuous layer of contractile processes runs *longitudinally.* When the muscle fibers contract equally, body or tentacles can be shortened rapidly, and then, on muscular relaxation, slowly extended again. If the muscular layer contracts more on one side than on the other, body or tentacles bend in the direction of greatest contraction.

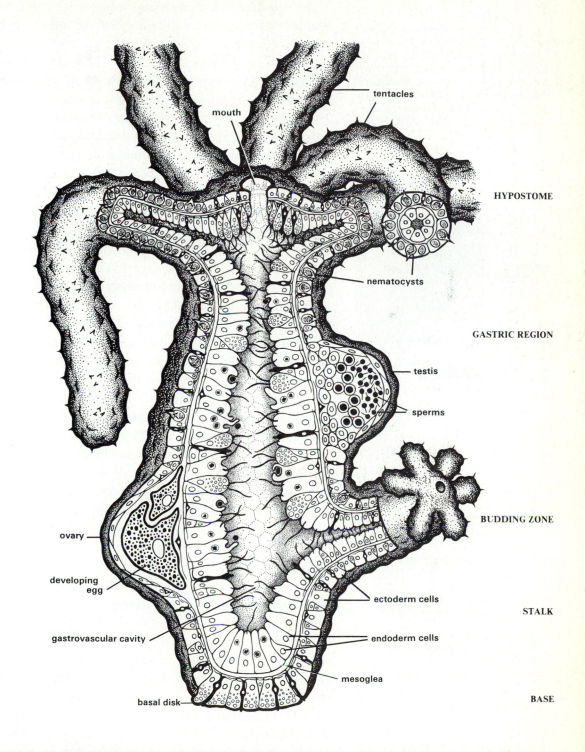

mouth

tentacles

HYPOSTOME

nematocysts

GASTRIC REGION

testis

sperms

BUDDING ZONE

ovary

developing
egg

ectoderm cells

STALK

gastrovascular cavity

endoderm cells

mesoglea

BASE

basal disk

Regions of the body are emphasized in this diagram of a hydra cut away in part. Cnidocytes and reproductive cells occur only in the **ectoderm.** Muscle fibers and nervous elements (shown in black) are most numerous and best developed there. The granule-filled cells of the basal disk secrete the sticky mucus by which the hydra attaches. In the **endoderm** mucus-secreting gland cells are numerous about the mouth, and enzyme-secreting gland cells in the gastric region, but the stalk region has none. The **mesoglea,** everywhere a thin sheet, is thickest at the base, decreases toward the mouth, and is scant in the tentacles.

Clusters of interstitial cells may often be seen, with all members of a cluster synchronously producing one type of nematocyst. In electron micrographs, cytoplasmic "bridges" are often seen between the cells in such clusters. The bridges may be channels of communication between cells that are differentiating in synchrony.

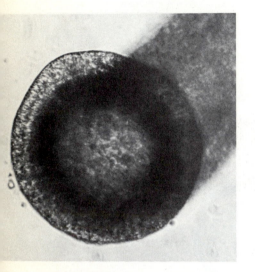

Basal disk seen from underneath in a hydra attached to a glass slide. The aboral pore is visible. (R.B.)

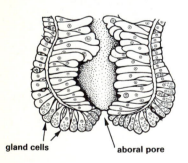

gland cells   aboral pore

An **aboral pore** (seen in longitudinal section of the base of a hydra) perforates the center of the basal disk, at least in some species. Its function is not yet clear. When open, it sometimes exudes a stream of fine particles, perhaps in part derived from the cells being constantly cast off at the basal end. (Modified after I.I. Kanaev)

The ectoderm contains many **interstitial cells,** so-called because they are seen tightly packed in the interstices between the tapering bases of other cells. These actively dividing cells are known to develop into sensory and nerve cells during normal cell replacement or regeneration of large parts of the body. Mostly, however, they are the source for the constant replenishment of cnidocytes. In capturing a small crustacean a hydra may lose 25 percent of the nematocysts on its tentacles. After the eversion of the nematocyst tube, the cnidocyte dies and is replaced, often within 2 days. Even unused nematocysts of tentacles and body are regularly replaced. Reproductive cells (eggs and sperms) also arise from interstitial cells, and are found only in the ectoderm.

The more sheltered **endoderm** has fewer cell types and functions. It is primarily a digestive lining, especially in the gastric region, where there are many tapering gland cells that secrete digestive enzymes. Still more numerous are the **nutritive-muscular cells** with ameboid free ends that engulf partly digested food particles. The basal processes of the nutritive-muscular cells (only one to a cell) run *circularly,* forming a circular muscle layer that, on contraction, decreases the diameter of the column. This muscle layer probably contributes less to body movements and more to the peristaltic muscular waves associated with digestion and circulation of food. Partial, constant contraction of the circular muscle layer keeps the fluid in the gastric cavity under slight pressure, so that the animal is supported by a hydrostatic skeleton. Sensory and nerve cells, and the interstitial cells that replace them, are sparse in the endoderm. Nematocysts are absent, but those ingested with prey or being discarded are often seen, partly digested, inside nutritive-muscular cells. Flagella on both glandular and nutritive-muscular cells help to keep the gastric fluids and food particles circulating about.

The **mesoglea**, as in *Gonionemus,* is apparently secreted by both cell layers, and is chemically related to and serves the same functions as the connective tissues of more complex animals. Under the high magnifications made possible with electron microscope preparations, it can be seen that the basal parts of the epithelial cells of both layers lie in intimate contact with the surface of the thin mesogleal sheet and send cytoplasmic extensions deep into the mesoglea, so strengthening their anchorage. These same preparations also show structural attachments between the epithelial portions of adjacent ectodermal cells, and bondings of the muscle processes where they

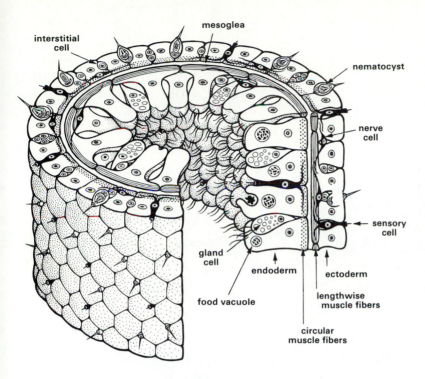

Interstitial cells are selectively killed when hydras are treated with certain chemicals (such as colchicine) or with gamma irradiation. The hydras continue to grow and to bud; and in time, the non-dividing cell types are diluted out, and the animals come to consist entirely of epithelial cells. They lack sensory and nerve cells, cnidocytes, gametes, and sometimes gland cells. Such hydras do not respond normally to stimuli and cannot feed by themselves; but if force-fed, they may be maintained indefinitely. From these experiments, it has been learned that the epithelial cells determine such characters as polyp size (which differs from strain to strain). On the other hand, interstitial cells appear to determine sex. When interstitial cells transplanted into these **"epithelial hydras"** develop into gametes, the sex of the gametes is that of the interstitial-cell donor, not of the individual in which they develop.

Diagrammatic slice through the gastric region of a hydra. A wedge-shaped piece has been removed to provide a longitudinal section of the 2 cell layers in addition to the cross-sectioned surface at the top of the slice.

In the **cross-section** the lengthwise muscle fibrils in the muscle processes of the ectoderm are shown as black dots. The muscle processes of the endoderm meet end-to-end, encircling the body. Sensory cells and nerve cells are shown in solid black.

In the **longitudinal section** at the right of the diagram the muscle processes run lengthwise in the ectoderm, and the circular processes of the endoderm are shown in section as black dots. One nutritive cell is engulfing a partly digested food particle. Four such cells contain food vacuoles in which finely fragmented food is undergoing digestion.

meet end-to-end. The structural attachments keep the delicate cells in place during active body movements. And the mesogleal contacts provide resisting surfaces against which muscle fibrils can pull when they shorten, thus exerting force or producing movement. All this reminds us that functional tissues are not just loose aggregations of cells of the same kind.

Growth and replacement of cells goes on continually in both ectoderm and endoderm. Cell division throughout the length of the animal adds new cells, while old ones are sloughed off at the tips of the tentacles and at the base. The epithelial cells of both layers maintain their numbers by division. But nervous elements and cnidocytes do not divide. They arise from interstitial cells.

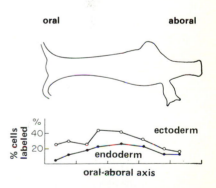

To locate and count dividing cells in a hydra, an experimenter injected into the gastric cavity a small amount of radioactive thymidine, a substance which is incorporated especially into the nuclei of dividing cells. Thymidine-labeled nuclei were scarce or absent in the tentacles and mouth region, increased in proportion along the gastric region, then declined somewhat toward the stalk, but nevertheless showed division of cells in all the regions of the body column. (Modified after R.D. Campbell)

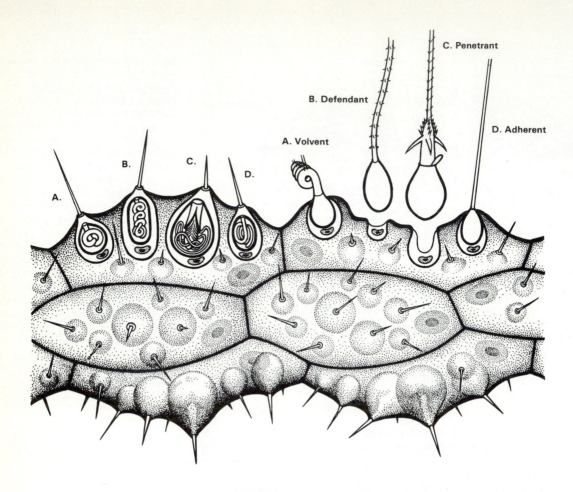

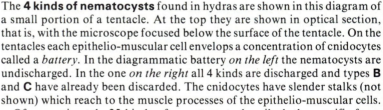

Several volvents are seen clinging to the bristles of a small crustacean. (After O. Toppe)

The **4 kinds of nematocysts** found in hydras are shown in this diagram of a small portion of a tentacle. At the top they are shown in optical section, that is, with the microscope focused below the surface of the tentacle. On the tentacles each epithelio-muscular cell envelops a concentration of cnidocytes called a *battery*. In the diagrammatic battery *on the left* the nematocysts are undischarged. In the one *on the right* all 4 kinds are discharged and types **B** and **C** have already been discarded. The cnidocytes have slender stalks (not shown) which reach to the muscle processes of the epithelio-muscular cells.

Of approximately 25 kinds of nematocysts described, the specific functions of only a few are known: their technical names (in parenthesis below) are based on structure alone.

A. Discharges to prey when chemical stimulation is accompanied by strong mechanical stimulation, as with the rough texture of a bristly crustacean or worm. Winds about bristles and protuberances, holding prey fast: **volvent** (desmoneme).

B. Discharges to disturbance other than prey. Probably effective in killing or repelling animals which are not accepted as food: **defendant** (holotrichous isorhiza).

C. Penetrates and paralyzes prey. The large barbs pierce hard coverings, opening the wound through which the long, barbed tube enters the soft tissue and releases toxin: **penetrant** (stenotele).

D. Anchors tentacles in locomotion. Discharges to prolonged contact with solid objects, not to brief contact as with passing prey, and is even inhibited by prey juices. Has barbs so minute as to be visible only in electron microscope preparations: **adherent** (atrichous isorhiza).

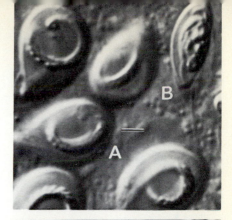

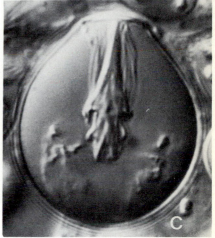

The **mechanism by which nematocysts are discharged,** after appropriate stimulation, is not definitely known. One hypothesis assumes a sudden increase in hydrostatic pressure due to rapid intake of water through the capsule wall. Another proposes that the capsule is constantly under tension and that the lid only needs to be released and opened for the tube to be discharged. A third idea is that fibers surrounding the capsule may contract and thus cause an increase in pressure within the capsule. Highspeed movies of discharging nematocysts demonstrate an initial volume increase, suggesting water uptake; this is followed by a volume decrease that is more rapid than could be accomplished by the fastest known contractile filaments and seems more likely to result from release of tension in the capsule wall.

Control of nematocyst discharge in cnidarians is also incompletely understood. Mechanical stimulation appears to be important, but chemical stimuli are often also involved. Although isolated nematocysts can be made to discharge, there is evidence that when the nematocyst is intact in the animal, discharge may be at least partly under nervous control. Some nerve cells appear to terminate on cnidocytes. And the threshold of discharge varies with the situation or with the physiological state of the animal. For example, an unfed animal will discharge nematocysts more readily than a satiated one. Physiological state could, however, act directly on the cnidocyte without first affecting the nerve cells. A better example is seen in certain sea anemones, in which discharge of tentacle nematocysts is inhibited by contact of the pedal disk with a specific substrate, mollusc shell.

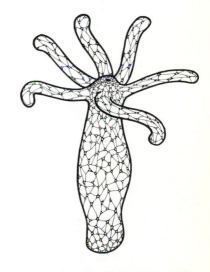

Undischarged nematocysts in a tentacle of a living hydra, lettered as on the facing page. Line = 1 μm. (R.D. Campbell)

The **nerve net** is more concentrated around the mouth and the base of a hydra than elsewhere, but there is no evidence of any specialized controlling group of nerve cells such as the nerve rings of the *Gonionemus* medusa. The nerve net is composed of separate nerve cells and permits diffuse conduction in all directions. A weak stimulus applied to the tip of one tentacle initiates few nerve impulses and results only in the contraction of the single tentacle stimulated. A very strong stimulus to a tentacle tip initiates many impulses and causes the whole animal to contract. The impulses travel much more slowly than in more complex animals; but the nervous mechanisms of a hydra serve well enough for the limited activities of a sedentary animal, enabling an organism composed of many thousands of cells to react as one integrated individual. Epithelial conduction may also play a role in hydras, as in medusas.

The **behavior** of hydras consists of a limited repertoire of responses with which the animals react to a variety of stimuli. Hydras of many species react to light and temperature. They tend to move toward a lighted area, where there are usually more food organisms, but away from a region of high temperature (above 25°C for most species). A hydra will move from the bottom of a dish to the top if the concentration of carbon dioxide rises above a certain level, but otherwise adult hydras do not migrate either up or down in response to gravity. However, in one species studied, young buds always migrated up a vertical surface, and they retained this response to gravity for up to three days after they detached from the parent. It may be that this temporary response helps to distribute young hydras.

The **nerve net,** drawn from a preparation stained with methylene blue, is composed of separate nerve cells. It is most concentrated around the mouth and base.

Although hydras need oxygen, they do not respond to a drop in the oxygen concentration of their surroundings. However, in their habitat in ponds or quiet streams, low oxygen is usually encountered together with high carbon dioxide, as near decaying bottom material. Hydras do respond to the high carbon dioxide content: they ascend toward the surface, where they find themselves in better oxygenated water.

This chemical control of hydra behavior is analogous to that of human breathing. The rate of breathing is not affected by a substantial drop in blood oxygen content. But if the carbon dioxide concentration of the blood increases, then the breathing rate increases, bringing more oxygen to the lungs.

Hydras have no specialized multicellular sensory structures like the statocysts of the medusa of *Gonionemus*. They probably receive information about the orientation of the body column from differential stretching of the muscle processes. For example, when a hydra is attached to a vertical or sloping surface, the pull of gravity will tend to curve the body downward, stretching the upper surface and compressing the lower. The animal maintains its characteristic posture, at right angles to the substrate, by greater contraction of the muscle processes on the upper surface.

The simplest method of **locomotion** in a hydra is a creeping on the slippery and muscularized basal disk. A more rapid, and the most common method, consists of a series of steps, often with quiet periods between, in which the hydra may make exploratory movements before proceeding. One hydra, watched for about 6 hours, took 37 steps, averaging one every 10 minutes. In an exaggerated form, especially under unfavorable conditions, the base swings over the mouth in a kind of somersaulting. In species in which the tentacles are 2 to 5 times longer than the body, a hydra can move by catching hold of some object with the extended tentacles, then loosening the base and contracting the tentacles until the body is pulled up to the object—as if the animal were "chinning" itself. Hydras have also been observed floating free in the water and being transported by water currents, sometimes buoyed up by a bubble of gas attached to the basal disk by a sticky secretion.

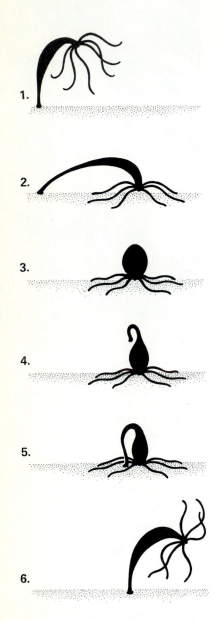

A hydra takes a "step." 1. Body extends and bends over. **2.** Tentacles are set down at a distance from the base. **3.** Column contracts to a knob. **4.** Body extends gradually and the base starts to bend. **5.** Base is set down somewhere close to the tentacles, on either side or between them. **6.** Body may straighten and explore the new site or bend over and initiate another "step." (Adapted from A. Trembley and from R.F. Ewer)

A hydra somersaulting. This is a rapid form of stepping. (Based on A. Trembley)

In **feeding,** the widely extended tentacles of a hydra snare a variety of small crustaceans, worms, tiny fishes, snails, and tadpoles, which in brushing even one tentacle may set off a barrage of nematocysts. The batteries of the tentacles consist mostly of the two nematocyst types directly involved in capture of prey—one or two of the large penetrants, surrounded by many of the small volvents—and a few of the other nematocyst types. The volvents have the longest cnidocils, but have a very high threshold for mechanical stimulation, so that they discharge only when chemical stimulation is combined with strong mechanical stimulation, as in contact with bristly crustacean prey. They wind tightly around any protuberances on the prey, holding it fast while it is numbed by the toxin exuded by the tubes of the penetrants. When enough penetrants have paralyzed the prey, the tentacles wrap around it and draw it towards the mouth, which opens widely to receive it. Food objects are swallowed by muscular contractions, aided by the secretion of mucus from the gland cells lining the area around the mouth.

Sometimes a hydra swallows its food so rapidly that it takes in one or more of its own tentacles, and a hydra was even observed to swallow its own base, which had a small crustacean attached to it. Fortunately, a hydra does not digest its own cells; and after a time the swallowed parts emerge apparently uninjured.

When the nematocyst tubes of a hydra penetrate its prey and inject their toxin, the broken cells of the wounded animal release into the water a chemical, glutathione (a peptide consisting of 3 amino acids). The response to glutathione has been most studied in *Hydra littoralis*. This hydra responds to glutathione with a typical feeding reaction: the tentacles begin to writhe and to curl toward the mouth, and the mouth opens widely to receive the prey. It has been found that the chemical alone will produce the entire sequence of behavior which characterizes the feeding reaction, and the presence of prey is unnecessary. The nutritional state of a hydra may modify its response. An unfed animal may respond for a longer time or to a lower concentration of the chemical than will a well-fed one.

Other cnidarians display feeding reactions in the presence of a variety of single amino acids which normally occur in the cells of any prey animal. Sometimes a combination of substances is active in sequence. One chemical may stimulate the tentacles to bend toward the mouth, while another may be necessary to induce swallowing.

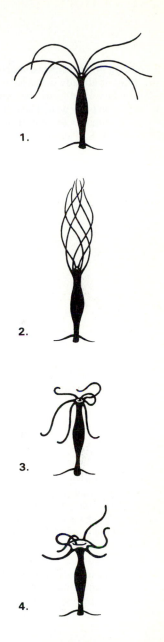

Reaction to glutathione. 1. Hydra with mouth closed and tentacles outstretched. **2.** Glutathione has been added to the water. Tentacles begin to writhe and to move inward toward the central vertical axis. **3.** The mouth opens and the tentacles bend toward the mouth. **4.** The mouth is opened wider, sometimes even turned inside out, as when feeding on large prey. The tentacles are in various stages of contraction, and one or more tentacle tips may be in the mouth. (Modified after H. M. Lenhoff)

Hydra eating an oligochete worm, part of which is already in the gastrovascular cavity and can be seen through the thin body wall. The raised rings on the tentacles are batteries of nematocysts. (P.S. Tice)

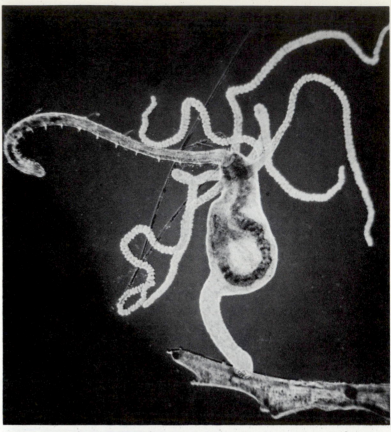

Digestion begins as soon as the worm is safely tucked away in the gastrovascular cavity. Then the hydra spreads its tentacles again and awaits a new victim. The long tentacles, radiating out from a central mouth and controlling the surrounding territory in all directions, are admirably suited to the needs of an animal that spends most of its time in one place waiting for prey to approach. (P.S. Tice)

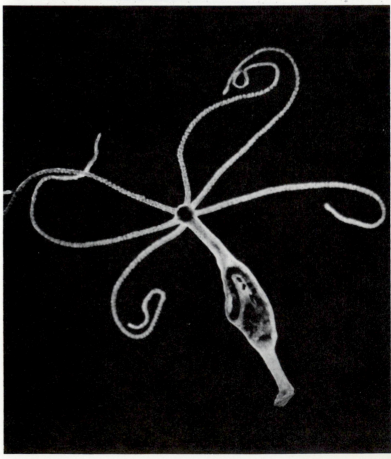

The behavior changes with the physiological condition of the animal. A well-fed hydra usually remains attached, with the tentacles quietly extended. At intervals the body and tentacles suddenly contract and then slowly extend in a new direction, increasing the amount of territory controlled by the animal. If a food organism does not appear after some time, the tentacles begin to wave "restlessly," and the body contracts and expands in a new direction more frequently. If food is still not forthcoming, the hydra may move off to new hunting grounds. While a light touch on the tentacles may cause the food-taking reaction, a stronger touch on the tentacles or the body, as well as shaking or jarring, causes contraction of the animal in varying degrees, depending upon the strength of the stimulus. Unless the hydra has been injured by the stimulus, the body and tentacles will soon again be extended slowly, and the "patient" life of trapping and digesting will be resumed.

Digestion begins in the interior cavity as *extracellular* breakdown of the prey. The gland cells of the gastric lining pour out enzymes (mostly protein- and fat-digesting) that reduce the prey to fragments small enough to be engulfed by the ameboid tips of the nutritive-muscular cells. In addition to active muscular waves in the body wall, flagella keep the fluid suspension of food particles constantly circulating. Once food particles are within the nutritive cells, enclosed within food vacuoles, *intracellular* enzymes complete the reduction of food to fine particles and eventually to single molecules which pass from cell to cell in the endoderm or through the mesoglea to the cells of the ectoderm.

Elimination of indigestible remnants of food organisms, such as the hard parts of small crustaceans, worms, or fishes, is through the mouth, which in all cnidarians is the only large opening of the gastrovascular cavity.

Respiratory exchange and the **excretion** of nitrogenous wastes presumably take place by diffusion. The thin cell layers are constantly exposed, inside and out, to circulating fluids. Excretions that diffuse into the gastrovascular fluid can easily be expelled through the mouth during contractions of the body.

Water regulation poses few problems for a marine cnidarian like *Gonionemus*. In freshwater animals as permeable as hydras, living in water with a salt content much lower than that of their tissues, water continually passes through the body surface by osmosis, and into the cells of the endoderm from the gastrovascular cavity in the process of feeding. However,

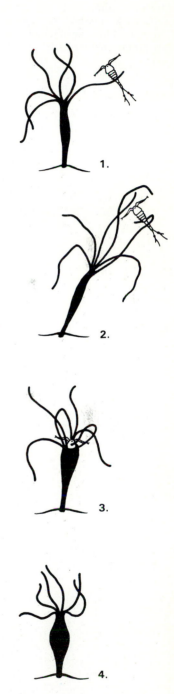

A hydra catches and eats a microscopic crustacean.

Freshwater medusas of the genus *Craspedacusta,* closely related to *Gonionemus,* have been shown to regulate water content as do polyps of *Hydra.*

Budding rate varies with temperature. The green hydra has its highest rate of budding at about 23°C (73°F).

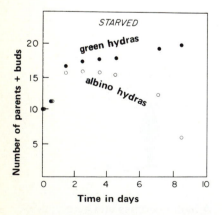

Budding rate and survival during starvation were measured in green hydras with and without symbiotic algas. The animals with algas survived and continued to bud for a longer time. (After L. Muscatine)

nothing like the contractile vacuoles of freshwater protozoans and freshwater sponges is observed in the cells of hydras. Instead, the endoderm of a hydra secretes salts into the gastrovascular cavity, thus drawing excess water from the tissues into the cavity. The accumulated water and salts are expelled through the mouth by periodic contractions of the body column. The hydra manages to maintain a sufficiency of essential salts by taking them up from the external environment, and from its food, and perhaps also by actively reabsorbing certain salts from the gastric fluid.

Accumulation of water in the gastrovascular cavity can be demonstrated by tying off the gastric region below the hypostome and above the stalk. The highly distensible and isolated gastric region soon swells until the body wall bursts, releasing the excess of accumulated water that can no longer be expelled through the mouth.

When well-fed and healthy, hydras undergo **asexual reproduction** by budding. The buds occur about one-third the length of the body up from the base. As in a polyp of *Gonionemus,* both ectodermal and endodermal layers hump up, forming a projection which elongates. At its base the gastrovascular cavity of the bud is continuous with that of the parent, and in this way the bud receives a supply of nourishment. It soon sprouts tentacles and a mouth at its outer end, and in 2 or 3 days the bud looks like a little hydra. It begins to feed on its own, and shortly after this it constricts off from the parent and takes up an independent life.

Because most hydras bud steadily as long as food and physical factors are favorable, observing the rate of bud production is an easy and reliable way of measuring growth and viability. For example, budding rates have been used in studying a green hydra, *Chlorohydra viridissima,* which normally harbors green algal cells in its endoderm. The plant cells give the hydra its color, and if an animal is caused to lose them, it becomes white. Groups of green and white individuals, kept in the light, were found to produce buds at the same rate when fed daily. But if not fed, the white hydras ceased to bud within 1-3 days and all died within another 6 days. Green individuals, however, continued budding for 7-9 days and survived another 7-10 days after that. The soluble nutrients which a hydra receives from the plant cells, although insufficient to support it indefinitely, can sustain it through an interval of starvation.

In studying animals that bud continuously, a question that comes to mind is whether or not such individuals ever age and die. When the rate of bud formation of a number of individually isolated hydras (called the parental generation) was compared with that of their youngest, that is, the most recent bud (the first generation), the youngest bud of this bud (the second generation), and so on through 12 generations, there was a definite decrease in the rate of budding of the parental generation with time. And individually identified hydras have been reported to cease budding and die after a little more than two years. However, clones of hydras, reproducing only asexually, have been maintained for decades without showing any signs of decreased vigor as a total population.

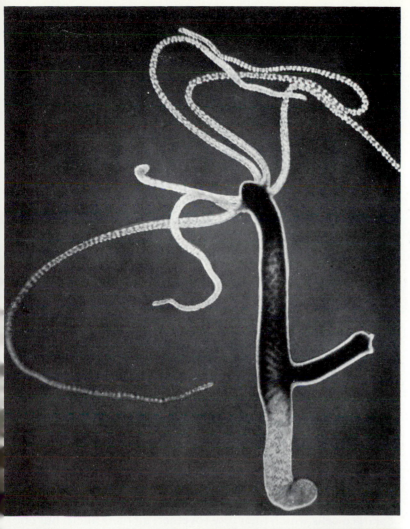

Hydra with young bud. The cavity of the bud is in direct communication with that of its parent, and when this photo was made, the food was seen washing back and forth. The lower third of the body shows no food content in the cavity. It serves chiefly as a stalk, which is more sharply set off from the trunk region in some species. Asexual reproduction of animals was first described in hydras by A. van Leeuwenhoek in 1701. *Hydra littoralis.* (P.S. Tice)

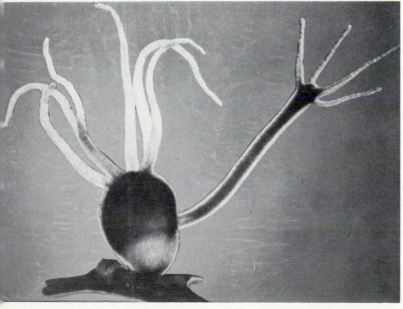

The next day the bud is larger and has well-developed tentacles. Its behavior is relatively independent of that of the parent. Here the parent is contracted, while the bud is extended. (P.S. Tice)

Regeneration would seem to be an easy matter for these simple animals, and hydras do, in fact, show a marked capacity for replacing tentacles or speedily repairing serious injuries. A supply of epithelial cells is provided by division of those already present; and the nondividing cell types—the nerve cells and cnidocytes—develop from interstitial cells. Even if a hydra is cut into a number of pieces, most of the pieces will grow the missing parts and will become complete and independent hydras.

Sexual reproduction, as well as asexual budding, is left to the polyp. Under suitable conditions, hydras will continue indefinitely producing buds which develop into new polyps, but they never produce buds which develop into anything even faintly resembling a medusa. There is no medusa stage in any phase of the life history. Instead, the polyp reproduces sexually at certain times of the year, generally in the fall or winter. In some species both male and female sex cells occur in the same individual, a hermaphrodite. In other species, the two sexes are always separate, but usually male and female individuals can be distinguished only when they are actively reproducing. In certain regions of the ectoderm sex cells suddenly start to grow rapidly, causing the body wall to bulge locally, and these simple bulges constitute the testes or the ovaries. In each **testis** the interstitial cells first enlarge and then divide a number of times to form many **sperms.** An **ovary** also contains many interstitial cells at the start, but a few of these cells begin to grow by engulfing neighboring ones. The result is several enlarged cells which then fuse together; all their nuclei but one degenerate. There finally remains a single large spherical **egg,** packed with food reserves that will later nourish the developing embryo. The ripe egg breaks through the covering ectoderm and projects with its free surface exposed to the water. Sperms, discharged from a testis, swim through the water, and one eventually fertilizes the egg.

Early development proceeds as in *Gonionemus*. Continued division results in a single-layered hollow **blastula.** In hydras the cells composing the single layer now divide, and some of the surface cells migrate inward so that cells accumulate in the interior cavity. This 2-layered **gastrula** drops from the parent and becomes fastened, by a sticky secretion, to the substrate. It is protected by a heavy membrane, or shell, secreted by the embryonic cells. Under favorable circumstances the young hydra may hatch from the shell after a week or more. In winter the developing egg may lie dormant until the following spring, when development is resumed. The young hydra hatches out by rupture of the shell.

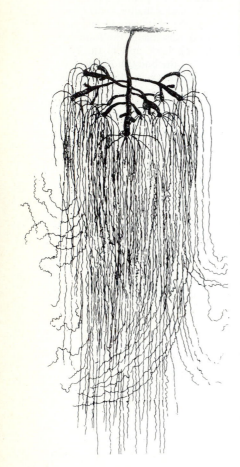

Buds of several generations may remain attached, as described by the Swiss naturalist A. Trembley in 1744. He also described, in hydras, the first successful grafting in animals, the first vital staining of tissues, and the first account of orientation to light of eyeless animals. (From A. Trembley, "Mémoires pour servir à l'Histoire d'un genre de Polypes d'eau douce, à bras en forme de cornes.")

The almost unlimited potentialities of hydras for restoring structural and functional integrity, after almost any experimental manipulation or mutilation, made Trembley express doubts about whether hydras should be classed as plants or animals.

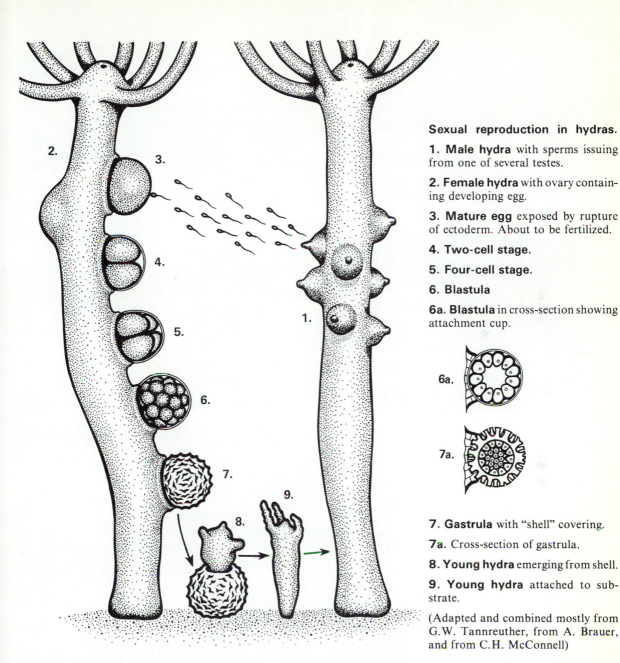

Sexual reproduction in hydras.

1. Male hydra with sperms issuing from one of several testes.

2. Female hydra with ovary containing developing egg.

3. Mature egg exposed by rupture of ectoderm. About to be fertilized.

4. Two-cell stage.

5. Four-cell stage.

6. Blastula

6a. Blastula in cross-section showing attachment cup.

7. Gastrula with "shell" covering.

7a. Cross-section of gastrula.

8. Young hydra emerging from shell.

9. Young hydra attached to substrate.

(Adapted and combined mostly from G.W. Tannreuther, from A. Brauer, and from C.H. McConnell)

It is not definitely known exactly what causes hydras to begin producing sex cells. Food supply may be involved; and another factor may be low temperature, since some species of hydras will produce ovaries and testes after being kept in a cold refrigerator for 2 or 3 weeks. Hydras begin to produce sex cells when the concentration of carbon dioxide increases, such as might occur naturally under conditions of stagnation and crowding. A hydra may also be induced to become sexually active by a small graft from another, already sexually active hydra, suggesting that hormones may be involved.

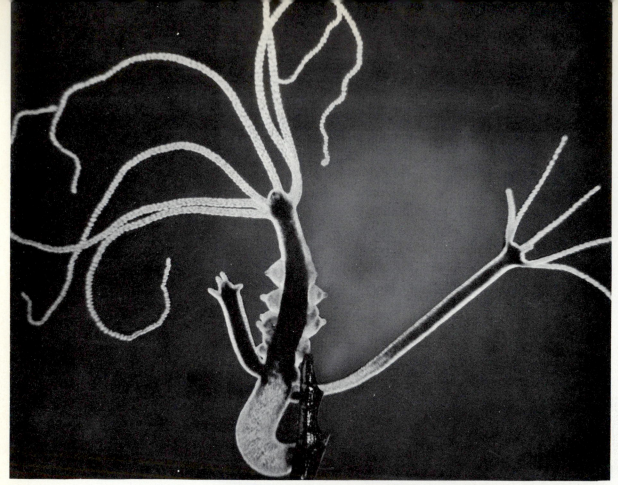

Male hydra *(above)* with rows of testes along the sides of the body, which also bears 2 buds. *Hydra littoralis.* (P.S. Tice)

Female hydra *(right)* with 2 eggs, one still immature and covered by the ectodermal epithelium and the other already extruded. Cnidarians are not noticeably dimorphic with respect to sex, but the sexes can often be distinguished by the color or form of the testes and ovaries during reproductive periods. *Hydra littoralis.* (P.S. Tice)

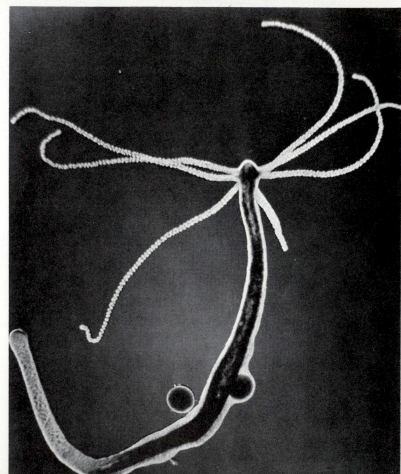

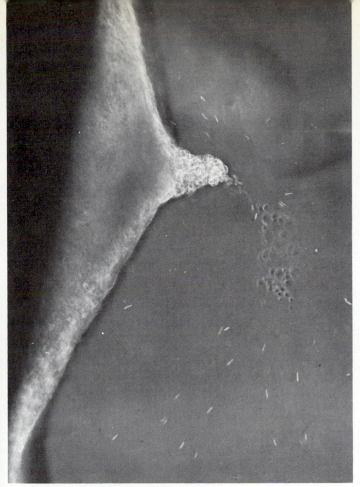

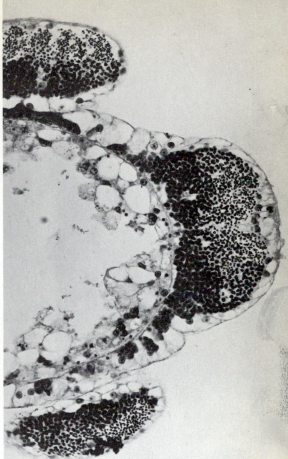

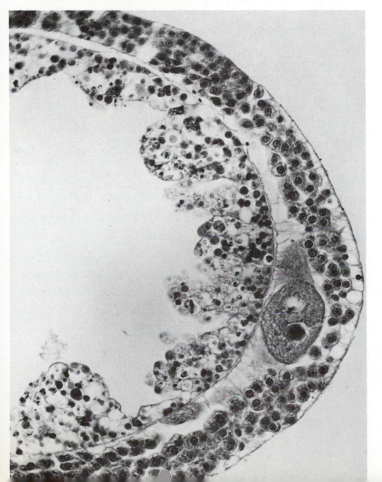

Spawning of sperms *(above, left)* from a testis of *Hydra littoralis*. The sperms, each bearing a single long flagellum, swim actively through the water. (P.S. Tice)

Three testes *(above, right)* are shown here in cross-section. The layered appearance of each testis reflects successive stages of maturity of the developing gametes from the base of the testis to the surface epithelium. (Stained preparation)

Immature egg *(left)* lies in the ectoderm. When the egg is ripe, it breaks through the surface ectoderm. If not fertilized within a brief period, the naked egg dies, but the hydra produces a succession of eggs. In green hydras the algal symbionts are transferred in the egg. (Stained preparation)

COLONY OF POLYPS—OBELIA

Branching colonial cnidarians known as hydroids often grow as delicate plantlike forms on seaweeds, rocks, and wharf pilings along seacoasts. Some of the commonest species are of the genus *Obelia,* colonies of which are usually several centimeters high. A colony arises by budding from a single polyp. In *Gonionemus* the buds separate as larvalike forms, and in *Hydra* they detach only after becoming well-developed, feeding polyps. In *Obelia* the buds fail to separate at all. After repeated budding, there results a branching, bushy growth, permanently fastened to some object and consisting of numerous polyps. The base of the colony is attached to the substrate by a network of horizontal tubes, or **stolons,** from which arise vertical uprights, or **stems.** At intervals along the stems, the polyps sprout on all sides. Each colony member, or **zooid** (zo'-oid), is to some extent subordinate to the colony as a whole—a feature common when clone members remain attached in colonies, but not when they detach and disperse.

Polyps, stems, and stolons are protected by a thin covering of tough protein and chitin, the **perisarc,** secreted by the ectoderm, which encloses all the stems and stolons and extends around each polyp as a transparent cup, or **theca,** shaped like a goblet. When irritated, the muscular polyp can withdraw into this cup, and the rapid contraction and slow expansion of the polyps are among the few movements that can be seen in a colony of *Obelia.* The stems are held erect and relatively rigid by the perisarc, but at certain points the continuous covering is interrupted by more flexible rings which allow for bending as the stems are swayed by water currents.

A polyp of *Obelia* is built on the same plan as a polyp of *Gonionemus* or *Hydra,* and consists of the same two cell layers, ectoderm and endoderm, held together by a thin sheet of mesoglea. The cell layers are composed of cell types similar to those already seen, and the polyp also feeds in the same way, capturing small prey by means of tentacles armed with nematocysts. The tentacles have a solid core of endodermal cells, as in a polyp of *Gonionemus.* But the stems and stolons are hollow, so that both cell layers and the gastrovascular cavity of every polyp are continuous with those of every other polyp in the colony. Food is partly digested in the cavity of the polyp which captures it, and the resulting suspension is circulated about through the stems and stolons by the beating of flagella of the digestive epithelium. Thus, food is distributed throughout the colony in thoroughly cooperative fashion, and digestion is completed in food vacuoles within the cells lining the gastrovascular cavity.

Hydroid of *Obelia.* Upright branch, 1 cm high. England. (D.P. Wilson)

Experiments have shown that distribution of the food is regulated so that if only one polyp on a branch is fed, the food is rapidly distributed to the other polyps; but if all of the polyps are fed, each retains more of the food for itself. Circulation of food is efficient enough so that even if the polyps on only one half of a colony are fed, those on the unfed half will live just as long.

Asexual reproduction by budding steadily increases the number of polyps in a colony, so that the entire colony may, after a time, consist of hundreds of zooids. If we examine such older colonies carefully, we see that the polyps are not all alike. Those already described have tentacles with which they catch prey, and may be called the **feeding polyps,** or gastrozooids. In some of the angles where the feeding polyps branch from a stem, there occur **reproductive polyps,** or gonozooids. These, having no mouth or tentacles and being unable to feed, are nourished through the activities of the feeding members of the colony. Each reproductive zooid is enclosed by a transparent, chitinized vase-shaped theca and consists of a stalk on which are borne little saucerlike medusa buds, the largest and most completely developed near the top, the smallest and least-developed near the base. The topmost one escapes through an opening at the upper end of the theca and swims off as a tiny, free medusa, with 8 statocysts. It gradually matures, adding more tentacles, and finally, gonads.

Sexual reproduction occurs in the mature medusas. Separate female and male medusas, produced by different polyp colonies, shed eggs and sperms into the seawater. Fertilization takes place, and the **zygote** develops into a **blastula** and then into a **gastrula,** with ectoderm and endoderm. The outer cells bear cilia, and their beating propels the little **planula** through the water. After swimming about for a time, it finally settles on rock or on seaweed, and by asexual budding produces a new colony of sessile polyps.

The occurrence of two or more forms in a single species is called **polymorphism,** and there are various kinds. Many other animals have different forms at different stages of the life history (for example, caterpillars and butterflies, chapter 24). Some besides cnidarians even have different sorts of zooids in a clonal colony (see bryozoans, chapter 26). But cnidarians have developed a unique mixture of polymorphic diversity. *Obelia* may be said to be polymorphic in two ways. First, the life history has both a short-lived medusa that feeds and grows briefly before it reproduces sexually and dies, and a long-lived polyp colony that feeds and grows by asexual budding, sometimes producing medusas season after season. Second, the colony has both a feeding polyp that snares and ingests prey, and a nonfeeding reproductive polyp that buds off medusas. Some kinds of cnidarians have colonies with still more kinds of polyps, such as two separate kinds that capture food or ingest it, or colonies with both polypoid and medusoid members.

The portion of a feeding polyp that includes the mouth, tentacles, and gastric cavity is called a **hydranth.**

Medusa of *Obelia* just released is only about 1 mm in diameter. This one was set free when a bit of an *Obelia* colony was placed in a drop of seawater on a slide. The number of tentacles increases as it grows. England. (D.P. Wilson)

What determines that different types of individuals are produced at different stages of a cnidarian life history is still incompletely understood by developmental biologists. Increased carbon dioxide and several other chemicals stimulate hydroid colonies of some species to begin producing reproductive polyps, or even cause feeding polyps to turn into reproductive polyps which bud off medusas. Grafts or extracts from reproductive polyps also transform feeding polyps in this way. Even medusa and polyp types of individuals are not absolutely distinct, since isolated or dissociated medusa tissue has been observed to regenerate in polyp form.

Despite the increased complexity of cnidarians, some show the same capacity for regeneration seen in sponges. The polyps of *Pennaria,* a colonial hydroid, have been forced through fine muslin, emerging as masses of dissociated cells, from which organized polyps were reconstituted.

Regeneration serves not only to repair mechanical damage but also to replace the colony after it regresses from seasonal or other environmental stresses.

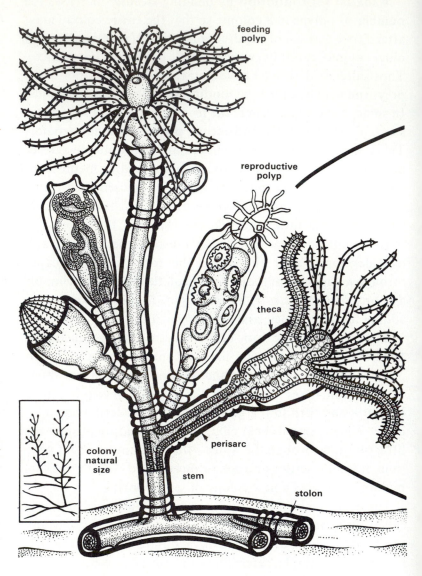

Life history of *Obelia.* One feeding polyp, part of the main stem, and one reproductive polyp (with developing medusa buds) are shown in longitudinal section.

THE CNIDARIANS show a level of structural complexity that goes significantly beyond that seen in sponges. They have **two well-developed layers of cells,** the ectoderm and the endoderm. Nervous elements coordinate the activities of groups of cells that work closely together in performing a single function, achieving what may be called the **tissue level of organization.** Unique devices, the **nematocysts,** used in

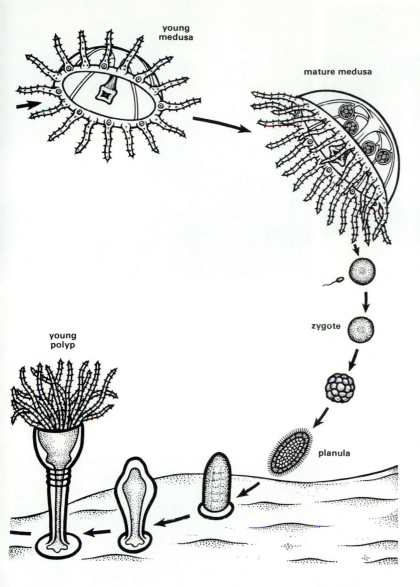

young
medusa

mature medusa

zygote

planula

young
polyp

The occurrence in cnidarian life histories of polypoid and medusoid forms (both diploid) is in no way similar to what is called "alternation of generations" in plants (in which there are haploid and diploid forms). The polyp probably arose as an immature stage and the medusa as the mature sexual adult, much as a caterpillar is a larval stage and the butterfly is the sexual form. If the caterpillar could develop mature sex organs, the butterfly stage could be dropped out of the life history. Such a process of sexual maturity of originally larval or juvenile forms, as has apparently happened in hydras, is known to occur in a variety of animals besides cnidarians, even in animals as complex as salamanders.

feeding, defense, and adhesion, are produced in no other phylum of animals. The variations among cnidarians result chiefly from a shifting emphasis in the evolution of the several groups—first upon the medusa stage, and then upon the polyp stage of the life history. Many of these variations are described in the next chapter.

Hydroids, illustrated here by erect polyp colonies of *Kirchenpaueria (=Plumularia) pinnata,* are members of the cnidarian class **HYDROZOA.** Most hydrozoan life cycles include both polypoid and medusoid phases. England. (D.P. Wilson)

Giant jellyfish of Antarctic waters (McMurdo Sound). The bell size is matched by certain Arctic medusas, but most members of the class SCYPHOZOA live in temperate and tropical waters and are closer in size to the diver's face mask. (P.K.Dayton)

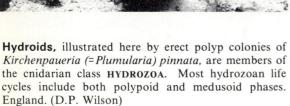

Sea anemone, *Bartholomea annulata,* quietly fishing for food with beaded tentacles gracefully outstretched. Sea anemones are included among the beautiful and diverse "flower-animals," the class **ANTHOZOA.** Florida. (R.B.)

Chapter 6

The Cnidarian Array

Hydrozoans Scyphozoans Anthozoans

THE more than 9000 cnidarian species have often been described as "the flowers of the animal kingdom." But they are almost entirely carnivorous in habit, and the flower-like shape of the polyps is an elegant design for collecting lively prey from all directions. On temperate shores fleshy sea anemones and delicate growths of colonial hydroids crowd tidepools and rock overhangs. Cold waters support vast numbers of the largest of the medusas and even tall branching coral colonies. Yet it is in tropical seas that the cnidarians really come into their own. There towering growths of colonial polyps sway with the currents, and massive banks of reef-forming stony corals occupy all suitable shores, dominating the lives of the other organisms as the trees in a forest dominate the other plants and the animals. Through every opening in the cnidarian thickets, there dart fishes and invertebrates of endless variety. If not avoiding their enemies by sheltering in coral crevices, they may be browsing on the delicate polyps or even chewing the hard calcareous skeletons for the animals that have burrowed into them. At night medusas by the thousands glow in the dark waters as waves or other disturbances stimulate them to luminesce. The scintillations of zoanthids and sea pens fade and light up anew as they react to the touch of wandering animals. Even small colonies of hydroids, among them species of *Obelia*, have been seen to luminesce.

In *Obelia geniculata* the luminescing cells, the photocytes, lie in the endodermal layer of the stolons, stems, and side branches, but not in the hydranths. Photocytes at a stimulated spot fire in synchrony, and from any point of stimulation the luminescence spreads in both directions.

A **trachymedusa,** *Liriope,* develops without any fixed polyp stage. It has a smooth margin and gonads suspended from the underside of the radial canals. (After A.G. Mayer)

The **actinula larva** of trachymedusas and narcomedusas undergoes metamorphosis into a medusa.

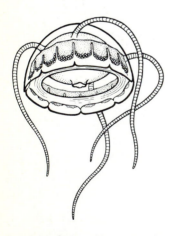

A **narcomedusa,** *Aeginopsis,* has a scalloped margin and gonads attached below the stomach pouches. There is no pendant manubrium. Narcomedusas often swim with the muscular tentacles extended forward (aborally) and actively seizing prey. (After H.B. Bigelow)

The diversity of living cnidarians is predominantly polypoid, and the largest and most complex of the three classes has no medusa stage at all. Yet it is generally believed that the ancestral cnidarian was a free-swimming medusa with no polyp stage. *Gonionemus, Hydra,* and *Obelia* served as examples of variations in cnidarian life histories. All belong to the least complex and supposedly the most primitive of cnidarian groupings, the class Hydrozoa.

HYDROZOANS

The class **Hydrozoa** ("hydralike animals") exhibits a bewildering spectrum of life histories that range from wholly medusoid to exclusively polypoid forms. All are small or moderate in size and more or less transparent and fragile. The medusas typically have a well-developed muscular **velum** that is important in propulsion through the water. The mesoglea, whether thin as in polyps or bulky as in the bells of medusas, contains no cells. The gastric cavity is a simple sac with no partitions. And the endoderm contains no nematocysts. These largely negative characters together distinguish hydrozoan medusas and polyps from the more complex forms of the other two cnidarian classes. All the truly freshwater cnidarians are found in this class. Besides the tiny hydras and freshwater medusas there is *Cordylophora,* a branching colonial hydrozoan that lives in rivers and brackish inlets.

Trachymedusas (order Trachymedusae) resemble the medusas of *Gonionemus* in structure but their life history comes closer to what is generally believed to be that of the ancestral cnidarians. Trachymedusas produce no attached polyp at all. The ciliated, free-swimming planula develops into another larval stage, the **actinula,** which has a rounded body with short tentacles and looks like a stalkless polyp. The actinula larva never settles to the bottom. Always free-swimming, it expands and flattens radially, with the area between mouth and tentacles pushing in to form the subumbrella during metamorphosis into the adult medusa. Some **narcomedusas** (order Narcomedusae) also follow this pattern, but in most the actinula buds other actinulas before metamorphosis. Lacking a polyp stage, both trachymedusas and narcomedusas are free of ties to shores, and most of them are found in the open ocean.

Limnomedusas (order Limnomedusae) include *Gonionemus* and the related freshwater genus *Craspedacusta.* The tiny round polyp of *Gonionemus* is like a fixed actinula, and it

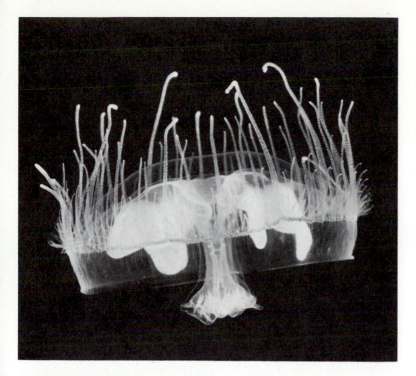

The **freshwater jellyfish** *Craspedacusta*, a limnomedusa, has the habit of swimming to the surface of the water and then coasting down with tentacles outspread and forming a trap. Diameter 2 cm. Ohio, eastern U.S.A. (R.B.)

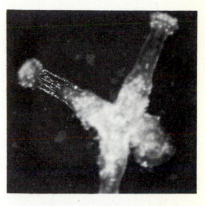

The **polyp** of *Craspedacusta* has no tentacles, but nematocyst batteries surround the mouth. It reproduces asexually, forming small colonies of connected polyps, larvalike buds that creep away to found new polyp colonies, and finally medusa buds. Before its relationship to the medusa was known, the polyp was named *Microhydra,* but now it carries the medusa's name. Height of one polyp 0.5 mm. Indiana, eastern U.S.A. (C.F. Lytle)

buds other polyps before producing medusas. But while the free-swimming actinula larva of the trachymedusas and narcomedusas metamorphoses directly to become an adult medusa, the polyp of limnomedusas remains always a polyp and produces medusas by budding. This pattern of development—with a polyp that does not develop into the adult medusa but gives rise to it by budding—allows the polyp stage a more or less permanent, independent existence, with the potential for indefinite asexual reproduction of both polyps and medusas by budding. All the rest of the hydrozoan orders share this pattern; their polyp stages are at least as prominent as their medusas and in many cases there are no free medusas at all.

The **hydroids,** which include *Obelia,* are commonly seen on rock overhangs, wooden pier pilings, seaweeds, and shells as white, pink, orange, or violet bushy growths of delicately branched polyp colonies. But there are solitary polyps also. The medusas of hydroids, collected from surface waters along with other plankters, are often hard to associate with their polyp colonies, so that the full life histories of many are still unknown, and the hydroid orders have acquired two sets of

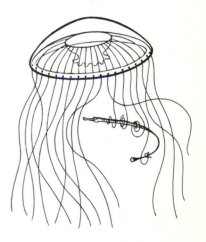

Leptomedusa, *Aequorea,* reaches 17 cm or more in diameter and is a giant among the medusas of thecate hydroids (most of them 2 or 3 cm in diameter or less). Also unusual are the multiple radial canals leading out from a very large stomach, all clearly visible through clear-as-glass mesoglea. *Aequorea* is shown here catching a pipefish. (Mod. after M. V. Lebour)

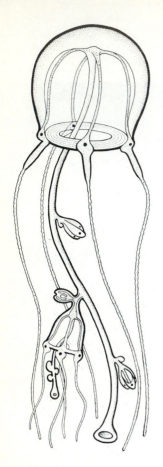

An **anthomedusa,** *Sarsia,* which reproduces asexually by budding off other medusas from the manubrium. The manubrium of different medusas varies in structure and function. In *Sarsia* the stomach is the swelling immediately inside the mouth opening. (After C. Chun)

names, one relating to the medusas and the other descriptive of the polyps.

Hydroids of the order Leptomedusae, or Thecata, such as *Obelia,* have medusas which are generally small and flat (*lepto* ="flat"). The hydranths, or feeding ends of the polyps, are enclosed in a chitinized cup, or **theca,** hence the alternative name. All the stems and stolons of the colony are supported and protected by a thin perisarc. In most thecate hydroids the reproductive polyps (gonozooids) produce reduced medusa buds (gonophores), which never develop into free medusas. The gonads develop and mature in these attached medusa buds. The gametes may be spawned into the seawater; or the eggs may be fertilized and develop within the gonophore, to be released as planulas. The life history is thus completed without any fully developed, free medusa stage.

Hydroids which have naked polyps and high, bell-shaped medusas comprise the order Anthomedusae, or Athecata ("without cups"). Though the hydranths lack enclosing thecas, the stems and stolons may be covered by a chitinized skeleton. The colonial athecate hydroids of the genus *Hydractinia* carry polymorphism of the colony one step further than we saw in *Obelia.* In *Hydractinia* there are several kinds of zooids: feeding polyps, reproductive polyps, and two kinds of stinging polyps. The stinging polyps (dactylozooids) are long and

Hydractinia, small portion of colony. *Left,* a reproductive polyp. *Center,* two slender stinging polyps. *Right,* a feeding polyp. (Adapted from G.J. Allman)

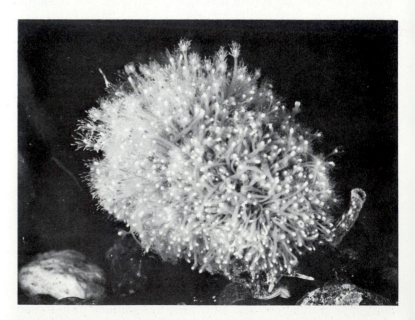

Hydroids, *Hydractinia,* **growing on the shell of a snail,** may enjoy advantages from the currents created by their host and freedom from the danger of being buried by sediments. The snail also benefits if potential predators are discouraged by its stinging camouflage. Misaki Bay, Japan. (R.B.)

slender and especially heavily armed with nematocysts. They may function in food capture and in the defense of the colony. In addition, the stolons become enlarged and heavily charged with nematocysts where they contact a competing colony of the same species. They grow toward the foreign colony and discharge nematocysts into its tissues. All parts of a colony of *Hydractinia* receive nourishment only through their connections with the feeding polyps. The reproductive polyps produce gonophores which release eggs or sperms into the water. The planula swims off and settles to the bottom, there giving rise to a new polyp colony.

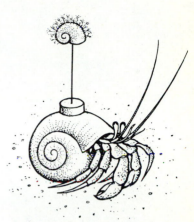

Settling of *Hydractinia echinata* planulas is selective. They may settle on a shell occupied by a hermit-crab, but not on a shell still inhabited by a snail or on an empty shell. However, the planulas will settle on an empty shell carried by a hermit-crab, suggesting that it is a suitable motion of the shell to which they respond, and not the occupant. (After C. Cazaux)

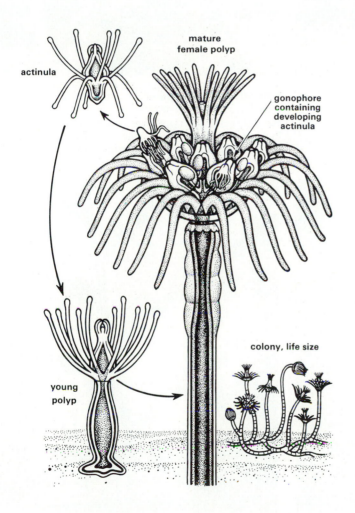

In the **large athecate hydroid** *Tubularia,* the eggs are produced and retained within the gonophores that encircle the body of the polyp between its 2 whorls of tentacles. Sperms released from a male polyp are attracted by specific chemicals in the eggs. The zygote develops into an actinula larva which is released, settles to the bottom, and develops into a fixed polyp. (Combined from G.J. Allman and from Pyefinch and Downing)

Although polyps appear to have realized the fairytale ambition of Peter Pan never to grow up, they nevertheless seem to grow old, and hydroids have been favored for research on the process of aging. The life span of a feeding zooid in a hydroid colony is usually only a few days. The zooid is then resorbed and its constituent materials used by the colony for continued growth and reproduction of new zooids. A zooid isolated intact from its colony will regress on schedule; but the life span of zooids may be experimentally increased by lowered temperature (which slows down metabolic processes) and by mechanical or radiation damage. Perhaps tissue repair and reorganization somehow function to maintain the zooid and delay its resorption.

Thecate hydroids, with the hydranths enclosed in cups of perisarc, are seen in all photos on this page. *Above,* a colony of *Thuiaria (= Sertularia) argentea* with hydranths fully extended from their cups as they feed. East coast of N. America. (R.S. Bailey)

Reproductive polyps of *Thuiaria argentea,* each enclosed in a clear vase of perisarc, are reduced to gamete-bearing stalks. The eggs are not discharged into the water but are fertilized in place. The dark oval bodies on the stalks seen here are developing planulas that will be set free. (R.S. Bailey)

Feather hydroid, *Aglaophenia,* is commonly found on beaches, where it is cast up along with seaweeds. The reproductive polyps bearing gonophores are clustered on special branches and protectively enclosed by sheaths (corbulas) that appear in the photo as dark elongate bodies. About ½ nat. size. Monterey Bay, Calif. (R.B.)

Plumelike colony results when branches are all in one plane and the protective cups are fused to the upper surfaces of the side branches, as in *Plumularia.* With tentacles fully extended, the polyps give small animals slim chance of slipping through. Bermuda. (R.B.)

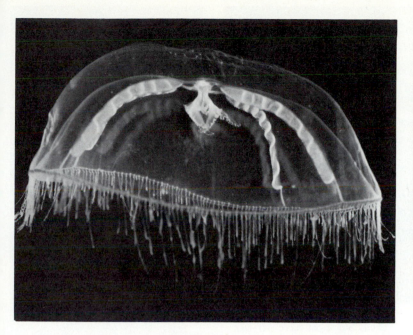

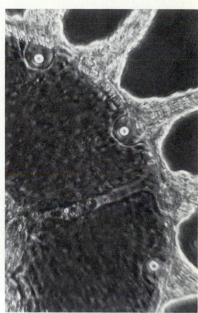

Leptomedusa is relaxed, *above,* and contracted, *below,* as the bell pulsates. The flattened bell, wider than high, is characteristic of leptomedusas. So are the gonads attached to the radial canals. The 4 radial canals are widened into pouches by extension of stomach tissue, and the wavy gonads can be seen on the upper part of the radial canals. The square mouth has 4 frilled lips. In the contracted medusa the wide velum has been pushed downward by the outrush of water from the contracting bell. The hydroid grows as single polyps from a stolon. The covering of perisarc is a lidded cylinder only 1 mm tall, but the hydranth can extend another millimeter. *Mitrocoma cellularia* is abundant in Puget Sound, where it reaches 10 cm in diameter. This one was from Monterey Bay, California. (R.B.)

Statocysts are characteristic of lepto-medusas (but not found in antho-medusas). In this photomicrograph of a portion of the bell margin of a newly released medusa of *Obelia,* we see 3 of the 8 statocysts and 7 of the 16 tentacles. As the medusa matures it grows from a diameter of 1 mm to 2.5 mm; it develops a gonad on each of the 4 radial canals, and as many as 84 tentacles. (R.B.)

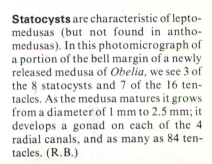

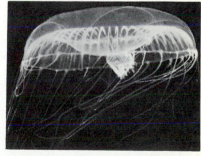

Aequorea (see drawing on p. 135) is large among mostly tiny leptome-dusas. It has especially bright, green luminescence when disturbed in darkness. The emission of light in this medusa starts with the activation of a protein (aequorin) by calcium. As the emission of light is directly related to the concentration of calcium ions, aequorin extracted from these me-dusas can be used to measure minute amounts of calcium in solution or even inside living cells. (R.B.)

Athecate hydroids may be completely naked, but most have perisarc covering the stolons and upright stems and stopping only at the bases of the hydranths. In *Clava squamata* (seen *above* in a cluster attached to a brown alga) each delicate pink polyp arises separately from the horizontal mat of stolons. The tentacles are strewn irregularly over the hydranth (a distribution considered to be primitive). As in most hydroids, the sexes are separate. But when male and female colonies of *Clava* were mixed and then squeezed through cheesecloth to separate the cells, the regenerating mass produced some hermaphroditic colonies. England. (D.P. Wilson)

Reproductive zooids, each with a cluster of 5 oval female gonophores are seen near the base of a side branch of *Eudendrium,* a colony with a bushy plantlike aspect. In this genus the sexes are very different in appearance, and male gonophores are long and narrow with 2 or more swellings along their length. The gonophores of *Eudendrium* are also unusual in starting out with mouth and tentacles. Later the mouth closes; the tentacles degenerate and are lost. The eggs of the female gonophores are fertilized in place, and the planulas that develop are set free. Florida. (R.B.)

Giant solitary hydroid polyp, *Lampra microrhiza,* about 50 cm high, is unusually large for a hydroid. Grapelike bunches of gonophores hang down from the hydranth between the 2 circlets of tentacles (see drawing of *Tubularia).* McMurdo Sound, Antarctica. 30 m deep. (P.K. Dayton)

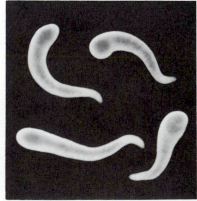

Planulas of *Eudendrium* are microscopic, pink, little ciliated larvas that swim actively with the broad end leading. After a time, the planula attaches by this end. The rear end of the planula (the free end when the egg was attached to the gonophore) becomes the mouth end of the polyp. Florida. (R.B.)

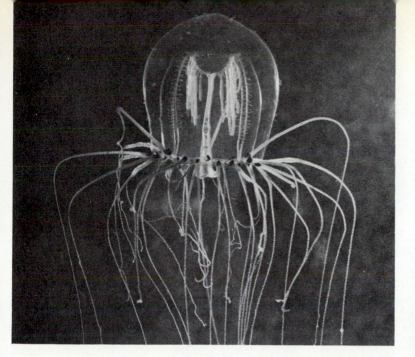

Anthomedusa, *Polyorchis,* is very large for the medusa of athecate (naked) hydroids. Along the Pacific coast of the USA, this genus reaches a bell height of 5 cm, a large size compared with most anthomedusas, some of them only 1 or 2 mm high. In other respects it conforms to the general pattern of this grouping, having a bell that is taller than wide, pigmented eyespots on the swellings at the bases of the tentacles, no statocysts, and gonads that hang from the manubrium (not from the radial canals, as in leptomedusas). *Polyorchis* is often seen from piers in harbors as it swims along by rhythmic pulsations, the long manubrium jerking and the stringlike gonads swaying with each pulse. It feeds on plankton and also on small crustaceans in submerged beds of eelgrass. The minute hydroid, only 1 mm high, has 2 whorls of tentacles. It has been found on rock scallops, attached by a spongy mass of stolons. It buds off 1 mm medusas that swim free. (R.B.)

Anthomedusa with clustered tentacles is *Nemopsis bachei,* with a bell about a centimeter high. Each cluster of about 16 tentacles (of 2 kinds) springs from a single tentacular bulb at the far end of each of the 4 radial canals. And each bulb bears a row of pigmented eyespots, one for each tentacle. From just above the mouth there spring 4 delicate branching oral tentacles with nematocyst batteries at the tips of the branches. Native to the Atlantic coast of the U.S., this species has spread to northern European waters, probably by attachment of the tiny hydroid colonies to the hulls of ships. Northwest Florida. (R.B.)

Leuckartiara is usually recognized by the jelly-filled projection at the apex of the bell, and by rudimentary tentacles intervening among the fully developed ones. It resembles other anthomedusas in having a high bell and pigmented eyespots on the swellings at the bases of the tentacles. Only 20 mm in widest diameter, it is a voracious feeder on copepods, crustacean larvas, tiny fishes, and even young squids. The naked hydroids of this medusa are less than 1 cm high and are usually unbranched. On the U.S. west coast they have often been found on the shells of certain living gastropods. (R.B.)

Colonies of *Millepora* grow in tropical or subtropical shallow waters, their leaflike or branching forms erected perpendicular to the direction of water flow. They may reach 60 cm in height, and are usually colored yellowish brown from the dinoflagellate symbionts in their tissues. Fiji. (J.S. Pearse)

Expanded stinging polyps of a branching millepore colony form a dense cover. Even when the polyps are retracted, contact of the skin with the colony can result in a fierce burning sensation that lasts for some time. Fiji. (J.S. Pearse)

The **hydrocorals** include two orders of polymorphic colonies of inconspicuous hydroidlike polyps which are remarkable among hydrozoans because they build rock-hard, massive skeletons of calcium carbonate.

All the **milleporine hydrocorals** (order Milleporina) belong to the genus *Millepora* ("thousand pores"), an important contributor to tropical coral reefs. The erect flattened or branching skeleton of the yellow-brown polyp colony is dotted with the many pores for which the genus is named. Each pore is the opening to a cuplike chamber, just below the surface of the skeleton, in which sits a polyp. Two kinds of naked polyps expand through these pores: the feeding polyps (gastrozooids) have knoblike tentacles surrounding a mouth; the slender stinging polyps (dactylozooids) have no mouth but are heavily armed with especially powerful nematocysts that give *Millepora* the common names "stinging coral" or "fire coral." The polyps are connected to each other by an epithelium which covers the surface of the skeleton, and by hollow strands of tissue which form a branching network within the skeleton. From these tissue strands, medusas are budded in special, rounded chambers which also open to the outside through pores. The tiny medusas have no velum, tentacles, or radial canals. They swim free for only a few hours before they shed their gametes and die. The planulas settle to the bottom and grow into new colonies of polyps.

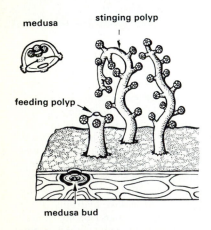

Millepora, small portion of a colony. The feeding polyp is only partially expanded; it may extend as far as the stinging polyps. (Polyps after H.N. Moseley; medusa after S.J. Hickson)

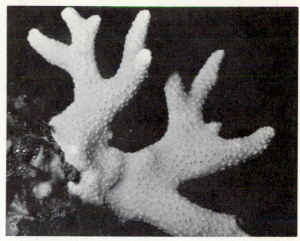

Skeleton of a millepore colony. *Above*, a portion of a millepore colony that has been cleaned and dried to show the multitude of pores. *Below*, closeup of a small portion of the same skeleton shows 2 sizes of pores. In the living colony the large holes are occupied by the plump feeding polyps and the small holes by the slender stinging polyps. Line is 1 mm long. (R.B.)

Temperate-water stylasterine, *Allopora californica*, is common subtidally in Carmel Bay, California. It grows as erect, branching colonies (usually pink, red, or purple) that superficially resemble the more familiar stony corals. Colonies established on steep or sloping rock surfaces may grow for up to 100 years. Those on the bottom are killed within a year by algal overgrowth or by sediments. (G.L. Ostarello)

A related species, *Allopora porphyra*, grows as a purple encrusting sheet on shaded surfaces protected from strong surf. It occurs from central California to British Columbia.

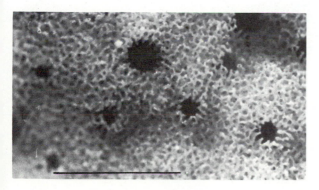

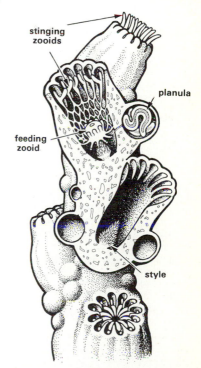

The **stylasterine hydrocorals** (order Stylasterina) are similar to the millepores, but occur mostly in colder waters and are often much more brightly colored in shades of pink, red, and purple. They differ from the millepores also in having a spine, or style, in the base of the cup in which each feeding polyp sits and in having different types of nematocysts. Each flower-shaped pore reveals a feeding polyp in its center surrounded by a circle of stinging polyps. There is no free medusa; the gonophores are budded from the network of tissue strands in special chambers that are often visible on the surface as rounded domes. The planula larva develops within the rounded chamber and must wriggle and squeeze through a narrow channel in the skeleton until it reaches the nearest opening and escapes. Shallow-water stylasterines that are uncovered at very low tides are usually of encrusting form. In deeper waters stylasterine colonies have slender branches.

Diagram of a **stylasterine hydrocoral,** part of which has been cut away. (Combined from H.N. Moseley and from H. Broch)

A. The medusa *Liriope* produces no attached polyp. Eggs or sperms are released from the gonads of the medusa into the water, and the zygote develops into a ciliated, free-swimming planula larva. The planula develops into another larval stage, the actinula, which in turn becomes a medusa.

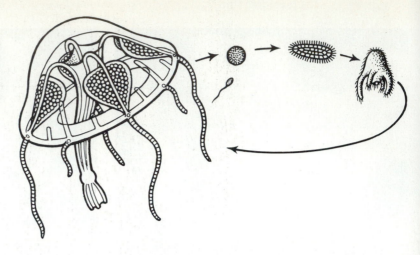

B. In the genus *Gonionemus* the planula larva settles to the bottom and metamorphoses into a minute feeding polyp. The polyp buds off small medusas that swim freely and mature into sexually reproductive adult medusas.

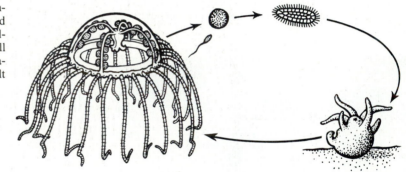

C. In the colonial hydroid *Obelia* the polyp stage is the more conspicuous. Certain of the polyps are specialized only for budding off small medusas of either sex. These mature into the sexually reproductive adult medusas.

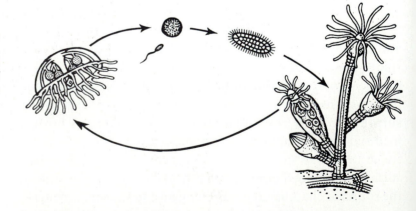

Life histories of hydrozoans display a spectrum that ranges from those with a medusa phase only and no polyp **(A)** to those with a polyp phase and no medusa **(F)**. But whether this represents the actual evolutionary sequence is speculative.

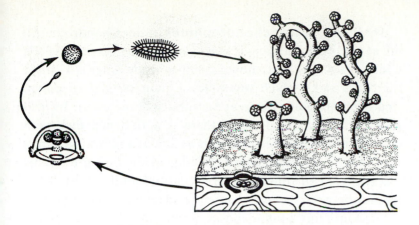

D. In the hydrocoral *Millepora* the emphasis is still more on the polypoid phase. From strands of tissue which form a branching network within the skeleton, medusas are budded in special rounded chambers which open to the outside. The tiny medusas have no velum, tentacles, or radial canals, and swim free for only a few hours before they shed their gametes and die. The planulas produced settle down and grow into new colonies of polyps.

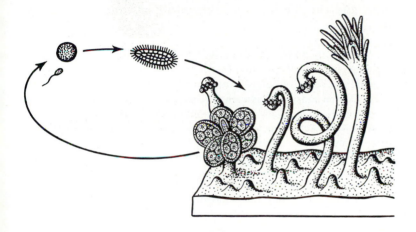

E. The colonial hydroid *Hydractinia* carries reduction of the medusa one step further. Reduced medusa buds (gonophores) are budded from the reproductive polyps but never develop into free medusas. Instead, the reduced medusa buds remain attached to the polyps and release eggs and sperms into the water. The planula larva swims off and settles to the bottom to produce a new polyp colony.

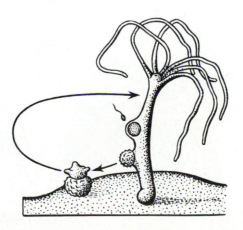

F. In *Hydra,* presumably evolved from marine cnidarians with more complex developmental sequences, the life history is ultimately simplified. The polyp produces eggs and sperms, and the zygote develops directly into a new polyp.

In all the life histories portrayed here, except for that of *Liriope,* there is a polyp stage that grows by asexual reproduction of additional polyps, which may separate (**B, F**) or remain connected as a colony (**C, D, E**). The multiple products of a single zygote, the polyps as well as the medusas they bud off, are members of a **clone.** Further growth of the clone through asexual reproduction by medusas is less common.

Although chondrophores were once grouped with siphonophores in the order Siphonophora, their similarities most likely reflect parallel adaptations to pelagic life, and they are probably not closely related.

The fragile, pelagic **chondrophores** (order Chondrophora) are at the other extreme of life-style from the massive, sessile hydrocorals. The best known member of the order is *Velella,* the "little sail," which resembles a giant tubularian polyp, floating mouth-side down at the surface of the water. Instead of a perisarc-covered stalk, there is a flattened disk with concentric air chambers that acts as a float and bears an erect chitinous "sail." All around the edge of the disk are hollow tentacles that surround the large central mouth. On stalks in the area between the central mouth and tentacles, medusa buds develop into tiny adult medusas, which are eventually freed to release their gametes. From the zygote and early embryo, through an actinula-like larva which in turn grows into a new sail-bearing polyp, all stages of development float freely in the water; and the life history takes place entirely in a pelagic environment.

Unlike the gonophores of tubularians, which are borne on simple stalks, the medusas of *Velella* are budded from what have been interpreted as individual reproductive polyps, since each has a mouth. The large central mouth has been considered as a single feeding polyp, lacking tentacles. And what appear to be tentacles at the rim have been thought of as a group of stinging polyps. According to this interpretation, each *Velella* represents a complex polymorphic colony. However, the alternative interpretation that chondrophores are single polyps related to tubularian polyps seems more convincingly supported. The two groups share corresponding features of polyp and medusan morphology, the actinula-like larva, and detailed similarities in neurophysiology and behavior.

Velella, a large polyp (often up to 8 cm in diameter) that floats at the surface of the water. A thin layer of tissue covers the float and sail. The front portion has been cut away to reveal the continuous gastrovascular cavity, which extends from the central gastric region into the hollow tentacles, the budding stalks, and a network of endodermal canals that surround the float and sail. The large central mouth and smaller mouths at the tips of the budding stalks all actively ingest food. (Modified after Delage and Hérouard)

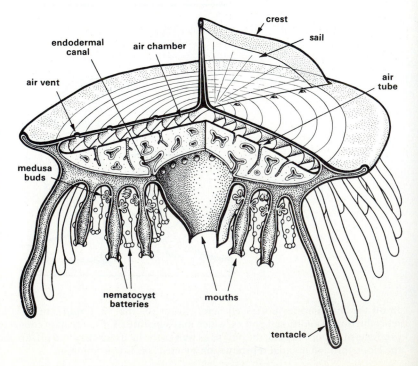

Velella is a familiar chondrophore of warm seas. These were from a California shore. The sail lies on a diagonal to the longer diameter of the disk. The animal can tack at an angle to the wind, sometimes as much as 63°, and thus avoids being stranded by all but the strongest onshore winds. The direction of the sail differs among individuals, and the two types, mirror images of each other, are sent in opposite directions by the wind. (R.B.)

Great windrows of *Velella* line warm Pacific and Mediterranean beaches when winds and waves bring them in by the millions. An Atlantic expedition once sailed through an aggregation of *Velella* that stretched for 260 km. Shown here are stranded velellas on the shore at Pacific Grove, California. (R.B.)

Porpita is 3.5 cm across the flat, circular, shiny disk. This chondrophore without a sail is found in the warm waters of the Mediterranean and of the Atlantic and Indian Oceans. Both *Porpita* and *Velella* are fed on by *Glaucus,* a nudibranch (see molluscs, chap. 15). It is seen here nibbling away at the tentacles (and later the soft underparts) of a Florida *Porpita.* (W. Stephens)

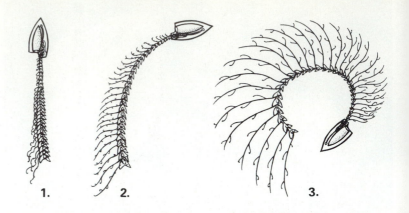

Siphonophore fishing *(Muggiaea).* **1.** With stem and tentacles contracted. **2.** The swimming-bell shoots forward, while the stem and tentacles relax, starting at the lower end. **3.** The drag of the stem and tentacles causes the swimming-bell to veer in an arc, and the tentacles spread centrifugally in the water, forming an efficient "fishing net." (After Mackie and Boag)

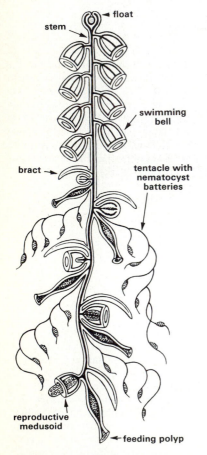

Diagram of a **siphonophore.** The gastrovascular cavity extends the length of the stem and is continuous from the float and swimming-bells through the feeding polyps, reproductive medusoids, tentacles, and bracts. (Mostly after C. Chun)

It is in the **siphonophores** (order Siphonophora) that we find the extremes to which colonial organization is carried. These complex floating colonies have not only more than one kind of polyp but also more than one kind of medusa. In addition to the sexual medusas (either free or attached), siphonophores may have numerous modified medusas, called **swimming-bells** or nectophores, which cannot feed or reproduce but serve only to propel the colony. Some siphonophores also have a gas-filled **float** and may regulate the amount of gas in the float so as to rise to the surface or sink below. Beneath the swimming-bells and/or float may hang a **stem** bearing repeated groups of budded zooids. Each group, or cormidium, usually includes a feeding polyp with a single, often branched tentacle, one or more gonophores, and sometimes a stinging polyp, all partially enclosed in a protective, gelatinous structure called a **bract.** Bracts have been variously interpreted as modified medusas, or polyps, or tentacles. In addition to protecting the other members, in some cases they beat actively and propel the colony. In some kinds of siphonophores, new cormidia are continually produced near the top of the stem, and the older cormidia at the bottom are set free to live independent of the parent colony. All this may suggest a loose degree of organization. But many siphonophores possess a well-developed colonial nervous system, and their individual zooids behave together in such highly coordinated fashion that it sometimes seems more appropriate to consider the colony as one complex organism.

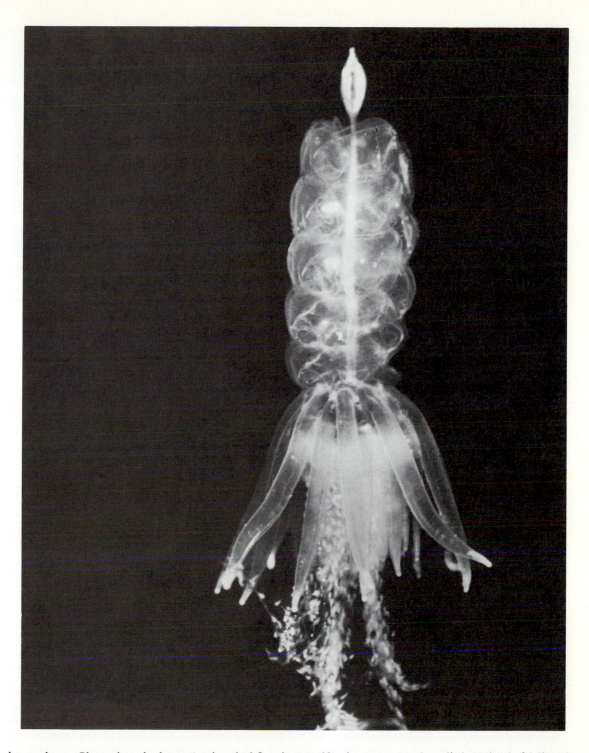

Siphonophore, *Physophora hydrostatica,* is only 6.5 cm long and has been compared to a little garland of delicately colored flowers. The gas-filled float at the apex and the 2 rows of compressed swimming-bells below the float are medusoids. The bells are highly muscular, and their pulsations can move othe colony along horizontally at 3 m/minute. Just below the bells is a crown of red-tinted dactylozooids, covered with nematocysts at their tips. These hang down over the crowded reproductive medusoids and feeding polyps that extend from the main stem as the colony moves along. If danger threatens the gonophores and polyps can be retracted into the protective shelter of the fingerlike dactylozooids. *Physophora* lives in warm seas but occasionally is carried northward by the Gulf Stream and can be found in the North Sea. Marine Laboratory, Santa Catalina Island, California. (R. Given)

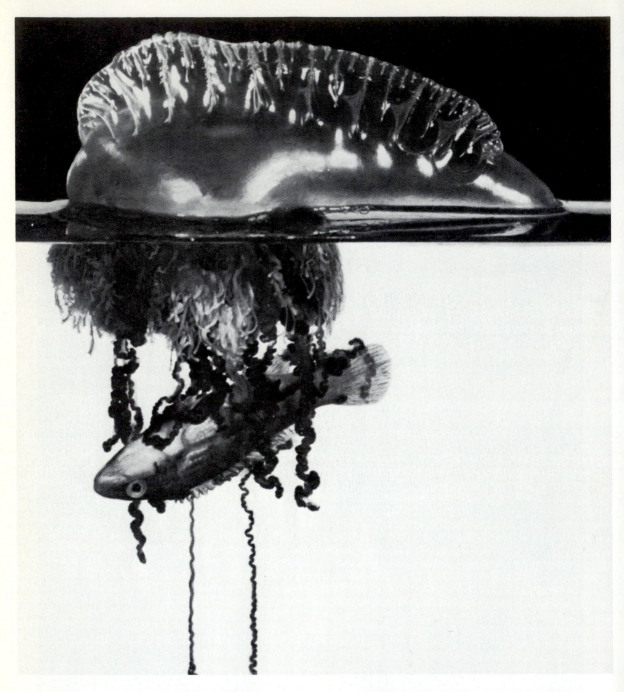

Physalia, the most notorious of siphonophore colonies, is often called the "Portuguese man-of-war." It has no swimming bells but is driven about by the action of the wind on its crested, gas-filled float. From the underside of the float hang several kinds of specialized polyps, clusters of attached medusas, and tangles of long tentacles that may reach a length of 20 meters or more. *Physalia* is armed with especially large nematocysts that can readily paralyze a fish. The vivid blue float is a familiar and beautiful sight on the surface of warm seas all over the world—but is not a welcome one to swimmers who know that the trailing tentacles can inflict serious and sometimes fatal injury. Some fatalities may result from drowning due to pain, or from anaphylactic shock in victims that have been stung earlier and have become allergic to the toxin. *Physalia* tentacles, even when torn from the colony, are capable of stinging—as they do when fishermen bring up gear entwined with tentacles, or sunbathers on a beach come in contact with dried tentacles in sand. The colony shown here (the float 18 cm long) has just caught a fish. England. (D.P. Wilson)

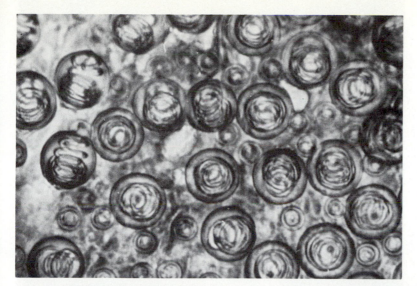

Nematocysts of *Physalia* fill the microscope field in this view of a small portion of the surface of a tentacle. There are several kinds of nematocysts, the largest ones clearly showing the coiled hollow thread within. Each of these large stinging capsules is enclosed within a cnidocyte that sends a projecting sensitive bristle up through the surface. At the basal end of the cnidocyte is a long stalk that reaches down to the mesoglea. The nematocysts of siphonophores are often large, and those of *Physalia* especially so. Bimini, Bahamas. (R.B.)

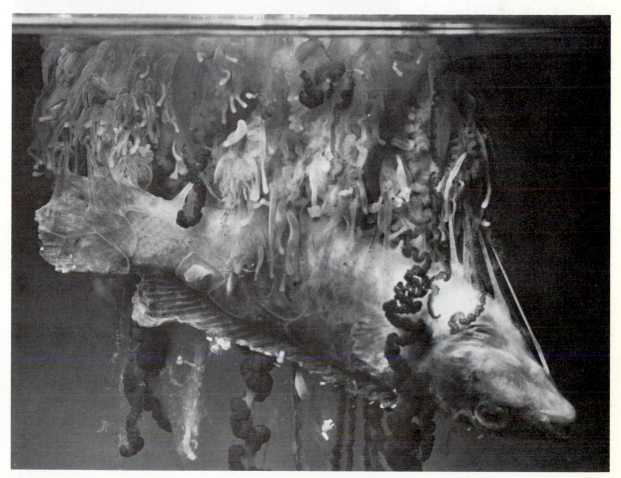

Physalia eating a fish. This photo was taken about 1½ hours after the one on the previous page. The dactylozooids which caught the 10 cm fish have now released it and are hanging below. The fish is held by the feeding polyps (gastrozooids), which can be seen stretching down, with their thin transparent lips spread over the surface and meeting edge to edge thereby enclosing the fish completely. From the feeding polyps digestive enzymes are poured onto the fish, disintegrating its substance so that it can be sucked up and later distributed to all members of the colony. England. (D.P. Wilson)

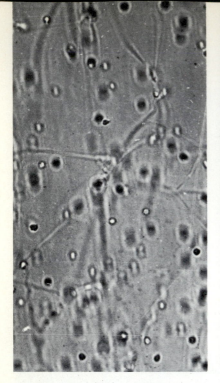

A bit of **mesoglea,** highly magnified, from the bell of *Aurelia.* The dots are cells. The streaks are fibers that strengthen the mesoglea. (R.B.)

Aurelia is easily recognized by the 4 horseshoe-shaped gonads, white in males and pink in females, the pattern of radiating gastrovascular canals, and the 4 transparent mouth lobes. Helgoland. West Germany. (R.B.)

scyphozoa = "cup animals"

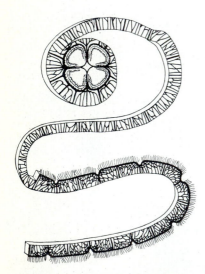

The **diffuseness of conduction in a nerve net** was demonstrated in 1875 by cutting the bell of *Aurelia* into a spiral strip, 2 to 3 cm wide and nearly a meter long, and removing all but one of the rhopalia and marginal nerve centers. Impulses initiated by the single remaining nerve center at one end of the strip were conducted the entire length of the strip, provoking a wave of contraction. (After G.J. Romanes)

SCYPHOZOANS

The class **Scyphozoa** (sy'-fo-zo'-a) includes the larger jellyfishes. All are marine and can be roughly distinguished from the hydrozoan medusas by their large size and by the absence of a velum. In contrast to the thin, noncellular mesoglea of hydrozoans, that of scyphozoans is usually thick and may contain cells.

Medusas of the genus *Aurelia* are among the commonest scyphozoans and occur in all seas. Large shoals of these shallow, saucer-shaped medusas can be seen drifting along together or swimming slowly by rhythmic contractions. They are usually about the size of dinner plates, ranging from less than 10 cm to about 25 cm across the bell. Exceptional individuals may reach a diameter of more than a meter.

The youngest *Aurelia* medusas **feed** on tiny fishes, but after some growth, shift to a diet of assorted animal plankton. At the end of a very short manubrium is a square mouth, the corners drawn out into 4 trailing **mouth lobes,** sometimes called oral arms. In the fold of each lobe runs a ciliated groove. Nematocysts in the lobes paralyze and entangle small planktonic animals, which are enveloped in mucus and swept up the grooves, through the mouth, into a spacious gastric cavity in the center of the bell. Plankton which collects in mucus on the exumbrellar and subumbrellar surfaces of the bell is swept by cilia toward the margin and picked up by the mouth lobes.

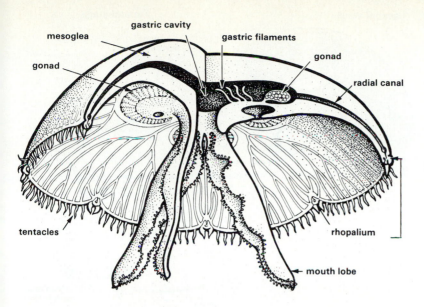

Diagram of a scyphozoan medusa, *Aurelia*.

In the complex radial system of canals, beating cilia maintain a definite pattern of **circulation,** moving fluids and food particles toward the margin in 8 unbranched canals and then back toward the gastric cavity through a series of branched canals. Currents in and out of the gastric cavity are also kept separate by a system of grooves. That no such complex circulatory system is known in hydrozoan medusas is probably related to their smaller size.

The **gastric cavity** is extended into 4 pouches, in which there are tentacle-like projections of the endoderm, called **gastric filaments**. The filaments are abundantly supplied with gland cells which secrete digestive enzymes. And they are also covered with nematocysts, which hold and paralyze prey that arrives in the pouches still alive and struggling. The presence of gastric filaments is another of the characters that distinguishes a scyphozoan from a hydrozoan medusa.

From the gastric cavity, partially digested food is moved through radial canals to the margin of the bell. As in hydrozoans, digestion is completed intracellularly throughout the endoderm. Cilia lining the entire gastrovascular cavity maintain a steady current of water. And this brings a constant supply of food and oxygen to, and removes wastes from, the internal parts of this large animal.

The numerous marginal **tentacles** are set closely together except where interrupted by 8 equally spaced notches. In each notch lies a complex of **sensory structures,** the rhopalium. The **nervous system** includes no marginal nerve rings as in hydrozoan medusas. Instead, nerve cells are concentrated in **marginal nerve centers** (marginal ganglia) near each rhopalium. These centers act as pacemakers for the swimming beat, and receive and integrate sensory information from the rhopalia. There are at least 2 **networks of nerve cells.** One extends over the whole exumbrella and subumbrella, the marginal tentacles, the manubrium, and the mouth lobes. It conducts nerve impulses relatively slowly and coordinates localized movements, as in feeding. The other is limited to the subumbrella, conducts impulses more rapidly, and coordinates the pulsations of the bell in swimming. The **swimming**

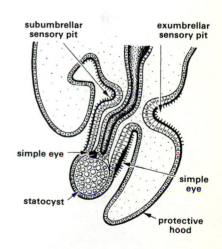

Complex of sensory structures, or rhopalium, of *Aurelia*, sectioned along one radius of the bell. The rhopalium consists of a statocyst with many statoliths, 2 pigmented simple eyes sensitive to light, and 2 sensory pits lined with cells that are thought to sense food or other chemicals in the water. (Combined from various sources)

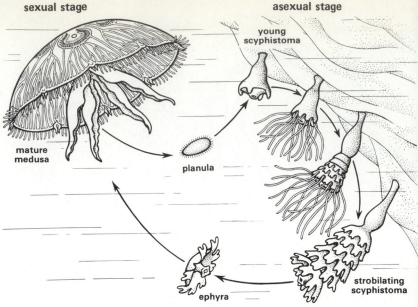

sexual stage   asexual stage

young scyphistoma

mature medusa

planula

strobilating scyphistoma

ephyra

Life history of *Aurelia.*

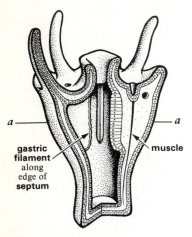

gastric filament along edge of **septum**

muscle

a —————— a

Diagram of a scyphozoan polyp, or scyphistoma, of *Aurelia*. The gastrovascular cavity is partially divided into 4 compartments, or **gastric pockets,** by 4 thin vertical partitions, or **septa.** Each septum consists of a supporting wall of mesoglea covered on both sides by endoderm. Protruding into the central cavity, the septa bear nematocyst-armed **gastric filaments** which serve, as in the adult, to hold and paralyze prey not wholly subdued by the tentacles and to secrete digestive enzymes. The basal disk is not well developed, and the polyp attaches and moves by means of stolons which grow out from the body column. (Combined from various sources)

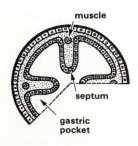

muscle

septum

gastric pocket

Cross-section of scyphozoan polyp in the diagram above, cut at level *a—a.*

muscles consist of circular and radial muscle fibers in the bases of ectodermal cells in the subumbrella.

During **sexual reproduction,** *Aurelia* can be recognized by the 4 horseshoe-shaped gonads, which show through the transparent umbrella. The testes and ovaries occur in separate individuals on the floor of the pouches of the gastric cavity. The **sperms** of a male medusa are discharged into the gastrovascular cavity and are shed to the outside through the mouth. The **eggs** are fertilized inside the female by sperms which enter with the feeding currents. After fertilization, the normal feeding currents are reversed for a short time, so that the young embryos are swept out of the mouth and lodge in special brood pouches in the folds of the mouth lobes, where they continue to develop into ciliated planula larvas.

The **planula** escapes and attaches to an overhanging rock ledge or other firm surface. There it grows into a small scyphozoan polyp, a **scyphistoma,** with long solid tentacles and a short stalk. The polyp feeds and stores nutrients, and may survive for many months, or even years, meanwhile budding off other small polyps like itself, usually from stolons. During cold seasons, fall to spring, it develops a series of horizontal constrictions, which gradually deepen until the polyp resembles a stack of saucers. One by one the "saucers" pinch off from the polyp and swim away as little 8-lobed medusas, **ephyras,** which gradually develop into adults. This method of budding off medusas by successive constrictions is called **strobilation;** it is characteristic of some of the most familiar scyphozoans but does not occur in the other 2 classes of cnidarians.

Strobilation of *Aurelia*. *Above:* Scyphistomas, most of them in various stages of strobilation, growing on the underside of a stone ledge. An ephyra, released only a few seconds before, is swimming away; 2½× natural size. *Below:* Three scyphistomas in successive stages of strobilation. As constrictions form, the polyp elongates and resorbs its tentacles; 2× natural size. Plymouth, England. (D.P. Wilson)

| February 24 | March 3 | March 10 | March 20 |

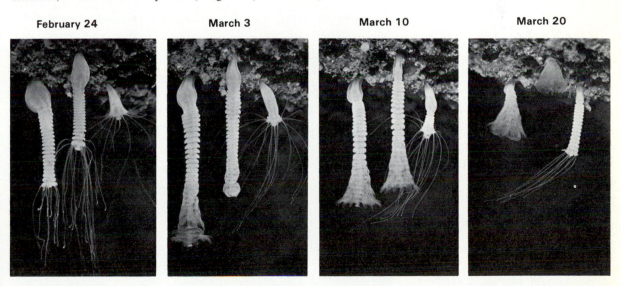

In *Aurelia* strobilating polyps release into the seawater a chemical, an organic iodine-containing compound, which causes other polyps to strobilate also. There is some evidence that this chemical is similar or identical to thyroxin, a hormone produced by the thyroid gland in humans and other vertebrates. In many vertebrates, thyroxin plays an important role in development, as in inducing tadpole larvas to metamorphose into adult frogs. If this hormone plays an analogous role in scyphozoans, the same may possibly prove true in other invertebrates which produce thyroxin but in which the hormone still has no known role.

Semeostome medusa, *Pelagia colorata,* is known only from the waters of the continental shelf of the Pacific coast of the USA. It has prominent purple streaks that radiate over the bell. The spiral coiling of the mouth lobes is distinctive to this Pacific species. This specimen from Monterey Bay, California, was about 30 cm in diameter, but some reach 80 cm, with mouth lobes several meters long. From time to time strong winds bring these medusas into bays but the members of the genus *Pelagia* have no fixed polyp and therefore are free of this tie to shores. The free-swimming planula develops directly into an ephyra, which then matures into an adult medusa. The better known *Pelagia noctiluca,* common in the Mediterranean, is often seen from ships at night, great numbers of the globular bells lighted up by greenish luminescence as they move through the water. (R.B.)

se-mee'-o-stome

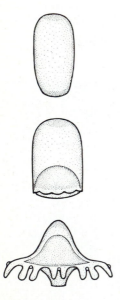

Development of *Pelagia noctiluca* from planula to an ephyra only 0.4 mm in diameter. The adult may reach 10 cm across the bell margin. (M. and C. Delap)

The 5 orders of scyphozoans are, as a group, less diverse than hydrozoans. The **semeostome medusas** (order Semaeostomeae) include the larger and more familiar scyphozoan jellyfishes, such as *Aurelia,* and most of them are relatively similar in appearance and life history.

Some members demand more caution in handling than others. Contact with *Aurelia* produces little sensation in human skin, but *Chrysaora* can deliver a painful sting, and even a small *Cyanea* can raise huge weals on the arms or legs. Fortunately, giant specimens of *Cyanea,* sometimes over 2.5 m across the bell with trailing tentacles 40 m long, occur in cold northern waters where few swimmers venture. Such huge masses of jelly are among the largest of the invertebrates.

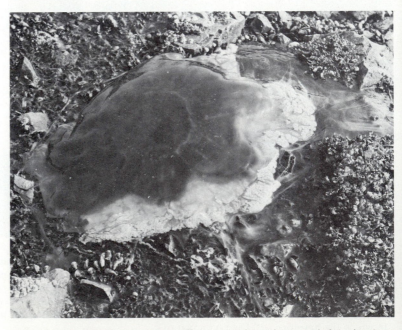

The **giant jellyfish,** *Cyanea capillata,* can still deliver a painful sting long after it is stranded on a beach. Maine, east coast of N. America. (R.B.)

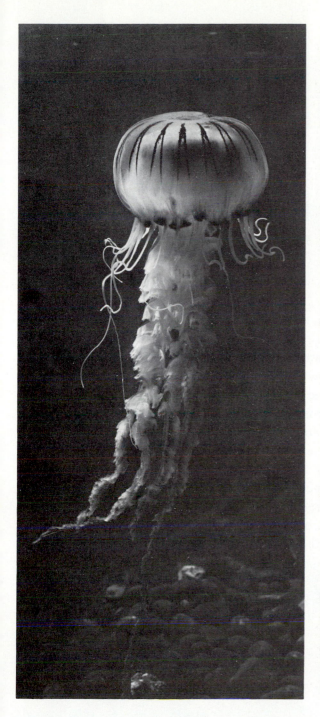

A medusa swims by alternately contracting and relaxing the bell. The bell is contracted, *left,* forcibly expelling the water from its concavity and so pushing the animal in the direction opposite to that in which the water is expelled. *Right,* the bell is relaxed, admitting water again. The "compass jellyfish," named for the V-shaped markings on its bell, is one of the commonest scyphozoan medusas. *Chrysaora hysoscella* occurs in great numbers, toward the end of summer, along the Atlantic coast of Europe; these photos were made in the Helgoland Aquarium, West Germany. Related species are found on North American coasts. In summer months the often painful stings of the "sea nettle," *Chrysaora cinquecirrha,* keep swimmers out of some favored East Coast resort areas whenever wind, weather, and water conditions combine to concentrate the jellyfishes. (F. Schensky)

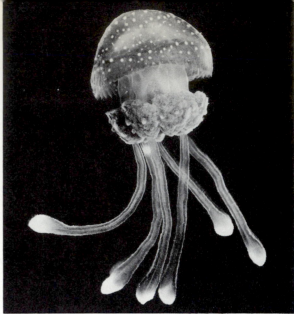

Cassiopeia is seen in its usual habit of lying mouth-up on sandy bottoms of shallow tropical bays, the mouth lobes exposed to food-bearing water currents. These medusas harbor in their tissues great numbers of photosynthetic dinoflagellates (zooxanthellas) which are thus exposed to the sun and supplement the nutrition of their hosts. Bimini, Caribbean Sea. (R.B.)

Mastigias also harbors zooxanthellas in its tissues. The presence of the dinoflagellates has been shown to be essential for strobilation of the polyp stage, even if it is well-fed; the exact role of the algas is not known. *Mastigias* is one of the rhizostomes with prominent clubshaped terminal appendages hanging from the mouth lobes. Hawaii. (R.B.)

The **rhizostome medusas** (order Rhizostomeae) inhabit mostly shallow tropical or subtropical waters. They are similar to semeostome medusas except that, early in the development of young rhizostomes, the 4 mouth lobes each branch as they grow out, to form a total of 8 lobes. The edges of the mouth and the grooves in the 8 mouth lobes then close together and fuse, forming a system of enclosed, branching canals with many tiny mouth-openings along the edges of the fused lobes. Without a large central mouth, and lacking tentacles around the rim of the bell, rhizostomes feed mostly on tiny organisms collected on the mouth lobes.

Rhizostoma, is a familiar sight in the Mediterranean and other warm Atlantic waters. In the Adriatic Sea, some 40,000 rhizostomas were once estimated in 1 km² of water. Naples Aquarium, Italy. (R.B.)

Beached rhizostome jellyfishes, *Catostylus,* lying in only a few centimeters of water and so numerous that it was hard to walk without stepping on them. They are not dangerous, but nematocysts of the mouth lobes can inflict a painful sting. Diameter 10 to 15 cm. Gulf of Thailand. (R.B.)

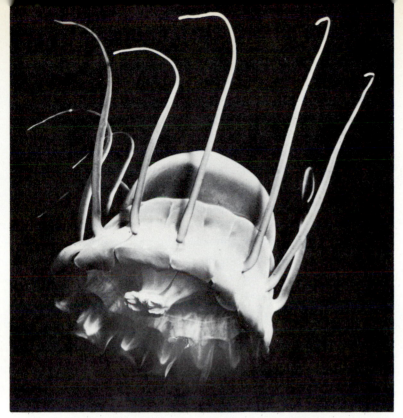

Coronate medusa, *Periphylla,* 30 cm in diameter, has a dome-shaped lavender-blue bell and is found in deep waters of all oceans. Occasionally it is seen at the surface, swimming with the tentacles streaming back from the bell margin. This one was seen, feeding on plankton, under the thick sea ice at McMurdo Sound, Antarctica. (G.A. Robilliard)

The **coronate medusas** (order Coronatae) are distinguished from others by a prominent horizontal groove which encircles the bell. Below this crowning (coronal) groove, the bell is further sculptured by a series of vertical grooves, each ending in the middle of a marginal lappet and making these delicately colored jellyfishes look like elaborately molded gelatin desserts. Unlike semeostomes and rhizostomes, adult coronate medusas retain the septa and gastric pockets of the polyp stage. The gastric filaments and gonads are located on the septa, instead of on the floor of the gastric cavity. Coronate medusas are typically deep-sea forms, but some may be seen at the surface in warmer waters, as *Nausithoe.* The scyphistomas of *Nausithoe* may be solitary or colonial, and their stalks are covered with chitinous perisarc, as in some hydroids.

Coronate medusa, *Nausithoe,* is common in shallow water in the Bahama-Florida area and in similar warm shallow waters in many parts of the world. (After A.G. Mayer)

Scyphistomas of a coronate medusa, *Nausithoe,* grow as a branching colony. These scyphistomas, found in the interior spaces of sponges and attached to other animals or to rocks, have long been known under the name *Stephanoscyphus.* They feed on plankton and finally strobilate, producing tiny ephyras that swim away and mature into adult medusas. Western Samoa. (K.J. Marschall)

A **cross-section through a cubo-medusa,** close to the base of the bell, shows how apt are the ordinal name and the common name, "box jelly." (After F.S. Conant)

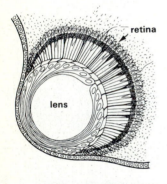

Eye of a cubomedusa, *Carybdea,* one of the more elaborate of several eyes in each rhopalium. There are roughly 11,000 sensory cells in the eye, and it is similar to the camera eyes of vertebrates. (Modified after E.W. Berger)

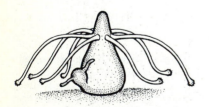

A **cubomedusan polyp,** *Tripedalia cystophora,* with its knobbed tentacles, looks more like a hydrozoan polyp than like a typical scyphozoan polyp. Polyp is only 1 mm high. (After Werner, Cutress, and Studebaker)

The cubomedusas (order Cubomedusae) might at first be taken for hydrozoan medusas. They are usually small (2 to 4 cm high although some may reach 25 cm), and the bell margin is simple, not scalloped. Also like hydrozoans are the 4 single tentacles or 4 tentacle clusters, the marginal nerve ring, and even a velarium, an inward bending of the bell at the margin, resembling and functioning as a velum, though anatomically different. Cubomedusas are so transparent that only careful examination reveals their septa, gastric pockets, and gastric filaments, which link them to other scyphozoans. The colorless bell is almost perfectly square, with a sensory complex (rhopalium) in the middle of each of the 4 flattened sides and the often colorful tentacles hanging from the corners. Each rhopalium bears a statocyst and the most remarkably complex eyes, sometimes of several types, seen in any cnidarian. In darkness a cubomedusa can detect the light from a match up to 1.5 m away and will turn toward it. Cubomedusas are strong swimmers, and feed mostly on shrimps and fishes.

The tiny cubomedusan polyp is unlike those of other scyphozoan orders. It has no gastric septa, pockets, or filaments; it is radially symmetrical instead of tetramerous; and it does not strobilate. After budding off new polyps asexually, it metamorphoses into a small 4-sided cubomedusa. The mixture of hydrozoan-like characters, especially in the polyp, with other scyphozoan characters in the medusa suggests that cubomedusas may represent an evolutionary link between hydrozoans and scyphozoans.

Cubomedusa, *Tripedalia cystophora,* with a bell only 2 cm high in the male shown here, is one of the many "sea wasps" or "fire medusas" that are much feared in shallow tropical and subtropical waters around the world. They give an extremely painful sting and it may leave long-lasting welts. *Tripedalia* is common around Puerto Rico, Jamaica, and other warm Atlantic waters. It is named for the cluster of 3 tentacles at each corner of the box-shaped bell and for the pedalia, the tough, bladelike, transparent, basal portions of the tentacles. The distal portions are contractile and ringed with nematocysts. Hanging from the stomach near the top of the bell is the slender manubrium. Midway in the bell is a ring of little oval testes, 2 on each side. Below the testes project 4 sensory organs, the rhopalia, one on each side. (R.B.)

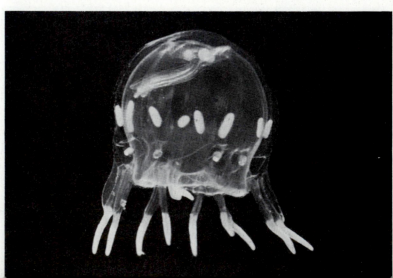

Australian box jellyfish, *Chironex fleckeri,* is one of the largest of cubomedusas, reaching 25 cm across the bell. And the sting of its nematocysts is responsible for more human deaths than is that of any other medusa. Despite its large size, *Chironex* is almost invisible in water because of its transparency when seen against a sandy bottom. Most victims are unaware of their peril until they feel the sudden searing pain. Where *Chironex* occurs along the tropical ocean shores of northeastern Australia it claims twice as many victims as do the sharks in those areas. The young medusas develop in sheltered waters of mangrove-lined estuaries. In December, when the monsoon rains come, the little medusas are flushed out into the ocean and spread along ocean beaches, growing rapidly to adult size. From December to April many popular beaches have to be closed when *Chironex* is detected. The majority of stings are not fatal, but survivors suffer intense pain and then are often left with ugly, whiplike scars. Extensive contact with tentacles of a large specimen may cause death, sometimes within 3 minutes, presumably of shock and heart failure. (K. Gillett)

Stauromedusas are sessile scyphozoans that live attached by a stalk to seaweeds, eelgrass, shells or stones in bays, sounds, or coastal waters, usually in colder seas. The knobs on the clustered tentacles are armed with nematocysts and catch tiny crustaceans and other small animals. The dark oval bodies in the notches between the marginal lobes are the rhopalioids described below. *At the left,* hanging by its stalk, is *Haliclystus. At the right,* seen in oral view, is *Craterolophus.* Plymouth, England. (D.P. Wilson)

The little oval marginal bodies in the notches between the tentacled marginal lobes of adult stauromedusas are called rhopalioids, but (unlike the rhopalia of other scyphozoan medusas) appear to have no special sensory function.

The **stauromedusas** (order Stauromedusae) differ from all other scyphozoans in appearance, life history, and habit. Never free-swimming, the adults are polyplike forms only a few centimeters across, with 8 outstretched clusters of knobbed tentacles. They live attached by an aboral stalk to seaweeds or rocks, catching small animal prey. Even the planula larva is unusual, for it has no cilia and creeps about like an inch-worm. The planula is capable of asexual reproduction, budding off others like itself. The young polyp does not strobilate, but instead metamorphoses directly into the adult form. During metamorphosis, the first 8 tentacles of the scyphistoma are reduced to tiny marginal knobs, and the adult tentacles are clustered on 8 armlike extensions of the rim of the umbrella. As in coronate medusas, the septa and gastric pockets of the scyphistoma are retained in the adult, and gonads develop on the septa. Because of the changes that occur during metamorphosis, the adult is usually considered a medusa rather than a persistent polyp, but its life style is certainly polypoid. Stauromedusas cannot swim, although some can move about and reattach in new places; others are permanently fixed.

The **Anthozoa** ("flower-animals") are marine polyps and have no medusa stage. This is by far the largest and most diverse of the 3 classes. Anthozoans are technically distinguished from hydrozoan and scyphozoan polyps by the fact that the surface ectoderm turns in at the mouth, the bulky mesoglea usually contains many cells, and the gastrovascular cavity is divided by larger numbers of nematocyst-bearing septa than in scyphozoan polyps. But superficially there is little difficulty in telling the large fleshy sea anemones or the calcium carbonate-secreting corals from most of the small fragile polyps of the other two classes.

The **sea anemones** (order Actiniaria) are large solitary polyps. The body of an anemone (a-nem'-o-nee) consists of a stout muscular cylindrical **column,** expanded at one end into a flat **oral disk** having at its center a **mouth** surrounded by several to many circlets of hollow **tentacles.** At the other end is a smooth muscular mucus-coated **basal disk** on which the anemone can creep about very slowly and by which it holds to rocks so tenaciously that one is likely to tear the animal in trying to pry it loose. In some anemones, the basal end forms an expansible bulb which anchors the polyp in sand or mud. And in one tropical family (the minyads), the basal end secretes a float, and the polyp hangs mouth-down at the surface of the water.

Anemones, though they are among the most complex of cnidarian polyps, lack the multicellular sensory structures and concentrated nerve centers common in the more active medusas. Nevertheless, several distinct conducting systems, including at least one extensive **nerve net,** receive information from **sensory cells** and serve several sets of specialized **muscles.** The usual feeding, opening, and closing movements of anemones are mostly slow; but if disturbed, an expanded anemone can contract very rapidly.

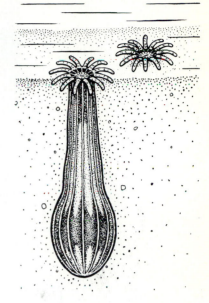

A **burrowing anemone,** *Halcampa,* with just its oral disk exposed on the surface of the sand, its body burrowed in and firmly anchored by the bulbous basal end. (After P.H. Gosse)

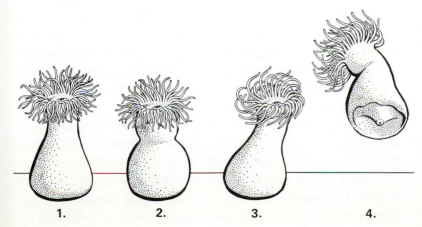

A few kinds of **anemones can swim** and do so on contact with certain sea stars and sea slugs. **1.** A relaxed expanded anemone, *Stomphia coccinea.* **2.** After contact with a sea star, the anemone withdraws slightly, and its sphincter muscle constricts. **3.** The anemone expands again and rotates its oral end in a whirling motion. **4.** The anemone detaches from the substrate and thrashes up into the water by vigorously bending and twisting. The anemone never moves far in any one direction by this inefficient method but may successfully escape a startled sea slug or a sluggish sea star. (Drawn from photos by Yentsch and Pierce)

1.   2.   3.   4.

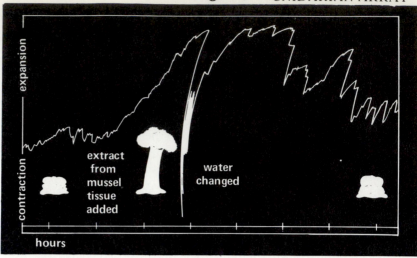

The slow **movements of anemones** are tedious or impossible to observe except by time-lapse photography or by mechanical recording devices. A kymograph traced the record shown here of an anemone, *Metridium,* expanding after a food extract (from mussel tissue) was added to the seawater, and contracting, over several hours, after the water containing the extract was replaced by clean seawater. (Modified after C.F.A. Pantin)

One may wonder how a nerve net can coordinate the many varied and graded responses observed in an anemone. One explanation is that a single nerve impulse does not cross the synapse between one nerve cell and the next, or between a nerve cell and a muscle cell, but it does produce a short-lived effect, perhaps by releasing a small amount of transmitter substance at the synapse. This then *facilitates* the passage of a closely following impulse. In a nerve net requiring such facilitation, the few impulses excited by a light touch would travel only a short distance in the net and produce only a local response. In the anemone *Calliactis,* impulses spread freely through the nerve net without facilitation, but a requirement for facilitation at nerve-muscle synapses prevents the anemone from responding unnecessarily to a light touch that excites only a single impulse, while insuring fast, symmetrical contraction when a vigorous poke excites many impulses following each other in quick succession. This simple arrangement allows the anemone a wide range of response because the *different frequencies of nerve impulses,* excited by lighter or heavier touches, determine which kind of muscles will respond and how fast they will contract.

The **preferred substrate** of the swimming anemone *Stomphia coccinea* is the surface of a shell. Here the tentacles of an anemone make chance contact with a mussel shell, and the inflated basal disk bends toward the shell until it touches. Then the anemone straightens up and slowly glides onto the shell. Only if an anemone is long established on another surface will it fail to transfer to an available shell. (Based on Ross and Sutton)

From the mouth a muscular, ectoderm-lined **pharynx** (stomodeum) leads into the gastrovascular cavity and is connected with the body wall by a series of vertical partitions, the **septa**. A septum is composed of two sheets of endoderm held together by an intervening layer of mesoglea. The septa increase the digestive surface of the cavity, making it possible for an anemone to digest rapidly a relatively large animal, such as a fish or crab. The free edges of the septa are expanded into convoluted thickenings, called **septal filaments,** which bear nematocysts and the gland cells that secrete digestive enzymes. The nematocysts may function to subdue struggling prey and to hold the digestive gland cells of the septal filaments closely appressed to the prey and enveloping the prey so completely as to keep the enzymes from being diluted by fluid in the gastric cavity. Enzymes injected by the nematocysts deep into the prey may also speed breakdown of the tissues and thus aid in digestion. Digestion begins in the gastric cavity and is completed within endodermal cells lining the cavity.

The pharynx is not cylindrical but flattened, and at one or both ends of its long axis is a groove, or **siphonoglyph,** lined with cilia that are much longer than those lining the rest of the pharynx. These long cilia beat downward, drawing a current of water into the gastrovascular cavity and providing the internal parts of the anemone with a steady supply of clean, oxygenated

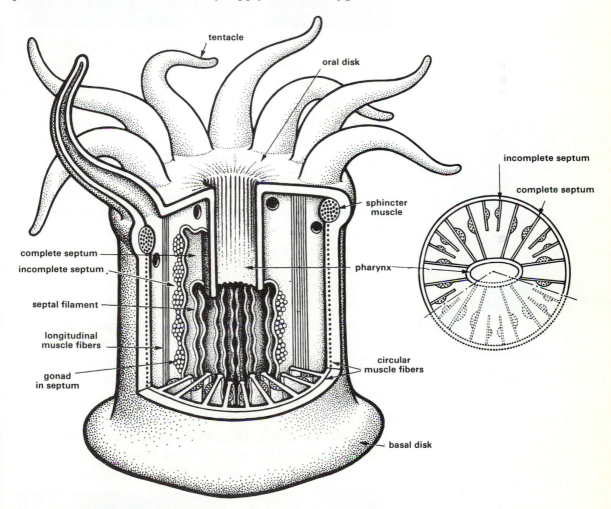

Diagram of a sea anemone cut away to show the large gastrovascular cavity divided by many **septa** (sometimes called "mesenteries"; in all animals except anthozoans, the term *mesentery* is restricted to folds of coelomic lining that support the viscera). The edges of the septa are thickened as **septal filaments,** which bear gland cells and nematocysts. Holes in the septa permit circulation between them. The strong **longitudinal retractor muscles** on the septa and the **circular sphincter** and **columnar muscles** are all endodermal. The ectoderm-lined, ciliated **pharynx** is characteristic of anthozoans.

Diagrammatic **cross-section** of a sea anemone shows the regular pattern of the septa and muscles, symmetrically arranged on either side of the elongate pharynx. (The dotted sector indicates the portion cut away in the 3-dimensional diagram.)

A **fully expanded sea anemone,** on the left, displays its several circles of tentacles and central elongated mouth, awaiting prey. The one on the right has just captured a small fish, which is held by the nematocysts on the tentacles while the anemone folds in the edge of the oral disk, bringing the fish toward the mouth. Helgoland, North Sea. (F. Schensky)

seawater. At the same time, the cilia lining the walls of the pharynx beat upward, creating an outgoing current of water that takes with it dissolved carbon dioxide and other metabolic by-products. During feeding, the cilia of the pharyngeal walls reverse their beat, and the food is swept down the pharynx and into the digestive cavity. A species of anemone is usually consistent in having either one or two siphonoglyphs. But some have individuals with different numbers, because asexual reproduction often leads to irregularities in symmetry.

Anemones may accomplish **asexual reproduction** by pulling apart into halves. In certain species, fragments of the basal disk are left behind as the animals move about, or the entire rim of the basal disk may separate and fragment, the fragments regenerating into tiny anemones.

In **sexual reproduction** the eggs or sperms form in the septa of the gastrovascular cavity. The sexes may be in separate female and male individuals or combined in hermaphroditic individuals. Ejected through the mouth, the egg is fertilized externally and develops into a planula, which finally settles in some rocky crevice and grows into a single anemone. Or, in some anemones, fertilization is internal, and the young may be brooded inside the parent or attached, presumably for protection, around the outside of the basal disk.

Anemone with juveniles around the base of the column is *Epiactis prolifera,* common on the U.S. Pacific coast, often on eelgrass. The tiny offspring are sheltered in brood pits, and when they reach a larger size will creep away and live on their own. (K. Sandved)

Anemone dividing. *Anthopleura elegantissima.* Large clones of these anemones, produced by repeated divisions, live in tight groups on intertidal rocks. Individuals within a clone become specialized for sexual reproduction or for defense, but to a lesser degree than do the connected polyps of a hydroid colony. Puget Sound, USA. (L. Francis)

Stinging filaments, the acontia, are present in some anemones and presumably serve them in defense. These long slender nematocyst-armed filaments are extensions of the free ends of the septal filaments and lie in a mass at the bottom of the digestive cavity. When sufficiently disturbed, the anemone pulls in its tentacled disk and then contracts the body column, forcing the acontia out through the mouth or through small openings in the body wall. This large *Metridium* was laid on its side to show the acontia and the withdrawn disk. (R.B.)

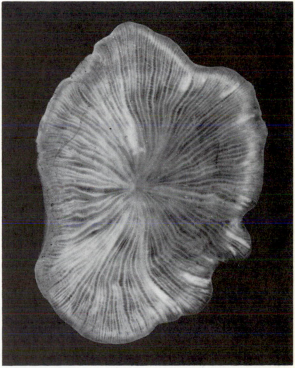

Oral disk of *Metridium* shows the elongated mouth and the light-colored siphonoglyph at one end. Both reflect the biradial symmetry of sea anemones (see chapter on ctenophores). (R.B.)

Basal disk of *Metridium* (seen through a glass wall of an aquarium) is marked by radiating lines where the septa are attached and reflects the basic radial symmetry of cnidarians. Maine. (R.B.)

Aggression in the anemone *Anthopleura elegantissima* takes place only between individuals that do not belong to the same asexually produced group (clone). This maintains a bare uninhabited zone between clones. The attack involves specialized organs bearing special nematocysts and is occasionally fatal. Monterey Bay, Calif. (L. Francis)

Two aggressive anemones, *Urticinopsis antarctica,* in a "head-on" confrontation. Although an instance was observed in which one anemone ate the other, this activity usually results only in separation of the two combatants and is thought to be a kind of territorial behavior. Antarctica. (P.K. Dayton)

Aggression in anemones is a factor in the formation of large clones. The subtidal anemone *Metridium,* which has mostly short delicate tentacles, develops long sweeper tentacles, which can recognize anemones of other clones and damage them. On the U.S. east coast the larger the clone of *Metridium,* the more likely it will survive attacks of the nudibranch *Aeolidia* (see chap. 15), which specializes in feeding on *Metridium;* scattered individuals are removed. *Metridium* develops a patchy distribution of large and long-established clones and complete absence in nearby areas.

Cnidarian eats cnidarian. A sea anemone, *Urticinopsis antarctica,* has captured a large semeostome medusa. Anemones of this kind have also been observed to cannibalize other anemones of the same species. Their most frequent prey are sea urchins, but the sea urchins sometimes camouflage themselves with shells and other debris covered with living hydroids. When an anemone contacts the stinging hydroids, it usually withdraws, and the sea urchin escapes. McMurdo Sound, Antarctica. (P.K. Dayton)

Corallimorphs (order Corallimorpharia) are like anemones in lacking a skeleton. But in most features they resemble stony corals, as in having knobbed tentacles and not having siphonoglyphs. The colonial habit of most is also coral-like. *Corynactis californica,* seen above, forms small sheetlike colonies on shaded rocks and ledges, concrete wharf pilings and various shells. The polyps measure 2.5 cm across disk and tentacles. When reproducing asexually *Corynactis* forms clones of a single color, but adjacent clones may differ. Colors are red, crimson, pink, purple, yellow, orange, brown, or buff, or almost white. This species is common in southern and central California, and other species of *Corynactis* are found around the world in temperate and warm waters. Large solitary corallimorphs occur only in the tropics. (R.B.)

Cup corals of the genus *Caryophyllia* are solitary corals of temperate waters, common off Mediterranean and northern European shores. The tips of their knobbed (capitate) tentacles catch the light, making each coral polyp look like a miniature display of fiber optics. Diameter of a contracted polyp 1 cm. England. (D.P. Wilson)

A coral colony arises by budding from a single individual polyp. White, pinkish, or orange colonies of *Astrangia* are found on both coasts of N. America. A colony may be over a meter across, but each polyp is less than 1 cm high. Woods Hole, Mass. (American Museum of Natural History)

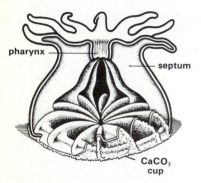

pharynx

septum

CaCO₃
cup

Young **stony coral polyp,** partly cut away to show the relation of the developing skeleton to the septa and to the ectoderm that secretes the skeleton. (After Pfurtscheller)

Skeleton of a solitary coral, *Paracyathus*. The delicate septa of endoderm and mesoglea project between the vertical calcareous skeletal plates. Diameter 1 cm. Monterey Bay, California. (R.B.)

The **stony corals**, with snowy white skeletons of calcium carbonate, are grouped under the order Scleractinia, a name that means "hard anemones" and is appropriate for anthozoans that build hard external skeletons with cups, or in some cases grooves, into which the small, anemone-like polyps can retract. From the wall of each cup or groove, a series of radially arranged vertical skeletal plates project inward, pushing up folds of tissue between the septa. The stony skeleton is constructed by the ectoderm and lies entirely outside and beneath the polyp. Almost all corals are permanently fixed in one place, their skeletons firmly cemented to a hard substrate. But a few kinds lie free on sand or mud.

Many kinds of solitary corals or small coral colonies grow in temperate waters along marine shores, and even the cold deep waters of the Norwegian fjords support great banks of a colonial branching coral, *Lophelia*. However, the great majority of corals are the species that construct tropical **coral reefs.** Small colonies of reef-building (hermatypic) species have been found growing in water which gets as cold as 9°C (48°F). But substantial coral reefs develop only in shallow tropical or subtropical waters and flourish best where the average annual sea temperature is above 23°C (73°F). Where the reef is exposed to strong wave action, the corals grow mostly as encrusting or rounded masses or with short branches. In the sheltered waters behind the reef front, or at depths below the turbulent wave zone, are found the taller corals with longer, more slender branches, or large flattened tablelike colonies supported by a slender central pedestal. In tropical latitudes coral reefs are absent from shores bathed by cold ocean currents (west coasts of Africa and South America) and from the mouths of great rivers, like the Amazon, which deposit large quantities of silt.

Acropora. Species of this genus are common and conspicuous members of tropical reef communities in both the Atlantic and the Indo-Pacific. Their forms vary from stout-branched rounded clumps, to large thin platelike structures, to great thickets of slender-branched "staghorn" types. *Above left,* a compact colony from an exposed reef on the north shore of New Caledonia. (R.B.) *Above right,* close-up of a slender branch with expanded polyps, each having many short tentacles and a single long one. Western Samoa. (K.J. Marschall)

Partly expanded colony of *Manicina* is responding to increasing light. Though closely related to brain corals, *Manicina* has the same niche in the Caribbean area that is occupied by the less closely related fungiid corals in the Indo-Pacific. Young stages of *Manicina* are attached by a stalk to shells or coral. Later they break free and sit on the sand. If turned over by shifting sand, they take in water, eject it forcibly through the mouths and so right themselves. Thus they can live in a sand habitat not open to other corals in the Atlantic. Bimini. (R.B.)

Large brain coral has contracted polyps in sunlit hours, expands and feeds at night. The sinuous skeletal grooves are lined with confluent polyps produced by budding. The many mouths lie along the bottoms of the valleys and are flanked on each side by continuous rows of tentacles. The large coral mass has been built up by a succession of polyps, perhaps over several hundred years. Jamaica. (V.B. Pearse)

Packets of gametes of both sexes are spawned into the water from coral polyps. Or the eggs may be fertilized in the gastrovascular cavities of the polyps and the zygotes develop there into planulas that are finally released. (K.J. Marschall)

Development from planula to coral polyp. The ciliated planula develops septa and a mouth while still free-swimming. It settles down on its aboral end, and tentacles begin to form around the mouth, while the base secretes the beginnings of the white calcium carbonate skeleton (here shown as black radiating spicules). As the polyp develops, it continues to add septa and tentacles in multiples of 6. (After J.E. Duerden)

Living mushroom corals, *Fungia, left,* with extended tentacles; *right,* with contracted tentacles. Hawaii. (R.B.)

Mushroom corals, *Fungia* and related genera, are single polyps which are budded and detached from a fixed, stalked stage. The free polyps are capable of righting themselves if overturned and can even roam about with surprising activity. (Modified after Delage and Hérouard)

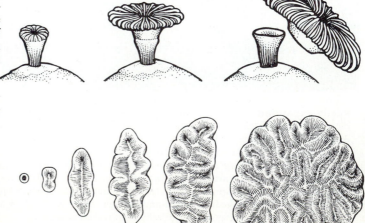

Development from coral polyp to colony of *Manicina* is illustrated here from a series of skeletons of increasing age. The original circular polyp elongates and then folds and branches. The colony, all from one planula, comes to consist of many mouths, each with its pharynx, lying in the valleys of the skeletal grooves and flanked on each side by continuous rows of tentacles. In a coral mass formed by fusion of more than one planula, each gives rise to a separate system of valleys. (After T.F. Goreau and N.I. Goreau)

Tropical, reef-building stony corals harbor **zooxanthellas,** single-celled, photosynthetic dinoflagellates. Living inside the endodermal cells of the coral, these symbionts are responsible for the rich golden-brown hue of many otherwise colorless corals. One may imagine possible advantages to both corals and zooxanthellas from this intimate relationship. The dinoflagellates sheltered within a polyp cannot sink into the dark depths of the sea or be eaten (unless the coral itself is eaten); and they have ready access to by-products of the coral's metabolism: materials containing phosphorus and nitrogen (in short supply in warm waters) and carbon dioxide. The coral may benefit from having these by-products recycled and, at the same time, from the oxygen and organic nutrients that are produced in photosynthesis and released by the zooxanthellas.

There is much experimental evidence for such exchanges of compounds of carbon and nitrogen between zooxanthellas and corals. But the relative importance of these interactions to both members of the symbiosis differs from place to place, according to the amount of light and nutrients available, and probably also in different coral species. For each situation, the nutritional relationship can be determined only by gathering quantitative information on the specific metabolic requirements of both kinds of organisms, the amounts of materials from various sources available to satisfy such requirements, and the degree to which each of these sources is actually available and used. Measurements of the metabolic rates of one coral species *(Stylophora pistillata)* and its contained zooxanthellas have indicated that over 95% of the carbon fixed by zooxanthellas is translocated to coral tissue. In shallow well-lit water, photosynthetic products of the dinoflagellates could more than satisfy all of the coral's energy needs. In the shade or in deeper waters with less light, the corals received only about 58% of the organic carbon necessary to fuel their respiratory requirements and appeared to need additional sources of energy for maintenance and growth. In both situations, zooplankton captured by the coral polyps may provide an essential source of nitrogen.

The rate at which corals add calcium carbonate to their skeletons gives an estimate of their overall growth rate and can be measured by using radioactive calcium or carbon. In many corals with zooxanthellas the skeletal deposition rate is several times faster in light (when the zooxanthellas photosynthesize) than in darkness. In corals without zooxanthellas the rate is slower and little affected by light.

Apparently because of the light requirements of the zooxanthellas, *active reef-building* by corals goes on only in relatively shallow waters, less than 30 m. Reef corals seldom grow at depths greater than about 90 m.

Like trees, corals lay down growth rings that form a record of past environmental conditions. Wide and dense rings in coral indicate times when conditions favored coral growth. *Porites* corals from the Great Barrier Reef, under UV radiation, show yellow-green fluorescent bands from fulvic acids in soils. The bands match rainfall and soil runoff records from the Australian mainland. Thus old *Porites* corals from around the world may provide a record of rainfall and soil runoff from centuries past.

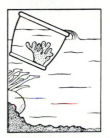

1. Loading vessel.

2. Decanting excess water.

3. Adding Ca⁴⁵.

4. Exposure on reef.

Measuring skeletal growth rate of a coral. A small colony is detached and placed in a vessel under water. The amount of seawater in the vessel is controlled, and a known amount of radioactive calcium (Ca⁴⁵) is added to the seawater. The vessel is then sealed and set back on the reef, so that disturbance to the coral is minimal and conditions of light and temperature are as authentic as possible. After a period of some hours, the coral is removed, and the amount of Ca⁴⁵ in its skeleton is determined. This gives an estimate of the total amount of calcium added to the skeleton by the coral during the experimental period; and the growth rate of the skeleton can then be calculated. (Modified after Goreau and Goreau)

Three main types of coral reefs are recognized. **Fringing reefs** border coasts closely or are separated from them at the most by a shallow narrow stretch of water. **Barrier reefs** also parallel coasts but are separated from them by a channel deep enough to accommodate large ships, and are many kilometers wide. **Atolls** are ring-shaped coral islands enclosing central lagoons, and thousands of them dot the tropical Pacific. Some are hundreds or thousands of kilometers from the nearest land, and their steep outer sides slope off into the depths of the ocean.

Charles Darwin reasoned that if an island, surrounded by a fringing reef, were to subside very slowly, so slowly that the reef could grow upward at about the same rate, the island would grow smaller and smaller, and the fringing reef would become separated from it by a wide, deep channel, finally becoming converted into a barrier reef. If this process were to continue, the island would finally disappear entirely beneath the surface of the water, and the rising barrier reef would become a ring-shaped island, or atoll. This theory is still the most widely accepted one, though changes in sea level during and after glacial periods, which Darwin did not know about, may also have played a role in shaping reefs. In some cases, an atoll may have been formed directly, without going through a fringing reef and a barrier reef stage, upon a submarine platform built up close to the surface by volcanic activity.

Although corals provide the framework of "coral reefs," they are not the sole builders. In some reefs calcareous algas are the major contributors, and the calcareous skeletons of a variety of protozoans and shelled invertebrates are also incorporated.

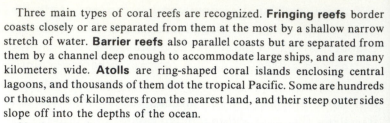

The **Great Barrier Reef** of Australia parallels the tropical northeast coast for more than 2000 km. It is as wide as 145 km and in some places is 120 m deep. It extends from 9°S latitude to as far south as 25°, a few degrees beyond the 23.5°S latitude that marks the southern limit of the tropics.

Fringing reef growing around an oceanic island.

Small barrier reef widely separated from subsided island.

An atoll. Accumulated debris builds up the islands of the ring.

Coral skeletons from the Great Barrier Reef of Australia have been dried and bleached in the sun and are lined up here to show some of the variety of form of tropical reef corals. The living corals show even more striking differences because the surface flesh varies in color and in the shapes of the polyps. The white skeletons are beautiful in themselves, and are sold as ornaments, but they give a poor impression of the exquisite beauty of living, expanded corals, delicately tinted in pastel shades of pink, violet, and yellow or brownish green. (W. Saville-Kent)

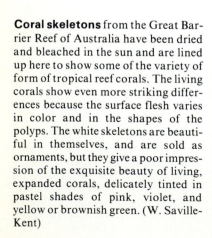

The sea anemones, stony corals, zoanthids, and several other anthozoan orders are commonly considered together as **hexacorallians** (subclass Hexacorallia, or Zoantharia) because their tentacles and septa are often, though not always, arranged in multiples of 6. However, the various hexacorallian orders differ greatly in many respects, not only in the possession or lack of a skeleton, and in the number and arrangement of septa, but especially in the developmental history that leads from egg to adult. Two "hexacorallian" orders, the cerianthids (tube anemones) and the antipatharians (black corals) are so different from the others that they probably should be assigned to two separate subclasses.

Zoanthids can be found in temperate waters, but most members of the order Zoanthidea are warm-water species, usually colonial, each polyp less than 10 mm in diameter. Some are solitary polyps, a few of large size. In the small colony seen above, the polyps arise from stolons that enclose canals connecting the gastrovascular cavities of the polyps. Zoanthids look like anemones, and like them lack a skeleton; but they differ in the way they form septa, in their lack of a basal disk, and in their mostly colonial habit. Santa Catalina Island, California. (R.B.)

Partly retracted zoanthid colony covers a vaselike sponge. Many zoanthids are symbiotic with sponges, gorgonians, bryozoans, crinoids, or hermit-crabs, so gaining a substrate and perhaps providing the host with a protective covering. In the type of colony shown here the polyps are embedded in a mat of epidermis-covered mesoglea, filled with ameboid cells and with canals that connect the polyps. Such a common flesh, or coenenchyme (seen-en-kyme), was seen also in colonial stony corals and forms the fleshy body of various other anthozoan colonies. Aquarium de Noumea, New Caledonia. (R.B.)

Black coral colony, also called thorny coral, takes its common names from the black or brown color and the spiny surface of the horny internal skeleton. The order name, Antipatharia ("remedy for suffering"), owes its origin to an ancient belief, still widely held in the tropics, that black coral has magical healing power. The coral skeleton is polished, cut into pieces, and then heated and bent into bracelets that are worn in vain hope of curing arthritis and other ailments. The tiny white polyps *(below left),* each with 6 tentacles, are interconnected by a thin layer of tissue that covers the whiplike or treelike axial skeleton. Australia. (K. Gillett)

Left. **Polyps,** 1 mm high, on a branch of black coral have 6 simple tentacles armed with nematocysts. They catch minute organisms, presumably those that settle down from surface waters. (K. Gillett)

Right. **Spines covering the axial skeleton** become evident when the living tissue is removed. (R.B.)

Large treelike black coral colony, its size indicated by comparison with the diver, is growing on a New Zealand ocean bottom. Extensive gathering of black coral branches in the Philippines and in Hawaii, for making jewelry, is seriously depleting stocks of this handsome hexacorallian. Most colonies are found in tropical or subtropical waters. (K. Tarlton)

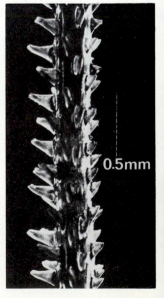

0.5mm

Tube anemone (order Ceriantharia) is one of the most graceful of anthozoans with its 2 whorls of long slender tentacles, one set surrounding the mouth and the other at the margin of the oral disk. The aboral end of a cerianthid is not a basal disk but is rounded and muscular and is used to burrow in sand or mud. The animal secretes around itself a tube of organic material encrusted with sand grains and other particles. In some species the tube may be almost 2 m long. Cerianthids lack strong septal retractor muscles (one of several ways that they differ in structure from actiniarian sea anemones) and have instead a well-developed layer of longitudinal muscles in the column. When disturbed, a cerianthid contracts these muscles and can disappear down its tube in a flash. Members of this order are solitary and mostly subtropical or tropical in distribution. Although their natural life span is not known, one specimen lived and grew for over 40 years in the Naples Aquarium, where this photo was taken. (R.B.)

The rest of the anthozoan orders comprise the **octocorallians** (subclass Octocorallia, or Alcyonaria), a much more homogeneous group, which includes such forms as soft corals and sea whips. The polyps of all octocorals are remarkably similar, having **8 pinnately branched tentacles** and **8 complete septa** that divide the gastrovascular cavity. The flattened pharynx always has one siphonoglyph lined with long cilia that maintain a current flowing into the gastrovascular cavity. Below the level of the pharynx, the free edges of 2 of the septa, those located directly opposite the siphonoglyph, bear heavily ciliated septal filaments whose cilia create a current flowing out of the gastrovascular cavity The free edges of the other 6 septa in most forms bear septal filaments with abundant gland cells that produce digestive enzymes.

Except for a few seldom-seen genera of apparently solitary polyps, all octocorals are colonial, and the body cavities of the polyps are connected with each other through **endodermal canals,** the solenia, which lack septa and which penetrate the whole colony, distributing food and other materials. Despite the uniformity of the 8-tentacled polyps throughout the group, the forms of the colonies and the **endoskeletons** they produce are strikingly diverse.

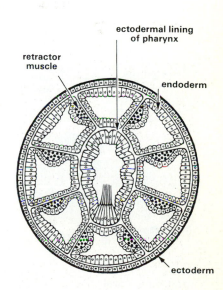

Diagrammatic **cross-section of the polyp of a sea whip** at the level of the pharynx shows the long cilia of the siphonoglyph. The 8 septa each bear a longitudinal retractor muscle. A sectioned polyp of any octocoral would show the same pattern. (Based on W.M. Chester)

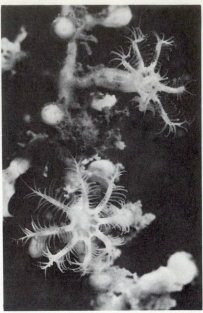

Stoloniferan polyps (order Stolonifera), only 2 mm high, and members of a colony 1 cm in diameter, are from the low intertidal zone of a central California shore. The polyps are all alike and arise separately from a creeping stolon attached to the substrate. The feathery tentacles, as in other octocorallians, are 8 in number. The feeding end of the polyp is thin-walled and delicate; the basal portion of the column wall is stiffened by closely arranged but separate calcareous spicules. The various kinds of stoloniferans grow mostly on shallow bottoms, temperate or tropical. (R.B.)

Telesto, typical of the order Telestacea, is a genus found only in the tropical Atlantic. Each main stem arises from a creeping base and is the elongated body of the axial polyp, its mouth and tentacles at the top. Its lateral polyps arise from the canals in the wall of the main stem. A hornlike external covering, secreted by the epidermis, is the only skeleton. (J. Rees)

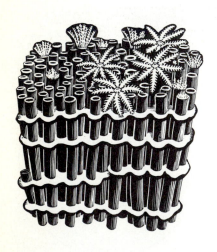

Organ-pipe coral, *Tubipora,* is a tropical stoloniferan. The calcium carbonate tubes, formed of fused spicules, are colored red, but the polyps are sage green. The lower levels of the tubes are abandoned by the coral polyps and become shelters for worms, crabs, and innumerable other small animals. (Modified after E. Haeckel)

In the simplest colonial types, the polyps are all alike, and since they arise singly from flat separate stolons or thin mats, they are called **stoloniferans** (order Stolonifera). Small stoloniferan colonies produce low growths, tightly adhering to rocks in shallow waters or even in the intermittently uncovered intertidal zone. The polyps and their stolons or mats are covered with ectoderm, and the bases of the polyps are connected by narrow hollow tubes of endoderm branching through the mesoglea. Cells in the mesoglea, which are probably derived from the ectoderm, produce an endoskeleton. In small, temperate-water colonies, such as *Clavularia,* the skeleton consists only of scattered **calcareous spicules.** A tropical stoloniferan, the organ-pipe coral, *Tubipora,* produces a far more substantial skeleton of vertical limestone tubes and cross-connecting platforms. Colonies of organ-pipe coral are organized basically like smaller stoloniferan colonies. The polyps arise from a basal mat, and as they grow upward they add similar connecting mats of stolons, from which new polyps may sprout. Colonies of organ-pipe coral may grow large enough to be important reef-builders.

The **soft corals**, or alcyonaceans (order Alcyonacea), grow as fleshy lobed or treelike colonies, in a wide variety of subdued colors. In the Indo-Pacific soft corals form a conspicuous and, in some places, the dominant feature of shallow sea bottoms. The polyps are embedded in a mass of fibrous mesoglea permeated by an abundant network of endodermal canals. Scattered cells throughout the mesoglea produce spicules that give the flabby mass some firmness. The bodies of the polyps are often very long, many reaching all the way to the base of the colony, others arising at intermediate levels; but much of their length is within the fleshy part of the colony (called the coenenchyme), and only the oral portions, with their encircling tentacles, emerge above the surface when the polyps expand. Some species of alcyonaceans, in addition to the usual **feeding polyps**, called autozooids, have a second kind of polyp, the **siphonozooid**. Siphonozooids often have reduced tentacles and septa but strongly developed siphono-glyphs, by which they drive currents of water through the canals of the colony. Not surprisingly, specialized circulatory siphonozooids are characteristic of large fleshy colonies, while in less massive colonies beating cilia in the endodermal canals suffice. There are no specialized reproductive polyps, and either the autozooids or siphonozooids may bear the gametes.

Masses of soft corals cover the bottom in shallow water areas along many tropical shores. The soft or leathery appearance can safely be confirmed by touch, since the nematocysts are small and innocuous, as in octocorals generally. One can only speculate that soft coral colonies may be protected from the nibbling of fishes by a disagreeable taste or by the presence of sharp calcareous spicules, which in some species can be quite prominent. Madang, New Guinea. (J.S. Pearse)

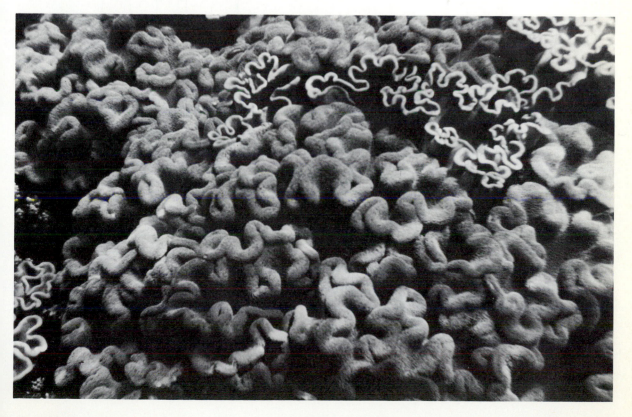

Temperate-water soft coral, common in European waters, is *Alcyonium digitatum*. The spongy, gelatinous lobes may rise as much as 20 cm from gravelly bottoms. White or flesh-colored, they suggest bloated fingers and are called "dead men's fingers" in England, and a less mentionable name in France. At the least disturbance the delicate polyps are pulled inside out, like the fingers of a glove, into the protection of the main mass. England. (D.P. Wilson)

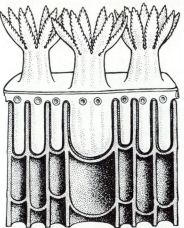

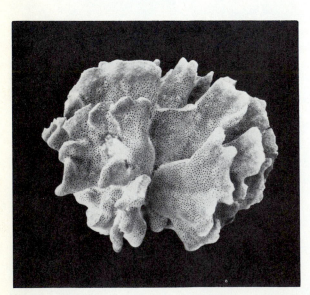

Skeleton of blue coral colony, *Heliopora coerulea.* The massive skeleton of calcium carbonate is blue in color—deep blue at its core. A large colony may reach 2 m in diameter. (R.B.)

Small portion of blue coral colony. *Heliopora,* the sole member of the order Coenothecalia, is found only in the warmest parts of the Indo-Pacific. The live colonies look chocolate-brown, a color imparted to the living flesh by its content of zooxanthellas. The brown flesh masks the blue of the skeleton. The 8-tentacled polyps are only 1 mm in diameter. They are united by a coenenchyme only 3 mm thick and filled with a network of endodermal tubes, the solenia (some shown in cross-section). This network connects with the polyps, and it sends down numerous slender, vertical, endodermal tubes that end blindly. The polyps and the vertical solenia sit in cavities formed by vertical skeletal partitions, and occur only in a single layer on the surface. As the colony grows, the polyps become more numerous, and the surface larger. (Adapted from H.N. Moseley and from G.C. Bourne).

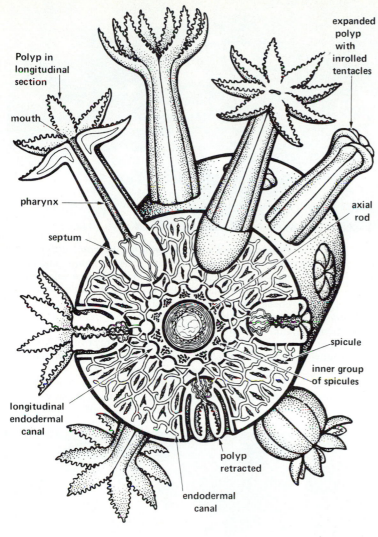

Diagram of a portion of a gorgonian colony. The stem of a single branch is shown in cross-section, and in 3 dimensions, with a few of its polyps both expanded and contracted. At the center of the stem, seen in section, is the main supporting skeleton, a flexible **axial rod** of gorgonin. Groups of small **calcareous spicules** encircle the axial rod, and a larger type is distributed throughout the tissue of the stem. (In many gorgonians the spicules extend onto the projecting portion of the polyp, providing internal support or a covering of overlapping scales, and the axial skeleton also contains spicules.) Just outside the smaller grouped spicules is a circle of large **endodermal canals** that run lengthwise in the stem. They distribute materials, received from the digestive cavities of the polyps, by way of smaller canals that ramify and anastomose throughout the fibrous, mesogleal, cellularized, spicule-filled mass (coenenchyme) of the stem. (Adapted from W.M. Chester)

Gorgonians, or horny corals (order Gorgonacea), have a skeletal axis, usually of a proteinaceous, flexible, hornlike material called gorgonin. Some live in temperate waters but most are subtropical or tropical. The brightly colored colonies grow permanently attached in a variety of shapes which have given them the common names "sea whip," "sea fan," and "sea plume." But even the same species may grow as a radial whiplike form in a spot with random turbulence and, only a meter away, as a flattened fanlike form, oriented perpendicular to a prevailing water current. Instead of a flexible horny axis, *Corallium,* the "precious red coral," has orange or blood-red calcareous spicules, fused into a solid skeletal core, which can be polished and made into jewelry.

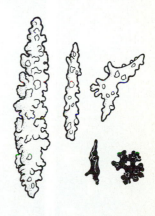

Calcareous spicules of *Pseudoplexaura,* a gorgonian colony. The large white spicules are scattered in the cellularized mesoglea among the endodermal canals. The smaller spicules, which occur in groups around the central core, are purple. (After W.M. Chester)

Gorgonians grow in all seas and in depths down to more than 4,000 m. They are most abundant in shallow warm waters. In Bermuda, the West Indies, the Bahamas, and Florida waters they often dominate the bottom seascape, their stout or slender branching stems, sometimes 2 or 3 m high, swaying gracefully with the currents. Because of their fixed habit and sturdiness, they serve as a substrate for many sessile invertebrates such as sponges, hydroids, bryozoans, brachiopods, and barnacles, in addition to small crabs, shrimps, and brittle stars that move about in the branches. Bahamas.

Portion of stem of sea whip, *Leptogorgia virgulata.* At the *left,* the stem shows polyps beginning to extend from the coenenchyme as the light increases. At the *right,* in higher magnification, the edge of the stem shows fully expanded polyps, gathering plankton with 8 pinnate tentacles. Two fully contracted polyps can be seen in the dark coenenchyme. *Leptogorgia* stems may be yellow, orange, red, or purple. Florida. (R.B.)

Precious red coral, *Corallium rubrum,* differs from other gorgonaceans in having no gorgonin. The red or pink skeletal axis is rigid and consists of a solid core of fused calcareous spicules. Investing this axis is a thin red coenenchyme strengthened by calcareous spicules, as in other gorgonians. The long white polyps gather plankton. The tiny circular ones scattered over the coenenchyme have no tentacles and serve only to circulate water through the colony. A colony may reach 0.5 m in height, with stems up to 4 cm in diameter. The living coral is stripped of its coenenchyme and worked into jewelry and ornaments in Mediterranean countries. A paler coral is collected and fashioned in Japan. (Naples Aquarium)

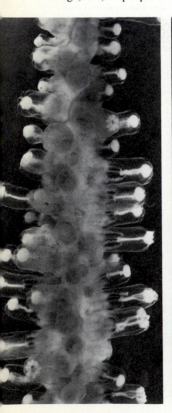
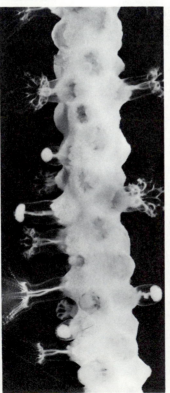

Temperate-water sea fan, *Eunicella verrucosa,* is a gorgonian that grows on European shallow-water bottoms. The name sea fan applies to gorgonians that have their branches in one plane. The dried colony retains its form, but soon loses its orange-pink color and turns a dull white, because the pigmentation is not in the spicules, as in most alcyonarians, but in carotenoid droplets in the living cells. England. (D.P. Wilson)

Latticed sea fan, *Gorgonia,* has cross-connections between the branches. The purple color is in the spicules, and the dried colony retains a purplish color. This is the common sea fan of subtropical west Atlantic waters. Bermuda. (R.B.)

Tropical sea fan, deep orange in color, is seen here in the Aquarium de Noumea, New Caledonia. (R.B.)

Sea plumes, with symmetrical horizontal branches, are abundant on Caribbean bottoms. (J.S. Pearse)

A fleshy **sea pen**, *Ptilosarcus,* sits with its stalk half-embedded in muddy sand. The many tiny feeding polyps are borne on each of the flattened side branches of the plumelike upper portion. They snatch animal plankton from the water currents that sweep past. Dense groups of these pale to deep-orange sea pens live subtidally along the west coast of N. America. Large specimens may be 50 to 60 cm long. (California State Fisheries)

Sea pansy, *Renilla,* is named for the purple color and reniform (kidney) shape of the fleshy apical polyp, which does not stand erect on its stalk but lies flat on the substrate, with secondary polyps only on the exposed surface. The tall, tentacled feeding zooids are covered over with a net of secreted mucus; small animals trapped by the mucus are stung and then swallowed. Dotted over the purple surface are tiny but conspicuous white rings, each a circle of spicules that surround a minute cluster of siphonozooids through which water flows into the interior channels. After circulating, the water is expelled through a larger single siphonozooid in the midline. *Renilla* luminesces at night. It is found in the Caribbean and along the southern parts of both coasts of the USA. (K. Sandved)

Pennatulaceans (order Pennatulacea) are all colonial and include the fleshy "sea pens" and "sea pansies" and various long slender colonies with wandlike stalks. Most are familiar from warm, shallow sea bottoms, but some of the long-stalked forms, up to a meter in height, come up in fishermen's nets in cold or temperate waters or have been photographed, at great depths, on antarctic bottoms. Pennatulaceans are the most complex octocorals. The fleshy colony consists of an elongated axial polyp which buds secondary polyps of at least two kinds, autozooids and siphonozooids, and sometimes even two kinds of siphonozooids, incurrent and excurrent, which circulate water through the colony. Support is provided by spicules and usually also a central horny axis. Yellow, orange, red, or purple spicules may lend their color to the colony. Pennatulaceans live on sand or soft mud bottoms, with the lower portion of the colony anchored in the substrate and the upper portion, bearing the polyps, held erect in the water. When suddenly disturbed, the portion of the colony bearing the polyps is swept by waves of bright blue, violet, green, or yellow luminescence and may expel water and contract down into the soft substrate.

CLASSIFICATION: Phylum CNIDARIA (COELENTERATA)

Class HYDROZOA

TRACHYLINE MEDUSAS (3 orders sometimes classed together as order Trachylina). Medusa predominant.

Order Trachymedusae. Medusas with smooth margin; gonads beneath radial canals. Free-swimming actinula larva; no attached stage. *Liriope, Geryonia.*

Order Narcomedusae. Medusas with scalloped margin; gonads beneath gastric pouches; no manubrium. Free-swimming actinula; young stages sometimes parasitic on other medusas. *Aeginopsis, Cunina, Pegantha.*

Order Limnomedusae. Medusas with smooth margin; gonads beneath radial canals. Small polyp stage. *Gonionemus, Olindias.* Freshwater forms: *Craspedacusta* (polyp *"Microhydra"*), *Limnocnida.*

HYDROIDS (2 orders sometimes classed together as order Hydroida). Solitary or colonial polyps predominant; with or without free medusa stage.

Order Leptomedusae (Thecata, Calyptoblastea). Colonial polyps with zooids, stems, and stolons enclosed by perisarc. Medusas mostly flat, with statocysts, sometimes also simple eyes; gonads on radial canals. *Obelia, Campanularia, Plumularia, Thuiaria, Sertularia, Aglaophenia, Aequorea, Eudendrium.*

Order Anthomedusae (Athecata, Gymnoblastea). Solitary or colonial polyps with hydranths naked, only stems and stolons sometimes enclosed in perisarc. Medusas mostly high and bell-shaped, with simple eyes; gonads on manubrium. *Hydra, Tubularia, Corymorpha, Cordylophora, Hydractinia, Pennaria, Sarsia, Nemopsis.*

HYDROCORALS (2 orders sometimes classed together as order Hydrocorallina). Polymorphic polyp colonies that construct massive skeletons of calcium carbonate.

Order Milleporina, stinging coral or fire coral. Tropical reef-builders. Tiny free medusas. *Millepora.*

Order Stylasterina. Feeding polyps with basal skeletal spine (style), each surrounded by circle of protective polyps. Embryos brooded in chambers to planula stage. *Allopora, Stylaster.*

Of the remaining 3 orders, the first 2 are probably most closely allied with the Anthomedusae; the affinities of the last are uncertain.

Order Chondrophora. Large pelagic solitary polyps that float at the sea surface. Tiny free medusas. Formerly classed with siphonophores; probably closely related to tubularia-like hydroids and sometimes classed under order Anthomedusae. *Velella, Porpita.*

Order Siphonophora. Swimming or floating polymorphic colonies, with both polypoid and medusoid members. *Physalia, Physophora, Muggiaea.*

Order Actinulida. Minute actinula-like forms found among grains of beach sand. *Halammohydra, Otohydra.*

Class SCYPHOZOA

Order Semaeostomeae. Medusas with 4 long frilly mouth lobes; scalloped margin usually with marginal tentacles; no septa or gastric pockets. Polyps with septa and gastric pockets; strobilation produces young medusas. Includes the most common large jellyfishes. *Aurelia, Chrysaora, Cyanea, Pelagia.*

Order Rhizostomeae. Medusas with mouth lobes branched and fused, closing off the central mouth; the lobes bear many small mouth-openings; scalloped margin without tentacles; no septa or gastric pockets. Polyps with septa and gastric pockets; strobilation produces young medusas. Mostly tropical, mid-sized jellyfishes. *Rhizostoma, Catostylus, Cassiopeia, Mastigias.*

Order Coronatae. Medusas with horizontal groove encircling the bell; scalloped margin with tentacles. Both medusas and polyps with septa and gastric pockets. Polyps grow in colonies, their stalks enclosed in perisarc; strobilation produces young medusas. Mostly small, deep-water jellyfishes. *Nausithoe, Linuche, Periphylla, Atolla.*

Order Cubomedusae, sea wasps or box jellies. Medusas with cuboidal bell, simple margin bearing 4 tentacles or tentacle clusters, velarium, septa and gastric pockets. Polyps without septa or gastric pockets; metamorphosis of polyp produces young medusa. Mostly tropical and small, but powerful stingers. *Carybdea, Chironex, Chiropsalmus, Tripedalia.*

Order Stauromedusae. No typical free-swimming medusa. Adult is sessile, attached by stalk to substrate; lacks sensory structures. Both adult and polyp with septa and gastric pockets. Metamorphosis of polyp produces adult form. Small, found mostly in shallow, cold waters. *Lucernaria, Haliclystus, Craterolophus.*

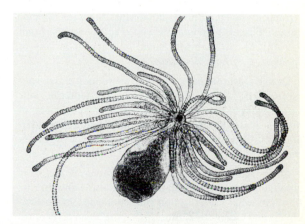

A **marine actinulid,** *Halammohydra schulzei,* from coarse sediment, 8 m deep, off Helgoland, North Sea. Body length, 0.3 mm. (G. Uhlig)

Class ANTHOZOA

Traditionally divided into 2 subclasses, Hexacorallia (Zoantharia) and Octocorallia (Alcyonaria). Hexacorals are a heterogeneous group, all with similar nematocysts but diverse structure, development, and symmetry (several orders show 6-parted symmetry but others do not). The tube anemones (Ceriantharia) and black corals (Antipatharia) are different enough that they are sometimes considered as two separate subclasses. Octocorals are a more homogeneous group, all with similar nematocysts, development, and 8-parted symmetry.

HEXACORALS

Order Actiniaria, the sea anemones. Solitary polyps without skeleton, one or more siphonoglyphs, aboral end usually a basal disk. Found at all latitudes and depths. *Halcampa, Metridium, Anthopleura, Stomphia, Actinia, Anemonia, Calliactis, Stoichactis.*

Order Ptychodactiaria. Very small group of anemone-like polyps. Arctic and antarctic species. *Ptychodactis, Dactylanthus.*

Order Corallimorpharia. Small group of solitary or colonial polyps that are anemone-like in lack of skeleton, coral-like in capitate (knobbed) tentacles, lack of siphonoglyphs, and other structural features. Tropical and temperate-water species. *Corynactis.*

Order Scleractinia (Madreporaria), the stony corals. Solitary or colonial polyps with massive calcium carbonate exoskeleton, no siphonoglyphs, usually capitate (knobbed) tentacles. Some temperate-water species; mostly tropical reef-builders. *Balanophyllia, Caryophyllia, Astrangia, Fungia* (mushroom coral), *Acropora, Pocillopora, Porites, Manicina.*

Order Zoanthidea. Solitary or colonial polyps; one siphonoglyph; no skeleton or basal disk. In colonial types, the polyps are connected by basal stolons or coenenchyme containing endodermal tubes, as in octocorals. Found in temperate and tropical waters, often growing on invertebrates such as other cnidarians or sponges. *Palythoa, Zoanthus.*

Order Ceriantharia, the tube anemones. Large solitary polyps with 2 whorls of slender tentacles, one siphonoglyph. Burrow in sand and secrete tubes of organic material encrusted with sand grains and other particles. Tropical and temperate-water species. *Cerianthus.*

Order Antipatharia, the black or thorny corals. Tiny colonial polyps with 2 siphonoglyphs, borne on a thorny, hornlike axial skeleton and interconnected by coenenchyme and endodermal tubes, as in gorgonian octocorals. Found mostly in deep, tropical waters. *Antipathes, Dendrobrachia.*

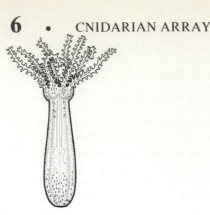

A **solitary octocorallian polyp,** *Hartea* (order Protoalcyonaria). (After P. Wright)

OCTOCORALS, all polyps with 8 pinnate tentacles, one siphonoglyph, 8 septa.

Order Protoalcyonaria. Polyps described as permanently solitary, producing buds that soon detach or no buds at all. With scattered calcareous spicules. Some biologists believe that such solitary polyps are only an early stage in the development of colonies of other octocorallian orders. *Hartea, Monoxenia.*

Order Stolonifera. Discrete polyps connected by basal stolons or mats. Skeleton of separate calcareous spicules, or spicules fused into tubes. Found in shallow temperate and tropical waters. *Clavularia, Cornularia, Tubipora* (organ-pipe coral).

Order Telestacea. Colonies of long axial polyps connected by basal stolons and bearing lateral polyps as side branches. Skeleton of calcareous spicules, sometimes loosely fused. Mostly deep-water species. *Telesto, Coelogorgia.*

Order Alcyonacea, the soft corals. Fleshy colonies with the lower parts of polyps embedded in main mass of coenenchyme, oral ends protruding; sometimes with 2 kinds of polyps. Skeleton of scattered calcareous spicules. Some cold- and temperate-water species, mostly tropical. *Alcyonium, Xenia, Sarcophyton.*

Order Coenothecalia, the blue coral. Polyps and connecting endodermal tubes with massive, blue, calcium carbonate endoskeleton. Found only in shallow waters of tropical Indo-Pacific. *Heliopora.*

Order Gorgonacea, the gorgonians; sea plumes, sea fans, sea whips. Polyps and connecting coenenchyme borne on axial skeleton of hornlike gorgonin and/or fused calcareous spicules; rarely 2 kinds of polyps. Some temperate species, mostly tropical. *Gorgonia, Eunicella, Leptogorgia, Corallium.*

Order Pennatulacea, the sea pens. Colony with single long axial polyp bearing lateral polyps on side branches; always several kinds of polyps. Skeleton of calcareous spicules, sometimes an axial rod. Found in all latitudes and depths. *Ptilosarcus, Pennatula, Veretillum, Stylatula, Umbellula, Renilla* (sea pansy).

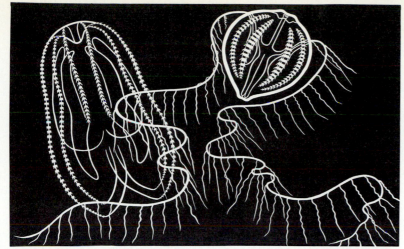

Chapter 7

Ctenophores

CTENOPHORES are mostly small, transparent, gelatinous, delicate animals that drift or swim slowly in all seas. They are especially abundant near shores, but also frequent the open seas, in surface or mid-waters. A few have been reported from depths of over 3,000 meters.

The name of the phylum **CTENOPHORA,** as well as the common name, "comb jelly," refers to the 8 vertical rows of **ciliary combs** that radiate over the surface of the animal like the meridians on a globe. Each comb row consists of a succession of little plates formed of large cilia fused at their bases like the teeth of a comb. In swimming, the combs stroke rapidly toward the pole opposite the mouth, and then recover more slowly to their relaxed position. In each row the combs beat in successive waves, like so many swimming paddles, gently propelling the animal through the water.

Some species flourish in bays or estuaries with only a third of full oceanic salinity, but there are no freshwater forms. Of the nearly 90 known species of ctenophores, most live in warm or temperate seas and only a few are restricted to the cold waters of polar latitudes or deep seas. The warm-water species (like many subtropical and tropical protozoans, sponges, and cnidarians) appear yellow-brown or faintly golden from the photosynthetic, modified dinoflagellate cells that live in their tissues. Ctenophores in cold waters are mostly colorless or have red pigments that lend a pinkish tinge.

Two genera are cosmopolitan, found from arctic to antarctic waters: the globular "sea gooseberry," *Pleurobrachia*, with comb rows and 2 long fringed tentacles, and the compressed, thimble-shaped *Beroe*, which has no tentacles. Distribution of ctenophores is closely tied to temperature changes, and many species migrate from north to south or from surface waters to deeper layers with the change of

Ctenophora (ten-ah'-fo-ra) from the Greek *ctene* = "comb" and *phora* = "bearers."

Ctenophores are the largest animals that move about by means of beating cilia. And the cilia of the combs, up to 2 mm in length, are the longest cilia known among organisms. The thousands of cilia in each comb are arranged in a transverse band and they beat in unison. The giant size of comb cilia provides unique opportunities for microsurgical intervention in the ciliary beat. Insights gained from experimental manipulation of the beating of ctenophore comb cilia are widely applicable among animals.

In the Black Sea *Pleurobrachia* is known to feed at the surface in the spring, and then descend gradually to about 35 meters below the surface as hot weather arrives. They also remain below throughout the winter, and in this way live always at a temperature close to 12°C.

There is some evidence, for northern waters, that a season of great ctenophore abundance is followed in subsequent years by smaller yields of oysters, cod, or herring.

Ctenophore light has been measured at a wavelength of 510 nm and at an intensity of about 0.2 μW/cm^2 at a distance of 1 cm.

An observer on the southeastern coast of Argentina has described *Beroe*-laden waters, dashing high on the rocks at night, as like a spectacular display of fireworks.

Pieces of ctenophore that include at least 4 combs will luminesce as long as they remain alive. If stimulated in the dark, even the early stages of developing embryos luminesce.

Sea gooseberries are stranded on beaches the world over, washed ashore during a storm or in wind-whipped waves. Seen on the wet sand as glistening balls, they give no hint of the live animal loping gracefully through the water, trailing its long, fringed tentacles. *Pleurobrachia bachei,* common along the Pacific coast of N. America. Life-size. (R.B.)

seasons. Calm weather brings the fragile comb jellies closer to the surface, and stormy weather or even small disturbances send them deeper again, away from the surface turbulence.

Being feeble swimmers, ctenophores are carried about by currents and tides and often accumulate in great wind-driven swarms. When so concentrated, their voracious feeding may decimate the surface plankton, removing large numbers of the fish fry important in the human food supply.

So transparent as to be almost invisible in the water, ctenophores are often overlooked. Even in favorable lighting the only clue to their presence, as one strives to see them from a small boat or boat pier, may be the rippling iridescence of the beating rows of combs as they diffract the light and produce a constant play of changing color. At night ctenophores become visible, when disturbed, as they react with a display of greenish **luminescence**, of an intensity said to exceed that measured in any other group. As they glide through the dark water, they flash along the walls of the 8 digestive canals that underlie the 8 rows of combs. If dipped up in a container with great care so as not to fragment the delicate bodies, and carried in a bucket of seawater to a lighted room, they gradually cease glowing. Then, if the room is darkened for a time, and the bucket gently shaken, they will luminesce again.

Ctenophoran structure and habits are well illustrated by *Pleurobrachia,* for the genus is a member of what is presumed to be the most primitive of the groups, the cydippid (sy-di′-pid) ctenophores. The surface of the ctenophore is covered with an epithelium of **ectoderm,** which turns in at the mouth and lines the pharynx. Beyond the pharynx, the stomach cavity and the canals that lead from it are lined throughout by a digestive epithelium of **endoderm.** Between the thin ectodermal and endodermal cell layers is a mass of gelatinous **mesoglea** that gives structural support to the delicate body. The mesoglea is penetrated by the gastrovascular chambers and canals. On opposite sides of the body are 2 highly extensible ectoderm-covered **tentacles,** which can be withdrawn into the **tentacle sheaths,** deep ectoderm-lined pouches that indent the mesoglea. The mesoglea is strengthened by connective-tissue fibers and by long muscle cells and contains many wandering ameboid cells. The muscle cells, derived from ameboid cells of the mesoglea, are long contractile cells, independent of the epithelial cells of the animal, and are of the *smooth muscle type*. They run longitudinally and circularly under the ectodermal epithelium, and there are denser concentrations of muscle cells in the core of the tentacles, surrounding the mouth, investing the pharyngeal walls, and around the aboral sensory area.

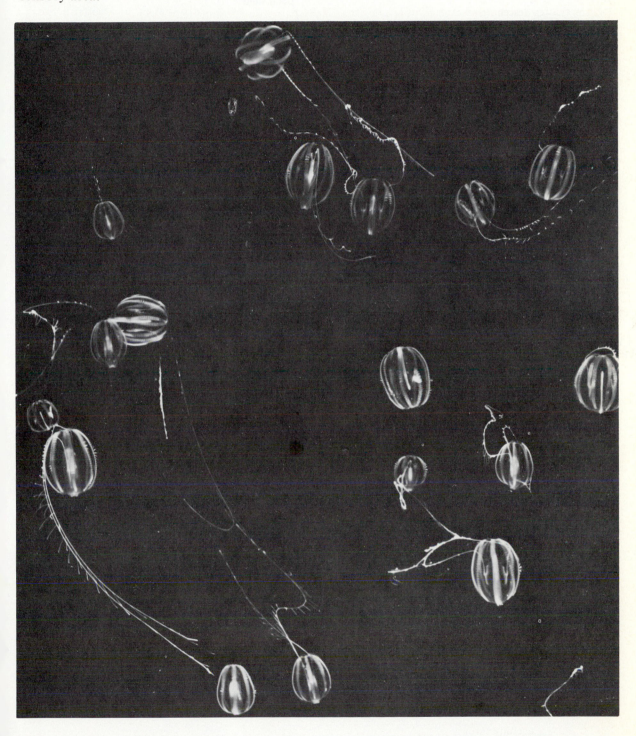

Swimming sea gooseberries, *Pleurobrachia,* shown about actual size. The largest usually seen are 18-20 mm in diameter. Under more natural conditions, the 2 long tentacles may be nearly 4 times the length of the longest shown here, and the sticky side branches are widely outspread as the tentacles sweep the water gathering food organisms. Helgoland, West Germany. (F. Schensky)

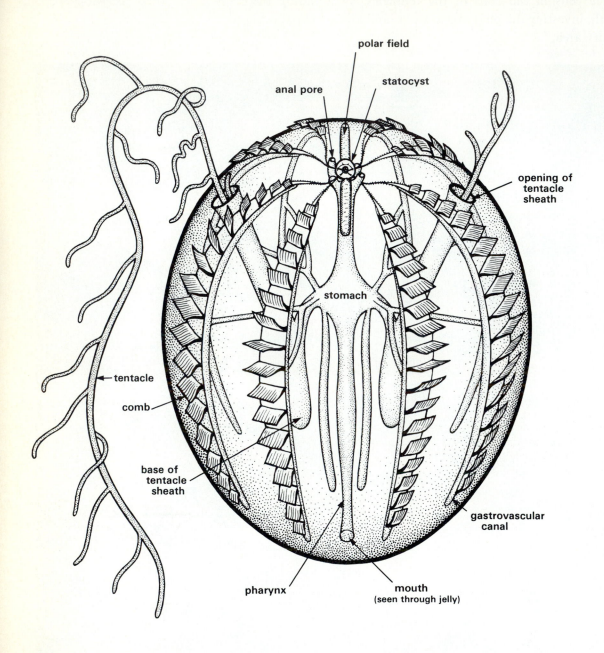

Diagram of a ctenophore, *Pleurobrachia.* The animal is tilted toward the viewer to show the prominent structures at the sensory pole. This shifts the mouth backward away from the viewer, but it can be seen through the transparent jelly. The gastrovascular chambers and canals penetrate the mesoglea, and the tentacle sheaths are indentations of the body surface. The 8 rows of ciliated combs are underlain by 8 gastrovascular canals.

The **symmetry** of ctenophores appears at first glance to be *radial,* with equally spaced comb rows radiating over the surface. But it is underlain by bilaterality even more marked than that we saw in the mouth and internal partitions of sea anemones and corals. In ctenophores, two-sidedness is conspicuous in the 2 tentacles and tentacle sheaths, the branching of the gastrovascular cavity, the elongation of the pharynx in a plane opposite that of the stomach and tentacles, and in the symmetry of some of the structures at the aboral pole. This *bilaterality* in ctenophore architecture begins at least as early as the 8-cell stage of the embryo, and it seems appropriate to speak of this group as having **biradial symmetry.**

Locomotion in *Pleurobrachia* is by the beating of the ciliary combs. A single comb consists of tall cells, each bearing about 50 cilia, so that a comb is made up of several thousand cilia that adhere and move as one. Since the power stroke is away from the mouth, the animal moves through the water with mouth end foremost, and when feeding or resting at the surface it hangs mouth end up. So we do not speak of upper and lower poles as they are shown in the diagrams, but refer to the mouth end as the **oral pole** and the opposite end as the **aboral pole** (or sensory pole.) This last has a concentration of nerve cells and sensory cells and a special balancing structure, the statocyst, which helps to keep the animal advantageously oriented in swimming, feeding, or reacting to changes in physical conditions.

The **statocyst** consists of a ciliated pit roofed over by a transparent dome derived from fused cilia. Within the pit is the **statolith,** a concretion of calcareous particles supported on 4 tufts of very large cilia. At the base of each tuft, called a **balancer,** there emerges a **ciliated groove,** which promptly divides into 2 grooves, and these run to the starting points of a pair of comb rows. Each beat of a tuft of balancer cilia can be seen to be transmitted as a wave of beating cilia along the grooves and then along a pair of comb rows.

The most aboral comb beats first, and the others successively along the row. Since some will be falling while others are rising, visible *waves of rising and falling combs* pass along the row from sensory area to mouth end. In any pair of rows leading from the 2 grooves of the same balancer the waves will pass along synchronously, but they will not be in unison with other pairs of comb rows.

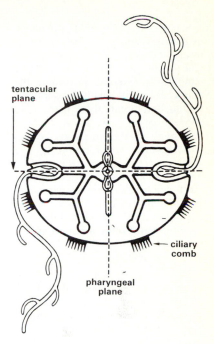

Biradial symmetry shown in a diagrammatic cross-section of *Pleurobrachia.* The dotted vertical line is the *pharyngeal plane,* in which pharynx and mouth are elongated. The dotted horizontal line is the *tentacular plane.* Pharyngeal halves are not equivalent to tentacular halves.

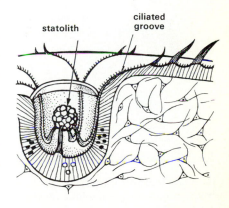

Statocyst of *Pleurobrachia.* The ciliated pit and its dome, the epithelium lining the pit, one of the ciliated grooves, and the first 3 combs of one row are shown in section. Nerve cells are seen in the mesoglea. The sectioned epithelium shows developing calcareous particles in the center and on each side a group of dark bodies thought to be photoreceptors. (Adapted from G.A. Horridge)

When the ctenophore strikes an object or is experimentally exposed to an irritating chemical, the effective stroke and the direction of the waves can be temporarily reversed, propelling the animal for a time with the statocyst leading and mouth at the rear.

Removal of the statocyst by an experimenter does not stop the beating of the combs. But it does result in lack of coordination between the 8 comb rows, erratic swimming, and inability to return to a vertical position if tilted, for this last depends on coordinated rates of beating in all 8 rows.

A cut through a ciliated groove causes a brief cessation of beating in all the rows. When the beating resumes, the row that has been severed from its ciliated groove beats independently of the other member of its pair.

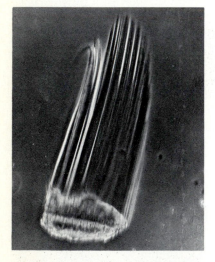

Isolated beating comb, detached from a fragmented ctenophore caught in a plankton net, is a common sight in marine plankton seen under the microscope. It will continue to beat as long as the epithelial cells, at the bases of the giant cilia, remain alive. As in ciliated cells generally, the beat is self-generating. The cells basal to the cilia are packed with large mitochondria, which presumably provide for the high energy requirements of moving so large an animal entirely by cilia. (R.B.)

The shifting weight of the statolith, as the ctenophore is tilted or tossed about by oscillations of the water, will press to a lesser or greater degree on each of the 4 balancers, accordingly slowing or speeding their rate of beating. This produces unequal numbers of ciliary waves in the uppermost or lowermost rows of combs in the horizontally swimming animal, or in the vertically positioned resting or feeding animal that is temporarily tilted by ripples in the water. Thus by differences in the rates of comb beats in the 4 quadrants of the animal, it returns to the vertical position, or it turns over, or it swims upward or downward, with ciliary beating coordinated by the statocyst. Which of these movements it will execute depends on information, from the environment, that modifies the response of the statocyst to gravity.

Behavioral responses of the whole animal, can be elicited by an adequate touch to almost any part of the body, by vibrations in the water, by unfavorable temperature changes, or by irritating chemicals. Any stimulus strong enough to abruptly stop the ciliary beating is accompanied by a sudden contraction of the tentacles, and then slow withdrawal, sometimes completely, into their sheaths. Such responses are invariably followed by downward swimming. Under natural conditions downward swimming carries the animal away from any unfavorable physical contact, from surface turbulence, and from the surface layers of water that sweep onto the shore. If the disturbance subsides, the tendency of the animal to swim upward reasserts itself. The tentacles reemerge, and though the sheaths are lined with cilia, extension appears to be essentially passive—mostly by muscular relaxation and by the force of water flowing past the delicate tentacles as the animal moves along.

When *Pleurobrachia* sinks passively downward, it experiences an increase in hydrostatic pressure, which then stimulates upward swimming. By this behavior, and by changes in buoyancy, ctenophores maintain themselves at particular, shallow depths where their planktonic prey abounds.

Localized responses during feeding, or when perceiving small ripples in the water, include the coilings or limited contractions of the tentacles, and opening or closing of the mouth. Another localized muscular response follows the slightest touch to the sensitive cilia that form the dome over the statocyst. The protruding statocyst is suddenly pulled down, by symmetrically attached muscles, indenting the surface.

The **ectodermal nerve net** underlying the surface epithelium can be made visible with certain dyes. The network functions much like that in cnidarians, with diffuse conduction in all directions from the point of stimulation. It probably is connected with the even more diffuse nerve network which pervades the mesoglea, among the muscle cells. Nervous impulses are transmitted from cell to cell across **synapses,** and there are no ganglia nor any controlling nerve center. However, there are regionally differentiated pathways in the general network, and some of these are seen as denser concentrations of nerve cells such as the **8 nerve strands** that underlie the comb rows.

Localized responses may be coordinated by specialized local pathways in the nerve net, and not by diffuse spreading.

Since under favorable natural conditions the animal tends to swim upward, and under unfavorable conditions reverses this tendency and swims actively downward, there must be sensory and nerve cells that receive and transmit information from the environment that in some way modifies the response of the statocyst cilia to gravity.

In *Beroe* the nerve net around the mouth is also more concentrated and forms an **oral ring,** visible on staining.

Sensory cells, with projecting bristles that increase sensitivity, occur over the surface of ctenophores, and the animals are sensitive to touch, vibrations, temperature, and chemicals. Of presumably sensory function are the 2 long ciliated depressions, the **polar fields,** that run, as extensions of the floor of the ciliated pit of the statocyst, onto the polar surface in opposite directions from the statocyst, and at right angles to the tentacular plane. The tall epithelium that lines the pit of the statocyst is completely ciliated, and many of these cells must be sensory, with the cilia acting as receptors. A group of cells near the origin of each of the 4 ciliated grooves have dark, lamellated bodies, derived from the membranes at the bases of cilia, that in structure resemble the photoreceptors of vertebrates. But experiments in which a swimming ctenophore *(Pleurobrachia)* was illuminated by a beam of light directed from below, from above, or from the side, showed no response in its swimming movements, at least, to the changes in light; all the responses noted were oriented to gravity.

Especially long sensory bristles, often 3 or 4 to a cell in some ctenophores, are abundant on the tentacles, the parts of the body most sensitive to touch or underwater vibrations. When stimulated, the tentacles shorten, ciliary beating stops, and the animal swims downward. But if the tentacles are cut off, a ctenophore responds in the same way to touch or vibrations, presumably by way of the body surface, through the nerve net to the statocyst, or by direct stimulation of the statocyst cilia. An isolated tentacle, floated in a dish of still water, will respond to vibrations.

There is more than one conduction system in ctenophores, as in many cnidarians. If excess magnesium ions are added to a dish of seawater containing a ctenophore *(Pleurobrachia)*, the sensory cells and structures, as well as the synaptic nerve nets of the ectoderm and mesoglea, become anesthetized. The animal does not react to external stimuli, as it normally would, by tentacle contraction, abrupt cessation of ciliary beating, or downward swimming. In contrast, the cilia in the grooves and the comb rows continue to beat, and at an even greater rate, once the braking action of the nerve net is removed. Coordination between comb rows is lost, but not coordination of waves within any row. Thus we see that ciliary beating in the normal animal is inhibited by the nerve net but maintained and coordinated within each row by a conducting system that does not show the usual attributes of nerve cell transmission. The nerve cells seen under the comb rows appear to be inhibitory ones related to the nerve net; they do not generate or coordinate the waves of beating along a comb row.

When a comb row is cut across, between 2 combs and through the underlying nervous layer, the wave of beating crosses the cut, and the combs continue to beat in coordination along the row. But if a comb is prevented from beating, or if it is trimmed off close to its base, the wave of beating does not pass the immobilized comb or stub, and the parts of the comb row on either side develop independent rhythms. These and other experiments suggest that in a ctenophore, coordination within a ciliary groove depends on non-nervous cell-to-cell conduction, while coordination within a comb row is by mechanical interaction between the combs. The combs are mechano-sensitive, and the movement of each comb sets up hydrodynamic drag forces in the water that trigger the beating of the next comb. In this way, within-row conduction coordinates the continuous beating of prolonged locomotory periods. But responses under nervous control, those mediated by the sensory receptors and the nerve net, permit a rapid change in direction of movement when physical conditions require it, or a sudden halt and quiescent posture in feeding. The same nervous pathways cannot both stop and initiate the same functions.

Sensory pole, freshly cut from the aboral end of a live *Beroe*, shows the centrally located statocyst, the two prominent **polar fields,** and the 8 ciliated grooves that run from the statocyst to the 8 severed rows of still beating combs. Each polar field is about 2 mm in length. Panacea, Florida. (R.B.)

In the actively **feeding** animal the 2 long trailing tentacles sweep a wide swath in the water. The muscular tentacles are fringed with side branches attached along one edge. These side branches are studded with adhesive cells, the **colloblasts,** which have sticky surfaces that adhere to such small animal food as eggs, larvas, copepod crustaceans, and fish fry. Struggling prey are also mechanically restrained by the coiling of the tentacles and their branches. As the tentacles are withdrawn, entrapped prey, embedded in mucus, are wiped off on the edges of the **mouth** and taken into the capacious **pharynx**. The walls of the pharynx are thick and folded, and it is here that digestive enzymes, with surprising speed, play their major role in the *extracellular* digestion that begins the process of reducing the sizeable prey to smaller particles that can be more completely digested later. The partly digested particles pass into the **stomach** and from there are distributed throughout the body by **gastrovascular canals** that end blindly. In the endodermal cells that line the canals digestion is completed by *intracellular enzymes*.

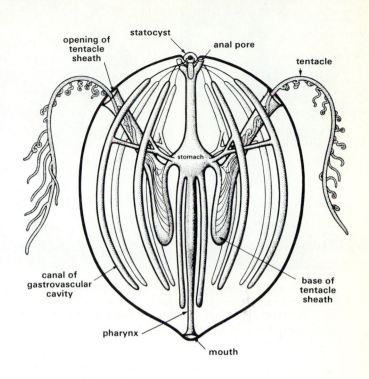

The gastrovascular cavity of *Pleurobrachia* (shown without the overlying comb rows). The pharynx appears narrow because its largest diameter is in the plane at right angles to the surface of the page; the stomach is widest in the plane of the page. Only small undigestible particles are eliminated from the anal pores; larger residues are ejected from the mouth, which is the only important opening of the gastrovascular cavity. The bilateral branching of the gastrovascular canals is a prominent feature of the biradial symmetry of the animal.

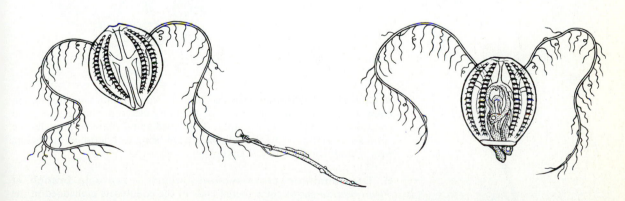

Pleurobrachia **catching a small pipefish.** (Adapted from M.V. Lebour)

Pleurobrachia **with pharynx crammed full of young herring.** (Adapted from M.V. Lebour)

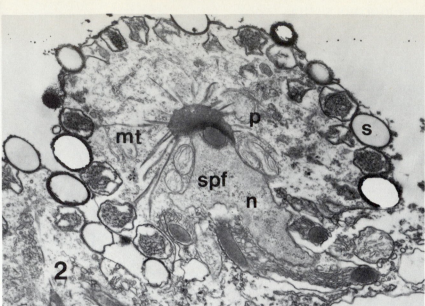

1. Colloblasts, covered solidly with adhesive spherules, crowd the surface of tentacle branches of *Pleurobrachia*. (Scanning electron micrograph, x2500 by P.V. Fankboner)

3. Diagrammatic interpretation of a longitudinal section through a colloblast from many electron micrographs of *Pleurobrachia*. The tip of the spiral filament is embedded in the muscle of the core of the tentacle branch, firmly anchored there by submicroscopic "rootlets." (From P.V. Fankboner)

2. Colloblast in longitudinal section. The hemispherical exposed end of the cell, which is covered with adhesive spherules, appears in section to be rimmed with a single row of spherules **(s),** some already spent and clear of content. Colloblasts are not pulled free at each contact, and can serve in food capture more than once, the spent spherules being replaced. Individual spherules are fastened, by bundles of microtubules **(mt)** to the head of the spiral filament **(spf)** at the plexus **(p),** the area of fusion of microtubules, spiral filament, and cell nucleus **(n).** The spiral filament coils several times around the elongated, tapering cell nucleus, and appears here as 4 separate oblique sections. (Transmission electron micrograph, x22,000; P.V. Fankboner)

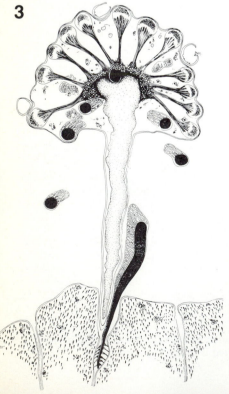

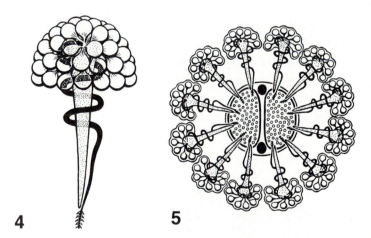

4. Colloblast in 3 dimensions has been drawn by combining information from light microscope observations, scanning electron micrographs, and transmission electron micrographs. The coiled spiral filament provides the springiness that prevents the colloblast from being torn loose by struggling prey. (Modified after P.V. Fankboner)

5. Diagrammatic cross-section through a **tentacle branch** of *Pleurobrachia* shows the tethered ends of the colloblasts embedded in the muscular core, which contains 2 large muscle bundles and 2 small nerve strands. (Based on P.V. Fankboner)

Regeneration of lost or damaged parts is speedily accomplished by all ctenophores which have been studied, and such repair probably goes on almost continuously in these extremely fragile animals. Whole new individuals can be regenerated even from small fragments. But this ability is used regularly for **asexual reproduction** only by a few of the creeping forms, in which small pieces separate from the edge of the body as the animal creeps about, and each fragment regenerates to form a new complete ctenophore.

Sexual reproduction occurs at two separate periods in the life of many ctenophores. The gonads may first develop when the individual is still a larval or a juvenile form. The eggs produced by larvas and juveniles tend to be small and few in number, and many fail to hatch, but a few offspring may result. The gonads then become inactive or even much reduced until the ctenophore reaches adult size and reproduces for the second time. *All ctenophores are hermaphrodites,* producing both eggs and sperms in the same individual. The gonads lie in the walls of some or all of the 8 meridional gastrovascular canals, both **ovaries** and **testes** in the same canal. The cells from which the **gametes** develop appear to be derived from the endoderm.

Most ctenophores *release eggs and sperms into the seawater* through pores in the epidermis above the gonads, and the **embryos** and **larvas** develop in the sea. The adult usually dies not long after spawning. A few of the creeping ctenophores brood their embryos internally. Ctenophores are not quite as ephemeral as they look, but probably they seldom live more than about a year.

Development is notably orderly and precise. As the zygote divides again and again, the contribution which each cell and its progeny will eventually make to the developing larva is determined. Hence this kind of division, or cleavage, is called **determinate cleavage,** and it results in **mosaic development,** in which the various parts of the adult are mapped out even in the early cleavage stages. If either of the first 2 cells, resulting from the first division of the zygote, is experimentally removed, the remaining one will produce an embryo with only 4 comb rows instead of the usual 8. One cell isolated from the 4-cell stage will produce an embryo with only 2 comb rows.

At the 8-cell stage, after the third division, the embryo consists of 2 rows of 4 cells each, with the 4 end cells slightly smaller and set at an angle, so that the rows seem to curve up at the ends. The *biradial symmetry* of this early embryo already reflects the final symmetry of the adult. The long axis of the cell rows is the future tentacular plane; a cut across the rows would mark the

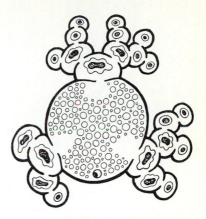

Egg with 3 nurse-cell clusters.

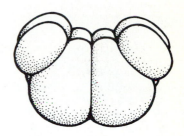

8-Cell cleavage stage.

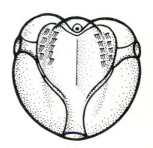

Developing **cydippid larva.**

Development of a ctenophore. Each potential egg divides until it is surrounded by "nurse cells" in 3 clusters. When growing nurse cells have increased several times in size, their cytoplasm flows into the egg. As they shrink the egg becomes large and filled with food reserves. The relationship between the symmetry of the egg, embryo, and larva has been experimentally established. The arrangement of nurse-cell clusters on the egg may also reflect the biradial symmetry of the adult. The position of the nurse cells with respect to this symmetry is still to be discovered, and the arrangement of clusters shown here is hypothetical. (After H. Pianka)

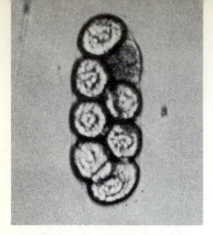

The long axis of this **8-cell embryo** will become the tentacular plane of the adult. The *biradial symmetry* of the adult is already established at this early stage. (R.B.)

future pharyngeal plane, with the upward curve directed toward the future sensory pole. At the 8-cell stage, only the 4 end cells can now produce comb rows, and as the cells of the embryo continue to divide, they acquire increasingly specific roles in the development of the various tissues and structures of the adult. The rigidly determinate development of the embryo contrasts sharply with the adult ctenophore's flexible powers of regeneration.

Nearly all ctenophores have a **cydippid larva,** which looks like a miniature version of adult individuals of the order Cydippida. In *Pleurobrachia,* as in other cydippids, the larva matures into an adult through only slight structural changes. In the other orders, however, the cydippid larva undergoes more or less dramatic changes as it grows into an adult that lacks the two long tentacles, or has large mouth lobes, or lacks combs, or lacks any kind of tentacles, or is very different in body shape from the globose cydippid larva. In the rest of this chapter, some of these diverse forms seen among the orders of ctenophores will be illustrated.

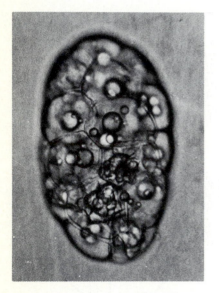

As **cell cleavage** proceeds past the elongated plate of 8 cells, these divide unequally, pinching off 8 smaller cells on the concave surface of the cell plate, the surface that will become the aboral pole. The smaller cells multiply, cover the aboral pole, and spread downward in a one-layered sheet, becoming the *epidermal covering* of the animal and producing the combs. The original large cells give rise to the layer of *endoderm* that forms the gastrovascular cavity. (R.B.)

The cydippids generally follow the description given for *Pleuro-brachia.* All have 2 long branched tentacles retractable into a pair of tentacle sheaths. The gastrovascular canals end blindly. *Mertensia* is a "sea walnut" familiar in the northern Atlantic and Pacific. It is often an "indicator species" of cold water; above a certain temperature it quickly dies. *Hormiphora* is pear-shaped and lives in warm waters.

A cydippid ctenophore. The waves in the beating combs are best seen on the left edge. The statocyst is visible at the aboral pole (here uppermost). The large mouth is open. The gonads are just visible in the gastrovascular canals. The 2 tentacles are only partly expanded. 2x actual size. Monterey Bay, Calif. (R.B.)

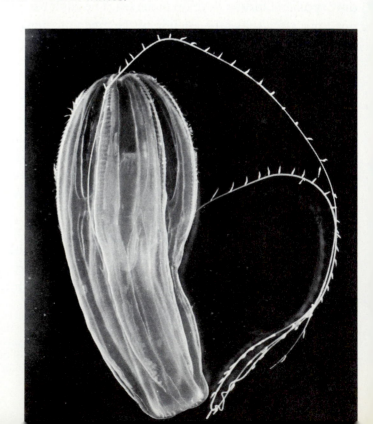

The **lobates** are more or less oval in outline and somewhat compressed. The comb rows are longer on the flattened surfaces, which are expanded at the oral end into muscular **mouth lobes** that waft small organisms toward the mouth and also help to enclose larger prey. In some species the mouth lobes are used in swimming. During metamorphosis the 2 long tentacles of the cydippid larva lose their tentacle sheaths, usually become reduced, and move down to lie one on each side of the mouth. **Tiny tentacles** lie along grooves that surround the mouth. Four **ciliated flaps** and the tentacles all join to entrap small food organisms and to sweep them, in mucous strands, into the mouth. The oral ends of the gastrovascular canals are joined.

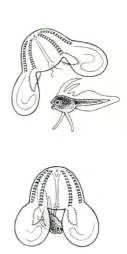

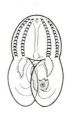

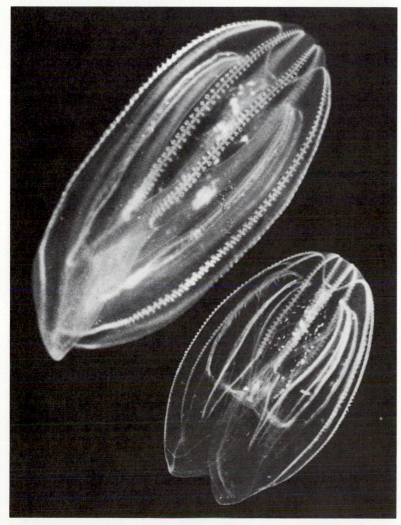

Lobate ctenophores common on the American East Coast from Cape Cod to South Carolina are *Mnemiopsis leidyi*. The mouths are at the left in both specimens. *Mnemiopsis leidyi* is noted for its great swarms in summer, and the brilliance of its luminescence when disturbed at night. Actual size. Woods Hole, Mass. (R.B.)

A lobate ctenophore (*Bolinopsis,* of cold Atlantic and Pacific waters) catching a young angler fish with small, tenacious tentacles around the mouth, and then enclosing the struggling fish between the mouth lobes until it is tucked inside. (Modified after M.V. Lebour)

Cestum ("girdle" or "belt") is encountered primarily in warm seas. *Cestum veneris,* up to 2 m long, is common in the N. Atlantic and Mediterranean. It is called "Venus's girdle" because of the beauty of the transparent, pale violet body that shimmers with blue or green iridescence in sunlight. At night it is luminescent. The similar but smaller *Velamen* also occurs in Florida waters. (Modified after C. Chun)

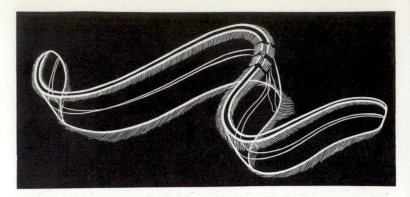

The **cestids** are ribbonlike ctenophores, transparent as glass. The young stage is a globular cydippid larva, but during development the tentacular plane becomes greatly compressed and the pharyngeal plane greatly elongated. The 8 comb rows come to lie in 2 lines along the aboral edge of the ribbonlike body, with the combs twisted so that their bases are almost parallel to the comb rows. Four of the comb rows, 2 on each side of the statocyst, have only a few combs. The other 4 comb rows consist of hundreds of combs and extend most of the length of the aboral edge. The 2 main tentacles are reduced but from their sheaths next to the mouth extend tufts of short tentacles, and 2 rows of short tentacles fringe the oral edge. When quietly feeding, the animal moves slowly through the water by beating its combs, but when disturbed it swims rapidly by undulations of the body.

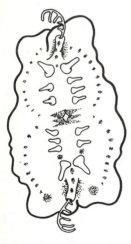

Coeloplana is a flattened, creeping platyctenid, elongated in the tentacular plane. The aboral surface shows the central statocyst, 2 tentacles just emerging from their sheaths, and rows of erect projections typical of this genus. There are no combs. (Modified after A. Krempf)

The **platyctenids** *(platy* = "flat," *ctenid* = "a form with combs")* are mostly very small, creeping or sedentary ctenophores with 2 main tentacles. They are so strongly flattened in the oral-aboral axis that statocyst and mouth are brought close together. Many are commensals that cling to particular hydroids or anthozoan colonies, or aquatic plants, or even to the spines of sea urchins. In those that have combs only in the larva, the resting or creeping adults can easily be mistaken for small, delicate marine flatworms—until the 2 long unmistakably ctenophoran tentacles suddenly emerge from their sheaths. Most are tropical or subtropical, but *Tjalfiella* is found in cold waters off Greenland living as a sedentary commensal on *Umbellula,* a pennatulacean. The platyctenid *Gastrodes* is the only ctenophore known to be parasitic. The planula-like stage bores into the test of a tunicate, emerges as a larval cydippid stage, settles to the bottom and loses its combs, then finally becomes a flattened creeping form with 2 long tentacles.

Coeloplana in *section* through the tentacular plane. Platyctenids have an everted pharynx, so spread out that its ciliated ectodermal lining forms the surface of the creeping ctenophore applied to the substrate. The opening that functions as a mouth is (anatomically) the connection between the everted and internal parts of the pharynx. (Modified after T. Komai)

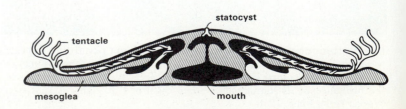

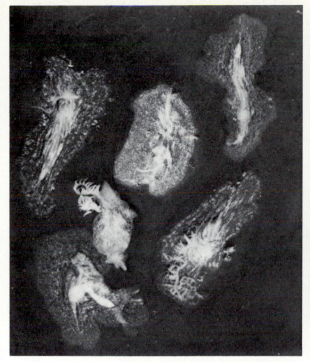 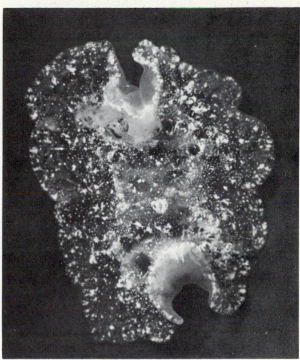

Specimens of **Coeloplana,** each about 1 cm long, collected from Sagami Bay, Japan, in April, from the surface of an alcyonacean colony, on which they formed an almost invisible mosaic pavement of hundreds of delicately pink platyctenids arranged edge-to-edge. They revealed themselves only when they thrust out the 2 tentacles. (R.B.)

Creeping platyctenid, *Vallicula,* without combs in the adult, is best known from Florida and Bermuda waters. It may reach 18 mm. By day it remains quiet and markedly flattened, its mottled pattern usually matching the natural background. Visible here is the aboral surface with the statocyst at its center, a number of dark surface papillas, and the tentacular edges upturned, a tentacle protruding. At night the animal humps up, creeps about and extends the long sticky tentacles, catching small crustaceans, etc. Length, 3 mm. Bimini. (R.B.)

Large platyctenid, *Lyrocteis flavopallidus,* has a body height of 11 cm. The tentacles, up to 70 cm long, arise from 2 chimneylike fused folds of the body. *Lyrocteis* is the first platyctenid reported from Antarctic waters (McMurdo Sound). Sedentary and usually found atop sponges, it can move 1 to 2 cm a day to reach advantageous feeding sites. The genus name refers to the lyre-shaped body and the species name to the straw-yellow color. The name is tentative, as it is based on a superficial similarity to *Lyrocteis imperatoris* of Sagami Bay, Japan; future examination may require it to be assigned to a different genus, or even a different family. (P.K. Dayton)

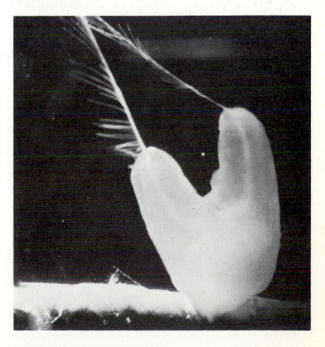

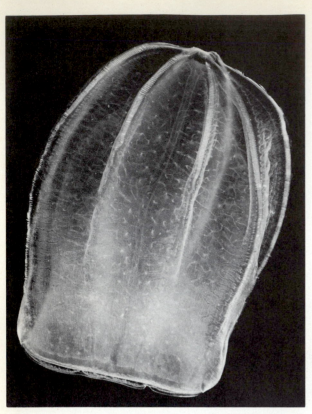

The beroids (be-ro′ids), named for the commonest and most widely distributed genus, are flattened and usually described as cone-shaped or thimble-shaped. Lacking tentacles in both larva and adult, they swallow prey directly, mostly other ctenophores, especially *Pleurobrachia.* The wide mouth and capacious pharynx stretch when swallowing larger prey, sometimes fishes larger than the normal outline of the ctenophore.

Beroe is pinkish or even reddish in cold seas, and milky in temperate waters. This specimen, from the warm waters of northwestern Florida, was golden in color from its contained zooxanthellas. The wide mouth stretches across the whole of the oral end (lower edge here). The branching pattern of the gastrovascular canals is visible, as are the rows of beating combs. Actual size. (R.B.)

THE GENERAL BODY PLAN of ctenophores must be said to be still at the **tissue level of organization,** for there is only minimal evidence of the combination of tissues into organs of the kind we see in more complex groups. Ctenophoran structure reminds us somewhat of that of medusas, and in many respects ctenophores are much like cnidarians; indeed, for a long time both groups were joined in the same phylum. But there are many differences to support their separation. The **biradial symmetry** of ctenophores goes beyond the beginnings of bilaterality seen in the cnidarians. The pattern of development is markedly different. The **ciliary combs** that propel ctenophores are unique in the animal kingdom, and so are the adhesive cells, the **colloblasts.**

If we must look for living animals that vaguely resemble what could have been an ancestral ctenophore, the most likely are certain trachyline medusas, the same group thought to be closest to the ancestral cnidarians. Searching in the opposite direction for links between the ctenophores and the flattened creeping flatworms of the next most complex phylum to be described, it is tempting to look at ctenophores like *Coeloplana,* which at a casual glance could be mistaken for delicate marine flatworms. This intriguing superficial similarity does not hold up under careful examination. Perhaps both ctenophores and flatworms were offshoots from some primitive planula-like stock that also gave rise to modern cnidarians.

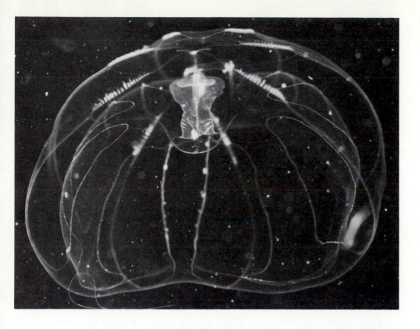

Medusa-shaped ctenophore. *Thalassocalyce inconstans* is too fragile to be caught intact in a plankton net and was only recently observed and collected by SCUBA divers, in warm shallow waters of the western N. Atlantic. Intermediate in structure between cydippid and lobate ctenophores, it may constitute a separate new order, Thalassocalycida. Up to 15 cm diam. (L.P. Madin, Woods Hole Oceanographic Institute)

CLASSIFICATION: Phylum CTENOPHORA.

Some prefer the name "Ctenaria," as consistent with Cnidaria. Others designate the Ctenophora and the Cnidaria as subphyla of the phylum Coelenterata. Two classes may be distinguished. The first, the class TENTACULATA, includes those with tentacles, long or short, and comprises the first 4 orders listed. In this class 2 additional, minor orders are sometimes recognized, intermediate in structure between cydippid and lobate ctenophores; each has only a single genus—*Ganesha* in the order Ganeshida and *Thalassocalyce* in the order Thalassocalycida (see photo on this page).

The second class, the class NUDA, includes only the beroids, which have no tentacles at any stage. However, some authorities on the ctenophores think that the loss of tentacles is not important enough to merit setting up a separate class for the beroids.

Order Cydippida (sy-di'-pi-da). Slightly flattened globular or pear-shaped forms. Two long branched tentacles retractable into tentacle sheaths. Gastrovascular canals end blindly. *Pleurobrachia, Mertensia, Hormiphora.*

Order Lobata. Oval in outline but flattened. Comb rows longer on flattened surfaces, which are expanded at mouth end into 2 large lobes. Mouth surrounded by 4 ciliated flaps, 2 main tentacles much reduced in size, and 2 grooves lined with tiny tentacles. *Mnemiopsis, Bolinopsis.*

Order Cestida. Ribbonlike and highly transparent, with comb rows that run along aboral edge of elongated body, and tiny tentacles along the oral edge. Swims by undulations of body as well as by combs. *Cestum, Velamen.*

Order Platyctenida (platy-ten'-i-da). Flattened creeping forms compressed in the oral-aboral axis. Two long fringed tentacles. Some have lost combs. *Ctenoplana, Coeloplana, Vallicula, Lyrocteis, Gastrodes, Tjalfiella.*

Order Beroida (be-ro'-i-da). Lack tentacles in both adult and cydippid larva. Cone-shaped in outline and flattened. Meridional digestive canals with many side branchings, forming a visible and characteristic pattern. Wide mouth swallows prey directly. Comb rows, according to species, reach halfway down body or almost to mouth. *Beroe.*

Chapter 8

Flatworm Body Plan

Bilateral symmetry,
Three layers of cells,
Organ-system level
of construction.
Regeneration.

IT TAKES MORE than a lively imagination to identify with an ameba, which has no fixed front or rear, or with a medusa, which has multiple eyes encircling an umbrella-shaped body. Even a ctenophore, with sense organs concentrated at one end of the body, creates confusion when it moves off with the sensory end trailing. Now we come at last to some invertebrates in which it is easy to tell head from tail, back from belly, and right side from left. These are the little freshwater planarians, worms named from the flatness of their elongated bodies. They have a definite head, with 2 large unblinking eyes and a pair of sensory lobes. We recognize at once the same head-dominated and two-sided symmetry we see in ourselves.

Freshwater planarians, along with marine and terrestrial ones, belong to the phylum **PLATYHELMINTHES** and to the class Turbellaria, which includes many other kinds of free-living flatworms and a few commensal or parasitic ones. The phylum also embraces 3 exclusively parasitic classes.

The body of planarians is clearly differentiated into front, or **anterior,** and rear, or **posterior,** ends. When the worm moves, one surface of the body always remains upward, while the other is kept against the substrate. The upper surface is termed **dorsal,** and differs from the lower, or **ventral** surface, which bears the mouth and most of the cilia.

A body form, usually elongate, in which there is a difference between anterior and posterior ends and between dorsal and ventral surfaces is said to have **bilateral symmetry.** The term *bilateral* refers not to these ends or surfaces but to the fact that in these animals the body structures are arranged symmetrically on the two sides with reference to a central plane, the *sagittal plane,* which runs from the middle of the head end to the middle of the tail end. The paired eyes and other sensory structures occur at equal distances to either side of the sagittal plane. Single structures are generally located in the mid-line and are bisected by the plane. Bilateral animals have

planar = "lying in one plane"

platy = "flat"
helminthes = "worms"

Turbellarians were named for the turbulence created in the water by the beating of their cilia.

dorsal = "back"
ventral = "belly"

Almost all details of the planarians described in this chapter are based on the genus *Dugesia.*

Members of this genus, being carnivorous like most turbellarians, can be lured to pieces of beef or beef liver placed in many springs or streams. Easily collected by the hundreds, these small worms (1 to 2 cm long) thrive in great numbers in large laboratory pans or individually in small glass dishes of standing water. And they have become widely used for classroom study and in research. Some of the other kinds of free-living flatworms are briefly treated in the next chapter.

right and *left* sides, while hydras, for instance, have no defined sides. A bilateral animal can be cut into two similar pieces only by one particular cut—along the sagittal plane. The two resulting pieces are not identical but are mirror-images of each other.

Bilateral symmetry is often imperfect in some degree, owing to the specialization of one side or part over another. Thus, in humans the right arm is usually larger and stronger than the left, and in the brain there is a speech center on the left side but none on the right. Much more marked asymmetries occur in other bilateral animals, such as the coiling of snails into a spiral (because of unequal growth of the two sides).

When bilateral animals become sedentary, they tend to evolve a modified symmetry which appears superficially like the radial symmetry of cnidarians. Examples are the sea stars, which are said to be secondarily radial.

The bilateral body form lends itself to "streamlining" and, with a head to direct movements, gives rise to many animals whose success depends on being able to move fast. Beginning with the flatworms, all more complex animals (unless secondarily modified) are bilaterally symmetrical.

For efficient **locomotion,** it is essential that one end should go first. In the planarians it is the head end, which bears the sense organs, that always ventures first into a new environment, while the rear end merely follows along. Such an animal would seem to be open to attack from the rear and from the sides, as compared with a cnidarian medusa or polyp, which can detect enemies and ward off their attacks on all sides. However, the concentration of sense organs at the front end enables the animal to detect danger ahead and so better avoid it. Also, specialization of anterior and posterior ends usually accompanies fast locomotion, and such animals can better pursue prey and escape from enemies than can radial animals.

Planarians do not swim freely through the water, but move in contact with a solid object or on the underside of the surface film. When a worm leaves the surface, it glides down attached to a thread of mucus. Planarians move about on a substrate in a characteristic slow, gliding fashion with the head bending from side to side as though it were testing the environment. The worms are propelled by muscular waves, not apparent to the naked eye, that pass from the head backward. If we prod the animals, they hurry away by conspicuous muscular waves. Their path is smoothed by secretion of mucus from a band of cells along each side of the ventral surface. All these movements result from two mechanisms. The gliding is muscular and probably also ciliary; the rapid movements are wholly muscular.

A **bilateral animal** is divisible into two equivalent pieces only along a single plane, the sagittal plane.

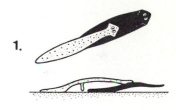

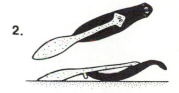

Predation of one planarian on another, as observed in the laboratory. (Stages 1 and 2 are shown in both top and side views.)
1. A spotted planarian, *Dugesia tigrina,* lunges and attaches its strongly adhesive head to a mid-dorsal region of a blackish planarian, *Cura foremanii.*
2. The tail of the spotted predator adheres tenaciously to the substrate, so that as the stretched worm shortens, it hauls in its prey.
3. The predator loops its body around the prey and begins to tear out and ingest pieces of tissue by means of its extended pharynx. (Modified after J.B. Best)

The relative importance of cilia and muscles in planarian locomotion can be tested. If the cilia are paralyzed, by treating the worm with lithium chloride, locomotion continues. If the muscles are paralyzed, by treating with magnesium chloride, locomotion stops.

The surface of a planarian consists of an epithelium, as in a hydra; but this epithelium is ciliated, particularly on the ventral side. Just beneath the epithelium are layers of muscle cells. The outer layer runs in a circular direction, and the inner layer in a longitudinal direction. Muscles also run, both vertically and obliquely, between dorsal and ventral surfaces and help to make possible all sorts of agile bending and twisting movements. The muscles are not part of the epithelial cells, as in hydras, but are independent muscle cells specialized for contraction. Also, they are not developed from the ectoderm or endoderm but arise in a different way.

The dorsoventral muscles are essential for maintaining the flatness of flatworms, an important consideration in animals that depend on diffusion from the surface to supply oxygen to the tissues of the whole body at an adequate partial pressure.

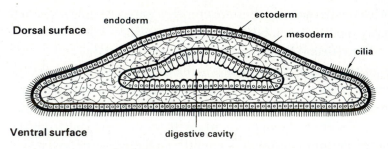

Cross-section through a planarian.

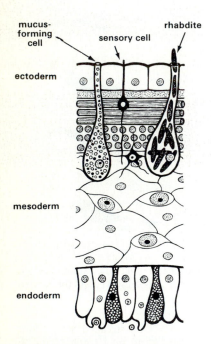

Cellular layers of the dorsal portion of a planarian as seen in cross-section. Many unicellular gland cells occur in the epidermis, but mostly they are sunk into the mesenchyme and have long necks, as in the mucus-forming and rhabdite-forming gland cells shown here. **Rhabdites** are packets of mucoid material and are thought to swell on contact with water, producing protective mucus.

Beginning with flatworms, all more complex animals have a mass of cells between the ectoderm and the endoderm, appropriately called the **mesoderm** ("middle skin"). This layer gives rise to muscles and to other structures which make possible an increasing complexity in animal activities. Like almost all characters of animals, the mesoderm does not appear suddenly in fully developed form. Its early beginnings are perhaps comparable with the ameboid mesenchyme cells in the mesohyl of sponges and the mesoglea of many cnidarians. We recognize it as mesoderm when, as in the flatworms and in all more complex animals, it is more massive than either ectoderm or endoderm and gives rise to definite structures, such as muscles and reproductive organs.

The largest two-layered animals are certain medusas; these have attained great size and some degree of body firmness by the secretion of great quantities of mesoglea in which may be found a sparse population of mesenchyme cells. In three-layered animals the mesenchyme has been increased from a scattered group of wandering ameboid cells to a cohesive tissue that gives firmness and bulk to the body.

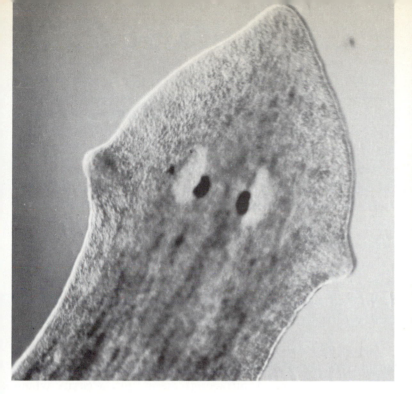

Head of planarian, *Dugesia tigrina,* shows bilateral symmetry. A pair of pigment-cup eyes and a pair of sensory lobes are arranged one on each side of the midline and are mirror images of each other. It is easy to imagine that the flatworm is returning one's gaze, but the eyes are not image-forming. (P.S. Tice)

The region between the outer protective epithelium, or epidermis, and the inner digestive epithelium (sometimes called the gastrodermis) is filled with various organs surrounded by mesenchyme in the form of ameboid cells, most of which are united into a network, although some are free and move about. Muscles run through the mesenchyme, and it also contains many gland cells which open to the surface and secrete mucus or sticky substances. The gland cells are largely derived from the ectoderm; but the sex organs, muscles, and mesenchyme are mesodermal.

The cnidarians were described as animals organized on the tissue level. From flatworms to humans, animals are constructed on a still more complex level of organization. Not only do cells work together to form tissues, but tissues of various kinds are closely associated to form single structures, called **organs,** each adapted for the efficient performance of some one function. For example, the planarian feeding organ, the pharynx, is an elongated tube covered by epithelium. Under the surface epithelium are several layers of muscle that run in different directions. The hollow of the pharyngeal tube is lined by a digestive epithelium that contains gland cells. The several kinds of tissues work together, coordinated by nervous tissue composed of both sensory cells and a nervous network.

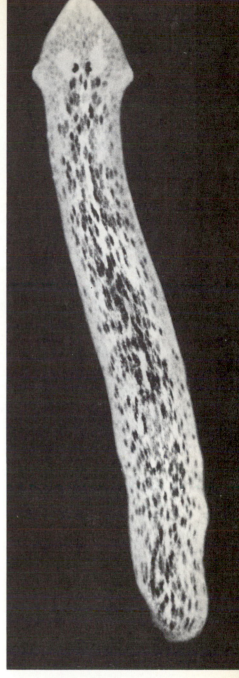

Spotted planarian, *Dugesia tigrina,* is spotted or streaked with brown pigments and may reach 18 mm in length. It is common across the U.S.A. on submerged vegetation or under stones in ponds, lakes, rivers, and streams. *D. dorotocephala,* shown in the figure that heads this chapter, is found in springs and spring-fed waters, is uniformly blackish, may reach 30 mm in length, and has a more pointed head with more pointed sensory lobes. (R. Buchsbaum)

Digestive system of a planarian.

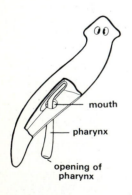

The **pharynx extends through the mouth** during feeding, as shown in this diagram with part of the body cut away.

Connective tissue binds the various layers together. An organ usually cooperates with other organs or parts in the performance of some life-activity, and such a group of structures devoted to one activity is termed an **organ system.** Thus, in vertebrates the stomach is part of the digestive system, and all the other organs of this system, such as the esophagus, liver, and intestine, are necessary for the proper performance of digestion. Complex animals are made up of a number of such organ-systems, as the digestive system, excretory system, circulatory system, nervous system, and so on. Flatworms do not have all of these, and the ones they do possess are not all very well developed; but they are the simplest phylum of animals built on the *organ-system level of construction.*

In the **digestive system** of the planarians the **mouth,** curiously enough, is not on the head but usually near the middle of the ventral surface. It opens into a cavity which contains the tubular, muscular **pharynx,** attached only at its anterior end. The pharynx can be greatly lengthened and is protruded from the mouth for some distance during feeding. Planarians feed on small live animals or on the dead bodies of larger animals. They can detect the presence of food from a considerable distance by means of sensory cells on the head. They move toward their food, mount upon it, and press it against the substrate by means of their muscular bodies. Struggling prey can be successfully held in this way, especially after they have become entangled in the slimy secretion from the worm. The pharynx is protruded posteriorly through the mouth and inserted into the prey. Enzymes secreted by the pharynx soften the prey tissue while sucking movements of the muscles of the pharynx tear the tissue into microscopic bits, which are then swallowed, along with the juices of the prey, by successive muscular waves of contraction that pass along the pharynx.

From the anterior attached end of the pharynx the rest of the digestive system extends as a branching **digestive cavity** throughout the interior of the animal. It consists of one anterior branch which runs forward and two posterior branches which pass backward, one on either side of the pharynx, to the posterior end. All three branches have numerous and fairly regularly spaced side branches, thus providing for the distribution of the food to all parts of the body.

Digestion of food in planarians takes place both extracellularly and intracellularly, as in cnidarians. The larger particles of food are broken down in the digestive cavity by enzymes released from secretory cells of the digestive epithelium. However, most of the food is already broken into small particles, by the suctorial action of the pharynx, before it enters the cavity, and is thus ready to be taken up by the digestive epithelial cells in ameboid fashion and incorporated into food vacuoles. The digested food is absorbed and presumably passes by diffusion throughout the tissues of the body. Some digestion may be completed within the fixed cells of the mesenchyme. There is only one opening to the digestive cavity; indigestible particles are eliminated through the mouth.

Experiments on a common species of planarian, *Dugesia dorotocephala,* which was fed on liver showed that, after a meal, all the ingested liver was taken into the digestive epithelial cells in about 8 hours, and that 3 to 5 days were required for the complete digestion of the food vacuoles so formed. Much of the food was found to be converted into fat, which was stored in the digestive epithelium.

Practically all animals can store food reserves upon which they draw in time of need. A small animal like an ameba stores very little and, unless it goes into the inactive encysted state, will die after about 2 weeks without food. Starved hydras survive much longer periods. But planarians are peculiarly adapted to go for many months unfed while remaining active. During this starvation period they use the food stored in the digestive epithelium, whole cells breaking down. Later they begin to digest other tissues, the reproductive organs usually going first. Externally one can observe only that the worms grow steadily smaller though retaining the same general appearance. A worm starved for 6 months may shrink from a length of 20 mm to one of 3 mm. Because of their ability to go for months without food, planarians make ideal household pets for busy or absent-minded people.

A new system composed of structures not found in any of the forms already studied is the **excretory system,** which lies in the mesenchyme. A network of fine tubules runs the length of the animal on each side and opens to the surface by minute pores. Numerous fine side branches from the tubules originate in the mesenchyme in tiny enlargements known as **flame bulbs.** Each flame bulb has a hollow center in which beats a tuft of flagella that suggests a flickering flame. The hollow center is continuous with the cavity of the tubules of the system, and a current of fluid moves along the tubules to the pores, aided by the beating of other tufts of flagella on the tubule walls. As with the contractile vacuoles of freshwater protozoans, the primary function of the flame bulb system in these freshwater flatworms is apparently to remove excess water from the tissues. The flow of water that constantly enters the tissues of the worm and is processed through the excretory system probably also flushes out some metabolic wastes.

Excretory system of a planarian consists of **protonephridia**, tubules with beating flagella at their closed inner ends.

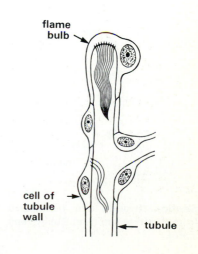

Diagram of **flame bulb** and flagellated tubule of *Dugesia tigrina.* The 30 to 90 flagella of the "flame" are closely apposed and coordinated in their beating. Water enters through the tubule wall surrounding the flame. (Adapted from J.A. McKanna)

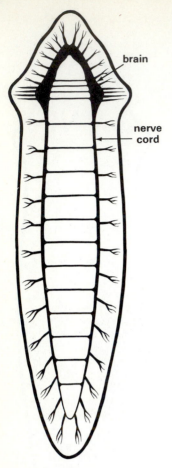

Nervous system. The brain and nerve cords are organized with the nerve cell bodies outermost, surrounding an inner core of nerve fibers. This arrangement is typical of the central nervous systems of all the more complex invertebrates.

Flatworms have a **central nervous system,** the kind of nervous system possessed and further centralized by most of the more complex animals. In planarians there is in the head a concentration of nervous tissue into a bilobed mass called the **brain.** From the brain two strandlike concentrations of nerve cells, the **nerve cords,** run backward through the mesenchyme near the ventral surface, giving off numerous side branches to the body margins. The two cords are connected with each other by many cross-strands like the rungs in a ladder. This type of system has been called the "ladder type" of nervous system. The brain and the two cords constitute the central nervous system, a kind of "main highway" for nerve impulses going from one end of the body to the other. If the two nerve cords are cut, muscular waves traveling from head to tail stop at the level of the cut, and the worm can no longer move as a coordinated whole. Thus the nerve cords appear to mediate the coordinated muscular movements of the body. The brain, on the other hand, is not necessary for the muscular coordination involved in locomotion, for a planarian deprived of its brain will still move along in coordinated fashion. But it moves more slowly, and responds only in a negative way to contact with objects, avoiding but never positively approaching any object. Thus the brain probably serves chiefly to maintain a certain level of excitability and spontaneity and also to receive sensory information from the sense organs and to process and transmit it to the rest of the body. The result is a much more closely knit behavior than is possible with the diffuse, noncentralized nerve net of the hydras, which lacks definite pathways and a coordinating center. Nerve nets do occur locally in planarians, for example, in the pharynx.

Another function of the ventral nerve cords is to inhibit spontaneous activity in the pharyngeal nerve net. If the cords are cut just in front of the pharynx so that this constant inhibition stops, the pharynx becomes so lively that it may twist loose and escape through the mouth, wriggling and swimming off on its own. Alone, it is fully capable of ingesting food, but cannot discriminate edible from inedible objects.

Nerve nets occur locally in the tissues of almost all the more complex animals. In humans, for example, a well-developed nerve net, connected with the central nervous system, serves the wall of the intestine.

Conditions in the external world are conveyed to the nervous system by **sensory cells,** slender elongated cells that lie, with their pointed ends projecting from the body surface, between the epithelial cells. Different ones are specialized to receive the stimuli of touch, water currents, and chemicals. Sensory cells are distributed all over the body surface, but in addition are concentrated in the head, especially in two

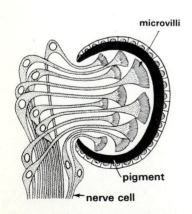

Section through eye. (Combined from R. Hesse and from Röhlich and Török)

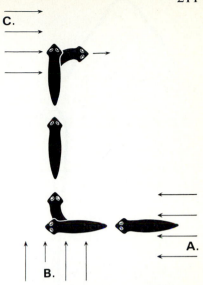

sensory lobes, or auricles, that project at the sides. If the auricles are cut off, the worm has difficulty finding food. The two **eyes** consist each of a bowl of black pigment filled with special sensory cells whose ends continue as nerves that enter the brain. The pigment shades the sensory cells from light in all directions but one, and so enables the animal to respond to the direction of the light. Unlike other regions of the ectoderm, that which is immediately above the eyes is unpigmented, and thus allows light to pass through to the sensory cells. Planarians whose eyes have been removed still react to light, but more slowly and less exactly than normal worms. This indicates that there must be some light-sensitive cells over the general body surface.

Sensitivity of the general body surface to light has been demonstrated in some other invertebrates, such as earthworms and centipedes, and even in some vertebrates, for example, tadpoles and newly-hatched pigeons.

By virtue of abundant sensory cells, specialized sense organs and a centralized, cephalized nervous system, planarians show more varied **behavior** and much more rapid and precise responses than do most cnidarians. Planarians avoid light and are generally found in dark places, under stones or leaves of water plants. If placed in a dish exposed to light, they immediately turn and move toward the darkest part of the dish. They are highly positive to contact and tend to keep the under surface of the body in contact with other objects. They respond to chemical substances in the water and quickly react to the presence of food by turning and moving directly toward it. That is why a piece of raw meat placed in a spring inhabited by plan-

Orientation to light. A light was turned on successively at **A, B,** and **C,** and the worm turned as shown in successive positions. (After W.H. Taliaferro)

Light from different directions (shown by arrows) illuminates different portions of each of the pigment-cup eyes of a planarian. This gives the animal information about the direction of light. (Modified after R. Hesse)

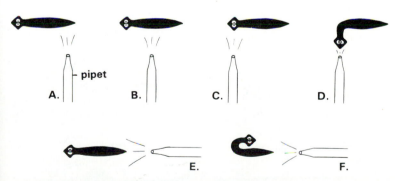

Reactions to water currents produced by a pipet. **A, B,** the current strikes the middle or rear of the body and there is no response. **C, D,** the current strikes the sensory lobe on the side of the head and the worm turns toward the current. **E, F,** the current from the rear passes along the sides of the body to the sensory lobes and the worm turns around toward the current. In nature these reactions orient the worm upstream. (Modified after I. Doflein)

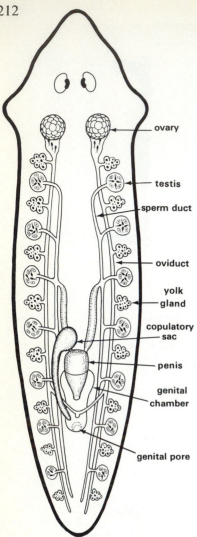

ovary

testis

sperm duct

oviduct

yolk gland

copulatory sac

penis

genital chamber

genital pore

The **reproductive system** of a planarian includes both male and female organs. The sperms are unusual in having 2 flagella. Moreover, these do not have the usual "9 + 2" pattern of microtubules that characterizes most cilia or flagella in planarians and other animals.

arians attracts hordes of worms, which glide upstream toward the food, guided by the meat juices in the current of water. Planarians react to water currents, and some species regularly move upstream against a current. They also respond to the agitation of the water produced by the animals upon which they prey.

Planarians have a highly complicated **reproductive system** for sexual reproduction. Ovaries and testes arise in the mesenchyme, and there is a system of tubules and chambers in which fertilization occurs, as well as complicated sex organs for the transfer of sperms. The animals are hermaphrodites, every individual having both male and female organs. After the breeding season, the reproductive system degenerates and is later regenerated at the beginning of the next sexual period.

When sexually mature, each worm has a pair of **ovaries** close behind the eyes. From each ovary a tube, the **oviduct,** runs backward near the ventral surface. Multiple **yolk glands,** consisting of clusters of yolk cells, lie along the oviduct, into which they open. There are numerous **testes** along the sides of the body. From each testis leads a delicate tube, and all these tubes unite on each side to form a prominent **sperm duct,** which runs backward alongside the oviduct. The sperm ducts, packed with sperms during the time of sexual maturity, connect with a muscular, protrusible organ called the **penis,** which is used for the transfer of sperms to another planarian. The penis projects into a chamber, the **genital chamber,** into which there also open the oviducts and a long-stalked sac called the **copulatory sac.** The genital chamber opens to the exterior by the **genital pore** on the ventral surface behind the mouth.

Cross-section through a sexually mature planarian. Note the gland cells opening to the ventral surface; they occur in two bands running the length of the body and secrete the mucus on which the worm glides.

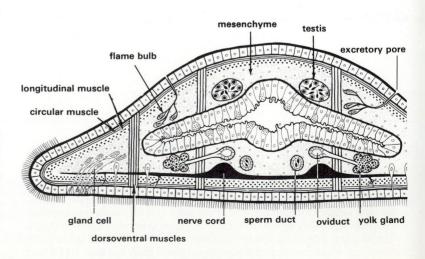

mesenchyme

testis

excretory pore

flame bulb

longitudinal muscle

circular muscle

gland cell

nerve cord

sperm duct

oviduct

yolk gland

dorsoventral muscles

Although each planarian contains a complete male and female sexual apparatus, self-fertilization does not occur; instead, two worms come together and oppose their ventral surfaces. The penis of each is protruded through the genital pore and deposits sperms in the copulatory sac of the partner. After copulation, the worms separate. The sperms soon leave the copulatory sac and travel up the oviducts until they reach the ovaries, where they fertilize the eggs as they are discharged. The fertilized eggs pass down the oviducts, and at the same time yolk cells are discharged from the yolk glands into the oviducts. When the eggs and yolk cells reach the genital chamber, they become surrounded by a shell to form an egg capsule. The eggs of most flatworms are peculiar in that food reserves do not occur in the eggs themselves but are kept in the yolk cells, which accompany the eggs. The capsules (each containing less than 10 eggs and thousands of yolk cells) are passed out through the genital pore and are fastened to objects in the water, often to the underside of stones, as shown in the drawing that heads this chapter. The eggs hatch in 2 or 3 weeks releasing minute juvenile worms, like their parents except that they lack a reproductive system.

Many planarians reproduce only sexually, but some multiply by **asexual reproduction**. In this process the worm, without any evident preliminary change, constricts at a region behind the pharynx, and the posterior zooid begins to behave as though it were rebelling against the domination of the anterior one. When the worm is gliding quietly along, the posterior zooid may suddenly grip the substrate and hold on, while the anterior one struggles to move forward. After several hours of this "tug-of-war," the anterior zooid finally breaks loose and moves off by itself. Both zooids regenerate the missing parts and become complete worms. Species which have this habit often go for long periods without sexual reproduction, and, in fact, some of them rarely develop sex organs.

Dugesia tigrina, the most common planarian in the U.S.A., breeds in the spring and summer. Sexuality can be induced in the laboratory by a rise in temperature after the worms have been at a low temperature for a period. It is often also necessary to coat with vaseline the surfaces of the pan in which the worms are cultured. This prevents them from getting a good hold on the substrate, so that asexual fission cannot occur. As the worms grow to a maximum size, the sex organs mature and produce eggs and sperms.

In some places the local strain remains permanently asexual; in others reproduction is exclusively sexual; and in still others the worms alternate between these two modes seasonally. The characteristic reproductive pattern of each strain is not simply limited by the local environment, but is apparently under genetic control, since it is maintained indefinitely in the laboratory. However, an investigator found that testes and copulatory structures did develop in the posterior two-thirds of an asexual individual which was grafted onto the anterior region of a sexual worm.

Planarian beginning asexual replication. It will divide, at the constriction, into 2 unequal pieces, and each will become a whole worm. *Dugesia tigrina.* (R.B.)

Asexual division. *Left,* just before division. *Right,* just after. The rear zooid will soon develop a head, pharynx, and other structures.

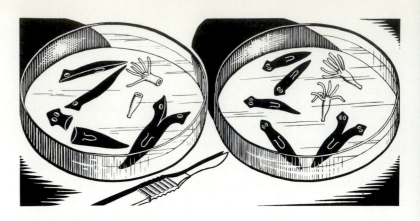

REGENERATION: NEW PARTS FROM OLD

REGENERATION, or the ability to renew tissues and repair damage, at least to some degree, is universal among animals. In humans, the epidermal cells of the whole body surface are continually sloughed and replaced; wounds of considerable size heal and broken bones grow together again; but a lost finger or toe cannot be regenerated. Lizards, on the other hand, can pinch off and replace the tail (although the new one lacks vertebras) and salamanders cans regrow an entire limb (but frogs cannot). An earthworm can replace its head, a sea star its arms, and a lobster a leg or an antenna, but in general, invertebrates with relatively complex organ systems show a correspondingly low capacity for regeneration. Protozoans, sponges, cnidarians, ctenophores, and planarians have greater powers of repair and reorganization, and some regularly reproduce asexually by fission or by pinching off small pieces. Within these groups, the members of some species regenerate readily after severe injury, while those of other species do not. Some kinds of freshwater planarians that regenerate complete worms from almost any piece are used extensively in researches on regeneration.

Regeneration depends upon the ability of uninjured cells to produce the kinds of cells destroyed. The more specialized the cells have become, the less able are they to produce cells different from themselves and so replace the missing parts. Planarians sequester, among the large, irregular cells of the mesenchyme, a supply of unspecialized cells. These smaller cells with large nuclei are called **regeneration cells,** or **neoblasts.** They can differentiate into all of the many different types of specialized planarian cells, and are distributed throughout the mesenchyme of the worm, especially adjacent to the brain and nerve cords. In a normal worm, they appear to be inactive, though they probably contribute to the regular replacement of cells in all body tissues.

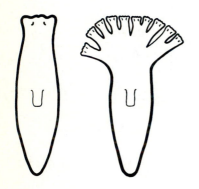

Ten-headed monster was the result of making many cuts in the head of a planarian *(Dendrocoelum)* and then repeatedly renewing the cuts so that the edges could not grow together again. (After J. Lus)

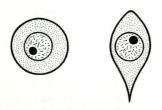

Regeneration cells have large nuclei and little cytoplasm. (Adapted from F. Stéphan-Dubois)

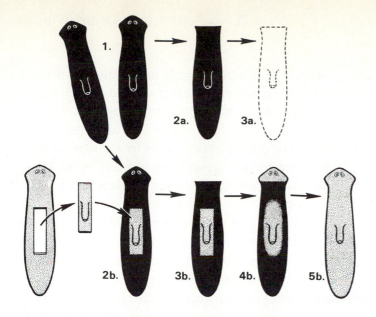

1. 2a. 3a. 2b. 3b. 4b. 5b.

Migration of neoblasts and their essential role in regeneration have been studied experimentally using X-ray and grafting techniques. Specimens of *Dugesia lugubris*, **1**, which normally regenerate readily, are exposed to X-rays. The irradiated worms, in which all the regeneration cells have selectively been killed by the X-rays, are divided into 2 groups. In one group, **2a**, the heads are cut off at once. These worms, **3a**, fail to regenerate and they die. In the other group, **2b**, each irradiated worm receives a graft from a healthy, nonirradiated worm (an individual of a different color, so that the graft tissue can be recognized). The heads of the irradiated worms with implanted grafts are then cut off. These worms, **3b**, regenerate heads, **4b**, colored like the graft. Microscopic examination shows that healthy regeneration cells from the graft have migrated into the wound and differentiated to form a head. After several months, the entire worm, **5b**, is colored like the graft, as its cells are gradually replaced by cells derived from the neoblasts of the graft. (Based on F. Stéphan-Dubois)

In a wounded worm, the activities of regeneration cells suddenly become conspicuous. In the tissues closest to the wound, these cells begin to migrate to the site of the damage within hours. They find the wound already closed; contraction of the surrounding muscles has pulled its edges close together and the remaining area of exposed interior tissue has been rapidly sealed over by a membrane. The arriving regeneration cells gather just under this membrane. Meanwhile, other regeneration cells farther away in the body have begun first to divide and then to migrate toward the wound. After several days, the accumulated cells form a small bulge, the **regeneration blastema,** at the site of the wound. The light-colored blastema contrasts sharply with the dark color of the old pigmented tissues at the edge of the wound; the regenerated portion will become pigmented later. Even if, for example, the wound has resulted in the loss of the entire front half of the worm, including the pharynx, the missing parts are restored. The cells of the blastema first differentiate to form the brain, then the eyes, then the pharynx, and other missing organs. The regenerated worm, at first strangely proportioned, eventually comes to look exactly like the original worm, though smaller.

How wounding stimulates the regeneration cells to divide and directs their migration is not known. Regeneration cells in glass culture dishes are attracted to bits of planarian tissue, suggesting that chemicals released from damaged cells may diffuse through the body and initiate migration and division. Another idea is that the nervous system transmits the stimulus. Special secretory cells in the brain and nerve cords have been found to multiply during regeneration, and it seems likely that these neurosecretory cells may initiate regeneration or control some other stage in the process.

In asexual reproduction, neurosecretory cells increase in number just before fission occurs. Apparently stimulated by the secretion of these cells, the regeneration cells in the fission zone at once increase their production of ribonucleic acid (RNA), needed for protein synthesis, and are thus prepared for their role in the regeneration that will follow fission.

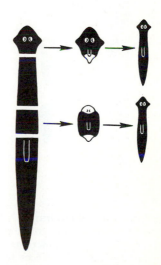

Remodeling to normal proportions is a complex process which continues even after regeneration of all the parts. It is one example of the wider category of normal developmental processes called **regulation.** (Modified after C.M. Child)

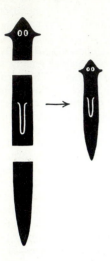

A regenerating piece retains its **polarity**—a head grows from the anterior end and a tail grows from the posterior end. (Based on C.M. Child)

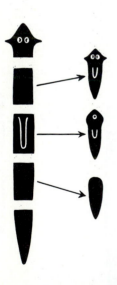

Capacity for regeneration decreases from the anterior to the posterior end in many planarians. In some species only the pieces from anterior regions are able to form a head, while those farther back effect repair but do not regenerate a head. In other species almost any small piece from any level can regenerate a complete worm; while in still others there is almost no capacity for regeneration at all. (Based on C.M. Child)

During regeneration the remaining undamaged structures of the body, and the new tissue mass produced by the migrating and dividing regeneration cells, must somehow interact to restore the original form of the worm in all details. Certain patterns observed during experiments on the regeneration of planarians characterize regeneration of many other kinds of simple animals also and seem to provide clues about how this interaction works.

In the first place, any piece of such animals usually retains the same **polarity** it had while in the whole animal, that is, the regenerated head grows out of the cut surface which faced the anterior end in the whole animal, and the regenerated tail grows out of the cut surface which faced the posterior end. This *anteroposterior differentiation* operates throughout the entire animal down to small portions.

Another generalization drawn from experiments is that the **capacity for regeneration** of the head is greatest near the anterior end and decreases toward the posterior end. Pieces from the anterior regions of planarians regenerate faster and form bigger and more normal heads than pieces from posterior regions, and there is a gradual change in these respects along the anteroposterior axis.

The head of a planarian is dominant over the rest of the body, and in general, any level controls the level posterior to it. One way in which this **dominance** can be demonstrated is by means of grafting. If a small bit of the head region of a planarian is grafted into a more posterior level of another individual, it will not only grow out into a head but will influence the adjacent tissues in some way so that a new pharynx, for example, may be formed in the body near this grafted head. If a head piece is grafted into a planarian and then the host's head is cut off, the grafted head may influence the anterior cut surface to regenerate a head without any brain or even to form a tail instead. In other words, grafts of head pieces reorganize the adjacent tissues into a whole worm in relation to themselves. Grafts from tail regions do not have these effects but are usually absorbed.

The dominance of the head over the rest of the body is limited by distance. If the animal grows to a sufficient length, its rear part may get beyond the range of dominance of the head. This happens in asexual reproduction when the rear part constricts off as a separate animal. That diminishing of control of the head over the body is the most important factor in asexual division is shown by the fact that separation of the rear part can be induced by cutting off the head.

Until the moment when a planarian starts to constrict, there is no external evidence of the physiological isolation of a posterior zooid. Yet its presence can be shown in experiments by appropriate cuttings. Pieces of worms taken from a region just behind the pharynx usually do not develop heads. But shortly behind this is a zone that almost always produces normal heads. This is the region where the worm constricts in asexual division. In a longer worm, it can be shown that there are 3 or even 4 developing zooids, one behind the other, each indicated by a region which shows an increased ability to produce worms with normal heads. In species of planarians that do not reproduce asexually, there is no evidence of such regions.

If the anterior end of a planarian is cut down the middle, and the two halves prevented from growing together again by renewing the wound several times, then each half will regenerate the missing parts and a two-headed planarian will result, with dominance divided equally between the two heads. If the cut goes back far enough, each head will influence the formation of its own pharynx.

All these facts indicate that there is some sort of gradation, in structure or processes or chemical substances related to regeneration, along the anteroposterior axis of a planarian, and we refer to this as the **anteroposterior gradient.** The basis of this gradient is not known. One hypothesis is that it may reflect an anteroposterior gradient in the *rate of general metabolic processes.* Such a gradient has been found in the rate of protein synthesis, for example. Another idea is that the *distribution of regeneration cells* may determine the gradient, since many species do have more regeneration cells in the head. However, the distributions of these cells do not always correlate well with the anteroposterior gradients. More widely accepted is the concept that the anteroposterior gradient may be based on the concentration gradients of *specific chemical inducers and inhibitors* of regeneration. There is much experimental evidence that such chemicals are produced in various organs and body regions, especially when they are actively regenerating. These

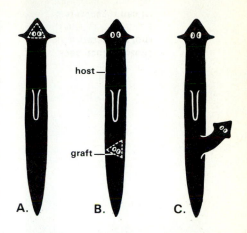

Grafting. A, a small piece, indicated by broken lines, is cut out of the head of the donor. **B,** the graft is placed in a wound made in the body of the host. **C,** the graft has grown into a small head. (Based on F.V. Santos)

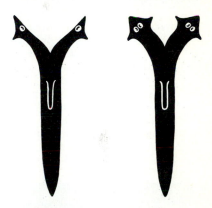

Two-headed planarian produced by a lengthwise cut through the middle of the head. Such monsters may split apart, and each then regenerates into a normal worm. (Based on C.M. Child)

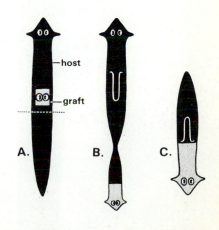

Reversal of anteroposterior axis by a graft. In this experiment host and donor were of two different species, and the tissues of each remained distinguishable as growth occurred. **A.** The host pharynx was removed and in its place was grafted a piece of the donor, including the eyes, part of the brain, and adjacent mesenchyme. One week later, the host tail was cut off at the level of the dotted line. **B.** A few weeks later, the graft has grown out as a small head and the host has developed a new pharynx. The 2 zooids are about to separate, as indicated by the constriction. **C.** The posterior zooid 74 days after grafting. A pharynx has formed in the original host tissue, but it is oriented in a direction opposite that of the old one. A tail has developed at the anterior end of the old host tissue where a head would be expected to develop if the graft were not present. The direction of beat of the host cilia has been reversed. (Based on J.A. Miller)

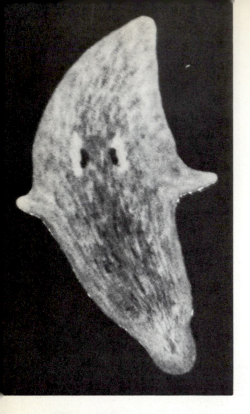

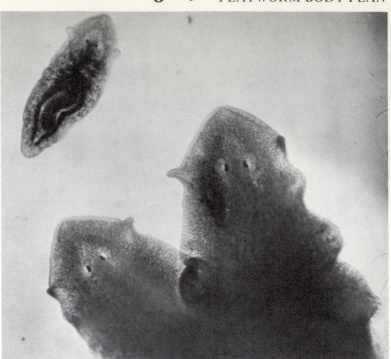

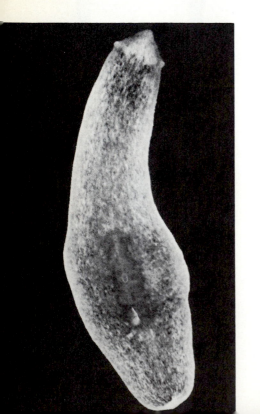

Regenerating pieces of *Dugesia tigrina* retain their original polarity and in time will regulate growth to produce worms of normal proportions. *Above,* a regenerating head piece. *Below,* a large transverse piece cut from the posterior half of the worm. Head and pharynx are partially regenerated. A tapering tail will develop later. (R.B.)

Tiny regenerated worm, only 4 mm long, is of the same species, *Dugesia dorotocephala,* as the two-headed monster. The small size and disproportionately wide body of the tiny planarian are a result of regeneration from a small transverse slice of a worm. Illinois, U.S.A. (R. Buchsbaum)

chemicals diffuse throughout the body, affecting regeneration according to their concentration, which is greatest closest to the source and decreases with distance.

For example, in the planarian *Polycelis nigra* it has been shown that a substance produced by the brain diffuses through the tissues and induces regeneration of the eyes, which in *Polycelis* occur in a row along the rim of the anterior portion of the body. In normal regeneration, the brain always regenerates before the eyes; and if the brain is repeatedly removed, or prevented from regenerating by local irradiation, the eyes never regenerate. However, if a suspension of ground-up planarian heads (head homogenate) is added to the water around an eyeless, brainless planarian, the eyes regenerate without regrowth of the brain. Presumably, the eye-inducing substance is present in the head homogenate, for tail homogenate is not effective. In another experiment, if the eye-bearing rim, with the eyes removed, is cut out and transplanted anywhere in the anterior part of the worm's body, the eyes regenerate. If it is transplanted close behind the brain, extra eyes regenerate. If it is transplanted into the tail end, no eyes regenerate, presumably because the concentration of active eye-inducer is too low at this distance from the head. If a brain is now grafted into the tail near the transplanted eye-bearing rim, the eye do regenerate. However, this brain-graft does not induce the nearby rim of the tail region to form eyes. This region apparently cannot form eyes even in the presence of the eye-inducing substance; only the anterior rim is competent to form eyes. Thus, in order for the formation of an organ to be induced, the appropriate *inducer* must encounter *competent tissue.*

Similar experiments also reveal the presence of *inhibitors* of regeneration. If a brain is grafted into the posterior part of a host worm, it will regenerate a complete head and an extra pharynx may even form near it, as described earlier. If a brain is grafted close to the head of the host worm, on the other hand, the graft usually fails to live and grow. However, if immediately after the grafting operation, the original head of the host is cut off, the graft may take the place of the original head. Sometimes the original head regenerates, with eyes, but with only a very small brain or none at all. Similarly a worm regenerating its head in the presence of head homogenate fails to develop a normal brain. Worms regenerating heads in the presence of homogenates of other body parts develop normally. It has been concluded that the brain of a planarian produces, in addition to a substance that induces eyes, a substance that inhibits the regeneration of other brains. Some experiments also suggest that the eyes may produce inhibitors that prevent the formation of additional eyes and in this way maintain their normal number and spacing. Both inducers and inhibitors of any planarian species are also active in other planarian species which regularly regenerate.

Further experiments and observations of the normal sequence of events in regeneration indicate a complex succession of such interactions. It is thought that, in the regeneration of an anterior portion, the brain must always differentiate first, from regeneration cells in contact with the nerve cords. The brain then induces the eyes, and other head structures, and the region in front of the pharynx. This region in turn induces the pharyngeal zone, which induces the pharynx and the region behind it. The brain also induces the gonads, which induce the copulatory apparatus. Production of specific inhibitory substances has been demonstrated only for the brain and pharyngeal zone, but it is hypothesized that each of the actively regenerating organs or regions produces a specific inhibitor which prevents its being regenerated again and again. Thus a sequence of balanced inductions and inhibitions would guide the orderly progression of the regeneration process, so that all the missing parts, and only the missing parts, are reconstructed, and a normal worm emerges.

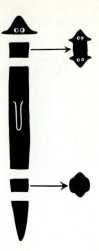

Short pieces may include such a small fraction of the anteroposterior gradient that there is no appreciable difference between the anterior and posterior cut surfaces. Short pieces cut from near the eyes in planarians regenerate a head at both ends. Short pieces from posterior regions regenerate a tail at both ends. (Based on C.M. Child)

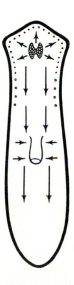

Representation of some of the **successive inductions and inhibitions** that are thought to control regeneration in planarians. *Polycelis nigra.* (After Wolff and Lender)

Degeneration in a poisonous solution begins at the head end. (Based on C.M. Child)

Degeneration after exposure to X-rays begins at head and tail ends. (Mod. after H.H. Strandskov)

Time-graded regeneration field of *Bdellocephala punctata* is shown by dots—the denser the dots, the faster the regeneration of a piece cut from that region. Regions behind the pharynx do not regenerate a head in this species. (After H.V. Brøndsted)

Comparable sequences of balanced inductions and inhibitions can be thought of as setting up gradients that direct development of animals in general. But one may inquire **how gradients get started**. It seems that they must arise early in development, under the direction of specific pattern-forming genes, in response to external factors. The position of the egg in the ovary is one such external condition. The egg is attached by one end to the ovary and is free at the other end. It is known for a good many eggs that this position determines, or at least is correlated with, the polarity of the egg, that is, which end of the egg is to become the anterior (or mouth end) of the future animal. The polarity is a property of the cytoplasmic structure of the egg and is based upon a gradient, a graduated difference of some kind. Once established, the polarity continues throughout embryonic development.

It has been found that many experiments on regenerating adult flatworms can be duplicated on developing eggs and embryos. By subjecting eggs and embryos to poisonous solutions of certain concentrations and at critical times, the development is greatly modified and all sorts of curious embryos are obtained. In general, because such solutions act most severely on the head, these embryos often have small heads, reduced eyes or eyes fused into one, and so on.

Such results suggest that some kind of gradient is an important factor in embryonic development, providing an underlying pattern which controls the orderly development of normal form and proportion. Thus we may think of the several kinds of symmetry as resulting from differences in the number of gradients which act in development. In the truly spherical animal all radii are alike and there is no one main axis of differentiation, though there is some difference between the interior of the cell and its exposed surface. Such an animal can show only a limited amount of differentiation. In an ameba all points on the surface are alike in that any point is capable of sending out a pseudopod, but the pseudopod proceeds in a definite direction. By means of chemical indicators it has been shown that in an actively moving ameba there is a definite, though constantly changing, physiological gradient—from the tip of the leading pseudopod to the opposite end of the animal. Radial and bilateral animals have more permanent axes of differentiation. In radial types the main axis (also called the "polar axis") is from mouth to base. In bilateral types there are, besides the main anteroposterior axis, two minor axes of differentiation, one extending from the mid-region toward both sides (the mediolateral gradient) and one

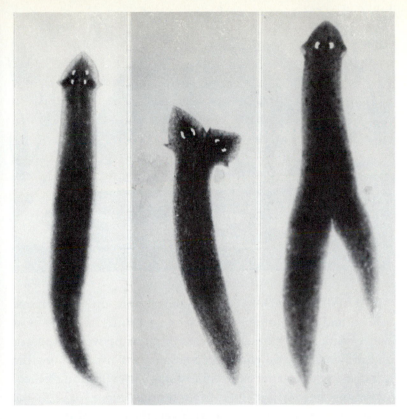

Grafting, within or between species, is a technique used in producing abnormal relationships that reveal much about normal growth. Host and donor are placed on trays of ice to slow their movements. Then a piece of tissue is cut out of the worm selected to be the host. In the resulting hole the experimenter places a piece of corresponding size and shape cut from a worm of the species selected to be the donor.

Left, a piece of head tissue containing eyes was cut from the donor worm, *D. tigrina,* and placed into the region just behind the eyes of the host worm, *D. dorotocephala.* The tissues of host and donor fused completely.

Center, the same operation produced a different result. The graft failed to unite with host tissue on all sides and induced the host to form a complete head with donor eyes but host characters (pointed head shape, pointed sensory lobes, and blackish pigmentation).

Right, a 2-tailed worm produced in 2 steps. A piece from the head of the donor worm (*D. tigrina*) was grafted into the tail region of the host worm (*D. dorotocephala*), inducing the host to grow a new pharynx oriented to the graft head. Then the host head and pharyngeal region were cut off. Instead of regenerating a head, the cut surface (now presumably dominated by the graft head) produced only a pharynx and tail. After this photo was made, the two tails fused along their inner margins. When this animal later divided asexually, the resulting worms had 4 eyes, 2 pharynxes and a double digestive cavity. (J.A. Miller)

extending from the ventral surface toward the dorsal surface (the ventrodorsal gradient). These two minor gradients are usually masked by the more prominent anteroposterior gradient. And in the more complex animals even the anteroposterior gradient is obscured by the complexity of the adult structure; it can be clearly shown only in embryos.

THE FLATWORMS, as illustrated by the planarians, differ from the cnidarians and ctenophores in the possession of many important characteristics that are shared by most of the more complex animals. Flatworms have *bilateral symmetry,* with differentiated anterior and posterior ends, and dorsal and ventral surfaces. They have a definite *head,* with a concentration of sense organs, and a central nervous system. And they have an extensively developed *third layer of cells,* the mesoderm. Either by itself, or in combination with ectoderm or endoderm, the mesoderm gives rise to *organs and organ systems.*

In addition to reproducing sexually, many planarians reproduce asexually, and this is often associated with a high capacity for *regeneration* of injured or lost body parts. Such animals lend themselves to experimental studies of regeneration, which have helped us to understand the general problems of animal form, growth, and development, and have been fruitful in suggesting new means of approach.

Chapter 9

Free-living and Parasitic Flatworms

Symbiosis: passengers, partners, and parasites

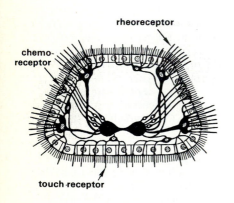

Section through the head of *Mesostoma* shows 3 of the types of sensory cells found in turbellarians. **Chemoreceptors** respond to CO_2 and various other dissolved substances, such as those diffusing from prey organisms; they are limited mostly to the head, where they often occur in pits or grooves. **Touch-receptors** respond to contact and are widely distributed over the whole surface, though concentrated on the head and the margins of the body. **Rheoreceptors** detect water currents and turbulence created by prey, and are distributed mostly along the margins of the head and body. (Modified after J. von Gelei)

THE MOST COMMON kind of relationship between one animal and another, or between an animal and a plant, is that of diner and dinner. The interactions between predator and prey or between herbivore and herb are often among the strongest selective pressures that shape the evolution of each. As rabbits become speedier, foxes become more foxy. As the crushing claws of crabs grow stronger, the shells of their snail prey grow thicker. Such changes are termed *co-evolution*.

When the association between members of two species becomes even more intimate, and lasts longer than a mealtime, co-evolution is intensified, and the form and habits of one or both associates may be entirely reshaped. Because almost every conceivable kind and degree of association is illustrated by flatworms, this chapter will focus on that aspect of their biology. Of the four classes in the phylum Platyhelminthes, three consist entirely of parasitic members—monogeneans, flukes, and tapeworms—and one includes members that span the full spectrum of relationships, the turbellarians.

TURBELLARIANS

LIKE the planarians described in the last chapter, most of the other 3,000 species in the class **Turbellaria** are free-living carnivores that prey on other small live animals or scavenge on larger dead ones. A few scrape algal films from rocks or eat other plant material. Although many turbellarians are sizeable, most are under 5 mm long and are best seen under the microscope. A majority of the species are marine.

Free-living habits require searching for food and exposure to the hazards of an external environment; and most turbellarians are liberally supplied with sensory cells and sense organs that guide their responses to light, gravity, and various mechanical and chemical stimuli. The sensory equipment of flatworms that live as parasites is generally less elaborate.

222

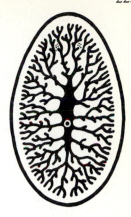

| **Acoel** (*Amphiscolops*) | **Rhabdocoel** (*Macrostomum*) | **Alloeocoel** (*Prorhynchus*) | **Triclad** (*Bdelloura*) | **Polyclad** (*Hoploplana*) |

Types of digestive cavities in turbellarians.

The orders of turbellarians are technically separated by details of the hermaphroditic reproductive system, but several groupings may roughly be recognized by the shape of the digestive system. Large to moderate-sized turbellarians with a 3-branched gut are **triclads** (or planarians), and those with a many-branched gut are **polyclads.** Smaller worms of several orders have a straight, unbranched gut (rhabdocoel type), and miscellaneous others have guts that are variously lobed or diverticulated (alloeocoel type). Among the smallest and simplest turbellarians are the **acoels,** which, as their name suggests, lack a digestive cavity; the mouth opens, sometimes through a pharynx, into a mass of loosely packed cells.

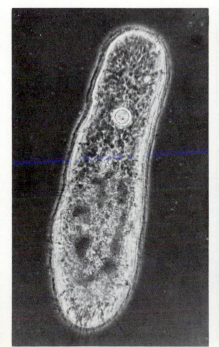

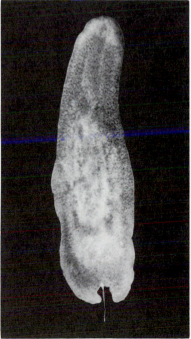

Acoel (*left*) with large statocyst and with oval outline is typical of many members of this group of turbellarians. The vesicle of the statocyst encloses a lithocyte (sometimes 2) that secretes a statolith in its center. A statocyst occurs in many acoels, and in some rhabdocoel and alloeocoel turbellarians; it lies close to or embedded in the brain. Acoels are almost all marine, measure less than a few millimeters long, and lack an excretory system. Most live among sand grains, eating a variety of tiny organisms and behaving much like the ciliated protozoans in the same interstitial habitat. Some acoels are pelagic. Monterey Bay, California. (R.B.)

Tailed acoel, *Polychoerus carmelensis,* from a central California tide pool. Reaching 6 mm in length, these worms are large for acoels and unusual in having 2 tail lobes and a trailing tail filament. They are conspicuous in their bright orange-red color and their habits, as they crawl about on the surface of the gravel or seaweeds in tide pools during low tides, descending into the gravel when the waves return. (R.B.)

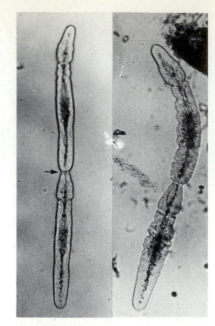

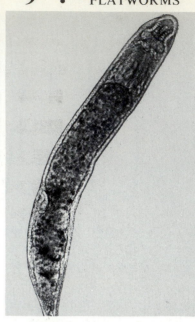

Asexual fission in *Microstomum* produces a chain of zooids. An indentation develops, then ciliated pits, then pharynx and brain. The first fission plane appears in the middle of the worm, and each of the 2 zooids thus formed later divides, producing a chain of 4. Successive fission planes can produce a chain of as many as 16 zooids in various stages of development. Besides *Microstomum* and others in the rhabdocoel order Macrostomida, members of the orders Catenulida (photos at right) and Tricladida reproduce asexually by fission. *Microstomum* includes hydras in its diet and retains the nematocysts in its epidermis. The nematocysts discharge when the worm encounters prey, presumably aiding in prey capture. (Modified after L. von Graff)

Catenulid rhabdocoel, *Catenula,* has the body divided by a circular groove into a tapered pre-oral lobe, containing the brain and statocyst, and a longer trunk, containing the digestive cavity and the reproductive organs when present. The worms are mostly seen as chains of 2 or 4 zooids, produced by asexual fission and still attached. Only rarely are they seen in a sexual state. *Left,* a chain of 2 zooids, with a constriction at the fission plane (arrow). *Right,* each of the 2 zooids shows a constriction, an external indication that 4 zooids are now developing. Lake Pymatuning, Pennsylvania. (R.B.)

Open ciliated pits, one at each side of the head, are prominent in *Stenostomum,* a catenulid that is among the most common of all microturbellarians in freshwaters. The pits are presumed to be chemoreceptors important in detecting food; if they are removed, the animal cannot detect food juices or orient to food. Lacking in *Stenostomum* are a statocyst (present in *Catenula*) and eyes (present in many rhabdocoels). The rear third of the worm is a developing zooid, with ciliated pits already visible. Lake Pymatuning, Pennsylvania. (R.B.)

Freshwater dalyellioid neorhabdocoel with a bulbous pharynx and an unbranched digestive system that occupies much of the body. As in most neorhabdocoels, there is a pair of prominent pigment-cup eyes, but there is no statocyst. Some species of *Dalyellia* harbor green symbionts that can be cultivated outside the body. Such host species are much larger (2 to 5 mm) than non-symbiotic species of *Dalyellia*. Lake Pymatuning, Pennsylvania. (R.B.)

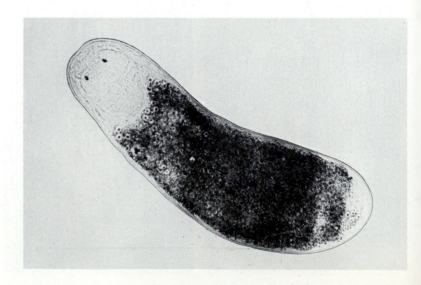

Cave flatworm, *Dendrocoelopsis americana,* a white triclad, reaches 18 mm in length and has a group of eyes on each side of the head. The flatworm pounces on small crustaceans and hydras, grasping them with a suckerlike adhesive organ at the center of the anterior margin. The prey is held against the substrate by the pharyngeal region as the flatworm feeds. This specimen was found in Missouri; the species can be abundant in caves and in cave springs in Missouri, Arkansas, and Oklahoma. Most of its white dendrocoelid relatives live in Eurasia. (R.B.)

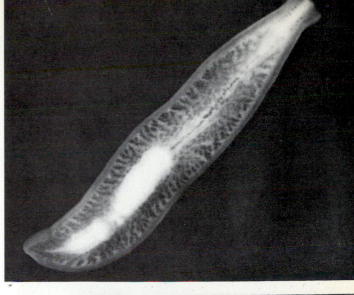

Garden planarian, *at right,* with a pointed head and stripes down the back is *Geoplana mexicana,* one of several species of land planarians imported into California with plant shipments. Here, in a Santa Cruz garden, it is taking shelter from the sun in a strawberry partly eaten by a slug. It emerges at night in search of prey, as does *Bipalium* (below). (R.B.)

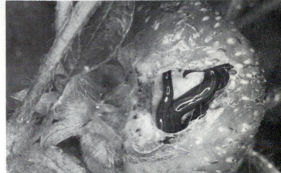

Large land planarian, *Bipalium kewense,* hanging from a leaf in a Hawaiian forest *(left)* and in closeup *(right).* This Asian species has been distributed around the world by commerce in tropical plants. Though well established in gardens in California and the southern U.S., it reproduces only by fragmentation there, never becoming sexually mature as does a related species. In temperate regions it lives mostly in greenhouses but the authors have found these worms, with 5 purple stripes on a yellowish background, under flowerpots in Pittsburgh and New York gardens. A large specimen reaches 25 cm or more; other land planarians range in length from about 1 to 60 cm. The halfmoon-shaped head is characteristic of the tropical family Bipalidae, with many species in the Indo-Malay region. (R.B.)

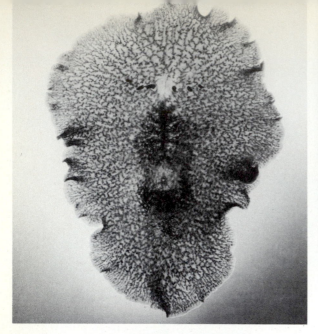

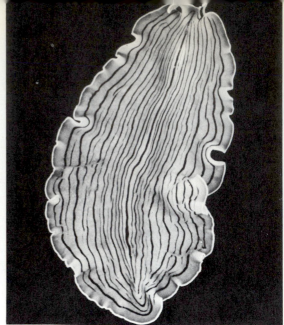

Polyclad, moving along on a piece of glass illuminated from below, is so thin that it displays all of its anatomy. In the dark central area are the pharynx and main branches of the digestive system and also the reproductive organs and openings. The anastomosing branches of the digestive cavity can be seen to penetrate the entire body. About a fourth of the way from the anterior margin is a row of darkly pigmented sensory tentacles spread out on either side of the brain (light area). Length, 3 cm. Misaki, Japan. (R.B.)

Polyclad laying eggs. Emerging from the worm is a continuous adhesive gelatinous string, secreted by special glands, and containing a row of eggs. The egg string is woven back and forth, forming a compact egg layer. NW Florida. (R.B.)

Striped polyclad, *Prostheceraeus vittatus,* from an English shore, is about 25 mm long. At the head end a pair of flaplike marginal tentacles bear multiple marginal eyes, and there are 2 clusters of eyes over the brain. The digestive system has many lateral anastomosing branches. Polyclads of this genus are common on cool shores of Europe and both U.S. coasts, but most of their cotylean relatives (polyclads with an adhesive disc behind the female genital pore) are found in warmer waters. Many of these are large and of striking form and color. (D.P. Wilson)

Müller's larva, named after its discoverer, is the free-swimming 8-lobed larva of many marine polyclads. This one was caught in the plankton off Bimini. The larva swims in the plankton for only a few days, then resorbs its ciliated lobes and flattens out into a young polyclad. Some polyclads and all other turbellarians have direct development, hatching as juvenile worms. (R.B.)

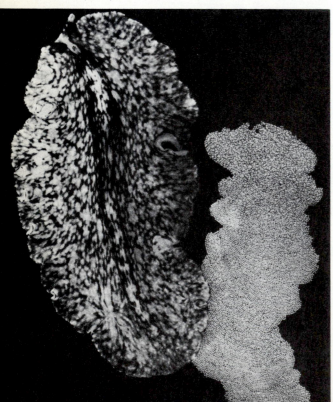

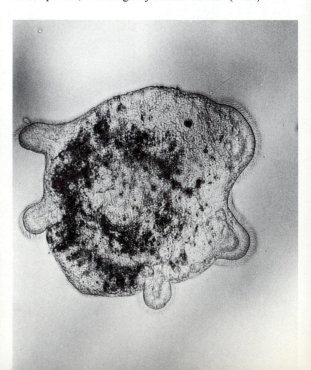

EXCEPT among the freshwater and land planarians, some members of all major turbellarian groups have given up the roving life-style for that of **symbiosis**—specific, regular, intimate, long-term association with other organisms, sometimes limited to another single species. The term symbiosis ("living together") is useful as a collective word for the several kinds of relationships (commensalism, mutualism, parasitism) that two organisms may share, or for designating an association in which the nature of the relationship is not known.

Any symbiotic relationship may be described as *facultative* if both members are also found separately; it is *obligate* for that member which is never found living independently. The smaller member, or *symbiont*, is described as an *ecto*symbiont or *endo*symbiont according to whether it resides on the outside or inside of the larger member, or *host*.

Commensalism is a relationship between two organisms in which the smaller (commensal) derives some benefit from the larger (host) without causing significant harm. The term implies sharing of food, but it is also used for relationships such as *phoresy,* in which the commensal is carried about on the host and the benefit derived is transportation, and *inquilinism,* in which the commensal occupies the nest or burrow of the host and the benefit derived is shelter. How a commensal relationship might develop is suggested by the behavior of small intertidal turbellarians *(Monocelis)* that are sometimes found during low tide inside barnacles or in the mantle cavity of limpets, protected in these moist shelters from dessication. During high tide the worms leave these temporary hosts and roam freely. It is easy to imagine that such a loose association might gradually become permanent, as in certain species of polyclads that live in the shells of hermit-crabs.

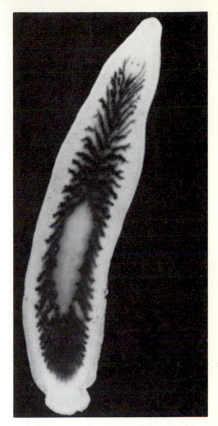

Commensal marine triclad of the genus *Bdelloura* lives on the gills of the horseshoe-crab *Limulus polyphemus* (see photo, chapter 22) on the Atlantic coast of N. America. The worms apparently receive transportation, aeration, and shelter, while feeding on small animals stirred up by movements of the host and on fragments of the host's food. *Bdelloura* sticks to its host by means of a marginal zone, supplied with adhesive gland cells and free of cilia and rhabdoids (see planarian cross-section, chapter 8). In ectocommensals such as *Bdelloura* the well-developed marginal zone is expanded into an adhesive disk at the posterior end. Length, 1 cm. (R.B.)

Commensal rhabdocoel, *Syndisyrinx franciscanus* (=*Syndesmis franciscana),* lives in the gut of sea urchins and eats ciliates that are also regular symbionts. This relationship has mutualistic possibilities (i.e., if the ciliates harm the host) but may also be partly parasitic, as when the worms enter the body cavity and ingest host cells. Echinoderms (chap. 27) are among the most common hosts for commensal and parasitic turbellarians, which occur in the gut or body cavity. Length, 2 to 3 mm. (R.B.)

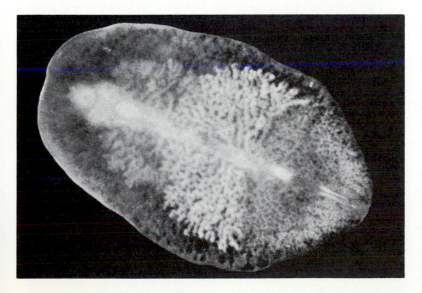

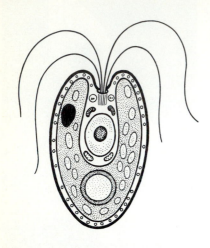

Free-living form of green flagellate, *Tetraselmis convolutae,* has 4 flagella, a cell wall, and an eyespot. Flagellates grown in laboratory culture may be introduced into young, white, laboratory-hatched worms, *Convoluta roscoffensis.* (After Parke and Manton)

Symbiotic form of green flagellate, *Tetraselmis convolutae,* in a transmission electron micrograph prepared from the tissues of the host acoel, *Convoluta roscoffensis,* has lost the flagella, cell wall, and eyespot. (J.L. Oschman)

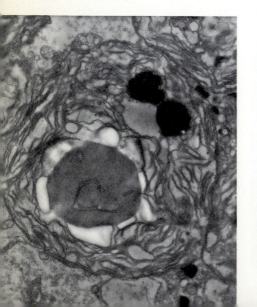

Commensalism sometimes evolves into **mutualism,** a relationship in which both members of the association benefit. Examples already mentioned include the intestinal flagellates of termites, and the unicellular algal symbionts of protozoans and cnidarians, especially reef-building corals. One of the most celebrated examples of mutualism involves the acoel turbellarian *Convoluta roscoffensis,* colored green by a unicellular photosynthetic symbiont, *Tetraselmis convolutae.* In certain areas of the Channel coast of France, the little green worms gather by the millions at low tide on the surface of sand beaches, in patches that look like splashes of green paint. The young worms are colorless when first hatched, but they promptly begin to feed on any available small organisms, and their first meal usually includes some of the green symbionts, handily left clinging to the egg case by the parent worm or found free-living in the environment. Once inside the worm, the flagellated symbionts lose their flagella, cell wall, and eyespot, and they begin to multiply beneath the epidermis of the host worm. Supplied by the worm with a moist sheltered place in the sunlight and with carbon dioxide and other nutrients (by-products of nitrogen and phosphorus metabolism), the photosynthetic symbionts produce oxygen and various organic compounds. A portion of the photosynthetic product is released by the symbionts to the host's tissues; and as the young worm matures, it ceases to feed and depends solely on nutrients supplied by the symbionts. The nutrients transferred are mostly fatty acids and sterols, which the worm cannot synthesize, and amino acids. Only late in life does the worm begin to digest its photosynthetic guests. The number of green cells declines, and eventually the worm dies.

The symbiosis of *Convoluta* and *Tetraselmis* is widely considered as a mutualism, and the relationship does appear to be essential to the worms, which never grow to maturity if uninfected with symbionts. However, the benefits to the green flagellates could be questioned. As they are found living free in the surrounding environment, they might be viewed not as partners, but as prisoners, doomed to remove the worm's nitrogen and phosphorus wastes and forced to give up part of their photosynthetic product.

If *Tetraselmis* is absent, other kinds of flagellates may become established as symbionts of *Convoluta,* but neither they nor the worms grow as well together. If *Tetraselmis* is then introduced, it replaces any other symbiont. Whether it has won a competition for a privileged place, or is merely the host's preferred victim, remains to be investigated.

To determine the most accurate view of the relationship would require analysis of the effects of symbiosis on the life history of the flagellates; of whether carbon dioxide and inorganic nutrients are limiting factors in the growth of the free-living flagellate population; of whether the symbiotic population might at times serve as a reservoir from which a free-living population, unable to survive extreme environmental fluctuations, could be re-established; and of other possible benefits and costs.

Restriction of habitat is notable for *Convoluta roscoffensis*. The worms occur only in patches or streaks where seawater oozing from the sand (from a previous high tide) continually runs seaward in rivulets covering the worms throughout the following low tide period. Roscoff, France. (R.B.)

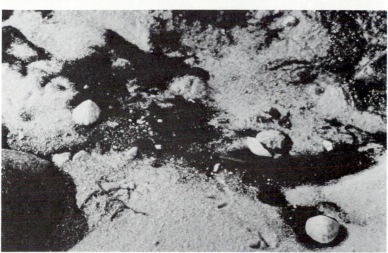

Dark green patches and streaks on the sand surface consist of millions of convolutas, densely crowded into the only areas with a constant cover of running seawater throughout low tide. Perhaps the worms are also kept together by chemical communication, as they excrete the volatile and pungent compound, trimethylamine. Dense aggregation by the worms probably protects them from excessive UV radiation and great changes in salinity during rains. Crowding probably also brings worms together for mating. Despite their negative response to gravity during daylight hours, the worms are very sensitive to vibrations so that anyone approaching patches of *Convoluta roscoffensis* must step lightly or the patches suddenly disappear. Roscoff, France. (R.B.)

Close-up of convolutas in a pan of sand and seawater in the laboratory reveals these small green, elongate, ciliated worms (3 to 5 mm long) gliding about in a film of secreted mucus. Tapping the pan causes the worms to descend suddenly into the sand. But undisturbed, in a lighted room, they rise to the surface again. For about a week after being brought into the laboratory, they retain their rhythmic behavior, rising to the surface at the same time that the tide goes out on the nearby beach outside, and retreating below the surface when the tide comes in again. In a few days, the rhythm is lost and the worms begin to rise with the sun and sink at sunset. Station Biologique de Roscoff, France. (R.B.)

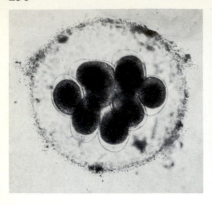
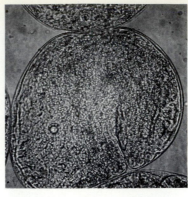

Left, **egg case** (diam. 0.6 mm) of *Convoluta roscoffensis.* Most contain 8 to 15 eggs. The cases, laid just under the surface of the sand, have green flagellates sticking to the mucilaginous surface. *Right,* **ready-to-hatch worm** is colorless and bent on itself within the egg membrane. (M. Parker)

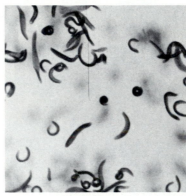

Survival of *Convoluta* in sealed glass tubes that contain seawater with a bubble of air. In the left-hand tube, kept in natural daylight and darkness for 7 days, the worms appear healthy and are swimming about. In the right-hand tube, kept in continuous darkness, the worms became degenerate in half the same time. Roscoff, France. (R.B.)

Orientation of *Convoluta roscoffensis* is seriously disturbed by the antibiotic streptomycin, added to the seawater as a means of obtaining worms free of bacteria, for preparation of tissue cultures. *Above left,* the worms swim about normally. *Above right,* after streptomycin has been added, the worms swim about in circles. (A common side-effect of streptomycin in humans is dizziness or disorientation and accompanying nausea. The mechanisms for these phenomena are not known.) Station Biologique de Roscoff, France. (R.B.)

Positive phototaxis of *Convoluta roscoffensis* can be seen by placing a dish of the worms next to a lighted window. The worms immediately start moving toward the light and form a dense, dark-green aggregation at the lighted side of the dish. (R.B.)

Algal-invertebrate symbioses are common among acoels, but none are known to be as intimate as that between *Convoluta roscoffensis* and *Tetraselmis*. More common is the type of symbiosis seen between the marine free-living acoel *Amphiscolops langerhansi* and its dinoflagellate partner *Amphidinium klebsii*. After the alga is ingested and lies among the mesenchyme cells close to the epidermis, it undergoes no visible changes: it still retains its covering theca and 2 flagella (it has no eye spot). In the laboratory an uninfected host can survive and reach adult size without the alga if fed enough. It will also undergo asexual fission, but it will not mature sexually. Perhaps the association involving *Amphiscolops* is of more recent origin than that described for *Convoluta*.

Commensalism sometimes evolves, not toward mutualism, but toward **parasitism,** a relationship in which the parasite benefits at significant cost to the host by taking nourishment directly from the host's digestive tract or tissues. One can see that a resident commensal could easily begin to take further advantage of its host through parasitism. This is almost certainly the history of the temnocephalid (rhabdocoel) turbellarian *Scutariella didactyla,* an ectoparasite that sucks the blood of its host shrimp but belongs to a group in which other members are ectocommensals that live on freshwater crustaceans, snails, and turtles, clinging to their hosts or moving about on them with leechlike movements by means of special adhesive tentacles and disks.

Another possible evolutionary route for an ectosymbiont which is in constant danger of being knocked off or picked off the host's surface is to shelter in host cavities. There protection is greater, and opportunities for taking advantage of the host increase. For such cases it is not always easy to discriminate between commensalism and parasitism. Some alloeocoel turbellarians *(Urastoma)* are found free-living and also in the mantle cavity of oysters and other bivalve molluscs. Although considered commensals, the worms edge toward parasitism when they become so abundant that the oysters are adversely affected. Further along the road to parasitism are rhabdocoels *(Triloborhynchus)* that live as endosymbionts in the digestive ceca of sea stars and feed on host tissue. Though the hosts are not conspicuously affected, the worms show several effects of their parasitic life style—reduced ciliation, loss of eyes and rhabdoids, and development of the rear end as a muscular, adhesive disk.

Symbiotic acoel, *Amphiscolops langerhansi.* (After L. Hyman)

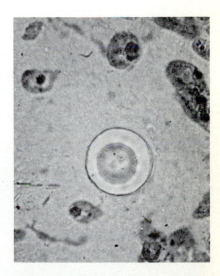

Statocyst of *Convoluta roscoffensis.* The statocyst is important in the worm's responses to gravity. (R.B.)

Few turbellarians have discernible impact on human affairs. Besides polyclads such as the oyster-leech, *Stylochus,* which damages commercial oyster beds, there are more welcome turbellarians that produce secretions lethal to mosquito larvas. In the rice fields of California, species of the rhabdocoel genus *Mesostoma* paralyze and then feed on the larvas of 2 species of mosquitos: *Culex tarsalis,* a vector of equine encephalitis, and *Anopheles freeborni,* a vector of malaria. Introducing the appropriate species of *Mesostoma* into malarial areas could help to reduce the incidence of malaria.

Polyclad feeding on bryozoan colony. *Below left,* the small polyclad *Hoploplana californica* on a bryozoan colony *(Celleporaria brunnea).* The flatworm so perfectly matches the bryozoan that its presence is revealed only by the clear areas around each of the 2 sensory tentacles on the head. *Below right,* the flatworm has been flipped over by the photographer to expose the pale ventral surface. Monterey Bay, California. (R.B.)

Less benign endoparasites are certain rhabdocoels *(Kronborgia)* in which the young stages bore through the exoskeleton of marine crustaceans and enter the host's body cavity. The worms lack a digestive tract and must absorb nutrients from the host's body fluids. The host fails to reproduce and eventually dies when the mature worms again break through the body wall and emerge to mate.

Although many commensal and parasitic turbellarians do not differ conspicuously from free-living forms, loss of the digestive tract has occurred in several. More common is reduction or loss of the eyes and of the cilia, rhabdoids, mucus glands, and pigments of the epidermis. Adhesive organs are extensively developed, and an outer syncytial layer is sometimes present. Relatively large numbers of eggs are produced. In many of these features, and in metabolic patterns also, these probably resemble the ancestors of the three parasitic classes of flatworms.

Parasitism may also evolve directly from a predatory relationship. The polyclad *Hoploplana californica* feeds on colonies of the bryozoan *Celleporaria brunnea* and is usually called a predator. However, the feeding is a sufficiently slow process that the bryozoan colonies are the habitat of the polyclad, which matches the bryozoans in color and texture and is found nowhere else, so that it could be considered an obligate ectoparasite.

Another polyclad that is a part-predator, part-parasite is the oyster-leech, *Stylochus frontalis,* of the southeast coast of the U.S., which is so thin it can slip in between the shell valves and eat an oyster bit by bit. Likewise, the free-swimming larva of the polyclad *Stylochus tripartatus* invades a barnacle and the tiny young worm is a parasite that can only nibble at the living host. By the time the worm has grown to a length of 5 cm, however, it is a predator that can open and clean out a large barnacle in short order.

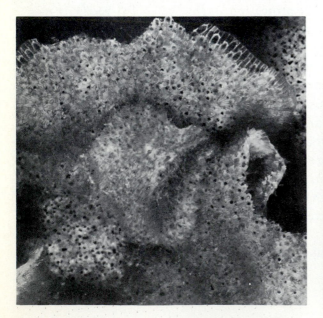
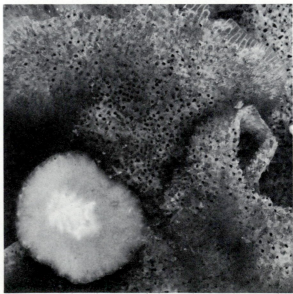

PARASITIC FLATWORMS

THE ADAPTATIONS to parasitism are often so dramatic that they obscure the basic flatworm body plan in the three parasitic classes. But most of the same features seen in planarians and other turbellarians are present, and trends seen in parasitic turbellarians are further developed. The body is constructed on a 3-layered plan, and a mouth in the anterior half of the body opens through a muscular pharynx into a variously branched digestive cavity, except in the tapeworms, which lack a gut entirely. An excretory system of flame bulbs and connecting tubules lies embedded in mesenchyme along with the other organ-systems. The sense organs and nervous system may be much reduced. The reproductive system is hermaphroditic, as in turbellarians, but its great volume and complexity leave no doubt about how these worms are allocating their resources—and those of their hosts.

MEMBERS of the class **Monogenea** are parasitic flatworms with a single host and a relatively simple life history. Most of the roughly 400 species live on the skin or gills of fishes; but other aquatic vertebrates (frogs, turtles, hippopotamuses) and invertebrates (squids and crustaceans) may also be hosts. The small worms (from less than a millimeter to 2 or 3 cm long) feed on epithelial cells, mucus, and blood, clinging to their slippery, fast-swimming hosts with a complex adhesive organ, or **haptor,** at each end of the body. The haptors of parasitic flatworms are often single or clustered suckers, aided by adhesive secretions, but these may be replaced or supplemented by a variety of hooks, clamps, and other structural means for holding on. Even these elaborate safety devices do not always insure against the fatal consequences of being separated from the host; and monogeneans of some species shelter in host cavities accessible from the outside (mouth, nasal passages, gill chambers, rectum, urinary tract).

Though hermaphroditic, monogeneans usually mate and fertilize each other's eggs. The zygote develops into a **ciliated larva** (oncomiracidium) with lensed eyes that are lost or reduced in the adult. On hatching, the larva swims freely in the water and is attracted by the mucus of the host skin. On contact with the host, the larva attaches with its posterior haptor and proceeds to shed its surface epithelium of ciliated cells. This startling metamorphosis, which is also stimulated by host mucus, takes only 30 seconds. The adult monogenean surface is a cytoplasmic syncytium containing mitochondria and other organelles but no nuclei.

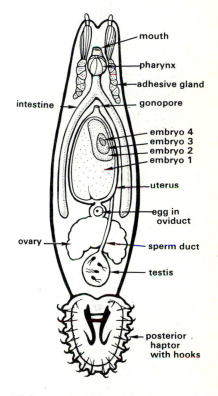

Viviparous monogenean, *Gyrodactylus,* broods its fertilized eggs internally. Each egg produces up to 4 embryos, one inside another, in different stages of development. The largest and most advanced, when released, will attach to the host and release the next embryo within a day or two, and so forth. Through such asexual reproduction of young stages, the population of worms can increase rapidly, although the scale is modest compared to the vast numbers of individuals produced by the several larval stages of flukes. There is no free-swimming larva; infestation spreads by contact between fishes and is a serious problem in fish ponds and hatcheries where fishes are crowded. Though each worm is less than a millimeter long, heavy infestations can be fatal to the host. (Combined from various sources)

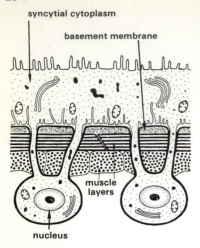

syncytial cytoplasm

basement membrane

muscle layers

nucleus

Tegument of monogeneans is organized like that of the other 2 parasitic classes of flatworms. (Modified after K.M. Lyons)

Polystoma from the mouth cavity of a turtle. (Modified after H.W. Stunkard)

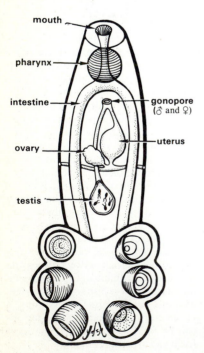

mouth

pharynx

intestine

gonopore (♂ and ♀)

ovary

uterus

testis

Such a surface, called a **tegument,** is present in members of all three parasitic classes. The tegumental surface is not ciliated but is covered with an organic coating layer (glycocalyx). The outer membrane has complex infoldings or bears many microvilli (microscopic fingerlike projections); these elaborations of the surface increase its area manyfold. Beneath the tegument are muscle layers, and beneath these, embedded in the mesenchyme, are the tegumental nuclei and most of the synthetic organelles, contained within cytoplasmic bodies that communicate with the tegument through narrow connections. In endoparasites, the tegument and its organic coat facilitate digestion and absorption of nutrients and protect the parasite against the digestive enzymes or immune defenses of the host. In the mostly ectoparasitic monogeneans, the tegument is mainly protective, but it is possible that these worms take up some nutrients through the tegument as well as through the mouth and digestive system.

The epidermis of certain parasitic (and even supposedly commensal) turbellarians is syncytial, like the tegument of the parasitic classes. Curiously, even some free-living turbellarians have epithelia covered with microvilli or nuclei in-sunk below the muscle layers. The functions of all these features are not yet known. Free-living and commensal turbellarians covered with microvilli may supplement their feeding by taking up dissolved nutrients from the environment; concentrations of organic material are often substantial in the interstitial environments in which many small turbellarians live. Moving among sand grains, turbellarians subject their delicate surfaces to constant physical abrasion. The removal of the nuclei and synthetic organelles from the vulnerable epidermis places this cellular machinery in a protected position from which it can regulate and supply frequent renewal of the surface. The apparent ease with which the flatworm epithelium is modified for uptake and for repair can be viewed as a *pre-adaptation* for parasitic life styles.

Adults of another species of *Polystoma* live in the urinary bladder of frogs. In winter when the frog hibernates, the worms are also dormant; in spring when the frog produces gametes and spawns, the worms release fertilized eggs which hatch into ciliated larvas just in time to infest the gills of the emerging tadpoles. When the tadpoles begin to metamorphose and resorb the gills, the worms migrate to the bladder. Young worms have been observed moving at night over the skin of the ventral surface of the host; the migration took only about a minute. The worms live for several years in the frog's bladder, synchronizing their reproduction with that of their host by means of host hormones. If an immature frog is injected with pituitary extract, the worms mature precociously and produce eggs.

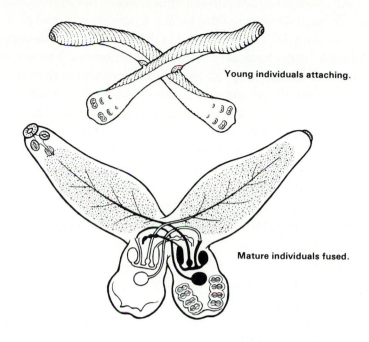

Young individuals attaching.

Mature individuals fused.

Mating for life is taken literally by monogeneans of the genus *Diplozoon,* in which pairs of hermaphroditic adults live fused together in a permanent state of mutual copulation. When two young individuals meet (on the gills of a fish host), they become attached firmly together, their tissues fuse, and their reproductive systems grow together so that each fertilizes the other. Unmated individuals fail to mature and soon die. (Combined from various sources)

TREMATODES

ALTHOUGH we usually know them as important human parasites, trematodes are primarily parasites of molluscs. Almost all members of the class **Trematoda** have a snail or clam as their first host. In the small subclass **Aspidogastrea,** a mollusc may be the only host. But the great majority of the 6,000 trematode species—the flukes, subclass **Digenea**—have evolved life histories that include a second (or third or fourth) host, which harbors the adult stage and which is almost invariably a vertebrate, sometimes a human.

FLUKES probably had ancestors with simple life histories involving a single host, as in monogeneans and in some aspidogastreans, but modern flukes lead complicated lives with *many asexual stages* and with *2 or more hosts.* A typical fluke has 3 hosts (sometimes 2 or 4). Two *intermediate hosts,* which harbor larval or juvenile stages, may be invertebrates or vertebrates; the *final host,* which harbors the sexually mature adult stage, is almost invariably a vertebrate.

Aspidogastrean, *Cotylaspis cokeri,* in ventral view. Aspidogastreans are distinguished by a large divided sucker or row of suckers along much of the ventral surface. The subclass Aspidogastrea of the class Trematoda includes about 40 freshwater and marine species, which are external or internal parasites of snails or clams. The larvas (cotylocidia) swim by means of tufts of long cilia until they locate a molluscan host; later stages of some species infest fishes and turtles. A single species may occupy a range of hosts, and some aspidogastreans can live for weeks without a host. These relatively unspecialized parasites are thought to resemble the ancestors of flukes. (Modified after H. W. Stunkard)

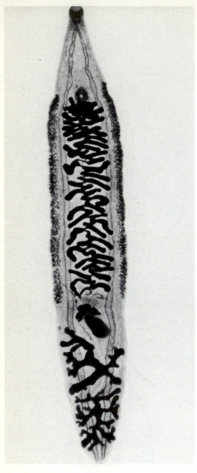

Adult of Chinese liver fluke. Most human infections with *Clonorchis sinensis* involve relatively few worms, and symptoms are mild or absent, but a long-lasting infestation with hundreds or thousands of worms results in severe liver disease and sometimes death. Flukes have been implicated in liver cancer. A simple control measure is thorough cooking of fish, which kills the encysted metacercarias. However, strong cultural preferences for raw fish, and sometimes a lack of fuel for cooking, combine to perpetuate the Chinese liver fluke as a serious health problem in many parts of the Orient. (R.E. Kuntz)

Life history of the Chinese liver fluke, *Clonorchis sinensis* (= *Opisthorchis sinensis*). (Based on E.C. Faust)

An example is the **Chinese liver fluke,** *Clonorchis sinensis,* common and widespread in China, Korea, Japan, and parts of southeast Asia. The **adult** lives in the liver and bile passages of humans, dogs, cats, and many wild carnivores. The worm is usually 1 to 2 cm long and has two suckers, one at the anterior end around the mouth and another a short distance back on the ventral surface. It feeds on epithelial tissue and blood, but may also absorb nutrients through the tegument. The reproductive system is hermaphroditic, and the female system includes a long coiled uterus, which contains the great numbers of eggs produced by these parasites. Fertilized eggs pass into the host intestine and out with the feces, which must get into freshwater (as they commonly do) if the eggs are to be found and eaten by a suitable freshwater snail, the *first intermediate host.* Each egg contains a developing ciliated larva, or **miracidium.** The egg opens within the snail's digestive tract and releases the miracidium, which burrows into the gut wall, loses its ciliated epithelium, and develops into a saclike **sporocyst.** This begins a phase of asexual reproduction in which many thousands of young may result from a single miracidium. Within the sporocyst there develop forms called **redias,** and each redia gives rise to a number of **cercarias.**

The cercaria has eyes and suckers and a muscular tail with which it can swim. It makes its way out of the snail and swims free in the water. Cercarias are short-lived and have only a

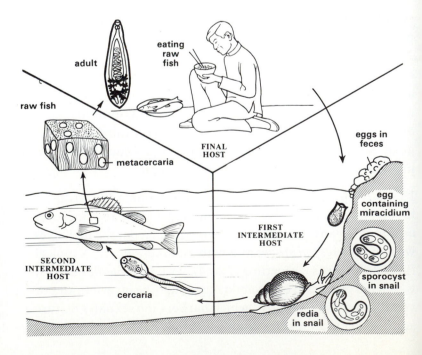

matter of hours to locate a *second intermediate host,* which may be a freshwater fish or crustacean. The cercaria alternately sinks and then, on contact with any surface or on stimulation by water movements or a shadow (such as a passing fish might make), it swims actively upward, increasing its chances of contacting a host. When contact is made, the cercaria attaches to the host with suckers, casts off its tail, and within minutes has digested its way through the skin. It then encysts in the second intermediate host as a **metacercaria,** a stage which can remain alive and infective for months or years.

The **adult** stage is reached in the *final host,* a human or other carnivorous mammal that eats an infected fish, so ingesting the encysted metacercarias. The cyst is digested and the young worms are released in the host's intestine. They migrate up the bile duct into the liver, where they may live for at least 8 years, releasing eggs. Any individual offspring has little chance of surviving to complete the complex life history, but the high risks are made up for by the large numbers of eggs produced over the life span of a successful adult and by the asexual multiplication of the juvenile stages.

Stages of a fluke, based mostly on *Clonorchis sinensis.* Each stage is generally a little more like the adult in structure. The **miracidium** and **sporocyst** have a flame-bulb excretory system but no mouth or digestive tract; they are nourished only by stored food and by nutrients absorbed through the tegument. The **redia** has an unbranched gut and can feed on host tissue; the excretory system is more extensive than in previous stages, and the redia is more active than a sporocyst. The **cercaria** has a well-developed excretory system and forked gut, but is short-lived and may not feed; eyes and a muscular tail help it contact a second intermediate host to which it attaches with suckers. In this host, it encysts as a **metacercaria.** When eaten by the final host, the fluke emerges from its cyst and grows to maturity. The **adult** feeds on host tissues and develops an elaborate hermaphroditic reproductive system. (Based mostly on Faust and Khaw, on Komiya and Tajimi, and on Yoshimura)

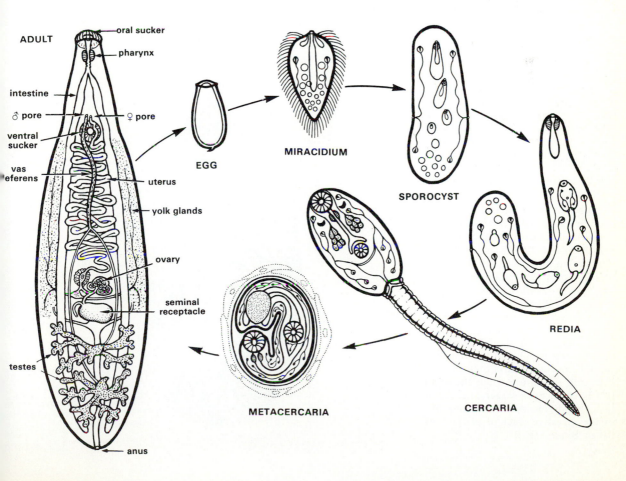

ADULT
oral sucker
pharynx
intestine
♂ pore
♀ pore
ventral sucker
vas deferens
uterus
yolk glands
ovary
seminal receptacle
testes
anus

EGG

MIRACIDIUM

SPOROCYST

REDIA

METACERCARIA

CERCARIA

Some of the many variations in fluke life histories can be seen in the **sheep liver fluke,** *Fasciola hepatica,* an important parasite of sheep, cattle, and other domestic stock. It inflicts severe, often fatal damage and results in large economic losses to stock raisers. It also infects a variety of wild herbivores and occasionally humans. The miracidium hatches from the egg in freshwater as a free-swimming larva that lives only 24 hours or less and so must quickly find and penetrate into a certain species of snail. The redias produced by the sporocyst give rise to more redias, which only then produce cercarias. Emerging from the snail, the cercarias encyst as metacercarias on aquatic vegetation or free in the water. There is no second intermediate host. A sheep or other final host ingests metacercarial cysts when grazing in poorly drained pasture or drinking water containing cysts. Humans may become infected by eating wild watercress.

Sheep liver fluke, *Fasciola hepatica,* is among the best studied of flukes, but still a serious problem for stock raisers. Snail poisons, or drugs given to sheep and cattle to kill the adult worms, are only partially effective. Improving pasture drainage is not always possible, and rotating animals among fields is tedious because the parasites remain alive for long periods within the snail host or encysted. This stained specimen shows the highly branched digestive system. Length, 2 cm. The sheep liver fluke is not closely related to the Chinese liver fluke, as can be surmised from their different appearance and life history. (A.C. Lonert)

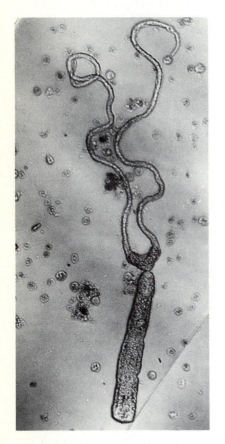

"Oxhead" cercaria from a marine clam has a forked tail that ends in 2 long slender branches. (A fancied resemblance of the tail branches to the horns of an ox gave rise to the names of the family Bucephalidae and the common genus *Bucephalus,* meaning "oxhead"). When released from the clam, the cercaria swims slowly in the water trailing the branches of the tail-fork, which (with luck) become entangled and caught on the fins of a small fish. The cercaria then encysts, emerging to mature when the small fish is eaten by a larger one. Adult bucephalids, less than 1 mm long, live in the digestive tract of predaceous fishes. Bimini. (R.B.)

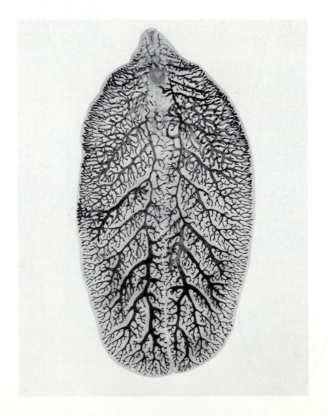

Animals commonly evolve special features that reduce the chances of their being eaten by a predator. But most trematodes must be eaten by an intermediate or final host in order to complete their life histories, and many have elaborate adaptations that increase the chances of predation. For example, some cercarias do not burrow into the intermediate host but sport bright colors and conspicuous behavior that attract a fish host to eat them. Once inside the intermediate host, cercarias of *Diplostomum* migrate specifically to the eyes or brain of the fish, damaging its vision or nervous system, so that it is more likely to fall prey to the parasites' final host, a fish-eating bird. One trematode *(Leucochloridium macrostomum)* develops in land snails that are probably not usually seen or eaten by the birds that serve as final hosts. However, the presence of the parasites changes the behavior of the normally photonegative snails so that they emerge into lighted, exposed areas. In addition, the sporocysts in the snail grow into large caterpillar-like bodies brightly banded with green, brown, and orange. They extend into the snail's tentacles and pulsate rapidly, so that they look (to human eyes, at least) well suited to attract the attention of an insect-eating bird. Birds that peck at the tentacles ingest the infective cercarias within the sporocysts.

BLOOD FLUKES called **schistosomes** affect an estimated 200 million people, mostly in tropical or subtropical Asia, Africa, and northeastern South America. Among parasitic diseases of humans, schistosomiasis (sometimes called bilharziasis, after a pioneering parasitologist) is today a world health problem second only to malaria. Victims are seldom killed outright, but are more often debilitated by years of continual discomfort and lack of energy, their lives eventually cut short by secondary illness.

The life history of schistosomes is a curiously modified and abbreviated one. The free-swimming miracidium infects a suitable freshwater snail and becomes a sporocyst, which then produces more sporocysts, and these produce cercarias. Although a redial stage is omitted, a single miracidium may give rise over a few months to over 200,000 cercarias. The cercaria, on emerging from the snail, does not burrow into a second intermediate host or encyst as a metacercaria, as expected. Instead, it burrows directly into the final host.

Because of their abbreviated life history, it is thought that schistosomes may be maturing precociously in what would formerly have been the second intermediate host and that the life history previously included a third and final host, which has been abandoned.

Whereas most adult trematodes live in the lumen of the host's digestive tract or organs connected with it, such as the lungs, schistosomes live within the blood vessels. Schistosomes are also unusual in having separate male and female individuals. In *Schistosoma mansoni,* as in the other 2 important species in humans, the male is shorter and stouter than the female, and the sides of his body curve under to form a canal, making the ventral surface look as though split along most of its length *(schistosoma* = "split body"). The longer, more slender female is held in the ventral canal of the male. Both worms feed on blood, primarily in the veins of the large intestine, and they also take in dissolved nutrients through the tegument. The female consumes about 10 times as much blood as the male, and she also receives nutrients from her mate through her tegument, devoting much of her intake to the production of eggs, about 300 per day for 20 years or more. From time to time, she leaves the male, her slender shape giving her access to the smallest veins of the host intestine, where she deposits her eggs. A sharp spine on the side of each egg helps lodge it in the vein. The egg must then pass through the wall of the vein and reach the lumen of the host intestine. Shed with the host's feces, the egg hatches when it reaches freshwater, and a new cycle begins.

Life history of *Schistosoma mansoni.* The egg **(1)** is shed from the human host with fecal matter. The egg hatches **(2)** in freshwater, and the free-swimming ciliated miracidium **(3)** is attracted to snails. The miracidium penetrates a snail, sheds its cilia, and turns into a sporocyst **(4)**. The original sporocyst produces more sporocysts **(5)**, which produce cercarias. The cercaria escapes into the water **(6)** and bores into the skin of the human host. The adult worms **(7)** live in the veins of the intestine for years. The species of snail identifies this scene as Puerto Rico. (Based partly on Faust and Hoffman)

Cercarias are not very discriminating, and many lose their lives burrowing into unsuitable hosts. Along some lakeshores and seacoasts, misguided cercarias of bird schistosomes attack human swimmers and penetrate the skin, causing itchy swellings, a condition known as "swimmer's itch." These unfortunate schistosomes quickly die, while the human victims may be rewarded for temporary discomfort with increased resistance to infection by dangerous human schistosomes.

In the other 2 important schistosomes of humans, the adults are a little larger and differ in some details from those of *Schistosoma mansoni.* In *S. haematobium,* the adult worms are associated mostly with the urinary system; the eggs, which have a terminal spine, are released into the bladder and escape with the urine. In *S. japonicum,* the adult worms are associated mostly with the small intestine; the eggs are small, with only a rudimentary spine, and they are easily distributed throughout the body, often causing damage to the central nervous system. Hence infection with only a few worms can be serious. *S. japonicum* can mature in many different mammals, whereas the other 2 species are limited to a few final hosts.

S. haematobium is found throughout Africa and Madagascar and in parts of the Arabian peninsula and the Middle East. *S. mansoni* has a more limited distribution in these regions, but was spread by the slave trade to northeastern S. America and to some of the Caribbean islands. (*S. haematobium* was certainly also present in slaves brought to the Americas, but failed to become established, probably because there was no suitable host snail.) *S. japonicum* occurs in scattered parts of Asia.

1. Adult male and female schistosomes, each with 2 suckers close to the anterior end, spend much of the time in intimate contact. Males average about 10 mm long, females about 16 mm. An unmated female does not grow to adult size and remains sexually immature. Nutrients and other substances transferred from male to female are necessary for the female to grow, mature, and remain in reproductive condition. Adult worms become coated with substances present in the host's blood and are thus protected against the host's immune system, which does not recognize them as foreign organisms. However, their presence sensitizes the host, and young schistosomes attempting to enter an already infested host are attacked by the immune system before they can acquire a protective coating. The degree of such immune resistance is variable among humans, and unfortunately, vaccines containing dead schistosome material or live schistosomes (weakened by radiation) have not yet been very successful.

2. Egg with young miracidium inside. How the eggs escape from the vein to the lumen of the intestine is still in question. Breakdown of the wall of the vein due to pressure of the eggs and their spines may be aided by digestive enzymes released from the eggs, or host tissue reactions may be entirely responsible for walling off the eggs and removing them from the path of blood flow. In any case, about 2 out of 3 eggs never do escape but are trapped in the wall of the intestine or are carried away by the blood to lodge in various other organs. The major ill effects of schistosomiasis are caused by these eggs. Each egg is a center of inflammation and eventually becomes surrounded by fibrous or calcified connective tissue. The walls of the intestine may become so scarred with connective tissue that eventually no eggs escape and tests other than checking for eggs must be used in diagnosis.

3. Miracidium hatching through a slit in the egg shell. Hatching is inhibited by high temperature, high osmotic pressure, and darkness. These conditions within the warm-blooded host help to prevent the miracidium from emerging prematurely. Dilution of the external medium is the most important factor that promotes hatching when feces containing the eggs are deposited or are washed into freshwater.

4. Free miracidium has long cilia for swimming. Miracidia swim toward snails, but the chemical attractants are not very specific, and the larvas will attempt to penetrate many species of snails unsuitable as hosts. Anterior glands of the larva produce secretions that digest the tissue of the snail and ease penetration. The beginnings of the digestive, nervous, and excretory systems are visible, as well as cells that will develop into the second "generation" of sporocysts.

5. Cercaria is covered with minute spines embedded in the tegument. The organ systems are further developed, and anterior and ventral suckers are present. Large glands that open anteriorly help the cercaria to escape through the body wall of the snail and later to penetrate the skin of the human host.

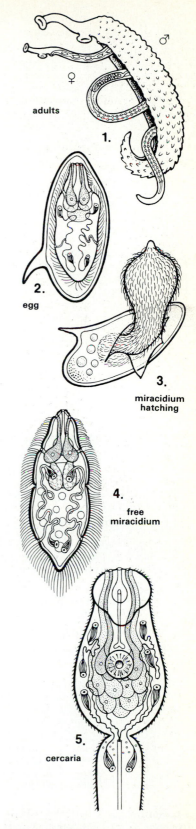

Stages of a human schistosome, *Schistosoma mansoni.* (Partly after Faust and Hoffman)

Control of schistosomiasis by treating victims with drugs may be successful, but not with people who are regularly exposed to reinfestation. Practical vaccines have not so far been developed. Sanitary disposal of human feces and urine, and education to reduce contact with contaminated water, are beyond the means of most of the countries in which the disease is widespread. Control of schistosomiasis has been most successful where several approaches were used simultaneously.

Most efforts have focused on snail control. Poisons are convenient, but few are both cheap and selective; if a poison kills more snail predators and competitors than snails, its use will *increase* snail populations. Some success has been achieved with biological control methods such as the introduction of predators (leeches, snails, ostracods, crayfishes, ducks and geese, and fishes) or parasites (protozoans, nematodes, flies). Certain other snails unsuitable as hosts compete directly with the snail or serve as decoys to some of the miracidia. Miracidia and cercarias may fall prey to hydrozoans, turbellarians, oligochetes, mosquito larvas, small crustaceans, and fishes. Another method has been to culture and distribute eggs of other trematodes (harmless to humans), particularly those with predaceous redial stages, which consume or outcompete the schistosome stages within the snail, while also damaging the snail population.

Making the environment unsuitable for snails (by draining ditches or increasing flow rates) is highly effective where practical. Unfortunately, development of irrigation for agriculture almost always favors host snails and increases schistosomiasis. This has long been a problem in Egypt, and since the building of the Aswan High Dam special efforts are being made to provide villagers with safe water for domestic use and to take other health measures. In the Gezira area of Sudan extensive irrigation has brought high rates of schistosomiasis. Control measures there include the introduction of mudfish that eat snails and of Chinese grass-carp that eat grass on which snails feed.

CESTODES

CESTODES, or tapeworms, are the dominant members of the unpopular class **Cestoidea,** which is entirely parasitic. The adults are almost all intestinal parasites of vertebrates, and few vertebrates are free of them. The young stages live in various tissues (but not the digestive tract) of invertebrates, especially arthropods, and also of vertebrates. The most conspicuous feature of all members of the class is that there is *no mouth or digestive cavity* at any stage of development. Another feature of most members is a body divided into *repeated units,* each with one or more complete reproductive systems. This reminds us of the chains of zooids produced by some turbellarians, and also of segmentation, a kind of organization that appears in a different form in annelids, arthropods, and some other phyla.

The class Cestoidea, like the class Trematoda, is usually divided into one minor subclass with members thought to resemble the ancestors of the group and one major subclass containing the rest. In the small subclass Cestodaria (see the end of this chapter), the body is not divided. This is true also of some members of the subclass Cestoda, but most have long bodies regularly marked by divisions, flattened and ribbonlike, and are well named the tapeworms.

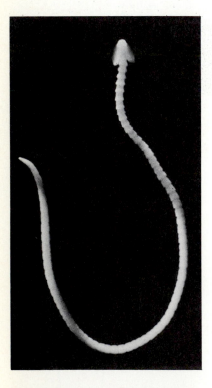

Tapeworm has a flattened, ribbonlike body divided into repeated units. Young specimen of *Parabothriocephalus sagitticeps* from a rockfish. (L.A. Jensen)

TAPEWORMS, subclass **Cestoda,** are typically whitish or yellowish. Of some 3,500 species, the smallest is only a millimeter long and the largest, in sperm whales, over 30 m. The adult holds fast to the intestinal wall of the final host with suckers, hooks, or other adhesive organs on the attached end, or **scolex.** Behind the scolex the body is divided into **sections,** or proglottids, each separated from the adjacent one by a transverse constriction and usually by a membranous partition. The sections are marked internally by regularly repeated features of the various systems, such as interruptions in some of the musculature and cross-connections between the main longitudinal nerve cords that run back through the worm from a small brain in the scolex. The nervous system is less well developed than in many free-living flatworms or in the more active parasitic groups, but similar in form. There are no eyes or other complex sensory organs, but the surface is supplied with sensory endings, especially numerous in the scolex and around the reproductive openings. The excretory system consists of pairs of longitudinal canals that receive fluid from branching tubules ending in flame-bulbs; in some tapeworms, cross-connections or pores opening to the outside occur in each section. Most conspicuous is the complete male and female reproductive system in each section; there may even be multiple sets of both male and female reproductive organs in each section. New sections are budded off from a neck region just behind the scolex, so that the anterior sections are generally smaller and less mature than those farther back.

The terms "segment" and "proglottid" are usually used interchangeably to refer to the *external divisions* of tapeworms, which in this book we call "sections." But in various tapeworms each section may contain from 1 to 14 sets of reproductive organs; in others several sections may enclose a single set; and in still others, no sections are present to reflect internal segmentation in the sense of single or paired repeated sets of organs along the body. "Segment" is usually restricted to this sense in the rest of biological literature, while "proglottid" is likewise restricted by some tapeworm specialists, who use "segment" to refer to an external division of a tapeworm.

In most bilaterally symmetrical animals, the end that goes first and that bears the mouth and sense organs such as eyes is considered to be the anterior end, whereas the end that trails along and bears the anus is the posterior. In tapeworms, which as adults hardly go anywhere and which have no digestive tract or complex sense organs, it is not obvious which end is which. Even the presence of nerve ganglia and concentration of sensory cells in the scolex does not settle the question, as concentrations of nerve cells sometimes occur at both ends of an animal. The development and behavior of young stages are most often cited as evidence, but different authorities reach opposite conclusions. Most consider the scolex to be the anterior end, but admit that the various criteria for distinguishing between dorsal and ventral surfaces are arbitrary.

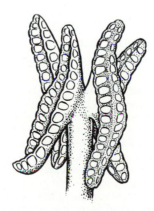

Scolexes may have only suckers and a crown of thorns, or shallow sucking grooves, or large leaflike sucking grooves combined with suckers. The various orders of cestodes have different types; shown here are two examples from the order Tetraphyllidea.

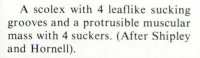

A scolex with 4 leaflike sucking grooves and a protrusible muscular mass with 4 suckers. (After Shipley and Hornell).

A scolex in which the leaflike structures are divided up by ridges into multiple little sucking grooves. (After E. Linton).

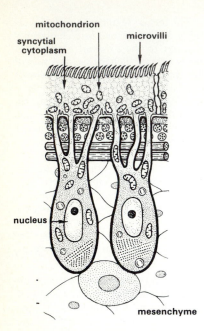

Tegument of a tapeworm, *Dipylidium caninum,* common in dogs and cats. The tegument of all 3 parasitic classes of flatworms has the same organization, differing only in details. The regular arrangement and specialized tips of the microvilli are tapeworm characters. (Modified after L.T. Threadgold)

As there is no mouth or digestive tract, tapeworms must obtain all their food through the **tegument** that covers the body. Surrounded as they are in the host's intestine by a handy supply of already digested food, they might seem to have no problems of nutrition. But every life style has its difficulties, and tapeworms face their share. First, they must compete for the available nutrients against the host's highly specialized digestive epithelium, while at the same time resisting attack by host enzymes and immune defenses. Tapeworms meet this challenge aggressively with a tegument that is remarkably similar in many ways to the vertebrate intestinal lining; both have the surface increased many fold by projecting microvilli and are covered with an organic coat (glycocalyx) that binds enzymes and nutrients. Nutrients that have been pre-digested by host enzymes are taken up at the surface of the tegument by passive diffusion or, mostly, by active transport. In addition, the tegument surface adsorbs most starch-digesting enzymes (amylases) and increases their activity, so that breakdown of starches to sugars is accelerated close to the tapeworm surface, where the products are most readily available to the worm. The simple sugar glucose is the main source of energy for tapeworms. Glucose not immediately metabolized is converted to the starch glycogen, stored, and used during periods of host starvation; glycogen reserves may amount to more than 50% of the dry weight of the worm. Host bile salts, and enzymes such as those that break down proteins and fats, are bound to the glycocalyx or tegumental membrane and are inhibited, which presumably helps to protect the tegument from digestion by the host. At the same time, the tapeworm's own surface-bound enzymes (peptidases and lipases) attack protein derivatives and fats, providing the supply of amino acids and fatty acids that the worm needs, as its synthetic capabilities are relatively limited.

Little is known about how the tegument surface resists digestion and immune attack by the host. Inhibition of host digestive enzymes bound at the tegumental surface is a possibility already mentioned. There is also evidence that host immunoglobins are bound and inactivated. Tapeworms may coat themselves with host proteins, as schistosomes do, or the glycocalyx may be resistant to attack. Even with such defenses, the tegument probably suffers repeated damage but maintains itself by continual repair and renewal of the surface membrane and glycocalyx.

A tapeworm's defenses may work in one host but not in another with a more vigorous or more readily mobilized immune response. A rat will harbor a few specimens of the tapeworm *Hymenolepis diminuta* indefinitely and will mobilize an effective immune attack only if infected with many worms. But any worm of this species that finds its way into a mouse will succumb after only a week or so. The tegument of mature sections is apparently resistant to attack, but the surface is severely damaged in the neck region of the worm, where new tegument is being rapidly added, and the body drops away. The scolex is unable to produce new sections in the immunized host; but if transplanted surgically to a mouse that has not previously been infected, the scolex quickly grows a new body. Only after some days does the new mouse acquire sufficient immunity to reject the worm.

A second challenge is the low oxygen and high carbon dioxide content of the tapeworm's habitat. Tapeworms (and other intestinal parasites), although they can use oxygen, have a primarily anaerobic metabolism; even when oxygen is available, they do not metabolize glucose completely to carbon dioxide and water, as most animals do. On the contrary, they fix carbon dioxide—high levels of which are necessary for normal rates of glucose uptake and for glycogen synthesis in tapeworms—and excrete organic acids which are intermediate breakdown products of aerobic metabolism. From a given amount of glucose, the anaerobic metabolism of tapeworms yields less energy than aerobic metabolism but more than the anaerobic metabolism of normally aerobic animals. By developing an anaerobic pathway in which carbon dioxide is fixed, tapeworms derive extra energy and turn a potential drawback of the intestinal habitat to their advantage. The basis for such biochemical creativity was probably inherited from free-living ancestors in anoxic mud substrates.

Tapeworms are not notably active animals, and most of their food and energy goes into impressively rapid growth rates and into the vast numbers of eggs necessary to counter the odds against completing the life cycle. The hermaphroditic **reproductive system** lies embedded in the mesenchyme and so dominates mature sections that tapeworms are sometimes described as nothing but bags of reproductive organs. The male system often develops first. It consists of numerous small testes, scattered in the mesenchyme and connected by fine tubules to a large coiled sperm duct, the end of which is modified as a muscular copulatory organ. The sperm duct leads into the genital chamber, which opens to the outside through a genital pore. Also opening into the genital chamber is the vagina, a female duct which receives sperms and conducts them to the oviduct, where the eggs are fertilized. Cross-fertilization can take place between mature sections of different worms when two or more are present in the same host, though it is not known how they locate each other. Self-fertilization also occurs, either within a segment or, when an animal is folded back on itself, between segments of the same worm.

Self-fertilization is an extreme form of inbreeding, which is commonly assumed to be rapidly deleterious. However, in studies of 2 species of *Hymenolepis,* there was no evidence of ill effects in worms self-fertilized for several generations, compared to cross-fertilized controls. *H. microstoma* self-fertilized for 14 generations without any indication that it could not be maintained indefinitely in this way.

Self-fertilization and other forms of inbreeding have been shown to be deleterious in some organisms that normally practice outbreeding and have pairs of dissimilar genes for many hereditary characteristics. Such organisms tend to accumulate large numbers of deleterious recessive genes, which are usually masked by functional dominant genes and are therefore tolerated. With inbreeding, similar pairs of such recessive genes turn up more often, and their deleterious potential is then expressed. On the other hand, in organisms that regularly self-fertilize and therefore have mostly pairs of similar genes, deleterious genes are regularly expressed and eliminated by natural selection.

From the ovary, commonly in the form of 2 large lobes, each egg passes into the oviduct and is moved along, past the opening of the vagina, where it is fertilized, and past the opening of the yolk duct, where it is joined by one or more yolk cells from the yolk glands. Then egg and yolk cells enter a bulbous region of the oviduct called the ootype, which shapes the egg capsule around them. The ootype is surrounded by unicellular glands (Mehlis's glands) which contribute a thin membrane thought to facilitate capsule formation; the yolk cells release globules of capsule-material, which line the ootype walls and coalesce to form the capsule. Some tapeworms do not form a thick capsule, but the thin membrane contributed by Mehlis's glands surrounds other protective layers. In any case, the worm must process huge numbers of eggs each day through the ootype, one by one, before they go on to accumulate in the large uterus. This "bottleneck" may be the key to understanding the selective pressures that have favored thousands of replicated reproductive systems over a single larger one.

Separate male and female individuals occur in a few species of several different orders of tapeworms. Presumably, they evolved from hermaphroditic ancestors, as in the schistosomes among trematodes.

Monogeneans and digeneans also form their eggs one by one through an ootype. However, monogeneans produce small numbers of eggs, and even digeneans produce relatively few eggs (compared to tapeworms), relying instead on extensive asexual reproduction by young stages.

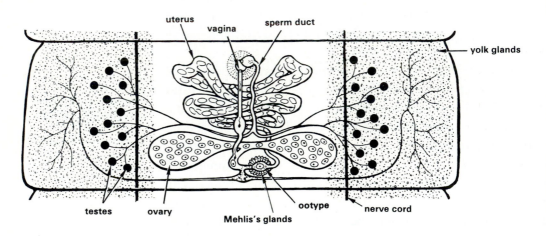

Section of a tapeworm, *Diphyllobothrium latum (=Dibothriocephalus latus),* showing the hermaphroditic reproductive system. (Combined from various sources)

A COMMON TYPE of tapeworm **life history** is one in which a crustacean and then a fish serve as intermediate hosts, with a fish or some other vertebrate being the final host of the adult. Transfer between hosts is entirely passive; each stage must be eaten by the next host. The **broad fish tapeworm,** *Diphyllobothrium latum* (order Pseudophyllidea), a large tapeworm of humans, illustrates such a life history. The adult, which reaches 10 to nearly 20 m in length, attaches to the intestinal wall of the final host with 2 adhesive grooves (bothria) on the scolex. Behind the scolex stretch 3,000 to 4,000 sections, of which many mature simultaneously, and the worm may shed a million eggs a day. The eggs leave the uterus through a uterine pore, their thick capsules tanned to protect them against digestion by host enzymes, and pass out with the host's feces. From time to time a strip of depleted sections detaches and is found in the feces.

Each egg capsule looks much like that of a trematode and contains a developing embryo and numerous yolk cells. It is sensitive to dessication, but if it reaches water, the embryo develops into a **6-hooked larva,** or oncosphere. On hatching, as a ciliated stage called a coracidium, the larva swims about until eaten by a suitable copepod (minute crustacean), the first intermediate host.

Inside the copepod's gut, the larva sheds its ciliated outer layer and uses its hooks to penetrate through the gut wall into the copepod's body cavity. There the larva metamorphoses

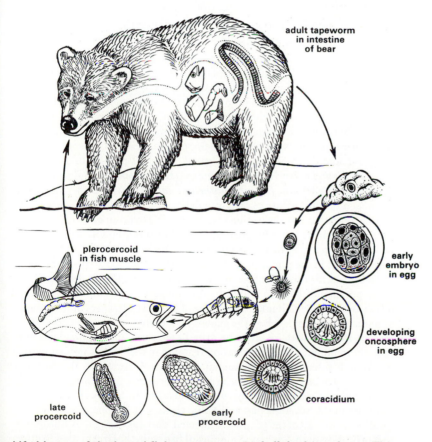

Life history of the broad fish tapeworm, *Diphyllobothrium latum.* The adult of this species occurs in a variety of fish-eating mammals, including humans. Pikes and other carnivorous freshwater fishes are intermediate hosts. Microscopic stages from early embryo in egg to procercoid are shown in circles. (Modified after various sources)

into a juvenile stage, called a **metacestode.** Metacestodes take many successive and different forms in various tapeworms, but in the broad fish tapeworm, the first metacestode stage in the copepod host is a tiny elongate worm (procercoid) with 6 hooks in a posterior disk; it resembles a monogenean. The metacestode absorbs nutrients from the copepod host and grows, but develops no further until the copepod is eaten by a fish and the worm is released into the fish's gut. Like the 6-hooked oncosphere, the metacestode promptly burrows through the gut wall into the tissues of the fish and there grows and develops into the next metacestode stage (plerocercoid); the hooked posterior end has been lost, and a scolex develops at the anterior end. If a small fish host is eaten by a larger fish, as often happens, the plerocercoid just re-establishes itself in the tissues of the new host; and it may pass through several levels of the food chain in this way until the fish host is eaten by any one of a variety of suitable mammals, including humans. In the intestine of the final host, the worm attaches by the scolex to the wall and grows into a mature tapeworm, with the body divided into sections.

Humans and other final hosts are infected with *Diphyllobothrium latum* by eating raw, undercooked, or inadequately smoked freshwater fish; ocean fish may transmit other species of *Diphyllobothrium.* An infected person may not even suspect the tapeworm is present until sections in the feces are noticed, but some people suffer mild symptoms similar to those associated with other tapeworms. Abdominal pain, nausea, weakness, loss of appetite, dizziness, and diarrhea may result partly from the physical bulk of these large worms and partly from their toxic excretory products. Moreover, a few people infected with *Diphyllobothrium latum* develop megaloblastic anemia, a serious disease caused by a deficiency of vitamin B_{12}, essential to normal digestion and absorption of nutrients and to production of red blood cells. The parasite takes up great quantities of this vitamin, which normal people can spare; but those who are subnormal in ability to absorb the vitamin may, in the presence of the tapeworm, absorb virtually none.

Treatment for this and other tapeworms has been difficult, because most drugs powerful enough to kill the worms are also hard on the host; an ideal drug would be one that takes advantage of known differences in the biology of parasite and host, such as the anaerobic metabolism of tapeworms. This may be the case with an effective drug now in use, niclosamide, which is thought to inhibit a phosphorylation reaction in the worm's mitochondria, but is well tolerated by the human host. After treatment, the worm is promptly expelled. Untreated, a large tapeworm may live as long as its host.

Diphyllobothrium latum occurs worldwide. It has been known for centuries in the Scandinavian and Baltic regions of Europe, where in some localities nearly all of the people are infected, and was brought by Europeans to North and South America. Shipping of fish from the Great Lakes to other parts of the U.S. and Canada has spread this parasite.

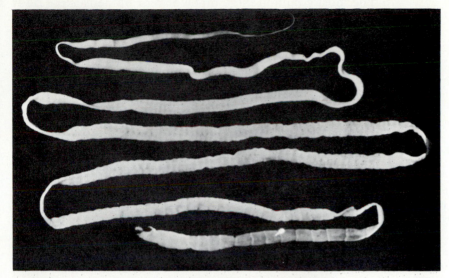

Diphyllobothrium from a cat.

Removal of tapeworm from a child is not a big event in some parts of the world. (L. Braithwaite)

MOST LIFE HISTORIES of tapeworms do not include a free-swimming coracidium; instead they have a non-ciliated 6-hooked larva that hatches in the digestive tract of the first intermediate host. Many tapeworms depend on host enzymes to digest the outer, untanned egg coverings and release the larva. Such eggs must, however, be protected from digestion in the intestine of the parent's host and cannot be freely shed into the intestinal lumen to pass out with the feces. Instead, the eggs are retained within the parent's uterus, which is a closed sac, and entire sections are shed from the posterior end of the worm. This pattern is seen in the large tapeworm order Cyclophyllidea that includes most of the important parasites of humans and domestic animals. In this group, there is usually only one intermediate host—a crustacean, sometimes a mollusc or annelid worm (if the final host is aquatic) and an insect, other terrestrial arthropod, or vertebrate (if the final host is not associated with water).

Scolex of the beef tapeworm, *Taenia saginata* (or *Taeniarhynchus saginatus),* order Cyclophyllidea. Humans are the only known final hosts.

Adults of the beef tapeworm often cause no symptoms in humans. But sometimes there is diarrhea, loss of weight, nervous symptoms, or rarely perforation of the intestinal wall.

The most common large tapeworm of humans is the **beef tapeworm,** *Taenia saginata,* which has 4 suckers on the knoblike scolex. The body, often 10 m long, consists of up to 2,000 sections that are progressively larger and more mature from the neck backwards. The male reproductive system matures first, then the female, then both regress; and fully mature sections at the posterior end of the worm devote most of their available space to the branched uterus full of developing embryos, about 80,000 per section. Each day an average of about 7 to 9 sections detach from the worm, one by one, and pass out with the feces or even crawl out the anus under their own power (causing a distinctive sensation that can only be imagined by those of us who have not experienced it). Once outside the human host the embryos are released through ruptures in the section. The egg capsule, although more prone to digestion than tanned ones, provides better protection against dessication.

The main intermediate host of the beef tapeworm is a domestic bovine of the genus *Bos* (usually referred to as a cow, although most beef cattle are male). The egg capsules may be carried to the cow on the hands of an infected person or by flies. Or the cow may acquire the embryos when it ingests vegetation, feed, or water contaminated with human feces.

1. Scolex of tapeworm *(Taenia pisiformis)* from a dog, showing suckers, hooks, and young sections. Diameter of scolex, 1.3 mm.

2. Immature section with male organs developed. Female organs are beginning to appear.

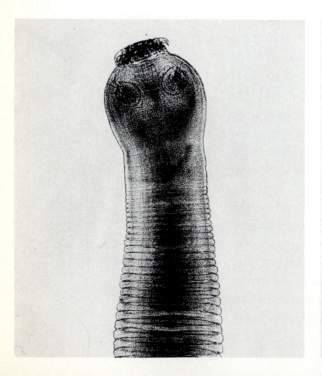

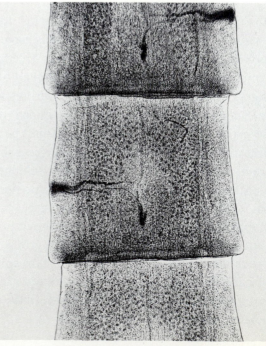

Freed of its protective coverings by the combined action of gastric and intestinal enzymes and bile, the larva bores through the intestinal wall of the cow with its hooks and penetration glands, enters blood or lymph vessels, and finally encysts in muscles or other tissues. There, over 2 to 3 months, it grows into a metacestode (called a cysticercus or bladderworm) in the form of a fluid-filled, ovoid bladder, from the inner wall of which the inverted scolex and neck develop. When a person eats raw or rare meat containing the bladders, gastric and intestinal digestive fluids cause the metacestode to evert the scolex. The worm attaches to the intestinal wall and matures to the adult stage.

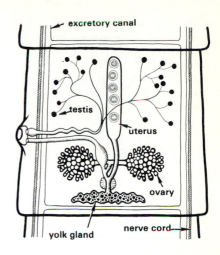

Section of beef tapeworm, *Taenia saginata.*

The bladders of the beef tapeworm occur most frequently in the muscles of the jaw and of the heart; these are the parts of the cow usually examined by meat inspectors. In the United States, meat inspection has greatly reduced the occurrence of this once common parasite; but not all meat is federally inspected and studies have shown that about one out of four infected cattle goes undetected. The bladders are about a centimeter long but can readily be overlooked. It is safest to avoid eating beef that has not been cooked thoroughly (to a temperature of at least 56°C throughout) or frozen at -10°C for 10 days. In many parts of Africa, Asia, and South America where sanitation is poor and beef may be prepared by roasting big pieces over an open fire, a large proportion of the human population is infected. In India, the Muslim population has a relatively high incidence of beef tapeworm, whereas Hindus, who follow religious restrictions against beef, are free of infection.

3. Mature section showing both male and female sex organs developed.

4. Ripe section is not much more than a sac containing the enlarged uterus filled with eggs. Width, 5 mm. (A.C. Lonert)

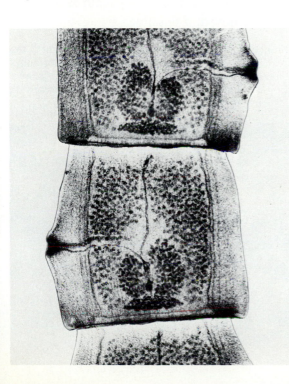

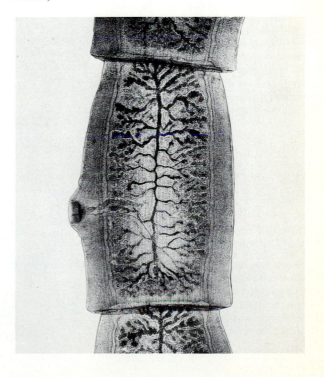

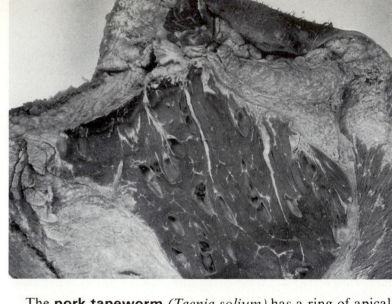

Measly beef, or beef containing bladders of the beef tapeworm, is now rarely found on the market, thanks to government meat inspection. Meat that is cooked until it has lost its red color is safe. (U.S. Army Medical Museum)

Pork tapeworm is rare among Muslims and Jews, who avoid the meat of the hog, and common in Christians in parts of Europe, Africa, the Middle East, and Mexico where pork is commonly eaten without thorough cooking.

The **pork tapeworm** *(Taenia solium)* has a ring of apical spines on the scolex in addition to 4 suckers, but otherwise resembles the beef tapeworm closely and has a similar life history, except that pigs usually serve as the intermediate host. This species is more dangerous, however, because the *metacestode bladders can also develop in humans.* A person may accidentally swallow the eggs, or eggs from an adult tapeworm may be released and hatch in the same host's digestive tract. If the metacestodes settle in muscles, no great harm results. Sometimes, however, they lodge and grow in the eyeball, interfering with vision. Metacestodes in the brain may result in convulsive seizures or other neurological disturbances and must be be surgically removed, if possible.

Bladder worms in brain of a 34-year-old woman, who was brought to the hospital with a history of seizures. These became more frequent until 3 days before her death. when convulsions set in every half-hour. Her brain (shown in longitudinal section) contains 100 to 150 bladders. (Specimen in Pathology Museum, University of Chicago)

6-hooked larva.

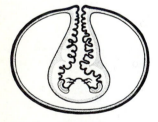

Inverted scolex of metacestode (cysticercus) in bladder.

Everted scolex of adult with hooks and 4 suckers.

Pork tapeworm, *Taenia solium.* In this and other members of the family Taeniidae, the developing scolex is inverted; most of the medically important tapeworms are taeniids. (Modified after various sources)

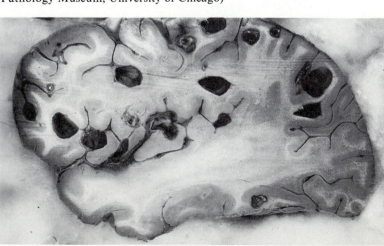

Even more lethal are the metacestodes of a related species, *Echinococcus granulosus,* a minute tapeworm that lives its adult life in dogs or other canines, causing little trouble. Humans become infected by drinking contaminated water or by allowing dogs to lick the face and hands. Because dogs clean the anal area by licking, the tongue of an infected dog is likely to carry tapeworm eggs. If a human or other potential intermediate host swallows the eggs, the larva that hatches migrates into the tissues, especially the liver, and develops into a metacestode in the form of a huge fluid-filled bladder. From the inner walls of the bladder grow smaller bladders, and within each of these are produced numerous scolexes. The whole structure is known as a hydatid cyst and may grow to the size of an orange or even a watermelon. This is one of the few examples of asexual reproduction in tapeworms (unless the budding of sections is considered as such). From the point of view of the tapeworm, development of the cysts in humans is unfortunate because human flesh is rarely eaten by dogs or other final hosts. The most frequent intermediate hosts are sheep, and the parasite is common wherever great herds of these and other domestic ungulates are raised, in North and South America, Africa, the Middle East, Central Asia, Australia, and New Zealand.

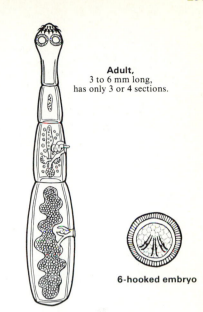

Adult,
3 to 6 mm long,
has only 3 or 4 sections.

6-hooked embryo

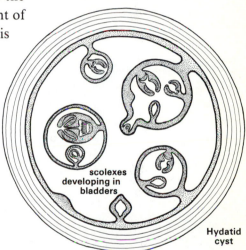

scolexes
developing in
bladders

Hydatid
cyst

Echinococcus granulosus. The adult is a midget among taeniids, but the metacestode stage (hydatid cyst) is huge. (Modified after various sources)

Echinococcus **cysts** in section of the liver of a person who accidentally swallowed the mature sections or eggs. The adult parasites live in the intestine of dogs. Actual size of cysts, 1 to 5 cm in diameter. (Specimen in Pathology Museum, University of Chicago)

The presence of hydatid cysts and other bladderworms is diagnosed by immunological tests, and they may be located by ultrasound and, if calcified, by X-rays. Treatment is by surgical removal but rupture of a large hydatid cyst, spontaneously or during surgery, releases great quantities of protein-aceous fluid that causes anaphylactic shock and often death; small bladders may also be released and dispersed in the body, to continue growth at new sites. To prevent this, the fluid is carefully withdrawn and a solution such as one containing silver nitrate is injected to kill the contents before removal of the cyst is attempted.

Other tapeworms which live as adults in non-human hosts may also occur as metacestodes in humans. Plerocercoid metacestodes of several species of *Diphyllobothrium* and the related genus *Spirometra* can be acquired by swallowing infected copepods in drinking water and by eating uncooked flesh of marine fishes, frogs, snakes, or other vertebrates that contain plerocercoids. The elongate metacestodes of these species cannot mature in a human but proceed as in any other intermediate host in the food chain; they penetrate the gut, enter various tissues, and continue as plerocercoids, growing to lengths of some centimeters. The oriental practices of eating raw snake meat to cure various ills, and applying a split frog to wounds or inflammations, seem as if designed to introduce plerocercoids into the human patient and do have this effect.

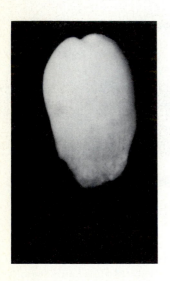

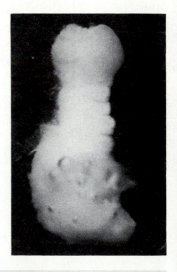

Bladderworms of beef tapeworm, *Taenia saginata,* normally excyst in the intestine of a human who has eaten raw or undercooked infected ("measly") beef. To learn about the factors that induce excystment, the bladderworms (shown here) were maintained in a glass tube, at 37°C, and treated with hydrochloric acid and pepsin to simulate conditions in the human stomach; then they were exposed to the enzyme trypsin (secreted into the intestine by the pancreas). *Left,* a bladderworm just beginning to evert. *Right,* the scolex and several sections are already everted. (R.B.)

Bladderworms in rat liver are the encapsulated metacestodes (cysticercus stage) of the tapeworm *Taenia taeniaeformis.* The adults live in the intestine of domestic and wild cats, doing little harm to their feline final hosts. Rats or mice swallow the eggs, and the larvas that hatch reach the liver and encyst there. This massive infestation was produced in a laboratory study. A natural case of rat cysticercosis would involve fewer bladderworms but could, nevertheless, cause serious damage and start cancerlike growth. (R.E. Kuntz)

MANY OTHER life history patterns occur among tape-worms. For example, the dwarf tapeworm, *Hymenolepis nana,* does not require an intermediate host. Fleas, grain beetles, and other insects sometimes serve as intermediate hosts, and the final hosts (rodents and humans) may be infected by swallowing these insects. However, infection more often results from directly swallowing the eggs (in contam-inated food). An egg swallowed by a human or rodent hatches, and the larva penetrates into the intestinal epithelium to develop into a tailed metacestode (cysticercoid), which then re-enters the digestive cavity to mature to the adult stage. This cycle may be repeated by eggs released within the intestine of an infected final host, leading to heavy infestations by thousands of worms. This species, which is perhaps the most common tapeworm of humans around the world, is found mostly in children, rarely in human adults, presumably because humans, like mice, are finally able to develop immune resistance to it.

Members of the genus *Archigetes* are likewise unusual in being able to mature from egg to adult in a single host, but in this case the host is a freshwater oligochete worm, and the tapeworm lives in the body cavity. These tapeworms are the only cestoideans that do not require a vertebrate host, although the adult stages are sometimes also found in fishes. For many other tapeworms, the full life histories are not known, and the biology of even the most common and best studied species of these parasites continues to offer questions and surprises.

In *Archigetes* and other caryophyllidean cestodes, the scolex is simple, the reproductive system is single, and the body is not divided into sections. Although usually considered to be progenic (precociously reproductive), these worms, and the spathebothriidean tapeworms, which have a series of reproductive systems but no external division into sections, suggest what primitive tapeworms may have been like.

In the small subclass **Cestodaria,** also, the body is not divided into repeated units and the anterior end lacks the distinctive haptors that characterize most tapeworms. The larvas bear resemblances to those of monogeneans, flukes, and tapeworms. Called lycophores, they are usually ciliated but shed the ciliated outer layer inside the host. Prominent groups of penetration glands open anteriorly, and the posterior end is armed with 10 hooks, whereas tapeworm larvas have 6. Adult cestodarians live in fishes and turtles. The subclass appears to offer glimpses of links between tapeworms and other flatworms, especially rhabdocoel turbellarians and monogeneans. But not all specialists agree that the two orders of cestodarians are directly related to tapeworms or even to each other.

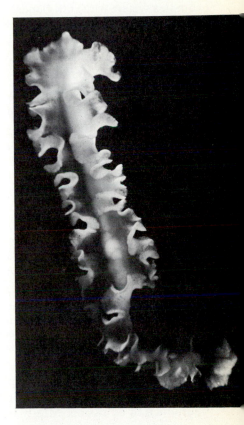

Cestodarian, *Gyrocotyle fimbriata,* lives in chimaerid fishes, *Chimaera* in the Atlantic and *Hydrolagus* in U.S. Pacific coast waters. (L.A. Jensen)

CLASSIFICATION: Phylum PLATYHELMINTHES

Class TURBELLARIA, mostly free-living forms, some commensal or parasitic. Great majority are marine, some freshwater or terrestrial. Almost all hermaphroditic.

Formerly separated according to the form of the gut, the orders are now separated more by reproductive characters. The number and arrangement of orders are in flux, and identification even to order is often a matter for specialists. Worms having a simple female reproductive system, yolk stored within the egg, and spiral cleavage of the egg (see chapter 16) are termed *archoophoran;* these characters are considered primitive. Worms having a complex system with yolk glands separate from the ovaries, yolk in accessory cells surrounding the egg, and modified cleavage patterns are termed *neo-ophoran;* these characters may have evolved independently in different orders.

Order Acoela. No digestive cavity; mouth or simple pharynx opens into solid mass of cells. No excretory system. Statocysts and sometimes eyes; 3-6 pairs of nerve cords. Archoophoran, no discrete gonads. Almost all marine and a few millimeters long, living among seaweeds or on soft bottoms from shallow to deep water; a few pelagic, or symbiotic with other invertebrates. *Polychoerus, Childia. Convoluta* and *Amphiscolops* with algal symbionts. *Oligochoerus,* freshwater. *Nemertoderma* and *Meara* with digestive cavity, sometimes placed in separate order.

The following 3 orders are often referred to as **rhabdocoels,** turbellarians with a straight, unbranched digestive cavity. Unbranched excretory system. With eyes and often ciliated sensory pits, rarely statocysts; usually one pair of ventral nerve cords. Mostly small, less than a few millimeters long.

Order Catenulida. Simple pharynx. Single median protonephridium. Archoophoran. Asexual reproduction by fission. Mostly freshwater, some marine. *Catenula, Stenostomum.*

Order Macrostomida. Simple pharynx. Paired protonephridia. Archoophoran. Asexual reproduction by fission. Freshwater or marine. *Macrostomum, Microstomum. Haplopharynx* with anterior protrusible adhesive organ (proboscis) and posterior anal opening; sometimes placed in separate order.

Order Neorhabdocoela. Bulbous pharynx. Paired protonephridia. Neoophoran; no asexual reproduction.

Suborder Dalyellioida. *Syndisyrinx, Syndesmis, Pterastericola,* commensal in echinoderms. *Fecampia* and *Kronborgia,* parasitic in marine crustaceans. *Dalyellia,* free-living in freshwater.

Suborder Typhloplanoida. Mostly freshwater, some terrestrial. *Mesostoma.*

Suborder Kalyptorhynchia. With anterior proboscis used in prey capture. Mostly marine, in shallow sandy bottoms. *Gyratrix.*

Suborder Temnocephalida. Ectocommensal on freshwater crustaceans, snails, and turtles. Cling to host with adhesive tentacles and disks. *Temnocephala, Scutariella.*

The following 3 orders are often referred to as **alloeocoels,** turbellarians with unbranched digestive cavity, straight or with many outpocketings. Pharynx variable. Pair of protonephridia, often branched. With or without eyes, ciliated sensory pits, and statocysts; usually 3-4 pairs of nerve cords. Neoophoran. Medium-sized, often one to several centimeters long.

Order Lecithoepitheliata. Simple or bulbous pharynx. Common in fresh and brackish waters, some terrestrial. *Prorhynchus.*

Order Prolecithophora, or Holocoela. Bulbous or plicate (folded, protrusible) pharynx. Marine or freshwater. *Proporoplana (= Plicastoma).*

Order Seriata, or Proseriata. Plicate pharynx. Mostly marine, some freshwater. *Monocelis.*

Order Tricladida, the planarians. Highly diverticulated, 3-branched digestive cavity. Plicate pharynx. Excretory system, a complex network with many pores. Eyes and other sensory structures, but no statocysts; usually 1-3 pairs of nerve cords. Neoophoran. Asexual reproduction by fission. Large, a few mm to 50 cm long. Marine: *Bdelloura,* commensal on horseshoe-crabs. *Procerodes.* Freshwater: *Planaria, Dugesia, Polycelis, Cura, Phagocata, Bdellocephala, Dendrocoelum, Dendrocoelopsis* (cave planarian). Terrestrial: *Bipalium, Geoplana.*

Order Polycladida. Highly diverticulated, many-branched digestve cavity. Plicate pharynx. Excretory system, a radiating network. Usually with multiple eyes and sensory tentacles, no statocysts; many radiating nerve cords. Archoophoran. Large, a few mm to 15 cm long, mostly broad and thin. Mostly marine and free-living on shallow bottoms, a few pelagic, a few commensal; rarely freshwater. *Stylochus, Leptoplana, Notoplana, Hoploplana, Thysanozoon, Pseudoceros, Prosthecéraeus.*

All members of the following 3 classes of flatworms are parasitic and, except in larval stages, are covered with a syncytial tegument.

Class MONOGENEA. With 1 host in life history, usually a fish, sometimes a frog, turtle, hippopotamus, squid, copepod. Mostly ectoparasitic. Free-swimming larva with hooks (oncomiracidium).

Subclass Monopisthocotylea, with 4 orders. Simple, hooked posterior haptor. Feed mostly on epidermal cells and mucus. *Gyrodactylus.*

Subclass Polyopisthocotylea, with 8 orders. Complex posterior haptor with hooks and multiple suckers. Feed mostly on blood. *Polystoma, Diplozoon.*

Class TREMATODA

Subclass Aspidogastrea (Aspidocotylea, Aspidobothria). Large divided sucker or row of suckers along ventral surface. Usually with 1 host in life history, a clam or snail; if 2 hosts, final host is a fish or turtle. Mostly endoparasites. Free-swimming larva without hooks (cotylocidium). *Aspidogaster, Cotylaspis, Stichocotyle.*

Subclass Digenea, flukes. Usually with 2 or 3 hosts in life history; first intermediate host is mollusc, final host is vertebrate. Endoparasites, mostly in digestive tract or associated organs, schistosomes in blood. Free-swimming larva without hooks (miracidium). Asexual reproduction by young stages in first intermediate host, usually followed by free-swimming cercarial stage. Adult mostly with 2 suckers.

SUPERORDER ANEPITHELIOCYSTIDIA
Order Strigeata. *Schistosoma,* blood flukes.
Order Echinostomata. *Fasciola,* liver flukes.

SUPERORDER EPITHELIOCYSTIDIA
Order Plagiorchiata. *Paragonimus,* lung flukes.
Order Opisthorchiata. *Clonorchis,* liver flukes.

Class CESTOIDEA. Endoparasites without mouth or digestive cavity. Usually with 2 or more hosts, adult in vertebrate intestine.

Subclass Cestodaria. Body not divided into sections, one set of reproductive organs, no scolex. Free-swimming larva with 10 hooks (lycophore).

Order Gyrocotylidea. In intestine of chimaeras (primitive cartilaginous fishes). Sometimes grouped with monogeneans. *Gyrocotyle.*

Order Amphilinidea. Young in crustacean intermediate hosts, adults mostly in bony fishes and turtles. *Amphilina.*

Subclass Cestoda, tapeworms. Body usually divided into sections, many sets of reproductive organs, scolex. Larva with 6 hooks (oncosphere), sometimes free-swimming (coracidium).

There are about a dozen orders of tapeworms, most of them small and restricted to either cartilaginous or bony fishes as final hosts. They have various reproductive arrangements and diverse adhesive organs on the scolex. All except two, order Caryophyllidea (in annelids and bony fishes) and order Spathebothriidea (in bony fishes), have the body divided into sections. Two major orders include species for which diverse vertebrates, including humans, are final hosts:

Order Pseudophyllidea. Life history usually with coracidium and oncosphere larva, encysted metacestode stages in 2 intermediate hosts (crustacean and fish), all 5 classes of vertebrates as final hosts. Includes largest tapeworms. Scolex with bothria; eggs are released before spent sections are shed. *Diphyllobothrium, Spirometra.*

Order Cyclophyllidea. Life history usually with oncosphere larva, encysted metacestode stages in 1 intermediate host (invertebrate or vertebrate), vertebrates except fishes as final hosts. Most tapeworms of birds and mammals are in this order. Scolex mostly with hooks and 4 suckers; sections containing eggs separate individually. *Taenia, Hymenolepis, Echinococcus.*

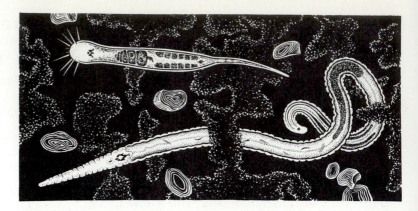

Chapter 10

Gnathostomulids

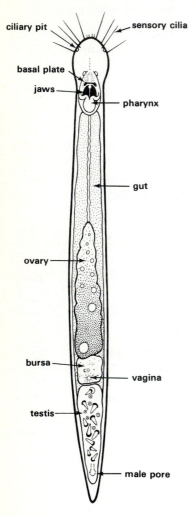

Diagram of a gnathostomulid, *Austrognatharia kirsteueri,* in dorsal view, showing the **pharynx** with **jaws** and **basal plate.** Long stiff **sensory cilia** and **ciliary pits** are borne on the head; the cilia that cover the entire body surface are not shown. The single **ovary** contains a series of eggs, progressively more mature posteriorly. Behind it a dorsal female pore, the **vagina,** opens into the **bursa,** where sperms are stored. The **testis** contains nonflagellated cone-shaped bodies, which may be giant sperms or sperm packets (spermatophores). The **male pore** opens ventrally. (After W. Sterrer)

[Labels on diagram: ciliary pit, sensory cilia, basal plate, jaws, pharynx, gut, ovary, bursa, vagina, testis, male pore]

THE PHYLUM **GNATHOSTOMULIDA** is one of only a few phyla whose members have escaped detection until the 20th century. And no wonder. They live on marine shores in fine-grained sand, but are most abundant, probably more so than any other metazoans, as one goes deeper into the sediment to the black anaerobic layer which even zoologists tend to avoid because its odor of hydrogen sulfide suggests that it will harbor little besides bacteria of decay. In collecting gnathostomulids, a sample of sand including the black layer is placed in a bucket with just enough seawater to cover the surface, then left to go from bad to worse. Only after some days or even months, when the odor of hydrogen sulfide has grown stronger and black spots have appeared at the sand surface, do the gnathostomulids emerge. In addition to being hard to coax from this improbable habitat, these elusive worms are semi-transparent and microscopic, ranging from 0.2 to 3 mm in length. Even if they are seen, moving among the sand grains and particles of debris, they may be passed over as minute turbellarians.

Gnathostomulids resemble turbellarians in being ciliated and elongate little hermaphroditic worms without an anus and without a body cavity. They were initially included in the Turbellaria, but as their distinctive features became recognized, they were removed first to a separate new class of flatworms and then to a new phylum. Their phylogenetic relationships remain uncertain. Fewer than 100 species have been described but many more are known, and the rate at which new ones are discovered in new locations suggests an eventual total close to 1,000.

the **bursa,** where sperms are stored. The **testis** contains nonflagellated cone-shaped bodies, which may be giant sperms or sperm packets (spermatophores). The **male pore** opens ventrally. (After W. Sterrer)

Austrognatharia is thought to belong to the most advanced group of gnathostomulids. Other kinds differ in lacking a permanent bursa and female pore and in having paired testes and sometimes a copulatory stylet. Two of these kinds are represented in the chapter heading: the long, slender, primitive *Haplognathia* and the stouter, tailed *Gnathostomula.*

A unique character of gnathostomulids is the **feeding apparatus** found in most species. It consists of a little pair of hard, toothed *jaws* and a central *basal plate*. With these, gnathostomulids scrape cyanobacteria ("blue-green algae"), diatoms, fungal filaments, bacteria, and other organic matter from the surfaces of sand grains. The epidermis is one-cell-layer thick and is **monociliated,** each epidermal cell bearing a single long cilium. **Sensory structures** on the head may include a halo of stiff sensory cilia, single or compound, and ciliary pits.

Gnathostomulids are further distinguished from flatworms by their scant mesenchyme and striated muscles, by a different type of protonephridia (with single flagella instead of with tufts of many flagella), and by their sperm structure.

When flagellated, gnathostomulid sperms have only a single flagellum with the usual 9 + 2 arrangement of microtubules. The flagella or flagellar axonemes of flatworm sperms almost always occur in pairs and have different microtubular patterns.

Although occasionally found on algal fronds or sea grasses, gnathostomulids are predominantly **interstitial,** living in the interstices of sand grains. Besides their small size, gnathostomulids share many characters with other interstitial animals. Elongate shape enables them to move through the channels among the sand grains, and the habit of adhering strongly to the surface of the grains prevents their being washed away. Below the surface layer of sand, there is little or no light, and eyes are often absent in interstitial animals. Finding a mate presents special problems in this habitat, and hermaphrodites have the advantage that any encounter with another sexually mature individual of the same species can result in copulation. In addition, gnathostomulids and other interstitial animals are often gregarious. Such tiny animals can produce only a few gametes. Gnathostomulids lay yolky eggs, one at a time, and direct development of the young, without any swimming larval stage, increases their chances of remaining in a suitable habitat.

The distribution of both young and adult gnathostomulids, in or near the organically-rich anaerobic black layer of sandy sediments, suggests that their respiration is largely or wholly anaerobic throughout the life history. They have not been kept alive in the laboratory, perhaps partly because they are exposed to too much oxygen; but when methods are discovered for keeping and observing them, there will be much to be learned about the biology of this group.

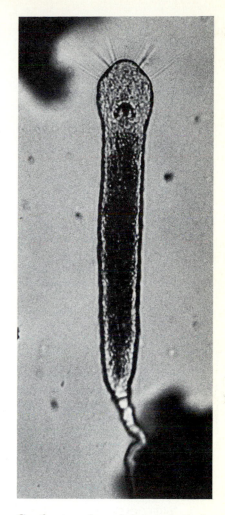

Gnathostomula axi from shallow water sediment at Bimini, Bahamas. (R.B.)

Gnathostomulids belong to the diverse assemblage of tiny animals referred to as **meiofauna,** those having dimensions roughly on the order of 1 mm. Many phyla of mostly larger animals have a few miniature meiofaunal members, but gnathostomulids fall entirely into this domain. Meiofaunal research has been actively pursued only during the last few decades and encompasses not only interstitial habitats but any others that harbor tiny animals.

Chapter 11

Nemerteans

*Digestive tract
with anus.
Circulatory system*

The name "bootlace worms" is sometimes given to very long, narrow kinds of nemerteans and the name "ribbon worms" to broad, flat ones. Sometimes these common names are applied generally to nemerteans (also called nemertines), but "proboscis worms" is more apt. Several versions of the phylum name are also current: Nemertea, Nemertinea, Nemertina, Nemertini, and Rhynchocoela. Most of these are derived from the name of a mythical Greek sea nymph, Nemertes.

The name Rhynchocoela refers to the cavity in which the proboscis lies. Any such body cavity lined by mesoderm is termed a *coelom* (see chapter 16). Nemerteans are nevertheless usually grouped with non-coelomate (acoelomate) animals, because their coelom is not a major body cavity (or widely thought to be reduced from a major body cavity—though this remains a possibility).

NEMERTEANS are elongate, either cylindrical or flattened worms. Nearly all of them are marine, but a few are terrestrial or in freshwaters. They are much like flatworms and would almost certainly lie submerged and unnoticed among the numerous groups of turbellarians had they not undergone some bold and innovative remodeling of the flatworm body plan along several lines. The distinctive and—literally—salient feature of all members of the phylum **NEMERTEA** (about 900 species) is a large **proboscis** that lies in a fluid-filled cavity above the gut. The proboscis can be everted through an opening at the anterior end of the body and is used in prey capture, defense, and locomotion. When a nemertean locates a prey organism, the proboscis is forcefully everted and quickly wrapped around the victim. The surface of the proboscis contains many gland cells, and the adhesive secretions of these cells, which rapidly entangle and secure the prey, are sometimes aided by paralyzing neurotoxins, also produced in the proboscis. In some nemerteans the proboscis is armed with one or more sharp, calcareous stylets that pierce the prey and help to introduce toxic secretions.

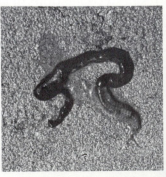

1. The dark-colored nemertean, *Paranemertes peregrina,* shoots out its yellow proboscis.

2. The proboscis coils around the segmented worm *(Platynereis bicanaliculata).*

3. The nemertean ingests its pre

Nemertean attacking a segmented worm. U.S. Pacific coast. (S.A. Stricker)

260

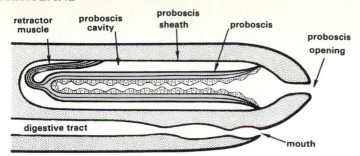

retractor muscle   proboscis cavity   proboscis sheath   proboscis   proboscis opening

digestive tract   mouth

Proboscis of a nemertean lies inside a muscular sheath just above the digestive tract and is firmly attached to the sheath at both ends—anteriorly, around its circumference; posteriorly, by a retractor muscle. Separate proboscis and mouth openings are present in most nemerteans that lack a stylet (orders Paleonemertea and Heteronemertea).

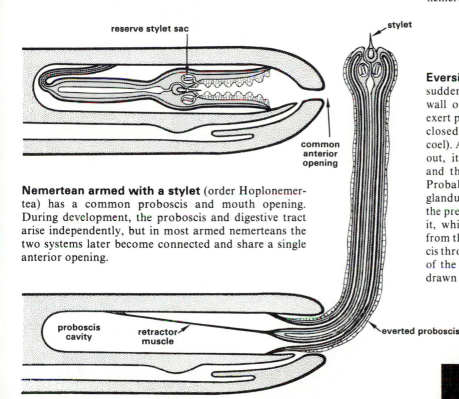

reserve stylet sac

common anterior opening

stylet

proboscis cavity   retractor muscle

everted proboscis

Nemertean armed with a stylet (order Hoplonemertea) has a common proboscis and mouth opening. During development, the proboscis and digestive tract arise independently, but in most armed nemerteans the two systems later become connected and share a single anterior opening.

Eversion of the proboscis is by sudden contraction of muscles in the wall of the proboscis sheath. These exert pressure through the fluid in the closed proboscis cavity (rhynchocoel). As the proboscis is turned inside out, its glandular lining is exposed and the stylet is brought to its tip. Probably, toxic secretions from the glandular epithelium are applied to the prey as the proboscis wraps about it, while additional toxin flows out from the posterior part of the proboscis through a narrow canal to one side of the stylet. The proboscis is withdrawn by the retractor muscle.

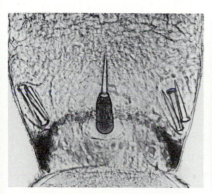

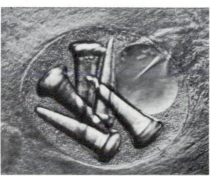

Stylet with its head embedded in the basis, a granular body that holds it in place. Flanking the central stylet are 2 reserve stylet sacs, each containing several stylets available to replace the central one if it is lost. *Tetrastemma.* Photomicrograph. (S.A. Stricker)

Reserve stylets are secreted in vacuoles within a single large cell in the center of the stylet sac. The shaft of the calcareous stylet forms around an organic core, visible here within a vacuole. *Zygonemertes virescens.* Nomarski optics. (S.A. Stricker)

Helical grooves mark this stout reserve stylet of *Paranemertes sanjuanensis.* Length, about 100μm. SEM. (S.A. Stricker)

The prey is usually swallowed whole through the anterior mouth (which may be separate from or combined with the proboscis opening, in different nemerteans). The **digestive tract** of nemerteans shows a major difference from the gastrovascular cavity of flatworms: it has 2 openings instead of one. In addition to the mouth, there is a posterior **anus** through which undigested residues are eliminated. This has the advantage that newly ingested food is not mixed in with partly digested food as it is in flatworms, and the digestive tract can become differentiated for sequential digestion and absorption along its length.

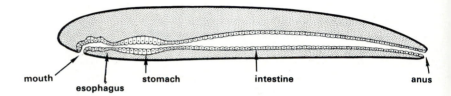

mouth   esophagus   stomach   intestine   anus

Digestive system of a nemertean.

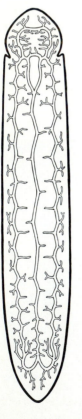

Circulatory system of a nemertean. (After O. Riepen)

In most nemerteans, the digestive tract has many lateral outpocketings, which are often long and branched. It is thus a gastrovascular system as in flatworms, serving both to process and to distribute nutrients. However, nemerteans have in addition a separate **circulatory system.** It consists of 2 lateral blood vessels running alongside the gut, usually a dorsal vessel, and sometimes an extensive system of smaller connecting vessels and sinuses. There is no heart and no constant direction of blood flow, but the contractile walls of the larger vessels help to circulate the blood. The functions of this simple vascular system are assumed to be similar to those of more complex ones in other animals: transport of nutrients, oxygen, metabolic wastes, hormones, and other substances from one part of the body to another. Evidence for such functions is inferred partly from the proliferation of blood vessels or sinuses that accompany or surround the gut and other structures. Extensive vascularization is generally associated with the **excretory system,** which consists of ducts leading from flame bulbs (protonephridia), much as in flatworms.

One kind of evidence for the repiratory function of the circulatory system is that the blood contains cells which sometimes bear yellow, green, orange, or red pigments, among them hemoglobin, an oxygen-binding pigment common in many animal groups. (Hemoglobin occurs even in a few protozoans and flatworms, which have no circulatory system.) A few relatively large, bulky nemerteans flush water in and out of the foregut, which is especially well vascularized and serves for respiratory exchange.

Most nemerteans are small or slender enough to require no specialized respiratory surfaces or structures. Sufficient respiratory exchange occurs through the general body surface. Nevertheless, it is the circulatory system, carrying oxygen from one end to the other, that permits these long worms to lie with most of the body safely buried in sediment or tucked away in stagnant crevices.

Contractile blood vessel, in cross-section, shows a lining epithelium and circular muscles in the walls. It has been suggested, on the basis of the lining epithelium and the arrangement of blood vessels, that they may be coelomic derivatives. (Modified after O. Riepen)

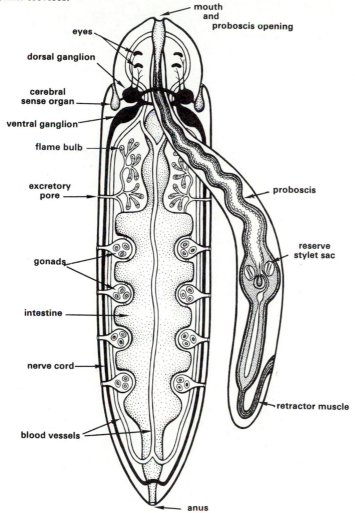

Diagram of a nemertean, in dorsal view, with the proboscis and its sheath drawn to one side. (Combined after various sources)

Nemerteans are not impressive hunters, possessing only relatively simple nervous and sensory equipment. The central **nervous system** is much like that of flatworms, with an anterior brain consisting of dorsal and ventral ganglia broadly connected around the proboscis apparatus. Two large lateral or ventral nerve cords extend from the ventral ganglia and run, with cross-connections, the length of the worm. In addition, there is a dorsal nerve in some nemerteans. The small, simple eyes also resemble those of flatworms. Cerebral organs that open on each side of the head through a ciliated canal are thought to be chemosensory (and may also have a neuro-endocrine function). Although some nemerteans can detect

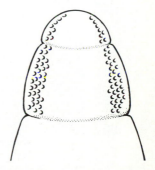

Multiple eyes distributed along each side of the anterior end of a hetero-nemertean. The number of eyes varies in different nemertean species from 2 to over 250. Closely set clusters of many eyes are usually associated with the detection of movement (as in the compound eyes of arthropods, see chapter 20). (After W.R. Coe)

General body structure of a nemertean shown in cross-section. (Based on various sources)

prey at a distance with chemoreceptors and move toward it, most seem to get along by blundering onto their food. Yet, once contact is made, nemerteans are formidable, and a soft-bodied and harmless-looking "proboscis worm" can quickly subdue and engulf an animal larger than itself. Besides feeding on live prey, including a variety of invertebrates and even fishes, nemerteans are sometimes scavengers.

The statement that a nemertean can swallow an animal larger than itself must be qualified because it is impossible to say just how large a nemertean is. These remarkably flexible worms can change in length by a factor of more than 5 times, their maximum extension being limited by crossed helixes of unstretchable fibers in the body wall. The longest nemertean on record is a specimen of *Lineus longissimus* that was estimated to measure 30 m at full extension. When the circular muscles of the body wall contract, the body elongates; when the longitudinal muscles contract, the body shortens. Local, alternating contractions of these body wall muscles, passing along the body, appear as waves of thinning and thickening. Such *peristaltic contractions* of the body wall serve to mix and move food in the digestive tract, which has no comparable muscle layers of its own, and to assist the flow of blood in the circulatory system. Peristalsis is also the basis of a kind of crawling that is the most common means of nemertean locomotion.

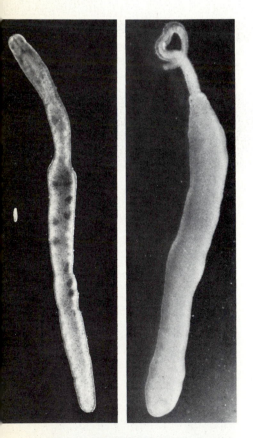

Freshwater nemertean, *Prostoma* (order Hoplonemertea). Members of this genus are small, reaching about 2 cm in length; they are found on aquatic vegetation or bottom debris in well-oxygenated ponds, lakes, and streams. Pennsylvania. (R.B.)

Young nemerteans, and adults when small, may move by gliding on mucus with the cilia of the epidermis, as flatworms do. As mentioned, the proboscis may also be used in locomotion. It is extended and attached to some surface, the body is drawn up to it, and the proboscis is released and extended again. Nemerteans burrow in much the same way, thrusting the proboscis into the substrate and anchoring it. Some nemerteans can swim, with dorsoventral muscles producing an undulating motion.

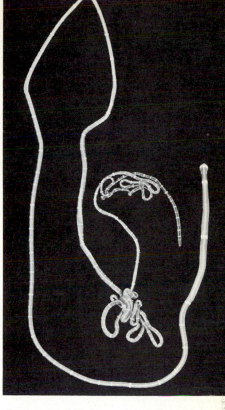

Uniform color can be conspicuous when it is bright red, vermilion, or orange-yellow, as in *Tubulanus polymorphus* (order Paleonemertea). This large nemertean, 3 m long when fully extended, hides under boulders in the low intertidal but comes out to feed. Females produce great numbers of eggs, which are easily teased out of fragments of the body and used in studies of fertilization and development. Oregon. (R.B.)

Marine nemertean *Amphiporus bimaculatus* is among those in which the proboscis is armed with one or more stylets (order Hoplonemertea). Up to 15 cm long. U.S. West Coast. (R.B.)

Patterns of lines or spots often decorate bright red, orange, yellow, green, or purple nemerteans, presumably making them conspicuous to predators. Combined with the presence of toxic substances in the tissues of the body wall, conspicuous colors and patterns may be a deterrent to a predator that has tried to eat such a nemertean and learned not to try again. *Tubulanus annulatus* (order Paleonemertea) is one of the few intertidal nemerteans known to feed during daylight hours when the tide is out and stranded prey is available. Drab nemerteans hide by day and emerge to feed only in darkness and when submerged. England. (D.P. Wilson)

Bootlace worm, *Lineus longissimus* (order Heteronemertea), has been seen to swallow a large segmented tubeworm. The nemertean pushed its head into the tube, stretched its mouth over the head of the tubeworm, and slowly drew the prey into its thin but very distensible body. England. (D.P. Wilson)

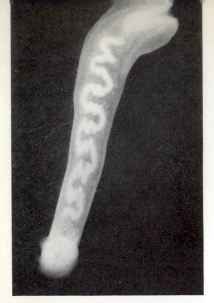

Parasitic nemertean, *Carcinonemertes,* shown on the egg mass of a crab. These parasites live on crabs of many species and feed on their developing eggs. They are one possible factor in the fluctuating abundance of the Dungeness crab, *Cancer magister,* an important commercial species on the U.S. West Coast. Shown here, among the crab eggs, are a nemertean, an egg-string, and a mucilaginous sheath in which eggs are laid and larvas hatch. Order Hoplonemertea. (R.B.)

Commensal nemertean *Malacobdella grossa* lives in many species of bivalves on American and European coasts, clinging to tissues in the mantle cavity with a posterior sucker. It filters plankton by means of ciliated papillas in the foregut. Larger plankters are caught by the proboscis. Order Bdellonemertea. (R.B.)

For worms as long as nemerteans, which inevitably break from time to time, **regeneration** is a useful capability. Like flatworms, most nemerteans have great powers of regeneration, and in fact some regularly fragment and regenerate as a means of **asexual reproduction.** Unlike flatworms, nemerteans have a rather simple **reproductive system,** usually with separate male and female individuals. A row of gonads lies along each side of the body, and the gametes are spawned to the outside from openings that break through when the gonads are ripe. In a few nemerteans fertilization and part or all of embryonic development take place within the ovaries; but usually eggs and sperms are spawned into the water or into a secreted mucoid mass or sac. Nemerteans that are solitary at other times may aggregate at spawning. It has been shown for some species that spawning is coordinated by substances (pheromones) which some individuals release into the seawater and which stimulate other individuals to spawn.

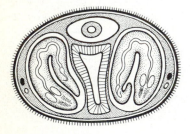

Internal development, the exception among nemerteans, occurs in a few aquatic species and in the terrestrial nemertean *Geonemertes agricola,* shown here. The cross-section reveals 2 young developing within the ovaries. (Modified after W.R. Coe)

Protective cyst around a regenerating fragment of *Lineus socialis.* Protective mucoid tubes or cysts are formed by many nemerteans. Some habitually live in such tubes; others secrete them only during egg-laying and brooding or during times of environmental stress. (After W.R. Coe)

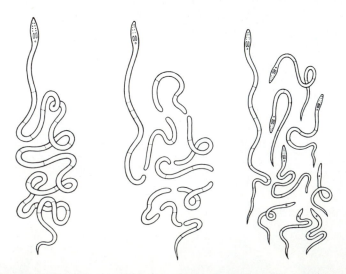

Fragmentation is a means of asexual reproduction among nemerteans in which a piece from any level can regenerate both anteriorly and posteriorly. In *Lineus socialis,* shown here, each of 6 to 10 or more pieces regenerates to produce a complete worm. (After W.R. Coe)

Embryonic development follows several different patterns among different nemertean species. A wormlike embryo may hatch and develop directly into an adult, or development may proceed through a free-swimming, feeding larva called a *pilidium,* or a nonfeeding larva, or a larva-like stage that never swims free. Metamorphosis of the pilidium involves a complex series of changes during which a new adult ectoderm is produced and the original outer epithelium is shed and often eaten.

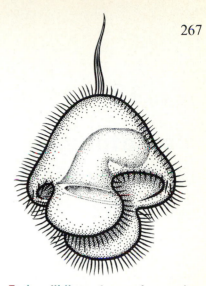

Early pilidium, larva of a marine nemertean, showing digestive tract with no anal opening. There is an apical sensory tuft. (Partly after C. Wilson)

THE MANY similarities to flatworms suggest that nemerteans arose from flatworm-like ancestors, and hints of nemertean characters can be found among living flatworms. Some rhabdocoels have an anterior protrusible proboscis, used in prey capture. Several flatworms (as well as some cnidarians, ctenophores, and gnathostomulids) have "anal pores," single or multiple openings of the digestive cavity to the outside, through which undigested residues are sometimes eliminated. However, nemerteans are among the simplest animals to have a regularly occurring, major, median exit to a one-way digestive tract. Such a device has presumably arisen many times, and the fact that most animals more complex than nemerteans have an anus, a coelomic cavity or cavities, and a circulatory system in no way implies that these structures are homologous in all or that nemerteans are either ancestral to or derived from these animals. No animals are clearly traceable from the nemertean line, although it has been proposed that nemertean ancestors gave rise to the vertebrates and other chordates (chapter 29), the chordate notochord (and ultimately the spinal column) being derived from the proboscis apparatus. Improbable as this might seem at first glance, the nemerteans merit a closer look.

Late pilidium with body of metamorphosing worm in center of the picture. (D.P. Wilson)

CLASSIFICATION: Phylum NEMERTEA

Class ANOPLA. Brain anterior to mouth, central nervous system within the body wall. Proboscis without stylets. Separate openings for mouth and proboscis.

Order Paleonemertea. *Tubulanus.*

Order Heteronemertea. Pilidium larva or other metamorphic development. *Lineus, Cerebratulus.*

Class ENOPLA. Brain posterior to mouth, central nervous system internal to body wall. Proboscis with or without stylets. Usually single anterior opening for digestive tract and proboscis.

Order Hoplonemertea. Proboscis with one or more stylets. *Amphiporus, Paranemertes. Carcinonemertes* (parasitic). *Geonemertes* (terrestrial). *Prostoma* (freshwater).

Order Bdellonemertea. Proboscis without stylets. Commensal in bivalve molluscs. *Malacobdella.*

Chapter 12

Nematodes
Nematomorphs
Acantho-
cephalans

Pseudocoelom

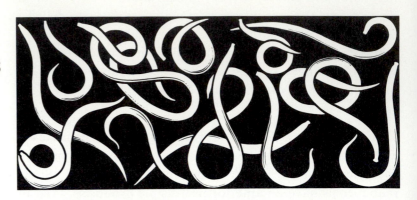

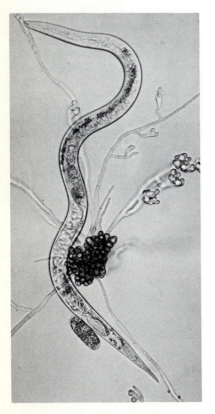

Free-living soil nematode, *Dity-lenchus myceliophagus,* about 0.7 mm long, is feeding on a fungal mycelium. Beside it is a single large egg, recently deposited. (C.C. Doncaster, Rothamsted Experimental Station, England)

ALMOST ANYTHING that one can say about roundworms as a group seems paradoxical. Though the phylum **NEMATODA** is now recognized as the second largest of all the animal phyla, and its members are more numerous than those of any other phylum of multicellular animals, many people have never seen nematodes. Even zoologists gave them scant attention until recent decades, and of an estimated half million species, only about 15,000 have been described and named so far. Though most nematodes are free-living and harmless to humans, mention of this group usually brings to mind the parasitic species.

Even the name Nematoda ("threadlike form") is somewhat misleading. The type of roundworm that is of even diameter throughout, and looks like a long white thread as it lies in loose coils among the muscles of vertebrates, is often seen by anyone who filets fish and by zoologists or their students who dissect vertebrates in the laboratory. But long, threadlike nematodes are a minority. Most are short and spindle-shaped, tapered at both ends. The smallest ones are less than a millimeter long; the largest known free-living ones, marine species, are only 5 cm long. The largest nematode known is a parasite from the placenta of a sperm whale; it measured almost 9 meters long and 2.5 cm wide.

These many discrepancies in our perception of nematodes are easily explained. Though the free-living kinds teem by the billions in all moist soils and in the bottom sediments of all bodies of water, marine or fresh, these are mostly microscopic or nearly so. They were not seen until after the invention of the microscope in the 17th century. Nor was anyone aware of the microscopic nematodes that parasitize almost all plants, invertebrates, and vertebrates, including humans.

The only roundworms that came to attention until recent centuries were the moderate-sized to large parasites of humans or of their livestock or pets. One could hardly overlook a pencil-sized worm *(Ascaris lumbricoides)* that emerged unexpectedly, pink and glistening, from one's nostril or from the opposite end of the body. Nor could the ancients remain unaware of a worm that revealed itself as a thin serpentine ridge, as long as one's arm, bulging under the skin. To remove such a "serpent" by winding it out, very slowly, onto a stick—without breaking the creature or causing infection—took practiced skill. And in biblical times the main business of surgeons in the Middle East was to remove the roundworm we now know as *Dracunculus medinensis* ("little serpent of Medina"). Such parasitic worms caused much human distress and often death. Early naturalists were more attracted to colorful butterflies and beetles, or to useful honeybees, and they considered disease-producing worms to be the concern only of medical practitioners. To this day a serpent wound around a staff continues to be the symbol of the medical profession. And though the treatment of parasitic worms in vertebrates is still the concern of physicians and veterinarians, the study of the biology of nematodes, both parasitic and free-living, has become the main interest of many biologists.

The biblical account of the "fiery serpents" that afflicted the Egyptians, and then the Israelites as they passed through the desert near the Red Sea, almost certainly refers to *Dracunculus medinensis*. The bronze serpent on a pole made by Moses, at God's command, was probably not a curative talisman, as implied in the biblical translations, but more likely served as an instructional device for showing how to remove the nematode by winding it on a stick.

Ascarids, *Ascaris lumbricoides,* are common parasites of the human intestine. Though large, they are not dangerous in small numbers. Heavy infections may involve hundreds or even thousands of worms.

Nematodes parasitize almost all animals and plants; and those that injure human health and food supplies continue to pose enormous problems. But we are gradually coming to

Marine nematodes (photos below and at top of facing page) display a diversity of shape and ornamentation. As in terrestrial soils, they are sometimes the most abundant metazoans in marine sediments, occurring at densities of up to 10 million individuals per square meter. Many live in associations with other animals or plants.

A **desmoscolecid,** *Desmoscolex,* with the bristles and annulate ornamentation characteristic of this order. Dredged from nearly 1,500 m deep in the North Sea. (G. Uhlig)

In dormancy and hibernation, familiar processes in animals, metabolic levels may drop to less than 1/100th normal. In dried juveniles of the stem and bulb nematode, *Ditylenchus dipsaci,* sensitive techniques of metabolic measurement, that could detect levels 1/10,000th of normal, reveal no signs of metabolism. *Ditylenchus* can survive drying for at least 23 years, a related nematode 39 years. Such capacity for returning to active life after rehydration takes on ominous possibilities when we consider nematodes parasitic on crops. Two other groups of microscopic metazoans, the rotifers and the tardigrades, are notable for their ability to lose almost all of the water content and survive. (See the discussion of **cryptobiosis** in the section on tardigrades in the next chapter.)

appreciate the benefits that nematodes can provide when we enlist their help in the biological control of insects or the study of fundamental biological problems. Because of small size, a limited number of cells, cell constancy, few chromosomes, and the ease with which certain microscopic species can be grown in large numbers in the laboratory on standardized media and under controlled conditions, nematodes have become important laboratory models for studies in genetics, in developmental biology, and in various aspects of metabolism. The biological processes involved in ageing, for example, are so fundamental to all living systems, that the results obtained with nematodes are believed to apply also to humans and to other less experimentally convenient organisms. Finally, the greatest contribution of nematodes to the human economy still remains to be assessed accurately by ecologists. As the most abundant and pervasive of multicellular animals that feed on the bacteria and fungi that grow in decaying plants and animals, nematodes play an important role in recycling the nutrients that keep biological cycles going.

As with species number, the **variety of habitats** occupied by nematodes is greater than that of all metazoan groups except the arthropods. Marine nematodes, readily distributed by water currents, are often represented by the same species from arctic to antarctic seas. They occur in the greatest densities in the highly organic muds near shores, and in shallow water they also live in sandy bottoms, on seaweeds or eelgrass, in the crevices of coral heads or barnacles, or in the gill chambers of crustaceans. They are found at great depths also, but being poor swimmers they are absent from open waters. Freshwater species are most abundant on the shores of lakes, but some are found even in the smallest running streams, transported there by wandering animals. Terrestrial nematodes are carried around the world by winds, by animals, and with plants in shipment. Many are cosmopolitan, occurring anywhere in the world that conditions provide enough moisture and food—some even in deserts or polar ice fields. Those that are highly tolerant of extremes of temperature, salinity, and pH, and that have stages resistant to drying, can survive in mud puddles that are cold and wet in winter, hot and dry in summer. One free-living species has been reported from moist soils in northern Europe, from mosses high in the Pamir mountains of central Asia, in the Baltic Sea, in saline waters of inland Germany, and in thermal springs of New Zealand.

Desmodorids from marine sediments. Two common families have a bulgy pharynx and "stilt bristles" on the ventral surface. The hollow bristles, supplied with adhesive glands, are attached to the substrate alternately with bristles on the head, in leechlike locomotion. *Above,* an epsilonematid, *Metepsilonema.* North Sea. (G. Uhlig). *Below,* a draconematid. Bimini. (R.B.)

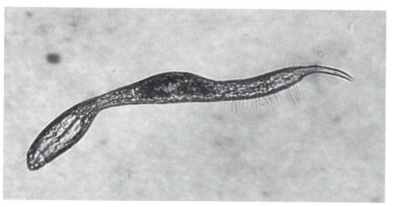

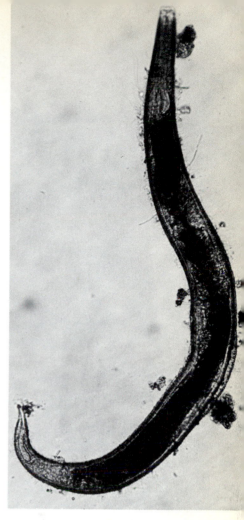

A **chromadorid** from Bimini. (R.B.)

Some species have very narrow niches. However, the one cited most often, *Panagrellus silusiae,* at first found only in the beer-soaked felt mats on which Germans set their beer mugs, has since been reported from natural habitats and is reared routinely in culture for laboratory studies. In the same family is *Turbatrix aceti,* the vinegar-eel, the first free-living nematode studied carefully by early microscopists.

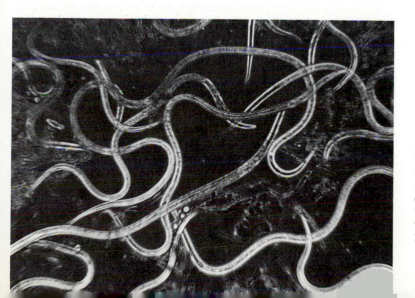

Vinegar-eels were formerly found in almost every household and restaurant in the organic material (mother of vinegar) at the bottom of vinegar containers. Transported there by tiny vinegar flies, the vinegar-eels tolerated the high acidity and fed on bacteria, yeasts, and molds. Commercial vinegars are now pasteurized, distilled, or filtered, and they lack nematodes; but vinegar-eels are still raised for classroom study, for various physiological experiments, and for testing of drugs used to kill the nematode parasites of vertebrates. *Turbatrix aceti,* about 2 mm long. (R.B.)

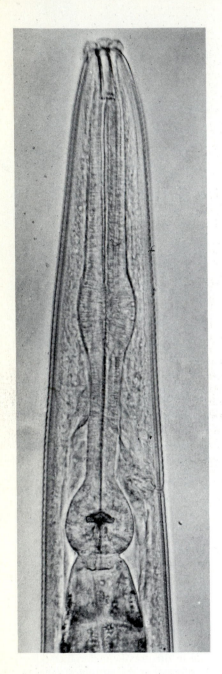

Pharynx of rhabditoid type in a nematode that feeds on bacteria and small organic particles, *Pelodera lambdiensis*. Diameter of pharynx, between the expanded bulbous portions, is about 15 μm. (C.C. Doncaster, Rothamsted Experimental Station)

The habitats of parasitic nematodes have been enormously expanded by human activities. Raising cattle or pigs in crowded quarters, or planting vast acreages to such crops as sugar beets or potatoes,have provided nematode parasites with convenient concentrations of food, where formerly their hosts were absent, or few in number and widely scattered.

The **feeding habits** of nematodes cannot be observed in natural settings. But when small free-living worms are brought into the laboratory and examined under the microscope they prove to be obliging beasts, almost as transparent as glass models. The feeding machinery of mouth cavity and sucking pharynx can be observed at work, often pumping food in so rapidly that it must be recorded by cinematography for later study, frame by frame. If the worms are examined promptly after they have been sieved from aquatic sediments or filtered out of soil mixed with water, the thin-walled intestine will reveal still undigested foods. According to habitat and mouth size, there may be recognizable bacteria, yeasts, fungal cells, algal cells, diatoms, organic bottom detritus, fragments of plants or animals, whole protozoans, and even whole metazoans such as rotifers, tardigrades, nematodes, and small earthworms. Sometimes these last are recognized only from small hard parts: the teeth or stylets of nematodes, the jaws of rotifers, or the bristles of earthworms.

Aquatic nematodes can be grown in nutrient fluid media and their nutritional needs determined; or they can be offered various live organisms. Soil nematodes may be raised on nutrient-agar seeded with bacteria. And when these worms multiply, predatory nematodes can be added to the dishes to prey on the bacterium-feeders.

By correlating direct observations of feeding, and careful examination of intestinal contents, with structural differences in mouth cavity armature and pharynx muscularity, nematologists have learned to make knowledgeable guesses about the feeding habit of almost any nematode from a known habitat. For example, a small mouth and a small tubular mouth cavity with a smooth cuticular lining, and 1 or 2 muscular enlargements of the pharyngeal wall, is associated with a feeder on bacteria and small soft particles either on muddy bottoms or in soils. In contrast, a large-mouthed worm with long teeth and abrasive denticles and/or sharp cutting plates, combined with a simple tubular pharynx of even diameter, is an active predator that swallows prey whole or tears it open and sucks up the contents. Plant-parasitic nematodes almost all have a piercing and sucking stylet for sucking out cell contents. But some predatory types also have a stylet; it injects digestive enzymes into the prey and then sucks out the pre-digested fluid content. Others have a solid stylet that can be driven into prey to make a hole through which fluid can be sucked up into the mouth. The nematode intestinal parasites of vertebrates have simple mouth cavities if they feed on intestinal content, but sharp teeth and cutting plates if they browse on intestinal lining.

NO ONE would want to be quoted as saying that if you've seen one nematode, you've seen them all. But it must be admitted that the external appearance and internal structure of these worms are remarkably similar for a group with such diverse habitats and feeding habits.

The elongate cylindrical body is not solid like that of a flatworm or nemertean, but contains a fluid-filled **body cavity** between the body wall and the tubular digestive tract. The body cavity plays a central role in the biology of nematodes. For one thing, it separates the distributive functions from the muscles, permitting the muscular body wall to contract strongly without the interference of multiple branches of gut, blood vessels, or excretory system as in a nemertean. Nutrients, oxygen, and wastes are dissolved and carried in the fluid of the cavity, which thus contains a complex mixture of dissolved organic materials, but no circulating cells. There are *no other circulatory or respiratory systems,* and the so-called excretory system probably functions primarily to maintain osmotic balance in the fluid of the body cavity. The fluid also serves importantly as a **hydrostatic skeleton** that is maintained under high pressure by the elastic and contractile elements of the body wall. In *Ascaris lumbricoides,* which is large enough to be convenient for many kinds of studies, the pressures measured were an order of magnitude higher than those on record for other invertebrates.

The term **pseudocoelom** is applied collectively to the body cavity of nematodes and to that of other invertebrate groups described in this and the next chapter.

The parasitic nematode *Ascaris lumbricoides* develops a hemoglobin that has an oxygen-binding capacity 2,500 times greater (in the body wall) and 10,000 times greater (in the coelomic fluid) than that of the hemoglobin of its pig host. Presumably *Ascaris* can thus obtain oxygen, even at the low oxygen tension in the intestine of the host.

The body cavity of nematodes is said to develop from the blastocoel, the cavity that first appears at the blastula stage, when the embryo is a hollow ball one-cell-layer thick. Any body cavity which develops in this way is called a pseudocoelom, in contrast with a coelom, a body cavity that develops within mesoderm. A pseudocoelom is not lined with the mesodermal epithelium that characterizes a typical coelom.

However, on the walls of the pseudocoelom of many nematodes, large mesodermal cells extend nucleated networks of membranes which cover the

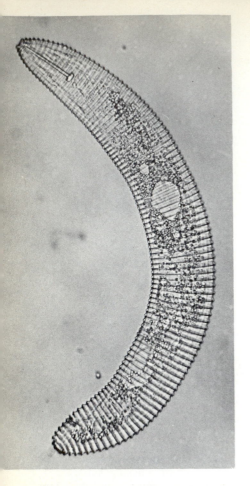

digestive tract and reproductive organs and which line the cavity. Body cavities of some other pseudocoelomate animals also have linings that are more substantial than those of some animals in well-established coelomate groups. Recent studies have revealed that the modes of development and of structure vary more within groups and overlap more between groups than had been believed. The significance of distinctions between the various kinds of body cavities is now in question.

The tapered body of a nematode is more blunt at the front end and usually quite pointed at the rear end. It is covered with a tough, proteinaceous **cuticle** consisting of several complex layers that include collagen fibers as a major component. The cuticle turns in and lines the digestive tract at both ends, often forming sharp teeth, a piercing stylet, or other feeding structures anteriorly and male mating equipment posteriorly; short portions of the excretory and reproductive ducts also have cuticular linings. The cuticle is molted 4 times during the life of most nematodes, although the first 1 or 2 molts occur within the egg shell in some species. Molting is not necessary to permit growth (as it is in arthropods, see chapter 20); juveniles grow between molts and growth often continues after the final molt to the adult stage. *Ascaris* grows from a length of about 2 cm to 40 cm or more following its final molt. However, the changes in cuticular structure and sense organs that occur at the molts may be important to different stages of the life history. For example, under unfavorable environmental conditions, juveniles of many nematodes enter a long-lived resistant stage (dauer stage) with a characteristic cuticle.

Ornamented cuticle with regular annulations helps this nematode secure a firm grip on the substrate as it moves. Note also the anterior stylet, which members of this genus use to pierce plant roots. *Criconemoides*, about 0.45 mm long. (C.C. Doncaster, Rothamsted Experimental Station)

Below left, **body wall** of a nematode consists of an outer collagenous **cuticle**, underlain by a cellular or syncytial **epidermis** (or **hypodermis, h**) and, beneath this, bands of **longitudinal muscles (m)**. In this longitudinal thin-section of a portion of the body wall, the cuticle shows several layers (vertical striations mark this as a dauer larva) and surface annulations separated by furrows that ring the body at regular intervals (about 1 μm) and provide added flexibility. *Below right*, cross section through the lateral line of a **nematode about to molt** shows the smooth juvenile cuticle, which will be shed, overlying the new adult cuticle, which has prominent longitudinal ridges. Transmission electron micrographs of *Caenorhabditis elegans*. (M. Kusch)

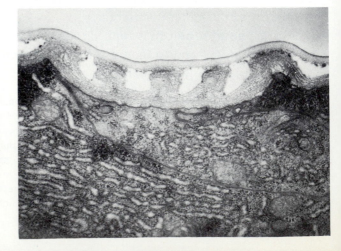

The cuticle is secreted by the underlying **epidermis** (also called hypodermis). The epidermis is usually cellular but, especially in large forms, may be a syncytium, with many nuclei in a continuous cytoplasmic layer undivided by individual cell membranes. The epidermis is thickened dorsally, ventrally, and laterally into **4 longitudinal ridges,** called cords, within which all of the nuclei are concentrated. *Ascaris,* with nuclei distributed in the epidermis, is exceptional. The large lateral ridges, and sometimes the smaller dorsal and ventral ones, may be visible externally as pale lines that run the length of the worm. Beneath the epidermis and between the epidermal ridges lie 4 bands of **longitudinal muscles,** which consist of elongate muscle cells that lie parallel to the long axis of the worm. Nematodes are peculiar in having no circular muscles in the body wall. The staggered arrangement of the thick and thin contractile filaments in the longitudinal muscles causes them to appear *obliquely striated.* Compared to the cross-striated muscles of many invertebrates and of vertebrates, the obliquely striated muscles of nematodes and of a number of other soft-bodied invertebrates are generally slower, but are capable of undergoing greater changes in length and of exerting greater stresses while expending less energy. These differences are related to the different arrangement, proportions, and numbers of the contractile filaments.

Locomotion depends on the opposing forces of the longitudinal muscles, elastic cuticle, and hydrostatic skeleton. The sinuous motion is eel-like (some nematodes are called "eelworms") but the range of movements is limited, partly by the lack of circular muscles and partly by the way that the nerves synchronize muscle contraction; both dorsal bands, or both ventral bands, always contract together, so that the body bends only in the dorsoventral plane. When the worms are free in the water their whiplike thrashing results in erratic locomotion, and on a flat surface the worms wriggle along on their sides. But when they are in soil, or aquatic sediment, or among the tissues or intestinal contents of a host, the solid particles provide surfaces against which to push, and the worms move along with speed and agility. In soil, movement is fastest when the nematode is 3 times as long as the diameter of the particles among which it moves.

Though most nematodes progress by whipping or by undulatory movements, some show other kinds of locomotion. One marine species, which curls its tail like a watch-spring, hops from one sand grain to the next on the tip of its tail. The tip adheres momentarily by means of one glandular secretion and then lets go by means of a second secretion.

Nematode moving in soil follows a sinusoidal track among the soil particles. Most soil nematodes are too small to displace many particles and are restricted to the channels that occur between particles. Observations of nematodes in graded soils show that locomotion is fastest when the channels are just slightly larger around than the worms are. Narrower channels act as barriers, for nematodes do not burrow well. Wider channels permit the worms to slip. (Modified after H.R. Wallace)

The straight **digestive tract** opens through a mouth at the anterior tip and an anus near the posterior. Against the continuous pressure of the fluid in the body cavity, the muscular cuticle-lined pharynx pumps food into the intestine, while a strong sphincter or valve prevents it from escaping at the rear. When the anus is opened by special muscles, the jet of intestinal contents from a large ascarid may shoot half a meter. The intestine is made up of a single layer of cells and typically lacks encircling muscles. Food is moved along within the intestine mostly by the ingestion of more food and by the general movements of the body. However, in some nematodes (oxyurids) a layer of muscle encircles the intestine and carries out peristaltic movements; in ascarids oblique muscles extend from the body wall to the intestine; and in other nematodes also, muscular elements may be associated with the intestine.

The digestive tract seems to play a major role in **excretion** of nitrogenous wastes; and the excretory system, as mentioned earlier, probably functions mainly in osmotic regulation. One or two lateral excretory canals, which run in or along the lateral epidermal ridges, join anteriorly to form a single duct that opens ventrally.

Triradiate lumen of the pharynx, seen in cross-section, is characteristic of nematodes and present in at least some members of about one-third of all phyla. Three nuclei can be seen, each corresponding to one sector of the pharynx. When the radial muscles of the pharynx relax, the lumen is closed (as here) by the elastic cuticle lining it. When the muscles contract, the cuticle stores energy, then releases it to open the lumen against the fluid pressure of the pseudocoelom. Analysis of mechanical models of 2-, 3-, and 4-rayed lumens suggest that the 3-rayed configuration places minimum demands on the muscles. Before its advantageous mechanical properties were appreciated, the triradiate configuration was thought to be a fortuitous structure unlikely to arise by convergent evolution and its presence in two groups was sometimes cited as evidence for an evolutionary relationship between them, an argument which can no longer be made with confidence. Transmission electron micrograph of a free-living bacterial feeder, *Caenorhabditis elegans*. Pharynx is about 12 μm in diameter. (M. Kusch)

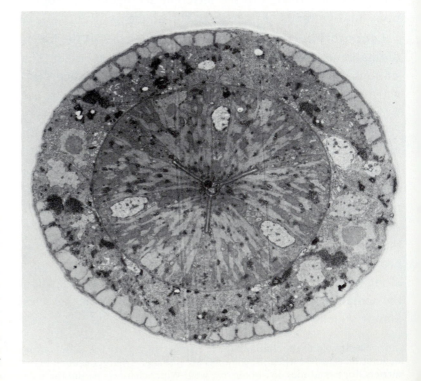

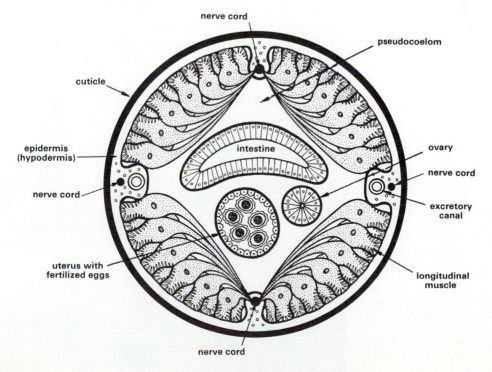

Diagram of a nematode, a male in lateral view. Near the posterior end, the cuticular spicules that aid in sperm transfer are shown protruding from the common (cloacal) opening of the sperm duct and digestive tract. In a female, the digestive tract is entirely separate from the reproductive system, which usually opens through a midventral slit in the central region of the body (see pinworm drawing). (Combined from various sources)

The **nervous system** consists of a nerve ring and ganglia, around the pharynx, from which nerves run both forward and backward. The nerve ring is the site of most sensory integration. There are usually 4 major nerve cords, running in or along the epidermal ridges and connected by crosswise commissures. The lateral nerve cords are sensory, the dorsal and ventral ones mainly motor. The small dorsal nerve cord consists mostly of processes from cell bodies located in the ventral nerve cord, which is larger and marked by ganglia. Nematodes are peculiar in that nerves do not branch out to the muscles; instead (see cross-section) long processes from the muscle cells extend to the midline and make contact with the dorsal and ventral nerve cords.

Cross-section of a nematode, a mature female.

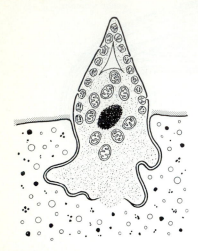

A variety of **sense organs** in the form of tactile bristles and papillas, and paired chemosensory pits, are concentrated especially at the anterior and posterior ends, but also around the genital openings. Structures that are probably simple photoreceptors have been found in some marine and freshwater nematodes, and some responses to light have been demonstrated, but structure and function have yet to be conclusively linked. Modified non-motile cilia in the mechano- and chemoreceptors are the only cilia found in nematodes, which have *no motile cilia or flagella,* even in their sperms (which are ameboid).

The **reproductive system** of a nematode lies coiled in the body cavity. In most species testes or ovaries occur in separate male or female individuals, which copulate; and fertilization is always internal. Mature nematodes, especially females, produce substances attractive to the opposite sex. The tubular gonads may be single or paired. In a male the sperms pass from the testis into a storage vesicle and later, at the time of copulation, into a long glandular sperm duct, which opens together with the digestive tract through a posterior common (cloacal) opening. Cuticular spicules, protruded from this opening during mating, aid in sperm transfer. In a female the eggs pass from the ovary through an oviduct into a uterus, which opens by way of a short vagina through a ventral slit, usually in the midregion of the body. The fertilized eggs sometimes begin development in the uterus, and may even hatch there, the young emerging from the parent as juvenile worms.

Ameboid sperm of *Ascaris lumbri-coides* entering the egg. The surface membranes of the two gametes will fuse and disappear, so that their cytoplasm becomes confluent; this has already happened in one region. Nematode sperms have no flagella but move within the female's uterus by means of pseudopods at the forward end. Nucleus and mitochondria occupy the central portion of the sperm. At the rear are a triangular refringent body (thought to contribute to ribosome synthesis) and numerous vesicles. (Based mostly on W.E. Foor)

Hookworms mating, *Necator americanus.* The male (8 mm) aligns himself perpendicular to the body of the female (11 mm) and attaches his posterior end over the female's vaginal opening, in the midregion of her body. The male's cuticular spicules probably serve to spread the opening; they have no duct through which sperms could pass but are solid. After sperm transfer, the worms separate. These adult hookworms were removed from the host's intestine, where mating occurs. (U.S. Army Medical Museum)

In a number of species, males make up a tiny fraction of the population or are unknown, and females produce offspring without mating, by parthenogenesis. Or, individuals that are morphologically female produce first sperms and then eggs in their gonads; these hermaphrodites may fertilize their own eggs or, if males are available, may mate and cross-fertilize the eggs. In some species, penetration by sperms of males or hermaphrodites is necessary for egg development, but the sperm nucleus never fuses with that of the egg, which develops parthenogenically. A mixture of reproductive processes seems to occur in some populations, and in certain parasitic species different modes are practiced by alternate generations. Most unusual are morphologically male individuals which produce gametes of both sexes; the eggs are fertilized and develop within the parent's reproductive system and the young worms that hatch must escape by penetrating the body wall.

That the **development** of nematodes is particularly convenient to study was as well recognized a hundred years ago as it is today. The transparent eggs and the small number of chromosomes made possible the first understanding of the cytological events of meiosis and fertilization. Not until 1883 was it shown, in an ascarid, that when the egg and sperm were joined, each contributed the same number of chromosomes to the resulting offspring. The pattern and number of resulting cells is relatively small and fixed, and by the time the young worm hatches, cell division is largely complete. Most growth is by cell enlargement. Therefore, it has been possible to trace the fate of individual cells as they differentiate to form the various organs of the worm. Ascarids and some other nematodes provide striking visible evidence of a basis for differentiation at the cellular level, as only those cells destined to become gametes (the germ line) retain the full complement of intact chromosomes, while the chromosomes of all other cells (the somatic line) proceed to lose material, presumably genetic material that is irrelevant to the progressively differentiating cell type.

Cell lineage studies have shown that a cell which is produced by a given series of divisions always develops in the same way. This rigid pattern, called *determinate development,* sharply differs from the more flexible situation of most other animals, which have *indeterminate development* with the differentiation of cells depending to a greater extent on their position within the embryo and the kinds of cells around them. Nematodes are especially suitable animals in which to investigate the effects of a local damage, or single genetic mutation, on a group of cells, and to try to determine the consequences for the development of structure and behavior of all stages of the worm. They are especially *un*suitable for studies of repair and regeneration, as they lack these capabilities. Asexual

Egg of *Ascaris lumbricoides,* with young worm developing inside, is the infective stage. It may survive dry and freezing conditions and remain alive for years. The egg covering is so resistant that brief exposure even to 10% formalin does not kill the embryo. The covering consists of 3 layers, including an outer protein coat secreted by the maternal uterus, a thick middle chitinous coat secreted by the egg, and a thin inner membrane. Longest diameter, 50 to 80 μm.

reproduction by processes such as fission or budding is, therefore, absent from this group, which has instead evolved a great variety of ways to reproduce by parthenogenesis and hermaphroditism with self-fertilization—modes of reproduction that are virtually asexual in the genetic sense of involving little or no genetic recombination.

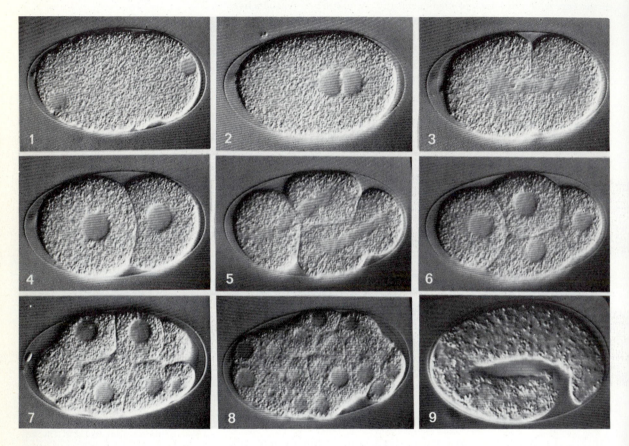

Development of *Caenorhabditis elegans*. In the study of the interactions between the various cells of the cleaving embryo researchers commonly take sequential photomicrographs during development. This is a sequence of an undisturbed embryo; but in some experiments, micromanipulators are used, for example, to destroy a particular cell at a particular stage and to determine how the death of the cell affects the subsequent development and the final form of the worm. (1) **Fertilized egg** with pronuclei at opposite ends. The symmetry of the embryo is already fixed at this stage: the female pronucleus is anterior, the male pronucleus is posterior. (2) **Pronuclei** about to fuse. (3) **Zygote** dividing. (4) **2 cells.** (5) **Two-cell stage** dividing. (6) **4 cells.** (7) **8 cells.** (8) **28 cells.** (9) **Young worm** before hatching. (E. Schierenberg)

Much recent research has been focused on a tiny free-living nematode species, *Caenorhabditis elegans*. Less than a millimeter long and easily reared on a diet of bacteria on agar plates, these hermaphroditic nematodes lend themselves to genetic analysis because they can be allowed to self-fertilize (quickly establishing a homogeneous line from a single mutant individual) or they can be cross-bred with males (for example, to test the effects of combining different mutant genes in the same individual). In addition, the total number of genes and chromosomes is small compared to most animals; of an estimated 2,000 genes, about 200 have already been identified and mapped on the 6 pairs of chromosomes (of which one pair normally determines sex: XX in hermaphrodites, XO in males).

Likewise, the total number of cells in the body is so small that their fates can be accounted for individually. It has been found, for example, that of 671 cells produced in the embryo, 113 are regularly destined to die before hatching, and 18 more die during juvenile stages. Studies of such "programmed cell death" have been viewed in the context of ageing in other animals, and these genetically-controlled events in early stages of nematodes broaden the perspective. In *C. elegans* about 350 of the cells of an adult belong to the nervous system, and since the number and arrangement are not much different in a giant nematode like *Ascaris lumbricoides,* the results of electrophysiological studies that are possible only on large worms can be applied with some confidence to the more detailed genetic and microstructural information available for the tiny ones.

Such comparative studies have indicated, for example, that acetylcholine (which serves as a chemical transmitter between nerve cells in many animals) is the transmitter at the junctions between excitatory nerves and muscles in these nematodes. Acetylcholine is also present in the epidermis, which may constantly release the transmitter directly onto the body wall muscles to maintain continuous contraction. Mutant strains of *C. elegans* which are deficient either in the enzyme that synthesizes acetylcholine (choline acetyltransferase) or in the enzyme that degrades it after release (acetylcholinesterase) cannot move normally. They are sluggish and uncoordinated or even paralyzed. Biochemical and genetic studies show that there is more than one form of the cholinesterase, controlled by 2 genes on 2 different chromosomes, and that locomotion is impaired only if both genes are defective. One of the mutants deficient in choline acetyltransferase is severely affected only as a young juvenile, and grows out of its uncoordinated behavior after the first molt, a time of active development and rearrangement within the nervous system. This kind of detailed, coordinated research holds promise for new insights into the genetic control of development.

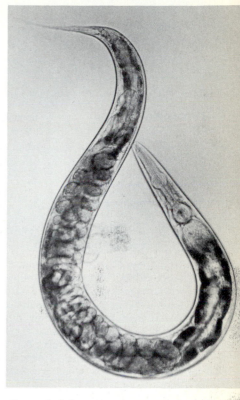

Caenorhabditis elegans. (R.B.)

THE IMPORTANCE OF **plant-parasitic nematodes** is not yet sufficiently appreciated, partly because they are small (most measure less than 1 mm) and less visibly destructive than the snails, slugs, and insects that come to mind as garden and agricultural pests. Only after the testing of the first pesticides effective against nematodes was the great impact of these worms revealed, as farmers saw yields increase by 20% to 200%. However, the usual problems of finding safe, effective nematocides have been encountered. Now it is known that many established agricultural practices such as crop rotation, fallowing, cultivation, addition of organic matter, and control of soil moisture and acidity are effective largely because they reduce populations of harmful nematodes. Planting a single crop year after year in the same soil allows nematode populations to build to devastating numbers, and rotation or fallowing is effective in preventing this only if the alternate crops do not support the target nematodes and if resistant stages of the nematodes do not persist. Added organic matter may reduce the worms by releasing organic acids toxic to nematodes as it breaks down in the soil, or by supporting growth of the many fungi that trap and digest nematodes.

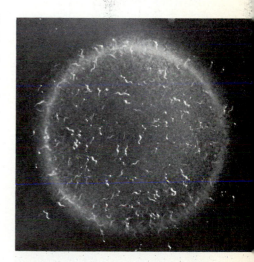

Laboratory culture of the nematode *Caenorhabditis elegans.* Feeding on a colony of bacteria reared on an agar plate, the small nematodes grow and multiply readily in small petri dishes. The availability of large numbers of animals of known breeding lines makes it possible to isolate uncommon mutants and to analyze their genetics and biochemistry. (R.B.)

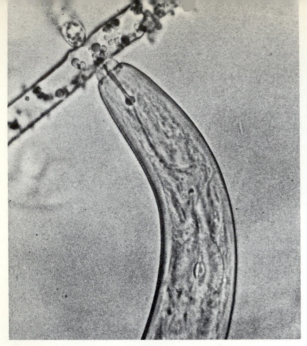

Cuticular stylet is used by all plant-parasitic nematodes to pierce cell walls. The stylet is hollow and through it the nematode ingests plant juices and cell contents or releases secretions. Some of these nematodes remain free and attack the plant from the outside; others enter the plant tissues. The worms feed on fungi, algas, mosses, and ferns as well as on vascular plants. This nematode *(Aphelenchoides blastophthorus,* order Aphelenchida) is piercing a hypha of a fungal mycelium. (C.C.Doncaster)

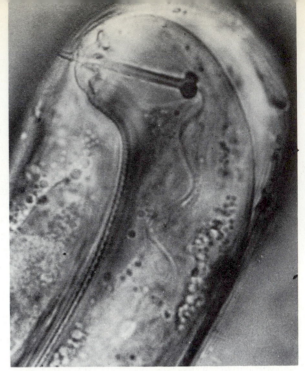

Cutting through the egg-shell with its stylet (about 0.014 mm long), this second-stage juvenile potato-cyst nematode *(Globodera rostochiensis)* has already molted once and is ready to seek out a host plant. It will molt 3 times more within the host tissue before maturing as a slender male or as a saclike female. (C.C. Doncaster)

Potato-cyst nematode, *Globodera* (or *Heterodera) rostochiensis,* is just one of many cyst-forming tylenchid nematodes, serious pests of many crops. The pale cysts are live female worms, their swollen bodies full of embryos, their head-ends still embedded in the root tissues and taking food from the plant. When the females die, their cuticle turns dark; the dark cysts drop off into the soil and the contained embryos remain viable for 10 years or more, ready to hatch when stimulated by material released from potato roots. Cysts, about 1 mm diam. (C.C. Doncaster)

Root-knot nematode, *Meloidogyne,* in a section of tomato root. The nematode secretes substances that induce the plant to form syncytial "giant cells" from which the worm feeds; several syncytia are visible here in the tissue in the core of the root. This mature female has become sedentary and her swollen body contains hundreds of eggs, eventually to be released into the soil. Root-knot nematodes, like many of the other important plant parasites, belong to the order Tylenchida. (C.C. Doncaster)

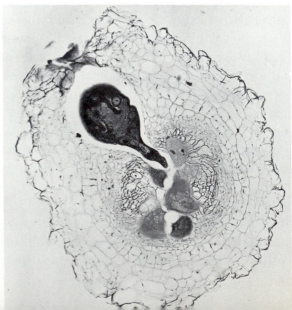

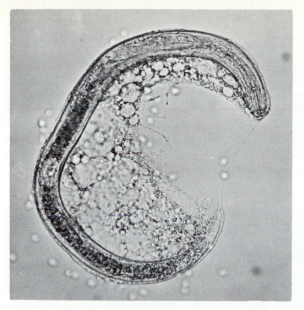 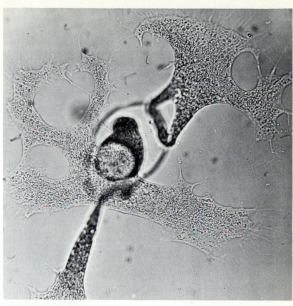

An ameba catches a nematode. The potatoes are safe from this second-stage juvenile potato-cyst nematode *(Globodera rostochiensis),* which has just been caught, *left,* by an ameba *(Theratromyxa weberi).* The ameba engulfs and surrounds the nematode with a digestive cyst, from which well-fed daughter amebas later emerge, *right.* (C.C. Doncaster)

Nematodes sometimes benefit agriculture by parasitizing insects; and researchers, growers, and companies are now cooperating to test and promote the use of nematodes for biological control of insect pests. For example, one species of nematode *(Deladenus siricidicola)* attacks both larvas and adults of a woodwasp that damages pine trees in Australia and New Zealand. The eggs of the wasps are damaged, but the adults are not killed and themselves spread the nematodes from tree to tree. Certain other nematode parasites *(Heterorhabditis, Neoaplectana, Steinernema)* enter the body cavity of insect hosts, introducing specific bacteria. When the bacteria have infected and killed the insect, the nematode feeds on the bacteria and perhaps also on the broken-down insect tissues. Such nematodes have proved promising as biological controls against a number of important caterpillar and beetle pests.

Insects that spread disease among humans and other animals are also targets for control by nematodes such as the mermithids, whose juveniles are parasites of insects or other invertebrates.

Romanomermis culicivorax, a mermithid nematode, attacks mosquitoes and blackflies, important vectors of many diseases (some of them caused by nematodes—see below). Large numbers of infective juvenile stages of *R. culicivorax* are laboratory-reared and then distributed (by spraying from the air, as with chemical pesticides) into bodies of water where the insects breed.

Nematodes are also preyed upon by a variety of small soil animals such as other nematodes, annelid worms, tardigrades, mites, and other arthropods. In addition, many fungi trap nematodes by means of sticky secretions or by forming rings that tighten around the worm as it crawls through; the fungal mycelium then penetrates the cuticle and digests and absorbs the tissues of the worm.

Using a sharp stylet, the juvenile mermithids penetrate into the body cavity of the insect larvas. When full grown, the mermithid emerges from its mosquito host, and the nonfeeding adult reproduces, so that under favorable circumstances a permanent population of nematodes is established and repeated spraying is unnecessary. The mosquitoes die after the parasites emerge. Development of this mermithid species is not completed in blackflies, which are killed during or shortly after penetration, so that repeated spraying is required for control. Still other nematodes, which are parasites of snails that harbor larval schistosomes (see chapter 9), are being investigated for use against the snails.

Mermithid nematode from soil in Pennsylvania. (R.B.)

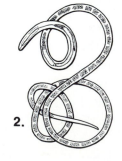
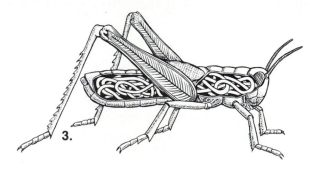

1.   **2.**   **3.**

Life history of a **mermithid nematode,** *Agamermis decaudata.* **1. Developing worm,** still enclosed by egg membrane. The young worm molts once before hatching into soil where the egg was laid. **2. Young juvenile,** 5 to 6 mm long, migrates to soil surface after an early summer rain and penetrates into a newly hatched grasshopper, using a sharp stylet (inherited from free-living predaceous ancestors). **3. Parasitic juvenile** in body cavity of grasshopper grows rapidly for 1 to 3 months, taking up nutrients through pores in the thin cuticle. Mermithids are among the most modified of nematode parasites, for they never feed through the mouth. The midregion of the digestive tract disconnects at both ends and serves as a storage organ for the nutrients that will sustain the worm throughout the rest of its life. In late summer the full-grown worm emerges from its host and enters the soil as a free-living but nonfeeding adult. It will overwinter in the soil, to reproduce the following summer. (After J.R. Christie)

THOSE NEMATODES that have become **parasites of vertebrates** are a minority and are concentrated mainly within a few of the many nematode orders. But even if we are so openly self-centered as to limit discussion to those species that occur commonly in humans and in our most popular pets, the subject is a large one and can be treated only briefly here. Humans are host to about 50 species of nematodes, of which about a dozen are major health problems, with cases numbering in the millions or even hundreds of millions.

A simple kind of life history is illustrated by *Trichuris trichiura*, a common cosmopolitan species that is estimated to infect 500 million people each year. Its members are called **whipworms,** because the slender anterior end and stouter posterior end look like the lash and stock of a whip. These nematodes live in the host's large intestine, probably feeding on blood. The eggs are shed with the host's feces and develop into infective embryos while free in water or soil. The warm, moist conditions of tropical or subtropical climates are necessary for their survival and development. Poor sanitation favors transmission to new human hosts, as the infective stage is the developing egg, which is ingested by the new host, usually in food or water contaminated with feces. The egg covering is weakened by host digestive enzymes, and the young worm hatches in the intestine and burrows into the intestinal wall, where it grows for a time. Mature males and females are moderate-sized worms, reaching as much as 50 mm in length. They live and mate in the intestinal lumen, and may survive for some years, but do little harm unless present in large numbers.

Related to whipworms are **trichina worms,** *Trichinella spiralis,* parasites in which some relatively small differences in life history result in drastic consequences for the unfortunate host. Adult trichina worms are smaller than whipworms, the females about 3 to 4 mm long, the males only half as long. They live only a month or so, in the wall of the intestine, and cause no symptoms or significant damage. The eggs, instead of being shed to the outside to reach other hosts, hatch within the reproductive system of the female worm, which thus gives birth directly to young worms, about 1,500 per female. The young worms, roughly 0.1 mm long, burrow through the intestinal wall into the blood and lymph vessels and are transported throughout the body of the host. They make their way into virtually all the organs, but especially into certain muscles. It is during this burrowing activity in the muscles that the host suffers excruciating pain, fever, and other disturbances; death results only occasionally, usually from respira-

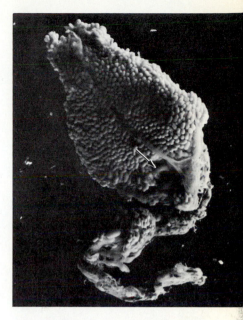

Nematodes in invertebrate host are of concern to doctors and veterinarians only as sources of infection to their vertebrate patients, the final hosts. But the parasites often cause serious damage to the intermediate host also, as shown here. Only the upper part of this gonad of a sea urchin (chapter 27) is normal; the shriveled remains of the lower portion contain 2 juvenile nematodes (*arrow* points to one worm broken free of its cyst). At Catalina Island, California, from 40 to 80% of sea urchins (*Centrostephanus coronatus*) were infected, and their reproduction was impaired, though they appeared otherwise healthy. The complete life history of these nematodes, *Echinocephalus pseudouncinatus* (order Spirurida), is not known; but the early stages of related nematodes are found in copepods (chapter 21), and sea urchins could become infected by accidentally swallowing copepods while feeding. The final host, which eats sea urchins, is the California horned shark. (J.S. Pearse)

In the USA about 1 in every 1,000 hogs butchered is infected with trichina worms. It is up to the consumer to cook pork adequately. About 90 cases of trichinosis occur each year from home-prepared smoked sausages.

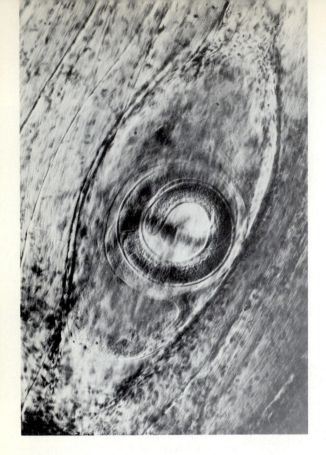

Trichina cyst in tongue muscle of a pig. As the cyst is only about 0.5 mm long, microscopic examination is necessary for meat inspection, and the U.S. government does not inspect pork. *Trichinella spiralis,* sectioned and stained preparation. (P.S. Tice)

Juvenile trichina worms freed from cysts. Several show stringy antibody-antigen precipitates extending from the mouth end. Work continues on the development of an immunological test that would reveal the presence of trichina cysts in pork reliably and economically. Photomicrograph. (R.F. Carter)

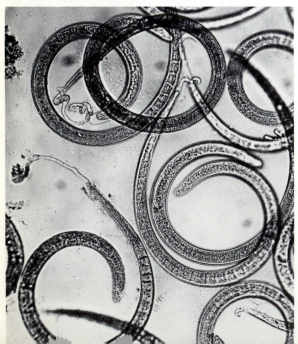

tory or heart failure. In the muscle, the worms grow to a length of about 1 mm, then encyst, curling up and becoming enclosed in a calcified wall. Within the cysts they may remain alive for many years but develop no further unless the host tissues (and cysts) are eaten by another suitable host. The parasites are thus transmitted directly from host to host. Humans usually become infected by eating insufficiently cooked pork, occasionally other meats such as bear, walrus, or wild boar. Pigs usually become infected when fed garbage containing scraps of infected pig flesh. Omnivorous scavenging by rats makes them common hosts also.

In contrast to the entirely parasitic life history of trichina worms, in which there is no stage that regularly occurs free in the environment, the life history of **threadworms,** members of the genus *Strongyloides,* includes one or more generations that are free-living and that alternate with parasitic phases of asexual reproduction. The adult, sexually-reproductive males and females live and mate in fecal matter or in the soil. They produce young that feed on bacteria, grow, and molt. After two molts, the juveniles may become infective to a variety of mammalian hosts; humans are most commonly parasitized by *S. stercoralis.* The juveniles penetrate the host's skin, enter the lymph and blood vessels, and are carried to the lungs, where they break through into the air passages. The worms then migrate or are coughed up from the trachea, are swallowed, and reach the small intestine, where they mature into parthenogenic females. The young produced asexually by these females may reinfect the same host, or pass out to infect new hosts, or become free-living males and females. This asexual phase of the life history may be compared to the multiplication of larval stages in trematodes such as the liver flukes. The parasite, having secured a host, exploits the host's resources to multiply asexually before embarking on the rest of its risky life history.

The flexible life history of threadworms, partly free-living and partly parasitic, suggests how nematodes with more completely parasitic life histories may have evolved. From this sort of life history, it is easy to derive, for example, that of the **hookworms**—parasites that afflict about one-quarter of all the people in the world. As in threadworms, juvenile hookworms that have completed their second molt in the soil are ready to infect a host by penetrating the skin, and they follow a similar path to reach the digestive tract, where they attach to the intestinal lining by means of teeth or plates in the mouth cavity and hold on while sucking in blood and tissue fluids. Sexually-reproductive males and females develop and produce great quantities of eggs (roughly 10,000 per female per day), which are shed in the host's feces. Under suitably warm moist conditions, the eggs hatch and develop into young juveniles, which are the only free-living stage in hookworms. Hookworms have been said to be among the most important causes of human ill-health worldwide, as the loss of blood makes the victim weak and anemic, and sharply reduces resistance to other afflictions. The most important species are *Necator americanus* in tropical Africa, Asia, and the Americas—responsible for 90% of all cases of hookworm—and *Ancylostoma duodenale* in Europe, North Africa, and Asia. Other species of hookworms infect cats, dogs, and other domestic and wild mammals.

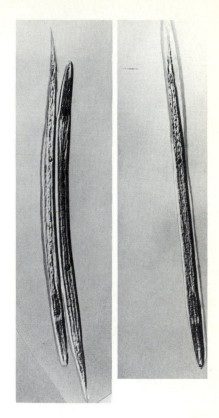

Juvenile hookworms in their third stage, following the second molt, are the infective stage found in soil. Length, 0.5 mm. *Necator americanus,* preserved specimens.

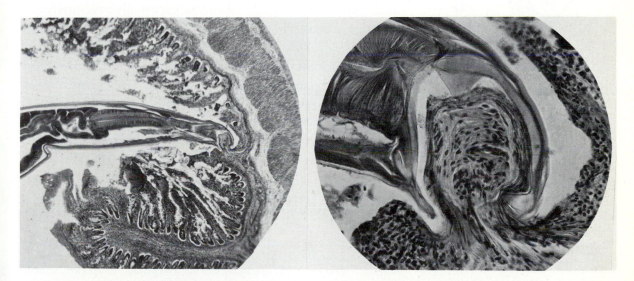

Hookworm attached to intestinal wall, in section of preserved material. *Left,* anterior portion of worm, including muscular sucking pharynx. Close-up, *right,* shows the worm firmly holding in its mouth cavity a small portion of the host's intestinal lining. When actively feeding, the worm sucks 120 to 200 times per minute. *Necator americanus* takes about 0.013 to 0.1 ml blood per worm per day, *Ancylostoma duodenale,* 0.15 to 0.25 ml. In a well-nourished adult, an infection with fewer than about 50 worms has little effect, but heavier infections produce measurable to severe anemia. The results are more serious in a person, especially a child, simultaneously suffering from some degree of malnutrition. (U.S. Army Medical Museum)

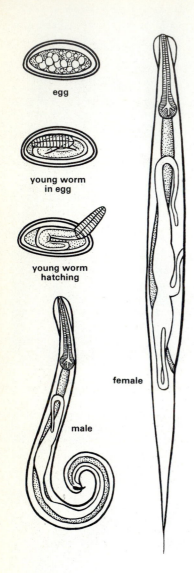

egg

young worm
in egg

young worm
hatching

female

male

Pinworm, *Enterobius vermicularis.*
(Mostly after R. Leuckart)

The life histories of other important human intestinal parasites share various combinations of features already described. **Pinworms** *(Enterobius vermicularis)* occur in at least 500 million people throughout the world, but seldom cause serious disease; they are especially common in children of North America and Europe, less so in the tropics. The mature females migrate out the host's anus and release thousands of eggs in the perianal area, causing severe itching and irritation. The female quickly dies, and the embryos in the eggs either hatch and crawl back in the anus, or lie dormant (often in bedlinens and household dust) until swallowed or inhaled, so the free-living phase is limited and involves no feeding stage.

This is even more the case in the **ascarids,** in which the only stage outside the host is the developing egg. Ascarids differ significantly from the other parasitic nematodes mentioned so far mainly in the giant size of the adults. Males of *Ascaris lumbricoides,* a human intestinal parasite, reach a length of 30 cm; females reach 40 cm and produce about 200,000 eggs per day. Infection occurs when a person swallows the eggs, usually in food or water contaminated with feces. The eggs, which already contain a developing worm, hatch in the digestive tract. Oddly, although they have already arrived at the site where they may live for many years as adults, the newly hatched juveniles migrate through the tissues much as was described for juvenile threadworms, eventually to return to the intestine. It seems likely that this behavior represents a vestige of an ancestral life history; in any case, as with many other intestinal nematode parasites, the greatest damage to the host occurs during this migratory phase. The adults feed on the intestinal contents, rather than on blood, and so do less harm than the tiny hookworms, but may obstruct the gut when present in large numbers (up to 5,000 worms per host have been found). Ascarids are major human parasites, about as common as hookworms, with over a billion cases estimated worldwide.

Infection with a related group, the **anisakids** *(Anisakis, Phocanema),* is a hazard of eating raw, salted, or pickled seafood, especially fish such as herring; freezing or cooking kills the worms. The juvenile worms are found in the tissues of many marine invertebrates and fishes; the adults mature in the stomach or intestine of marine mammals and other fish-eating vertebrates. Anisakids do not mature in humans, but the juvenile worms may cause pain and tumorlike growths, mimicking ulcers or gastric carcinoma.

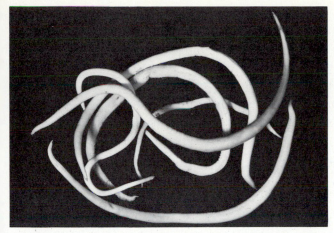

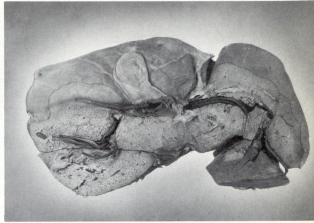

Horse ascarids, up to 18 cm long, were expelled by a 3-year-old horse after medication. This species, *Parascaris equorum,* is common in young horses, which ingest the eggs when feeding from the ground in horse pastures, paddocks, or stalls; older horses develop a degree of resistance. (R.B.)

Ascarids do relatively little harm as intestinal parasites, unless present in numbers, but they occasionally cause trouble by wandering into the stomach, appendix, pancreatic duct, or even up the esophagus and out the nose. This human liver cut away to reveal several ascarids that entered through the bile duct. (U.S. Army Med. Mus.)

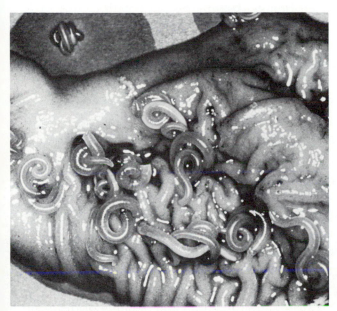

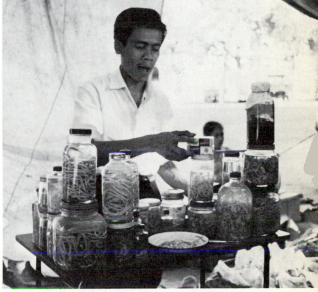

Nematodes in pets are sometimes a health hazard to their human owners. Shown here are nematodes in the stomach of a cat, a condition to be expected unless the pet is regularly given preventive medication. Some species of dog and cat nematodes *(Toxocara)* also readily infect humans, especially children. The eggs require some time to develop to an infective stage and are usually acquired by ingesting dirt, rather than by direct contact with an infected animal.

Worms create jobs. Displaying jars full of nematodes and other parasites to stimulate sales, this man had a thriving business selling worm medicine in an open market in Bangkok, Thailand. The combination of warm, moist climate and many low-income communities with poor sanitation gives the tropical regions more than their share of parasites. (R.B.)

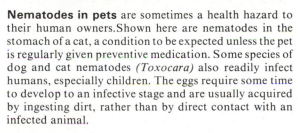

Juvenile guinea worms in a cope-pod, an intermediate host. (After Martini)

In regions of Africa and Asia where guinea worms are common, a large part of the population is incapacitated during part of the year. The appearance of the itching, burning blisters is accompanied by nausea, vomiting, diarrhea, and dizziness. The traditional method of treatment is to remove the worm by slowly and painfully winding it out on a stick. This method is often successful, but if the worm breaks, bacterial infection usually ensues, and loss of the limb or even death may eventually result. Supplementary treatment with anti-inflammatory drugs and antibiotics may be effective. An alternative is surgical removal of the worm. Control measures for this disease include filtering all drinking water and minimizing contact of infected people with communal water supplies.

Microfilaria, shown with 3 red blood cells for scale. This distinctive young stage of filarial nematodes moves actively while still ensheathed in an egg membrane and has only a rudimentary gut. (Modified after E.C. Faust)

A final group of nematode parasites that afflict humans have life histories with an *invertebrate intermediate host.* The intermediate host is some sort of arthropod (chapter 20), and the final human host acquires the nematode by accidentally ingesting an infected arthropod or being bitten by one.

An example is the **guinea worm** (*Dracunculus medinensis*), one of the more serious discomforts of life in many warm parts of the world. The only free-living stages are the young juveniles, which swim about in freshwater until ingested by a copepod (*Cyclops,* chapter 21). In the body cavity of this small crustacean, they develop to a stage infective to humans. When a person drinks unfiltered water containing an infected copepod, the worms are introduced into their final host. They then penetrate the human intestinal wall, grow, and finally mate in the connective tissue. The male worms, inconspicuous and only a few centimeters long, die after mating. The females grow to about a meter in length, a millimeter in diameter. When full of developing embryos, they migrate to tissues just under the skin, usually of the legs or feet. A painful blister forms and then breaks, leaving an ulcer with a hole in its center. When this ulcer is suddenly plunged into cold water (as when the human host enters a stream to wash clothes or to fish), large numbers of tiny juvenile worms are released into the water.

An important group of parasites with arthropod intermediate hosts are the **filarial nematodes,** which are obligate parasites at all stages of the life history and have no free-living stages. Over 300 million people (as well as domestic animals), mostly in tropical areas around the world, suffer from infection with these long slender worms, which are transmitted by mosquitoes, biting flies, fleas, and ticks. The most widespread causes of filarial disease, or filariasis, are species of *Wuchereria* and *Brugia.* When a mosquito bites a potential human host, the infective stage of the worm enters through the wound and into the lymphatic vessels, where they may live for many years, causing inflammation and swelling of the arms, legs, scrotum, or breasts. When mature, the worms reproduce in the lymph nodes and release tiny young, called **microfilarias,** which circulate in the lymph and blood. Mosquitoes become infected by sucking the blood of a human and ingesting the microfilarias, which develop in the insect host to a stage which can infect another human host when the mosquito bites again. The parasites are more dangerous to their insect hosts than to humans; many mosquitoes die or suffer such severe muscle damage that they cannot fly.

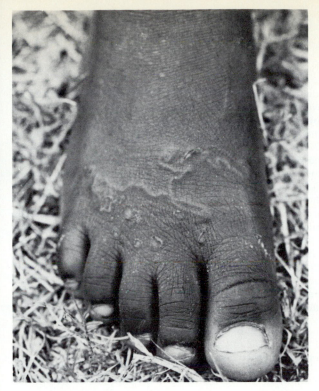

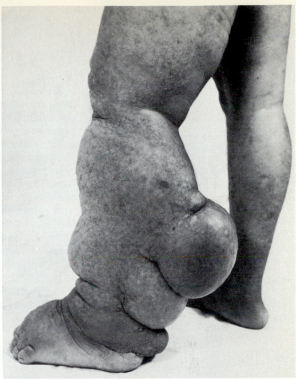

Guinea worm, *Dracunculus medinensis*, beneath the skin of the foot. Ghana, west Africa. (L. Pittman)

Elephantiasis is a relatively uncommon outcome of filariasis. Lymphatic blockage and growth of fibrous tissue result in grotesque enlargement of the limbs or other affected parts. This condition is probably in part an allergic reaction, and it develops only in a minority of people after repeated and prolonged infection with species of *Wuchereria* or *Brugia*. Western Samoa. (W.J. Vermeulen)

In *Onchocerca volvulus,* transmitted by blackflies, the adults live and reproduce in large nodules under the skin, which reach as much as 5 cm in diameter but do little real harm. It is the young microfilarias that are most troublesome to the host, causing dermatitis and, more serious, producing blindness when they invade tissues of the eyes. Onchocerciasis, also known as "river blindness" because it occurs mostly near rivers and streams where blackflies breed, afflicts 30 million or more people in Africa and in Central and South America. The life history is similar to that of other filarial nematodes.

The **symptoms** caused by many nematode parasites are not specific and are often attributed to other causes; diagnosis usually depends on finding and identifying eggs, microfilarias, or other stages. Intestinal nematodes may cause no symptoms, or there may be abdominal discomfort, nausea, diarrhea, or anemia, with intestinal obstruction in cases of heavy infection. Young stages that migrate through the tissues do the most harm, causing fever, severe pain, irreversible damage to muscles and other tissues, pulmonary and cardiac disturbances, and blood disorders. Extra-intestinal types such as guinea worms and filarial nematodes produce different syndromes.

Treatment with a variety of drugs is now available for most cases of nematode parasites, but is not a satisfactory solution if the patient is cured only to become promptly infected again. There are no vaccines against nematodes, and the acquisition of **immunity** following infection is poorly understood; it seems to depend partly on the type of nematode and partly on differences among individual hosts. Many people seem able to develop immune defenses against ascarids and pinworms, which are far more common in children than in adults, but there is no evidence for resistance against guinea worms, which occur repeatedly in the same individuals.

The only filarial nematodes familiar to many people in the U.S. are the **heartworms** (*Dirofilaria immitis*) that infect the family dog. Heartworms are distributed widely around the world and are transmitted by many species of mosquitoes. When a dog is bitten, the infective stages develop in various tissues for 2 to 3 months, then migrate to the heart, where they grow to adult length, about 15 cm in males, up to 35 cm in females. Dogs can usually tolerate the light infections they may acquire between regular checkups; but the worms live for years, and if large numbers are allowed to accumulate, they irritate the heart lining and obstruct blood flow to the lungs, causing respiratory symptoms and eventually weight loss and weakness. The microfilarias that circulate in the blood also cause damage and serve as a source of further infections.

Prevention depends on knowledge of life histories and on breaking a link in the chain of events leading to infection. For intestinal parasites that spread through fecal contamination, personal and community sanitation is the key, but is beyond the economic means of great numbers of people around the world. Preventive measures are sometimes most effective on an individual level, as when the infective stages can be avoided by simple precautions, such as sufficiently cooking pork to kill encysted *Trichinella*. Where the infective stages live in soil and enter through the skin, as in hookworms, wearing shoes can sharply reduce chances of infection. Where they occur in the water, as with guinea worms, filtering or boiling all drinking water is recommended. In the case of filariasis transmitted by biting insects, individual measures such as wearing protective clothing are best combined with vector control on a regional level. In addition to conventional chemical insecticides, mermithid nematodes which are parasites of mosquitoes and blackflies have been used as biological control agents against these insect vectors of filarial disease. Also, filarial nematodes themselves destroy many of their insect hosts, as already mentioned.

THE ADAPTABILITY of the nematode design for living is the most astonishing thing about the group. The blueprint consists essentially of a feeding and digestive tube surrounded by a tubular body wall—the two cylinders separated by a body cavity filled with fluid. This incompressible fluid under high pressure acts as a hydrostatic skeleton, giving turgidity to an otherwise soft worm and enabling it to move by undulations through soils or sediments. Modifications of mouth armature and pharynx musculature enable the feeding equipment to serve the needs of an estimated half-million species occupying almost every habitat on earth. Whether aquatic or terrestrial— free-living forms, microscopic or giant worms, or parasites of plants or animals—most of them are so much alike in external appearance and in internal structure as to make it difficult to tell one species from another.

Nematodes seem pre-adapted to parasitism by their generalized body plan and by their genetic plasticity. Also, as was noted for platyhelminths, the apparent ease with which many free-living nematodes adopt oxygen-poor environments may ease their entry into the similar conditions that parasites encounter within the bodies of hosts. Sometimes free-living species, parasitic ones, and ones on the way to becoming parasitic, occur in one genus, so that nematodes have been referred to as "animal weeds," ready to move into almost any niche that opens up. The rapid takeover of land habitats by flowering plants co-evolving with insects, and both groups exploited by land vertebrates, may be the evolutionary setting that opened a whole spectrum of new feeding opportunities, thus accelerating nematode speciation.

"If all the matter in the universe except the nematodes were swept away, our world would still be dimly recognizable, and if, as disembodied spirits, we could then investigate it, we should find its mountains, hills, vales, rivers, lakes, and oceans represented by a film of nematodes. The location of towns would be decipherable, since for every massing of human beings there would be a corresponding massing of certain nematodes. Trees would still stand in ghostly rows representing our streets and highways. The location of the various plants and animals would still be decipherable, and, had we sufficient knowledge, in many cases even their species could be determined by the examination of their erstwhile nematode parasites." (N.A. Cobb)

CLASSIFICATION: Phylum NEMATODA

Many of the orders of nematodes are not agreed upon by nematologists. The great diversity of classification schemes in this group reflects the difficulty of sorting out relationships among nematodes.

Class ADENOPHOREA (APHASMIDIA). Includes most nematodes free-living in marine and freshwater sediments; feed mostly on microorganisms, detritus.

SUBCLASS CHROMADORIA (TORQUENTIA). Mostly in marine sediments.

Order Chromadorida. Marine and freshwater sediments, soil, hotsprings.

Order Monhysterida. Marine and freshwater sediments, soil.

Order Araeolaimida. Marine and freshwater sediments, soil.

Order Desmodorida. Marine sediments. *Chaetosoma* (draconematid), *Metepsilonema.*

Order Desmoscolecida. Mostly in marine sediments. *Desmoscolex.*

SUBCLASS ENOPLIA (PENETRANTIA). Mostly in soil or freshwater sediments, some parasites of plants or animals.

Order Enoplida. Free-living, mostly in marine sediments, some freshwater or soil; feed on microorganisms and minute animals.

Order Mononchida. Mostly free-living; feed with teeth on minute animals and microorganisms.

Order Dorylaimida. Free-living in freshwater sediments and soil; feed with stylet on minute animals and microorganisms. Some parasites of plants. Either or both of the two following orders (as well as dioctophymatids) are sometimes included as dorylaimids, sometimes combined in a single order Stichosomida.

Order Mermithida. Parasites of invertebrates; juvenile uses stylet to penetrate host. *Romanomermis, Agamermis.*

Order Trichurida (Trichocephalida, Trichinellida). Parasites of vertebrates. Whipworms, *Trichuris;* trichina worms, *Trichinella.*

Class SECERNENTEA (PHASMIDIA). Includes some free-living and most of the parasitic nematodes, almost all terrestrial.

SUBCLASS RHABDITIA, free-living and parasitic.

Order Rhabditida. Mostly free-living in soil; feed mostly on microorganisms. Variety of associations with plants and animals. *Caenorhabditis, Panagrellus, Turbatrix, Neoaplectana, Steinernema, Pelodera.* Threadworms, *Strongyloides.*

Order Strongylida. Parasites of vertebrates. Hookworms: *Necator, Ancylostoma.*

Order Ascaridida. Parasites mostly of vertebrates. Ascarids: *Ascaris, Toxocara.* Anisakids: *Anisakis, Phocanema.* Pinworms, often placed in a separate order Oxyurida: *Enterobius.*

SUBCLASS SPIRURIA, all parasitic.

Order Drilonematida. Parasites of annelids and other invertebrates.

Order Spirurida. Invertebrate intermediate hosts, vertebrate final hosts. *Gongylonema, Physaloptera, Spirocerca, Tetrameres, Thelazia, Echinocephalus.* Giant kidney worm, *Dioctophyma,* traditionally placed close to trichurids, reclassified with spirurids by some nematologists.

Order Filariida, filarial worms. Arthropod intermediate hosts, vertebrate final hosts. *Wuchereria, Brugia, Loa, Onchocerca, Dirofilaria.*

Order Camallanida. Copepod intermediate hosts, vertebrate final hosts. Guinea worm, *Dracunculus.*

SUBCLASS DIPLOGASTERIA, includes most plant parasites.

Order Diplogasterida. Free-living in soil, a few parasitic in insects.

Order Tylenchida. Parasites of plants and invertebrates. With stylet. *Ditylenchus, Criconemoides, Heterodera, Globodera, Meloidogyne, Deladenus.*

Order Aphelenchida. Parasites of plants and insects, or predators on minute animals. With stylet. *Aphelenchoides.*

NEMATOMORPHS

LONG WIRY WORMS have often startled people who happened to look into drinking troughs or other stock watering places. As the worms appear suddenly at full size, and come in about the same range of colors and lengths as the hairs of horses' tails, people believed they arose from horsehairs that had fallen into the water and called them "horsehair worms." Experimental attempts, by early naturalists, to produce them by soaking horsehairs in water invariably failed. The abrupt appearance of the full-grown worms in various bodies of freshwater was explained only after determined observers had begun to understand their curious life history. The issue of their origin from horsehairs is no longer controversial (at least among biologists), but controversies about the biology and relationships of these enigmatic worms continue to spark discussion and make them attractive objects of study. Over 230 species of horsehair worms, members of the phylum **NEMATOMORPHA,** have now been described.

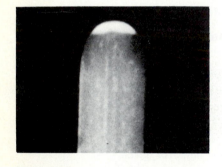

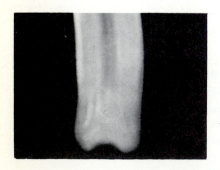

Rounded, white-tipped **anterior end,** *above,* and rounded or lobed **posterior end,** *below,* of a nematomorph help to distinguish it from a nematode, with tapered ends. (R.B.)

Nematomorphs resemble nematodes in many ways. As in nematodes, the slender elongate body of nematomorphs is circular in cross-section and covered by a thick collagenous cuticle that is molted; the body wall contains no circular muscles but only longitudinal ones; circulatory and respiratory systems are absent; and motile cilia are lacking, even in the sperms. Unlike nematodes, nematomorphs do not show constancy of cell number, and the pseudocoelom usually becomes largely filled with mesenchyme. Also, there are no lateral epidermal ridges and no excretory system; only a single ventral nerve cord runs backward from the anterior nerve ring; and the genital ducts of females as well as males open into a cloaca.

Gordian worm is a common name for a nematomorph and is derived from the habit of coiling into complicated loops, suggesting the "Gordian knot" of Greek mythology. Central California. (R.B.)

Nematomorphs are notably similar to mermithid nematodes in their way of life. Like mermithids, juvenile nematomorphs live in the body cavity of a host, usually an insect, and take up nutrients through their body surface. The digestive tract is rudimentary, and often there is not even a mouth. The full-grown worms emerge when the host approaches freshwater, and they spend the rest of their lives in water or moist soil. Observations suggest that the worm, when ready to emerge, influences its insect host to seek water, but no one knows how the parasite does this, or how it senses when the insect has reached a suitably moist spot. Emergence of the worms is quickly followed by the death of the host.

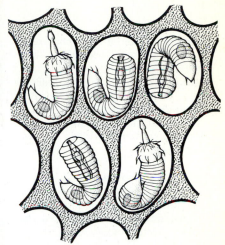

Larval nematomorphs in an egg mass. After hatching, the larvas can remain free-living and active for only a few days. If not ingested by a host, they encyst in the water and on vegetation; or, if ingested by a host that will not support their development, they encyst in the tissues of the host and resume development later if eaten by a suitable host. Nematomorph larvas have been found in a variety of aquatic and terrestrial insects, in other arthropods (myriapods and arachnids), in trematodes, snails, and annelid worms, and in fishes and tadpoles. Juvenile nematomorphs, however, develop in a more limited range of hosts and are most commonly found in grasshoppers, crickets, beetles, and other large terrestrial insects, which presumably ingest the active or encysted larvas when drinking or feeding near the water's edge; or in predators such as mantids, which acquire the parasites by eating insect prey containing encysted nematomorph larvas. Members of a marine genus of nematomorphs live in crustaceans. A few aberrant occurrences in humans have been reported. (After A. Schepotieff)

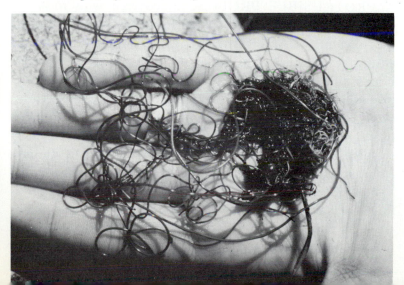

Mass of adult nematomorphs, *Paragordius varius,* taken from freshwater in Illinois. (C.J. Swanson)

The **phylum Nematomorpha** is divided into 2 classes or orders:

Gordioidea: freshwater, includes the great majority of species, mostly in aquatic and terrestrial insects; pseudocoelom filled with mesenchyme; single ventral epidermal ridge. *Gordius, Chordodes.*

Nectonematoidea: marine, one genus, in decapod crustaceans; pseudocoelom not occluded with mesenchyme; dorsal and ventral epidermal ridges. *Nectonema.*

Although the adults are free-living in terms of habitat, they do not feed and depend entirely on nutrients obtained during the parasitic phase. These nutrients may sustain them for months, as their activity is sluggish. Their bodies are rather stiff and wiry, and their behavior consists mostly of slow writhing or undulating movements. Males are usually smaller and more active than females. After mating, the females lay eggs in long gelatinous strings that are wound around water plants. The eggs hatch into distinctive larvas with a spiny eversible proboscis armed with 3 stylets. When ingested by a suitable host, they use the proboscis to make their way from the host's intestine into the body cavity, where they grow to adult size and form.

ACANTHOCEPHALANS

A PROTRUSIBLE PROBOSCIS, covered with rows of stout, recurved spines, is the most characteristic structure of the phylum **ACANTHOCEPHALA** ("spiny-headed"). As adults, acanthocephalans live in the digestive tract of vertebrates, firmly attached to the intestinal wall by the spiny proboscis. The worms have no mouth or digestive tract at any stage, and nutrients are taken in from the host's intestine through the syncytial epithelium of the body surface, as in a cestode. Tubular branched canals extend into the epidermis from surface pores and expand the area for assimilation by 20 to 60 times, even more than in cestodes. The body cavity is a fluid-filled pseudocoelom; and through it stretches a thick strand of tissue called the ligament, which may be a vestige of the digestive tract. From the ligament are suspended 1 or 2 ligament sacs, which enclose smaller fluid-filled cavities, and within these are the testes or ovaries. In some species flame-bulb excretory organs (protonephridia) empty into the reproductive ducts, which open at the rear end of the body. The overall body organization is therefore quite simple, befitting an animal that spends its adult life soaking up nutrients in the guts of others. Nevertheless, as already seen with parasitic flatworms and nematodes, there are formidable challenges to living successfully in such an environment.

Reproduction, and transit during the life cycle from one host to another, pose major problems. Unlike flatworms, which are hermaphroditic, acanthocephalans have separate sexes. In males, sperm ducts lead from the testes to the outside, where they open into a cuplike chamber (bursa) that is everted over the female opening during copulation. After sperm transfer, cement from male accessory glands seals the female opening, preventing the female from mating again with another male until the cement dissolves some days later.

Occasionally males also seal the reproductive openings of other males with this cement, and not because they cannot discriminate between the sexes, as there is no evidence for sperm transfer to other males. The more worms to which a male can apply his cement, the greater is his competitive advantage, provided he avoids having any applied to him. For a male engaged in such a game of tag, there is an urgency to mate as soon as possible, which may explain why males mature and remain at a much smaller size than females. As in most parasites, the females produce great numbers of eggs, for which large size is an advantage.

Within the female, balls of ovarian tissue float free in the ligament sacs or pseudocoelom. Sperms fertilize the ripe eggs and development begins within the female's body cavity. Those eggs that contain embryos are selected by a special sorting mechanism, extruded through the female opening into the host's intestine, and shed to the outside with the host's feces.

The hooked larva, called an acanthor, hatches only when the egg is eaten by an intermediate

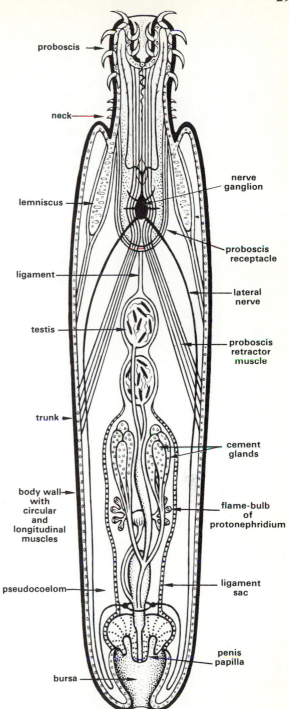

Diagram of an acanthocephalan, a male in dorsal view. A system of "lacunar canals" in the epidermis may serve as a circulatory system; there is no special respiratory system. These canals extend into the muscles and also into a pair of spongy saclike bodies, the lemnisci; they seem to serve as fluid reservoirs for the proboscis. A nerve ganglion in the proboscis receptacle is considered ventral. Males have paired genital and bursal nerve ganglia, not found in females. Flagella are present in the flame-bulbs of the protonephridia and in the tails of sperms.

host, usually a crustacean or an insect, sometimes a millipede. The larva bores through the arthropod's intestine into the body cavity and there develops through several acanthella stages into a juvenile worm ready to infect a final host. Further development depends on the intermediate host being eaten by a suitable vertebrate; fishes and birds are the most common hosts, but amphibians, reptiles, and mammals also entertain acanthocephalan guests. In some cases the intermediate host is eaten by a *transport host,* usually a small fish, within which the juvenile hatches, bores through the intestinal wall, and encysts. The juvenile can survive passage through several transport hosts, being eaten and freed from its cyst, then re-encysting in one transport host after another—ready to resume development to the adult stage if and when the transport host is eaten by a suitable *final host.*

Some of the best known acanthocephalans are *Moniliformis* in rats, acquired from cockroaches, and *Macracanthorhynchus* in pigs, acquired from beetle larvas. Both are occasionally found in humans, who become infected by accidentally swallowing insects, but do not suffer serious disease unless a large number of worms is present. Humans also can become infected as transport hosts by eating raw fish that contains encysted juveniles, but the re-encysted juveniles in human tissues are rarely noticed, and they are probably doomed because their human host is unlikely to be eaten by the fish or seal that serves as their proper final host.

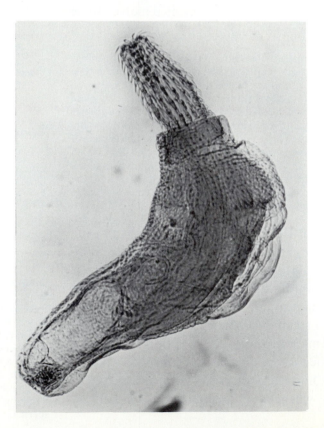

Acanthocephalan. Stained preparation.

Acanthocephalans are difficult to relate to other inverte-brates. The embryology, hooked larva, syncytial surface, lack of a gut, and many other features are reminiscent of parasitic flatworms, especially cestodes, but most experts now believe that these are convergences. Electron microscopic studies indicate that the "cuticle" is not an extracellular secreted structure but a dense layer inside the epidermal plasma membrane; it most closely resembles the so-called cuticle of a pseudocoelomate group described in the next chapter, the rotifers. Among other characters that seem to connect acantho-cephalans with various other pseudocoelomates are the con-stant number of nuclei, and the ligament and ligament sacs (see chapter 13). In addition, acanthocephalans show strong similarities to burrowing predatory marine worms called priapulans, which are themselves of uncertain affinities (chap-ter 18). Some experts have concluded that acanthocephalans have no clear affinity with any living phylum known.

THE WORMS, soft elongate backboneless animals, were all classified together by early taxonomists in a group simply called Vermes ("worms"), regardless of how they were organ-ized inside. As knowledge of worms accumulated, it became clear that many worms have very different body plans with little in common except being vermiform. So the worms have been divided into a number of groups, and they now comprise more than half the commonly recognized animal phyla.

However, many of the worms do fall into groupings of phyla within which the members have much in common. One such major grouping, sometimes considered a phylum, is the **Aschelminthes,** (ask-hel-min′-theez). It comprises most of the animals in this chapter and the next, and sometimes the priapulans (see chapter 18). Tardigrades (chapter 13) and sometimes acanthocephalans are excluded.

Aschelminths are bilaterally symmetrical and have a pseudo-coelomate body cavity between the body wall and gut (but in some groups, such as nematomorphs, there is not really much of a cavity). The name refers to the pseudocoelom (ascos = "cavity," helminth = "worm"), and it distinguishes the aschel-minths from the solid-bodied flatworms, gnathostomulids, and nemerteans but not from other worms with body cavities. Typically aschelminths are small, often microscopic, and have a simple digestive tract with mouth and anus (some of the parasitic worms discussed in this chapter being notable exceptions). A highly differentiated pharynx, such as in nematodes, is often present. The nervous system includes an

The phylum Acanthocephala is divid-ed into 3 orders or classes, which differ in various technical features (the location of the main lacunar canals, whether the ligament sacs per-sist in adult females, whether there are multiple cement glands or a single syncytial one in males, etc.).

Archiacanthocephala. Adults para-sitic in terrestrial birds and mammals; intermediate hosts are insects or myria-pods. Protonephridia present in one family, absent in all other acantho-cephalans.

Palaeacanthocephala. Adults para-sitic in fishes, amphibians, reptiles, aquatic birds, marine mammals; inter-mediate hosts are crustaceans.

Eoacanthocephala. Adults para-sitic mostly in fishes, a few in amphib-ians or reptiles; intermediate hosts are aquatic insects or crustaceans.

Parasitic worms of different kinds (especially trematodes, cestodes, nema-todes, and acanthocephalans) are often grouped by parasitologists and referred to as helminths ("worms"); this is not a formal taxon but an assemblage of animals which share common habitats and problems and which have convergently evolved some strong similarities in structure, physi-ology, and life history.

anterior brain or nerve ring around the gut, giving off lateral or ventral nerve cords. Respiratory and circulatory systems are absent, which is what one would expect of minute organisms; substances are moved about in the fluid of the body cavity or simply diffuse the minute distances between gut, muscles, and other tissues. The muscles are usually in separate bundles rather than in the regular circular and longitudinal muscle layers of most other worms. Excretory and water-regulatory systems with protonephridia like those in flatworms occur in some groups. The sexes are usually separate and the reproductive systems are relatively simple. Excretory and reproductive ducts may open into the posterior intestine, which is then a cloaca. Development is determinate with a pattern resembling spiral cleavage (see chapter 16), and there tends to be little capacity for regeneration. At least in some of the phyla, the number of cells (or of nuclei in syncytia) is relatively small and constant.

Whether these rather general characters indicate close evolutionary relationships among the aschelminths is debatable, but because it is at least convenient to discuss and think about the aschelminth groups as an assemblage, we have placed those with large parasitic members together in this chapter and those of the microscopic world in the next.

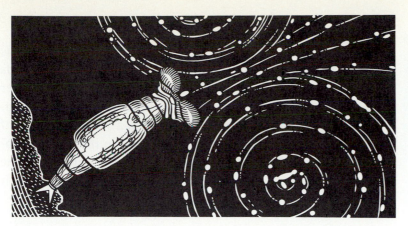

Chapter 13

Microscopic Metazoans
Rotifers
Gastrotrichs
Kinorhynchs
Loriciferans
Tardigrades

THE ANIMALS presented in this chapter—the micro-metazoans—live their lives on a miniature scale. They may be visible to the naked eye as tiny specks. But to be appreciated, they must be observed under the microscope, among the many kinds of protozoans and other protists that are their prey, and sometimes their predators. Micrometazoans, though they live in a protistan community, are solid members of the animal kingdom, with multicellular organ systems, a pseudocoelomate body cavity, and other aschelminth traits described in the last chapter.

ROTIFERS

ROTIFERS are among the smallest of metazoans. Though a few giants reach lengths of 2 or 3 mm, most measure only 100 to 500 μm (0.1 to 0.5 mm), and the smallest are only 40 μm long. These tiny, incessantly active animals are abundant, widespread, and easily found in almost any body of freshwater, except where rushing currents sweep them away. Close to 2,000 species have been described, of which relatively few can tolerate seawater and fewer still are exclusively marine.

The commonest rotifers can be recognized by the presence of an anterior crown of cilia, the **corona,** which serves as the chief organ of locomotion and also as the means of bringing particles of food to the mouth. In these forms the beating of the cilia, which are arranged around the edge of a pair of disk-shaped coronal lobes, gives the appearance of revolving wheels—hence the name of the phylum, **ROTIFERA,** which means "wheel-bearers."

Rotifers vary in shape from wormlike bottom-dwellers, or flowerlike sessile types, to rotund forms that float near the surface. In many species the body is elongated and is roughly distinguishable into 3 regions: a **head** which bears the mouth and ciliated corona; a main central **trunk**; and a tapering tail

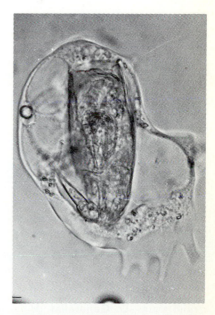

Rotifer engulfed by an ameba is smaller than its protozoan predator. The rotifer, with its pointed tail folded against itself, is enclosed in a food vacuole and is undergoing digestion. (R.B.)

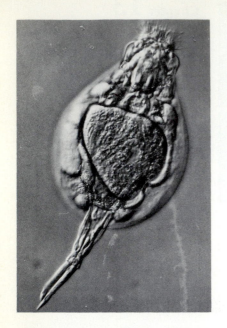

Marine rotifer, 0.2 mm long, is from coarse sediment at a water depth of 8 m. Helgoland, West Germany. (G. Uhlig)

portion quaintly named the **foot**. At the end of the foot are the "toes," pointed projections from which open cement glands that secrete an adhesive material used to anchor the rotifer during feeding. In the bdelloid rotifers (the ones most commonly seen in pondwater) the toes aid in a second method of locomotion in which the animal proceeds in inchworm fashion. It stretches out, attaches by an adhesive secretion at the front end, releases the toes, and contracts the body; then it fastens the toes again, stretches, and so on.

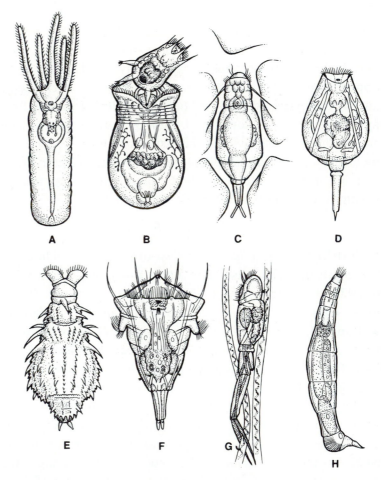

A variety of rotifers from many habitats. (After various sources)

A. Sessile rotifer, *Collotheca judayi,* with long slender coronal lobes used to trap prey; the body is surrounded by a gelatinous sheath. **B. Planktonic predator,** *Asplanchna,* lacks a foot. It is shown ingesting a smaller planktonic rotifer, *Brachionus.* The prey grows spiny projections in response to a substance released by *Asplanchna,* an effective defense only against smaller individuals of the predator. **C. Interstitial rotifer,** *Bryceella tenella,* among sand grains. **D. Generalist,** *Monostyla,* a widespread genus with many species. Suspension-feeders found near shores in both fresh and marine waters; on various solid substrates and in sand; in hot springs; in low-oxygen sediments; in acid waters; and in terrestrial habitats among mosses. The surface of the trunk is a stiff, protective lorica. **E. Moss dweller,** *Macrotrachela multispinosus,* able to revive after extreme dessication. **F. Marine planktonic rotifer,** *Synchaeta vorax,* feeds on phytoplankton. The spines and inflated shape may slow sinking and discourage some predators. **G. Slender-bodied algal dweller,** *Scaridium longicaudum,* a notommatid rotifer, lives among algal filaments. Many notommatid rotifers suck the contents of algal cells. **H. Endoparasite,** *Albertia gigantea,* with reduced corona, lives in the intestine and coelom of earthworms.

The body is transparent and flexible. In some species, the **stiffened surface** is divided into sections, and head and foot can be telescoped into the trunk. The "cuticle" seen and described by light microscopists in rotifers was assumed to be a secreted, extracellular (extrasyncytial) structure. More recently studies with the electron microscope have revealed that it consists of two separate and different components, both of which are sometimes referred to in the literature as "cuticle." *Outside* the outer plasma membrane of the syncytial epidermis, an extracellular organic layer is present in at least some rotifers and may be present in others reported to lack it, as it is difficult to preserve with fixatives. Of unknown function in rotifers, it consists of granular and/or fibrillar organic material and seems comparable to the organic coat (glycocalyx) of parasitic flatworms and acanthocephalans. Just *inside* the outer plasma membrane of the syncytial epidermis, an intracellular dense cytoplasmic layer has been demonstrated in all rotifers examined. It presumably gives form and support to the body and serves as a place of attachment (by means of fine fibrils running through the epidermis) for underlying muscles. Rotifers do not have any substantial external cuticle such as is found in nematodes and nematomorphs, and so do not molt.

Bdelloid rotifers with corona extended. The cilia on the 2 lobes of the corona are beating steadily and draw food particles to the mouth. The jaws of the mastax are of the ramate type (see next page) in bdelloid rotifers. (R.B.)

Bdelloid rotifer with corona withdrawn and the anterior part of the body partly telescoped into the trunk. At the anterior tip there are 2 prominent red eyes. (R.B.)

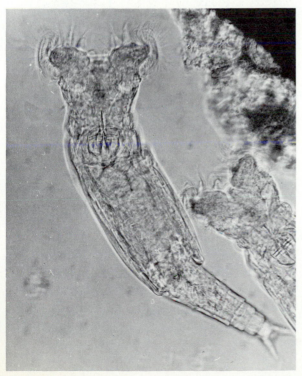

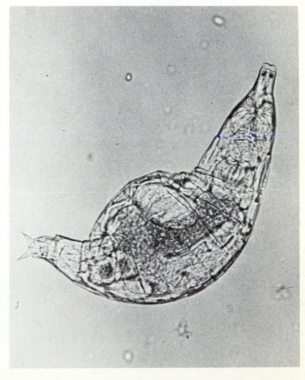

The epidermis of rotifers, surprisingly, most closely resembles that of acanthocephalans. Both groups have the surface covered with pores that lead, through a dense cytoplasmic layer, into channels within the epidermis. But the channels of rotifers end in short sacs and are probably involved in the secretion of the organic coat; whereas those of acanthocephalans are long and branched and represent an enormous expansion of the surface, a feature associated with absorption of nutrients by these parasites from the surrounding medium in the host's intestine.

Most rotifers are suspension feeders. When feeding, they attach by the foot to an aquatic plant or other substrate, and the rapid beating of the crown of cilia draws a food-bearing current of water toward the mouth. Protozoans, algal cells, and organic particles are directed into a food groove that lies between two opposing rows of cilia. From there the food is swept through the mouth into a **muscular pharynx,** or mastax, which contains a chewing or grinding apparatus consisting of little hard **jaws,** operated by muscles. In predaceous rotifers, long pincerlike jaws can be extended through the mouth and used, like a forceps, for grasping prey. Usually the pharynx leads into a straight digestive tract that opens dorsally at the junction of trunk and foot. In some species the digestive tract ends in a blind stomach, and undigested residues are regurgitated through the mouth.

Shapes of pharyngeal jaws of rotifers vary with mode of feeding. The jaws, or trophi, are the most distinctive structures of rotifers and are used by taxonomists to distinguish one species from another. (After various sources)

A. Malleate type, common in many rotifers and considered least specialized, serves primarily for chewing or grinding small organisms or particles of detritus, but may also help to grasp prey; *Proales.* **B. Virgate type,** characteristic of planktonic predators, grasps plant or animal prey while contents are sucked out and provides surface for attachment of muscles that effect suction; *Synchaeta.* **C. Ramate type,** found in bdelloid rotifers, is reduced to stout pieces bearing parallel ridges for grinding the food, which consists of small organisms and detrital particles. **D. Forcipate type,** with elongated elements, is protruded through the mouth to seize and tear prey; *Dicranophorus.* **E. Incudate type,** another version of a forceps-like structure used to seize prey; *Asplanchna.*

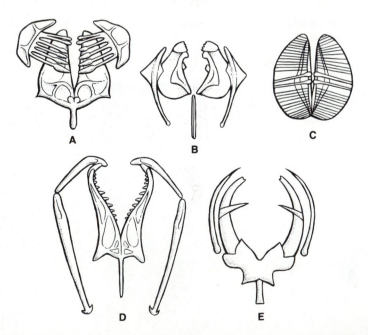

Like most other aschelminths, rotifers have a fluid-filled pseudocoelom between body wall and digestive tract, and there are no respiratory or circulatory systems. Discrete bundles of muscles are coordinated by a simple **nervous system** that centers about a brain in the anterior end, from which run 2 main ventral nerve cords. Many rotifers have simple eyes and sensory projections. The **excretory system,** which serves mainly to regulate water content, comprises a pair of protonephridia, each with multiple flame bulbs. In most rotifers, the terminal portion of each excretory tube opens into a bladder that pulsates, ejecting its contents periodically. The excretory system discharges into the posterior part of the intestine, the cloaca. The **reproductive system** also opens into the cloaca. The sexes are separate, and there is either a single gonad or a pair, a character that divides the two classes of rotifers. The eggs are usually deposited on a substrate or attached to the mother's body; a few species brood them internally and give birth to live young.

Most tissues of rotifers are not divided into distinct cells but, like some of the tissues of flatworms and nematodes, consist of **syncytia.** Cell membranes are present in embryonic stages but later disappear. It is also a striking fact that the number of cells of the late embryo, or the number of nuclei of the adult, is constant (about 1,000) for each individual of a species; and, further, each nucleus occupies a constant position, so that all the nuclei of a rotifer can be numbered and mapped. Such cell or nuclear constancy also appears to some degree in nematodes, acanthocephalans, and tardigrades (see end of chapter).

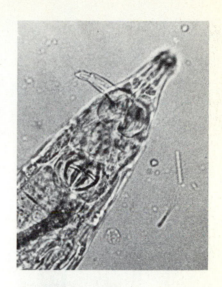

Antenna of a bdelloid rotifer is a moveable fingerlike projection tipped with sensory bristles. It extends from the midline of the dorsal surface of the trunk. In many rotifers there is a pair of similar antennas, one on each side, near the head or on the trunk. (R.B.)

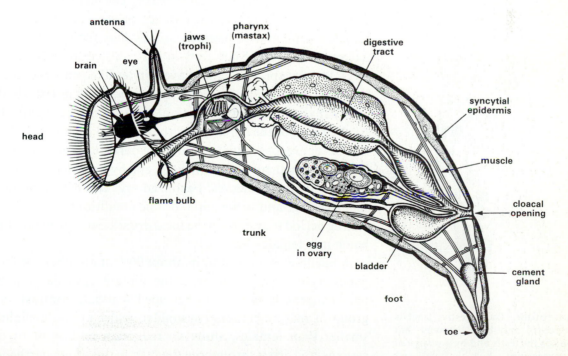

Rotifer, showing general structure in diagrammatic lateral view. Only a few of the nuclei are shown. (Combined from various sources.)

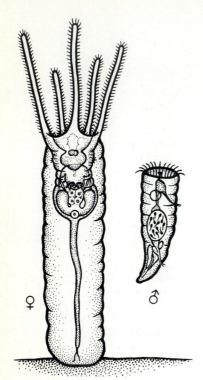

Sexual dimorphism. Male rotifers are smaller and often simplified. In *Collotheca* the adult females are sessile and the males swim freely. (After W.T. Edmondson)

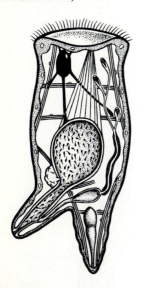

Male rotifer. The digestive tract is represented only by a ligament that supports the testis and by a vestige of the cloaca into which the testis and excretory system open. (After Delage and Hérouard)

THE EXTRAORDINARY **life histories** of most rotifers can be understood only in terms of the constantly changing environment that they face in small or temporary bodies of freshwater. During most of the year, populations of typical rotifers consist entirely of females. These give rise to other females by way of eggs that develop without being fertilized, a process called **parthenogenesis.** Parthenogenesis is accomplished by various organisms in many different ways, with different genetic consequences. In rotifers, during most of the year, eggs are produced without meiosis, so that they are diploid, like other cells of the body. The female offspring are also diploid and (except for mutations) are genetically identical with the mother, so that this is a kind of asexual reproduction, in the sense that no genetic recombination occurs.

In another kind of parthenogenesis, a haploid egg is produced by meiosis but then becomes diploid by fusion of its nucleus with another haploid nucleus, usually one of those arising during meiosis or at the first cleavage. In this case, recombination of genetic material between chromosomes may occur and the offspring are not identical with the mother or with each other, although there is less genetic diversity than in typical sexual reproduction, when mating occurs between two unrelated individuals.

During brief periods of the year, some female rotifers of some species produce haploid eggs. These eggs, if unfertilized, develop by parthenogenesis into males. The males are haploid, are smaller than females (in some species only 1/10th the female's size), and sometimes entirely lack the digestive and excretory systems. Such individuals can live for only a few days. In most species, males have a stylet-like penis and inject their sperms through the female's body wall and into the body cavity (hypodermic insemination). The sperms fuse with the haploid eggs of the female; and the fertilized diploid eggs laid thereafter are distinguished from the parthenogenic ones by a hard thick shell, often ornamented, and by a greater content of energy-rich food reserves. Fertilized eggs can withstand drying, freezing, and other unfavorable conditions, and after a resting period (commonly the local dry season or winter), they hatch into females.

In bdelloid rotifers, and in about 90% of all other rotifers, males have never been seen; the diploid eggs develop by parthenogenesis and always become females. In contrast, in a group of marine rotifers (seisonids), males are only slightly smaller than females, show no reduction or loss of organ systems, and occur throughout the year; in this group parthenogenesis is unknown.

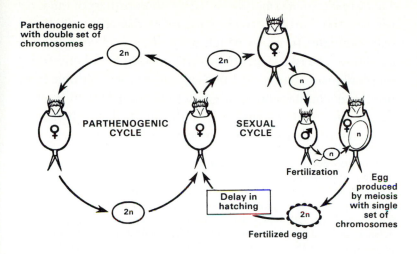

Parthenogenic egg
with double set of
chromosomes

2n

PARTHENOGENIC
CYCLE

2n

2n

2n

n

SEXUAL
CYCLE

Fertilization

♂

n

n

n

Egg
produced
by meiosis
with single
set of
chromosomes

Delay in
hatching

2n

Fertilized egg

Parthenogenic and sexual cycles of *Euchlanis dilatata,* typical of the minority of monogonont rotifers in which males have been observed. Reproduction is parthenogenic most of the time, sexual only during brief periods. (Modified after C.E. King)

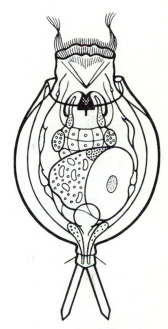

By parthenogenic reproduction, a single freshwater rotifer could theoretically populate a small lake in about 2 months. This calculation is based on individuals that have a lifespan of 10 days, mature by the third day, and produce one offspring per day thereafter; it takes into account the normal mortality of these short-lived animals. The final population density calculated for a lake 1 km by 0.5 km by 2 m deep is 1,000 rotifers per liter, a figure comparable to naturally occurring densities. The total population at this time would be 10^{12} (a million million) genetically identical individuals (barring mutations).

All of these individuals would be competing with one another for the same food resources if it were not for another trick that has evolved in some rotifers. In the largely predaceous genus *Asplanchna,* genetically identical females can develop into forms so different in size and shape that they were at first described as separate species. Laboratory studies have shown that the frequency of the different forms depends upon diet. Those that hatch in the spring from fertilized eggs are small rounded (saccate) forms that prey mostly on protozoans or on smaller kinds of rotifers. As the season progresses and green algal unicells multiply in the water, so do large herbivorous zooplankters, potential prey for *Asplanchna.* The green cells (like those of higher plants) produce a compound called tocopherol, or vitamin E, which animals cannot produce. As tocopherol increases in the diet of the saccate rotifers, they begin to give birth to young that grow to much larger sizes and are able to feed on the bigger prey that are now available.

Resting egg of a rotifer is a diploid, fertilized, female-producing egg with a hard, thick shell that protects the embryo during unfavorable conditions. (After H. Miller)

This larger (cruciform) type, if fed on other rotifers of the same genus or on certain crustaceans, gives rise to a still larger (campanulate) form, to which prey of even greater size are available. The two larger forms can no longer feed effectively on small protozoan prey, but as all three types continue to be produced simultaneously, a greater selection of food is available to the clone as a whole. In addition, the larger forms are protected by their size from some predators.

With the development of larger forms in *Asplanchna,* cannibalism becomes a problem, especially for the small saccate females, which are readily devoured by larger forms of their own or other clones. In addition, the smaller males are in danger of being eaten when they approach hungry females to mate. The males bear extensible projections that appear to be defenses against cannibalism, rather than against predation by other animals, because their size is related to the specific appetites of their own females. In one clone examined, the largest females voraciously attacked the males, which had well-developed defensive projections and were rarely captured or ingested. In another clone, the large females seldom attempted to eat the males, which had only rudimentary defensive projections; these same females did not hestitate to attack either sex of the first clone.

In *Asplanchna,* sexual reproduction occurs only in the larger forms. Thus the presence of tocopherol—which is a fairly reliable indicator of a time when food is abundant and the population of rotifers is rapidly growing—is also used by these rotifers to trigger sexual reproduction under conditions that are favorable, both for the chances of males to encounter females and for the females to have sufficient resources to produce the energy-rich resting eggs. Other species of rotifers use different indicators of an auspicious time for sexual reproduction. Some respond to an unidentified product that accumulates in the water when the rotifer population increases to sufficient densities. Others, presumably living in more regularly predictable habitats, time their reproduction by daylength.

Another way in which some rotifers have evolved to cope with the often extreme conditions in terrestrial habitats or small bodies of freshwater is by developing the ability to withstand **dessication.** In an almost completely dried state, rotifers with a normal lifespan of only a few days or weeks may survive for years. As soon as they are again immersed in water, they swim about and feed actively. Because of this capacity, rotifers can live in places that are only temporarily wet, such as roof gutters, cemetery urns, rock crevices, among moss, and in similar habitats. When the water evaporates, the body contracts to a minimum volume and loses most of its water content. Sometimes the animal itself dies but its contained eggs

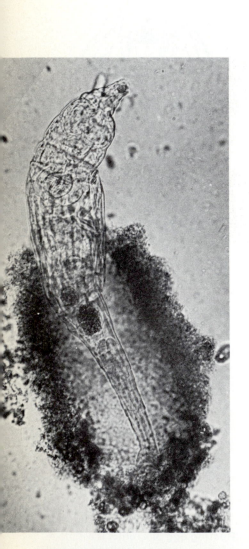

Gelatinous case, secreted by certain bdelloid rotifers, serves as a receptacle for the deposit of eggs. (R.B.)

survive until moisture returns. Because of their small size and their capacity for withstanding temporary drying, certain rotifers have been distributed the world over, chiefly by wind and by birds. If environmental conditions are similar, a lake in Africa is likely to contain some of the same species of rotifers as a lake in North America. The same is true for some other small animals that can withstand drying, such as nematodes and tardigrades (see end of this chapter).

Colony of rotifers lives permanently fixed to submerged vegetation in freshwater ponds. Its members, all females, are attached by an adhesive secretion from the foot and arranged so regularly that the colony is nearly spherical (diameter, 4 mm). The eggs, produced synchronously, are attached to the mother's body; and the juveniles that hatch after 3 to 4 days, although they can crawl and swim, are tethered to the parent colony by threads of adhesive

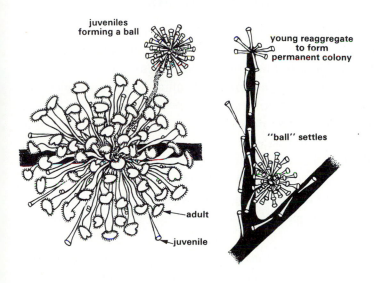

material. Within hours of hatching, the juveniles aggregate into a ball which is at first bound to the parent colony by a thick strand formed of their individual tethers intertwined. The ball is jerked and twisted by the combined movements of the active juveniles, and it soon breaks free and swims away. The juveniles have red eyespots (later lost) and direct the ball consistently toward brightly lighted areas, where the organisms on which these rotifers feed are most abundant. When darkness falls, the ball of juveniles prepares to form a permanent colony, testing any substrate that they contact by moving their coronas along its surface. When they have located a suitable plant, they begin to disperse out of the ball, continually secreting adhesive threads from the tip of the foot as they crawl along, exploring the surface. Using these as guidelines to locate each other, they finally reaggregate to form a permanent colony. Males, produced by the colony only at certain times, are free-swimming and resemble the juveniles. *Sinantherina* (or *Synantherina*). (After F.M. Surface)

CLASSIFICATION: Phylum ROTIFERA (or ROTATORIA)

Class DIGONONTA, with 2 gonads.

Order Seisonidea. Marine, commensal on crustaceans *(Nebalia)*. Sexual reproduction only, no parthenogenesis; males constantly present and well developed. *Seison.*

Order Bdelloidea. Leechlike rotifers that can creep and swim; among the commonest animals in freshwater. Also live in mosses and other habitats subject to drying, and most can withstand prolonged dessication. Parthenogenic reproduction only, no males known. *Philodina, Macrotrachela, Rotaria, Habrotrocha* (freshwater). *Zelinkiella* (marine).

Class MONOGONONTA, with single gonad.
Includes 90% of species. Sexual reproduction (in at least some species) and parthenogenesis; males known in a minority (about 10%) of the species, always smaller than females and with some systems reduced. Some secrete gelatinous sheaths or construct tubes.

Order Ploima. Includes majority of freshwater species. Most creep and swim along edges of freshwaters; some species, especially predaceous types, swim freely away from substrates; some symbiotic forms. *Notommata, Scaridium, Proales, Bryceella, Albertia, Dicranophorus, Synchaeta* (many brackish and marine species), *Asplanchna, Brachionus, Euchlanis, Monostyla.*

Order Flosculariacea. Many sessile as adults, colonial or solitary. *Sinantherina, Trochosphaera.*

Order Collothecacea. Usually sessile and solitary. Large lobed or tentaculate corona, often funnel-shaped. *Collotheca.*

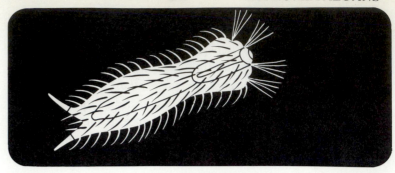

GASTROTRICHS

gastro = "belly"
tricha = "hair"

ALMOST ANY aquatic debris that contains rotifers will also contain a few members of the small phylum **GASTRO-TRICHA.** These minute colorless metazoans are about the size of rotifers but have no crown of cilia. They move over the substrate, and sometimes swim, by means of cilia on the ventral surface. The dorsal **cuticle** is often elaborated into scales, spines, or bristles, and at first glance gastrotrichs can be confused with ciliated protozoans. But after a few moments of watching a microscope field full of busy ciliates that zip about, one learns to distinguish gastrotrichs by their usually bristly appearance, typically forked tail, and unhurried, graceful gliding.

   Sensory structures may include tufts of long cilia and bristles, especially on the head; ciliary pits or small tentacles; and occasionally, red pigmented spots, in the brain, that are thought to be photoreceptors. The **digestive system** is a straight tube with a muscular pharynx. Bacterial, protozoan, and algal cells, along with bits of organic detritus, are ingested by the sucking action of the pharynx or by ciliary currents. The anus opens near the posterior tip of the body or, in the commonly seen freshwater species, at the base of the forked tail end. **Cement glands** occur in the tail fork or along the body and secrete adhesive materials that serve in anchoring the animal to solid objects or sometimes as an aid in inchwormlike creeping. Paired **protonephridia** (each with flagella borne singly rather than in "flame" tufts) have been found widely in freshwater gastrotrichs, rarely in marine ones. There is no body cavity (though spaces between the organs are probably pseudocoelomic). As in other aschelminths, there are no special circulatory or respiratory systems.

   About half of the roughly 500 described species of gastro-trichs are marine, living interstitially in sand, and half are freshwater forms. Most of the marine species **reproduce sexually** and are *hermaphroditic;* each individual functions as both male and female, either simultaneously or sequentially.

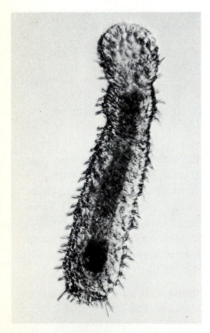

Marine gastrotrich shows adhesive tubules along both sides and the rear. *Diplodasys,* 300 μm long, from depth of 250 m, NE Atlantic. (G. Uhlig)

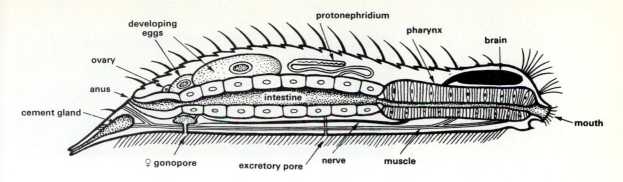

Diagram of a freshwater gastrotrich, lateral view. From the **brain,** two lateral nerves extend along the body. **Muscle bands** are shown only ventrally, where they are most prominent; the muscular system is similar to that of rotifers. The **digestive system** resembles that of nematodes. In the parthenogenic form shown here, there is a **female reproductive system** only. The eggs originate in a pair of posterior ovaries and lie between the gut and body wall within a delicate oviduct. They may be released through a ventral posterior pore or by rupture of the body wall. In hermaphroditic forms, a male system is also present, usually consisting of paired anterior testes with ducts leading to a ventral male pore. Each **protonephridium** opens through a ventral pore. (Partly after C. Zelinka)

All or almost all of the freshwater species **reproduce asexually** by **parthenogenesis,** females laying unfertilized eggs that hatch into females. Such asexually produced "generations" may succeed each other at intervals of only 2 or 3 days, or the animals may lay resting eggs, which can withstand unfavorable conditions for long periods. The eggs of gastrotrichs are attached to sand grains or to vegetation, and development is direct.

Gastrotrichs are among the smallest metazoans, many measuring less than 0.1 mm and few exceeding 1 mm in length. Their eggs, in proportion to body size, are among the largest.

CLASSIFICATION: Phylum GASTROTRICHA

Order Macrodasyida. Typically strap-shaped with many adhesive glands and tubules along body. Hermaphroditic. All marine or brackish. *Macrodasys, Diplodasys, Turbanella, Thaumastoderma.*

Order Chaetonotida. Typically bottle-shaped with adhesive glands and tubules in forked tail end. Almost all freshwater species are parthenogenic females; marine species are parthenogenic or hermaphroditic. Mostly freshwater; some marine species. *Chaetonotus, Lepidodermella, Neogossea, Xenotrichula.*

A few parthenogenic freshwater forms have, in addition to a functional female reproductive system, a reduced male system. Bodies thought to be sperms have been found in some species, but there is as yet no evidence for either self- or cross-fertilization. No purely male individuals have ever been reported.

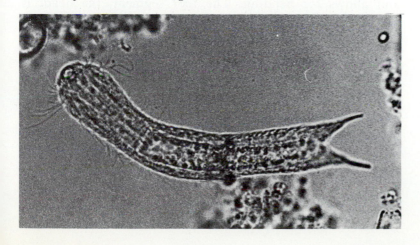

Freshwater gastrotrich feeding on organic debris; clearly visible are the sensory bristles on the head, cilia at the sides of the body, the digestive tract, and the tail fork. (M.B.)

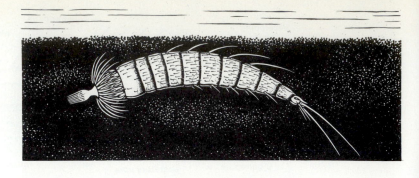

KINORHYNCHS

IN THE MUD of marine or estuarine shores live minute spiny animals that seem to be prisoners within their own armor of cuticle. Having no locomotory appendages or external cilia, and unable to swim, they burrow through the mud, pulling themselves along by a sort of snout. The phylum takes its name, **KINORHYNCHA** (kino = "motion," *rhynch* = "snout") from this method of locomotion.

Only about 130 species of kinorhynchs have been described, but the total number has been estimated at perhaps 500. They have been found in scattered locations around the world, and more will undoubtedly be discovered with further searching. But kinorhynchs are less than a millimeter long, and finding them is not so easy. One must skim off the upper 1 or 2 cm of muddy sediment, where kinorhynchs mostly occur, add some clean seawater, and bubble air through the mud slurry (a bicycle pump is handy). Kinorhynchs and other interstitial inhabitants with hydrophobic (non-wettable) surfaces are caught in the bubbles and brought to the surface film; from there they can be lifted off with a sheet of paper, rinsed into a dish, and examined under the microscope.

Between each of the 13 **body segments,** there is a narrow zone of thin flexible **cuticle** that separates stiff cuticular plates and allows for bending. The first segment, the head, is extended during locomotion and anchored in the mud by a ruff of **long recurved spines,** while the trunk is pulled up. A **protrusible mouth cone,** with the mouth at its tip encircled by spines, can be withdrawn into the head, which in turn can be withdrawn into the neck (the second segment), and in some species head and neck can be withdrawn into the trunk. With all these parts fully extended, a kinorhynch feeds on diatoms, other microorganisms, and organic particles in the mud, sucking in food with its **muscular pharynx.** Foregut and hindgut are lined with cuticle, which is periodically molted, along with the cuticle of the body surface, by growing juveniles. The anus opens at the posterior end of the trunk.

kye'-no-rink

The discoverer of these little animals regarded the section bearing the long backwardly-directed spines as the neck, and suggested the name echinodere *(echino* = "spiny"; *dere* = "neck"). Some zoologists use the alternate names Echinodera or Echinoderida for kinorhynchs.

The concept of segmentation, as it is usually defined in connection with development of the coelom in annelids and arthropods (see chapters 16 and 20), is different when applied to pseudocoelomate animals. Although the kinorhynch body is organized as a series of repeated units (involving the cuticle, muscles, and nervous system), some biologists prefer to call these units *zonites* rather than segments.

The general structure is typical of that seen in other aschelminths, with a spacious **pseudocoelom** between body wall and straight digestive tract. From an anterior nerve ring leads a midventral nerve cord with a ganglion in each segment. And the muscles are also segmentally arranged. Two flame-bulb protonephridia open on the eleventh segment, and two saclike gonads (testes or ovaries in separate but similar male or female individuals) send ducts to open on the last segment. In these and other micrometazoans, the sperms and eggs are comparable in size to those of larger animals and appear relatively huge within the tiny adults. Mating has never been observed in kinorhynchs, but some of the males possess special spines which they may use as do nematodes. Young kinorhynchs have been observed to hatch by forcefully everting the spiny head and rupturing the egg membrane. The newly hatched juveniles had only 11 segments and lacked mature patterns of spination (important in recognizing species) but resembled adults.

The **phylum Kinorhyncha** is divided into 2 orders.

Order Cyclorhagida, which includes the most common kinorhynchs and the most species. Generally small kinorhynchs with the body circular or oval in cross-section. *Echinoderes.*

Order Homalorhagida, in which individuals are generally larger. Body triangular in cross-section, with a sharply arched dorsal cuticular plate and 2 or 3 flat ventral plates in each segment. *Pycnophyes.*

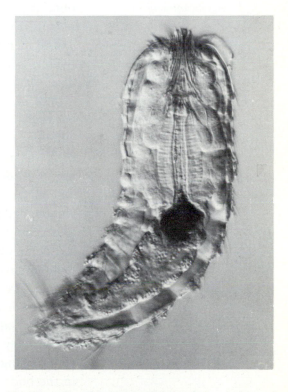

Kinorhynch with head extended in surface view. During locomotion, the prominent ruff of long recurved spines anchor the head, while the trunk is pulled forward. Then, dorsal and lateral spines on the cuticle of the trunk anchor the animal and prevent its slipping backward as it again extends the head. Adhesive tubules, as in gastrotrichs, may also anchor the animal to sand grains. 1,500 m. Great Meteor Bank, Iberian Sea. (G. Uhlig)

Kinorhynch with head withdrawn seen in optical section, with the microscope focused at the very center of the animal. When the head is withdrawn, a circle of plates on the neck segment close together over the front end. One can see the folded spines of the withdrawn head, the cavity and the muscular wall of the pharynx, and the stomach-intestine. 1,100 m. Josephine Bank, N.E. Atlantic. (G. Uhlig)

LORICIFERANS

THE NEWEST PHYLUM, described in 1983, results not from some technical taxonomic rearrangement but from the exciting discovery of some previously unknown animals. They are very small, the adults of the first species measuring only 235 μm long, and they live hidden away in the interstices of clean coarse marine sediments (sand or shelly gravel). Nevertheless, as many small species are already known from this habitat, the new animals come as a surprise. Their discoverer suggests that any specimens collected earlier were confused with rotifers or other known groups, and that few were found in any case because the usual extraction procedure does not loosen their tight hold on the sediment particles. His own success was partly accidental, as is often the case with scientific discoveries. Pressed by lack of time on his last day during a visit to the marine laboratory at Roscoff, France, he skipped over his established collection techniques and hastily washed his marine sediment sample with freshwater. Although he had obtained a few larval specimens earlier, this time he was rewarded with a complete series of larval and adult stages.

Although new species and genera are regularly described in many phyla, the establishment of a new family or order is an event, and new classes are rare. Only 2 other phyla of newly discovered animals have been described in the 20th century, the Pogonophora and Gnathostomulida. Those who consider the aschelminths as a phylum will regard the loriciferans as a new class—still a significant addition.

The members of the new phylum **LORICIFERA** ("corset bearers") have a variety of aschelminth characters, and their discovery seems to make the aschelminths a more cohesive group. The vase-shaped **lorica,** in which most of the body is encased and for which the group is named, is constructed of 6 cuticular plates and looks much like that of some rotifers (and of larval priapulans, see chapter 18). The **mouth cone** and **spiny head,** or introvert, which can be withdrawn and extended, resemble those of kinorhynchs. The mouth is surrounded by a circle of 8 or 9 **stylets,** and 2 accessory stylets appear to protrude into the digestive tract just in front of the muscular pharynx; these remind us of the stylets of larval nematomorphs (or of tardigrades, see end of this chapter). The simple digestive tract has one pair of glands and ends in a slightly dorsal anus. The nervous system comprises a relatively large dorsal brain, a circle of ganglia around the mouth, presumably associated with the spines, and two or more ventral ganglia; the details are still unknown. Paired gonads lie in the body cavity of separate male and female individuals. Smaller **larval stages** are found in at least two discrete size classes, as are empty loricas, indicating that growth and development are by molting. The larvas have 3 pairs of ventral locomotory spines and a pair of leaflike posterior "toes." They climb slowly over sand grains using these ventral spines and the tips of their toes, or swim by paddling with the toes.

Specimens have already been found at widely scattered locations (on both sides of the N. Atlantic and in the Coral Sea), always in similar marine sediments, indicating that these virtually unknown animals are distributed worldwide. If so, biologists the world over can look forward to the challenge of trying to find out how loriciferans feed, reproduce, and otherwise behave, how they develop, how they interact with other organisms, and whatever else may be learned from these tiny newly-met members of the animal kingdom.

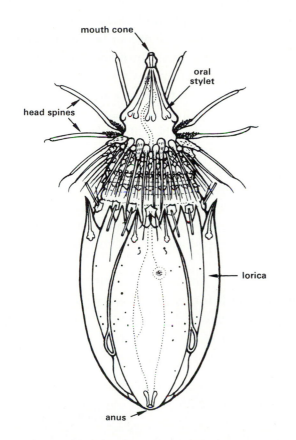

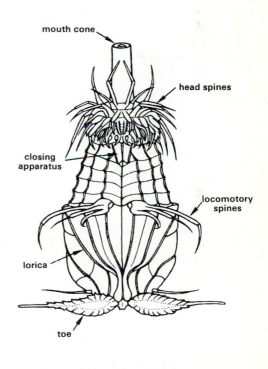

Adult loriciferan, ventral view of a female that is the holotype, the primary specimen which forms the basis for describing the first genus and species, *Nanaloricus mysticus,* family Nanaloricidae, order Nanaloricida, and phylum Loricifera. Collected from shelly gravel substrate near Roscoff, France, at 25 to 30 m depth in 1982. (After R.M. Kristensen)

Higgins larva of loriciferans has only 4 main plates in the lorica. Visible in this ventral view are the mouth cone, abnormally protruded due to the freshwater extraction procedure; the tousled head spines; the closing apparatus, which folds over the head (introvert) when it is withdrawn; the locomotory spines; and the terminal "toes." The ensemble somehow makes the larva look like an appealing if somewhat seedy character from a cartoon strip. (After R.M. Kristensen)

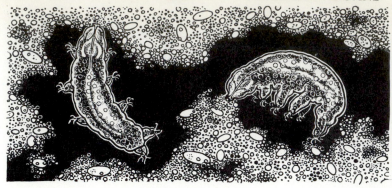

TARDIGRADES

NO ONE who looks through a microscope at tardigrades can fail to be enchanted by these plump-bodied "water-bears," slowly clambering about on plant debris or lumbering along the bottom on 4 pairs of stubby legs. The name of the phylum **TARDIGRADA** refers to the slow pace of water-bears, which are mostly herbivorous and seem never in a hurry to run after or away from anything. Tardigrades are all aquatic but cannot swim and, paradoxically, most of the 550 species are found on land—in tiny pockets of water trapped by mosses and lichens or in other temporarily wetted places such as roof-gutters or cemetery urns. In these out-of-the-way habitats, there are few enemies, though tardigrades do fall prey to large amebas and to nematodes. Tardigrades can also be found on submerged plants or bottom debris in freshwater ponds and among sand grains on ocean beaches.

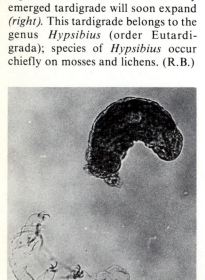

The body is completely covered by a **cuticle** which is delicate in the more aquatic forms and thickened into plates of armor in moss-dwellers. The cuticle is periodically molted. The head bears 2 eyespots but is not set off from the trunk, and sensory bristles may jut from various points on the body.

Tardigrade molting. *Above,* emerging from the old cuticle. *Below,* newly emerged tardigrade will soon expand *(right).* This tardigrade belongs to the genus *Hypsibius* (order Eutardigrada); species of *Hypsibius* occur chiefly on mosses and lichens. (R.B.)

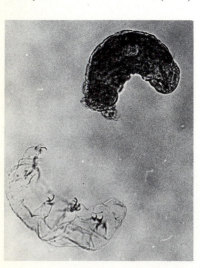

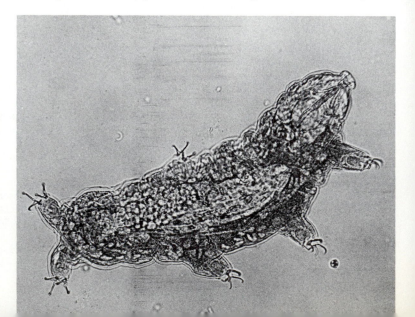

Tardigrades are all small (0.05 to 1.2 mm in length), and several features of their **internal anatomy,** or lack of it, can be related to their size. As in other micrometazoans, there are no circulatory or respiratory systems, but as the animal moves about, fluid in the body cavity is irregularly mixed, distributing food and bathing the various organs. Respiratory exchange takes place through the general body surface. An excretory system is also lacking in some tardigrades, except for certain cells in the epidermis and gut lining that have excretory functions. Other tardigrades have 3 small glands that open into the hindgut and are thought to be excretory.

Projecting into the digestive tract and capable of being protruded from the mouth are 2 sharp **stylets,** secreted by paired glands. Tardigrades use the stylets to puncture the cell walls of mosses and algal filaments or the bodies of small animals such as rotifers and nematodes and occasionally other tardigrades. They then suck out the liquid contents with a muscular pharynx.

The **reproductive system,** with separate sexes, includes a single dorsal gonad. Eggs are released at the female's molt and are often fertilized and protected within the shed cuticle; internal fertilization is also known. In some moss-dwelling species, males are rare or have never been observed, and parthenogenic reproduction presumably occurs. The young hatch from the egg by rupturing the shell with stylets and claws. Except for being smaller, they are like adults and have close to the full adult number of cells; growth is mostly by cell enlargement, as in nematodes, acanthocephalans, and rotifers.

IT IS COMMON for plant seeds and animal eggs to survive periods of drought and freezing and then germinate or hatch out, sometimes years later, when conditions are again favorable. But tardigrades are among the few organisms that can interrupt their normal activities at almost any stage of life and enter a state of **cryptobiosis** ("hidden life"), in which all activity ceases and metabolism declines to almost undetectable levels. In this state tardigrades can survive conditions that would quickly kill an active individual: extremes of temperature from less than 1° above absolute zero to over 150°C; atmospheres with low concentractions of oxygen, high concentrations of CO_2 or H_2S, or even an almost total vacuum; immersion in alcohol and other toxic chemicals; and irradiation with large doses of ultraviolet or X rays. Tardigrades may resume normal life after up to 5 or more years of cryptobiosis, and it has been estimated that repeated cryptobiotic periods can extend a life span of only a year or so to 60 to 70 years.

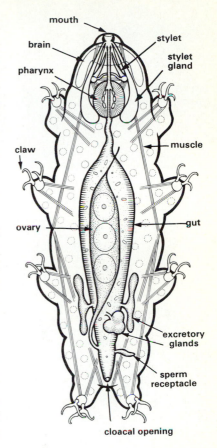

Internal anatomy of a tardigrade, in diagrammatic dorsal view. The 2 stylets and the glands that secrete them lie partly beneath the brain (shown transparent). The presence of excretory glands and gonoduct opening into the gut are marks of the tardigrade order Eutardigrada.

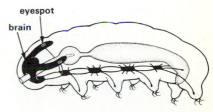

The central nervous system of tardigrades is arthropodlike. The relatively large dorsal brain bears 2 dorsal eyespots. Paired ventral nerve cords connect 5 double ganglia, an anterior ganglion plus one for each pair of legs. (Based on E. Marcus)

Claws of *Macrobiotus echinogenitus*. Scanning EM. (D.O. Gross)

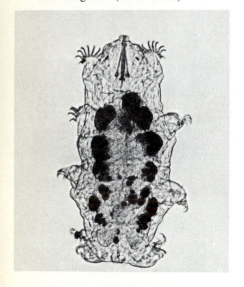

Marine tardigrade *Echiniscoides sigismundi* is only 0.2 mm. It has on each appendage a row of similar claws. Note the 2 eyes, stylets, and food-filled gut. This heterotardigrade was found, at a depth of 2 m, on the sea bottom off Helgoland, W. Germany. (G. Uhlig)

Toed tardigrade *Tanarctus,* only 0.1 mm in length, has long, branching sensory bristles. It was dredged from the sea bottom (N.E. Atlantic) at a depth just over 600 m. (G. Uhlig)

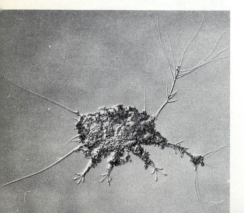

Cryptobiosis is most commonly induced by evaporation of the small body of water in which a tardigrade lives. The animal contracts into a barrel-shape, shielding highly permeable surfaces of its body from contact with air, and slowly dries out. Its water content drops over several days from about 85% to only 2 to 3%. Slow drying is important because it allows time for synthesis of the sugar trehalose, a substance that protects cell membranes during drying. Anesthetized animals, which could not retract the limbs and assume the barrel-shape, lost water 1,000 times faster, and all died. When returned to water, a cryptobiotic tardigrade expands and resumes activity within minutes or hours, the recovery time depending on the duration of suspended activity. Cryptobiosis may also be induced by freezing temperatures, high osmotic pressures, and low concentrations of oxygen.

Tardigrades are found primarily in small temporary pockets of water, not because they actively seek out such habitats (these slow-moving animals depend on being passively carried by wind or water to new homes), but because their capacity for cryptobiosis allows them to survive where few others can. In permanent bodies of water such as ponds or lakes, tardigrades are probably quickly eliminated by the greater variety and larger numbers of predators and competitors. As one might predict, the other animals which are commonly cited as prey or predators of tardigrades, and which share their precarious habitats, are also ones known to be capable of cryptobiosis—protozoans, rotifers, and nematodes.

ALTHOUGH tardigrades are here placed close to the aschelminths, their affinities are far from clear. They resemble arthropods in some ways (e.g., the series of paired legs and the associated nerve cords and ganglia, see chapter 20), but they resemble aschelminths in other ways (e.g., the feeding apparatus). Tardigrade characters such as the absence of cilia and the molted cuticle are shared by both nematodes and arthropods. Although the cuticles and muscles of nematodes and arthropods show distinctive differences, fine-structural and biochemical studies in tardigrades have yielded inconclusive results. In any case these characters may tell us less about affinities than about the particular adaptations of tardigrades. It seems best at the moment not to focus on the phylogenetic puzzle presented by water-bears, but to enjoy them for what they show us about the microworld in which they live.

The **phylum Tardigrada** is divided into 3 orders. The **order Heterotardigrada** includes marine members such as *Tanarctus* and *Batillipes* (which cling to sand grains with adhesive toes) and *Echiniscoides,* as well as freshwater and moss-dwelling types. The **order Eutardigrada** includes the well-studied genera *Macrobiotus* and *Hypsibius* and others occurring chiefly on terrestrial plants or in freshwater. The **order Mesotardigrada** has only one genus, *Thermozodium,* found near Japanese hot springs.

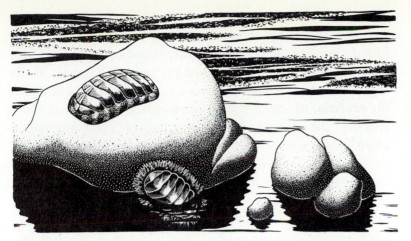

Chapter **14**

Mollusc Body Plan

Chitons

THE SECOND LARGEST of all invertebrate groups—and the one with the largest array of marine forms—is the phylum **MOLLUSCA.** The name means "soft-bodied"; and because of their soft, fleshy bodies, the molluscs, more than any other invertebrates, are widely used as food by humans. Some of the better-known molluscs are snails and slugs, clams and oysters, squids and octopuses.

Despite the differences in the external appearance of a snail, a clam, and a squid, their body plan is fundamentally the same and is distinct from those of all other invertebrate phyla. The basic features of molluscs are much modified, and some are even lost, in a highly specialized animal like a clam. They are less changed—from what we think was the condition of primitive molluscan ancestors—in the chitons.

The **chitons** (ky′-tons), class **Polyplacophora,** are readily recognized by their armor of 8 shell valves that arch over much of the convex dorsal surface. They are sedentary or slow-moving marine animals that graze on the low algal growths and on the animals that encrust rocks. Most live in the intertidal zone of rocky shores or in the subtidal shallows, but one family of chitons has been dredged from depths down to 4,000 m. When disturbed, chitons clamp down upon the rock so tenaciously with their powerful muscles that it takes much persistence, and often a knife, to pry them loose. Such an enterprise is seldom undertaken these days by any but biologists interested in chiton structure or behavior. In leaner times, however, Amerindians of the Pacific coast of North America often satisfied their hunger with the tough texture and fishy flavor of the giant chiton, *Cryptochiton,* which reaches 33 cm in length.

The chiton body is bilaterally symmetrical. At the anterior end is a small, ill-defined head, without eyes or other major

Americans have long used the spelling *mollusk* for a member of the phylum Mollusca. In recent years there has developed a strong trend to adopt the British spelling *mollusc,* which has the advantage of consistency with the phylum name and with the adjective *molluscan.*

poly = "many"
placo = "shell"
phora = "bearers"

There are only about 650 living species of chitons, but the group has lasted relatively unchanged for over 500 million years.

319

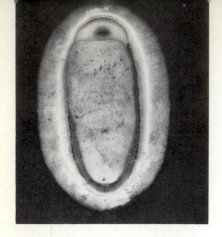

Ventral surface of a chiton is a large fleshy foot. Anterior to this is a head bearing the mouth. Surrounding the head and the foot is the tough marginal mantle, called the **girdle** in chitons; in the groove between foot and girdle on each side is a row of gills. Length, 7 cm. Bermuda. (R.B.)

The **shell valves of a chiton** were obtained by putting a dead animal on the lawn and allowing ants to clean off the flesh. If a chiton dies in the sea, hermit-crabs do the scavenging. Having 8 valves is stable; there are fossil chitons with 8 valves that are 400 million years old. The shell is covered with a pigmented organic layer, part of which may wear off, and a thicker underlying calcareous layer. Actual size. Bermuda. (R.B.)

sensory organs. The ventral surface is largely taken up by a broad, flat, muscular creeping **foot,** abundantly supplied with mucus glands. The **visceral mass,** containing most of the organs, lies dorsal to the foot and is covered by a layer of tissue, the **mantle,** which extends down as a fold alongside the foot and over the head. On its upper surface the mantle secretes a **shell,** which consists of 8 separate, mostly calcareous plates, overlapping from front to rear. The mantle fold creates a sheltered space, the **mantle cavity,** between itself and the foot, and in this groove is suspended a row of gills on each side of the foot. The kind of gills found in chitons and most other molluscs are called **ctenidia.** Each ctenidium consists of a series of flat thin-walled leaflets through which respiratory exchange occurs. There are from 6 to about 90 pairs of gills, the number varying slightly among individuals and markedly among species. Beating cilia on the gills, and on the roof and walls of the mantle groove, create water currents that enter the mantle groove near the anterior end, wherever the animal lifts the mantle edge slightly to provide an inlet. The water flows along the groove lateral to the gills, passes inward between the gill leaflets, continues along the foot, and finally exits at the posterior end. In the roof of the mantle cavity are sensory cells that presumably respond to changes in the content of dissolved materials or of sediment in the water.

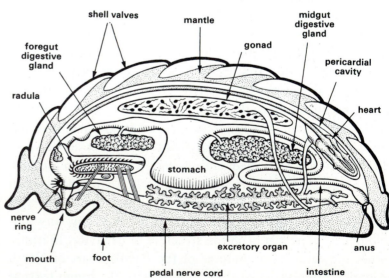

Longitudinal aspect of a chiton, showing mostly the organs of the midline, but also the left member of the paired digestive glands and excretory organs. The two rows of gills, one in each mantle groove, and the lateral nerve cords are not shown. They are represented in the cross-section.

The **digestive system** is a tube extending from the mouth at the anterior end to the anus at the posterior end of the animal. At the junction of the mouth cavity with the muscular pharynx, the **radula** protrudes from the long tubular sheath in which it lies. Unique to molluscs, the radula is a chitinous ribbon covered with many rows of hard, recurved teeth. The teeth of some chitons contain iron, part of it as magnetite, the only known example of biological production of this common mineral. When feeding, a chiton opens its mouth and protrudes the taste-organs from a pouch beneath the radula. Then, if food is detected, the cartilaginous mass that supports the radula is alternately protruded and withdrawn, with a rocking motion; and as the teeth of the radula move over the surface of rocks or seaweeds, small fragments of food are rasped off and ingested. Glands in the mouth cavity supply lubricating mucus, and the bits of food are formed into a mucous string that passes through the pharynx and esophagus into the stomach. Two pairs of digestive glands pour enzymes into the gut. The foregut digestive glands supply starch-splitting enzymes at the junction of pharynx and esophagus, and the midgut digestive glands secrete enzymes for the digestion of proteins, fats, and other substances in the stomach. Digestion continues as the food passes through the intestine, and consolidated fecal pellets passing out at the anus are swept away by the water currents leaving the mantle cavity.

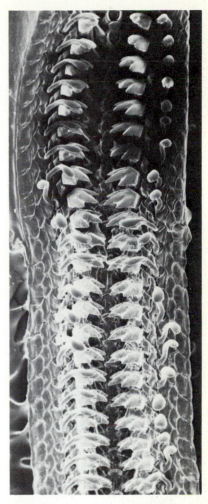

Radula of a chiton, *Nuttallina*. The radula of this chiton (about 30 mm long) was 12 mm long and 1.2 mm wide, with about 50 rows of teeth. Worn or broken teeth at the anterior end of the radula are replaced by new ones secreted at the posterior end within the radular sac. Scanning electron micrograph. (D. Eernisse)

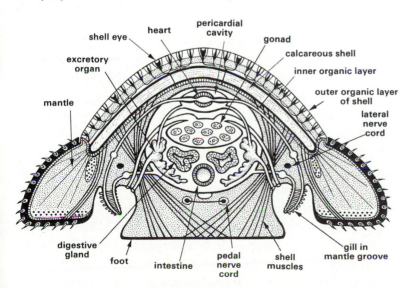

Cross-sectional aspect of a chiton shows the position of one of the numerous pairs of gills, the relationships of mantle and mantle cavity, and the vessels that carry blood from the visceral cavity to the gills and thence to the heart to be redistributed. Note also the sense organs and simple eyes that perforate the shell. The gonad, shown here below the heart, actually lies anterior to it. The cavities between viscera are blood spaces.

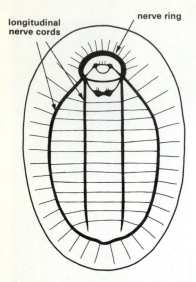

Nervous system of a chiton showing only the largest ganglia and the main branches.

The terms *excretory organ, nephridium, kidney,* and *renal organ* are sometimes used interchangeably. Such names usually imply the excretion of nitrogenous wastes; but in many animals these organs also serve in regulation of the water and salt concentration of the body fluids (osmoregulation), and this may even be their principal function.

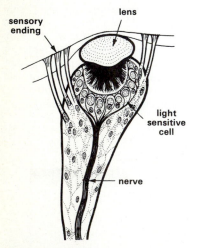

Shell eye flanked by sensory endings at the shell surface. (After M. Novikoff)

The **circulatory system** consists of a specialized pumping organ, the **heart,** a few closed vessels, and extensively branched **open blood spaces,** or sinuses, which are surrounded by connective tissue but lack a lining epithelium. Through these sinuses, blood passes to all parts of the body, then through the gills (where it picks up oxygen and discharges carbon dioxide), and finally back again to the heart. The pericardial cavity is a **coelomic cavity,** a body cavity lined with mesoderm (as are also the cavities inside the gonad and excretory organs). The cavities surrounding the visceral organs are blood spaces. The blood contains cells and dissolved hemocyanin, a blue copper-containing protein that carries oxygen.

Two **excretory organs** lie in a large blood sinus. Lined with a glandular epithelium, they extract nitrogenous wastes from the blood and regulate its ionic content. Coelomic fluid from the pericardial cavity passes through the excretory organs, which resorb salts and other materials. Just before the excretory organs open to the exterior, each has a bladderlike enlargement where, presumably, outgoing fluid accumulates temporarily. The waste material is discharged into the excurrent flow along the foot, near the posterior end.

The major elements of the **nervous system** include an anterior ring of nervous tissue around the gut, connected with two pairs of longitudinal nerve cords. It is a "ladder type" of nervous system, not very different from those of flatworms or nemerteans. As mentioned earlier, there are many patches of sensory cells, especially in the mantle cavity, on the gills, and in the floor of the mouth (buccal cavity). In the outer, organic layer of the shell valves, from the base to the surface, run narrow canals containing strands of cells with sensory and secretory endings at the shell surface. In some chitons, the ends of the larger strands may develop into **simple eyes,** and thousands of these tiny eyes dot the surfaces of the shell.

The **reproductive system** consists of a gonad, located dorsally and in front of the heart, with a pair of ducts that open into the mantle cavity near the posterior end. The sexes are almost always separate. Sperms and eggs are shed into the sea; or the eggs may be retained in the female's mantle cavity, where they are fertilized by sperms entering with the respiratory currents and are brooded until the embryos become well-developed young chitons.

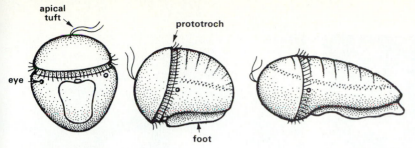

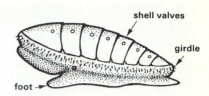

Development of a chiton.
1. Ventral and side views of a swimming trochophore show the foot and the beginnings of the shell valves.

2. While still swimming, the larva begins to elongate and flatten. The divisions between all 8 shell valves have become visible.

3. Settled larva has lost the prototroch; the shell valves are more distinct and the form is becoming adult. (After H. Heath)

Development of the fertilized egg leads to a free-swimming larva, the **trochophore** (similar to that of the marine annelids discussed in chapters 16 and 17). The typical trochophore is top-shaped, and has around its middle a prominent band of cilia (the prototroch) that serves as a locomotor organ and also brings food to the mouth. Though equipped with a mouth and intestine, most chiton trochophores live on the yolk provided in the egg and do not feed. At the top pole of the larva is a group of sensory cells connected to a tuft of long cilia, and just below the prototroch is a pair of simple eyes. Prototroch and sensory tuft disappear after the larva settles, and the rounded larval form becomes progressively flattened as the young chiton takes up its creeping way of life on a rock.

Egg of *Lepidochitona hartwegii.* Chiton eggs are commonly covered by a sculptured hull, shaped by accessory cells in the ovary. Diameter 230 μm. SEM. (J.S. Pearse)

The evolutionary history of modern chitons is just as long as that of the more specialized molluscan classes, and with time the chitons have also become specialized in some ways. The possession of 8 separate shell valves is unique to chitons and probably not a primitive molluscan feature.

Many chitons have homing sites to which they return during the day from their nightly or high tide feeding forays. The flexible girdle permits a tight fit on rock of uneven contour, minimizing water loss and dislodgement by predators or by surf. This species, 5 cm long, has the girdle covered with calcareous spines. Sagami Bay, Japan. (R.B.)

Sperm of *Ceratozona squalida.* Most chiton sperms are unusual in having a long anterior projection, which contains a tightly wound extension of the nucleus. The head, from the tip of the anterior process, to the flagellar attachment (at bottom of photo), is 8 μm long. SEM (J.S. Pearse)

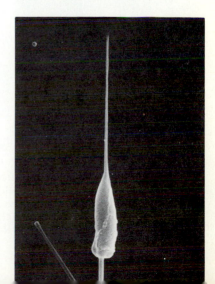

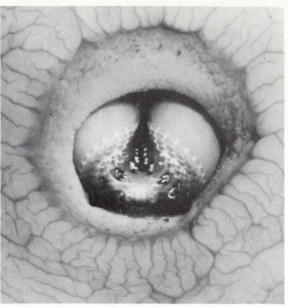

Radular tooth markings made by a chiton on alga-encrusted rock. Santa Cruz, California. (R.B.)

Radula of *Cryptochiton* photographed against the glass of an aquarium. As in other chitons, there are many transverse rows of 17 teeth—one central tooth flanked by 8 teeth on each side. At the very front the radula is unfolding as it scrapes the glass. Monterey Bay, Calif. (R.B.)

A chiton rolls up when detached from a rock—about the only defensive trick these animals have when dislodged from the substrate that normally shields the soft parts underneath. At the *left* the ventral edges of the mantle are pulled aside to show the two mantle grooves, in each of which lies a narrow row of gills. *Cryptochiton stelleri* is the largest species of chiton known; it may reach 33 cm long. The 8 shell valves, embedded beneath the dorsal surface, are not visible. Monterey Bay, California. (R.B.)

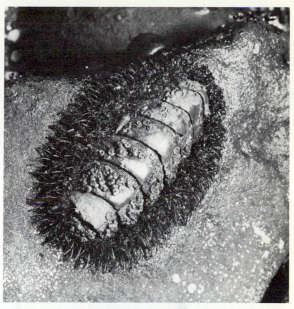

Predatory chiton, *Placiphorella velata,* may browse on encrusting algas or sponges, but it is unusual among chitons in also capturing worms or small crustaceans (even small crabs) by clamping down on them with its "head flap," a very large anterior extension of the mantle. In awaiting its prey, it lifts the front end and supports itself on tentacles that fringe the mantle fold surrounding the head. Length 5 cm. Monterey Bay, California. (R.B.)

The **hairy chiton,** *Mopalia muscosa,* is more tolerant of light than are many chitons of the American Pacific coast. On foggy days it may be seen in the open, but mostly it spends the daylight hours under rocks. This one is on an overturned stone. The girdle is thickly beset with coarse flexible bristles, which shelter many small animals. The eroded shell is encrusted with various small shelled invertebrates. Length, 3 cm. Oregon. (R.B.)

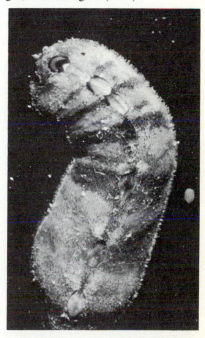

THOUGH LESS COMMON than their more conspicuous and more economically valuable relatives, the chitons are described here because they display the **molluscan body plan** in a more basic form. The body consists of a head and ventral muscular foot, a dorsal visceral mass, and a fleshy mantle which secretes the protective shell. A unique molluscan character is the *radula.* Not all molluscs have a radula, but nothing like it is found anywhere else in the animal kingdom. A shell is not peculiar to molluscs, for hard protective coverings have been developed by various groups; but many molluscs have taken advantage of this protective device in ways that allow them unusual freedom in their activities. The success of the molluscan body plan may be measured by the great number of living species of molluscs, besides a very large number of fossil species. How the various kinds of molluscs have adapted the same body plan to their specialized and different ways of life is the subject of the next chapter.

Cryptoplax is notable for its elongated shape and its widely spaced shell valves, which are only slightly exposed. The separation of the valves permits unusual flexibility. About natural size. Western Samoa. (R.B.)

Snails with helically coiled shells are the most numerous and most familiar molluscs. They comprise a majority of the class Gastropoda, the largest, by far, of the 7 classes of molluscs. This pair, *Caracolla caracolla,* are mating on a tree trunk in a tropical forest. Like all land snails (and slugs) they are hermaphroditic. After a mutual exchange of sperms, both will lay eggs. El Yunque, Puerto Rico. (R.B.)

Marine snail feeding on clam is a common interaction involving members of the 2 largest classes of molluscs, the Gastropoda and the Bivalvia. The large gastropod (17 cm) is *Murex fulvescens,* with a coiled shell bearing stout spines. Its powerful foot muscles are pulling apart the 2 valves of the southern hard-shell clam, *Mercenaria campechiensis,* opening the way for insertion of the snail's proboscis. NW Florida. (R.B.)

Sea slug, *Phidiana (= Hermissenda) crassicornis,* is one of many gastropods that have given up a shell for increased mobility. It relies for protection from predators on chemical and other defenses. The many respiratory projections along the back are tipped with stinging nematocysts derived from feeding mostly on cnidarians. This species frequents mud flats, rocky pools, and pier pilings from Alaska to Baja California. *Phidiana* and other sea slugs are good subjects for studying behavior and learning on a cellular level. Behavioral responses can be pinned down to particular identifiable cells in the head ganglia, something not yet possible in the trillion-cell human brain. (R.B.)

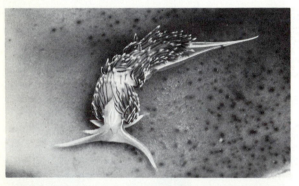

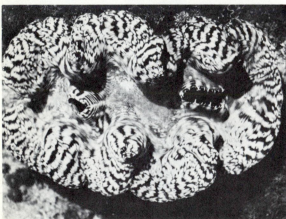

Giant clam, *Tridacna gigas,* is the largest of bivalves. Species of *Tridacna* live in or among coral masses in the reefs of the Indo-Pacific. The fused mantle edges, greatly thickened and often vividly colored, protrude from the shell gape, exposing their extensive surface to sunlight. Algal symbionts (zooxanthellas) in the mantle tissues use clam wastes and supply their host with a steady supply of nutrients. The clam also feeds in the usual bivalve manner—straining plankton from the water that passes through the gill filaments, propelled by beating cilia. (K.B. Sandved)

Common octopus of the Mediterranean, *Octopus vulgaris,* is also familiar in warm waters of the U.S. East Coast. Octopuses have only a trace of a shell, but this is not true of other members of the exclusively marine class Cephalopoda. *Nautilus* has a heavy, chambered shell. Cuttles and squids have a shell remnant buried in the mantle. Naples Aquarium, Italy. (R.B.)

Chapter 15

Molluscan Radiation

Gastropods
Bivalves
Scaphopods
Cephalopods
Aplaco-
phorans
Monoplaco-
phorans

THE MOLLUSCAN BODY PLAN is a winning combination, and chitons have survived with their version of it, scarcely changed, for some 500 million years. But molluscs could not have radiated out into so many varied species by competing with each other for a limited number of rocky grazing sites on marine bottoms. Instead, they have adapted to many new life-styles, some swimming freely in the oceans chasing prey, others climbing to the tops of trees in tropical forests to scrape algas and lichens from the bark. Clams and oysters, enclosed within two protective shell valves, lack head and radula and use the gills for straining microscopic food out of the water. Squids have sacrificed a heavy protective shell for greater mobility in the active pursuit of prey.

Though 80 percent of molluscs are under 5 cm in length, there is a wide spectrum of sizes. At the lower extreme are minute snails and clams less than a millimeter long. At the other extreme are giant squids that measure 18 meters from tip to tip and exceed in length all other invertebrates and all vertebrates except the larger whales. Giant squids and giant clams are the heavyweights among invertebrates. And the jet-propelled squids also hold the record for speed among aquatic invertebrates. But it is their relatives the octopuses that have the distinction of being the invertebrates most capable of learning.

Of the 7 classes of molluscs, 5 have never made their way out of the marine habitat, and together these last account for only a few percent of molluscan species. The rest of this large phylum is grouped into two really large classes: gastropods and bivalves.

Of an estimated total of nearly 110,000 species of living molluscs, about 90,000 are gastropods (snails and slugs). Of these 43,000 are marine, some 35,000 are terrestrial, and more than 12,000 are freshwater. Bivalve species number about 15,000, of which 13,000 are marine and only 2,000 freshwater. Unless otherwise indicated, aquatic molluscs shown in photos in this chapter are marine.

GASTROPODS

Shells of gastropods, like those of other molluscs, have an outer thin organic covering of tanned protein, the *periostracum,* which may wear off. Under this are thick layers of calcium carbonate laid down in a small amount of proteinaceous matrix.

Relatively few gastropods have large, shapely, or colorful shells deemed worthy of collecting. A majority are small snails with coiled, drab shells noticed only by molluscan specialists. In combining complex structure with small size, tiny snails have a competitive advantage over larger complex invertebrates, or less complex small ones, in occupying innumerable *niches* in water and on land. Minute land snails are more widely distributed than larger ones, being readily carried, even to oceanic islands, in the plumage of birds, on floating debris, and by winds.

The gastropods, class **Gastropoda,** comprise by far the largest and most varied of the molluscan groupings. Roughly 5 out of every 6 molluscs are gastropods. The name of the class means "belly-footed," an attempt to describe a group in which the broad muscular foot occupies most of the under side. The sluglike gastropods, with no shells or only embedded remnants of shells, vary more in body shape and in internal structure than do the shelled snails. Both snails and slugs come in creeping, swimming, and floating types. And gastropods as a group carry to the greatest extremes the propensity of their phylum for radiating out into strikingly diverse forms adapted to different habitats and to many ways of life. Many are creeping grazers like the chitons, but there are predatory types with biting jaws in addition to the radula, and parasitic snails that have no radula and feed with a piercing proboscis. It seems impossible to say anything about gastropods without immediately noting exceptions and variations.

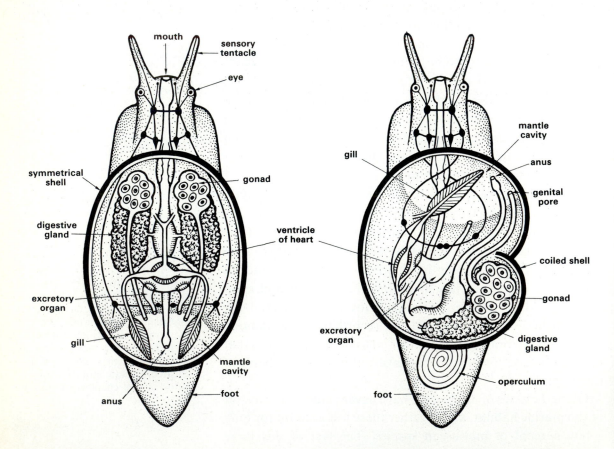

Construct of a gastropod as it would be if torsion and asymmetrical growth did not occur in early development.

Diagram of prosobranch snail in dorsal view, showing torsion and asymmetry of the organ systems.

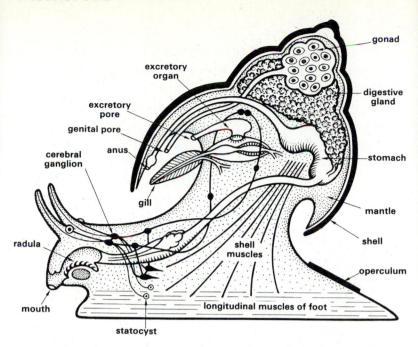

Marine snail in side view. The helically coiled shell identifies it as a snail, and the "gill in front" as a **prosobranch** snail. The bipectinate gill (filaments on both sides of a central axis) is characteristic of primitive prosobranchs; the more highly evolved ones have filaments on only one side.

Most gastropods show all of the basic molluscan features; but they are now, or have been in their evolutionary past, highly modified in a dramatic way: the mantle, shell, and visceral mass are reoriented so that the mantle cavity with all its discharges lies above the head. The advantages of this arrangement are not clear, and neither is its evolutionary history. However, the manner in which it is brought about during larval development has been observed and described in some species of gastropods.

Prosobranch snail, *Littorina irrorata,* the marsh periwinkle, climbing a stem of marsh grass as the tide comes in. At about 15 cm above the water surface at high tide level it will attach by a mucous secretion and close the shell opening with the operculum. This puts it beyond the reach of fishes and larger predatory snails that arrive with the incoming tide. However, its chief predator, the blue crab, *Callinectes sapidus,* can swim to the surface and reach up 7 cm to grab, with both claws, any snail on a stem too short to allow it to climb higher. At low tide the snail descends to the exposed mud surface to browse on diatoms, algal filaments, and plant debris. N.W. Florida. (R.B.)

The whelk *Buccinum undatum* is a carnivore, grasping its prey with the large muscular foot and then attacking it with a long extensible proboscis which has the radula at its tip. (The organ seen protruding from the edge of the shell is not the proboscis but the siphon, a tubular prolongation of the mantle for directing water to the gill.) It uses its proboscis to bore a hole through the hard armor of a recently dead crab or lobster. But, in attacking a scallop, it simply waits until the valves are agape, and sticks the edge of its own shell between the open valves to prevent them from closing. Then it inserts the proboscis and rasps away the soft parts. Shell, 12 cm long. Plymouth, England. (D.P. Wilson)

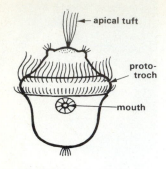

Early trochophore in ventral view.

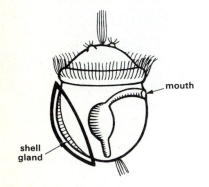

Later trochophore in side view.

Veliger before torsion.

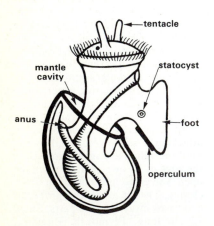

Veliger after torsion. (Mostly after K. Heider)

At an early stage of development, a molluscan embryo becomes a **trochophore,** with an equatorial girdle of cilia. Only the more primitive gastropods have a free-swimming larva of the trochophore type. Most forms pass through the trochophore stage while still confined within the egg capsule. As development continues, the girdle of cilia becomes expanded into a *velum,* often with large ciliated lobes, which serves in swimming and in bringing food to the mouth. This second, usually free-swimming larval stage with the expanded ciliated zone is called a **veliger** and is unique to molluscs. It occurs in the gastropods and in the next two classes described in this chapter, the bivalves and the scaphopods. The development of most of the organs and systems of the adult gastropod is well begun in the veliger. The few strictly larval organs —the velum, some excretory cells (not protonephridia), a larval heart—are lost at metamorphosis. After the larva settles to the bottom, the velum is resorbed, discarded, or devoured by its owner. Some gastropods pass through even the veliger stage within the egg capsule and emerge as juveniles. In land and freshwater gastropods development of the eggs is modified, and usually there is no recognizable veliger stage. In some cases the eggs are not laid at all but develop within the body of the parent.

The veliger at first has an anterior mouth, a posterior anus, a dorsal shell, and a ventral foot. As development proceeds, the digestive tract becomes bent back upon itself so that the anus lies near the mouth. This approximation of mouth and anus occurs also in some other classes of molluscs. But what follows is peculiar to gastropods, a process called **torsion**. While the head and foot remain stationary, the visceral mass, mantle, and shell are in effect rotated through an angle of 180°, so that the anus and mantle cavity are shifted around the right side of the body and upward, and finally come to lie above the head. All of the visceral and mantle organs are reversed right-to-left (but remain top-side-up), and the loop of the central nervous system that serves the viscera and mantle is twisted into a figure-8. In some cases, at least the first stages of torsion occur in as little as a few minutes and appear to result from the contraction of a single large muscle running from the right side of the shell to the left side of the head and foot. Torsion may be completed or may proceed entirely by a much slower process of asymmetrical growth.

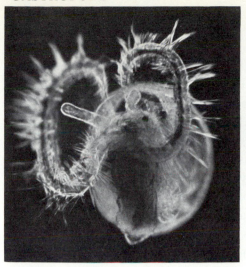

Veliger of *Rissoa sarsi,* a tiny marine snail, has two large velar lobes. This mostly ventral view shows the sole of the elongated foot, and the tentacles and eyes on the head. The visceral mass is enclosed by the larval shell.

Veliger of *Nassarius incrassatus* is notable for the size of the 4 velar lobes that serve for swimming and food-collecting. The adult is a scavenging carnivorous snail. It lays its eggs in tiny egg capsules, each separately attached to the substrate. England. (Both veliger photos, D.P. Wilson)

The **helical coiling** of the shell that distinguishes most gastropods also results from asymmetrical growth, but this process is separate from torsion. A helically coiled shell provides a more compact and stable arrangement, for a visceral mass of a given size, than could a shell in the shape of a straight cone or even a spiral shell in one plane. The great majority of snail shells are in the form of a right-handed coil, the tip of the spire projecting backward and toward the right side of the animal, and the widest portion of the chamber of the shell on the animal's left. This leaves less room for the organs on the right side of the mantle cavity; and the absence, in most marine snails, of the right gill, the right kidney, and the right auricle of the heart, is thought to be related to helical coiling. It has also been suggested that a single gill, rather than two, provides a more advantageous pattern of water currents in the flushing of fecal, excretory, and genital discharges. In fact, in most marine snails the single remaining gill is further reduced to a half-gill with filaments on only one side. Those few species with two gills have shells vented by special slits or holes, for example, key-hole limpets and abalones.

Giant keyhole limpet, *Megathura,* uncovered at low tide. The black mantle almost covers the shell, which has a central opening for escape of the respiratory current. Length, 13 cm. Monterey Bay, Calif. (R.B.)

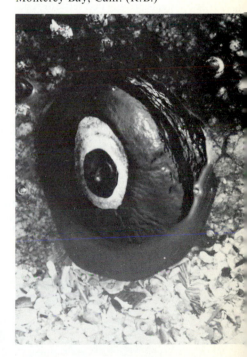

Left, **small keyhole limpets,** mostly 2-3 cm long, occur in warm seas. These were on the Pacific coast of Panama. A related species, *Fissurella volcano,* with a pink shell, extends from Panama to N. California. (R.B.)

We sneer at a snail's pace, but compared to chitons, most snails and other gastropods lead a fast life. Their more active habits are reflected in the prominent **head** and in the complex **nervous system** with well-defined ganglia, highly concentrated, or even fused, in the most active forms. The **sensory structures** are also more abundant and versatile. Like chitons, gastropods usually have sensory equipment in the mantle cavity. The most prominent type in aquatic forms is a raised, ciliated sensory area called an **osphradium,** usually located near the base of the gill, a strategic site for monitoring chemical substances, and perhaps the sediment content, in water entering the mantle cavity. Active carnivorous snails have an especially large osphradium and can detect prey at some distance. Most gastropods have a pair of eyes and sensory tentacles on the head, and a pair of statocysts embedded in the foot. When a snail retreats into its shell, the head with its array of sensitive structures is withdrawn first, the foot follows, and the shell opening is blocked by a flat, horny or calcareous door, the **operculum,** secreted and securely held on the dorsal side of the trailing portion of the foot.

The generalized picture of gastropods that we have given so far mostly portrays **prosobranchs** ("gill in front"). This group includes most marine gastropods, and a few land and freshwater species, typically shelled snails in which the viscera, nervous system, and large anterior mantle cavity and gills clearly show torsion. Other gastropod groups demonstrate the enormous flexibility of the basic molluscan plan, elaborating or eliminating what seem to be among its most characteristic features.

Separate sexes are characteristic of prosobranchs and occur in the herbivorous turban snail, *Tegula.* Attached to this turban snail are 3 smaller slipper shells, *Crepidula,* which feed by collecting small organisms in the mucus of the gill. Slipper shells are unusual among prosobranchs in being *protandric hermaphrodites,* developing first as males and then becoming females. The duration of the male phase can be influenced by secretions from other individuals. Young that settle on rocks or on other surfaces have a brief male phase. Young that settle on or near a large female *Crepidula* remain longer as males. Monterey Bay, Calif. (R.B.)

All photos on this and the next 5 pages are of **prosobranchs.**

Abalone, *Haliotis,* in dorsal view. The flattened spiral shell consists mostly of the last whorl. Along its outer edge a row of holes permits escape of the respiratory current. Wastes are discharged beneath one of the holes. As the animal grows, the smaller holes are sealed and new holes develop. Misaki, Sagami Bay, Japan. (R.B.)

Ventral view reveals the large muscular foot on which the abalone glides about, grazing on seaweeds, at depths from the intertidal to 50 m. A pair of large sensory tentacles are borne on the head and the mantle edge is drawn out into many small sensory tentacles. Length, 7 cm. (R.B.)

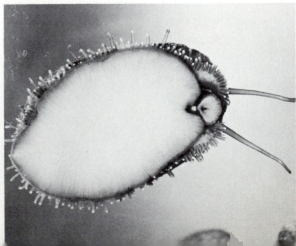

Limpets are gastropods that have a simple caplike shell. In this acmaeid limpet, the underside shows the head with two sensory tentacles, the large muscular creeping foot, the enveloping mantle edged with sensory projections, and to one side of the head, the single (left) gill. Mount Desert Island, Maine. (R.B.)

Limpet among mussel patches is *Patella vulgata,* the common European limpet. It clamps down tightly when the tide is out. When immersed it patrols rock surfaces, grazing on diatoms and other small organisms that settle on rock. This results in bare patches among seaweeds, barnacles, or (as shown here) mussels. Diameter of base, 3.5 cm. West Coast, England. (R.B.)

Homing of limpets can be seen by marking individuals and checking their position at intervals. Each limpet *(Patella)* leaves a "scar" in the rock from chemical dissolution and radular scraping. England. (R.B.)

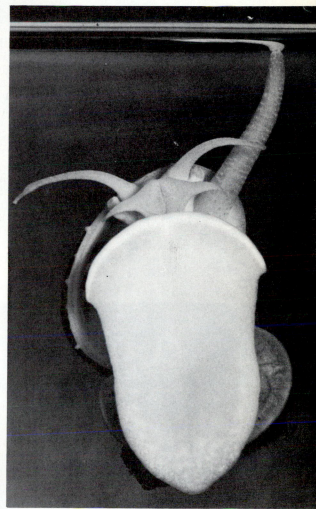

Freshwater prosobranch, *Ampullarius,* taking air through its long siphon. The mantle cavity is partitioned into an air-filled lung and a water-filled gill chamber. Thus this snail (from S. American swamps) can live in stagnant waters and is a popular choice for aquariums. Length, 3 cm. (R.B.)

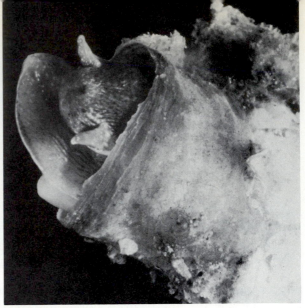

Vermetids, as juveniles, are motile snails with coiled shells. When they find a suitable site, they cement the shell to rock or other shells and become permanently sessile. Further growth produces a tubular shell, sometimes loosely coiled, often straight at the free end. Vermetids feed by secreting sheets or strands of mucus that trap small organisms and other food particles. Cluster of *Petaloconchus montereyensis* (left) has 2-mm openings. Two individuals have head and foot protruding; others show only the operculum, which is shed and regenerated from time to time. *Serpulorbis squamigerus* (right), with an 8-mm tube opening, has no operculum. Vermetid tubes can be distinguished by their shiny white lining from worm tubes, which have a dull lining. A small tube worm, with feathery tentacles protruding, can be seen coiled around the tube of a retracted vermetid at bottom of photo at left. Monterey Bay, California. (R. Buchsbaum)

Heteropod, *Carinaria,* is a pelagic predator and swims upside down with the swimming fin upward and the caplike transparent shell, enclosing the viscera, hanging below. A row of gills protrudes from the shell. The head *(at left)* is bent upward with the mouth at its tip. Behind the mouth are a pair of short pointed tentacles, and near them the brain. The pedal ganglia are visible just below and to the left of the fin. Nerves, muscle fibers, and the digestive tract are revealed by the transparency of the body. About natural size. Monterey Bay, Calif. (R.B.)

Juvenile *Caecum* still has spiral shell, 0.5 mm long. As the animal grows, the spire is cast off, leaving only a curved tube, sealed at its apex and with prominent rings. Tentacles and foot (with operculum) protrude. From coarse sediment in water 8 m deep. Helgoland, W. Germany. (G. Uhlig)

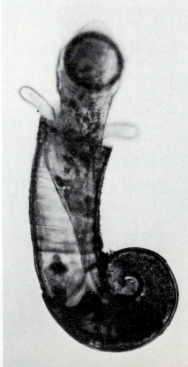

Violet snail, *Janthina (=Ianthina) janthina,* is pelagic in warm seas. It floats upside-down at the surface, with the foot attached to a raft of air bubbles trapped in dried mucus. The snail feeds mostly on *Velella,* browsing on the soft tissues beneath the float of this pelagic cnidarian. After storms the fragile, usually damaged, violet shells are cast up on many warm shores. Some related species attach their egg capsules to the bubble raft, but *J. janthina* is ovoviviparous and ejects small brownish packets that release fully-formed veligers. About 2× natural size. England. (D.P. Wilson)

Moon snail, *Polinices,* produces the rubbery "sand collars" seen on beaches and shallow bottoms. Sand becomes cemented into the gelatinous egg sheet, which the foot shapes into an open ring. When the eggs hatch, thousands of veligers are released. The adult snail is carnivorous and uses the radula and a shell-softening secretion to drill neat, round, countersunk holes through the shells of various bivalves, then sucks out the contents. Delaware Bay, USA. (R. Buchsbaum)

Cowrie, *Cypraea,* about 10 cm, starts out with a spiral shell. Later the lip of the largest whorl curls in, leaving only a slitlike opening in the flattened underside. Cowrie shells have a high shine because the mantle completely covers the shell most of the time and adds lustrous shell layers over the whole surface (covering over the tiny spire). Here the mantle is only partly extended, revealing the shell. Cowries occur in Florida and California, and in European waters, but most are tropical. Hawaii. (R.B.)

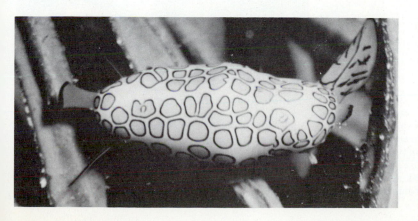

Flamingo tongue, *Cyphoma gibbosum,* common in the West Indies and Florida, creeps along branches of sea whips, browsing on the living polyps. Males stake out territorial claims and fend off invading males. The creamy shell, 25 mm long, is a collector's item but is less beautiful than the pinkish-orange spotted mantle that envelops it. The species name means "humpbacked" and refers to the transverse ridge across the shell. Lerner Marine Laboratory, Bimini, West Indies. (R.B.)

Commensal snail, *Cochliolepis parasitica,* lives under the scales of the large scale worm *Polydontes lupina* in warm U.S. waters from N. Carolina to Texas. Despite its species name, it is not a parasite but one of the few gastropods that live as commensals on other invertebrates (most commensal molluscs are bivalves). This small snail (4 mm in diameter) has a translucent shell, revealing the reddish body and large cream-colored gonad that occupies most of the spire. It feeds on diatoms and organic particles that the host worm brings into its tube in the feeding and respiratory currents. NW Florida. (R.B.)

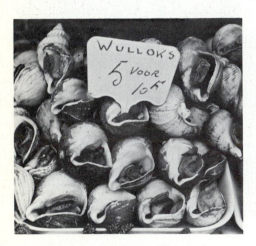

Left, **whelks** *(Buccinum)* for sale in the Netherlands. *Right,* the **pink conch,** *Strombus gigas,* may reach a length of 30 cm and weigh over 2 kg. Overcollected in the Florida Keys, it is still abundant in the Caribbean and used in chowder. The shells have a flaring, pink lip and are sold as souvenirs. Here a conch collector is removing the animals from their shells. Bimini, West Indies. (R.B.)

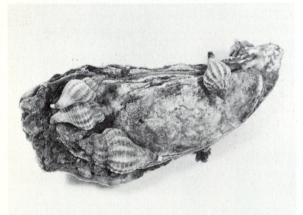

Oyster drills, *Urosalpinx cinerea,* on an oyster. The snail uses both the radula and the acid secretion of a protrusible boring gland to drill holes in bivalve shells. The work is slow and a full week may be spent drilling into a 10-cm oyster before the radula can rasp out the soft tissues. It is not surprising that *Urosalpinx* prefers young, thin-shelled oysters, but this habit of taking many small meals makes the snail a serious pest of commercial oyster beds. 2-3 cm. Beaufort, N.C. (R.B.)

Banded tulip, *Fasciolaria hunteria,* feeds on bivalves by holding the valves firmly with the fleshy foot and slipping the edge of its own shell between the slightly gaping valves, so that they cannot close, then inserting its proboscis to feed at leisure. The tulip favors oysters but will pass them up if *Urosalpinx,* the oyster drill, is abundant. In such cases, an oyster bed may benefit from the presence of banded tulips. Shell length 8 cm. Caribbean and S.E. and Gulf coasts of USA. (R.B.)

Left-handed whelk, *Busycon contrarium,* 25 cm long, differs from most *Busycon* species of the eastern U.S. and Gulf of Mexico, which are dextrally coiled. Here the operculum is visible on the outstretched muscular foot, as the overturned animal attempts to right itself. *Busycon* forages on bivalves, which it finds by sensing the outgoing current of the buried prey. At right, the **egg string** (diam. 25 mm) is a series of leathery flattened cases containing developing young. NW Florida. (R.B.)

Snails laying egg capsules that protect developing young. *Nucella lapillus* lays about 500 to 1000 eggs per capsule, of which more than 95% are "nurse eggs" that are eaten by the few viable young. Only 20 or so young snails emerge from each capsule. Cornwall, England. (R.B.)

Olive snail burying a dead fish before beginning to feed on it. An "olive" has a torpedo-shaped shell streamlined for burrowing in sand with only the upright siphon protruding. *Oliva sayana.* 5 cm. N.W. Florida. (R.B.)

Cone capturing a live fish. *Conus striatus* lies buried in sandy bottoms throughout the Indo-Pacific. It emerges in seconds when it detects chemical cues from a fish. **1.** The fish touches the tip of the proboscis. **2.** The single hollow radular tooth is harpooned into the fish from the proboscis, impaling the fish and injecting a paralyzing venom. **3.** The proboscis contracts, drawing the fish into the mouth. The tooth is replaced from a reserve sac. No one should pick up a cone by the mouth end; the sting of some species may be fatal. Hawaii. (A.J. Kohn)

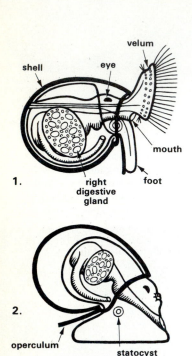

1.

2.

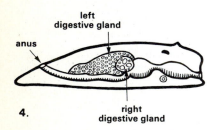

3.

4.

Metamorphosis of an opistho-branch, *Phestilla,* an eolid nudi-branch. **1.** A swimming veliger with velum, shell and operculum, and dor-sal mantle cavity. **2.** The larva at-taches by the foot, the velum is lost, and the larva detaches from its shell and operculum. **3.** The shell and operculum are cast off, and the vis-ceral mass is pulled down and flat-tened against the foot. **4.** The viscera are further straightened as the young animal continues to flatten and elon-gate. (Modified after Bonar and Hadfield)

There is a tendency among many groups of gastropods, including some prosobranchs, toward reduction or even com-plete loss of the shell, which for all its advantages is a handicap to active locomotion. This is often accompanied by loss or reduction of the mantle cavity and gills, and by detorsion and uncoiling of the visceral mass. These trends characterize the diverse groups of gastropods classed as **opisthobranchs** ("gills behind"). Most of these are slow, shell-less creepers known as sea slugs, which have perhaps the most spectacular colors and elaborate forms of any animals in the sea. Opistho-branch veligers have a shell (with a distinctive left-handed twist), which is abandoned at metamorphosis, except by those few types that bear shells as adults. The larval body shows varying degrees of torsion, with the mantle cavity more or less dorsal and anterior, and the anus opening anteriorly on the right side. The mode and degree of detorsion are quite different in the various opisthobranchs. It is not a simple reversal of torsion, but in at least some cases proceeds by a similar combination of muscular contractions and asymmetrical growth. The mantle cavity and gills, if any, and the anus tend to shift posteriorly along the right side. The characteristic twisting of the nervous system, which is one of the clearest indications of torsion in prosobranchs, can be seen in only a few opisthobranchs. In some others, the nervous system shows evidence of untwisting, but in most the ganglia have become so closely concentrated in the head that they can give no clue to the degree of torsion or detorsion that has occurred.

Nudibranchs ("naked gills") have no shell or mantle cavity and bear respira-tory projections on the dorsal surface. In eolid nudibranchs like this one, the dorsal respiratory projections are cerata, fingerlike structures that contain branches of the digestive gland and cover the dorsal surface *(left)*. The tips of the cerata are constricted off as sacs filled with nematocysts from cnidarian prey. The ventral view *(right)* shows the animal laying a spiral egg string, characteristic of nudibranchs. Beaufort, N.C. Length, 1 cm. (R.B.)

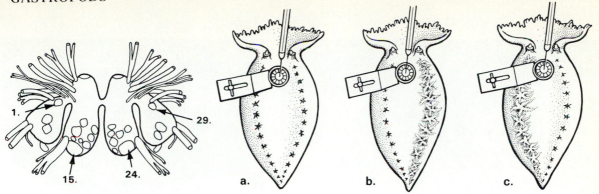

Single nerve cells can trigger discrete behaviors in the nudibranch *Tritonia*. The 3 anteriormost pairs of major ganglia are concentrated (see *above, left*) and have many large, distinctively pigmented nerve cells which can be consistently, individually identified in different specimens. By attaching electrodes, it is possible to record the responses in these cells after various kinds of sensory stimulation of the animal, or to stimulate the cells electrically and then observe the behavioral responses of the animal. Finally, one can construct a map of the nervous connections between cells and of the nerve fibers that lead out from the ganglia to many parts of the body. For example, (see *above, right*) stimulation of certain cells elicits withdrawal of the gills, which in an undisturbed animal expand from the dorsal surface in two rows. Stimulation of cells No. 15 and 24 elicits withdrawal of all the gills **(a)**; cell No. 1 causes withdrawal of the gills on the left **(b)**; and cell No. 29 causes withdrawal of the gills on the right **(c)**. Even a single stimulus applied to certain single cells may result in a complex sequence of escape responses. (Modified from A.O.D. Willows)

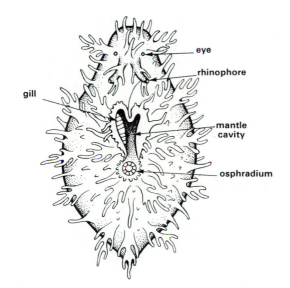

Sea hare, *Bursatella leachii,* an opisthobranch without a shell, from the west coast of Florida. *Above,* **free-swimming larva,** at 15 days, has a coiled shell. *Left,* **adult** (about 14 cm long) has lost the shell. The body has many projections which, with mottled yellow-brown coloring, may help to conceal the animal as it moves about on the bottom feeding on eelgrass and seaweeds. (After L.M. Henry)

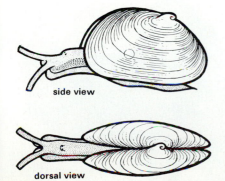

Bivalved gastropod, *Berthelinia limax,* is about 10 mm long, and is green in color from feeding on the alga *Caulerpa.* As in other sacoglossan opisthobranchs, the radula bears only a single file of teeth. The algal wall is punctured with a single curving tooth and the fluid contents sucked out. The juvenile *Berthelinia* hatches with a helically coiled shell and an operculum. This shell becomes the left valve of the adult shell, retaining the spire. As the animal grows, the left valve, which encloses most of the body, adds new shell on the two sides separately, producing the bivalved form. The origin of the bivalved shell is independent of that in the class Bivalvia. (After S. Kawaguti)

Sea hare, *Aplysia,* is used extensively in experimental studies of behavior, reproduction, and development. In the creeping animal, two flaps, extensions from the foot, are folded over the dorsal surface *(left),* and their undulations help to ventilate the underlying mantle cavity. The flaps are spread open in the swimming animal *(right),* revealing the mantle, which covers the visceral mass and the single gill on the right side. Most aplysiids hatch as veligers with a sinistrally coiled shell and an operculum. The pelagic veliger filter-feeds on phytoplankton and grows. After about a month, the larva is ready to settle and metamorphose, and will do so promptly if it encounters the proper algal species. Each aplysiid species settles preferentially on only one or a few algal species, usually the same ones that the adults graze on. The colors of sea hares tend to match the marine plants they eat. This individual was dark brown with lighter mottlings; those of other species may be yellow or olive with dark circles or other markings. After metamorphosis, the shell grows as a flat visorlike projection, coiling dextrally, and the operculum is cast off. In some species, the shell is also eventually lost. Length, 15 cm. Florida gulf coast. (R.B.)

Ragged sea hare, *Bursatella leachii pleii,* like many other aplysiids, releases a purple ink when disturbed. Like virtually all other opisthobranchs, sea hares are hermaphroditic. They aggregate and mate reciprocally in pairs or in long copulatory chains, each inseminating the one in front. Then each lays millions of eggs in gelatinous strings. Florida gulf coast. (R.B.)

Opisthobranch with spiral external shell, *Akera bullata.* The foot is extended into 2 large flaps with which the animal swims gracefully, especially during the spawning season in spring. Most of the time it glides about on European mud flats feeding on seaweeds and plant detritus. Length, about 4 cm. England. (D.P. Wilson)

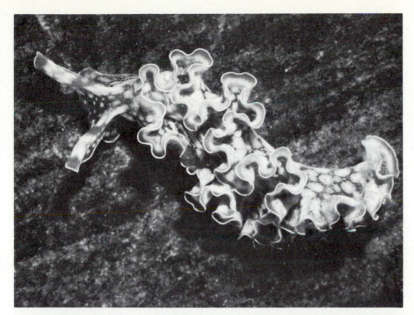

Sacoglossan feeds by sucking out the contents of algal cells. In some sacoglossans, the algal chloroplasts are taken up into cells of the digestive gland and continue to carry on photosynthesis. Products released by the chloroplasts are transported within the animal and used by various tissues. If kept without food, animals in the light lose weight more slowly and live longer than animals in the dark. Chloroplasts probably do not replicate in the animal host and remain functional in most sacoglossans from a few hours to a few weeks (in one species, 3 months), so they must be replaced regularly with freshly ingested ones. *Tridachia crispata.* Length, 6 cm. Caribbean. (L. Harris)

Behavioral studies on the opisthobranch *Pleurobranchaea* have been aimed at such questions as how an animal makes "choices." This carnivore will usually feed rather than right itself, mate, or "sleep," but it will not feed while laying eggs. A hormone from the central nervous system evidently suppresses feeding during egg-laying, but how feeding suppresses other behaviors is still unknown. As behaviors are often controlled by a few nerve cells (see *Tritonia,* p. 339), and as there is evidence that these animals can learn, experimenters also hope to find out how the nervous system is changed during learning. (R.B.)

Pteropods ("wing-footed") are opisthobranchs that swim in the open ocean by flapping a pair of wing-like extensions of the foot. When numerous they provide food for whales. The shelled pteropods seen here are probably not closely related to naked ones. Left, *Cuvierina,* 8 mm long. (W. Beebe). Right, *Corolla spectabilis.* (N.R. Swanberg)

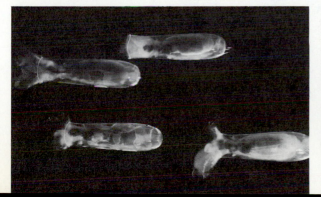

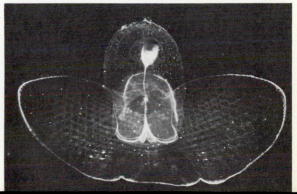

Gray sea slug, *Aeolidia papillosa,* is common on European and on both N. American coasts. Most eolid nudibranchs feed on hydroids, but *Aeolidia* and other members of the family to which it belongs feed mostly on sea anemones. A common prey is the dark red anemone *Actinia,* shown here on the upper left (expanded) and upper right (contracted). As the nudibranch rasps its flesh, the irritated anemone may move away, at 5 cm/hr, by muscular waves in the pedal disk. The nudibranch, at 500 cm/hr, can easily follow. Then the anemone may detach and inflate the column with water. Biting such a loose and inflated anemone is like trying to bite a balloon, and the slug soon turns its attention to another anemone. In some coastal areas *Aeolidia* feeds on the subtidal sea anemone *Metridium,* and is most successful with scattered anemones. This contributes to a patchy distribution of *Metridium,* each patch consisting of a dense aggregation in which individuals are better protected from *Aeolidia.* In its own defense against predators, such as fishes, *Aeolidia* may be protected by the camouflage of its form and coloring, by cnidarian nematocysts accumulated in the ceratal tips, and by acid and mucous secretions of gland cells in the cerata. Length, 7 cm. Plymouth, England. (D.P. Wilson)

Pelagic nudibranchs, *Glaucus,* feeding on a disk-shaped chondrophore, *Porpita* (related to *Velella,* on which *Glaucus* also feeds). Two of the blue and white nudibranchs are suspended, foot uppermost, at the surface and have already nibbled away most of the radiating tentacles of the cnidarian prey. The cerata of a third *Glaucus,* feeding on the soft tissues underneath, protrude from beneath the shining, circular float. *Glaucus* stores prey nematocysts in the cerata; they provide an effective defense, especially when the nudibranch has fed on *Physalia.* Species of *Glaucus* occur in all warm seas. Florida. (W.M. Stephens)

Nudibranch feeding on cerianthid does not kill; it only bites off some of the tentacles. These are regenerated, thus restocking the larder of the nudibranch. *Left,* the nudibranch (about 10 cm long) climbs up the tube, seizes a tentacle with the jaws, and then yanks hard, causing the cerianthid, as it retreats, to pull its attacker down with it *(right).* Both disappear from sight. The nudibranch remains in the tube, feeding on tentacles, for from 20 minutes to 2.5 hours. No matter how long it stays, damage involves no more than about 10 tentacles. Nudibranch: *Dendronotus iris.* Cerianthid: *Pachycerianthus fimbriatus.* Monterey Bay, Calif. (D.R. Wobber)

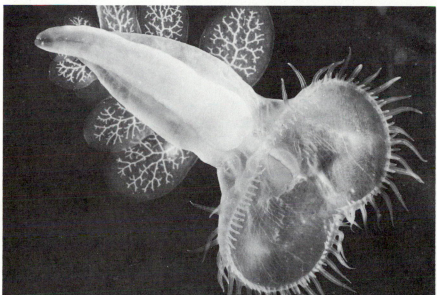

Hooded nudibranch, *Melibe leonina,* with large flattened cerata, swims with foot up (as here) when disturbed or when moving about. Mostly it clings by its foot to eelgrass (Puget Sound) or to subtidal kelp (Monterey Bay). It has no radula and feeds on plankton, sweeping the water with its capacious hood. The tentacles fringing the hood strain small crustaceans from the water expelled as the hood contracts. Alaska to Gulf of Calif. (R.B.)

White-lined nudibranch, *Dirona albolineata,* is translucent grayish white, with flattened and pointed cerata outlined in opaque white. It feeds on small snails, and also ascidians and bryozoans. Size ranges from 1 to 6 cm. Found from San Diego to Alaska and occurs also in Japan. (R.B.)

Dorid nudibranch, *Archidoris britannica* (seen in side view), is called the "sea lemon" from its color. It eats yellow sponges; but a related red species rejects yellow sponges and eats red ones. In dorids the anus opens posteriorly in the middle of the back and is encircled by frilly external gills. Length, 8 cm. Plymouth, England. (R.B.)

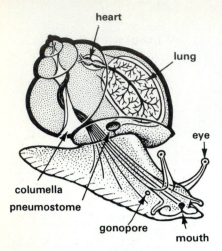

heart

lung

eye

columella

pneumostome

gonopore

mouth

Terrestrial pulmonate snail with shell drawn as if transparent to reveal the highly vascularized respiratory sac and its connection with the pneumostome through which air enters and leaves. The retractor muscles of head and foot are firmly anchored on the columella, the central axis of the shell. The small gonopore near the head serves for both exchange of sperms and egg laying in these hermaphroditic animals. If touched, the eyes are withdrawn to safety by inversion of the tips of the tentacles.

Pulmonates are gastropods that lack gills and are air-breathers. Most live on land—the familiar terrestrial snails and slugs—but some are freshwater snails and a few are marine intertidal limpets. They can be superficially distinguished from prosobranchs by the lack of an operculum. The front edge of the mantle is fused to the neck of the animal, so that the mantle cavity is an enclosed "lung," filled with air and open to the outside only by a small pore, the *pneumostome*. The walls of the lung, supplied with an abundant network of blood sinuses and often raised into folds, present a large moist surface for respiratory exchange. In land forms the pneumostome permits renewal of the air but limits water loss. Freshwater and marine types keep the pore closed while submerged, periodically climbing to the water surface and opening the pore to admit air. The anus and excretory pore usually lie just inside the pneumostome. Pulmonates are all hermaphroditic. Freshwater and land pulmonates lack free larval stages in their development. The mantle cavity and visceral mass show torsion, but the major ganglia of the nervous system, as in the opisthobranchs, are concentrated in the head and their short connectives are not twisted.

Radular teeth of *Stenotrema,* a land snail. The teeth point backward in the animal. Scanning EM. (G.A. Solem, Field Museum of Natural History)

Tooth marks made by radula of *Helix aspersa* scraping mildew off a leaf in a California garden. Leaf is 2 cm wide. (R.B.)

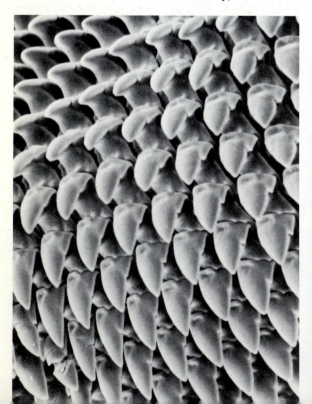

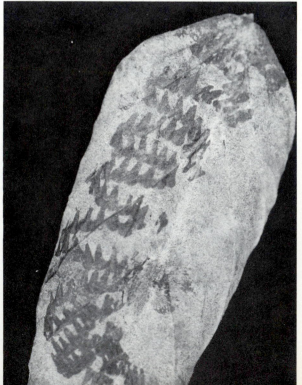

Tree snail, *Liguus fasciatus roseatus*, lives in trees, feeding on algas and lichens that grow on the bark. It descends to the ground to deposit its eggs in the leaf mold. After about 8 months minute snails emerge. During the dry cool winter, the snail attaches itself firmly to a tree by a mucous secretion that hardens and prevents drying. With the return of rain, the cement softens and the snail begins active life again. Pulmonates have no operculum to keep out small insects or mites, but this species secretes a fluid that repels ants. Shell, 4 cm long. Royal Palm State Park, Florida. (R.B.)

Desert snail from the Sahara survives by the same strategies found in many snails of other deserts: white color, retreat to crevices in the rocks, dormancy most days of the year. Desert snails lay few eggs. (R.B.)

Giant African snail, *Achatina fulica*, may reach almost 30 cm, the shell, 20 cm. In their native E. Africa these snails eat mostly decaying vegetation. But introduced into S. Asia, Indonesia, and many Pacific islands, they have become destructive agricultural pests. They are also major vectors of parasitic worms that cause severe, even fatal infections in humans. When first introduced into a new area, the snail populations undergo explosive growth and reach alarming densities, but then appear to decline spontaneously. Hastily introduced predatory snails, intended to control *Achatina*, have instead endangered the native snails of the islands. Moen Island, Truk, Caroline Isl. (R.B.)

Terrestrial slug is a pulmonate that has lost the shell or has only a shell remnant buried in the mantle. The pneumostome can be seen on the right side. *Ariolimax columbianus* reaches a length of 20 cm in the humid coast regions around Puget Sound, Washington. (R.B.)

Garden snail, *Helix aspersa,* a common European pulmonate, may be eaten when the larger and more favored *Helix pomatia,* of French cuisine, is scarce. It has become established in climates as varied as those of eastern Australia, hot moist Malaysia, and hot dry Arizona. Known as the "curse of California," it is no joke to gardeners coping with an animal freed of its native parasites and predators. It emerges from shelters to feed at night or after rains, and estivates during dry periods, attaching to the substrate by a thick mucous seal. (R.B.)

Mating slugs, *Limax maximus,* hang suspended in mid-air from a long cord of thick mucus attached to a branch of a tree or shrub. The courtship that precedes this spectacular mating is also no small event. It may last 2.5 hours, during which the hermaphroditic slugs trail each other in a circle, constantly licking each other and swallowing the mucus. Suddenly they wind their bodies together spirally and hang, heads downward. They unroll their long penises and twine the free ends into a tight knot covered by large umbrella-shaped folds of tissue. In this photo the slugs are in the midst of exchanging spermatophores. Later each slug will fertilize its eggs with the sperms received during mating and will lay a batch of eggs. (D. Norkett)

Planorbid snail has whorls in one plane. This freshwater pulmonate comes to the surface to take air into the lung. Certain planorbids harbor larval stages of *Schistosoma mansoni.* Diam. 20 mm. California. (R.B.)

BIVALVES

IN THE CLASS **Bivalvia,** the "two-valved" molluscs, the body is laterally compressed, and the right and left sides are covered by a pair of shell valves secreted by the two-lobed mantle that encloses head, foot, and viscera. This class includes clams, mussels, oysters, scallops, and some less familiar groups. Many are edible and from prehistoric times have formed an important part of the human food supply, as well as a sizeable part of the diet of our food fishes. A few bivalves are valued for the quality of the pearls they secrete or for the lustrous "mother-of-pearl" lining the inner surfaces of the two valves. Much more important in the human economy is the role of bivalves in recycling vast amounts of organic material and in straining harmful bacteria from polluted waters. For bivalves, with a few exceptions, are filter-feeders. They have no radula.

Most bivalves are clams that can move about slowly by expansions and contractions of the large muscular foot. Mussels seldom move; they live most of their lives attached by threads secreted by the small, narrow foot. And oysters do not move at all, being cemented by one valve to rock or to other oyster shells. Only a few bivalves, such as the scallops, which swim about briefly by clapping the shell valves, move fast enough to flee predators. The others depend for protection upon burrowing habits and the shelter of sturdy shells. The head of a bivalve consists of little else but a mouth flanked by feeding appendages, the palps. Sensory structures are largely concentrated in the mantle edges, which are directly exposed to the external environment and bear sensory projections or in some cases rows of eyes.

Most bivalves are **suspension feeders,** sitting quietly in one place and setting up ciliary currents that flush great quantities of food-laden water through the mantle cavity. The perforated and thickly ciliated **gills** are greatly enlarged and folded, and they serve as sieves for straining out the microscopic plants, bacteria, and organic particles on which the animal feeds. Of the many kinds of filter-feeding mechanisms among invertebrates, the bivalve gill is perhaps the most elegant device for concentrating the billions of microscopic plant cells that proliferate in sunlit waters and form the base of aquatic food chains. The gill also *sorts* the material collected from the water, separating the small particles of food from the mostly larger particles of sediment. Further sorting may take place on the palps. Other bivalves, the **deposit feeders,**

Burrowing clam, beginning to dig in, shows both shell valves, the horny ligament, and the long, partially contracted siphons at the posterior end. Calif. (R.B.)

Attached bivalve, the edible mussel, *Mytilus.* It lives fastened to rocks or to other mussels by strong *byssus threads* secreted by a gland in the foot. These mussels usually live in large clusters and in strong surf. West Coast, England. (R.B.)

Sessile bivalve, horse-hoof clam, *Hippopus,* lies with ligament down. It has a rudimentary foot and never moves. About 30 cm long, and one of the largest of clams, it is dwarfed only by the giant clam, *Tridacna.* Australia. (O. Webb)

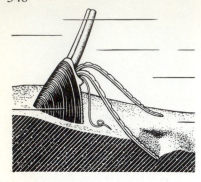

Primitive bivalve, *Yoldia limatula,* lives below low tide mark, in soft mud, and embedded at a steep angle with the posterior end protruding into the overlying water and the *siphons* nearly vertical. Below the two siphons protrude two long *palp-appendages,* extensions of the large palps that flank the mouth. Each palp-appendage is folded to produce a longitudinal ciliated groove, and when the tip of the appendage is inserted in the mud a steady stream of foraminiferans, ostracods, small lamellibranchs and gastropods, and other smaller organisms pass between the palps and into the mouth. The gills are suspended from muscular membranes and are moved up and down, pumping water into the incurrent siphon and out the excurrent one. A long, unpaired, retractile, sensory tentacle emerges near the base of the siphons and lies, usually coiled, on the surface of the mud. It is sensitive to touch and jarring, as are also the sensory projections along the edge of the mantle. (After G.A. Drew)

Late veliger of wood-boring bivalve, *Lyrodus pedicellatus,* with shell and foot well developed. Scanning EM. 0.4 mm diam. (C.B. Calloway)

gather the dead cells, undigested food, and organic particles that settle to the bottom from surface waters and mingle with bottom deposits.

Some of the more primitive (protobranch) deposit feeders use slender folded extensions of the palps to reach out onto the substrate and pick up small particles; the gills are primarily respiratory and do little food-collecting or sorting. A few protobranchs appear to scoop up bottom material by muscular movements of the mantle and foot, sorting the particles first on the gills, then on the palps. In other deposit feeders, the incurrent siphon (a long tubular extension of the mantle) sweeps the surface of the substrate like a vacuum cleaner, sucking up organic particles, which are sorted mostly by the palps. Another mode of feeding is practiced by certain deep-water bivalves which act as carnivores or scavengers. They are called septibranchs because the gills form muscular partitions or septa, that divide the mantle cavity into dorsal and ventral chambers. Up-and-down movements of the septa suck into the ventral cavity small live animals or bits of decaying animal matter and organic debris, which are then swallowed.

Specialized modes of feeding restrict the bivalves to aquatic habitats and prevent them from exploiting the many herbivorous and carnivorous niches open to bottom-living gastropods. However, on shallow, uniform sandy bottoms, where most food grows suspended in the overlying water, individual bivalves often outnumber the gastropods by 5 to 1, or even by 30 to 1.

Marine bivalves include over 13,000 species, about 85 percent of all bivalve species. They range in size from minute clams less than a millimeter long to the giant clams of tropical waters, with shells over 1.5 m long and weighing 270 kg. The massive bodies of such clams have never been weighed, but their volume has been estimated to be 2 billion times that of the smallest clams. Most marine bivalves live on shallow shores, where the water is enriched by erosion of materials from the land and by upwelling of nutrients from nearby depths. But some have been dredged from depths of 10 km.

Eggs and sperms are usually shed into the seawater, where fertilization takes place, a chancy event in open water if spawning were at random. However, in those species which have been carefully studied, it is found that the sexes stimulate each other to spawn simultaneously when sex products expelled by one sex enter the feeding current of the other. Usually it is the males that spawn first, stimulating egg release. In some species the eggs are retained until fertilized internally, a strategy that greatly increases the chances that sperms will reach the eggs. Internal fertilization may be followed by immediate release of the zygotes, which develop into free-swimming **trochophores** and then **veligers** with ciliated velar lobes that gather microscopic food and direct it to the mouth. (Remnants of the velum may later take part in the development of the ciliated feeding palps that guide food into the adult mouth.) In other cases incubation may continue within the gill chambers of the female until the developing embryos are released as juvenile clams, bypassing the free larval stages and sharply increasing the likelihood of survival of the young.

Except for a few species of mussels, **freshwater bivalves** are all clams, and there are about 2,000 species of them. There is hardly a permanent body of freshwater anywhere, no matter how small or isolated, sometimes even an arctic pool or a hot spring, that does not have in it the tiny "fingernail clams." These are less than 15 mm long when full-grown, and the minute young clams are easily carried from one body of water to another in the mud that clings to flying insects or to the feet of birds. Also favoring wide distribution is hermaphroditic reproduction, with self-fertilization. About half of the clams in freshwaters are of closely related families that are often grouped as "unionids." These are found in lakes, large ponds, and streams, but especially in the larger rivers, where in highly oxygenated shallow rapids they thickly carpet the bottom. Because of their large size, up to 25 cm, and their dependence on fishes as hosts for the parasitic young stage, unionids are never found in small isolated ponds. The large size and relatively unspecialized structure of unionids has made them convenient for study. The description that follows is based on some of the common freshwater unionids, but applies in a general way (except for the parasitic life cycle) to most bivalves.

The shell consists of multiple layers. A dark, horny outer layer, or **periostracum,** protects the calcareous shell from being dissolved by acids in the water. It is thin and usually eroded from the older parts of the shell, such as the umbo. A middle, or **prismatic layer** consists of crystals of calcium carbonate arranged perpendicularly to the surface of the shell and laid down in an organic matrix (mostly conchidin). The innermost, pearly or **nacreous layers** consist of tablets of calcium carbonate, alternating with thin sheets of organic material, laid down parallel to the surface of the shell. The first two layers are secreted only by the edge of the mantle, and show prominent concentric markings of discontinuous growth. The inner layer is laid down by the whole surface of the mantle and has a smooth, often lustrous surface. The ligament is also secreted in three layers by the mantle but is not heavily calcified like the shell, remaining flexible and elastic.

Pearls are secreted by the mantle as protection against foreign bodies, usually the larval stages of parasitic flatworms.

Edge of shell and mantle, sectioned to show the layered structure of the shell and the 3 folds of the mantle edge. The outermost fold secretes the periostracum and prismatic layer of the shell. The middle fold is sensory and the inner fold is muscular. Where mantle edges are fused locally to form siphons, these may derive from the inner folds *(Macoma),* from inner and middle folds *(Mercenaria),* or from all 3 folds *(Mya, Tresus)* and have a dark sheath of periostracum. (Modified after Taylor, Kennedy, and Hall)

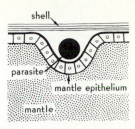

Formation of a pearl. 1. A parasite (round black body) lodges between the shell and the mantle epithelium.

2. The parasite is enclosed in a sac formed by the epithelium, which secretes thin, concentric layers of pearly substance.

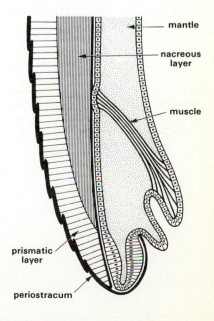

3. Pearl surrounds parasite and prevents it from harming the clam. (Based on F. Haas)

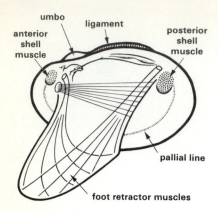

Right shell valve of a clam, showing the main muscles. The large anterior and posterior adductor muscles, shown cut, close the shell. Attached beside the anterior adductor, with fibers running around the upper part of the foot, are the protractor muscles, which help to extend the foot. Anterior and posterior sets of retractor muscles fan out into the foot and provide the force that draws the clam forward when it is digging through the substrate. The muscles that retract the mantle edge are attached along the pallial line.

In unionid clams the anterior and ventral margins of the mantle lobes are free of each other, and this makes for slow and inefficient burrowing as compared with that of a rapid burrower like *Ensis,* in which these mantle edges are fused except for a slit through which the foot protrudes. This improves the hydraulics involved, providing a more powerful jet of water for loosening the sand into which the shell will penetrate.

The rate at which extensions of the foot occur varies with the species. The razor clam *Ensis arcuatus* has been clocked at 90 extensions per minute and the soft-shelled clam *Mya arenaria* at only 1 per minute.

Right and left sides of the **shell** are fastened to each other dorsally by an elastic **ligament.** The gape of the shell valves is ventral. Near one end of the ligament on each valve is an elevated knob, the **umbo.** The end of the animal nearer the umbo is the anterior end. At the opposite or posterior end are the openings through which currents of water enter and leave the clam.

The umbo lies within the oldest part of each shell valve. As the animal grows, the mantle secretes successive layers of shell, each projecting beyond the last one laid down. This results in a series of concentric growth lines which mark the external surface of the shell and represent the successive outlines of its margin.

The shell valves are held agape by the elasticity of the ligament, and closed by the contraction of two large **muscles,** one anterior and one posterior. Near these are smaller muscles which extend and retract the foot; and attached to the shell close to its margin is a row of small muscles which retract the edge of the mantle. When one shell valve is lifted back (after cutting the muscles), its inner surface will show scars representing the former attachments of all these muscles. Also, it will be seen that the dorsal margin of each valve has long ridges and irregular toothlike projections, the **hinge teeth,** that fit into grooves or pits in the opposite valve. This interlocking arrangement fits the two valves together and prevents their slipping past each other when the animal moves.

During **locomotion** the animal burrows through the sand or mud like an animated plowshare. First, the pointed foot is extended forward into the sand, with the blood sinuses acting as a hydrostatic skeleton around which the muscles of the foot contract. The shell, slightly agape ventrally, anchors the clam in the sand. Then, the two large shell muscles further close the valves, which forces blood into the foot, anchoring it

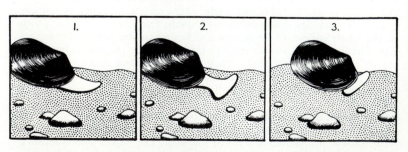

A moving clam, half buried in the sand, extends its foot ...   the tip of the foot swells and acts as an anchor ...   the muscles of the foot contract, drawing the clam forward.

by its swollen tip, and also expels water from the mantle
cavity, loosening the sand around the shell. Finally, the muscles of the foot contract and draw the clam forward. Such a
slowly moving animal, with a shell that is heavy and cumber-
some to carry about, could hardly run down its prey.

The large visceral mass lies dorsally, most of it between the
two large muscles that close the shell. The **mantle** drapes the
visceral mass and extends ventrally as two mantle lobes, one
apposed to each shell valve. The large space between the mantle lobes is the **mantle cavity.** At the posterior end the lobes
are thickened locally and form the openings for the entrance
and exit of water. The rims of these openings, bearing sensory
structures, extend out just beyond the margin of the shell
when the valves are agape. A current of water passes into the
mantle cavity through the ventral, or **incurrent opening,**
through the sievelike **gills,** and out through the dorsal, or
excurrent opening.

Unionid clam lies obliquely in the
sand or mud, with posterior end
tilted up and exposed above the substrate. The mantle edges that form
the incurrent and excurrent openings
barely protrude from the shell, and
this limits such clams to shallow burrowing habits. Many have heavy
shells that compensate for this exposure to predators; muskrats and
raccoons prefer the thinner-shelled
unionids, and fishes prefer the small
sphaeriids (fingernail clams). Some
sphaeriids burrow deeply in soft substrates, as they have siphons, tubelike
extensions of the mantle edges that
have at their tips the incurrent and
excurrent openings. Long siphons,
either separate, or united in a sheath
(see photos of *Mya, Tresus),* have
enabled many of the marine clams to
exploit a great variety of bottom
habitats.

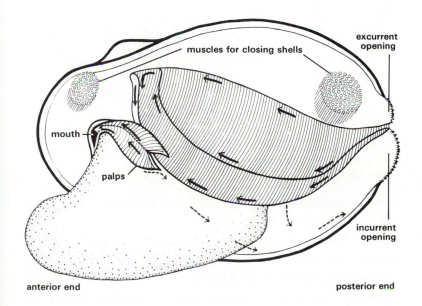

muscles for closing shells

excurrent
opening

mouth

palps

incurrent
opening

anterior end   posterior end

The **feeding mechanism** is ciliary;
food particles, caught on the surface
of the gills, are carried to the mouth as
shown by the *solid arrows.* Rejected
particles are removed from gills and
palps, as shown by *dotted arrows.*
(The left valve and mantle lobe have
been removed.)

Food particles left on the surfaces of the gills by the steady
stream of water are sorted, mostly by their small size, from silt
and other undesirable materials during their passage to the
mouth. Heavy particles of sand or mud drop from the surface
of the gill to the edge of the mantle, are carried backward by
cilia on the mantle, and are expelled posteriorly. The lighter
particles become entangled in mucus secreted by the gills and
are carried, always by beating cilia, to the edges of the gills and
then forward to meet the ciliary tracts on the **palps,** ridged

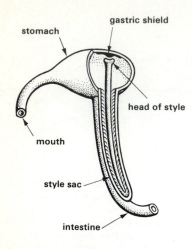

The **crystalline style** of a clam lies in a ciliated sac, and the beating of the cilia in the style sac rotates the style and moves it forward so that its upper end is constantly rubbed against a tough shield of cuticle on the stomach wall. At the base of the rotating style, enzymes and protein-aceous matrix, secreted by certain cells in the style sac, are incorpo-rated into the style, later to be worn away at the upper end of the style and mixed with the stomach contents.

A crystalline style is also present in some gastropods, especially in ciliary feeders on unicellular plants.

folds on each side of the mouth. Further sorting occurs here, and the larger particles are carried to the tips of the palps and then dropped off into the mantle cavity, from which they are removed.

The cilia of the palps direct selected materials, entangled in strings of mucus, into the mouth, which lies between the two lips, or ridges, that connect the palps. There is no radula, nor could it be of any use to an animal that feeds in this way. The food-laden strings of mucus are drawn through the esophagus

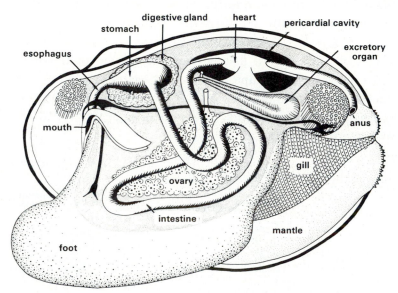

The **digestive system** and other organs are shown here in a clam in which the left mantle lobe, the left gill, and part of the foot are cut away. Excretory organs and gonads open into the mantle cavity.

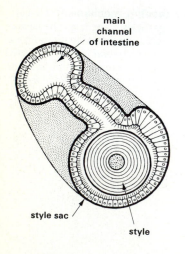

Cross-section of intestine reveals that the style consists of concentric layers of secretion added successively by the style sac in which it lies. (After T.C. Nelson)

into the stomach by being wound around a gelatinous rotating rod, the **crystalline style,** that lies in a sac off the intestine and projects into the stomach. At the same time, the food is mixed with extracellular digestive enzymes set free in the stomach by the slow dissolution of the style tip. Surrounding the stomach, and connected to it by ducts, is a large **digestive gland,** the main organ of digestion and absorption. Partly digested food from the stomach enters the digestive gland, where cells readily ingest and further break down the solid particles inside food vacuoles. As we might expect in animals that eat only finely divided food, digestion is mostly *intracellular.* Undigested residues from the digestive gland and stomach are directed into the intestine by ciliary channels, separate from those that carry incoming food. The intestine then offers a last chance for both extracellular and intracellular digestion. The anus opens near the excurrent opening, and the feces are carried away in the outgoing current.

The **gills** are perforated by microscopic **pores**. The beating of the cilia draws water through the incurrent opening into the mantle cavity, and then through the pores, leaving suspended particles on the surface of the gills. Within the gills the water flows up the **water tubes,** vertical channels formed by partitions that subdivide the cavity between the inner and outer walls of each gill fold. The water tubes open dorsally into the **dorsal gill passages,** which run one above each gill fold and open posteriorly near the excurrent opening.

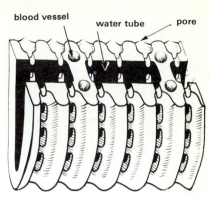

Small portion of a gill enlarged to show how the partitions, in which the blood vessels run, divide the space between the inner and outer walls of the gill fold into vertical water tubes which communicate with the mantle cavity through microscopic pores.

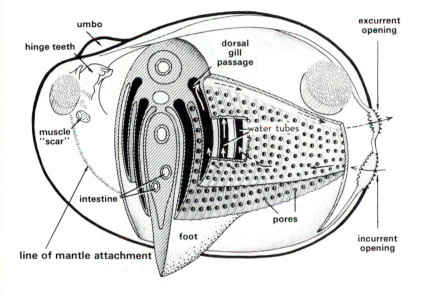

The direction of the **respiratory currents** of a clam are shown by arrows. The anterior part of the clam has been removed. A piece of one gill has also been cut away to show that the space between the inner and outer walls of each gill fold is divided into vertical *water tubes.* The *pores* of the gill are microscopic and are shown here greatly enlarged. The relations of the water tubes to the *dorsal gill passages* can be seen in the sectioned surface. (The intestine appears more than once in the section because it coils back-and-forth in the foot.)

The **circulatory system** consists of a heart with two auricles and a ventricle, blood vessels, and blood sinuses. The heart lies in the pericardial cavity, which is lined with an epithelium and filled with fluid. Heart and pericardium encircle the intestine so closely that the intestine appears to run through the heart, prompting the saying that "the way to a clam's heart is through its stomach." The heart pumps blood both forward and backward through the arteries. Anteriorly, the arteries supply the viscera and foot. Posteriorly, they supply the mantle. Many of the arteries lead into irregular blood sinuses in the tissues. The blood sinuses in the foot and mantle sometimes function as a fluid skeleton, as well as

Gills are visible through the translucent shell of a young scallop, *Patinopecten yessoensis,* 92 days old. The visceral mass is dark. The mantle edges are rimmed with sensory tentacles that alternate with tiny pigmented eyes. Shell length 3.4 mm. South Korea. (Pilae Kang, Institute of Fisheries of South Korea)

distributing the blood. In these clams, as in most bivalves, there is no oxygen-carrying pigment, and the blood is colorless but contains ameboid cells. Some clams, more often those that occupy poorly aerated habitats, have hemoglobin contained within cells.

Osmoregulation in most marine molluscs is poor. The total salt concentration of the blood changes with that of the surrounding seawater, although the concentration of particular ions may be regulated. Marine bivalves maintain their blood concentration temporarily, as during a rain, by keeping their valves tightly closed. In freshwater molluscs, the blood is maintained at a concentration higher than that of the external medium by resorbing most of the salts from the large quantities of water constantly flushed out through the excretory organs and by taking in salts with the food. Salts are probably also actively absorbed through parts of the body surface.

With water unlimited, bivalves and other aquatic molluscs excrete nitrogenous waste in large part as ammonia, which is toxic but can be readily washed away in copious urine and from vascularized surfaces. Land snails, however, must conserve water. They produce nitrogenous excretions in solid form (mostly insoluble uric acid and other purines, less toxic than ammonia) and expel them at long intervals, a convenient arrangement for animals that may be dormant during some seasons.

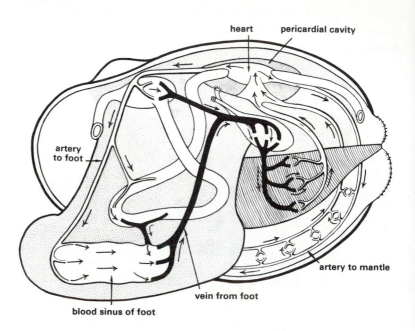

The **circulatory system** of clams is an *open* system with irregular channels and sinuses in the tissues, supplied and drained by blood vessels. The filling of the large sinus in the foot helps to produce the swelling of the tip of the foot in locomotion. The heart is not powerful enough to force blood into the foot sinus. The pressure required to do this, and to expel water from the mantle cavity in a jet strong enough to push aside sand, is provided by the large shell muscles that bring the 2 valves together. Thus the bivalved shell is not only a protective device but also an important part of a clam's equipment for burrowing in soft bottoms.

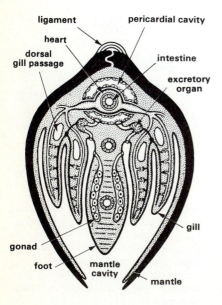

The **excretory organs** lie, one on each side, just below the pericardial cavity. Each is like a tube bent back on itself, with the two parts lying parallel and one above the other. The lower part, with glandular walls, connects at its anterior end with the fluid-filled pericardial cavity. This glandular part extracts waste products of metabolism from the blood and reabsorbs salts from the pericardial fluid. At its posterior end it is continuous with the upper part, a thin-walled bladder which opens anteriorly into a dorsal gill passage. The wall of the bladder is ciliated and maintains an outgoing current.

Cross-section of a bivalve showing relations of shell, mantle, foot, gills, and visceral organs. Gonads may be paired as shown, or fused along the middle. Gonads and excretory organs open into the mantle cavity, either by way of the dorsal gill passages, as described in the text, or into the angles between the inner gill folds and foot, as shown in this diagram. The heart wraps around the intestine.

The **nervous system** is relatively simple, not surprising in sluggish animals that live with the anterior end buried in the mud. The head is rudimentary and without sensory structures. Sensory and integrative functions, both centered in the head in gastropods, are in bivalves divided between two cen-

Simple eyes are present in some bivalve veligers and may persist in the adult at the anterior base of the gill on each side, just above the cerebral ganglia, which innervate the eyes. In the mussel *Mytilus edulis* the eyes are located beneath the special translucent "windows" in the shell.

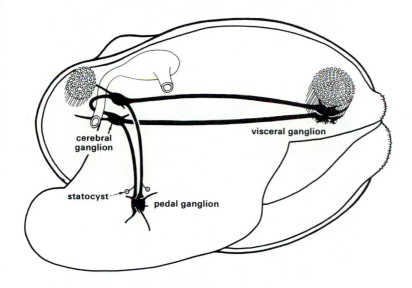

The **nervous system** of clams has a pair of ganglia for each main muscle mass of the body. The *cerebral ganglia* lie on each side of the mouth, joined by a commissure that runs around the esophagus. They send nerves to the palps, statocysts, anterior shell muscle, and anterior mantle. From the cerebral ganglia connectives run ventrally to the *pedal ganglia,* which supply the muscles of the foot. Two connectives also run from the cerebral ganglia to the *visceral ganglia,* which send nerves to the digestive tract, heart, gills, excretory organs, gonads, posterior shell muscle, and posterior mantle.

ters, the cerebral and visceral ganglia. Each of these represents fusions of ganglia which are separate in primitive gastropods. The cerebral and visceral ganglia are associated with the two large shell muscles, and a pair of pedal ganglia with the muscles of the foot. Like the ganglia, the **sense organs** are distributed in the body, not concentrated in the head. Near the pedal ganglia is a pair of statocysts, lined with sensory cells and containing a calcium carbonate concretion. A patch of yellow epithelial and sensory cells lies on the visceral ganglia and is thought to be sensitive to chemicals in the water that flows through the mantle cavity. The mantle has scattered sensory cells, which are most abundant on the small projections along the edges of the mantle at the openings for the water currents. They probably respond to light and touch. When a clam is irritated, the foot and mantle edges are withdrawn, and the two valves close very tightly—or as we say, "clam up."

Statocyst of a small bivalve, *Sphaerium notatum,* resembles that of unionids. An outer fibrous capsule surrounds the cellular layer that is in contact with a small spherical calcareous concretion, 19 μm in diam. The nerve that emerges through the capsule wall does not pass to the nearby pedal ganglion but joins the pedal-cerebral connective and carries its sensory messages to the cerebral ganglion. (Modified after C.R. Monk)

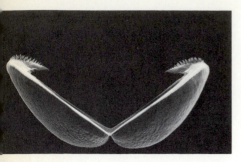

Shell of glochidium of *Anodonta cataracta* from Halfway Pond, Plymouth, Mass., USA. Length of each valve, 0.4 mm; SEM. (C.B. Calloway)

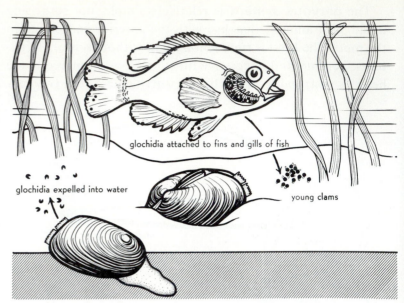

glochidia attached to fins and gills of fish

glochidia expelled into water

young clams

Life cycle of a freshwater clam. In some clams the passing of a fish stimulates the female to eject a cloud of glochidia directly at the fish. In other species the freed glochidia are stimulated to increased clapping of the shell valves when a fish is near. The glochidium clamps its valves tightly into the tissues of its host and in some way stimulates them to grow around it, thus forming the so-called "blackheads" of the fish. After 3-12 weeks of parasitic life, the juvenile clam falls off and becomes independent. In this figure the gill covering of the fish has been cut away to expose the parasitized gills. (Based on Lefevre and Curtis)

Migration into fresh waters from the large, stable, and interconnected oceans poses serious problems for invertebrates. They must make osmotic adjustments to the sharply lowered salt concentration. They must modify or give up free-swimming larval stages that could be swept downstream and out to sea. And they must solve problems of dispersal to bodies of water that do not connect. For hundreds of millions of years freshwater bivalves have met all these challenges, especially in the large rivers of eastern North America, where there are vast stretches of "riffles" or "shoals," made by fast-running shallow waters passing over sandy or gravelly bottoms. Some 40 to 50% of all freshwater molluscan species evolved in this highly oxygenated habitat, and many of them are not found elsewhere. During the last 100 years silting, industrial wastes, and pesticides from agricultural runoff have polluted these pristine rivers. But the greatest damage has been done by building of dams that change fast-running rivers into stagnat waters low in oxygen and food organisms. **Extinction of freshwater bivalves** has occurred on a massive scale and continues as more and more dams are built. As dams and irrigation come to S. America, Asia, and Africa, extinctions will follow there, but on a smaller scale, as these continents never had so magnificent a fauna as that of eastern N. America.

The **reproductive system** consists of a pair of lobulated gonads, surrounding the coils of the intestine that lie in the foot and opening near the external pores of the bladder into the dorsal gill passages. The sexes are separate. The male sheds sperms from the testes into the outgoing water current. They enter the female through the incurrent opening, pass through the pores on the surface of the gills, and reach the interior of the gills, where the eggs are held and fertilization effected. The zygotes develop within the gills to bivalved larvas called **glochidia** (singular, *glochidium,* "point of an arrow"). Tremendous numbers of glochidia are produced and expelled into the water, where they slowly sink to the bottom, and most of them die. To develop further, they must, within a few days, become attached to a fish, on either the fins or the gills (depending on the species of clam) and live as parasites until they have become juvenile clams.

In some freshwater clams there is no parasitic stage; the juvenile clams develop in spaces within the gills of the mother. Marine bivalves, as mentioned earlier, practice various strategies, and often have a free-swimming trochophore followed by a veliger.

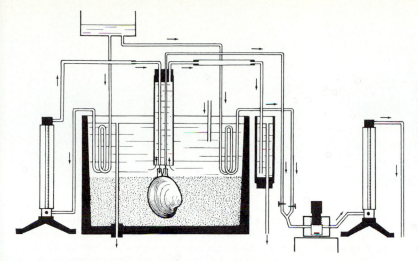

The hardshell clam, *Mercenaria mercenaria,* is also known as the Venus clam from its former generic name, *Venus*. When small, it may be called the cherrystone clam. It lives deeply buried, with the long slender siphons extending to the surface of sand or mud. When dug up at low tide, however, the siphons are withdrawn and the shell tightly closed for a period, after which it extends the foot and digs in at the rate of about 16 forward probes a minute.

Measuring the pumping rate of a clam. Since the water that flows through the mantle cavity supplies both food and oxygen, estimating the pumping rate is a first step to understanding a bivalve's nutrition and respiration. In the experiment illustrated above, using the hard-shelled marine clam, *Mercenaria mercenaria,* pumping rate and oxygen consumption were determined simultaneously and found to be directly related but highly variable. The rate may be related to nutritional needs at one time and to respiratory requirements at another, and is sensitive to light, temperature, salinity, oxygen, water currents, and the kinds and concentrations of particles and chemicals in the water. Some species show rhythms in pumping rate which are tidal, daily, or monthly. Despite all these variable factors, it appears that certain types of bivalves, for example, oysters, mussels, and scallops, which do not have long siphons, consistently pump faster than others. Large oysters in warm water have been estimated to pump up to 40 liters per hour at peak periods of activity. (After Hamwi and Haskin)

Gaper clam, *Tresus nuttallii,* one of the largest clams of the Pacific coast of the U.S., has a shell that reaches 20 cm. It lives deeply buried with long siphons united and reaching to the surface of the sand in bays. People concerned chiefly with their own digestion find "gapers" good eating. Biologists interested in the digestive process of bivalves find these large clams good subjects for experimentation. This specimen (prepared and studied by T. L. Patterson at Hopkins Marine Station) is anesthetized with ether; a rubber tube, with a balloon tied to its end, has been pushed into the incurrent siphon, across the mantle cavity, through the mouth into the stomach. The tube is connected with a device for recording changes in pressure exerted on the balloon by contractions of the clam's stomach. Monterey Bay, California. (R.B.)

Oysters are sessile as adults and have no foot. They live cemented by the left valve to rocks or other oyster shells. *Crassostrea virginica,* edible oyster of the E. Coast of the U.S., has a deeply cupped shell that rises off the bottom and enables this oyster to withstand turbid water, as in this S. Carolina mud flat. (R.B.)

Cockle, *Clinocardium nuttallii,* lives in mud flats in bays and estuaries from the Bering Sea to Baja California. It burrows shallowly, having incurrent and excurrent openings that are only short extensions of the mantle edge. Basket cockles are flavorful but tough, and not abundant. Oregon. (R.B.)

Razor clam of the Atlantic coast of N. America, *Ensis directus,* stands vertical in the sand at low-water mark, but to dig it up requires an alert shoveller. Even when captured it can spring from one's hand, land meters away, and disappear into the sand at a rate of about 90 forward movements of the foot per minute. It has a fine flavor, but is seldom seen in the markets. Delaware Bay. (R.B.)

Jackknife clam, *Tagelus plebeius,* up to 9 cm long, is found from Cape Cod to Brazil in mud or muddy sand flats. It lives vertically in its deep burrow and extends the siphons (longer than the shell) to the surface of the substrate for suspension feeding. The outgoing current attracts *Busycon,* which favors *Tagelus* over oysters or hard-shell clams. Human preference is the opposite. (R.B.)

Pacific razor clam, *Siliqua patula,* lies buried in ocean beaches with strong surf, from Alaska to California. Dug up at low tide and laid on the sand covered by water, as here, it can bury itself in seconds. The one in the foreground has just done so, leaving only the siphons protruding. It makes excellent eating but there are times in summer when it can be toxic. Length up to 15 cm. Coos Bay, Oregon. (R.B.)

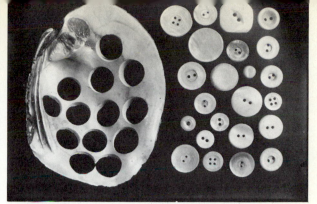

Pearls from freshwater clams, like those from marine bivalves, are mostly irregular; but occasional valuable ones, either spherical or baroque, are found while collecting and processing freshwater unionid clams into **pearl buttons.** Thousands of tons of shells, mostly from the Mississippi River Valley, were once used annually for buttons. Now most buttons are plastic, but pellets of shell are shipped from the U.S. to Japan and used in the production of cultured pearls in marine oysters. (C. Clarke)

Many molluscs produce pearls, but few pearls are valuable, and most commercial ones come from marine "pearl oysters," species of *Pinctada.* About 1000 pearl oysters must be collected by divers to find one valuable pearl. Shown *below, left,* are highly valued black pearls from a black variety of *Pinctada* in the Gulf of Mexico, and a dull, non-commercial pearl in place in the hardshell clam, *Mercenaria. Below, right,* Mother-of-pearl covers a fish that lodged between the shell and mantle. (Amer. Mus. Nat. Hist.)

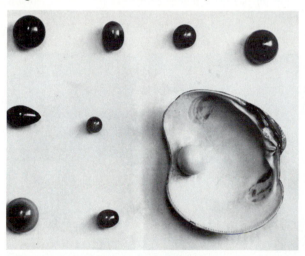
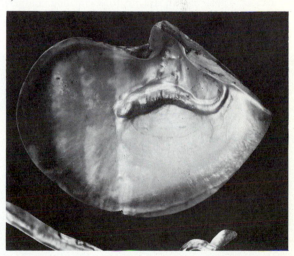

Culturing of pearls is carried out in Japan in the "pearl oyster," *Pinctada,* by inserting into the deeper layers of the mantle a shell pellet wrapped in a tiny piece of the outer surface of the mantle of another oyster. *Left,* the pellet is inserted with a forceps into a 3-year-old mature oyster. *Right,* the operated oysters are hung on racks that raise them above the reach of sea stars or oyster drills. Covered at high tide, they can be pulled up for cleaning. Nacreous material is secreted by the oyster at the rate of 0.09 mm a year. It takes up to 7 years to produce a valuable pearl. (R.B.)

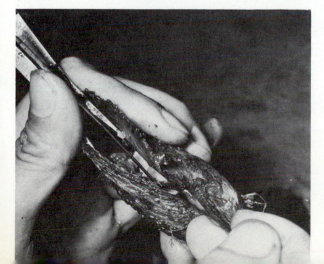
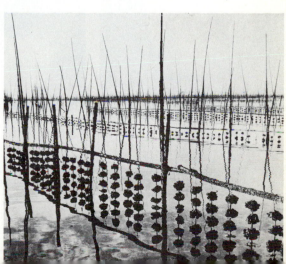

Hardshell clam, *Mercenaria mercenaria,* often called by the Amerindian name, "quahog," is gathered by the ton on the U.S. East Coast. The shell is heavy, its inner lining non-lustrous and tinged with purple. This panful of clams, destined for a steamed clam dinner, was subjected all day to a steady stream of clear seawater to encourage the clams to flush out the sand that they contained when collected. Those shown are up to 8 cm long. Larger, tougher specimens are used for chowder. Lewes, Delaware Bay. (R.B.)

Softshell clam, *Mya arenaria,* is a favorite edible clam of the U.S. East Coast and has been transplanted to the Pacific Coast. It is also called the "long-necked clam" because of the long siphons in a common sheath. The leathery periostracal covering over the sheath must be slipped off before the clam is eaten. *Mya* buries to a depth of 30 cm or more in compact sand or mud, but if dug up has little chance of burrowing so deeply again before it is taken by predators. Shell, about 5 cm. Solomons, Maryland, Chesapeake Bay. (R.B.)

Bent-nosed clam, *Macoma nasuta,* with long incurrent siphon that extends above the surface and sweeps the surface mud, sucking in food particles. The short excurrent siphon empties into the mud. Shell, 5 cm. Monterey Bay, Calif. (RB)

Coquinas, *Donax variabilis,* have short siphons and live barely covered by the surface of the sand. They can migrate up or down the beach with the tides. The flattened, vari-colored shells have earned the name "butterfly-shells" and serve decorative purposes. When abundant, coquinas may be gathered for soup; and in Italy a species is eaten raw. Those shown are in a tray of water and have siphons and foot protruding. Shell, 15 mm. Beaufort, N. Carolina. (R.B.)

Wood-boring bivalve, *Teredo navalis.* Only the anterior end, with the shell valves surrounding the suckerlike foot, is exposed in this split piling. Plymouth, England. (D.P. Wilson)

Preserved teredos in thoroughly riddled wood. As the shell valves are rotated, their razor-sharp ridges rasp the wood. Wood particles are carried to the mouth, and 80% of their cellulose content digested. Symbiotic nitrogen-fixing bacteria supply cellulolytic enzymes and supplement the nitrogen-poor wood diet. These bivalves with greatly elongated siphons, popularly called "shipworms," do great damage to wooden ships and pilings on all coasts. (R.B.)

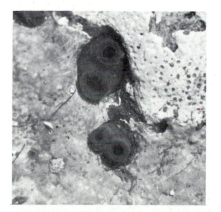

Left, **rock-boring clams** live embedded in rock with only the siphon tips protruding. *Right,* **3 rock-borers in split rock.** Only one at right is fully exposed, revealing anterior end with roughened shell valves that do the burrowing. When numerous they do damage to concrete harbor works and even to the lead sheathing of submarine telephone cables. (R.B.)

Scallops, *Chlamys opercularis,* swim about by rapid jerks in the direction opposite to that in which water is expelled from the mantle cavity by sudden clapping of the valves. They live in sand but swim in escaping from predators, such as this sea star *(Asterias rubens).* Plymouth, England. (D.P. Wilson)

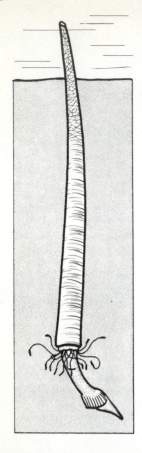

SCAPHOPODS

A SMALL and homogeneous class of molluscs is the class **Scaphopoda** ("spade-footed"). Scaphopods are all marine, and there are about 350 living species. These elongated animals, enclosed in tapering tubular shells, burrow in sandy bottoms, sometimes near shores but more often in deeper waters to 3,000 m.

On some beaches scaphopod shells are cast up in great numbers, and long ago, mistaken for fish teeth, they came to be called tooth shells. Scaphopods like *Dentalium* and *Antalis,* in which the marked taper and curvature of the shell remind one of an elephant's tusk, are often called tusk shells.

Scaphopods are bilaterally symmetrical and resemble bivalves more than they do any other group of molluscs. The larva of a scaphopod has two mantle lobes but these fuse along the ventral line, producing a tubular mantle that secretes a tubular shell, open at both ends. (In some species fusion is incomplete, and there are small slits or a long cleft along the ventral surface of the shell). Length of shell varies with species from less than 5 mm to nearly 15 cm.

A scaphopod lies buried at an oblique angle, with only the narrow posterior tip of mantle and shell protruding and providing both entrance and exit for the respiratory current. At the buried wider end are an inconspicuous proboscis-like head and a muscular protrusible foot. There are no eyes, but there is a pair of statocysts in the foot.

In **feeding,** the long filamentous **tentacles** (captacula) are extended by an inflow of blood and/or by ciliary gliding and then retracted by longitudinal muscle fibers in their walls. There may be more than 100 tentacles, and each is expanded at its tip into a flattened, ciliated bulb which contains a minute ganglion and sensory receptors. As the tentacle bulbs explore the water-filled spaces among the sand grains, they find, select, and gather food particles that are then passed to the mouth either by ciliated tracts or by contraction of the tentacles. *Dentalium entalis* is notable for specializing in foraminiferans. The proboscis may often be seen bulging with a solid mass of forams, but it seldom contains other particles or organisms except for occasional large diatoms or larval bivalves. The **radula** is much larger, in comparison to other digestive structures, than in any other mollusc examined, and it is presumed that this outsized radula must aid in crushing foram shells.

There are no gills; and though the entire mantle can absorb oxygen to some extent, there is a specialized **respiratory area,**

Tooth shells, the largest 3 cm long, in shell debris on sandy beach at Songkla, Thailand. (R.B.)

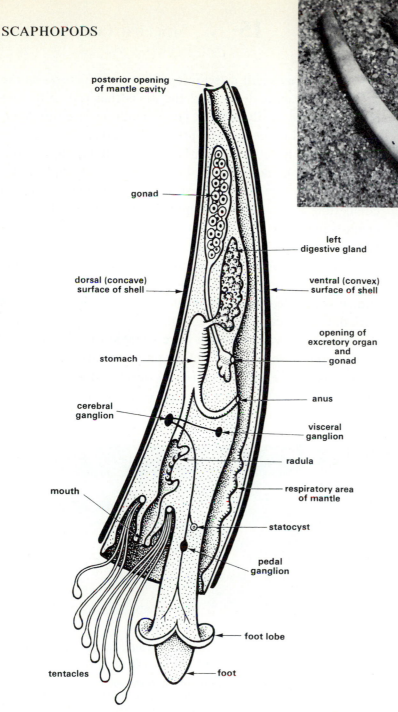

Four scaphopods, *Dentalium,* laid out in a water-filled pan of sand. The one at the far right has partially extended the foot. The one at the far left has the foot fully extended, with the tip and its lobes firmly anchored in the sand. Roscoff, France. (R.B.)

posterior opening of mantle cavity

gonad

left digestive gland

dorsal (concave) surface of shell

ventral (convex) surface of shell

opening of excretory organ and gonad

stomach

anus

cerebral ganglion

visceral ganglion

radula

mouth

respiratory area of mantle

statocyst

pedal ganglion

foot lobe

tentacles

foot

Scaphopod structures as seen from the left side. Based mostly on *Dentalium (Antalis) entalis,* which has a bilobed foot.

just anterior to the anus, in which the mantle surface is greatly increased by being thrown up into folds which have especially powerful cilia. A pair of **excretory organs** open into the mantle cavity on either side near the anus; and a single **gonad** discharges its gametes into the right excretory organ. The sexes are separate and the eggs, shed singly, develop into free-swimming trochophores and veligers.

In animals as small as scaphopods, such structures as the heart, the pericardial sac, and the ducts that run to the excretory organs, easily seen in larger molluscs, are tiny, delicate, and hard to see, even in prepared sections. Investigators who claim to have seen them have not been able to demonstrate them convincingly to others. Even if we accept the accounts of a contractile blood sinus near the anus, its role in circulation could hardly compare with the role played by changes in hydraulic pressure that result from powerful, periodic (every 10 to 12 minutes) contractions of the foot. Water enters the posterior mantle opening gradually, due to steady ciliary beating and slow extension of the foot at the anterior end. It is expelled suddenly through the same opening by the periodic contractions of the foot.

Unlike the bivalves, which have adapted the gills for filter feeding and have radiated out into a great variety of niches, scaphopods have lost the gills and developed a type of tentacular feeding that limits the diet of the animals and their evolutionary diversification.

1. Scaphopod beginning to dig in after having been laid on the surface of the sand. The muscular foot has been extended and thrust into the sand.

2. The foot is engorged at its tip by an inflow of blood into the foot sinuses, and thus firmly anchored in the sand. Foot lobes *(arrow)* help to anchor foot.

3. Scaphopod is pulled into the sand by contraction of foot muscles. Repeated extensions and contractions of the foot pull the animal along in the sand.

4. In normal position in the sand the scaphopod lies buried obliquely, with only the narrow posterior tip protruding into the water above. (R.B.)

CEPHALOPODS

THE MOST HIGHLY ORGANIZED molluscs are the nautiluses, squids, and octopuses—all marine and members of the class **Cephalopoda.** The name means "head-footed," for in these animals the foot is closely associated with the head. As in gastropods, all degrees of reduction of the shell can be found. While the nautiluses have a large, externally coiled shell, the squids and octopuses have only vestiges of a shell.

Squids are among the mostly highly developed invertebrates. Certain of their structures will be described here to illustrate the ways in which they have adapted the molluscan body plan to a fast-moving, predatory life comparable with that of fishes.

Giant squid, *Architeuthis,* is best known from strandings in Newfoundland, northern Europe, and New Zealand. It probably lives normally in the cold depths and dies, from lack of oxygen, at temperatures above 10°C. It forms an important part of the diet of sperm whales, as we know from examination of whale stomachs. The specimen in the photo, stranded at Ranheim, Norway in 1954, is relatively small: only 9.24 m in total length. The largest stranded specimens have measured up to 18.3 m in total length and 5.2 m in mantle length. A squid of such size could weigh 500 kg or more. (E. Sivertsen)

An authentic "sea-monster" was a giant squid encountered by the French steamship *Alecton,* near the Canary Islands, in 1861. The event is described in a French account, written by the ship's commander and read before the Academy of Sciences in Paris: "... After several encounters which permitted only of its being struck by ten or so musket-balls, I succeeded in coming alongside it, close enough to throw a harpoon as well as a noose ... a violent movement of the animal disengaged the harpoon; the part of the tail where the rope was wrapped around broke, and we brought on board only a fragment weighing about twenty kilograms. ... the giant squid ... seems to measure 15 to 18 feet to the parrot-beaked head, enveloped by eight arms 5 to 6 feet long. Its aspect is frightful, its color a brick-red. ... Officers and seamen begged me to have a boat lowered and to snare the animal again, ... but I feared that in this close encounter the monster might throw its long arms, equipped with suckers, over the sides of the boat, capsize it, and perhaps choke several seamen in its formidable whips charged with electrical emanations. I did not think that I ought to expose the lives of my men to satisfy a feeling of curiosity, even for science, and despite the fever of excitement that accompanies such a chase, I had to abandon the mutilated animal ..." (Adapted from an old engraving in L. Figuier, based on a drawing made by one of the officers on the *Alecton.*)

Opalescent squid, *Loligo opalescens,* 20 cm in length, is the common squid along the Pacific Coast of N. America. Contraction or expansion of pigment cells provides rapid color changes as the squid blanches when excited or as it moves from light sandy bottom to darker surroundings. Squids swim in schools and feed on crustaceans, fishes, and worms. (R.B.)

Sucker marks, 25 mm or more in diameter, on the skin of a whale, tell of an encounter between this largest of vertebrates and the largest of invertebrates, a giant squid. (L.L. Robbins)

The foot of a squid forms the **funnel,** a conical muscular tube that projects from under the head. Ten **sucker-bearing arms** surround the mouth. When the animal is swimming, the arms are pressed together and aid in steering. Two slender tentacular arms, longer than the rest and with suckers only on the expanded tips, can shoot forward to seize prey and then withdraw it toward the mouth. There the prey is held firmly by 8 shorter arms, which are equipped with suckers from base to tip, while two strong piercing **jaws** help to introduce a fatal toxin. Large bites of food are swallowed so rapidly that the **radula,** which is quite small in squids, probably plays a minor role.

A squid relies for protection not on a heavy shell but on its ability to leave the scene of danger in a hurry. The vestigial **shell** is thin and chitinous, and lies buried under the mantle of the upper surface. The **mantle** is thick and muscular and has taken on some of the protective functions which in other molluscs are served by the shell. The mantle is also the chief swimming organ. At the rear end, its upper surface is extended into a pair of triangular **fins** that stabilize the animal horizontally and can be adjusted to change the direction of movement or undulated to move the animal slowly. At the front, the mantle ends in a free edge which surrounds the neck between the head and visceral mass.

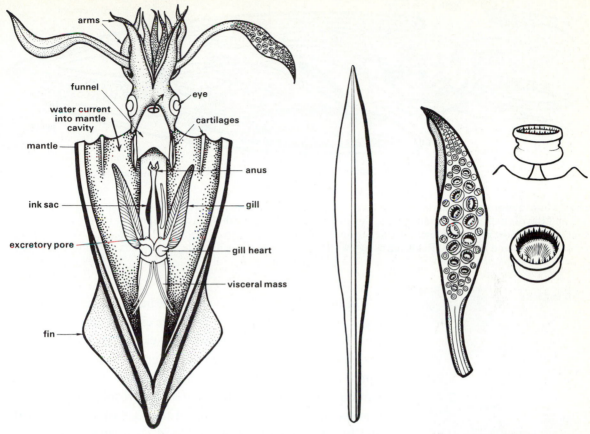

Ventral view of male squid with mantle slit open to show some of the principal organs.

The thin chitinous **shell,** or *pen,* lies buried in the upper surface of the mantle.

Tip of a **tentacular arm** of a squid, showing **suckers,** each attached by a muscular stalk and lined by a horny ring.

Three cartilaginous ridges on the mantle edge interlock with grooves on the visceral mass and funnel. When the mantle is relaxed, water enters the mantle cavity around the free edge; and when the mantle contracts, the edge is tightly sealed and water is forced out through an opening in the funnel. Thus water is rapidly flushed through the mantle cavity, constantly renewing the supply of oxygen and carrying away respiratory, excretory, and digestive wastes. When the squid is excited, the mantle is contracted strongly, forcibly expelling water from the funnel and moving the animal by jet-propulsion. When the tip of the funnel is bent backward, the squid darts quickly forward to seize its prey. When the tip of the funnel is directed forward, the animal shoots backward like a torpedo, and this is its usual behavior in escape. When attacked, it may emit a cloud of inky material from a special **ink sac** which opens into the funnel through the anus. The ink cloud serves as a dark object that may distract a predator while the squid goes off in another direction, or as a concealing "smoke screen," and it also may interfere with the chemoreceptors of a predator.

Three rows of **teeth from the radula** of a squid.

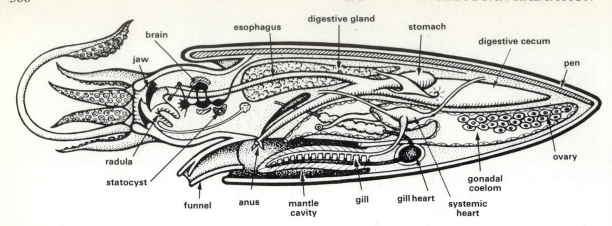

Longitudinal aspect of a female squid. Flattened mesodermal epithelium lines the cavity surrounding the systemic heart and the cavities of the gonad and the excretory organs; together these cavities represent the coelom. A duct drains fluid from the pericardial cavity to the coelom of the excretory organ on each side. The ovary sheds eggs into the gonadal coelom; and the eggs, after being coated with layers of jelly (from accessory sex glands not shown here), are released into the mantle cavity.

The long axis of a squid is dorso-ventral, instead of anteroposterior as in most bilateral animals. To compare the body with that of a snail, one would have to place the squid so that the head and foot were down and the pointed end up, as in the large squid held here by Ed Ricketts. The upper surface of a swimming squid is structurally anterior and dorsal. The under surface is structurally posterior and dorsal. Thus, a squid usually swims with the ventral surface forward and the dorsal surface hindmost. Monterey, Calif. (R.B.)

The active life of squids would not be possible with the slow respiratory currents produced by the ciliated gills of clams, or with the very different respiratory exchange mechanism of snails. **Circulation of water** through the mantle cavity of a squid is provided not by ciliary currents but by vigorous, regular contractions of the mantle. And respiratory exchange is speeded by the large surface of the extensively folded **gills** and by special features of the blood chemistry.

Squids live mostly in the open sea, where the cold water contains much O_2. They use large amounts of O_2 to maintain their active lifestyle, and their blood chemistry is specialized in several ways to provide O_2 to their tissues quickly. First, compared to blood from other molluscs such as snails, squid blood can carry more O_2 because it contains a higher concentration of the blue respiratory pigment hemocyanin, which picks up O_2 in the gills, carries it in the blood, and releases it to the tissues. Second, squid hemocyanins have a low O_2 "affinity," binding O_2 especially loosely and releasing it even when the O_2 content of the surrounding tissues is already relatively high, so that up to 90% of the O_2 carried by hemocyanin can be released to the tissues. Third, the O_2 affinity is even lower when the surrounding CO_2 concentration is increased, or the pH is decreased, as it tends to be in active muscles, so that more O_2 is released when it is most needed. This convenient effect of CO_2, while found in many blood pigments, including human hemoglobin, is more pronounced in squid hemocyanin than in any other blood pigment. However, the same low O_2 affinity that facilitates release of large amounts of O_2 to the tissues, strictly limits the conditions under which squid hemocyanin will pick up O_2 in the gills. The water in the mantle cavity must contain a great quantity of O_2 and very little CO_2, as it usually does in the open sea. A freshly-caught squid, placed in a bucket of freshly-dipped seawater, will die in a dismayingly short time, often less than an hour.

How then can a bottom-living snail survive in a similar bucket of seawater for days? Snail hemocyanins have a high affinity for O_2, releasing it only when the O_2 content of the surrounding tissues is exceedingly low. Also, the effect of CO_2 on O_2 affinity is much less, or in some cases even reversed. Therefore, the blood is able to pick up O_2 even when the O_2 content of the water in the mantle cavity is low and the CO_2 content is high, as it often is in stagnant bottom habitats. However, the O_2 and CO_2 levels in the tissues must remain low and so must the activity levels of the snail.

Thus the blood chemistry of squids permits their active lifestyle but imposes strict limits on where they can live, while that of snails lets them wander even onto stagnant muddy bottoms but keeps them to a snail's pace. Most of the bivalves, with even more limited activities, make do with no oxygen-carrying blood pigment at all.

The **circulatory system** is much more highly developed than that of other molluscs and provides for the rapid distribution of oxygen through the tissues. The blood flows within a continuous system of vessels, which are lined throughout with an epithelium—not in irregular unlined spaces among the tissues, as in clams and other molluscs—so that both the rate of flow and the blood pressure are comparatively high in squids. The tissues and gills are permeated with networks of very small vessels, the **capillaries,** through the thin walls of which respiratory exchanges take place rapidly. There are separate pumping mechanisms for blood going through the gills and that going out to the tissues. The deoxygenated blood returning from the tissues enters two **gill hearts,** each of which pumps blood through one gill. This gives the blood a fresh impetus, so that it passes through the gills at higher speed and pressure. Freshly oxygenated blood from the gills enters a single **systemic heart,** from which it is pumped out again to the tissues.

The **nervous system** of cephalopods is highly developed, with far greater numbers of nerve cells than in any other invertebrates—in sharp contrast with that of the slow-moving clams. The large brain of squids encircles the esophagus and lies between the eyes. The brain consists of several pairs of enlarged ganglia all fused together. It has multiple centers of nervous control, which in many other invertebrates are lodged in separate ganglia spread out over the animal. Besides an olfactory organ and a pair of statocysts, a squid has two large **image-forming eyes.** They are remarkably like human eyes but are developed in quite a different way. When two similar structures having a similar function appear in two distantly related groups, so that there is no possibility of a common ancestor which could have possessed such a structure, then the structures must have evolved independently. Thus the similarities in the eyes of squids and the eyes of vertebrates are said to have arisen by **convergent evolution.**

Effective jet propulsion in a squid requires that the muscles of the mantle contract rapidly and simultaneously on all sides. This means that the nerve impulses that trigger contraction must travel especially fast, and the mantle muscles are supplied by a special system of *giant nerve fibers*. It has long been known that the speed of conduction of a nerve impulse increases with increase in the diameter of the nerve fiber. In the squid *Loligo* the giant fibers are up to almost a millimeter in diameter and conduct impulses at relatively

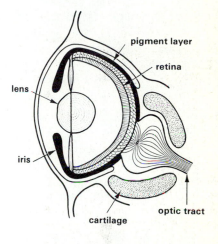

The **eye of a cephalopod,** as well as that of humans and other vertebrates, is called a "camera eye" because it is structured on the same principle as a camera. It consists of a dark chamber to which light is admitted only through the *pupil,* an opening in the *iris* (or, in a camera, the diaphragm). Behind this opening is a *lens* which focuses the light on the *retina* (or, in a camera, on a light-sensitive film). The lens of the cephalopod eye is adjustable for near and far vision, and both iris and retina accommodate to bright and dim light. The nerve fibers from the retina cross as they enter the *optic lobe;* and so do some of the fibers of the *optic tract,* which thus sends information to the optic lobe on the opposite side of the head, as we know from the fact that something learned through the eye on one side will be retained even after removal of the optic lobe on that side. In addition to conveying information to other parts of the brain, the optic tract carries incoming nerve fibers that control the muscles of the eye.

great speeds of up to 20 m/sec. It has been estimated that the giant fiber system gets the animal moving in about half the time that would be taken by nerve fibers of ordinary size. Moreover the largest, fastest giant fibers go to the farthest points on the mantle and the somewhat smaller, slower ones to nearer points, their sizes being precisely graded so that the muscles on all sides of the mantle are stimulated to contract simultaneously. Giant fibers are found wherever fast responses occur among invertebrates of many phyla, but those of squids have been studied most extensively and have supplied much of the basis of our knowledge about the mechanisms of nervous conduction.

Another example of convergence between squids and vertebrates is the development of **internal cartilaginous supports.** Squids have a number of internal cartilages which support muscles and form interlocking surfaces, but most interesting in this connection is the large cartilage which encloses and protects the brain, reminding us of the vertebrate brain case. The squids, perhaps more than any other invertebrates, have evolved along the same lines followed by the fast-moving predatory aquatic vertebrates: large size, streamlined shape, rapid locomotion, internal skeleton, very effective respiratory and circulatory systems, large brain, and highly developed sense organs.

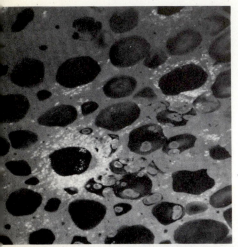

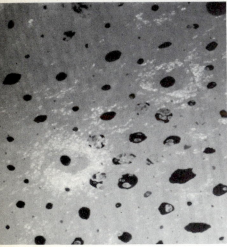

Color changes in squids and other cephalopods are produced by **chromatophores,** pigment cells distributed in the epidermis. Each cell contains an elastic sac filled with pigment and is attached to a radiating set of separate muscle cells under nervous control. *Above:* As a cephalopod flushes from a pale to a deep color, the muscle fibers of each chromatophore contract and the pigment cell is stretched and flattened to cover up to 10 times the area, showing off its content of pigment. *Below:* As the animal pales, the muscle fibers relax, and the elasticity of the pigment sac restores the cell to a smaller size. Shown here are chromatophores in the skin of the mantle of a squid, *Loligo opalescens.* ×10. (R.B.)

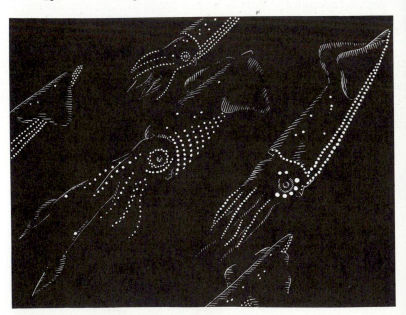

Luminous squids. (After C. Chun)

Bioluminescence, or the production of light by living organisms, is a widespread phenomenon found in some members of about half the phyla of animals. Its distribution follows no special evolutionary lines, but accompanies certain ways of life. Among freshwater animals, luminescence is known only in a few insects and in the New Zealand limpet, *Latia,* which produces a luminous secretion. But animal light is extremely common among marine forms, particularly cnidarians, ctenophores, squids (but rarely octopuses), and fishes.

Fishes and squids, members of the two phyla which show the best development of eyes, also have the most highly developed light-producing organs.

In some squids these organs are amazingly complex, having cells which produce the light, lens tissue, reflector cells, a pigment layer, and a fold of skin that can act like a blackout curtain or be lifted to let the light shine through. The adaptive value of luminescence to squids is not clear. The light may serve to deter or confuse predators, or to attract prey, or to help keep schools of squids together.

The chemical basis of light-production in most luminescent organisms studied involves the interaction of two substances, which retain their activity when extracted from luminous tissues. One is a heat-stable compound called *luciferin*. The other is an enzyme, *luciferase,* readily destroyed by heat. The two may be obtained separately in hot and cold aqueous extracts and when mixed in a test tube, in the presence of oxygen, they produce light. The mixture continues to luminesce until the luciferin has been completely oxidized.

In most cnidarians and ctenophores the luminescence appears to work in a different way, not requiring oxygen and involving only a single substance, a "photoprotein" which produces light in the presence of calcium (see *Aequorea* in chapter 6). However, chemical analysis of the components in both kinds of systems suggests a common basis; the photoprotein probably represents a stable complex of luciferin, luciferase, and oxygen, ready to be activated by calcium.

The luminescent flashes of animals occur mostly in response to mechanical disturbance and are under nervous control. The luciferin and luciferase, or photoprotein and calcium, must be kept separate, ready to be mixed on demand. The light of some squids and fishes is produced by symbiotic bacteria that live in the animal's light organs; the bacteria luminesce continuously, and the animal controls light emission by means of a shutter.

Reproduction in squids, as in all cephalopods, involves separate sexes. One of the male's arms, called the hectocotylus, is modified for transferring sperms to the female. The sperms are enclosed in intricate packets, the spermatophores. A school of spawning *Loligo pealei,* develops a social hierarchy. Males defend their chosen females by fighting sham battles, rushing at each other but seldom actually touching. Things may get out of hand, and the loser retreats minus the tip of an arm. The ability to acquire a receptive female, and to maintain exclusive conjugal rights, appears to be based on body size, aggressiveness, and eternal vigilance. After and between matings, the female deposits gelatinous capsules of eggs. Tattered and emaciated, both members of the pair die.

Mating squids, *Loligo pealei,* shown here in the side-by-side position (after G.A. Drew). The male wraps his arms firmly around the female and inserts his hectocotylized (left ventral) arm into her mantle cavity, holding his bundle of spermatophores in place until they ejaculate. The entire act takes no more than 10 seconds. The string of fertilized, jelly-coated eggs is extruded through the female's funnel, and she attaches it to a stone or shell or to the egg clusters of other females. Females of *L. opalescens* have together produced deep masses 12 m across of cigar-shaped egg capsules.

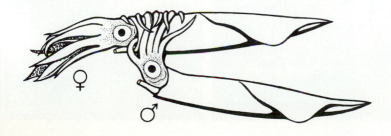

Spermatophores of *Loligo opalescens.* The external transparent tunic encloses a coiled ejaculatory apparatus, a small rounded cement gland, and a long opaque sperm mass. At mating, the male releases a bundle of 20 to 40 spermatophores through his funnel, picks them up with his hectocotylus, and transfers them to the female. He must act quickly, for in seawater the spermatophores rapidly take up fluid, and elastic and hydrostatic forces combine to effect ejaculation. Spermatophores of *Loligo* are only about a centimeter long, but those of the giant octopus *Octopus dofleini* measure more than a meter. (R.B.)

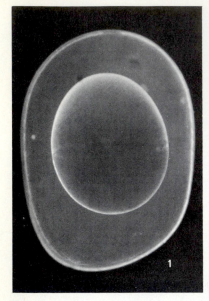

Development of a squid, *Loligo pealei.* The large, yolky eggs of squids and other cephalopods do not follow the complete, spiral cleavage pattern typical of other molluscs. Early cleavage is limited to a thin superficial disk of cytoplasm at one end ("animal pole") of the egg. **1.** The cells continue to divide and spread until they enclose all of the yolk in a sac. **2.** The embryo (shown with egg membrane removed) takes form at the animal pole. Small rounded arm buds encircle the base of the large ovoid yolk sac. Behind them are the relatively enormous eyes, at the ends of thick protrusions from the head region. **3.** As the embryo grows using its store of yolk, the yolk sac shrinks. The embryo is now easily recognizable as a young squid. Darkly pigmented chromatophores dot its surface. Its eyes are literally "bigger than its stomach." **4.** By the time the yolk is almost gone, the embryo is about 3 mm long, and its head and body begin to show more adult proportions. **5.** There is no free larval stage. At hatching from the egg membrane, the juvenile squid, with prominent funnel, is not yet adult in size, proportions, or behavior and becomes so only gradually as it grows and matures. Duke University Marine Lab., Beaufort N.C. (R.B.)

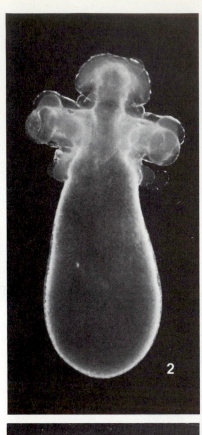

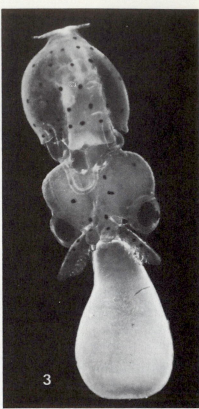

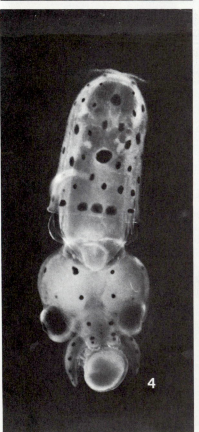

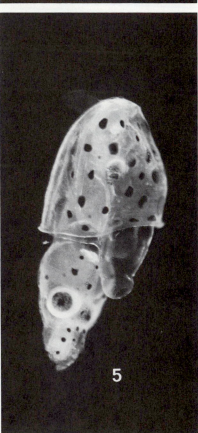

The closest relatives of the squids are the **cuttles,** or cuttle-fishes, which like squids have 8 short arms and 2 long slender tentacular arms. Instead of the thin chitinous pen of squids, cuttles have a substantial, shield-shaped, chitinized but mostly calcareous internal shell. The shell lies buried under the mantle of the upper surface and acts not only as a skeleton but also as a variable buoyancy tank. Tiny gas-filled chambers in the rigid shell give the cuttle a degree of buoyancy which is practically independent of depth. Buoyancy can be decreased by osmotically filling some of the chambers with liquid, thus displacing some of the gas, or increased by removing liquid, allowing expansion of the gas. *Sepia* does this daily, as it sinks to bury itself in the sand by day and emerges to hunt for small fishes and crustaceans at night.

Most cephalopods can change color, but cuttles display especially dazzling patterns. In less than a second they can switch from a striped swimming pattern (that appears, to human eyes at least, like ripples in the water) to a pale dotted beige that makes cuttles hard to see as they hover low over sand, or to dark and light patches that blend with a background of multicolored pebbles. When disturbed they may turn on a variety of patterns, one after another: flickering stripes, or ghostly pallor, or large black spots like eyes, all of which we presume could frighten or confuse a predator. Males have a special pattern of display.

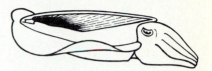

The **shell** of *Sepia* occupies 9 % of the animal's total volume (compared with 5% for the gas-filled swim bladder of a fish). The shell is partitioned into about 100 separate chambers by thin sheets of calcified chitin laid down one below the other. The lower tips of the chambers form a permeable surface through which shell and blood exchange fluid. About 10 of the largest, most forward chambers are kept largely gas-filled and are most important in maintaining *buoyancy*. Since they occupy the mid-region of the body they make possible the characteristic and effortless hovering posture from which cuttles may shoot out the tentacles to grab passing fishes. The smaller chambers at the rear are mostly fluid-filled and more important in determining *orientation*. When they lose liquid the rear rises and the head is tilted down, a posture convenient for hunting bottom crabs and prawns. (After Denton and Gilpin-Brown)

Striped swimming pattern of *Sepia officinalis,* the cuttle of shallow waters in the Mediterranean and eastern Atlantic. The ink sac provides a rich brown pigment, sepia, prized by artists. The calcareous internal shell is the "cuttlebone" given to pet birds as a calcium supplement. Naples. (R.B.)

Cuttle catching a prawn. *Above,* on seeing prey, the cuttle *(Sepia officinalis)* comes at once "to attention," turning first the eyes, then the body, towards the prey. Brilliant color changes, shimmering over back and arms, may distract the prey from the parting of the arms and the gradual extension of the two long tentacles. *Below,* as the cuttle advances within range, the tentacles are suddenly shot forward to seize the prey. Cuttles may lie in wait or they may hunt actively over the bottom, with the funnel blowing jets of water down to expose small shrimps or crabs buried in the sand. They also feed on fishes or even on smaller cuttles. The stimulus to attack must be entirely visual; a shrimp enclosed in a sealed glass tube is promptly attacked, as is the reflection of a shrimp on the glass wall of an aquarium. Plymouth Aquarium, England. (D.P. Wilson)

Cuttle attaching an egg to a stick in a tank at the marine laboratory in Den Helder, Netherlands. Each egg emerges singly from the funnel and is fertilized as it passes over the sperm receptacle in the membrane surrounding the female's mouth. The egg is covered with glandular secretion (that later hardens) and is stained black with ink from the ink sac. The female uses her large ventral arms to twine the flexible stalk of the egg around the vertical stick, and when the stalk hardens, the egg remains firmly affixed. In the ocean, eggs are usually attached to fixed marine plants or other solid objects. The animal shown here laid many eggs, sharing the stick and taking turns with another laying female. When the young hatch from the egg, they are about a centimeter long and squirt ink if handled. (R.B.)

Some kinds of cuttles live in deep water. *Spirula* can be netted in tropical or subtropical waters, down to 500 m or more. Buried within the mantle of this cuttle is a delicate coiled calcareous shell with gas-filled chambers. *Spirula* can move swiftly this way or that by directed jets from the funnel. When disturbed, it can withdraw head and arms completely into the mantle cavity, so that only the tough slippery mantle is exposed. The animals are thought to live only about 20 months, and when they die, the shells float to the surface and are cast ashore all over the world, making these cuttles seem familiar and much more widely distributed than they are.

Octopuses differ from squids and cuttles in having only 8 arms, all equally long and agile, with rows of suckers from base to tip. The rounded, compact body has only a trace of a shell and is so flexible that it assumes a streamlined form when the octopus swims squid-like by jets from the funnel. The animal seems to flow as it half-swims, half-crawls over rocky bottoms by graceful movements of its webbed arms, slipping easily through small openings between the rocks or into narrow crevices in search of shelter or of prey, usually crabs and other crustaceans, bivalves, worms, or fishes. Most bottom-living octopuses reside in a particular hole in the rocks; or, in most sandy areas, in a shelter that they build from scattered rocks. Discarded glass bottles are quickly claimed as octopus homes. Unfortunately for octopuses, these snug homes are no defense against some of the most enthusiastic octopus predators, moray eels. Courtship often involves elaborate color displays, and the male uses one arm, specialized for this function, to transfer a packet of sperms into the mantle cavity of the female. The eggs are fertilized internally, and the female attaches them to the ceiling of her home. Eating little or nothing, she keeps watch over the eggs, guarding and cleaning them constantly for up to several months until they hatch. At about this time the female dies.

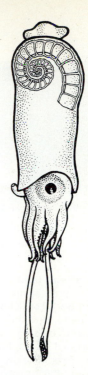

Spirula, **a pelagic cuttle.** Buoyed up by the shell, the animal hangs head down in the water. At the aboral end, from between two rounded fins, a light organ shines upward. About natural size.

Octopus design on a Cretan vase, from about 1500 B.C., is one of the first pictures of living invertebrates.

Octopuses mating. The male has his hectocotylized 3rd right arm inserted into the female's mantle cavity. Copulation may last for an hour or more, while multiple spermatophores are carried by peristalsis along a groove in the arm. The male generally matures at a smaller size and risks being eaten by any female larger than himself if he neglects to court her. A smaller female may be unceremoniously approached and inseminated. (Based on J. Meisenheimer and on Wells and Wells)

Octopus swimming. The "lesser octopus" of European shores, *Eledone cirrhosa,* is seen here with the funnel directed to the right and the animal propelled to the left. Length, 20 cm. Plymouth, England. (D.P. Wilson)

Octopus eating a crab. *Eledone cirrhosa* pins a crab against aquarium glass. The powerful beak bites into the crab, helping to introduce an immobilizing toxin. The lair of an octopus may be revealed by the litter of discarded skeletons of crabs and other prey. Certain Pacific octopuses, at least, are known to use the radula to drill holes in gastropod and bivalve shells through which to inject toxin. Plymouth, England. (D.P. Wilson)

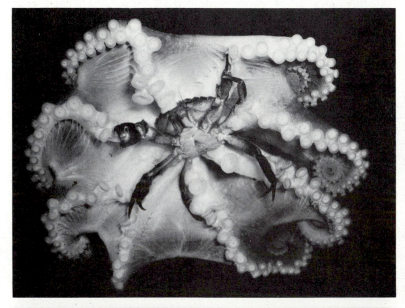

Egg string of *Octopus vulgaris* was one of many hanging in festoons from overhanging rock. Each egg was 2 to 3 mm long. England. (D.P. Wilson)

Common octopus of Europe, *Octopus vulgaris.* Marine Institute, Split, Yugoslavia. (R.B.)

1. **Dwarf octopus,** *Octopus joubini,* here climbing on the glass of an aquarium, is only 5 to 7 cm long, including the tentacles. When at home in the Gulf of Mexico it lives in empty gastropod shells, using bits of shell or gravel to close the shell opening. At night the octopus emerges to hunt for small crabs. (This species is indeed dwarfed by *Octopus vulgaris,* which may attain an arm span of 2 m in warm waters. A very large specimen of the Pacific giant octopus, *Octopus dofleini,* has been reported to reach a span of 9 m and weigh over 90 kg.) 2. **Female brooding eggs** in an empty cockle shell. At any threat her arms pull the valves closed. She will stay with her eggs constantly, neither feeding nor accepting proferred food. She protects and cleans the eggs until they hatch in 30 days. After brooding, she dies.

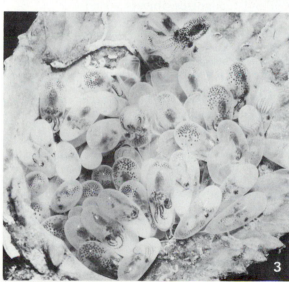

3. **The eggs** are attached in a single layer to the inner surfaces of both valves of the cockle shell. They are fewer but much larger (6 to 8 mm) than the eggs of *O. vulgaris.* As the female laid the eggs in successive batches, many stages are present at once. 4. **A young octopus,** just emerged, is seen at the top. Others have already left behind empty egg membranes, and the remainder are in various stages of development. 5. **Just-hatched young** of *O. joubini* is well developed, nourished by the generous supply of yolk available in the large egg. The remains of the yolk sac are still visible and provide a brief supply of food for the little octopus that is already able to crawl about on the bottom and will soon find food for itself. It does not undergo a planktonic period, as do the less-developed hatchlings of *O. vulgaris.* (R. Buchsbaum)

Sitting on the optic lobes, but neurally connected elsewhere in the brain, are a pair of small, spherical secretory bodies, called the optic glands. When these were removed from females of *Octopus hummelincki* after the animals had spawned eggs and were caring for them, the females stopped brooding the eggs and began to eat regularly again. Without optic glands, both males and females exceeded average weight and life span. The optic glands, the only endocrine glands identified in octopuses, control the maturation of gonads, sexual behavior, feeding, growth, and time of death.

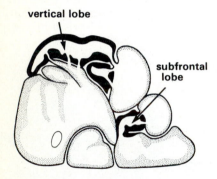

The **octopus brain,** with its 30 or so interconnected lobes, is among the best-mapped of all invertebrate brains. The functions of many of the parts have been established by observing changes in behavior after stimulating a given part electrically, or after experimental surgery. The functions of the human brain have been localized in much the same way, in people who have suffered localized head injuries or tumors.

In this supra-esophageal portion of an octopus brain, sectioned slightly to one side of the midline, the black designates areas consisting of millions of very small nerve cells. Damage to these small-cell areas has no effect on movement or posture, but is specifically disruptive to learning. Tactile learning is found to be associated most importantly with the subfrontal lobes and visual learning with the enormous optic lobes (which project out to each side and do not appear here). Lesions in the vertical lobe affect both learning systems. (Simplified after J.Z. Young)

Studies of **learning** in octopuses *(Octopus)* and cuttles *(Sepia)* have shown these animals to be the most apt subjects among invertebrates. They learn quickly and can remember for periods up to several weeks. During a learning experiment, for example, an octopus may be allowed to attack and eat crabs that are presented on a square background, but given a mild electrical shock when it is presented with a crab on a rectangular background. After only a few trials, the octopus no longer attacks in the presence of the rectangle, but attacks the square even in the absence of a crab. Substituting backgrounds of varying shapes, patterns, colors, and positions has shown that octopuses can discriminate some subtle differences between objects but fail to separate others that seem obvious to us. An octopus can barely tell a square from a circle, but it can distinguish a square resting on one side from a square resting on a corner, or a 4 cm square from an 8 cm square, even if the larger one is at twice the distance. Comparison of many such selected pairs helps experimenters to discover how an octopus perceives and recognizes objects visually.

Octopuses learn by touch as well as by sight, exploring objects with the sensitive suckers of the arms. They discriminate in this way between rough and smooth surfaces, or even between objects with different proportions of rough surfaces, but cannot discriminate by size or shape unless there are textural cues, such as sharp curves or corners. They never learn to discriminate by weight. Although the arms handle a light object with ease and strain at a heavy one, evidently no information about these different movements is fed back to the learning centers of the brain.

Visual and tactile learning by octopuses occur in separate parts of the brain, as shown by localized operations, which affect one kind of learning or the other. These reveal also that certain parts of the brain play specific roles in learning and memory, while others are involved primarily in muscular control (as in humans and other vertebrates). It has not proved possible, however, to destroy a specific memory by cutting out a small part of the appropriate area of the brain. Rather, all memories are impaired by such a cut, regardless of the exact location; the greater the amount of tissue damaged or removed, the greater the general loss of memory, again as in vertebrates.

A cut that separates the right and left learning centers of the brain prevents the usual transfer of learning between them, so that a "split brain" octopus must be taught independently on the arms of both right and left sides to recognize an object by touch. In a normal, intact brain, transfer of learning between the halves does occur but only slowly, on a time scale of hours. This suggests that it must involve something other than simple nervous conduction between the two sides.

One visible structural change, which may be associated with learning as a cephalopod matures, is especially dramatic in *Sepia*. Young cuttles are slower to learn than adults and remember less. They depend to a greater extent on genetically determined behavior, and the parts of the brain important in learning and memory are correspondingly small. However, within a few months these parts increase to over half again their original size relative to the rest of the brain. This seems to be associated with a switch to increasing dependence on learned behavior.

Bathypelagic octopod, *Amphitretus pelagicus,* with deeply webbed arms and telescopic eyes on stalks. The body is covered with a loose sheath of transparent jelly. Although found in the tropical Indo-Pacific, these octopods do not swim in bright, warm waters; they live in cold, dark depths below 500 m or so, although small individuals caught in lesser depths suggest that adults may swim upward to breed. This specimen (among the largest recorded) was caught in a trawl west of Seychelles, Indian Ocean. Length, 30 cm. (V.B. Pearse)

"Paper nautilus," *Argonauta argo.* Freshly caught female, placed on sand, is guarding eggs in the shell. The web of the first pair of arms, usually spread over the shell, is here mostly withdrawn. Crustaceans or fishes that chance to touch the web are grasped by the fourth arm. Shell, 17 cm. Skiathos, Aegean Sea. (D.P. Wilson)

Some kinds of octopods are pelagic, spending their whole lives swimming in mid-ocean, often at great depths. These brood their eggs by retaining them internally or holding them in their arms. The female argonaut, or "paper nautilus," *Argonauta,* secretes from the web between her two uppermost arms a fragile, lightly calcified "shell" in which she sits and lays her eggs. This shell is not homologous with the mantle-secreted shells of other molluscs. The male argonaut, about 2 cm long, is dwarfed by the 20 cm long female. He secretes no shell but often shares that of his mate.

The large and elaborate mating arm of the male argonaut breaks off inside the female's mantle cavity during mating and remains there, moving actively about under its own power for some time; but to our disappointment, the oft-told story that it swims freely in the ocean, finding and entering the female entirely on its own, is without foundation. The first biologists who found the detached male arm inside the female thought it was a parasitic worm, and it was named *Hectocotylus.* Later fanciful accounts described it as a whole male cephalopod. Now, the specialized mating arm of any octopus or squid is called the hectocotylus or is said to be hectocotylized.

The "vampire squid," *Vampyroteuthis infernalis,* is hauled up occasionally in trawls from about 2000 m. It is like octopuses in having 8 arms, but resembles squids in having numerous light organs and in bearing large fins on its deep purple mantle. Two slender filamentous tentacles, probably not homologous with the arms of squids and octopuses, may be retracted into pockets just outside the ring of webbed arms or, fully extended, reach well beyond the arms. Because of these and other peculiarities, *Vampyroteuthis* is placed, at present, in a separate order, the Vampyromorpha.

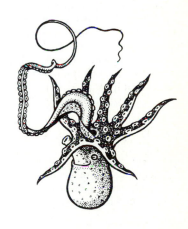

Male *Argonauta,* only 1-2 cm long, is unusual among cephalopods in differing markedly from the female in size and form. The hectocotylized third left arm may be 10 times as long as the mantle. (After H. Müller)

A pelagic octopod, *Tremoctopus violaceus,* is found with fragments of the tentacles of the siphonophore, *Physalia* ("Portuguese-man-of-war") attached to the suckers of the 4 dorsal arms. These suckers appear to be modified to hold the cnidarian tissue. The powerful nematocysts are probably used in both food capture and defense.

Two small octopuses, lightly sautéed in oil and served with a herbed oil and vinegar sauce, made a memorable luncheon in Venice, Italy. (M. Buchsbaum)

Blue-ringed octopus, *Hapalochlaena maculosa,* is common on Australian shores. Shy and long considered to be harmless, it was often handled by divers and by students in Australian classrooms. It rarely bites, and only in the early 1960s was it realized that the bite could be rapidly fatal. It is the only octopus known to inflict fatal bites on humans. The venom is in the salivary glands and enters through the wound made by the bite. (K. Gillett)

Dried squids, pressed wafter-thin between a pair of rollers, hang in rows from a rack in Bangkok, Thailand. They may be toasted over a fire for an extra fee. One squid provides a mid-afternoon snack that is nutritionally far superior to the American equivalent—a candy bar. (R.B.)

Preparing squids to sauté in oil. The pen of one has been pulled from the upper surface of the mantle and lies on the left. Chitinous jaws, head, and viscera are also discarded. Only the mantle and tentacles are eaten. *Loligo opalescens.* Monterey Bay, California. (R.B.)

The heavy coiled shell of a **nautilus** looks at first glance better suited as the home of a bottom-crawling snail than a swimming cephalopod. The few species of *Nautilus* in the tropical Indo-Pacific are the only living cephalopods with an external shell, although thousands of shelled fossil species of nautiloids and ammonoids have been found (see chapter on phylogeny). Nautiluses differ also from all other living cephalopods in having two pairs of gills, of excretory organs, and of auricles, instead of one pair, and in having "pin-hole camera" eyes with no lens. Instead of 8 or 10 large arms, nautiluses have about 90 short slender tentacles, all of them retractable into sheaths and lacking suckers but strongly adhesive. A unique fleshy hood over the head blocks the opening of the shell when the animal withdraws inside. As a nautilus grows, it seals off successive chambers in its shell by secreting pearly calcareous partitions across the whorls of the shell. The animal occupies only the outermost, largest chamber. The chambers are filled with gas and liquid, the amounts of each being regulated by a thin strand of mantle tissue that runs in a porous calcareous tube through the centers of the partitions and older chambers, so that both animal and shell are close to neutral buoyancy at a considerable range of depths, down to about 500 m. The few observations that have been made on nautiluses in the ocean suggest that they rest during the day in coral crevices, attached by their tentacles, and swim out to feed at night on fishes and crabs.

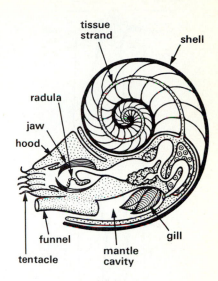

Diagram of *Nautilus*, the only living cephalopod with an external shell.

The egg capsules of *Nautilus* have not yet been found in nature, but in 1982 some females collected off Palau and maintained in aquariums began to deposit capsules on rocks and aquarium walls. Development of the embryos, as in other cephalopods, is modified by the mass of yolk that supports direct development, without larval stages. The shell develops very early.

Chambered nautilus, *Nautilus pompilius,* trapped below 60 m in Philippine waters but shown here swimming in the Vancouver Aquarium. Eye, numerous tentacles, and speckled hood are visible. Nautiluses feed mostly on crustaceans but will accept a proferred fish. (R.B.)

X-radiograph of *Nautilus macromphalus* shows the curved partitions separating the gas-filled chambers of the shell, the calcareous tube that connects the chambers, and the living animal occupying the largest, outermost chamber. Noumea, New Caledonia. (R. Cátala)

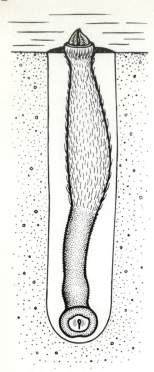

Burrowing aplacophoran (order Chaetodermatoidea) lies head down in its burrow. The gills protrude at the surface. (Based on H. Heath)

Solenogaster (order Neomenioidea) has a midventral groove. It climbs here, head up, on a hydroid colony. (Based on H. Heath)

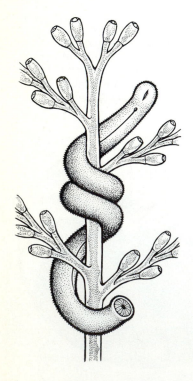

APLACOPHORANS

THE SMALL CLASS **Aplacophora** ("bearers of no shells") is entirely marine and includes drab, little wormlike animals, most of them about 25 mm long and lacking in any obvious external features that would place them in the same phylum with the large, swift, and intelligent cephalopods. The presence of a **radula** stamps them as molluscs. The whole surface of the elongated or plump body is studded with calcareous spicules, as in the girdle of some chitons, but there is no shell. Although there is a mouth at the anterior end, there is no distinct head. The anus opens into a small cavity at the posterior end.

Of the roughly 250 species, only a few live in shallow waters; most are deep-water forms and are seen only on dredging. In the smaller of the two groups that compose the class, the members burrow in muddy bottoms, feeding on unicellular organisms and organic debris. Not surprisingly in a group with these feeding habits, some species have a rudimentary radula. The animals lie head down in vertical burrows. A pair of typically molluscan featherlike gills protrude from the cavity around the anus.

The larger of the two groups consists of aplacophorans which creep along the mud surface or are found entwined in the branches of particular species of hydroids or soft corals or other colonial cnidarians, rasping off the polyps. They creep slowly on their hosts, on a mucous trail. From the cavity at the posterior end there protrude thin respiratory folds. In the ventral surface there is a longitudinal groove, or channel, from which they are called **solenogasters** (*solen* = "channel," *gaster* = "stomach"), a name often used for all aplacophorans.

According to one interpretation of aplacophoran structure, a fold of tissue in the ventral groove is the foot; the posterior space and the narrow spaces on either side of the foot represent the mantle cavity; and the entire spicule-studded surface of the body is the mantle. Another view is that aplacophorans have no foot or mantle at all.

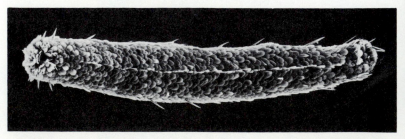

Interstitial solenogaster *Meiomenia swedmarki* shows the ventral groove that characterizes this group of aplacophorans. The surface is covered with flattened overlapping calcareous spicules and some spinelike ones. The minute aplacophoran, only 2 mm long, was collected at Friday Harbor, Puget Sound, Washington. SEM. (M.P. Morse)

Some aspects of the anatomy of aplacophorans are similar to those of chitons, especially the nervous system with two pairs of longitudinal nerve cords. Aplacophoran development, like that of many other molluscs, includes a trochophore stage. In general, the many peculiarities of structure make it difficult to relate aplacophorans to other groups of molluscs and to decide whether, of all the groups, they are the closest and most similar to primitive molluscs or are yet another later modification of the molluscan body plan.

MONOPLACOPHORANS

THE DANISH SHIP *Galathea,* dredging at about 3,600 m in an abyssal Pacific trench off Costa Rica in 1952, brought up 10 medium-sized "limpets" with broad low-domed shells and some very peculiar features. They were carefully preserved and set aside for later study. Back in Copenhagen, a closer look revealed that they were not gastropod limpets at all, but members of another class, the **Monoplacophora,** a group known only from fossils and believed to be extinct since the Devonian period about 350 million years ago. The discovery caused a sensation among biologists around the world. The first specimens were named *Neopilina galatheae,* and later several other species of *Neopilina* and another genus, *Vema,* were found in various places, but almost always in deep ocean waters, where the animals were feeding on unicellular organisms and organic material in the bottom mud. A tiny species of *Vema* has been found living on rock at depths of only a few hundred meters, on the continental shelf off southern California.

Univalve shell of *Neopilina galatheae* suggests that of a limpet. It is almost round, with the apex near the front margin. The thin fragile shell consists of the same 3 layers as in gastropods and bivalves. (After Lemche and Wingstrand)

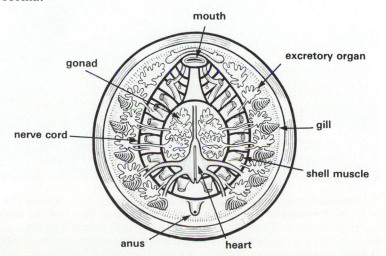

Diagram of monoplacophoran labels: mouth, gonad, excretory organ, nerve cord, gill, shell muscle, anus, heart

Diagram of monoplacophoran shows the repeated organs. (After Lemche and Wingstrand)

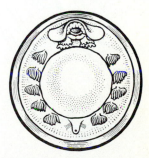

Undersurface of *Neopilina galatheae* reveals a bilaterally symmetrical animal with an anterior mouth and a posterior anus. On each side of the fleshy foot are 5 gills (or 6 in some species) with muscular attachments. The beating of the gills presumably aids respiratory exchange. (After Lemche and Wingstrand)

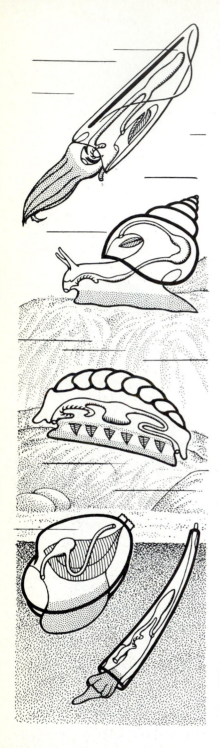

The discovery of *Neopilina* was exciting not only because it added a new class to the list of living molluscs, but also because the structure of these primitive, non-torted molluscs turned out to be remarkable for having multiple pairs of the elements of the organ systems: 10 pairs of cross-connectives in the ladder-type nervous system, 7 or 8 pairs of shell muscles (attaching the shell to the foot), 5 or 6 pairs of excretory organs, 5 or 6 pairs of gills, 2 pairs of gonads, and a heart with 2 pairs of auricles and a pair of ventricles. This multiplicity of parts in monoplacophorans is suggestive of *segmentation* found in annelid worms, the phylum described in the next chapter. Along with the many developmental similarities of molluscs and annelids, repetitive parts in monoplacophorans are thought by some to add weight to the evidence for closely linked origins of the two phyla.

IN COMPARING chitons, gastropods, bivalves, cephalopods, and other molluscs, we have seen that the fundamental body plan of an animal may become so modified in adaptation to a special way of life that many of its structures reflect the kind of life it leads rather than its relationship to its more typical relatives. The basic blueprint of the molluscan body plan lends itself to extraordinary modification and is expressed in a wide spectrum of ecological niches. This is often described by the phrase **adaptive radiation.** One or more of the most distinctive elements of the plan have been dropped out altogether in some molluscan evolutionary pathways, while in others these same features have been greatly elaborated and specialized. The result in either case is impressive numbers of species and great abundance of individuals that turn over enormous quantities of organic material and play significant roles in aquatic and land ecosystems.

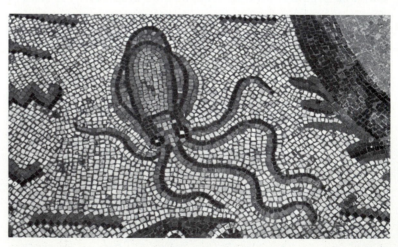

Divergent ways of life in the 5 major classes of molluscs: free-swimming cephalopods, mostly bottom-living gastropods and chitons, and burrowing bivalves and scaphopods.

Cephalopod image in a mosaic floor of a 1st century A.D., Roman villa in Sicily. It resembles a small species of octopus, with fins, that has been described for the Mediterranean. More than any other invertebrate phylum, living molluscs or their shells have played important roles in human cultures as food, as ornamentation, as monetary exchange, as symbolic images, and in religious ceremonies. (R.B.)

CLASSIFICATION: Phylum MOLLUSCA

Class APLACOPHORA (sometimes grouped with Polyplacophora in class Amphineura). Wormlike, without well-differentiated head, foot, shell, or mantle; spicules embedded in body surface. Marine bottoms at moderate to abyssal depths.

Order Neomenioidea, the solenogasters. With midventral longitudinal "foot" groove; respiratory folds at posterior end. Feed with radula on other benthic invertebrates, especially colonial cnidarians. Hermaphroditic. *Neomenia, Meiomenia.*

Order Chaetodermatoidea. Without ventral groove; with featherlike gills at posterior end. Burrow in mud, eating unicellular organisms and organic particles; radula may be rudimentary or absent. Separate sexes. *Chaetoderma.*

Class POLYPLACOPHORA, chitons. Rudimentary head without eyes or tentacles; broad creeping foot; shell consists of 8 overlapping plates along dorsal surface of body; many pairs of gills. All marine, mostly grazers of intertidal habitats. *Chiton, Cryptoplax, Katharina, Mopalia, Cryptochiton* (giant chiton), *Placiphorella, Tonicella.*

Class MONOPLACOPHORA. Limpetlike, but showing no torsion. Bilaterally symmetrical with repetition of parts along the body. Mostly extinct; only a few living species at moderate to abyssal depths. *Neopilina, Vema.*

Class GASTROPODA, the snails, limpets, and slugs. Generally well-developed head with eyes and tentacles; broad creeping foot; coiled or cone-shaped shell or no shell. Showing some degree of torsion and asymmetry.

Most modern authorities view the traditional subclasses listed below as patterns of organization rather than as natural groups with common origin.

Subclass PROSOBRANCHIA. Typically snails (with coiled shell and operculum) or limpets, showing clear torsion. Sexes usually separate. Mostly marine, a few freshwater and terrestrial.

Order Archaeogastropoda, mostly herbivorous, radula with many teeth, bipectinate gill or gills (filaments on both sides of gill axis): *Megathura* (keyhole limpet); *Haliotis* (abalone); *Acmaea* and *Patella* (limpets); *Calliostoma, Tegula.*

Order Mesogastropoda, herbivorous, carnivorous, and filter-feeding, radula with fewer teeth, monopectinate gill (filaments on only one side of gill axis). *Ampullarius* (freshwater prosobranch); *Littorina* (periwinkle); *Rissoa; Caecum; Petaloconchus* and *Serpulorbis* (vermetids); *Janthina; Crepidula* and *Crepipatella* (slipper snails); *Polinices* (moon snail); *Carinaria* (heteropod); *Lambis* (spider conch); *Strombus* (conch); *Cypraea* (cowrie); *Cassis* (helmet snail).

Order Neogastropoda. Mostly carnivorous, narrow radula with very few teeth, monopectinate gill. *Nucella* (dog winkle); *Urosalpinx* (oyster drill); *Buccinum* and *Busycon* (whelks); *Fasciolaria* (tulip snail); *Ilyanassa* (mud snail); *Conus* (cone); *Oliva* and *Olivella* (olives).

Subclass OPISTHOBRANCHIA. Typically sea slugs, mostly without shells, showing detorsion. Almost all hermaphroditic. Almost all marine.

Order Cephalaspidea, *Bulla, Navanax.*

Order Anaspidea. *Aplysia* and *Bursatella* (sea hares), *Akera.*

Order Thecosomata, shelled pteropods. *Cuvierina.*

Order Gymnosomata, naked pteropods. *Clione.*

Order Sacoglossa. *Berthelinia* (bivalved gastropod), *Tridachia, Elysia, Placobranchus.*

Order Notaspidea. *Tylodina, Pleurobranchaea.*

Order Nudibranchia. *Tritonia, Melibe, Dendronotus, Aeolidia, Phestilla, Glaucus, Phidiana, Hopkinsia, Archidoris, Dirona.*

Subclass PULMONATA. Mostly snails (without operculum) and slugs, showing some torsion, with mantle cavity modified for air-breathing. Hermaphroditic. Mostly terrestrial and freshwater, a few marine.

Order Basommatophora. Aquatic pulmonates: *Lymnaea, Planorbis, Siphonaria.*

Order Stylommatophora. Terrestrial pulmonates. *Helix* (garden snail); *Achatina* (giant African snail); *Liguus* (tree snail); *Limax, Ariolimax* (slugs).

Cephalaspidean *Navanax (= Chelidonura) inermis* has no shell. A voracious predator, it seeks out *Bulla* (a shelled cephalaspidean) and nudibranchs, by following their mucous trails. Catalina Island, Calif. (R.B.)

Class BIVALVIA (PELECYPODA, LAMEL-LIBRANCHIA), clams, mussels, oysters, scallops, file shells, fan shells, etc. Two shell valves, right and left, secreted by large two-lobed mantle. Narrow foot usually used in digging or secreting byssal threads. Rudimentary head without radula or specialized sense organs. Sexes usually separate. Mostly filter-feeders. Burrow in sand, mud, wood, rock, or live attached to rock, mostly in shallow waters, marine and freshwater.

The first two subclasses (sometimes called "protobranchs") include clams with several features considered primitive. **Protobranch** gills, small and not folded, with separate filaments, much like the gills of gastropods. No crystalline style. Flattened sole on foot. Marine, subtidal.

Subclass PALAEOTAXODONTA. Deposit-feeders that collect and sort bottom material with the palps. *Nucula, Yoldia.*

Subclass CRYPTODONTA. Deposit-feeders that collect bottom material with the mantle and foot, sorting on gills and palps. Some with reduced gut or no gut are apparently nourished by symbiotic sulfur-oxidizing bacteria in the gills and/or by uptake of dissolved organic materials. *Solemya.*

The next two subclasses include most bivalves. They feed with large **lamellibranch** gills, the filaments elongate and folded (forming a W-shape in cross-section). Adjacent filaments are bound into sheets by patches of interdigitating cilia (**filibranch** type) or by fusions or bridges of tissue (**eulamellibranch** type).

Subclass PTERIOMORPHA. Most live attached to surfaces, anchored by byssus threads or cemented by one shell valve. Foot modified to produce byssus threads, or rudimentary, or absent. Filibranch or eulamellibranch gills. Some fusion of edges of mantle lobes. Mostly marine. *Arca, Anadara* (ark shells); *Mytilus* and *Modiolus* (mussels); *Pinctada* (pearl oyster); *Pinna* (fan shell); *Pecten, Aequipecten,* and *Chlamys* (scallops); *Lima* (file shell); *Anomia* (jingle shell); *Placuna* (window pane oyster); *Ostrea* and *Crassostrea* (oysters).

Subclass HETERODONTA. Includes most of the familiar clams. Burrowing foot. Eulamellibranch gills. Edges of mantle lobes extensively fused; usually with elongate siphons. Marine: *Clinocardium* (cockle); *Tridacna* (giant clam), *Hippopus* (horse-hoof clam); *Mercenaria, Macoma; Donax* (wedge clam); *Ensis, Siliqua* and *Tagelus* (razor clams); *Tresus* (gaper clam); *Mya* (long-necked clam); *Pholas* (piddock), *Teredo* (shipworm). Freshwater: *Unio, Elliptio, Anodonta, Lampsilis, Margaritifera* (unionids); *Pisidium* and *Sphaerium* (fingernail clams).

Subclass ANOMALODESMATA. Some members with eulamellibranch gills: *Lyonsia.* Others with **septibranch** gills, modified as muscular partitions: *Cuspidaria, Poromya.*

Class SCAPHOPODA, tooth or tusk shells. Digging foot; tubular shell and mantle; rudimentary head bearing capitate feeding tentacles; no gills. Separate sexes. Sandy marine bottoms. *Dentalium, Antalis, Cadulus.*

Class CEPHALOPODA, the nautiluses, squids, octopuses. Large well-developed head with complex eyes and a cluster of tentacles around the mouth; funnel derived from the foot; shell mostly small or absent. Sexes separate. Active predators. All marine.

Subclass NAUTILOIDEA (Tetrabranchiata), the nautiluses. Mantle secretes a calcareous external coiled shell with gas-filled chambers. About 90 small, slender tentacles without suckers. Two pairs of gills, of excretory organs, and of auricles. Mostly extinct; one living genus in deep tropical Indo-Pacific waters. *Nautilus.*

Subclass COLEOIDEA (Dibranchiata). Muscular mantle covers body; shell internal or absent. Either 8 or 10 arms with suckers. One pair of gills, of excretory organs, and of auricles.

Order Decapoda, the cuttles (or cuttlefishes) and squids. With 8 short arms and 2 long tentacles; mantle bearing fins. Cuttles (suborder Sepioidea) with flat calcareous shell, on or near bottom in shallow waters: *Sepia, Sepiola;* or, with fragile coiled shell, bathypelagic: *Spirula.* Squids (suborder Teuthoidea) with flat thin chitinous shell ("pen"); pelagic in coastal or oceanic, shallow or deep waters worldwide: *Loligo, Architeuthis* (giant squid), *Cranchia.*

Order Vampyromorpha, "vampire squid." With 8 webbed arms and 2 filamentous tentacles (not homologous with tentacles of decapods). Deep purple mantle bearing large fins. Deep waters, 800-2,500 m. *Vampyroteuthis infernalis.*

Order Octopoda, the octopuses. With 8 long arms, fully webbed in pelagic and deep-water forms. Trace of shell, usually no fins. Shallow benthic: *Octopus, Eledone.* Shallow pelagic: *Argonauta* ("paper nautilus"). Bathypelagic: *Amphitretus.* Deep benthic: *Cirroteuthis, Cirrothauma.*

Chapter 16

Annelid Body Plan

Nereids and Earthworms

Segmentation

THE ANNELIDS familiar to most people are the useful, garden-variety earthworms and the notorious leeches. But the annelid body plan is better introduced by less specialized worms, such as the **nereids** (nir′-ee-ids).

The Nereids of ancient Greek mythology were sea nymphs, usually represented in female human form. Their invertebrate namesakes are mostly marine worms that can swim gracefully through the water, their slender bodies gently undulating from side to side. However, they spend most of their time on the bottom, safely hidden under stones, in burrows in the sand or mud, in mucous tubes built among blades of seaweeds or sea grasses, or even in shells occupied by hermit-crabs. When they emerge from these refuges, they are easy targets for hungry birds, fishes, crustaceans, and certain nemertean worms that specialize in eating nereids. Some nereids have managed to invade brackish sounds and estuaries, and a few live in freshwater lakes. In the Indo-Pacific, where humidity is high and where freshwater or marine shores adjoin damp soils and mangrove swamps, nereids may be found in moist soil, in damp litter, or even in the axils of coconut palms.

The most noticeable feature of the nereids is the ringing of the body, which is not merely external but involves many of the internal structures as well. The nereids and their relatives, along with the earthworms and the leeches, comprise the phylum **ANNELIDA** ("ringed forms"). The ringed condition is more often known as **segmentation**, or metamerism, and each ring is called a **segment,** or metamere.

Except for the head and the posterior tip, a nereid consists of externally similar segments. They have on each side a projecting appendage, or **parapod** ("side foot"), consisting of flattened fleshy lobes supported by chitinous rods (acicula). From the parapods there protrude bundles of chitinous **bristles,** characteristic of the annelidan class to which the nereids belong, the **Polychaeta** ("many bristles"). The bristles are

An annelid bristle is often dignified with the Latin term *seta* (plural *setae*) or the latinized Greek term *chaeta* (*chaetae*), and the scientific name of the class to which the nereids belong is Polychaeta. But there is no gain in precision in using these terms for annelid bristles as the terms are not specific to annelids, being used also for bristles of many kinds of animals and plants, as well as for a slender mouthpart of some insects.

Cross-section of nereid showing the protruding bundles of bristles and the chitinous rods that support the parapods.

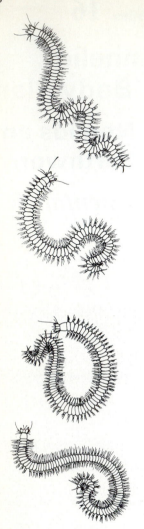

sharp, and probably protect the animal as well as enable it to obtain a hold on the smooth walls of its burrow. When a nereid is crawling slowly over the bottom, the parapods move like little legs and the bristles help them to grip the substrate at each step.

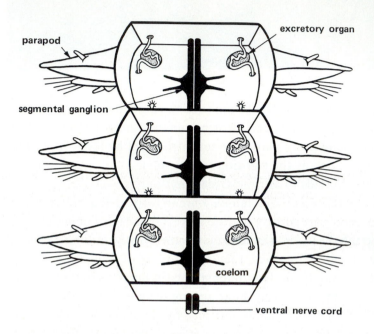

Repetition of parts in the annelid body plan.

The cuticle of annelids is tough and contains nearly unstretchable protein fibers. These are arranged in alternate layers of left and right-handed helixes, forming a cylindrical trellis that can be deformed and restored as the worm shortens and lengthens in swimming and bending.

The muscles of the body wall of annelids are obliquely-striated. Muscles of this type, with a structure intermediate between those of smooth and cross-striated muscles, are characteristic of the body wall of nematodes and are also found in molluscs and a few other groups. It has been inferred from this distribution that obliquely-striated muscle is particularly well suited to maintaining and controlling pressure in the hydrostatic skeleton of soft-bodied animals.

The outer covering of a nereid is a tough but flexible collagenous **cuticle** secreted by the epidermis. Beneath the epidermis is a layer of circular muscles, then a layer of longitudinal muscles, and finally a thin lining of mesoderm cells (the coelomic lining discussed below). Together these various layers constitute a definite **body wall**. They run the length of the worm and are interrupted by the segmental partitions, or **septa** (singular, *septum*). The longitudinal muscles occur as pairs of dorsal and ventral blocks. The circular muscles form continuous bands around each segment except where they pass out into the parapods as the parapodial muscles. Bundles of oblique muscles run in each segment from the midventral line to the parapods; they move the parapods back and forth. When a worm is in its burrow, waves of contraction pass along the body as the dorsal and ventral blocks of longitudinal muscle in each segment contract alternately. This produces gentle dorsoventral undulations of the body which create a current through the burrow, bringing the worm food particles or chemical stimuli from nearby food and constantly renewing the water for respiratory exchange.

When the worm is rapidly crawling or swimming, the waves of contraction are phased so that the dorsal and ventral blocks *on each side* contract alternately, and the body undulates from side to side as the parapods paddle slowly back and forth.

Nereids and other polychetes are the only animals in which the waves of muscular swimming contractions pass from tail to head, instead of from head to tail. This has led to the conclusion that nereids would swim backward, if it were not that the paddling of the parapods overcomes the presumed backward thrust of the body. However, calculations from an analysis of the fluid dynamics suggest that the parapods modify the flow of water over the worm such that the undulations of the body, even though they proceed from tail to head, actually combine with the paddling of the parapods to drive the worm forward in the water.

Feeding follows many styles. Nereids may be predatory carnivores or browsing herbivores, or scavengers on both plant and animal remains, or deposit feeders, or filter feeders. Having few defenses when they emerge to search for food, they extend only the anterior part of the body from the burrow, seize a piece of food with two strong jaws, and quickly drag it back into the safety of the burrow. If a nereid detects a piece of food too far from the burrow to reach, it may dig beneath the surface and emerge again closer to the food. These behaviors reduce the risk of being caught in the open by a predator.

Nereid moving along a sandy bottom. Muscular waves pass from tail to head. N.W. Florida. (R.B.)

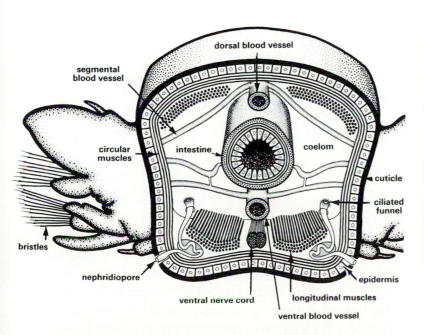

Cross-sectional aspect of a nereid with prominent circular and longitudinal muscles in both body wall and intestinal wall. The excretory organs, or *nephridia*, pass through the septum and their ciliated funnels open into the coelom of the segment anterior to the one shown. The coelomic lining is omitted.

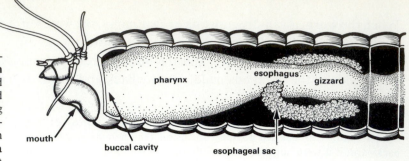

In the polychete body the **head** consists of 2 main parts: the **prostomium** ("region in front of the mouth") and the **peristomium** ("region around the mouth"). Behind the head is a long trunk consisting of mostly similar segments that bear parapods and contain muscles, nephridia, gonads, and a portion of the digestive tract. A short posterior region, the **pygidium,** bears the anus and often sensory projections. New segments develop in front of the pygidium. Prostomium, peristomium, and pygidium are not considered to be segmental (although in nereids and most other polychetes, one or two segments are incorporated into the peristomium during development).

Digestive system of a nereid is specialized only at the anterior end. The septa of this region are reduced to muscular strands that suspend the gut.

To describe one segment of a nereid is to describe nearly the whole worm. Only the **digestive system** shows much differentiation from the anterior to the posterior ends. The mouth leads into a thin-walled buccal cavity lined with cuticle bearing numerous dark brown, sharp denticles. The buccal cavity opens into a thick-walled muscular pharynx, on the inner walls of which are the two large jaws already mentioned. Buccal cavity and pharynx together form an eversible **proboscis;** and when the proboscis is turned inside out and extended through the mouth, the jaws grasp food, which is then swallowed by withdrawing the proboscis. Behind the pharynx the digestive

Head of nereid, *at left,* shows sensory projections and 4 eyes. *At right,* the proboscis is protruded inside out, exposing the jaws of the pharynx and the hard denticles on the cuticle that lines the buccal tube. Extension of the proboscis is only in small part aided by protractor muscles. Most of the forward thrust comes from contraction of circular muscles in the body wall as they exert pressure on the incompressible coelomic fluid. Strong retractor muscles that run from body wall to pharynx pull the proboscis in again.

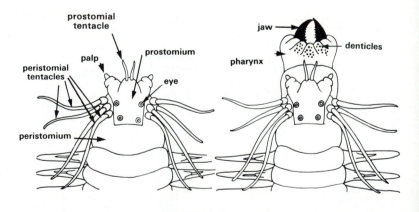

The hard parts of nereids such as the bristles and jaws consist mainly of organic substances (polysaccaride and protein), but probably owe their hardness at least partly to small but important inorganic components., The chitinous bristles contain calcium, magnesium, and iron. The jaws of nereids are not chitinous, but are cuticular derivatives containing significant quantities of zinc. In other polychetes, the jaws have been found to contain copper, calcium, or magnesium.

tube narrows to an esophagus, which runs through several segments and into which open a pair of glandular pouches, the esophageal sacs, which probably secrete digestive enzymes. Depending on the feeding habits of the species, the esophagus may be followed by a thick-walled gizzard or stomach, or open directly into a long, straight intestine that runs the length of the body to the anus at the posterior tip. In the wall of the digestive tract *peristaltic contractions* of the thick muscle layers produce a succession of rhythmic waves of constriction which push the food along independently of movements of the body wall.

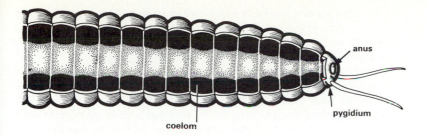

Posterior portion of digestive tube. The intestine, which extends from behind the gizzard throughout the length of the worm (most of which is omitted here), is relatively unspecialized.

Between the digestive tube and the body wall of a nereid is a definite space, a body cavity called the **coelom** (see'-lom), which is lined completely by a sheet of mesoderm cells, the *coelomic lining,* or peritoneum. One important advantage of the coelom is that it separates the gut from the body wall. As already mentioned, this allows the muscles of the digestive tube to push food along independently of the movements of the body. It also permits a freer play of the body-wall muscles. The coelom is filled with fluid and functions as a *hydrostatic skeleton* around which the muscles of the body wall can contract. The coelomic fluid contains ameboid cells and many dissolved substances; it bathes all of the internal organs and supplements the role of the circulatory system, although it has no direct connection with that system. The coelom also plays a role in excretion and in reproduction.

At the anterior end of the worm, the coelom is not divided by septa and provides space for the movements of the pharynx and for the development of the fluid pressure which aids in everting the pharynx. Behind the esophagus the coelom is partitioned by the septa into a series of chambers that correspond to the external segmentation.

The coelom arises by the formation of a pair of spaces in the embryonic mesoderm of each segment of the body. These spaces enlarge and give rise to a series of paired coelomic sacs. The inner walls of these sacs envelop the digestive tube; and where they meet in the mid-line, they form a double layer of coelomic lining, the *mesentery,* which supports the gut above and below. The anterior and posterior walls of the coelomic sacs form the septa. In the nereids, and in many other annelids, the mesentery below the digestive tube is present only during early development. It disappears in the adult, and right and left coelomic spaces are confluent below the digestive tube.

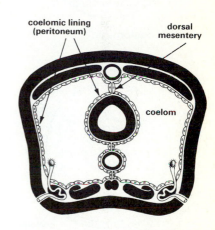

Cross-section of annelid emphasizing the coelom and its mesodermal lining.

The completeness of each septum, and the extent of the septa along the body, differ among the various polychetes; but even where septa are complete they usually have some perforations that allow limited communication between segmental chambers. The completeness of the mesodermal lining of the coelom also varies among polychetes, especially smaller forms, in some of which there is no recognizable coelom.

Acoelomate.

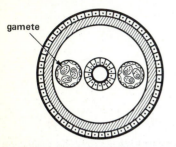

Pseudocoelomate.

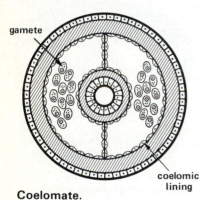

Coelomate.

Cross-sections of **3 body types** seen in invertebrate groups. Coelomates are probably derived from acoelomates; there is no evidence that they have passed through a pseudocoelomate stage.

Segmental blood vessels carry the blood in a loop around each half of each segment. Blood flows from the median dorsal vessel downward along the wall of the intestine in an extensive bed of capillaries (drawn much simplified). Collected again ventrally into larger vessels, some of the blood flows posteriorly in the median ventral vessel but most of it flows laterally into further capillary beds in the excretory organs, body wall muscles, and parapods, and finally back to the median dorsal vessel. (Partly after P.A. Nicoll)

The presence of a coelom is considered of such importance that animals are often divided into two large groups, those with a coelom *(coelomates)* and those without it *(acoelomates* and *pseudocoelomates),* categories which correspond roughly with what is meant by the "higher" and "lower" invertebrates. A space between the digestive tube and the body wall occurs in many of the phyla we have studied already. But in such groups as the nematodes and rotifers it has no definite mesodermal lining and is therefore not considered to be a coelom, although it serves many of the same functions. In molluscs the coelom is limited, for the most part, to the cavity surrounding the heart and the cavities of the gonads and excretory organs. Nearly all groups to be described hereafter have a coelom.

The general type of body structure seen in a nereid—with a muscular body wall separated from a muscular digestive tract by a space lined with mesoderm—occurs in all vertebrates. In humans the coelom is divided into an abdominal cavity, a cavity surrounding the heart, and two cavities which contain the lungs. When the coelomic lining, or peritoneum, becomes infected, as from a ruptured appendix, the serious condition that results in the abdominal coelom is known as "peritonitis."

In the **circulatory system** of the nereids, the blood flows entirely in closed vessels which send branches into all parts of the animal. The largest vessels are a *median dorsal vessel,* which runs just above the digestive tract and carries blood anteriorly, and a *median ventral vessel,* which runs just beneath the digestive tract and carries blood posteriorly. In each segment, some blood flows from the dorsal vessel into *segmental vessels* which branch further in the walls of the gut and other organs into intricate networks of very fine vessels, the *capillaries.* The thin walls of the capillaries are composed of a single layer of flattened cells that permit rapid exchange of dissolved substances. Capillary beds are especially extensive in the walls of the intestine, where nutrients enter the blood to be distributed to other tissues, and in the parapods and body wall, where respiratory exchanges occur and some nitrogenous

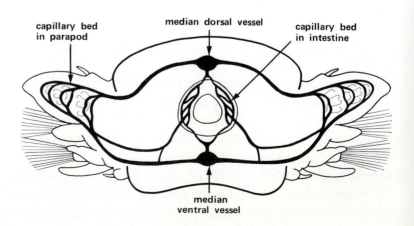

wastes leave the body. From the segmental vessels blood returns to both the median longitudinal vessels, but the segmental circulation is largely independent of the longitudinal flow. If, for example, the median dorsal vessel is tied off at any point along the worm, the segmental circulation continues scarcely disturbed for hours.

Extensive branching does not of itself make a good circulatory system, for, as the name implies, the blood must be in constant circulation. In the nereids there is no heart but most of the larger vessels and even certain capillaries are contractile, having muscle cells arranged circularly in their walls. Rhythmic peristaltic waves of contraction run along the median dorsal vessel and the lateral segmental vessels, maintaining the general pattern of flow. However, contractions of the various vessels are neither vigorous nor mutually coordinated, and the sluggish flow is not entirely consistent in direction, especially in the smaller vessels and capillaries.

Besides a good circulatory system to distribute the oxygen after it has entered the blood, large and active animals require an extensive **respiratory surface** which is freely exposed to oxygen, either of the air or dissolved in water. In the nereids the amount of body surface exposed to the seawater is enormously increased by the thin, flattened parapods, within each of which is an extensive network of capillaries. The capillary beds of the parapods and of the dorsal and ventral body walls lie very close to the surface, facilitating respiratory exchange. The oxygen-carrying capacity of the blood is increased by the presence of hemoglobin, which is dissolved in the blood instead of being contained within blood cells, as it is in some invertebrates and in vertebrates.

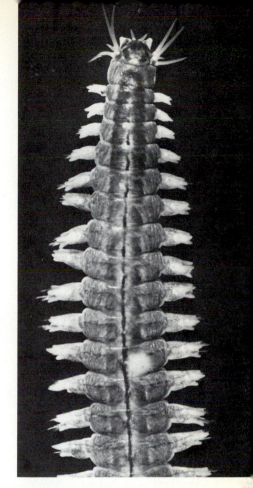

Neanthes diversicolor, anterior end. Through the translucent cuticle are visible the dorsal blood vessel and the vascular networks of body wall and parapods. This species is noted for its adaptability to brackish water and has been the subject of many studies on osmoregulation in nereids. Plymouth, England. (D.P. Wilson)

Most polychetes have red, iron-containing hemoglobin dissolved in the blood as nereids do. In others a similar iron-containing greenish pigment called chlorocruorin occurs in solution in the blood, alone or together with hemoglobin. Hemoglobin may be present in the coelomic fluid as well, either dissolved or in cells. Certain species of polychetes even have hemoglobin in the muscles of the body wall, but very small species usually have no oxygen-binding pigment at all.

The **excretory system** is segmentally arranged. Paired excretory organs, or **nephridia** (singular, *nephridium),* lie on the floor of nearly every segment. Each organ consists of a tubule coiled throughout most of its length, the coils compacted within a mass of connective tissue. At the inner end the tubule opens by a ciliated funnel into the coelomic chamber just anterior to that in which the main body of the organ lies.

In most marine invertebrates, the chief barrier to invasion of fresh waters is osmotic stress, not so much on the adults as on young stages in which efficient osmoregulatory mechanisms are not yet developed. The young are also at higher risk of washing out to sea. *Neanthes limnicola* of the N. American west coast lives in brackish streams and even in freshwaters. The worm is hermaphroditic (rare in polychetes) and internally self-fertilizing; and the developing young are sheltered in the coelom, protected from osmotic stress.

Development of an annelid nephridium. 1. The most posterior and youngest segment shows an enlarged cell, the nephridioblast, lying between apposed layers of mesodermal lining that form the septum. **2.** Nephridioblast divides. **3.** Cord of cells forms a ciliated canal that opens through the septum anteriorly. At its posterior end the canal meets an invagination of the ventral epidermis. **4.** Fully developed nephridium opens into the coelom through a ciliated funnel and carries excretory wastes to an external opening, the nephridiopore. (Modified after E.S. Goodrich)

Nereids have the type of nephridium diagrammed here, called a **metanephridium**. In some polychetes, the nephridial canal does not open internally through a funnel but is closed and contains beating flagella, much as in the flame bulbs of flatworms; this type is called a **protonephridium**. In most polychetes, a combination of nephridial cells and cells of the coelomic lining form a **nephromixium** which looks like the metanephridium of nereids, but usually serves both as an excretory organ and as a duct for the release of gametes.

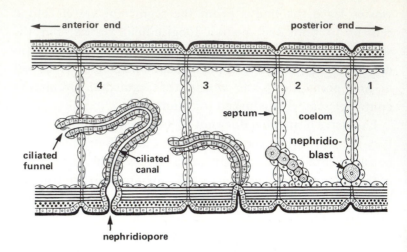

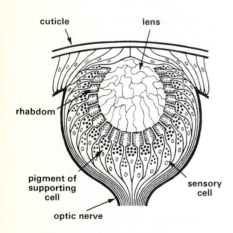

Eye of a nereid. Pigment shields the light-sensitive rhabdoms of the sensory cells; only rays passing through the pupil reach them, and the eye may analyze the direction of light. Movements of the retinal cells change the size of the pupil in bright and dim light. The central "lens," of uncertain function, consists of transparent extensions of the pigmented supporting cells; compare eye of honey bee, chap. 20. (Based mostly on Fischer and Brökelmann)

At the outer end the tube opens by an external pore at the base of the parapod. Cilia line the tubule, and their beating sweeps coelomic fluid into the ciliated funnel and out through the pore. As the fluid passes through the tubule, certain salts and other reusable materials are extracted from it, and waste materials from the blood may be secreted into it. Since nitrogenous waste is mostly in the form of ammonia, which diffuses readily, diffusion from the capillary beds of the parapods and body wall through the body surface probably accounts for a major part of nitrogenous excretion. Both the proportions and the total concentration of salts in the coelomic fluid are generally similar to those of the surrounding seawater, and also to those of the fluid leaving the nephridia. However, when seawater is diluted beyond a certain level, the nephridia become more active in resorbing salts and excrete greater volumes of dilute fluid. Thus they conserve salts and dispose of excess water. The body surface also helps to regulate water and salt balance in diluted seawater by actively taking up salts and decreasing in permeability to water and to salts. All nereids can tolerate some dilution of the body fluids, but relatively few species can osmoregulate well enough to survive extreme dilution of the surrounding seawater, as in estuaries.

The **nervous system** of nereids is more centralized and complex than those of flatworms and nemerteans but less so than those of molluscs and arthropods. The head bears two pairs of eyes and several pairs of projections and pits which are sensitive to touch and to food and other chemicals in the water. All these sense organs at the head end send sensory information to the large **dorsal ganglion** ("brain"), which lies above the pharynx and connects, by a ring of nervous tissue passing around the foregut, with another large ganglion, the sub-pharyngeal, or **first ventral ganglion.** From this first ventral ganglion the **ventral nerve cord** runs posteriorly along the length of the worm, and in each segment the cord enlarges into a **segmental ganglion,** which gives off nerves to the muscles of the body wall and parapods. Special controlling functions are shared between the dorsal ganglion and first ventral ganglion, but muscular coordination is performed by the nerve cord and segmental ganglia. If the nerve cord is cut midway in a nereid, the segments posterior to the cut can still crawl and swim in coordinated fashion but are not synchronized with the front half of the worm.

If the dorsal ganglion is removed, the animal is deprived of much *sensory input* but can still move in a coordinated way and in fact, it moves about more than usual. If it meets some obstacle, it does not withdraw and go off in a new direction but persists in its unsuccessful forward movements. This very unadaptive kind of behavior suggests that in the normal nereid the dorsal ganglion has an important function, that of *inhibition* of movement in response to certain stimuli. Removal of the dorsal ganglion prevents a nereid from *learning* the way through a maze but does not destroy previous learning nor prevent simpler kinds of learning if the training is sufficiently intensive.

In planarians spontaneous activity is reduced or eliminated after removal of the brain, but in nereids this happens only after removal of the first ventral ganglion, which is apparently responsible for general *excitation,* including spontaneous behavior and muscle tone. Thus, although the dorsal ganglion is often loosely referred to as the "brain." it should share this title with the first ventral ganglion.

The **reproductive system** of nereids is structurally simple, and as in polychetes generally, the sexes are separate. Breeding usually occurs at some definite season. Certain cells in the coelomic lining, segregated from the general mass of meso-derm cells early in development, begin to divide and give rise to the gametes. Floating freely in the coelom, the gametes are nourished by dissolved nutrients and perhaps also by cells in the coelomic fluid. In female nereids the eggs usually are shed by rupture of the body wall, but the sperms of the males may escape through special openings in the last segment. In many other polychetes both eggs and sperms are shed through the excretory organs or through special ducts opening from the coelom to the exterior.

Many nereids have a modified sexual stage, the **epitoke,** which swims toward the surface to spawn. At spawning the sexes are attracted to each other; and as the females burst and shed their eggs into the seawater, the males discharge their sperms. After the gametes are shed, the worms usually die.

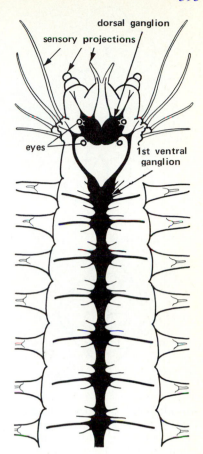

Nervous system of a nereid has 4 pairs of segmental nerves branching from each segmental ganglion, although annelids typically have 3 pairs. The ventral nerve cord and segmental ganglia also arise as paired structures, but fuse during development in nereids and in almost all other annelids.

An **epitoke** often looks so different from the immature stage of the worm that it can scarcely be recognized as belonging to the same species. The eyes may be enlarged, the sensory projections modified, and the body differentiated into two distinct parts: an anterior region with unmodified parapods, and a posterior region with parapods that have enormous and well-vascularized lobes and flattened, oar-shaped bristles, effective in swimming. The muscles are also modified for swimming; the gut and body wall are partially broken down; and all available materials and structures serve the production and eventual release of the gametes.

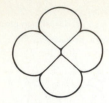

1. 4-cell stage

2. 8-cell stage

3. 16-cell stage

Spiral cleavage pattern is characteristic of dividing annelid eggs. It results from tilting of the mitotic spindle toward the less yolky pole of the cell and in many polychetes is evident as early as the 4-cell stage. At each cleavage the division products are slightly displaced so that they do not lie directly one above the other. In successive divisions the displacement is alternately clockwise or counterclockwise (as shown by the arrows). In eggs that are large (over 100 μm) and very yolky the cell divisions produce pairs of unequal size. In small polychete eggs divisions are equal.

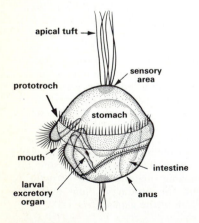

Trochophore of a polychete. (After R. Woltereck)

Hormones that control regeneration and reproduction are produced by the dorsal ganglion, or "brain," in nereids. Brain hormone is necessary for regeneration, for if the brain is removed, amputated segments will not be regenerated unless the same or another brain is implanted into the body. On the other hand, brain hormone inhibits sexual maturation. Since the amount of brain hormone declines with age, young worms are sexually immature and regenerate readily; whereas older worms are full of gametes but can scarcely regenerate at all. Removal of the brain from an adult worm is immediately followed by rapid development of the gametes and metamorphosis to the sexual form. However, removal of the brain from a young worm will result in abnormalities in the gametes and in sexual metamorphosis. Evidently a minimum level of brain hormone is necessary for early stages of gamete formation and preparation for metamorphosis.

What keeps this complex system in balance is still not fully understood. There is some evidence that hormonal levels may be affected by external factors such as temperature or light, and by internal factors such as feedback from gametes in the coelom. All of the hormonal influences, both positive and inhibitory, could be explained by a single hormone present in decreasing concentration and exerting different effects on structures at successive stages of development. Once we know more about the chemical nature of brain hormone, we may be able to discover how many active substances are actually involved, and whether they are produced by neurosecretory cells or by non-nervous cells in the brain.

Embryonic development of annelids follows several patterns that characterize a grouping of phyla called the **protostomes.** The name of the group refers to the fact that the *first* ("proto") external opening into the embryo, the blastopore, becomes the *mouth* ("stome") of the larva and then of the adult. In addition, protostomes have **spiral cleavage;** mesoderm that originates from a pair of primitive mesoderm cells; and a **schizocoel** ("split cavity"), a coelom that arises from spaces within the mesoderm. Finally, protostomes share **determinate cleavage,** in which the partitioning of the egg produces a mosaic of cells with specific and different potentialities. If a cell is separated from the embryo it develops only those parts it would have produced had it remained in an intact embryo.

The embryo of some polychetes, including some of the nereids, develops into a ciliated larva called a **trochophore.** Its most characteristic structure is a ciliated band about the equator, the *prototroch,* which serves as the chief organ of locomotion and also directs a food-bearing current toward the mouth, which lies just below. At the upper pole is a group of sensory cells from which arises a tuft of cilia. Some trochophores have a tuft of cilia at the lower pole also. Internally the most complex trochophores have a complete digestive tract with esophagus, large bulbous stomach, and a short intestine that opens through an anus at the lower pole. The paired larval excretory organs are **protonephridia,** tubules with beating flagella at their closed inner ends. They open near the anus.

Complex trochophores that swim and feed in the plankton and have a fully differentiated digestive tract generally belong to species that produce many small eggs containing little yolk. Nereids and most other polychetes have larger, yolkier eggs and a simpler trochophore larva that depends on the store of yolk; feeding and differentiation of the internal organs do not occur until a later stage of development (see young stages of nereid on the following page). Or, there may be no free trochophore larva; the trochophore stage is passed within the egg, and the animal that hatches is a later, segmented larval stage or even a juvenile worm. Free-swimming, feeding trochophores are present in relatively few polychetes but are thought to represent the more primitive condition.

The trochophore larva is of considerable theoretical importance because the same type occurs in several phyla. Few animals seem further apart in adult structure than a segmented worm and a snail. Yet their early stages of development are almost identical, cell for cell; and the trochophores that result are similar in many respects. Beyond the trochophore stage, however, marked differences begin to appear and the adults are very unlike. The close relationship thought to exist between these two phyla would never have been suspected except for the similarities of their embryonic development.

The most remarkable of the developmental resemblances that link the annelids and molluscs is the origin of the mesoderm in the two groups. The early stages of development of certain embryos have been followed so closely that each cell has been numbered and mapped. As a result of this extremely painstaking kind of work it is possible to trace the "cell lineage" of any portion of the early embryo. The adult mesoderm comes from a single cell (the "4d" cell), which arises in the same way in both annelids and molluscs. This cell divides into a pair, the *primitive mesoderm cells,* that lie one on each side of the region that will become the posterior end of the adult. These two cells give rise to two bands of mesoderm in annelids, or clusters of mesoderm cells in molluscs, which finally become hollowed out to form the coelom of the adult.

The development of the larval polychete into the adult worm proceeds with the formation of bristles on the elongating lower region and with constriction into segments. The ciliated bands disappear, and the upper part of the trochophore becomes the head. When the young worm settles to the bottom and takes up its adult mode of life, it continues to grow by addition of new segments just in front of the posterior tip, or pygidium.

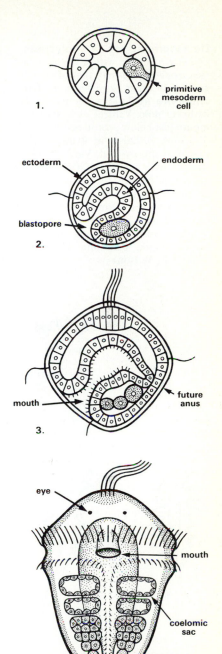

Development of mesoderm in a polychete. Blastula (1), gastrula (2), and young trochophore (3) are seen in section and from the left side. The older trochophore (4) is represented as the whole larva viewed from its oral surface. The chains of small cells budded off from the primitive mesoderm cells are breaking up into blocks which will become hollowed out to form coelomic sacs. (**1, 2,** and **3,** based on B. Hatschek; **4,** based on K. Heider and on D.T. Anderson).The mode of gastrulation shown here is typical of polychetes with small eggs containing little yolk. In polychetes with large, yolky eggs (e.g., nereids) the early endoderm cells do not invaginate but become internal as they are overgrown by the smaller and more rapidly dividing ectoderm cells.

Development of a polychete, *Ophelia bicornis,* with a less specialized trochophore than that of nereids and many other polychetes. **1. Trochophore at 2 days,** elongating at its lower pole. Ventral view, showing mouth just below prototroch. Length, 0.13 mm. **2. Larva at 4 days,** with first segments evident. View from left side, showing eye-spot and developing parapods. Length, 0.16 mm. **3. Young worm at 19 days,** after metamorphosis. Dorsal view, showing head with eyes, 3 segments with bristles, and developing digestive tract. Length, 0.33 mm. (Modified after D.P. Wilson)

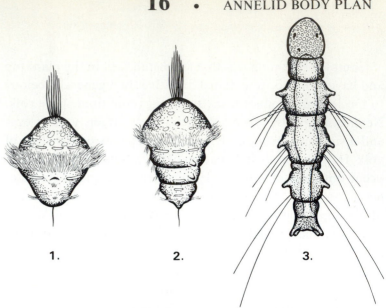

1.   **2.**   **3.**

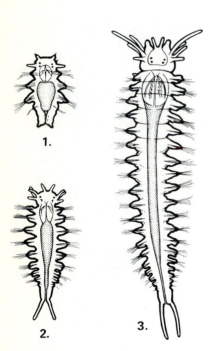

1.

2.   **3.**

Segments are added as a young polychete grows, and in these early stages of *Neanthes diversicolor* the number of segments is directly proportional to the length of the worm. Meanwhile, differentiation of the various parts proceeds. Note here the development of the head and tail appendages and of the jaws and digestive tract. **1.** Larva with 4 segments, 4 weeks old, 0.6 mm long. **2.** Larva with 9 segments, 7 weeks old, 1.5 mm. long. **3.** Young worm with 19 segments, 10 weeks old, 3.6 mm long. (Modified after R.P. Dales)

Some pelagic polychete larvas metamorphose in open water, and the young worms then settle to the bottom. Others, like the larvas of *Ophelia bicornis,* sink to the bottom before metamorphosis and use sensory cues to examine the chemical and physical characteristics of the substrate, *delaying metamorphosis* up to several weeks until a suitable substrate is encountered. In careful investigations by D.P. Wilson (see photo below) the larvas were found to be attracted to quartz sand grains of smooth and rounded shape, fairly uniform in size, covered with the usual bacterial coating of the sand in the natural habitat of adult worms. The larvas were repelled by sand grains of smaller size, or greater angularity, or without the bacterial coating. If the larvas fail to find suitable bottom sand they eventually die.

Unmetamorphosed larva of *Ophelia bicornis,* searching for a favorable site, has attached to a grain of "suitable" sand from a natural habitat of the adult worms, the Bullhill Bank at Exmouth, Devon, England. (D.P. Wilson)

EARTHWORMS

THE EARTHWORM caught by the early bird is no early worm but one that stayed out too late, for earthworms are nocturnal animals, emerging only at night and retreating underground in the morning. Even at night they usually do not leave their burrows but protrude the anterior part of the body in search of the seeds, leaves, and other parts of plants on which they feed while the posterior end maintains a firm hold on the burrow. Such retiring habits have probably contributed to the marked success of these animals which are quite helpless above ground. Since earthworms are adapted to living on so abundant and widely distributed a food as the decaying organic matter of the soil, it is not surprising that they occur in countless numbers in moist soils all over the world. Ever since Darwin made their activities the object of a careful study and wrote that "it may be doubted if there are any other animals which have played such an important part in the history of the world as these lowly organized creatures," it has been recognized that the work of earthworms is of tremendous significance to agriculture.

Earthworms spend most of their time swallowing earth underground and depositing it on the surface around the mouths of their burrows in the form of the "castings" familiar to everyone. In loose soil the burrow may be excavated simply by pushing the earth away on all sides, but in compact ground the soil must actually be swallowed. In moist and rainy weather the worms live near the surface, often doubled up on themselves so that either mouth or anus can be protruded. But in cold weather they plug the opening of the burrow and retreat into its deepest part, usually an enlarged chamber where one or several worms, rolled up together into a mucus-covered ball, pass the winter. In hot weather, also, they live far from the surface, as much as 3 m deep.

There are about 1800 species of earthworms, but details of this account are based on the common earthworm *Lumbricus terrestris,* called the "lawn worm" in England and the "night crawler" in the U.S. It reaches 30 cm, a medium size for earthworms. Other species range from less than a millimeter to giants more than 3 m long. The number of segments is usually not fixed and ranges from a minimum of 7 segments in certain minute worms to 500-600 in the giants. *Lumbricus terrestris* may have about 180 segments.

Lumbricus terrestris is one of a group of earthworm species that people from northern Europe have unwittingly carried around the world in soil accompanying plants or used as ballast in ships. In the large cities of eastern N. America, southern S. America, Australia, and New Zealand, the only earthworms to be found are of species from northern Europe that have flourished and have driven out the native species. In 1966 a survey of 60 earthworm species in the United States and Canada revealed that 37 (or 61%) were introduced from Europe. In the United States, *L. terrestris* is now found even in the highest remote areas of the Rocky Mountains.

Lumbricus is favored in laboratory classes because of its large size, but in the U.S. smaller earthworms, such as *Eisenia foetida* and species of *Allolobophora,* are more common.

The swallowing of earth is not alone a means of digging burrows. The soil passed through the digestive tract contains organic materials of various kinds: seeds, decaying plants, the eggs or larvas of animals, and the live or dead bodies of small animals. These organic components provide nourishment to earthworms, while the main bulk of the soil passes through. When leaves are abundant on the surface, the worms drag them into the burrows, and few castings are thrown up. When few leaves are taken in as food, the amount of castings increases.

The *effects of earthworms on the soil* are many. Both the castings, which become mixed with the soil, and the open channels created by burrowing ease the downgrowth of roots and enhance the fertility of the soil by increasing aeration and drainage. The thorough grinding of the soil in the gizzard of the worm is a most effective kind of soil "cultivation." The leaves pulled into the ground by earthworms are only partially digested, and their remains are thoroughly mixed with the castings. This organic material is then further broken down by microorganisms of the soil, releasing nutrients in a form available for absorption by plants. In this way earthworms have helped to produce the fertile humus that covers the land everywhere except in dry or otherwise unfavorable regions.

Thus when earthworms are present in the soil, agricultural productivity is generally higher, and in some cases greater crop yields have been achieved by introducing earthworms into soils where they were absent. Unfortunately, agriculture does not similarly benefit earthworms, which usually decrease in abundance when grassland is plowed and put under cultivation and when heavy doses of pesticides are used. Decline in earthworm populations is less when as much plant material as possible is left on the land and little or no nitrogen fertilizer is used.

The quantity of earth brought up from below and deposited on the surface has been estimated to be as high as 90 metric tons per hectare (40 tons per acre) in temperate regions, or if spread out uniformly, more than a centimeter per year. Estimates range much higher in the tropics. The layers of soil are thus thoroughly mixed, seeds are covered and so enabled to germinate, and stones and other objects on the surface become buried. In this way ruins of ancient buildings have been covered and so preserved, but many small and subtle features of archeological sites have been destroyed.

Certain pesticides to which earthworms are relatively resistant accumulate in their tissues, sometimes reaching concentrations 10 times or even 100 times those in the soil. Small mammals and birds that eat great numbers of earthworms concentrate the pesticides further and often die.

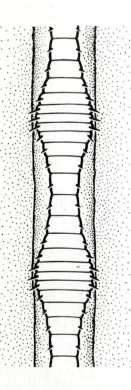

Bristles are extended or retracted as **muscular waves** pass down the body from head to tail. Midportion of a worm crawling upward in its burrow.

Externally an earthworm differs from a nereid in its *adaptations to a subterranean life.* As in other burrowing animals, the body is streamlined and has no prominent sense organs on the head or any projecting appendages on the body which would interfere with easy passage through the soil. On each segment are 4 pairs of chitinous **bristles,** which protrude from 4 pairs of small sacs in the body wall and are extended or retracted by special muscles. The bristles are used to anchor the worm firmly in its burrow, as can be readily discovered by trying to pull one out. But their main function is assisting in **locomotion.** As the worm works its way along, there pass down the body successive peristaltic waves of thickening and

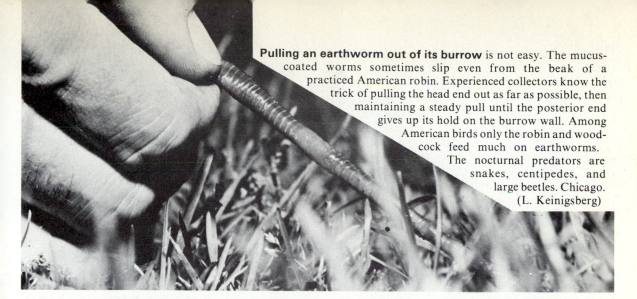

Pulling an earthworm out of its burrow is not easy. The mucus-coated worms sometimes slip even from the beak of a practiced American robin. Experienced collectors know the trick of pulling the head end out as far as possible, then maintaining a steady pull until the posterior end gives up its hold on the burrow wall. Among American birds only the robin and wood-cock feed much on earthworms. The nocturnal predators are snakes, centipedes, and large beetles. Chicago. (L. Keinigsberg)

thinning (7-10 per minute), as the longitudinal and circular muscles contract alternately. At each place where the body bulges out at a given moment, the bristles are extended and grip the burrow walls. The thinner segments in front of the bulge are pushed forward by contraction of their circular muscles, and those behind are pulled forward by contraction of their longitudinal muscles. As the wave passes on, the extended bristles are retracted, and the segments of the bulge thin out and are moved ahead in turn, each segment thus taking 2 to 3 cm "steps." The coelomic fluid acts as a hydrostatic skeleton. Because the septa between segments prevent coelomic fluid from flowing from one part of the worm to another as the muscular waves pass, the force of the contracting muscles is effectively localized and transmitted through hydrostatic pressure to the walls of each segment.

The lack of prominent **sense organs** on the head does not mean that an earthworm is insensitive to stimuli. There is a concentration of sensory cells at the anterior end, especially on the *prostomium,* the anteriormost tip of the body in front of the mouth. And, as in a nereid, cells sensitive to light, touch, and chemicals occur also among the epithelial cells of the epidermis along the body.

The absence of well-developed eyes, in an animal belonging to a phylum in which such eyes are common, is not unusual. Many animals that live in relative darkness have rudimentary eyes or no eyes at all; examples are burrowing forms such as earthworms or moles, cave animals such as certain crayfishes and fishes, and nocturnal forms such as many beetles. On the other hand, many other nocturnal or deep-sea animals have enormous and specially developed eyes to capture what little light there is.

The *light-sensitive cells* of earthworms are most abundant on the regions most frequently exposed to light, especially the prostomium, but also the dorsal surface of the anterior and posterior ends. Distributed all over the body are groups of from 35 to 45 cells, each with a hairlike process which projects through the cuticle covering the surface. These may be responsible for the sensitivity of earthworms to *touch,* to change in *temperature,* or to salts and certain other *chemicals.* The surface is notably sensitive to *acids.*

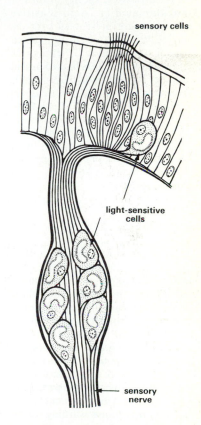

sensory cells

light-sensitive cells

sensory nerve

Sensory cells of the prostomium of *Lumbricus terrestris.* Those with clear vesicles lined by microvilli are light-sensitive; they occur singly among the bases of the epidermal cells or in clusters below the epidermis. The tall slender sensory cells with sensory hairs projecting through pores in the cuticle occur in small groups, about 1000 to a typical midbody segment and more densely at anterior and posterior ends. (Modified after W. N. Hess)

Taste cells probably occur on the prostomium and in the mouth and pharynx, as the worms seem to show definite food "preferences." All earthworms tested prefer manure or lettuce to tree leaves, and even will select green leaves over brown tanned ones from the same tree. The sense of *smell* is very feeble; and the worms are unresponsive to *sound,* which requires a complicated receiving apparatus. More important for a subterranean animal is the ability to detect *vibrations* transmitted through solid objects, and to these, earthworms are extremely responsive. It is said that one way to collect earthworms is to drive a stake into the ground and then move it back and forth, setting up vibrations in the ground which cause the worms to emerge from their burrows. There are no statocysts, and it has been suggested that the sense of *gravity* may depend on certain cells in the muscle layers that respond to stretch or tension.

Earthworms will not burrow into soil with a pH below a certain level, which varies from species to species. This may partly account for the distribution of earthworm species in soils of particular acidity. Acid-sensitive nerve fibers are present all over the body. Electrical recordings taken from certain single fibers show activity when the skin is bathed in acid solutions. The threshold of response is remarkably sharp. In *Lumbricus terrestris* the fibers respond at pH 4.1 to 4.3, and the worms will not burrow in soils of pH below 4.1. For *Allolobophora longa* pH 4.5 is the lower limit. The pH-sensitive fibers do not respond to various other chemical solutions tested.

The **central nervous system** is essentially the same as that described for nereids. A bilobed dorsal ganglion (brain, suprapharyngeal ganglion) lying above the pharynx connects by two nerves with the first ventral ganglion (subpharyngeal ganglion), lying below the pharynx. These two ganglia send nerves to the sensitive anterior segments and are considered to be the "higher centers." The dorsal ganglion is thought to direct the movements of the body in response to sensations of light and touch; and, as in nereids, it has important inhibitory functions, for if it is removed the worms move continuously.

Copious secretion of mucus follows exposure to various substances, or to handling, pinching, or electric shock in *Lumbricus terrestris*. The mucus secreted after such stimulation causes avoiding behavior in other members of this species, as well as in the secreting individual if it encounters its own mucus at a later time. Such mucus, which is not water soluble and which tends to retain its aversive properties even after several months, could act as an *alarm pheromone,* warning worms away from a spot where a noxious encounter occurred earlier.

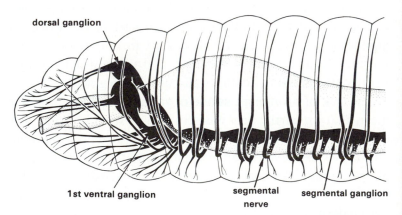

Nervous system at the anterior end of an earthworm. The digestive system is drawn as if transparent. (Modified after W.N. Hess)

Do earthworms feel pain? Immunohistochemical tests of sections of the brain and ventral ganglia contain substances similar or identical to the "natural opiates" enkephalin and beta-endorphin, pain-blocking substances produced by humans and other mammals. The production of these pain-blockers by earthworms suggests that they too may feel pain.

After removal of the first ventral ganglion, the worms no longer eat, and they cannot burrow in normal fashion. From the first ventral ganglion the double nerve cord runs to the posterior end of the body, enlarging in each segment to a double segmental ganglion from which 3 pairs of segmental nerves branch off. Each segmental ganglion serves its own segment and portions of the segments on either side of it, receiving impulses from sensory cells and sending impulses that result in contraction of the muscles.

The ganglia coordinate the impulses so that the longitudinal muscles relax while the circular muscles contract, or vice versa. Without this arrangement the two sets of muscles might only counteract each other's activities, and no movement would result. The smooth muscular waves that pass down the body in the ordinary creeping movements of the earthworm are not controlled by the large ganglia at the anterior end, for almost any sizable piece of an earthworm will creep along as well as a whole worm. The coordination is achieved partly through impulses relayed from one segment to another by nerve cells in the cord and partly through mechanical stimulation of successive segments. We infer this from the fact that coordinated peristalsis continues after cutting of the nerve cord or the body wall, but not both. As the nerve fibers are small, and as there is a certain amount of delay involved in the transfer from cell to cell in this chainlike succession of connecting fibers, the impulses travel slowly, at about 0.5 m/sec.

Besides the ordinary creeping or burrowing movements, earthworms can suddenly contract the whole body in response to strong stimulation of any region. If the anterior end is extended from the burrow and receives some unfavorable kind of stimulus, the longitudinal muscles contract as a whole, and the worm disappears into its burrow almost instantly. Such a response requires very rapid nervous transmission, and we do find certain "giant fibers" in the ventral nerve cord which pass throughout the length of the cord. The speed of transmission in these giant fibers varies with the temperature and has been measured at up to 45 m/sec. The speed is less in the giant fibers of nereids, up to 5 m/sec, but may be almost 10 m/sec in some other polychetes. These figures seem low when compared with the rate of nervous conduction in human motor nerves, in which impulses travel at about 100 m/sec.

The **digestive system** is differentiated into a number of regions, each with a special function. Food enters the mouth, is swallowed by the action of the muscular pharynx, and then passes through the narrow esophagus, which bears on each side the **calciferous glands.** These glands excrete calcium carbonate concretions into the esophagus, the calcium being derived from salts present in the food and the carbonate from metabolic carbon dioxide. If these glands are removed, and the worms placed under conditions of high atmospheric carbon dioxide such as may be routine in earthworm burrows, the body fluids become acid and high in calcium. The calciferous glands thus must have the dual function of maintaining the body's acid-base balance and regulating calcium content.

The esophagus leads into a large thin-walled sac, the **crop,** which apparently serves primarily for temporary storage, as the food undergoes little change and does not remain there very long. Behind the crop is another sac, the **gizzard,** with heavy muscular walls which (aided by mineral particles and very small stones swallowed by the worm) grind the food thoroughly. From the gizzard, the food passes through the intestine, which continues as a nearly uniform tube to the anus, where undigested residues pass out. Digestion is largely or

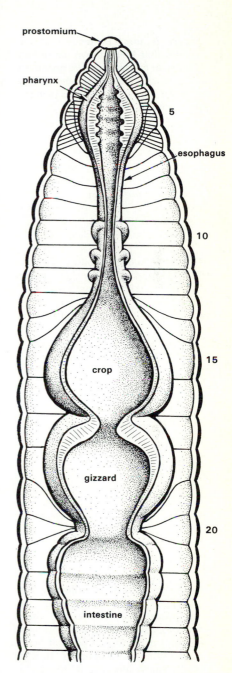

Anterior portion of digestive system of an earthworm. The esophagus bears a pair of small rounded sacs in segment 10 and paired calciferous glands in segments 11 and 12.

There is some evidence that the enzymes that break down chitin and cellulose in earthworms are produced by the worms themselves and not by symbiotic microorganisms as in many other animals.

entirely extracellular, and the pharynx, crop, gizzard, and intestine all secrete digestive enzymes, but the bulk of digestion occurs in the intestine. The roof of the intestine dips downward as a ridge or fold, the *typhlosole,* which increases the digestive surface. The intestinal wall is abundantly supplied with blood vessels, and the digested food is distributed to the rest of the body by the circulatory system.

The **circulatory system** is similar to that of the nereids. A median contractile dorsal vessel, which lies on the digestive tube and accompanies it from one end of the body to the other, is the main collecting vessel. In it the blood flows forward, propelled by rhythmic peristaltic waves. A median noncontractile ventral vessel, suspended from the digestive tube by the ventral mesentery, is the main distributing vessel. In it the blood flows backward and out into segmental branches which supply the various organs. In almost every segment the blood flows from the ventral to the dorsal vessel through capillary beds of the digestive tract, body wall, nephridia, and nerve cord. In the region of the esophagus (segments 7-11), the direction of flow is reversed: the blood flows directly from the dorsal to the ventral vessel through 5 pairs of enlarged muscular transverse vessels, the **hearts,** which pump the blood through the ventral vessel. Valves in the dorsal vessel and hearts prevent the blood from backing up during irregular contractions. Muscular contractions of the body wall and gut also assist the flow of blood. Hemoglobin dissolved in the blood carries 40% of the oxygen; the rest of the oxygen is simply dissolved in the blood plasma without being bound to any blood pigment.

Anterior part of circulatory system of an earthworm, showing only the principal blood vessels. Blood flows forward in the dorsal vessel and downward through 5 pairs of hearts surrounding the esophagus. In front of the hearts blood flows forward in the dorsal and ventral vessels to the head; from the head it flows backward in the subneural vessel. Behind the hearts blood flows backward in the ventral vessel and out into the segmental branches.

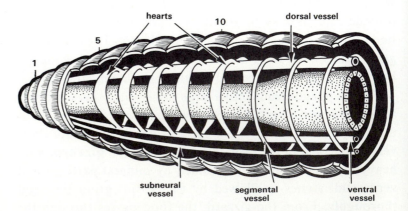

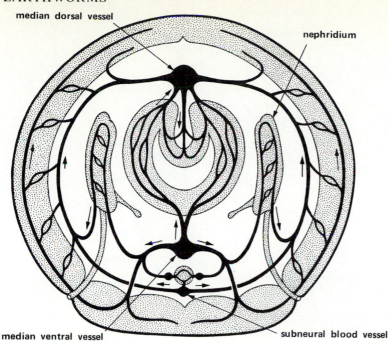

median dorsal vessel

nephridium

median ventral vessel

subneural blood vessel

Segmental circulation of an earthworm, showing only the principal vessels, in cross-sectional aspect. In each segment the ventral vessel sends branches to capillary beds (drawn much simplified) of the digestive tract, body wall, and nephridia, and to the lateral neural vessels, which in turn supply the nerve cord. The subneural vessel drains the nerve cord. Each segment is drained by paired segmental vessels that run to the dorsal vessel.

Earthworms are terrestrial animals, but they have not really solved the problems of land life. They have merely evaded them by restricting their activities to a burrowing life in damp soil, by emerging only at night, when evaporation to the air is low, and by retreating deep underground during hot, dry weather. Animals well adapted for land life have a heavy impermeable skin or cuticular covering which prevents excessive drying, but it also prevents respiratory exchange through the surface. In such animals oxygen reaches the internal tissues by means of special respiratory devices, such as lungs. Earthworms, on the other hand, carry out **respiratory exchange** in the same way as their aquatic ancestors. That is how they can live for months, completely submerged in water, yet will die if dried for a time. The collagenous cuticle of earthworms is extemely thin (7 μm) and must be kept moist so that respiratory exchange can occur by diffusion through the general body surface, which is underlain by capillary networks. Moistening of the surface is accomplished by mucous glands in the epidermis and also by the coelomic fluid which issues from *dorsal pores* located in the mid-dorsal line in the grooves between segments.

The **excretory system** consists of a series of excretory organs, or *nephridia,* one pair in every segment (except the first 3 and the last). As in nereids, each nephridium really occupies two segments, because it opens externally by a pore on the ventral surface and internally by a ciliated funnel which lies in the coelom of the segment anterior to the one containing the

Because respiratory exchange in earthworms occurs only through the body surface, growth in size is limited by the ratio of surface to volume, and worms increase in size mostly by elongation of the cylindrical body. Taking into account the partial pressures of oxygen inside and outside the worm body, the measured distance between the surface and superficial blood vessels, and the average of the measured metabolic rates of various tropical earthworms, it has been calculated that a worm diameter of about 30 mm is the upper limit for size. Some of the largest of the giant earthworms measure almost 30 mm when moving along on the ground surface. A similar calculation suggests that an earthworm *without* a circulatory system or a coelom could not exceed a few mm in diameter.

The epidermal mucus-secreting cells of *Lumbricus terrestris* occur over the full length of the body. The mucus serves not only in respiratory exchange, as already mentioned, but also as a lubricant that eases passage through the burrow. Also, in binding soil particles together, it prevents the walls of the burrow from collapsing.

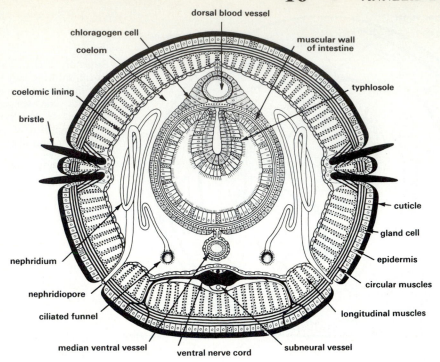

dorsal blood vessel

chloragogen cell

coelom

muscular wall
of intestine

coelomic lining

typhlosole

bristle

cuticle

gland cell

epidermis

nephridium

circular muscles

nephridiopore

longitudinal muscles

ciliated funnel

median ventral vessel

ventral nerve cord

subneural vessel

Cross-section of an earthworm. Only 2 of the 4 pairs of bristles are shown, with the muscles that move them. The ciliated funnels and the external openings of the excretory organs lie in successive segments, as described in the text. A layer of circular muscle and a layer of longitudinal muscle surround the coelomic cavity, exerting pressure on the fluid skeleton provided by the coelomic cavity and its contained fluid. The cross-section of the ventral nerve cord reveals 3 giant fibers lying dorsal to the others. The intestine has circular and longitudinal muscles also, and is sheathed with chloragogen cells presumed to be modified coelomic lining. The thickness of the cuticle is exaggerated; it is only 3-7 μm thick.

If foreign particles, such as India ink or bacteria, are injected into the body cavity of an earthworm, and the content of the coelom then examined, the particles soon appear within ameboid cells of the coelomic fluid. After a long interval these particle-laden phagocytic cells have all disappeared from the coelomic fluid. Some of them can be seen to be incorporated into rounded "brown bodies," along with chloragogen granules, old bristles, dead nematodes, and the cysts of gregarines. The brown bodies accumulate in the coelom throughout the life of the worm, especially at the posterior end. Other phagocytes make their way among the muscle fibers, depositing their pigmented content in the body wall.

body of the nephridium and its external pore. The urine contains ammonia, urea, and perhaps other nitrogenous wastes. Earthworms regulate their water and salt balance to a limited extent but are more noteworthy for their tolerance of extreme changes in water content. When an earthworm is taken from moist soil, about 85 percent of its weight is water. Submerged in water, the worm will gain up to 15 percent of its original body weight, or placed in a dry environment, will survive a loss of 60 percent or more of its weight.

The nephridia are not the only means of excretion in earthworms. About half the total nitrogen excreted is in the mucus secreted by the epidermis. In addition, the coelomic lining surrounding the intestine and the main blood vessels is modified into special **chloragogen cells,** the functions of which are not fully understood. They may participate in nitrogen metabolism and excretion, as well as storage and metabolism of starches and fats. Chloragogen cells are rich in iron, and the heme of hemoglobin could be synthesized there. The chloragogen cells, or at least the greenish-yellow granules that they accumulate, are released into the coelom, where they are engulfed by ameboid cells.

Lumbricus has brown and red pigments in the body wall, and the red protoporphyrin pigment is photodynamic, sensitizing the tissues to light, particularly the ultraviolet, which is harmful to earthworms. Brief exposure to strong sunlight causes complete paralysis in some worms, and longer exposure is fatal. This may explain the death of many of the earthworms seen lying in shallow puddles after a rain. They have not been drowned, as many people suppose, for earthworms can live completely submerged in water. However, during a rain the burrows may become blocked or filled with water that has filtered down through the soil and therefore contains very little oxygen. Lack of oxygen (or high levels of carbon dioxide or carbonic acid) may cause some of the worms to come to the surface, where they are injured by light and after a time can scarcely crawl. The rain puddles may afford some protection from ultraviolet radiation. Many of the dead worms seen after a rain were probably sick beforehand as the result of heavy infestation with parasites; their death has only been hastened by the rain.

The organ systems of earthworms described up to this point have shown little or no increase in division of labor among segments, as compared with nereids. The nephridia are identical in all segments in which they occur, and the central nervous system is practically the same as that of nereids. The circulatory and digestive systems of earthworms show some increase in specialization. However, the greatest differentiation among earthworm segments is found in the reproductive system.

The complexity of the **reproductive system** is an adaptation to life on land, where the naked gametes cannot simply be discharged to the exterior, as in marine polychetes, but must in some way be protected from drying and other adverse conditions during the development of the young. Earthworms, unlike most nereids, are *hermaphroditic,* each individual having a complete male and female sexual apparatus. As pointed out earlier, hermaphroditism is thought to be an adaptation to any mode of life which provides relatively few contacts between individuals.

The **sex organs** are located in the anterior end of the worm in particular segments. At some distance behind the sex organs is a swollen ring, the **clitellum,** formed by the thickening of the surface epithelium, which contains great numbers of gland cells. These produce secretions which play an important role in mating and in protecting the embryos.

Ant's-eye view of an earthworm with the anterior end lifted reveals the mouth and 4 **rows of bristles,** which can be felt by drawing the worm between the fingers from tail end to head. In the region of the **clitellum,** thickening of the body wall obscures external segmentation and bristles. The position of the clitellum is definite for each species of earthworm. In *Lumbricus terrestris* it extends over segments 31 or 32 to 37.

Annelids synthesize hemoglobin, as do humans, by combining protoporphyrin with iron and then with the protein globin. In earthworms, protoporphyrin accumulates and is deposited in the body wall. Free porphyrins are the precursors of such important compound porphyrins as chlorophylls, hemoglobins, and the cytochromes involved in the respiration of all living cells (except anaerobic bacteria). But free porphyrins are photodynamic, and in the presence of light and O_2, they sensitize cells to oxidations that deactivate enzymes, denature proteins, and disrupt membranes.

Any high level of free porphyrins in living tissues is called porphyria, and must be considered potentially damaging even when it is characteristic of a species. Porphyria is not common among animals because in most environments, there is strong selective pressure against it. However, earthworms that do not ordinarily emerge from their burrows during sunlit hours, or polychetes that remain in tubes, escape such selection. When for any reason they are exposed to strong sunlight they are seriously affected and often die.

Human porphyria ranges from mild to severe, or fatal, and is caused by hereditary defects in the formation of one or more of the succession of enzymes that convert protoporphyrins into heme, from which hemoproteins such as hemoglobin are made. The symptoms of porphyria usually start at an early age and may include redness, swelling, and scarring of the skin, anemia, liver abnormalities, dark urine, gallstones, discoloration of teeth, and other problems. One kind of porphyria, not hereditary, is caused by toxic agents such as hexachlorobenzene.

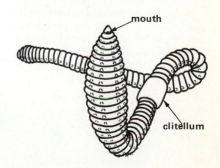

Among the North American earthworms of the family Lumbricidae, about half reproduce primarily or entirely by **parthenogenesis,** egg development without fertilization. Parthenogenesis seems to be correlated with type of habitat. It occurs in species that can tolerate a variety of marginal but widespread soil and habitat types; they can thus invade small, patchy habitats, subject to change with time, such as decaying logs, forest floor litter, stream beds or other wet areas, or shallow soil layers. All these are unstable habitats, and selective pressures are almost entirely from *physical* components of the environment such as temperature or pH. The occurrence of these worms is too scattered and too unpredictable to permit long-term *biological* challenges (by predators or parasites or competitors) that can be met by evolving variations in genotype. By replicating copies of itself through parthenogenesis, a single individual of such a species can rapidly exploit an unoccupied habitat. Once such a clone is established it really takes off, physical conditions permitting.

Why then do parthenogens not replace all the more slowly multiplying sexual species? Presumably it is because worm species able to dig deeply into the soil have the advantage of a habitat with great spatial continuity and with physical stability over long periods of time. Such species thrive in relatively constant numbers that provide a predictable resource attractive to both predators and parasites. In such a constant habitat an unvarying clone of worms could be wiped out by competitors, parasites, or predators. Perhaps only sexual reproduction, which regularly produces new genotypes, some of them superior to competitors or resistant to co-evolving parasites and predators, can avoid excessive mortality and localized extinction. *Lumbricus terrestris* lives in deep soil layers and is a sexual species.

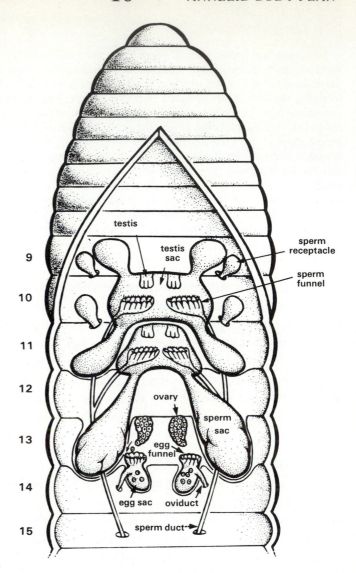

The **reproductive system** of earthworms is highly specialized. The testis sacs are drawn as if they were transparent, so that the testes and sperm funnels, located within them, can be seen. The large sperm sacs in segment 12 press down on the coelomic partition, so that they appear to lie in segment 13. This diagram is based on the structure of *Lumbricus terrestris*.

The male sex cells are formed in two pairs of *testes*, located in segments 10 and 11, and each pair is enclosed within a *testis sac* that communicates with the *sperm sac* in which the sex cells undergo further development. The mature sperms pass back into the testis sacs, into the *sperm funnels*, and through the *sperm ducts* to the two male genital openings on the ventral surface of segment 15. Two pairs of small sacs, the *sperm receptacles,* in segments 9 and 10, open through pores to the ventral surface. During mating they receive sperms from the other partner. The eggs are formed in a pair of *ovaries* in segment 13. When nearly fully developed, they are shed from the free end of the ovary into the *egg funnels* situated on the posterior face of segment 13. These funnels lead into two pouches, the *egg sacs,* where the eggs complete maturation and are stored. When ripe, they are discharged through the *oviducts*, which open by 2 minute pores on the ventral surface of segment 14.

Copulating earthworms, *L. terrestris,* are partially extended from burrows only 8 cm apart in a grass lawn, on a moist night. It was not love at first sight. The worms were foraging among grass and clover when their anterior ends touched. This contact sent both worms into sudden retreat into their burrows, presumably as fast as their giant neurons could deliver the message to the muscles. Then the worms re-emerged, and after another contact assumed the position shown here and described in the text. Chicago. (L. Keinigsberg)

The **mating process** is no simple shedding of the gametes, as in most marine polychetes. At the sexual season, spring to early fall, when the ground is wet following a rain, the worms may emerge at night and travel some distance over the surface before they mate; where abundant they merely protrude the anterior end and mate with a worm in an adjoining burrow. The two worms appose the ventral surfaces of their anterior ends, the heads pointing in opposite directions. The clitellum of one worm is opposite segments 9-11 of the other, and this is the region of most intimate attachment. Mucus is secreted until each worm becomes enclosed in a tubular mucous "slime tube," which extends from segment 9 to the posterior edge of the clitellum. When the sperms issue from the male genital openings in segment 15, they are carried backward in longitudinal grooves (which are converted into tubes by the presence of the mucous sheath) to the openings of the sperm receptacles on segments 9 and 10 of the mating partner. Then the worms separate; egg laying and fertilization occur later.

At the start of **egg-laying,** the gland cells of the clitellum produce a ring of secreted material which the worm moves forward over its body. As the ring passes the openings of the oviducts (segment 14) it receives several ripe eggs; and then, as it passes the openings of the sperm receptacles (segments 9 and 10), it receives sperms which were deposited there previously by another worm during mating. Fertilization of the eggs takes place within the ring, which finally slips past the anterior tip of the worm and becomes closed at both ends to form a sealed capsule, sometimes called a "cocoon." Within the capsule, which lies in the soil, the zygote develops directly into a young worm which then escapes.

Both head and tail can be regenerated, within limits. The extent of **regeneration** depends on the species, as well as on the position of the wound and on the size of the fragment that remains. Generally, the larger the number of segments lost,

In *Lumbricus terrestris* the egg capsule is only 6 to 8 mm long and lemon-shaped. Several eggs are deposited in each capsule, but only one egg develops into a worm, the others serving as nurse cells. In some species several worms may emerge from one capsule.

Regenerating rear end of a giant earthworm from montane forest north of Nono, Ecuador. The slender regenerating portion develops all the segments it can regenerate before it begins to grow in diameter. (R.B.)

the shorter the regenerated worm will be, until a point is reached where regeneration fails altogether. As in planarians, there is a gradient of regenerative capability from front to rear, and nervous elements are important. At a wounded end, regeneration begins only with the first segment in which the ventral nerve cord is intact, and hormones from secretory cells in the large anterior ganglia have been implicated in both regeneration and reproduction, as in nereids.

There have been many ideas about the *evolutionary origins of coelomic cavities and of segmentation.* It has been variously suggested that coelomic cavities arose by the separation of pockets of the gastrovascular cavity or gut, by enlargement of the cavities of the gonads or the excretory organs, or by the development of entirely new cavities within the mesoderm. All of these conjectures find support in morphological and embryological evidence from certain coelomate phyla but meet with difficulties when applied to others. Thus it seems unlikely that we shall soon agree on a single explanation for the origin of coeloms. Probably coeloms have arisen more than once and in different ways in several evolutionary lines.

The same may be said of the evolution of segmentation. One hypothesis is that segmentation arose by incomplete fission as in planarians, and that each segment represents a subindividual which was produced by asexual budding but failed to differentiate fully and to detach. Other suggestions are that crosswise septa developed in wormlike animals that already had repeated elements of all the organ systems strung out along the body, as in some flatworms and nemerteans; that septa first divided the coelom and repetition of parts followed; or that the septa of cnidarian polyps became modified as the septa of segmented coelomates. There are other ideas as well, but the evidence for or against them is even scantier than with coelomic origins, and we cannot expect that either of these evolutionary puzzles will be easily resolved.

ALMOST any large fluid-filled space in an animal can serve as a *hydrostatic skeleton* if muscles are suitably arranged around it. The gastrovascular cavities of cnidarians and flatworms and the body cavity of the nematodes and other pseudo-coelomates act partly as fluid skeletons. And it is commonly believed that the original and principal function of the coelom in annelids and many other coelomate phyla is hydrostatic, with the other mechanical and physiological functions evolving secondarily. Transmission of pressure by the fluid-filled coelom increases the effective speed and power of muscular action in locomotion and in other movements of the body. Division of the coelom by septa is thought to be especially important in localizing forces along the body during peristaltic burrowing, as in earthworms, or in maintaining turgor in the parapods of polychetes. In some of the very small or relatively sedentary polychetes, the septa are usually few or incomplete, and we shall see in the next chapter that in leeches, which move in a different way, the coelom is much reduced and not segmented.

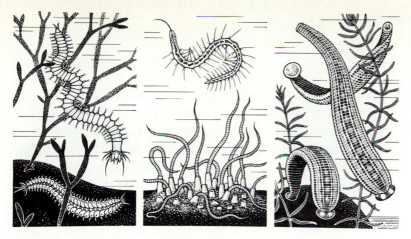

Chapter 17

A Diversity of Annelids

Polychetes
Oligochetes
Hirudineans

THE PHYLUM ANNELIDA embraces about 13,000 species of elongated and usually cylindrical worms with segmented bodies. All 3 of the annelid classes are found in the oceans, in freshwaters, and on land. But the many-bristled polychetes are mostly marine; the few-bristled oligochetes are mostly terrestrial earthworms; and the leeches, which lack bristles, live mostly in freshwaters. The drawings above represent only aquatic members. At the left are two marine polychetes, a nereid swimming and a scale worm crawling. In the center are two freshwater oligochetes, *Stylaria* swimming near the surface and *Tubifex* burrowing in great numbers in the stagnant bottom where the oxygen content is low. At the right are looping and swimming freshwater medicinal leeches, *Hirudo medicinalis*.

POLYCHETES

BRISTLE WORMS comprise the largest and most diverse class of annelids, the class **Polychaeta,** named from the "many bristles" that protrude from the parapods in discrete bundles. The class includes the nereids and a great variety of other marine forms that are sessile, sedentary, crawling, or pelagic. Some live in brackish and fresh waters, and a few species are terrestrial. Many of the marine forms share the shells or burrows of other invertebrates or cling to the bodies of their hosts as commensals, snatching bits of their food. Parasitic polychetes are rare.

Polychetes are among the most common animals of seashores, but as long as they remain under stones or in burrows in mud or sand, they are seldom seen except by biologists intrigued by their varied forms and habits—or by people who dig them up to use as fishing bait. It is only when marine polychetes swarm at the surface during spawning, near shores on warm summer nights, that they attract much attention, especially when the worms are luminescent or become visible as

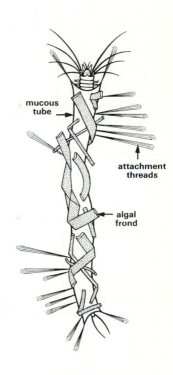

Nereid in tube is *Platynereis dumeri-lii,* removed from its attachment among mussels in the Bassin d'Archachon, France, and observed in a glass dish in the laboratory. The building of this new tube took about an hour. Specialized glandular parapods on several anterior segments secrete the threads which form the framework of the tube and then put them in place. The threads are cemented together with adhesive secretion also used to fasten the short bundles of threads that anchor the tube to the substrate. Bits of algal frond placed in the dish were incorporated into the tube, strengthening and perhaps camouflaging it. At night the worm abandons the tube to feed. (After J.M. Daly)

mucous tube

attachment threads

algal frond

they collect in great numbers below lights on wharves or boats.

Although a few polychetes copulate, most release their gametes into the sea. Some species produce epitokes, as described for nereids, their swarming habits providing for the simultaneous release of eggs and sperms from many closely gathered worms and insuring the fertilization of the greatest possible number of eggs or sperms of each individual. In addition to the precise timing of swarming, polychetes have other devices which bring about the simultaneous extrusion of eggs and sperms. In the family of nereids the discharge of sperms is often set off by a secretion from the spawning female. A member of the syllid family, *Odontosyllis enopla*, attracts attention in Bermuda because the meeting of the sexes involves the exchange of light signals. The worms come to the surface to spawn each month a few days after the full moon at about an hour after sunset. The female appears at the surface first and circles about, emitting along its length a bright greenish glow. The smaller male then darts rapidly toward the female, emitting flashes of light as it goes. Commonly several males gather around a single female and swim in a tight circle as both sexes shed their gametes. The eggs, surrounded by a cloud of luminous secretion, are fertilized in the water. The luminescence of *Odontosyllis* is bright enough to be easily visible to people watching from a beach or boat.

In some polychetes, only the posterior part of the worm forms gametes and acts as an epitoke. After undergoing changes in form and color, this modified sexual portion breaks off, rises to the surface and swims about, shedding the eggs or sperms. The unmodified anterior part remains on the bottom and regenerates a posterior end. Other species bud off, from the posterior end, large numbers of gamete-filled epitokes, each with a head and swimming parapods.

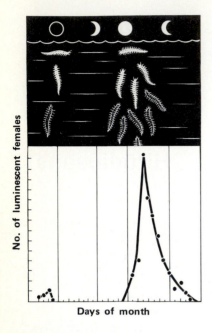

Swarming of *Odontosyllis enopla* is regulated by the lunar cycle. A few worms can be seen on about half the nights of the month, but there is a major peak near full moon. A lesser peak comes with the new moon. (Based on Markert, Markert, and Vertrees)

No. of luminescent females

Days of month

Male *Autolytus* budding off a **chain of epitokes,** the oldest one at the posterior end. Budding occurs in both sexes, producing male or female epitokes which break free. Not all species of *Autolytus* produce chains of buds. Some bud off epitokes one at a time. Others do not bud at all; the whole individual metamorphoses into an epitoke. (Based on Y.K. Okada, on G. Thorson, and on L. Gidholm)

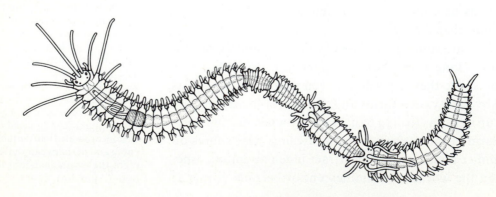

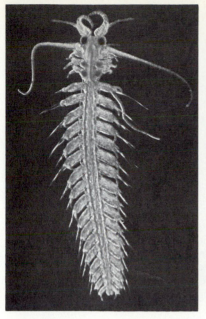

Epitokes of *Autolytus,* a genus of syllid polychetes. Each is 3 mm long. Millions of these epitokes swarm in the surface plankton spawning sperms or eggs.

Left, the **male epitoke** with big eyes, modified head appendages, and greatly enlarged parapods, modifications for active swimming.

Right, the **female epitoke** has simpler head appendages than the male. The eggs are extruded and fertilized during swarming in the plankton, and she broods them in a secreted ventral sac until they hatch. Plymouth, England. (D.P. Wilson)

Pacific palolo worm, *Eunice viridis,* (=*Palola siciliensis),* shelters in crevices in dead coral reefs, lining its burrow with mucus. The immature worms are negative to all but the dimmest light and emerge to feed only at night. This species is noted in Samoa and Fiji for swarming mostly on a single night of the year, usually the 7th night after the first full moon that follows the autumnal equinox. Because it could be predicted so accurately, the day of the "big rising" of the palolos marked the beginning of the Samoan New Year in pre-missionary times. Related species in the Caribbean and Mediterranean have less highly synchronized swarming. W. Samoa. (K.J. Marschall)

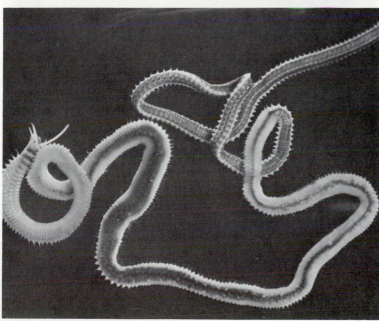

Specialization of body regions is conspicuous in palolo worms, which may be 40 cm long and have as many as 1000 segments. The anterior, photonegative portion, which does not develop gametes, includes the head plus some segments that lack gills, and a longer midregion in which the dorsal surfaces of the parapods bear simple gill filaments. The posterior region, the prospective epitoke, metamorphoses during the last weeks before spawning, when the gametes are maturing; the segments are longer and narrower and have a row of dark eyes along the ventral midline. The separated epitoke is positive to light. W. Samoa. (K.J. Marschall)

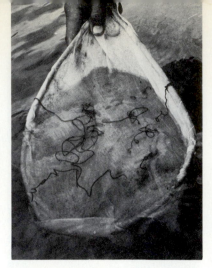

Gathering of palolos during hours of dawn. During the night epitokes rise to the moonlit surface by the millions after breaking off from the anterior portions of the worms, and the water has been compared to vermicelli soup. The Samoans gather the wriggling epitokes by the netful—before the light has stimulated the synchronous discharge of eggs and sperms. The worms will be pooled for a great communal feast, but hungry gatherers can be seen sneaking worms into their mouths as they work. Some worms from this gathering were served to two of the authors as part of a many-course invertebrate dinner. The lightly fried worms retained their green color and had a pleasant flavor. W. Samoa. (K.J. Marschall)

Male and female epitokes. The female epitokes appear to be nothing but egg-filled cylinders with a row of pigmented simple eyes along the midventral line. The sperm-packed male epitokes are the ones with more conspicuous eyes. W. Samoa. (K.J. Marschall)

Anterior atokous part of palolo worm retreats into burrow after separation of epitoke and begins to regenerate a new posterior region, which, newly laden with gametes, will break off on the corresponding day next year. W. Samoa. (K.J. Marschall)

THE LIFESTYLES of some polychete families are so specialized that the worms are strictly limited to a particular mode of feeding and dwelling. The nereids are more opportunistic, their members following a wide scope of habits, as mentioned earlier. The various families of polychetes that pursue similar ways of life tend to look generally alike and to share certain characteristics. However, specific structural and developmental features often argue against close relationships. Trying to figure out the convergences and divergences of the polychetes still baffles zoologists, and the modes of life briefly described below do not necessarily coincide with natural taxonomic groupings.

Crawling polychetes, many of which can also swim, have a well-differentiated head with prominent sense organs, an eversible pharynx bearing horny jaws or teeth, and well-developed parapods on each of their numerous and similar segments. Many of these worms, for example some of the nereids, build temporary burrows or tubes but can leave them to feed. If they depart for a new feeding site, they quickly dig in or build a new home. Others find some safety from predators under rocks, among seaweeds, in mussel beds, or around other sessile invertebrates. Most crawling polychetes are themselves predators, seizing small animals with the sharp jaws of the everted pharynx and swallowing them whole. They prey on small crustaceans, molluscs, and other animals, including other annelids. Some feed on seaweeds, or scavenge for a living, or process detritus-laden mud, while others mix their diets freely.

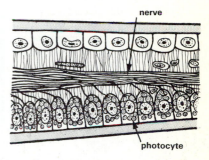

Luminescent scale of a polynoid, *Acholoe astericola,* may be automotized (severed by the worm itself) in response to attack by a predator or irritation by an experimenter. The rhythmically flashing scale could distract the predator while the worm escapes. (After J.A.C. Nicol)

Section through a luminescent scale of *Acholoe astericola.* Unpigmented epidermal cells at the upper surface allow light to shine through from the luminescing **photocytes,** modified epidermal cells. Luminescence is under nervous control, and rhythmic flashes proceed in both directions from the point where the worm is touched. But if the animal is quickly cut in half, only the posterior half luminesces. Perhaps it acts as a "lure," while the dark anterior half escapes and regenerates. (After J.A.C. Nicol)

Below left, a **scale worm** from Monterey Bay, California. Scale worms (polynoids) have 2 rows of exposed scales that cover all or most of the dorsal surface. *Below right,* an **aphroditid,** *Aphrodita aculeata,* 15 cm long, from waters near Plymouth, England. Several species of *Aphrodita* occur off both coasts of the U.S. This massive polychete is called "the sea mouse" because of the "furry" covering of densely matted long, fine bristles that obscure rows of scales. It ploughs through soft bottoms, feeding on various invertebrates, including large active polychetes as well as sedentary ones. Sand and mud are kept out of the respiratory current by the feltlike dorsal covering, as well as by the longer and brilliantly iridescent bristles that line the sides of the body. The animal protrudes only the posterior end, which provides entrance for a respiratory current that flows forward under the rows of scales. (R.B.)

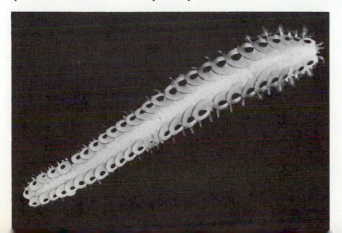

Phyllodocid polychete, the paddleworm, *Phyllodoce lamelligera,* may reach 30 cm in length. The mid-nineteenth century English naturalist, Philip H. Gosse, described it as "... looking like a centipede, but of a bright green colour, ...lithely crawling and turning among the seaweeds ... difficult to get hold of, from their length and slipperiness." Gosse's writings excited so much interest in living invertebrates that many readers set up marine aquariums in their London homes, and in stocking them decimated some of the most beautiful stretches of the rocky Devonshire coast. Similar destruction of fragile intertidal habitats is now being inflicted on U.S. shores by well-meaning admirers of invertebrates. England. (D.P. Wilson)

Lysaretid polychete, *Lysarete brasiliensis,* is bright orange and shimmers, with iridescence of the cuticle, when dug from its burrow and exposed to light. The irritated animal secreted copious amounts of mucus, to which sand adhered, so that the photograph was made only with some difficulty. The head, at the bottom of this picture, is already digging into the sandy substrate. Gulf of Mexico, NW Florida. (R.B.)

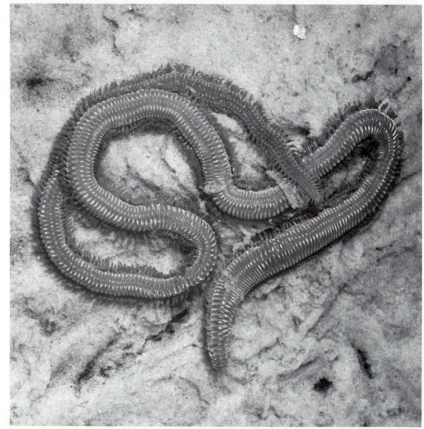

The **pelagic polychetes** are permanently free-swimming and resemble in some respects the swimming epitokes of the various bottom-living species. Large flattened parapods make them effective swimmers. Like many other pelagic invertebrates, these polychetes tend to be transparent, colorless, and luminescent, and some have enormous camera-type eyes with well-developed lenses. They prey actively on a variety of fast-swimming invertebrates and tiny fishes.

Burrowing polychetes worm their way through sand or mud, partly by peristaltic contractions, as earthworms do, and often also by forceful thrusts of an eversible pharynx. They tend to be streamlined, the head end lacking sensory projections and the parapods small. Some also feed like earthworms, eating their way through the substrate, digesting what they can and eliminating the rest. Others are carnivores, devouring whatever small animals they encounter as they actively burrow. Still others tend to dig a burrow and remain in one place. Most of these sedentary types feed on the rich organic deposits in the surface layers of the bottom, and many have long feeding tentacles near the head end that extend out over or through the substrate and are often all that is visible of the buried worm.

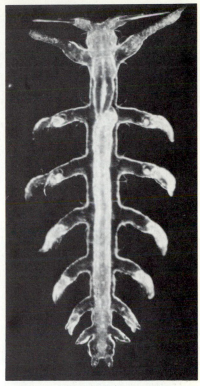

Pelagic polychete, *Tomopteris helgolandica,* is of glasslike transparency as it swims in graceful undulations at the surface. This young specimen is only 2 mm long, but as it matures will reach 1 cm, and have about twice as many parapods. The bilobed parapods, which lack bristles, will be much more expanded, and the long projections at the front may reach half the length of the body or more. Together, these extensions of the body surface increase the resistance to sinking. When threatened, *Tomopteris* rolls up into a ball and sinks to a safer level. Brought up in a tow-net at night, it may be brilliantly luminescent from light-organs between the 2 lobes of each parapod. England. (D.P. Wilson)

Terebellid, *Neoamphitrite,* lives in a mucus-lined burrow in mud and extends feeding tentacles in all directions over the surface. Tentacles are extended by ciliary creeping on the grooved surface, and retracted by longitudinal muscles. When the grooved surface is turned up, particles are moved along by cilia. The tentacles are then wiped across the lips where further sorting occurs as food is conveyed into the mouth. Behind the light-colored tentacles is a dense cluster of branching, deep red, respiratory filaments. NW Florida. (R.B.)

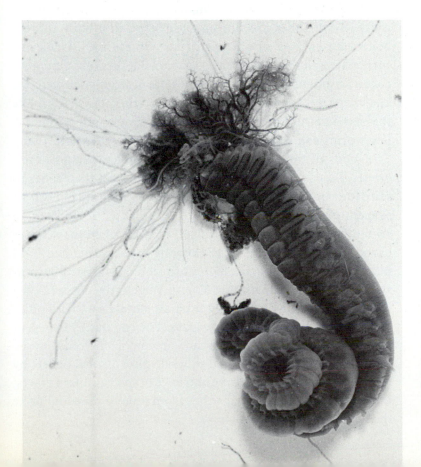

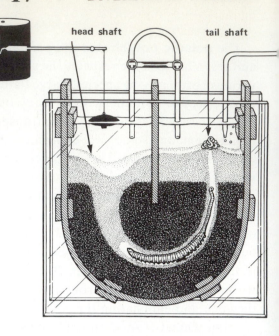

Castings of lugworm, *Arenicola marina,* are easily recognized on muddy sand flats at low tide by castings thrown up at the open posterior end and by the depression in the sand above the head end. Roscoff, France. (R.B.)

Recording activity cycles of *Arenicola marina* reveals that feeding occurs at about 7-minute intervals. The worm lies head-down in its mucus-lined burrow and creates a feeding current that draws suspended organisms and particles down into the sand, which is then ingested. At 30-40 minute intervals the kymograph records bursts of irrigation movements that bring freshly oxygenated water to the gills along the body, and defecation movements in which the worm rises tail-first to the surface to deposit castings. Marine Biological Laboratory, Plymouth, England. (Modified after G.P. Wells)

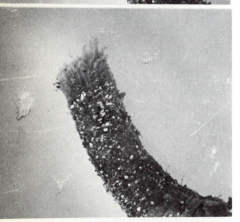

Shadow reflex of a sabellid in response to sudden decrease in light, as from the shadow of a predator. Withdrawal of tentacles, in a fraction of a second, is mediated by 2 giant axons. Monterey Bay, Calif. (R.B.)

Tube-dwelling polychetes are in many ways similar to the sedentary burrowing types; but instead of living in a simple burrow, they build a discrete, sometimes permanent tube. Many of the tubes are built of mucus and sand, some as loose irregular structures. Others, built by many worms living side by side, form great honeycombed mounds over intertidal rocks. Some tubes are fibrous or papery, and built entirely below the surface of the substrate. Others are of calcium carbonate and are cemented to rock. Still others are elaborate constructions with bits of shell and sand of strictly selected size and shape methodically layered in an overlapping pattern like the tiles on a roof or precisely fitted into a smooth mosaic. A few worms can move about, carrying their tubes with them. The foods of tube-dwelling polychetes are as varied as their homes. Some species emerge from the shelter of their tubes to seize prey or pieces of floating seaweed. Others are deposit or suspension feeders, and some of these (e.g., spionids) can switch from deposit to suspension feeding when local water currents increase in velocity, carrying more suspended food particles, and change back to deposit feeding when the current slows. Most polychetes that live in the confined space of a burrow or tube must regularly circulate water past their bodies to obtain oxygen and remove wastes. They do this by means of parapods, cilia, or peristaltic movements of the body. Some that live head-end-up in blind tubes dispose of fecal pellets by transporting them to the tube entrance in special ciliary tracts; others turn around in their tubes from time to time and poke the hind end out to defecate.

Tentacular eyes, seen here as dark spots on the main axis of the feathery tentacles, are characteristic of sabellids. Like their close relatives, the serpulids and the spirorbids (see following pages), the sabellids are sessile suspension feeders. All three families show the "shadow reflex," but only sabellids have tentacular eyes, so the exact function of these eyes needs further study. Monterey Bay, Calif. (N. Burnett)

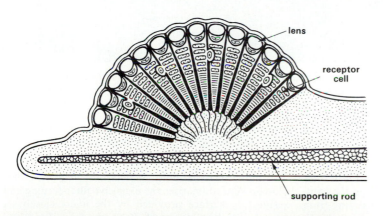

Longitudinal section of **tentacular eye of sabellid** *Megalomma (= Branchiomma) vesiculosum,* shows a basic structure that suggests the compound eyes of insects and other arthropods. Each tentacle has an eye near its tip, and the section reveals also the cartilaginous supporting rod that stiffens the tentacle. Each unit of the worm eye has one receptor cell, capped by a lens cell. The units are shielded from each other by pigment cells. If the eyes are removed, the worm still shows the shadow reflex, withdrawing into the tube in response to any sudden decrease in light intensity with or without eyes. *Megalomma* can become habituated to sudden flicking off of the laboratory light, and cease to respond. But such habituated animals will still respond to *moving* shadows. This suggests that sensitivity to a moving shadow is a separate sensory process, and there is also evidence that such eyes are especially suited to the detection of movement. (Combined from R. Hesse and from Krasne and Lawrence)

Tentacular crown of sabellid is a food-gathering net of outspread featherlike tentacles covered with beating cilia that convey microscopic food particles to the central mouth and provide an extensive respiratory surface. Monterey Bay, Calif. (R.B.)

Symmetrical crown of *Sabella pavonina* consists of 2 apposed half-circles of stiffly held tentacles, each supported by a cartilaginous rod that runs the length of the pinnate tentacle. Cilia on the tentacles create the feeding currents that move water between the small side branches, trapping microscopic organisms and other particles in mucus and then conveying them, in a groove along the length of each tentacle to its base, where further sorting occurs. The largest particles are discarded, and only the small sizes directed into the mouth. Selected sand may be stored in special sacs and later incorporated along with mud particles into the secreted mucus that forms the wall of the tube. Plymouth, England. (D.P. Wilson)

Division of the crown of sabellids, with 2 groups of tentacles arising bilaterally, is conspicuous in the twin-fan worm, *Bispira volutacornis,* here protruding from tube built in an opening in a rock. England. (D.P. Wilson)

Tentacle tips of sabellid, *Sabellastarte magnifica.* This closeup view dramatizes the extensive filter-feeding and respiratory surface provided by the many closely-set branches of the tentacles. (K.B. Sandved)

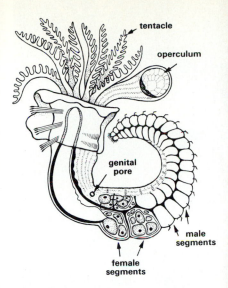

Spirorbid aggregation on the brown alga *Fucus*. Roscoff, France. (R.B.)

The larvas of *Spirorbis spirorbis* actively search for suitable habitats and tend to settle where other members of their species are already attached, commonly on rock or on *Fucus* at low-tide level. Were they not guided there by the presence of others they might attach, during high tide, in the upper shore levels and dry out at low tide. When larvas of two species of spirorbids were mixed in glass dishes containing paired pieces of slate, each slate with only one species of spirorbid attached, 90% of the larvas settled down on the slate that had members of their own species. The tendency to attach near other members of the species increases up to an optimum of about 10 per cm². Larvas arriving in areas of greater density move on to less crowded areas nearby, where they space themselves out evenly.

Spirorbis spirorbis taken from its calcareous tube. The first 2 sexual segments are female and the rest are male. There is a genital pore at the anterior end of the first female segment. (After King, Bailey, and Babbage)

As the eggs are spawned into the animal's tube, sperms are simultaneously released from a muscular sac just inside the mouth. This odd location, and the fact that immature worms have sperms in the sac, suggest that the sperms are collected, from many contributing neighbors, in the feeding current. This insures abundant cross-fertilization and a large batch of young that are brooded in a transparent envelope in the tube and released as late trochophores.

In many species of spirorbids, the embryos are brooded not in the tube but in a special chamber in the modified operculum.

Serpulid, *Serpula vermicularis,* has a feathery (pinnate) crown of tentacles like that of sabellids. But the tube it secretes is hard, white, and calcareous. It is coiled only slightly, in its lower part, where it is attached to hard substrates such as rock, shell, or wood. One of the tentacles is highly modified into a funnel-shaped stopper that seals off the tube opening when the crown is withdrawn. The closely related spirorbids are often included in the same family with the serpulids. They also have a crown of branched tentacles, a calcareous tube, and an operculum. The small, tightly spiralled tube is fastened to hard substrates or to seaweeds. *Serpula vermicularis* is cosmopolitan. (D. Wobber)

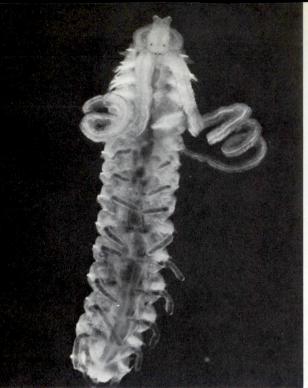

Spionids are burrowing and tube-dwelling polychetes of small size (10 to 150 mm). On soft bottoms species of spionids often outnumber those of all other polychete families. The 2 species shown (removed from their mucoid tubes) were collected in San Francisco Bay. Both are native to the U.S. Atlantic coast and were probably accidentally brought to the West Coast in transporting oysters. At left, *Polydora ligni,* in ventral view of the anterior end. The 2 long grooved and ciliated palps can be extended from the tube to gather suspended food. At right, *Streblospio benedicti,* in side view. It is perhaps unique among spionids in brooding the developing young in dorsal pouches on the middle segments. (R.B.)

Highly specialized tube dweller, *Chaetopterus variopedatus,* has many modifications of the parapods that serve in feeding. Near the middle of the body 3 large "fans" propel water through the burrow. Behind the head, 2 winglike extensions continuously secrete (during feeding) a mucous bag in which food particles are trapped. At the rear end of the bag the food-laden mucus is constantly rolled up into a pellet, and when this reaches a certain size the fans stop beating and the pellet is carried forward by cilia to the mouth. *Chaetopterus* emits a luminous secretion, but how this can benefit a worm that never leaves its tube is not clear. England. (D.P. Wilson)

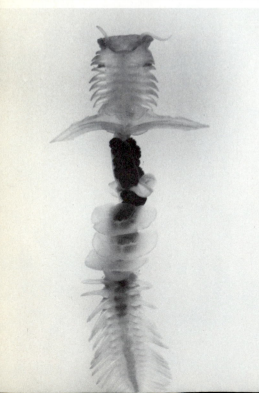
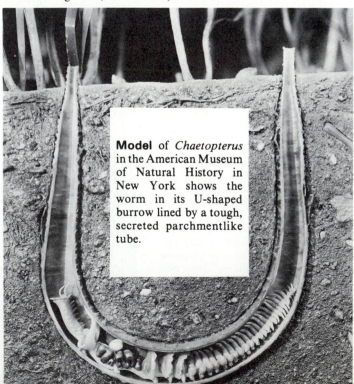

Model of *Chaetopterus* in the American Museum of Natural History in New York shows the worm in its U-shaped burrow lined by a tough, secreted parchmentlike tube.

Reef-building polychetes *Sabellaria alveolata* seen at low tide on exposed rock surrounded at the base by sand. The massive colonies, composed of thousands or even millions of fused tubes, grow by settlement of young worms and do best near low water mark, where wave action when the tide comes in is strong enough to stir up the calcareous shell sand grains used in tube building, but not so strong as to destroy the colony. Eventually a great storm and/or aging of the deeper portions of the tubes will result in destruction of even the largest colony. It will then be replaced by new batches of young worms that settle on the remains of the old. Duckpool, North Cornwall, England. (R.B.)

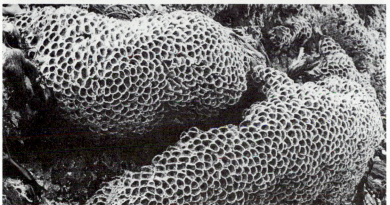

Closeup of smaller colonies shows why *Sabellaria* is often called the "honeycomb worm." The building, destruction, and resettlement of the reefs at Duckpool were studied from 1961 to 1975 by D.P. Wilson. Laboratory studies demonstrated that the worms are gregarious and delay metamorphosis until they make direct physical contact with the cement secreted by members of their species in tube building. Thus they tend to settle on top of old *Sabellaria* tubes, assuring them of a site that in the past supported flourishing colonies. England. (D.P. Wilson)

Head of sabellariid is protruded when the tide is in and the sticky, ciliated tentacles gather food particles. At low tide the tube opening is closed by an operculum consisting of 2 fans of wide and flattened blade-like bristles. Sabellariids should not be confused with sabellids. Calif. (R.F. Ames)

Cirratulids are mostly burrowers in soft bottoms, but *Dodecaceria fewkesi*, shown here, builds calcareous tubes and lives in dense colonies produced asexually. The tubes form extensive reefs on rocks. Each worm extends a pair of dark, thick palps and 3 pairs of tentacles. Monterey Bay, (N. Burnett)

Cone-shaped tube is constructed by *Pectinaria* from carefully sorted flat sand grains arranged in a neat mosaic. The worm feeds on detritus below the surface, where it is safer from predators, but where supplies of detritus soon give out, so that *Pectinaria* must move about, carrying its tube along. Length 4 cm. NW Florida. (R.B.)

Below, **head end of archiannelid,** *Protodriloides chaetifer,* seen under the microscope. The entire worm is shown on the next page. The head bears a pair of sensory tentacles but no eyes. The slitlike mouth leads to a strongly outlined pharyngeal bulb. (R.B.)

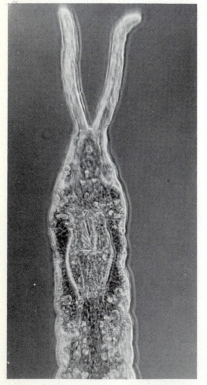

Onuphid polychete, *Diopatra,* when removed from the tough membranous tube in which it lives, promptly automotized the posterior part of the body. The anterior portion shown here is mostly obscured by sand adhering to mucus secreted by the irritated worm, but 2 pairs of a series of spirally-branching gills, borne dorsally on the parapods, show clearly. On intertidal bottoms the meter-long tubes are imbedded in the substrate, with a "chimney" projecting above the sand surface. The predatory worm which reaches out of the tube opening to catch its prey is only 30 cm long. NW Florida. (R.B.)

Interstitial polychetes—those that live in the tiny interstices between sand grains—feed mainly on single-celled algas, bacteria, and bits of detritus. The larger kinds are found in coarse sands, where the spaces between the grains are largest, and the smallest in fine sands with smaller spaces, but few of these worms much exceed 2 to 3 mm in length. Many of them look for the most part like typical members of their various families, except for their extremely small size. Certain others look like larval or juvenile polychetes or superficially resemble the many other minute, elongate animals that share the interstitial habitat. They may lack parapods, bristles, and distinct segmentation, and glide among the sand grains by means of cilia. Because of their simple structure and their lack of (or rudimentary version of) typically annelidan features, some biologists believe that these are the most primitive living annelids. Thus they are called *archiannelids* and sometimes placed in a separate class, along with slightly larger relatives from different habitats. However, it seems more likely that they are polychetes in which simple form is related to small size and interstitial habitat, as it is in many other phyla.

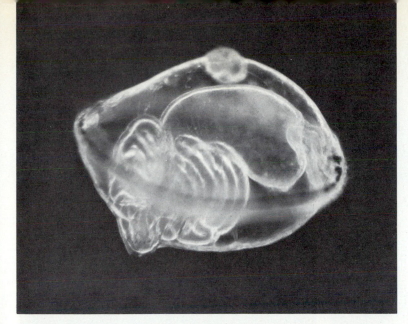

Segmented larva (metatrochophore) of *Polygordius lacteus,* an interstitial archiannelid. This larva, from the North Sea, is called an "endolarva" because the developing segments are folded up inside the larval body, 0.67 mm in diameter. The mouth is at the right, just below the equatorial band of cilia, and it leads into the bulbous stomach. In the Mediterranean the same stage of this species is called an "exolarva" because the rudimentary segments trail behind the broad presegmental portion. *Polygordius* has no parapods or bristles, and the segmentation is internal only. Adults live in silt or sand, and are among the largest of the archiannelids, sometimes reaching 12 cm. Except for their much larger size and lack of bristles, they resemble the tiny *Protodriloides,* being long and slender and having 2 tentacles on the head. England. (D.P. Wilson)

Interstitial syllid from deep bottom sediments in the North Sea, 1 mm long, is one of many species of polychetes that compose about a third of all the invertebrates collected from soft sea bottoms. The delicate dorsal cirri of the parapods, conspicuous even in this ventral view, extend out far beyond the bristle bundles of the ventral lobes. (G. Uhlig)

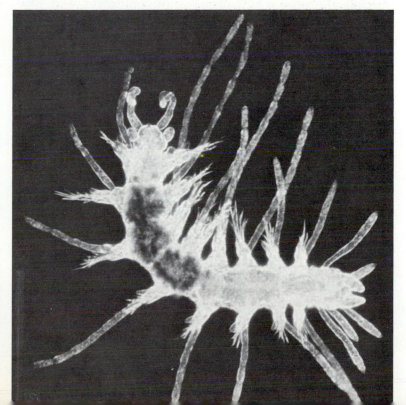

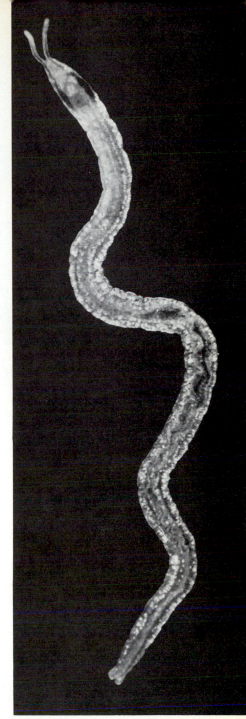

Archiannelid, *Protodriloides chaetifer,* is found among sand grains on intertidal bottoms. This young one, only 3 mm long, is from Monterey Bay, Calif. It has no parapods, but the segments bear, on each side, a pair of very short, hooked bristles. The worm moves by a ciliated band along the ventral surface, and can cling tenaciously to any solid object by means of 2 bands of adhesive glands on the bilobed posterior end. Adults may be 13 mm long and have 50 segments. This species is also widely distributed in Europe. (R.B.)

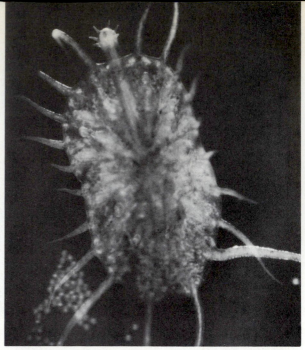

Hesionid polychete, *Ophiodromus (=Podarke) puget-tensis,* commensal on sea stars, especially *Patiria miniata,* crawls about on the oral surface of the host, apparently attracted there by chemosensory means. The worms are also free-living on muddy bottoms. Monterey Bay, California. (R.B.)

Myzostome with extended proboscis. The small round objects are eggs. A hundred or more myzostomes may scurry along the arms of a single crinoid, taking food gathered by the crinoid and camouflaged by matching the host's color. *Myzostoma.* Length, 1 mm. Lizard Island, Great Barrier Reef, Australia. (D.L. Zmarzly)

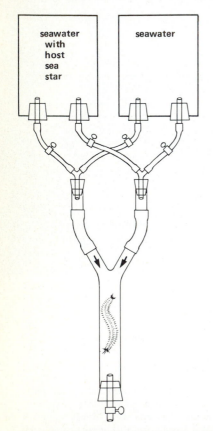

seawater with host sea star

seawater

Symbiotic polychetes are generally crawling types, related to free-living worms that inhabit rock crevices and creep out to pick up bits of detritus. It is not hard to imagine how such polychetes might gradually develop the habit of commensalism, living in some other animal's burrow. The burrows of mud shrimps, of sea cucumbers, of echiuran worms, and especially the tubes or burrows of various polychetes, regularly shelter commensal polychetes. Others live in the shells of hermit crabs and creep out to snatch extra bits of the host's food. Still others cling to the bodies of sponges, crabs, sea stars, and sea urchins. One species of scale worm, *Arctonoe vittata,* lives in the mantle cavity of the giant chiton and of the keyhole limpet, *Diodora,* on the U.S. Pacific coast. The worm has been observed to nip at the tube feet of predatory sea stars crawling over the host limpet. As the sea stars appear to be driven off, the worm-limpet association may be mutualistic. Parasitic polychetes are rare, but a few are found in the bodies of other polychete worms, one in echiuran worms, one on the gills of eels. One whole group of polychetes, the *myzostomids* (sometimes considered a separate class of annelids), are commensals or parasites on or in the bodies of echinoderms, especially crinoids.

Chemical attraction of commensal to its host. In 95% of trials in a Y-tube, the commensal scale worm *Arctonoe fragilis* ascended the arm with seawater containing its most common host, the sea star *Evasterias troschelii,* rather than the arm with plain seawater or with seawater containing another sea star, *Pisaster ochraceous,* rarely host to *Arctonoe.* Friday Harbor Laboratories, University of Washington, Puget Sound. (Modified after D. Davenport)

Commensal nereid, *Nereis fucata*, lives as a juvenile in a mucus-lined tube in the sand. Later it adopts as a host any of several species of hermit-crabs, entering the gastropod shell which the hermit-crab has itself adopted. *Above,* a hermit-crab, *Pagurus bernhardus* and its commensal worm have been removed from the shell of the gastropod *Buccinum* and have accepted a glass model of a gastropod shell. This reveals the behavior of the nereid, which occupies the smallest whorls of the shell. As soon as the hermit-crab starts eating, *below,* the worm comes to the opening of the shell and snatches a morsel from its host. Marine Biological Laboratory, Plymouth, England. (R.B.)

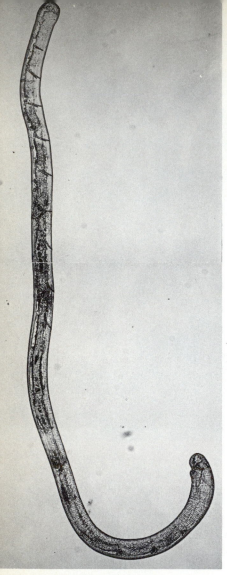

OLIGOCHETES

THE SECOND LARGEST group of annelids is the class **Oligochaeta** ("few bristles"), of which the majority are earthworms. The rest are mostly small or minute worms that occur in soils or in freshwaters, and there are also some marine forms. A hydra can be seen eating a freshwater oligochete in a photograph in chapter 5.

Oligochetes differ from polychetes in several respects. There are no parapods, and the bristles are fewer, small, and emerge from pits in the body wall. Whereas most polychetes have separate sexes and the sex cells are produced from the coelomic lining in numerous segments, the oligochetes are hermaphroditic and the sex cells are produced in special organs which occur only in certain segments, as was described for earthworms. Like other land and freshwater animals, oligochetes have no free larval stage comparable with that of marine polychetes, and the fertilized eggs are enclosed in a secreted capsule, where they develop directly into juvenile worms.

Those few polychetes that have managed to invade freshwaters show reproductive adaptations similar to those of oligochetes: hermaphroditism, copulation (instead of simple shedding of gametes), and large eggs with direct development, during which the embryos are protected from the external environment. In the few marine oligochetes, on the other hand, reproduction and development follow the pattern of the land and freshwater forms. So it is presumed that the reproductive habits of marine oligochetes are retained from freshwater ancestors and may be lost with time. The same can be said of marine members of the third annelidan class, the leeches, which are also primarily freshwater and land forms.

Marine oligochete, only 3 mm long, is from coarse sandy bottom at a depth of 30 m. Pairs of segmental bristles are visible. Whether marine or freshwater, aquatic oligochetes are in general much smaller than the terrestrial forms. Certain species are less than a millimeter long and have as few as 7 segments. Helgoland, West Germany. (G. Uhlig)

Freshwater oligochetes, *Tubifex tubifex,* in mass laboratory culture. These worms, about 4 cm long, are reddish in color from the respiratory pigment, erythrocruorin, dissolved in their blood. They tolerate low oxygen content, as in stagnant ponds, the bottom of deep lakes, or rivers polluted by sewage. A high concentration of *Tubifex* is an indicator of pollution. The worms lie with heads down and rear ends projecting from tubes and waving vigorously, so increasing aeration. *Tubifex* is sold in pet shops as live food for aquarium animals. (R.B.)

Giant earthworms. Terrestrial oligochetes are generally larger than aquatic ones. The greatest length is attained by Australian earthworms, but there are also giants in South Africa and Sri Lanka, and those in the photos above were from the Andean region of South America. *Left,* a giant earthworm from a montane forest near Nono, Ecuador. Such worms are sometimes seen crossing roads in the humid hours at dawn and dusk, even in the dry season. *Right,* a giant earthworm, *Rhinodrilus fafner,* from the humid montane forest near Cali, Columbia, here compared with a fully extended *Lumbricus terrestris.* The giant is mostly contracted and limp, after being handled, but the anterior tip is just beginning to dig in. The anterior segments are visibly turgid with coelomic fluid that provides the "fluid skeleton" against which the muscles can act. (R.B.)

Pulling an Australian giant from its burrow may cause it to squirt milky coelomic fluid from the dorsal pores in jets 30 cm high. Normally the screted fluid lubricates the worm's passage through its burrow and protects against soil bacteria and small parasites. The worms can be located by the gurgling sounds they make underground. The Kookaburra is the only bird known to feed on these tough earthworms. In the photo is *Megascolides australis,* the largest earthworm species, reported to exceed 3 m in length. It is long known from the Gippsland area of southeastern Australia, but there are other Australian giants in Queensland and New South Wales. (Globe photo)

Three egg capsules of *Rhinodrilus fafner.* The longest measured 6 cm. Such capsules contain 1 to 3 worms and when recently formed contain milky albuminous fluid, but as the worms develop they apparently digest the fluid and occupy the space. The young worms hatch in the large burrow occupied by the parent. (R.B.)

Marine tubificid from shallow muddy bottom at Beaufort, North Carolina. Such small, threadlike, reddish members of the family Tubificidae are common on both marine and freshwater bottoms (see *Tubifex* earlier). As they tolerate high levels of heavy metals in the sediments in which they live, the tubificids can be used as monitoring tools in field surveys of heavy metals and perhaps as indicators of heavy-metal pollution. (R.B.)

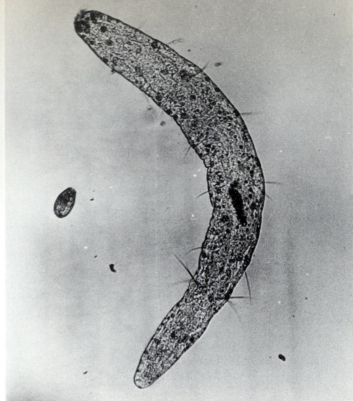

Freshwater oligochete, *Aeolosoma,* is only a few mm long and is often seen when looking through the microscope at bottom debris containing protozoans and minute metazoans. It glides along like a little turbellarian, propelled by cilia on the underside of the prostomium; but it is immediately recognizable as an oligochete by the spaced bundles of long bristles that reveal the segmentation. Internally the septa are represented only by fine muscular strands. Members of the family Aeolosomatidae reproduce mostly by asexual fission. (R.B.)

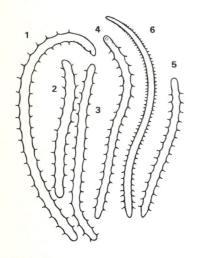

Fragmentation, followed by regeneration of the fragments into complete worms, is the regular mode of asexual reproduction in the South American freshwater oligochete *Allonais paraguayensis*. After reaching a length of 30 to 40 mm the worm spontaneously breaks up into several pieces, usually 6 to 8. In the drawing the 6 pieces are numbered according to their position in the body before fragmentation. In a few months this worm could produce several thousand descendants. Related species of the family Naididae also reproduce asexually, but in doing so form a chain of 2 or more zooids, each of which develops a fully differentiated head and posterior end *before* it breaks away from the chain. (After L.H. Hyman)

Enchytraeids range in length from 1 to 50 mm, but most are minute and live in moist soils, feeding on algal cells, bacteria, fungi, and organic matter. In acid soils, these oligochetes may occur 2.5 million to the m^2, and they take over the niches occupied by earthworms in more alkaline soils. Enchytraeids are also abundant in sewage beds and on the shores of lakes and oceans. Those called "snow worms," are found living in mountain snow banks. (R.B.)

LEECHES

THE LEECHES, class **Hirudinea,** are predominantly a fresh-water group and are found in lakes, sluggish streams, ponds marshes, or open ditches. They also occur in all seas and are widespread in moist soils, and on vegetation in tropical forests. In the U.S.A. leeches are especially abundant in the northern half of the country, and all species are considered to be aquatic, but some wander into gardens or agricultural fields, feeding on small invertebrates, especially earthworms. Some of these leeches pose minor problems for humans, and in hot climates both aquatic and land leeches may be serious health hazards.

Leeches do not crawl or burrow as do other annelids, and they lack both parapods and bristles. Although aquatic species can swim, leeches mostly move along with stepping or looping movements, much as does a hydra or a measuring-worm. With each step they alternately attach and release two suckers, a small one at the anterior end surrounding the mouth and a larger one at the posterior end. This type of locomotion does not depend on a spacious fluid skeleton, and the body of leeches is nearly solid, the coelom being largely filled with connective tissue and limited to a system of narrow sinuses and canals. In many leeches the coelomic canals replace the circulatory system. The segments are not separated by septa internally, nor do they correspond to the external rings, of which there are usually several per segment. However, the number of segments is constant and said to be 33 or 34, according to one's interpretation of observed structures.

Despite all their modification, the leeches are closely related to the oligochetes and are probably derived from them but have had less success in meeting terrestrial stresses. The best adapted for land life are the land leeches comprising the family Haemadipsidae; in these, the suckers are kept moist enough to function by the secretions of the first and last pairs of nephridia. Like oligochetes, leeches are hermaphroditic and have a clitellum which secretes the egg capsules. There is no character of leeches which is not present in at least some degree in some oligochete. For example, members of one oligochete family (the branchiobdellids) live on the gills of crayfishes, attached by suckers, and have a constant number of segments, as do leeches. Various oligochetes lack bristles or septa or a large coelom. On the other hand, *Acanthobdella,* once classified as an oligochete, is now considered a leech although it has bristles on a few of its 30 segments and a relatively large coelom divided by septa. Thus some biologists prefer to combine oligochetes and leeches in a single class **Clitellata.**

Changes in shape of a single leech, *Placobdella,* suggest the muscularity and flexibility of leech bodies. (From H. Nachtrieb)

In 18th century Europe, leeches were esteemed for their medical uses, and the subtle colorings and patterns of some species were copied onto fabrics for women's clothes. Today leeches have a poor "public relations image," but they are still used to a limited extent in medical practice, are sold as fishbait, and serve increasingly in biological research.

Septa are present in the embryo but are later lost. In the adult, segments are indicated by the positions of the ganglia, which number 34 in all; but the first ganglion innervates the prostomium, which in other annelids is not considered a segment. Most accounts list 6 ganglia concentrated in the head, 21 distributed along the trunk, and 7 concentrated in the posterior sucker.

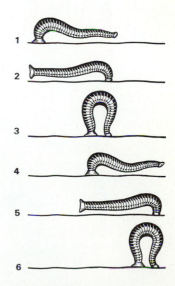

Leech looping is shown taking 2 steps. Aquatic leeches can swim by undulations of the body.

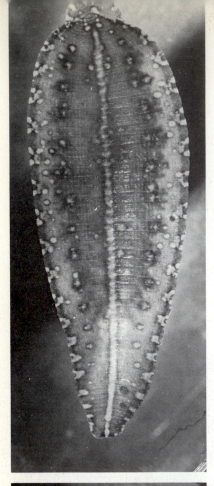

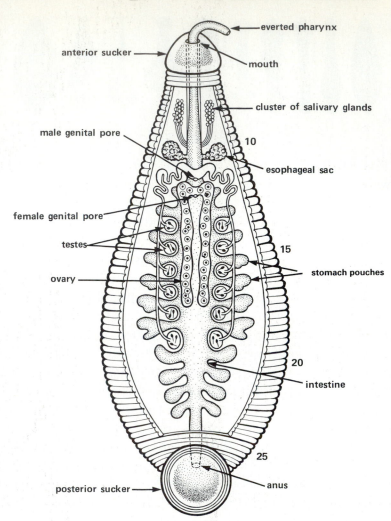

— everted pharynx

anterior sucker — — mouth

— cluster of salivary glands

male genital pore — 10

— esophageal sac

female genital pore —

testes — 15

ovary — — stomach pouches

20 — intestine

25

posterior sucker — — anus

Structure of leech, *Placobdella,* with part of the ventral body wall re-moved. Leeches of this type suck blood by means of an eversible pharynx. They have no jaws like those of the medicinal leech, *Hirudo.* The salivary glands are unicellular and their ducts enter the pharynx at its base but open near its tip. The two esophageal sacs harbor symbiotic bacteria thought to aid in digesting blood. The anus opens dorsally at the base of the posterior sucker, as in other leeches. Only the digestive and reproductive systems are shown. (Based partly on E.E. Hemingway)

Freshwater leech, *Placobdella parasitica,* lives free and feeds on worms and other small animals during the breeding season when carrying the eggs, and then the young leeches, attached to its ventral surface. At other times it sucks blood from the bases of the hind legs of freshwater turtles. Removed to an aquarium, the leech in the photo has attached itself to the glass by the large posterior sucker and hangs head down. **Dorsal surface,** *above left,* shows the external ringing of the body; except in the two end regions of the body, there are 3 external rings to each internal segment. The ground color is deep olive green, and the stripes and spots are yellow, orange, and green. Large specimens are 8-10 cm long. **Ventral surface,** *left,* reveals much of the internal anatomy and should be compared with the diagram above. The blood-filled branching stomach pouches can be seen through the body wall as dark patches. On each side of the midline are 6 testes and one ovary with its storage sac. In the midline, two small bright circles mark the position of the male genital opening and, just behind it, the female opening. Tennessee, U.S.A. (R. Buchsbaum)

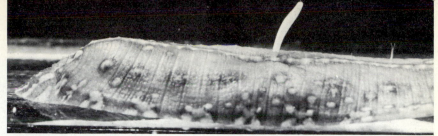

Transfer of sperms in *Placobdella parasitica,* as in many other leeches that lack an eversible penis, is by hypodermic injection through the skin of the mating partner. This leech has just received a spermatophore; it was injected into the dorsal surface and is seen projecting vertically. The spermatophore is packed with sperms, but at its lower tip contains granular material that cytolyzes the skin of the recipient. The walls of the spermatophore shrink on contact with water, forcing the sperms through the skin and into the underlying coelomic spaces in which the ovaries lie. Leeches are hermaphroditic and protandrous, developing first as males and later producing female sex organs. (R.B.)

Marine leeches. *Oceanobdella blennii,* ectoparasitic on the shore fish *Blennius pholis.* These leeches hatch from capsules in December, enter the opercular cavity of the fish, suck host blood, and grow. Two of those shown here are protruding from under the edge of the operculum. Usually, at this stage, they would have moved out and attached under the pectoral fin of the host. Here that position is already occupied by 2 large leeches in which the clitellum can be seen as a narrowed, whitish area. In April and May the mature leeches lay 20 to 30 capsules (each with one egg) on the underside of boulders. Then most adults die. North Wales. (R.N. Gibson)

The leeches that people are most likely to encounter are, naturally, those that suck human blood; but there are many other types as well. Some feed on the blood of aquatic vertebrates or on the soft tissues and body fluids of various invertebrates; others are scavengers or predators on small invertebrates such as earthworms, insects, slugs, and snails, swallowing them whole. Leeches are generally not fixed in their habits; even those which are specialized blood-suckers as adults may act as predators when young, and mostly predaceous species may take a blood meal if they get a chance. In their development of clinging suckers, leeches remind us of the parasitic flatworms. On the other hand, most leeches attach only temporarily to each blood-donor and need to move about to locate victims; they have eyes and are less modified than flukes or tapeworms. Few leeches are really parasites in the sense of prolonged living and feeding on a host.

Just as blood-sucking tsetse flies transmit trypanosomes to humans and other animals (chapter 2), leeches are intermediate hosts for certain species of trypanosomes and coccidians that infect fishes, amphibians, and reptiles.

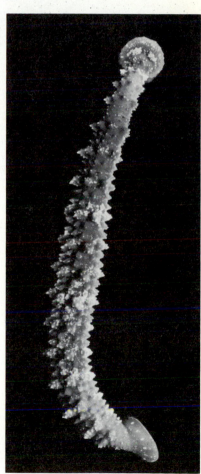

Marine leech, *Pontobdella macrothela,* a greenish worm brought up in a net haul with guitar fishes and presumed to have been attached to one of them. This specimen is from the Gulf of Mexico; the species has also been reported from the Caribbean Sea. (R.B.)

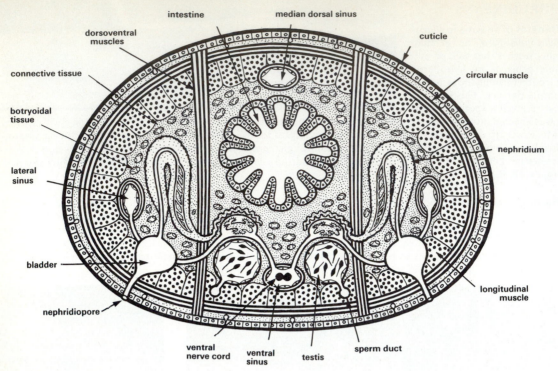

Cross-section of *Hirudo*. The main body cavity is obliterated by the growth of connective tissue. Remnants of the coelom are 4 longitudinal sinuses in which fluid circulates. The 2 lateral sinuses have muscular walls and drive the fluid forward and through branches in each segment. It then flows backward in the dorsal and ventral sinuses. The coelomic fluid contains dissolved hemoglobin and ameboid cells. A network of capillary channels (botryoidal tissue) is lined by globular cells full of brown pigment; these cells probably correspond to the chloragogen cells of earthworms. Leeches have fixed numbers of organs, and *Hirudo* has 17 pairs of nephridia. All obtain excretory products from the coelomic sinus system. But the funnels are separated off and are part of a capsule enclosed within a sinus. This rests on a testis and is overlain by part of the nephridium. The capsule produces the ameboid cells of the coelomic sinus system, and these cells are wafted out into the circulation by multiple small ciliated funnels on the capsule. The ventral nerve cord and its ganglia lie within the ventral sinus. (Based partly on A.G. Bourne)

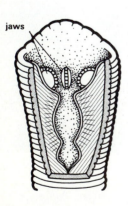

Head of *Hirudo* opened ventrally and cut away to show the muscles of the sucking pharynx and the 3 sawlike jaws that inflict a Y-shaped wound. Each jaw has about 50 teeth. (Modified after Pfurtscheller)

In sucking blood, a leech attaches to some animal by the posterior sucker, applies the anterior sucker to the skin, and makes a wound, often with the aid of little saw-toothed jaws inside the mouth. It fills its digestive tract with blood, sometimes taking an amount equivalent to 10 times its own weight, and then drops off, remaining torpid while digesting the meal. The salivary glands of leeches manufacture a substance called *hirudin,* which prevents coagulation of the blood while the leech is taking its meal as well as afterwards, during the long period that the blood is stored and slowly digested. Leeches may also inject an anesthetic and a substance that causes the small blood vessels to dilate, so that the wound bleeds for a long time after the leech has detached itself. Large blood meals are few and far between, but the digestive tract has lateral pouches which hold enough blood to last for months. During this time the blood is slowly broken down by symbiotic bacteria, into a form that can then be further digested by the enzymes of the leech. The symbiotic bacteria also inhibit the growth of other bacteria that might attack the store of blood.

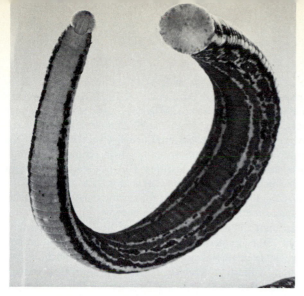

Medicinal leech of Europe, *Hirudo medicinalis*, now rare in nature because of over-collecting and destruction of habitat, has been introduced into the northeastern U.S., but is uncommon. This one was purchased in a Chicago drugstore, where there is still a demand for leeches to remove the blood around "black eyes." The upturned ventral surface (toward the left) shows mouth and anterior sucker; the dorsal surface (green with reddish-brown stripes) and the posterior sucker are seen toward the right. This leech was 12 cm long when extended in swimming. (R.B.)

Land leeches occur mostly in tropical and subtropical wet forests. They are most abundant in Southeast Asia and Sri Lanka, where they cause chronic blood loss in plantation workers. These little leeches, only 25 mm long when they attach, have sharp jaws and make a wound so painlessly that busy workers may be unaware of leeches hanging from their ankles like bunches of grapes. *Below left,* a leech from Malaya is attached to a leaf by its posterior sucker and is making searching movements with the anterior end. It has sensed the vibrations made by the footsteps of a mammal with a camera. Even if prey stops moving, alerted leeches rapidly converge on prey. *Below right,* the leech has attached to human skin and is sucking blood from the ankle of R.B.

Horse leech, *Haemopis marmorata,* about 16 cm, is common in America and occasionally takes blood from horses or humans that enter ponds. Like the European horse leech, *Haemopis sanguisuga,* it is misnamed. Members of this genus are all carnivores and scavengers, swallowing their small invertebrate prey whole, because they have weak jaws with few or no teeth. In the U.S. this genus is the only one with members that can leave the water. *H. marmorata* spends much time on muddy shores and also wanders inland, feeding mostly on earthworms. Even more terrestrial is *H. grandis,* the largest of N. American leeches, reaching 30 cm. It feeds in gardens and farm soil a kilometer or more from water. (C. Clarke)

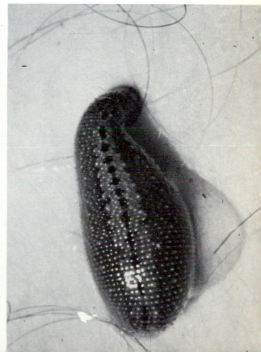

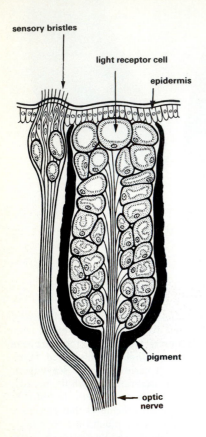

sensory bristles

light receptor cell

epidermis

pigment

optic nerve

Eye and other sensory cells of *Hirudo* are similar to those of earthworms. (Modified after R. Hesse)

The medical use of leeches goes back at least to the recommendation of Galen, a Greek physician and writer born in 130 A.D. By the 16th century doctors in England were often referred to as "leeches." And in the single year 1846, near the peak of bloodletting, France used 25 million leeches, many cultivated in France or imported from eastern Europe. Over the course of the 19th century, France imported a billion leeches from Hungary. *Hirudo medicinalis* is now scarce throughout Europe. This is a source of serious concern, for the leeches are still in demand, not so much for bleeding as for the use of their hirudin as an anticoagulant for certain patients with heart disease or undergoing surgery. *Hirudo* has also become valuable to neurophysiologists as an experimental animal. Each of its segmental ganglia contains only about 350 nerve cells, and these are unusually constant and individually recognizable in their sizes, shapes, positions, branching patterns, and connections with other nerve cells. Exactly 14 of these cells have been shown to provide the entire epidermis of the segment with sensitivity to mechanical stimulation. Each cell is sensitive either to light touch, or to prolonged pressure, or to noxious stimulation such as pinching or scratching, and each always serves the same, particular area of the epidermis, connecting directly to a motor nerve cell which in turn stimulates a certain portion of longitudinal muscle. Mechanical stimulation of the skin thus results in local shortening, and the entire reflex pathway is known and can be studied. Corresponding motor nerve cells on both sides of the ganglion are linked so that their activity is synchronous and shortening is symmetrical. The seemingly complex movements of swimming are also found to be controlled by only a few cells in each segment. The leech nervous system provides relatively simple experimental material for the study of fundamental nervous interactions.

Because *Hirudo* is now rare in nature and difficult to cultivate in the laboratory, biologists are looking to *Haementeria ghilianii,* a giant South American leech that sucks blood from mammals. It measures 50 cm long, breeds well in the laboratory, and produces large eggs useful for studies of nervous development.

Macrobdella, in the northern U.S. and Canada, and *Philobdella,* in the southern states of the U.S., are the only common native American leeches that regularly take human blood. *Macrobdella decora* was once used extensively for bloodletting and sometimes called the American medicinal leech.

SEGMENTATION seems to offer the same general possibilities as the dividing-up of an organism into cells, namely, the potential for the different segments to become specialized for different functions. In the nereids the segments are practically all alike, and this is thought to be the primitive condition. Even in this family, however, the beginnings of *specialization of body regions* are seen in the sexual epitokes, in which the segments of the posterior portion of the body become modified. In other polychete families, especially tube-dwelling types, the segments of different regions along the body have become further differentiated in both external and internal structures. In the earthworms and other oligochetes, specialization is most conspicuous in the hermaphroditic reproductive system; and in the leeches there is additional modification of segments at each end to form the suckers. In other segmented animals there are various degrees of specialization, some of which are extreme, as we shall see in the arthropods.

CLASSIFICATION: Phylum ANNELIDA

CLASS POLYCHAETA, the bristle worms. Typically with bundles of bristles borne on parapods; spacious coelom; variable number of segments, separated by septa. Sexes almost always separate. Mostly marine; a few freshwater and terrestrial.

Groupings within the class vary greatly. Many experts now believe that the traditional division into two subclasses is probably artificial, and that grouping the families into 10 or so orders better expresses their natural relationships. As both schemes are commonly encountered at present, they are combined here. Some of the major orders and a few of the more than 80 families are given.

Subclass ERRANTIA. Typically active crawling or swimming worms with large parapods, prominent sensory structures on the head, often jaws for seizing animal prey or pieces of seaweeds and with large numbers of similar segments.

Order Phyllodocida. Family Phyllodocidae: *Phyllodoce*. Family Nereidae: *Nereis, Neanthes, Platynereis*. Family Hesionidae: *Ophiodromus*. Family Syllidae: *Autolytus, Odontosyllis*. Family Polynoidae (scale worms): *Arctonoe, Acholoe*. Family Aphroditidae: *Aphrodita*. Family Tomopteridae: *Tomopteris*. Family Glyceridae: *Glycera*. Family Nephtyidae: *Nephtys*.

Order Eunicida. Family Eunicidae: *Eunice* (palolo), *Marphysa*. Family Onuphidae: *Diopatra*. Family Lysaretidae: *Lysarete*.

Subclass SEDENTARIA. Typically sedentary burrowing or tube worms with small parapods, without prominent sensory structures or jaws; usually deposit or filter feeders, with fewer segments, often differentiated into body regions.

Order Spionida. Family Spionidae: *Polydora, Streblospio*. Family Chaetopteridae: *Chaetopterus*. Family Cirratulidae: *Dodecaceria, Cirriformia*.

Order Capitellida. Family Capitellidae: *Capitella*. Family Arenicolidae: *Arenicola*.

Order Opheliida. Family Opheliidae: *Ophelia*.

Order Terebellida. Family Sabellariidae: *Sabellaria, Phragmatopoma*. Family Pectinariidae: *Pectinaria*. Family Terebellidae: *Amphitrite, Neoamphitrite, Terebella, Pista*.

Order Sabellida. Family Sabellidae: *Sabella, Megalomma (=Branchiomma), Myxicola, Bispira*. Family Serpulidae: *Serpula, Hydroides, Pomatoceros*. Family Spirorbidae: *Spirorbis*.

The next 2 groups are of uncertain taxonomic position.

Order Archiannelida. Mostly tiny, interstitial forms, often ciliated and lacking parapods, bristles, and distinct segmentation. Sometimes placed in a separate class. Family Protodrilidae: *Protodriloides*. Family Polygordiidae: *Polygordius*. Family Dinophilidae: *Dinophilus*.

Order Myzostomida. Small, scalelike commensals or parasites on or in the bodies of echinoderms. Sometimes placed in a separate class. Family Myzostomidae: *Myzostoma*.

CLASS OLIGOCHAETA, including the earthworms. Typically with 4 or 8 bristles on each segment, sometimes more; spacious coelom; variable number of segments, separated by septa. Hermaphroditic, with a clitellum that secretes egg capsules. Terrestrial, freshwater, and marine. Oligochete orders are separated on the basis of details of the bristles and reproductive organs, and are omitted here as unmeaningful to the general student. Earthworms: *Lumbricus, Eisenia, Allolobophora, Pheretima*, and giant earthworms *Rhinodrilus, Thamnodrilus*, and *Megascolides*. Freshwater: *Tubifex, Dero, Stylaria, Aeolosoma, Allonais*.

CLASS HIRUDINEA, the leeches. With suckers; bristles typically lacking; coelom reduced; small, fixed number of segments, usually 33 (or 34), not separated by septa. Hermaphroditic, with a clitellum that secretes egg capsules.

Order Acanthobdellae. Differs from other orders in having a few bristles, a relatively spacious coelom, some septa, and only 30 segments. One genus, parasitic on freshwater fishes: *Acanthobdella*.

Order Rhynchobdellae. Jawless, with an eversible pharynx (proboscis) that penetrates tissues of aquatic invertebrates and vertebrates to suck blood and other body fluids. Freshwater: *Glossiphonia, Placobdella*. Marine: *Pontobdella, Branchellion*.

Order Gnathobdellae. Bloodsuckers of vertebrates; with jaws and sharp teeth that pierce the skin. Or carnivores, with weak jaws and blunt teeth, that feed on small invertebrates or scavenge on dead flesh. Freshwater: *Hirudo, Macrobdella, Philobdella* (bloodsuckers); *Haemopis* (carnivore). Terrestrial: *Haemadipsa* (bloodsucker).

Order Pharyngobdellae. Jawless carnivores that prey on small invertebrates. Freshwater: *Erpobdella*. Terrestrial: *Trocheta*.

Chapter 18

More Worms

Echiurans
Sipunculans
Pogonophores
Priapulans

Annelid Satellites

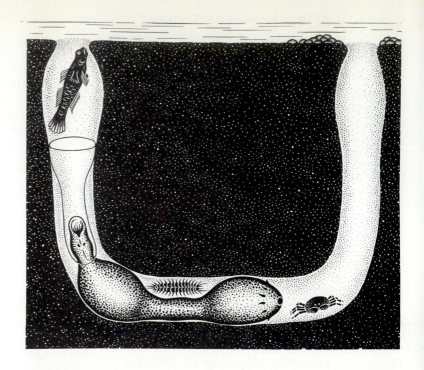

The echiuran shown in the drawing above (based on Fisher and Mac-Ginitie) is *Urechis caupo,* which has a very short proboscis and feeds in a way unusual among echiurans. *Urechis* secretes a funnel-shaped tube of mucus that filters out tiny organisms and particles from the respiratory and feeding current that flows through the burrow. The current is produced by peristaltic waves that pass along the muscular body. At intervals the worm swallows the mucus net, with its content of food, and secretes a new one. Large particles of food that drop from the net as it is swallowed are quickly snatched by small commensals that share the burrow. One or more fishes (a species of goby), a pea crab, and a reddish scale worm are the 3 most frequent commensals. The plump body and the hospitality afforded by *Urechis* to its guests have earned this echiuran the common name "fat innkeeper."

S ATELLITES may have a common origin with the star or planet that they orbit (all having come from a single mass that broke apart or having condensed from the same cloud of matter). Or, two bodies traveling through space may converge and pass so close that the smaller is captured by the gravitational field of the larger one. In this chapter 4 small phyla of burrowing worms are grouped as metaphorical satellites, because their similarities to annelids may result from common origins or from convergence.

ECHIURANS

THE MEMBERS of the phylum **ECHIURA** are among the most inoffensive and defenseless of animals. They burrow in sand and mud, or live protected in rock crevices, snail shells, and other secure shelters. The soft, sausage-shaped body is covered only with a delicate cuticle, and the animal keeps to the safety of its burrow or other shelter even when feeding. Most echiurans feed by sweeping organic detrital particles into the scoop-shaped proboscis, which is a richly ciliated extension of the head anterior to the mouth. In some species, it is many times longer than the rest of the body. The proboscis is extended over the substrate first in one direction, then another, as the feeding echiuran neatly clears detritus from the vicinity of its burrow.

There are about 130 species, all marine and distributed worldwide from shallow to abyssal depths. Individuals are from 1 to 40 cm long, not including the proboscis.

438

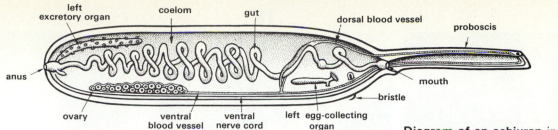

left excretory organ — coelom — gut — dorsal blood vessel — proboscis

anus — ovary — ventral blood vessel — ventral nerve cord — left egg-collecting organ — bristle — mouth

The muscular body wall surrounds a large coelom in which the coils of the long gut are suspended. There is usually a simple closed circulatory system, a ventral nerve cord with a loop into the proboscis, and a pair of excretory organs that open into the gut near the anus, at the posterior end of the worm. Respiratory exchange takes place through the body wall. The eggs or sperms do not complete their development in the single ventral gonad but are released into the coelom, where they float freely. When mature, they are collected by special organs (from 1 to hundreds of pairs in different species), which appear to have been derived from annelidlike nephridia. The gametes are spawned into the sea, where fertilization occurs and where the free-swimming larvas develop.

An exception is the genus *Bonellia,* in which fertilization is internal, as the males are minute non-feeding forms, 1 to 3 mm long, that live inside the egg-collecting organs of the females. Sex is not determined until after the young settle. A young individual that settles alone on unoccupied substrate almost always grows into a female. An individual that settles in territory already occupied by mature females usually attaches to the proboscis of a female and is influenced by substances in the proboscis to develop into a male.

The early development of echiurans proclaims their affinities with annelids, as the pattern of cleavage is cell-for-cell like that of any polychete, and the echiuran larva is a trochophore. Later development, however, diverges from the annelid pattern. Despite irregular, transient suggestions of mesodermal bands and nerve ganglia in some species, echiurans do not become segmented.

The **proboscis** of echiurans, primarily a feeding organ, also provides a large surface for respiratory exchange. The body of this small echiuran was only about 1 cm across. Central Atlantic coast, U.S.A. (R.B.)

Diagram of an echiuran in lateral view, cut lengthwise just to the right of the midline. The excretory organs open to the coelom through hundreds of tiny ciliated funnels. (Modified after Delage and Hérouard)

In *Urechis* there is no circulatory system. The hind gut is regularly flushed with seawater, and respiratory exchange occurs (through the thin wall of the hindgut) between the gut lumen and coelom. Cells containing hemoglobin circulate in the coelomic fluid, which is kept mixed by cilia and by frequent peristalsis of the gut and body wall. *Urechis* is also unusual in lacking a discrete gonad.

Bonellia viridis, a green echiuran, gathers its diet of algal cells with a long forked proboscis, which may reach a length of 3.5 m in a worm only 10 to 12 cm long. The toxic skin of *Bonellia* gives some protection against predators. Banyuls, France. (R.B.)

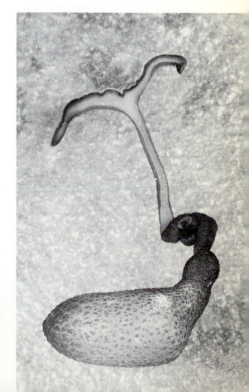

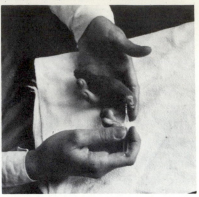

Urechis caupo lives in sandy or muddy bottoms on the U.S. West Coast and can be dug up at low tides. (R.B.)

Mud is shoveled away until the pink, cylindrical worm is exposed in its burrow. (R.B.)

Eggs or sperms are easily collected from *Urechis* by inserting a glass pipet into the genital openings. (R.B.)

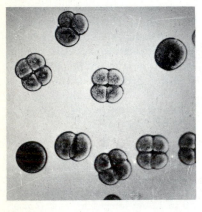

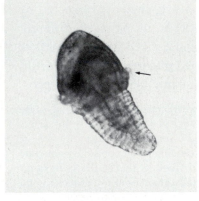

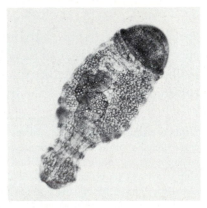

Zygotes divide to form 2-cell, then 4-cell embryos practically indistinguishable from the early embryos of many other animals. Early events of fertilization and development (sperm entry, membrane elevation, changes in oxygen consumption, enzyme activation, RNA and protein synthesis, etc.) have been extensively studied in *Urechis caupo*. (R.B.)

Free-swimming larva of *Urechis* near the time of metamorphosis. Arrow points to the cilia of prototroch. Length, 2 mm. (M. Paul)

Newly metamorphosed juvenile of *Urechis caupo*. What appears to be segments are rings of ectodermal mucous glands, not internal segmentation. Length, 3 mm. (M. Paul)

Chitinous bristles like those of annelids occur in echiurans—2 bristles near the mouth and sometimes also 1 or 2 circles of bristles near the anus. *Urechis caupo* (shown here) uses the bristles in digging and cleaning its burrow. Length, 10 cm. Elkhorn Slough, central California. (R.B.)

The phylum **ECHIURA** is divided into 3 orders. Order **Echiuroinea:** *Echiurus, Bonellia*. Order **Xenopneusta:** *Urechis*. Order **Heteromyota:** *Ikeda*.

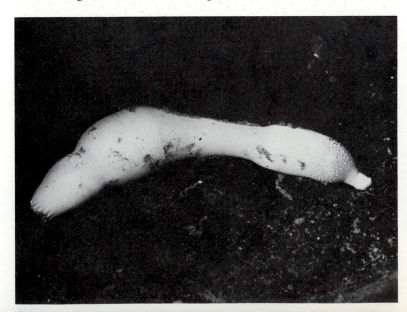

SIPUNCULANS

LIKE echiurans, sipunculans are unsegmented, coelomate, burrowing worms that live on particulate organic matter. The phylum **SIPUNCULA** includes about 320 species, found in oceans throughout the world at all depths and latitudes. Some sipunculans burrow actively and feed in the manner of earthworms, swallowing large quantities of sand or mud and digesting the contained organic matter. Many live in more permanent burrows or rock crevices and gather food particles into the mouth by means of a bouquet of frilly ciliated tentacles that also provide surface for respiratory exchange. Sipunculans burrow with the anterior portion of the body, which can be withdrawn for protection inside the trunk and is called the introvert, a term appropriate to these shy animals. When retracted, the introvert adds an anterior bulge to the already bulging posterior trunk, earning for sipunculans the common name "peanut worms."

Branching, ciliated tentacles of *Themiste* extend from the calcareous tube in which this sipunculan has taken permanent refuge. Santa Catalina Island, California. (R.B.)

Long slender introvert that can quickly be inverted into the plump trunk is characteristic of sipunculans. A circlet of short, unbranched tentacles extend from the tip of the introvert. *Phascolosoma* reaches a length of 12 cm. Monterey Bay, California. (R.B.)

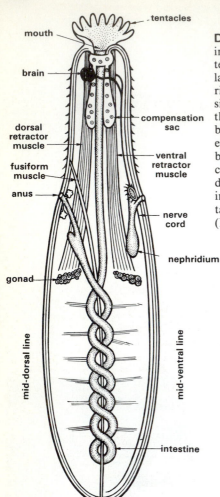

tentacles
mouth
brain
dorsal retractor muscle
fusiform muscle
anus
compensation sac
ventral retractor muscle
nerve cord
nephridium
gonad
mid-dorsal line
mid-ventral line
intestine

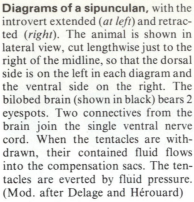

Diagrams of a sipunculan, with the introvert extended (*at left*) and retracted (*right*). The animal is shown in lateral view, cut lengthwise just to the right of the midline, so that the dorsal side is on the left in each diagram and the ventral side on the right. The bilobed brain (shown in black) bears 2 eyespots. Two connectives from the brain join the single ventral nerve cord. When the tentacles are withdrawn, their contained fluid flows into the compensation sacs. The tentacles are everted by fluid pressure. (Mod. after Delage and Hérouard)

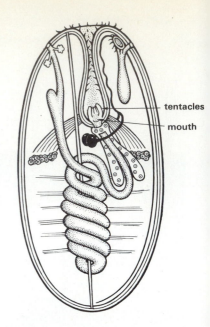

tentacles
mouth

The introvert is withdrawn by retractor muscles that insert on the midregion of the trunk. It is everted by pressure exerted through the fluid-filled coelom by the contraction of the body wall muscles of the trunk. Suspended in the coelom is a long gut that spirals back on itself and opens through an anterior dorsal anus. There is no circulatory system, but floating free in the coelomic fluid are cells containing the respiratory pigment hemerythrin, as well as developing gametes and excretory cells; other excretory cells lie in the coelomic lining. A pair of annelidlike nephridia (sometimes one is lost) open ventrally. They regulate volume and serve as the outlets for gametes, which are spawned into the sea.

As in annelids and echiurans, the fertilized egg undergoes spiral cleavage, and a free-swimming trochophore larva usually develops. In many sipunculans, the trochophore is followed by the pelagosphera, a large rounded larval stage propelled by a band of cilia that forms behind the prototroch. The pelagosphera may feed on plankton or subsist on stored yolk, and eventually settles to the bottom. Besides reproducing sexually, a few sipunculans reproduce asexually by unequal division of the body or by budding.

Sipunculus nudus is a large pink sipunculan, about 25 cm long. Sipunculans range from 1 cm to more than 60 cm. Brittany, France. (R.B.)

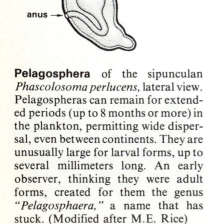

apical tuft
eyespot
prototroch
metatroch
mouth
anus

Pelagosphera of the sipunculan *Phascolosoma perlucens,* lateral view. Pelagospheras can remain for extended periods (up to 8 months or more) in the plankton, permitting wide dispersal, even between continents. They are unusually large for larval forms, up to several millimeters long. An early observer, thinking they were adult forms, created for them the genus *"Pelagosphaera,"* a name that has stuck. (Modified after M.E. Rice)

POGONOPHORES

THE PHYLUM **POGONOPHORA** ("beard bearers") was named for the thick tuft of long tentacles that distinguishes this group of about 120 species. Some beard worms have scraggly beards of only 1 or 2 relatively large tentacles, but the fine tentacles in a luxuriantly bearded species may number to about 200 (or more than 200,000 in the ones called vestimentiferans). The slender and delicate body is protected within a secreted chitinous tube, which the worm adds to at both ends as it grows, so that the tube consists of a series of rings or slightly funnel-shaped pieces. The tubes are embedded in soft marine sediments, mostly in cold, deep waters, and are usually less than a millimeter in diameter. Collected by deep dredging, the tubes look at first glance like detrital plant fibers and it is not surprising that the Pogonophora was one of only a few phyla to escape description until the 20th century.

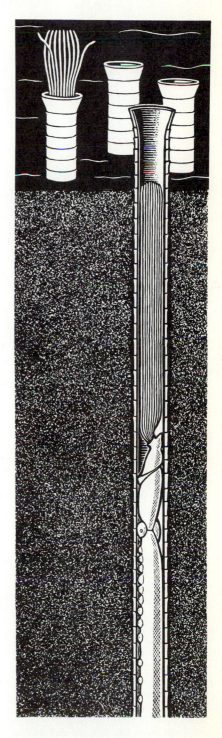

The oddest thing about pogonophores is that they have *no mouth or digestive tract* at any stage, a condition almost unique among free-living animals, although known in some parasites which take up dissolved nutrients through their surfaces. Experiments suggest that pogonophores could nourish themselves in much the same way, taking up sugars and amino acids from the surrounding seawater. In addition pogonophores harbor endosymbiotic bacteria in a richly vascularized tissue (trophosome) in the posterior part of the body; and there is evidence that the symbionts fix carbon dioxide into organic compounds that nourish both themselves and the pogonophores. The bacterial symbionts alone could sustain pogonophores in the nutrient-poor habitats where they are commonly found.

One kind of evidence that the bacteria may be a source of nutrition for pogonophores is the presence of enzymes involved in carbon dioxide fixation. The enzymes are localized mainly in the trophosome region and are of a type not found in animals but common in bacteria. Further evidence is the unusual ratio of stable carbon isotopes in pogonophoran tissues and tubes, which suggests that most of their food does not come from ordinary plant or animal material; bacterial fixation of carbon dioxide would account well for the unusual isotope ratio.

Besides pogonophores, the only entire phyla of free-living animals that lack a mouth and digestive cavity are the sponges and placozoans, in which individual cells ingest food particles. Exceptional free-living gutless species in other phyla include interstitial catenulid turbellarians and nematodes, oligochetes and archiannelid polychetes, protobranch bivalves, and concentricycloid echinoderms. Many internal parasites lack a gut.

The absence of a digestive tract at any stage of development makes it difficult to define the orientation of the body of pogonophores and to relate their structure to those of other groups. The end of the worm's body that points up and bears the tentacles is generally accepted as the anterior end, but which side is dorsal or ventral depends on phylogenetic interpretations. If pogonophores are considered to be related to other invertebrates in the chordate line, then the nerve cord runs along the dorsal side. If they are considered to be related to other invertebrates in the annelid-arthropod line, then the nerve cord runs along the ventral side. The latter view has received strong support from the stunning discovery of a short, segmented posterior end with chitinous bristles, looking like a tacked-on piece from an annelid. The posterior end breaks off easily from these fragile worms and had probably been lost from previously collected specimens, but its presence has now been confirmed in a number of species. It is thought to function in burrowing.

Another phylogenetic clue comes from embryological findings that cleavage follows a modified spiral pattern, and formation of the coelom is also annelid-like. The embryos and larvas are usually protected within the female parent's tube, but they may undergo a brief free-swimming period before settlement. Once settled in a tube, a pogonophore probably never leaves it. Males may produce free sperms or package them into spermatophores, but it is not known how sperms in either form reach the females, or how fertilization proceeds. These and other pogonophoran puzzles remain to be solved.

The members of the phylum **POGONOPHORA** fall into 2 distinct groups. Various specialists consider their many differences to be at the class or subphylum (or even phylum) level. The great majority of species are small, slender pogonophores of the **subphylum Perviata, class Frenulata,** divided into 2 orders: **order Athecanephria:** *Sieboglinum, Oligobrachia;* and **order Thecanephria:** *Lamellisabella, Heptabrachia.* The minority are giant pogonophores, described only in the last 2 decades and placed in the **subphylum Obturata, class Afrenulata, order Vestimentifera:** *Lamellibrachia, Riftia.*

Diagram of a pogonophore, much shortened, showing the main regions of the body: a short forepart bearing the tentacles, a long trunk, and a short segmented posterior end. Glandular regions in the forepart and trunk secrete tube materials, adhesives, and mucus. The trunk is marked by glandular papillas, thought to anchor the worm in its tube, and by ciliary bands. The segmented posterior end (opisthosoma) of pogonophores living in erect tubes in soft substrates probably projects from the bottom end of the tube. Its primary function is almost certainly that of burrowing. (After J.D. George and E.C. Southward)

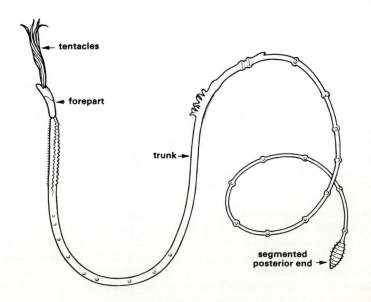

Pogonophoran tubes look like slender straws. Only the tops of these tubes project above the surface of the substrate. (Southward and Southward)

Pogonophore removed from its tube. This and other members of the genus *Siboglinum* have only one tentacle. (Southward and Southward)

Vestimentiferans are giant pogonophores with rather stiff tubes that stand up well above the substrate surface. Among typical (perviate) pogonophores, the largest tubes are a millimeter or so in diameter and the maximum number of tentacles is about 200, but the large vestimentiferan *Riftia pachyptila* may have tubes almost 40 mm in diameter and over 1.5 m long, and its bright red tentacles number over 200,000. *Riftia* was photographed and collected by means of a submersible at a depth of about 2,500 m, on the Galapagos Rift, where dense populations of these worms live around vents of warm (up to 23°C), anoxic water, rich in hydrogen sulfide. Like other pogonophores, *Riftia* is probably nourished at least partly by the symbiotic bacteria found in great numbers in the loose tissue (trophosome) which amounts to as much as 60% of the worm's wet weight. Physiological and enzymatic studies of *Riftia* suggest that the bacteria oxidize hydrogen sulfide to sulfate, and reduce carbon dioxide to organic compounds, which nourish both the symbionts and the pogonophoran host. The worm's bright-red blood carries oxygen, bound to dissolved hemoglobin, and also sulfide, tightly bound to another protein; chemical binding keeps these two substances from reacting together in an unproductive fashion before they are delivered to the symbionts and protects the tissues against the toxic hydrogen sulfide. The vent communities are among the few on earth that do not depend (except for oxygen) on solar energy. (M.L. Jones) (See color photo, p. 817)

PRIAPULANS

A DOZEN or so species of distinctively shaped worms are recognized as members of the exclusive phylum **PRIAPULA**. Living buried in sediments, priapulans are found from the intertidal to considerable depths, mostly in cold waters. They range in size from large specimens of *Priapulus,* 20 cm in length, to tiny interstitial forms such as *Tubiluchus.* The anterior part of the body is an introvert, which can be withdrawn into the longer, posterior trunk and which serves in burrowing, as in sipunculans. The muscular body is covered with a thin cuticle of chitin and protein. Circular ridges bearing papillas or spines give the trunk a superficially segmented appearance, and longitudinal ridges mark the introvert. Spines surrounding the mouth and continuing into the pharynx are not visible in a resting priapulan but are everted to capture prey, mostly soft-bodied annelids or other priapulans encountered in the sediment and swallowed whole.

Pentagons of spines surround the mouth area (about 0.1 mm across) and extend into the pharynx. SEM of a priapulan larva. (C.B. Calloway)

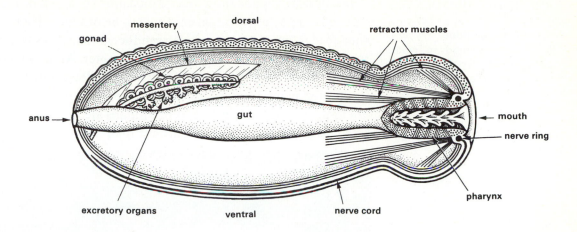

Diagram of a priapulan in lateral view, cut lengthwise just to the right of the midline. Shown in black are the cut ends of the nerve ring around the pharynx and the median ventral nerve cord, which lies in the epidermis. The pharynx has been cut open to reveal its spiny lining. The left gonad, the excretory organs, and their duct are shown suspended by a mesentery from the lateral body wall. The retractor muscles of the introvert also attach to the body wall. (Adapted from Delage and Hérouard)

The straight gut ends in a posterior anus and is suspended in a large fluid-filled body cavity, which serves as a hydrostatic skeleton. Without knowledge of its development, the cavity cannot be identified as either a pseudocoelom or a coelom; it has a membranous lining, no peritoneal epithelium. Floating in the fluid are a variety of cells, including some containing the respiratory pigment hemerythrin. In *Priapulus* the body cavity extends into one or two lobed appendages attached at the posterior end of the trunk. These are thought to be accessory respiratory structures, but the worm survives their removal without obvious effects and they are not present in all species. There are no other special respiratory or circulatory structures.

Priapus, the Greek god of fertility, is typically represented with a regal and ready copulatory organ. His namesakes have nothing comparable. The sexes are separate but usually not distinguishable. A pair of gonads is suspended in the body cavity and each shares a duct with the excretory organs (protonephridia). The ducts open near the anus, and at least in most priapulans the gametes are spawned into the sea. The fertilized eggs sink, and the larvas develop in the sediment. In larval priapulans the cuticle of the trunk is relatively thick and rigid, forming a protective case, or lorica. The cuticle is repeatedly molted throughout life as a priapulan grows.

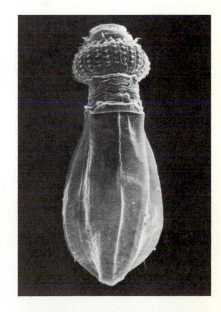

Priapulan larva with introvert extended. By retracting the introvert, the larva can withdraw into its protective lorica, formed by the thick cuticle of the trunk. SEM. *Priapulopsis bicaudatus.* Length 1 mm. Grand Banks, Newfoundland, Canada. (C.B. Calloway)

The small phylum **PRIAPULA** includes 3 families. The larger priapulans are in the **Priapulidae,** *Priapulus, Priapulopsis,* found in cold waters. The minute interstitial species are in the **Tubiluchidae,** *Tubiluchus,* found in shallow waters of the tropical and subtropical western Atlantic, and **Maccabeidae,** *Maccabeus,* found in deeper waters (60 to 550 m) in the Mediterranean.

Because priapulans show an odd mixture of characters, their relationships to other phyla continue to be obscure. They are similar in many ways to various pseudocoelomates, as well as to some of the coelomate worms in this chapter. As mentioned above, the nature of the body cavity must be determined from its development. The development is poorly known and does not appear to be annelid-like. If priapulans are indeed annelid satellites at all, their orbit is a distant one.

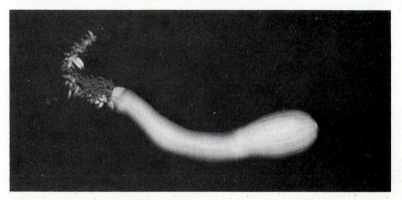

Large priapulan, *Priapulus caudatus,* with barrel-shaped introvert, ringed trunk, and lobed tail. This ventral view reveals the nerve cord. Preserved specimen, 4 cm long excluding tail. New Brunswick, Canada. (C.B. Calloway)

Interstitial priapulan, *Tubiluchus corallicola,* is exceptional in many ways besides its conspicuous tail (mostly cut off in this photo). It lives in calcareous sediments in warm waters and is a detritus feeder. The dimorphism of the sexes and the shape of the sperms suggest that mating and internal fertilization may occur in this species. The rows of ventral bristles, and the specialized structures associated with the genital area, mark this specimen as an adult male. Length 1 mm excluding tail. Bermuda. SEM. (C.B. Calloway)

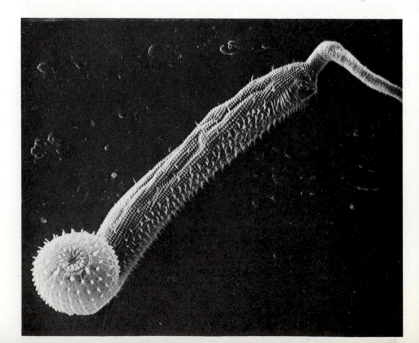

Chapter 19

Onychophorans

A Missing Link

IF WE COULD FIND an animal clearly intermediate in structure between two modern phyla, we would have good evidence that the two phyla are closely related. That such an animal has never been found is not surprising. Indeed, it would be more remarkable if the very form which at some remote time in the past gave rise to two stocks, now represented by two modern phyla, had also persisted unchanged through the ages. Fossils show that certain forms have remained unchanged for very long periods of time, but none is so old that it traces back to the time before all of the modern phyla had evolved. Therefore, we often speak of these missing ancestral forms as "missing links."

Animals that come as close as any to being the "missing link" between two phyla are the members of the small phylum **ONYCHOPHORA,** the name of which refers to the curved claws on the feet. Onychophorans are rare animals, found in moist places under logs and leaf litter only in restricted areas of some tropical or south temperate regions. The scattered distribution of the onychophorans, with 70 or so species in widely separated parts of the world, suggests that the group is an old one, having evolved before the ancient landmasses broke up into the present arrangement of continents.

Onychophorans look much like caterpillars, up to 15 cm long, with soft velvety skin and many pairs of legs. While some of their structures are like those of arthropods (to be described in the next chapter), onychophorans also have similarities to annelids and, of course, some special features of their own. As already pointed out, neither the onychophorans as we find them now, nor any other living animals, could be ancestral to any group as old as a phylum; but there is little doubt that onychophorans have descended from a line which branched off close to the primitive annelid-arthropod stocks.

on-i-kah'-for-a

The name onychophoran, meaning "claw-bearer," makes these mild animals sound rather fierce and would apply as well or better to crabs, eagles, and tigers. The common names of onychophorans are more descriptive: "velvet-worm" from the distinctive velvety texture of the skin, or "walking-worm" from the unusual combination of walking legs with a worm-like body, or "peripatus" from the first-named and widely known genus *Peripatus* (which means "the walker" or "the wanderer").

Onychophoran mother and her young. Found on a forest floor in Panama; soon after, the mother gave birth to two young, but one ate the other. When disturbed, both mother and young shot out sticky slime (visible on the leaf). Some species can eject a stream of slime with enough force to strike a target 50 cm away. The fluid slime becomes sticky on contact with air, then quickly hardens, forming a web of adhesive threads that effectively entangle prey or attacker. Young onychophorans can walk and feed immediately after birth, but stay close to the mother for several days. This large female is about 12 cm long. Male onychophorans are smaller than females, just one example among many animals in which a different size proves to be most advantageous to each sex. *Macroperipatus geayi.* (R.B.)

Unlike typical annelids and arthropods, onychophorans show no external segmentation, except for the **many pairs of legs,** each corresponding to a segment of the body. The legs differ from arthropod legs in that they are not jointed, but they end in claws which superficially resemble the claws of insects.

The outer covering is a **cuticle;** it is thin and flexible like that of annelids, but is chitinous and is structured somewhat like that of arthropods. It is covered with microscopic papillas that give it a velvety texture and that prevent the cuticle from being readily wetted by water. The cuticle and the underlying epidermis that secretes it are much folded, so that when the animal sets out on a walk, it may extend to more than twice its resting length. Onychophorans can also squeeze through holes much smaller than their body diameter, for they have no hard skeletal parts. Beneath the epidermis is a substantial layer of connective tissue that contributes firmness to the body wall. Continuous layers of muscles contract around the fluid of the body cavity, keeping the body turgid, as in the hydrostatic skeleton of annelids. (The body wall of arthropods is quite different, having a heavy articulated cuticle that serves as an exoskeleton and is acted upon by discrete bundles of muscle.) Onychophoran muscle fibers are categorized as smooth and are like neither the obliquely-striated (helical) fibers of annelids nor the cross-striated fibers of arthropods.

An onychophoran usually comes out at night and feels its way about by means of two sensory **antennas** on the head. There are also two small eyes, which show similarities to those of both annelids and arthropods. Onychophorans prey on insects and other small animals, cutting into the food with a pair of chitinous **jaws** that resemble the claws on the legs. The sickle-shaped jaws work alternately with a slashing motion, unlike arthropod or vertebrate jaws that work together and cut against each other. When startled or attacked, the animal ejects a stream of sticky fluid from a pair of glands that open on two stubby appendages, the **slime papillas.** The head is delimited by these 3 pairs of appendages (antennas, jaws, slime papillas) but is not otherwise set off from the long trunk.

The internal anatomy is a mixture of annelidlike and arthropodlike structures. The **digestive tract** is simple, without ceca, and both ends (foregut and hindgut) are lined with cuticle, as in arthropods. The **circulatory system** is open and like that of arthropods. A single long contractile dorsal vessel, the heart, extends the length of the body. On leaving the heart, the blood flows into large sinuses in the tissues and finally collects in the space surrounding the heart. It reenters the heart

The legs are soft and stumpy and very unlike those of arthropods, but the curved claws are quite arthropod-like. *Macroperipatus geayi.* Panama. (R.B.)

The mouth, on the underside of the head, holds the prey while the little jaws slice into it. Digestive secretions from a pair of salivary glands liquefy prey tissues which are then sucked out. New Zealand. (R.B.)

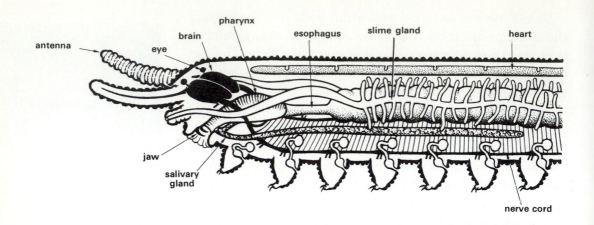

antenna — eye — brain — pharynx — esophagus — slime gland — heart

jaw — salivary gland — nerve cord

Anatomy of an onychophoran. (Combined from various sources)

Tracheal tubes lined with thin chitinous cuticle supply air throughout the body. The tracheal systems in onychophorans and in various terrestrial arthropods, although evolved independently by each group, do not arise out of entirely different structures. The selective pressures imposed by terrestrial life seem to have readily evoked this type of adaptation from the combination of characters, especially the chitinous cuticle, shared by all the groups.
(Modified after K.C. Schneider)

through a series of openings, one pair in almost every segment of the body. Together the blood sinuses constitute a **hemocoel,** as is typical of arthropods. But the most arthropodlike character of all is the respiratory system, consisting of air-filled **tracheal tubes** that open from the external surface and extend throughout the body, piping air directly to the tissues. Although such structures occur nowhere else in the animal kingdom except in terrestrial arthropods, they are thought to have arisen independently in the onychophorans and in various groups of terrestrial arthropods.

The tracheal tubes of an onychophoran differ from those of most arthropods in several important respects. In arthropods there are relatively few openings to the outside, and they usually have closing mechanisms. The openings lead into large tubes that branch repeatedly, the branches decreasing in size and ramifying throughout the body. In onychophorans a large bundle of unbranched tubes arises directly from each external opening. The external openings are necessarily numerous and scattered over the body; and as they lack closing mechanisms, the loss of water through this system of exposed air tubes is considerable. Experiments designed to test water loss under comparable conditions showed that an onychophoran loses water about as fast as an earthworm, 20 times as fast as a millipede, 40 times as fast as a smooth-skinned caterpillar, and 80 times as fast as a cockroach. Although the thin cuticle of onychophorans is relatively permeable to water, moisture is probably lost mostly through the tracheal system. Water is replaced by fluids in prey, by drinking, and by uptake through thin-walled vesicles (near the bases of the legs) that can be everted and pressed against moist substrates.

The central **nervous system** is more diffuse than in annelids and arthropods, lacking well-defined segmental ganglia. From the paired dorsal ganglia (brain) in the head run 2 widely separated ventral nerve cords which are only slightly thickened in each segment and which are connected by numerous fine commissures.

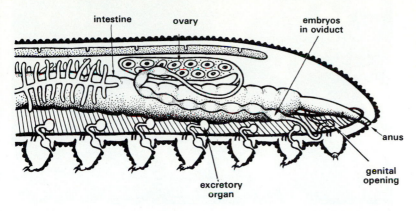

intestine   ovary   embryos in oviduct

anus

genital opening

excretory organ

At the anterior and posterior ends of the animal, in segments that have no excretory organs, the coelomic ducts (from which these organs are derived) give rise instead to the slime glands and to the reproductive ducts.

The **excretory system** consists of segmentally arranged pairs of coiled tubes that resemble the nephridia of annelids, but develop somewhat differently. The inner end of each tube opens into a small coelomic sac, from which fluid is moved into and along the tube by beating cilia, as in annelids. This last is the most unarthropodlike character, for motile cilia occur nowhere in arthropod systems (except in some arthropod sperms). The fluid, containing salts and small amounts of nitrogenous wastes, accumulates in a small bladder and is discharged at intervals to the outside through a pore at the base of the adjacent leg. The major site of nitrogenous excretion is the gut; the midgut lining excretes uric acid, which is eliminated with the feces. The total amount of water lost through both types of excretion is small.

The **reproductive system** is of coelomic origin and, like the excretory system, is ciliated. The sexes are separate. In some species, the male simply introduces the sperms (either free or enclosed in a spermatophore) into the female's genital opening. In others, the male attaches spermatophores to the outside of the female's body; wherever a spermatophore is attached, ameboid blood cells of the female make a hole through the body wall and cuticle. Sperms enter and swim through the blood sinuses, finally penetrating the ovaries and fertilizing the eggs. Only a few kinds of onychophorans lay their eggs. In most the embryos develop in the oviducts and are nourished either by yolk in the egg or by nutrients supplied continuously by the mother. In some species there is even a placentalike connection between maternal and embryonic tissues. The young are born looking like miniature adults. As they grow, their nongrowing cuticle becomes too small, and they must regularly **molt,** shedding and replacing the cuticle, as do arthropods.

Most aquatic animals dispose of nitrogenous waste as ammonia, which diffuses readily through permeable surfaces and requires no special excretory organs. The fact that it is toxic and must be rapidly diluted poses no problem for an animal living in water. Land animals, on the other hand, need to conserve water, and many do so partly by converting nitrogenous wastes to uric acid, which is not toxic and is relatively insoluble in water. Thus it can be excreted in semifluid or solid form with very little accompanying water. Excretion of uric acid as a conservation measure serves a wide variety of terrestrial animals besides onychophorans, such as snails, centipedes, millipedes, insects, lizards, and birds. Many other land animals, such as earthworms, spiders and their relatives, and mammals, have evolved different excretory products.

Two extensible antennas projecting from the head, and a simple eye at the base of each, provide sensory information to the brain. The antennas are covered with sensory projections, which also occur, less densely, on the body and legs. *Macroperipatus geayi.* Panama. (R.B.)

The **phylum Onychophora** is not divided into multiple classes or orders but considered to consist of 2 families.

Members of the **Peripatidae** are tropical in distribution, occurring in central America and the Caribbean, in central Africa, and in and around the Malay Peninsula. Members of the **Peripatopsidae** are known mostly from temperate climates, being distributed around the Southern Hemisphere in Chile, South Africa, New Guinea, Australia, and New Zealand. This gap between equatorial and southern groups is thought to result from a separation by desert conditions in the distant past. Then, following the breakup of Gondwanaland, onychophorans became separated by oceans as the continents drifted apart.

ANIMALS with characters resembling those of two or more otherwise discrete groups exist at every taxonomic level and present difficulties in classification. Onychophorans in particular continue to be controversial, though they are rarely grouped with annelids today. Some put them with arthropods, but doing so stretches the definition of arthropods so thin that it loses much distinctiveness. Although onychophorans have emerged from millions of years of evolution with a body plan uniquely their own, many of their features do suggest what the ancestors of at least some arthropods might have been like.

Polychete (Annelida) **Onychophoran** **Millipede** (Arthropoda)

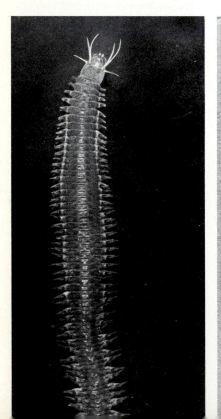

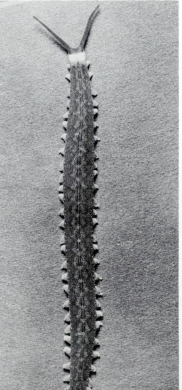

Chapter 20

Arthropod Body Plan

Jointed armor

THE ANIMAL GROUPS which are said to have attained the greatest "biological success" are those that have the largest numbers of species and of individuals, occupy the widest stretches of territory and the greatest variety of habitats, consume the largest amounts and the most diverse kinds of foods, and are most capable of defending themselves against their enemies. By these standards the phylum that occupies first place among all the animals (invertebrate and vertebrate) is the phylum **ARTHROPODA.** Of the more than a million known species of living animals, at least 3 out of 4 are arthropods. On the other hand, the arthropods may be better understood not as a single phylum of animals to be compared with the other phyla but rather, like the protozoans, as a type of organization.

arthro = "jointed"
poda = "feet"

Because they are so abundant and relatively conspicuous, many arthropods are familiar. The **crustaceans** are mostly aquatic, and they include both freshwater and marine forms such as crayfishes, lobsters, crabs, shrimps, and barnacles. The largely terrestrial **arachnids** (scorpions, spiders, mites) have marine relatives, the **merostomes** (horseshoe-crabs), and may be related also to the **pycnogonids** (sea-spiders). Finally there are the **myriapods** (centipedes, millipedes) and, by far the largest group of all, the **insects**.

Of the more than 900,000 described species of living arthropods, there are roughly 35,000 crustaceans, 70,000 arachnids, 11,000 myriapods, and over 800,000 insects. However, because many arthropods are small and easily overlooked, the number of living species is probably several times higher than the number of those already described.

Some arthropods do untold damage, destroying crops, undermining wooden buildings, and transmitting diseases. Others are prized by humans for providing food or some valuable service such as pollination of fruit trees. Still others, which also provide valuable services, usually go unappreciated, if not despised, such as spiders and centipedes that devour many field and household pests, or the insects that help to recycle carrion.

456

Insects destroy crops. The Colorado potato beetle, *Leptinotarsa decem-lineata,* has a history that exemplifies how human agriculture, which provides massive concentrations of the same plant, converts ordinary insects, minding their own small businesses, into destructive pests that create economic havoc. This beetle was a local species in the Rocky Mountain region of the U.S.A. and not abundant as long as it fed on its wild and scattered food, the buffalo bur, *Solanum rostratum.* When the West was settled, and potato crops were introduced into the home territory of the beetle it took to the new food with unabating zeal and soon spread to the Atlantic Coast and then to European countries. In the 19th century the beetle decimated European potato crops, causing widespread damage and resulting in heavy emigration to the U.S.A. (U.S. Bur. Entomology)

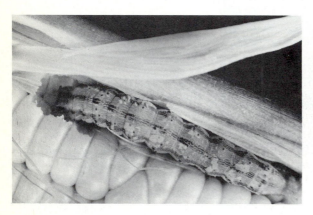

The corn-ear-worm is another important competitor for the human food supply. This 3-cm caterpillar is the larval stage of a noctuid moth, *Heliothis zea.* Consumers of sweetcorn in the U.S. see it only too often in the tender tips of the ears. It also feeds on cotton bolls, tomatoes, and other crops. (P.S. Tice)

Arthropods spread diseases of humans and domestic stock. The most common vectors are mosquitoes and other flies, lice, fleas, biting bugs, and ticks. Shown here is a mosquito of the genus *Anopheles,* a vector of malaria, with its mouthparts piercing human skin. (Science Service)

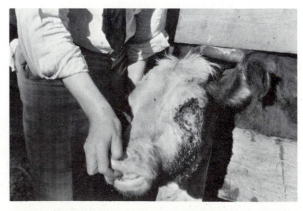

Damage by crustaceans runs into many millions of dollars annually around the world. Barnacles foul ship bottoms. Various crabs and shrimplike crustaceans are serious pests in oyster beds and rice fields, and land crabs in southern Florida attack tomato crops. Marine isopods burrow into wood, cement, and stone harborworks. *Limnoria quadripunctata,* the tiny marine isopod shown here, is only about 3 to 4 mm long, but when it occurs by the millions, at 20 per cm³ of wood, it can destroy the pilings of a wharf in less than 2 months. (C.A. Kofoid)

Parasitic fly larvas do serious injury to many mammals. The eye of this steer has been damaged by American screw-worms, *Cochliomyia hominivorax,* which infest livestock and stockhandlers from the southern USA to Argentina. The screw-worm fly lays its eggs in open wounds or in the nostrils, ears, and eyes. (U.S. Bureau of Entomology)

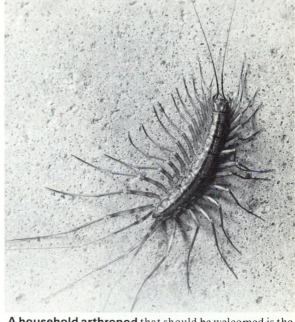

Many insects pollinate plants as they gather pollen and nectar from flowers. Among these are the honey bee, *Apis mellifera,* shown here, and several other kinds of bees, as well as beetles, flies, moths, and thrips. Honey bees are many times more valuable for cross-pollination in orchards and fields than they are for their honey and wax. Yields of fruits and legumes have been tripled by providing adequate numbers of bees. Unfortunately, the sprays used to kill pests on fruit trees also poison the honey bees and other economically valuable insects. (RB)

A household arthropod that should be welcomed is the 3-cm long house centipede, *Scutigera coleoptrata,* which preys on cockroaches, flies, and other household pests. Its bite is not dangerous to humans. Large tropical centipedes can reach lengths of over 25 cm and their poisonous bite can make a person ill. (C. Clarke)

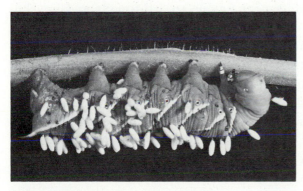

Insects that parasitize other insects without sparing the life of the host are not typical parasites but have the effect of predators. Called parasitoids, they comprise about 12% of all insects. Shown here are cocoons of a braconid (related to wasps) attached to a sphinx-moth caterpillar (tomato worm). The eggs were laid beneath the skin of the caterpillar; the larvas fed on host tissues, and have emerged through the skin to pupate. The exhausted caterpillar will soon die. Parasitoids serve as an important check on the insects that harm agricultural crops. (R.B.)

Caterpillars of the silkworm moth, *Bombyx mori,* native to China, are now cultivated in various warm countries, mainly China, Japan, India, and the Mediterranean area. Cared for, and fed on mulberry leaves, the caterpillars grow rapidly, attaining a length of 75 mm or more. At pupation they spin a cocoon with a continuous silk thread that may be more than 1,200 m long. The pupas are killed and dried before the moth can emerge, and the silk thread is unwound. About 5,000 cocoons provide one kg of silk. (C. Clarke) (See p. 647 for photograph of the silkworm moth.)

Arthropod body plan is relatively unspecialized in this centipede, *Lithobius forficatus.* (P.S. Tice)

In arthropods there are *no motile cilia or flagella,* except for the tails of some arthropod sperms. Nonmotile, modified cilia have been found in certain sensory structures.

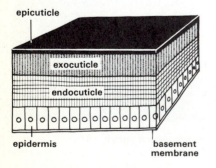

Body wall of an arthropod. The *epicuticle* is thin and non-chitinous. *Exocuticle and endocuticle* are a composite of chitin and proteins. The *epidermis* secretes the cuticle and is underlain by a *basement membrane.*

THE ARTHROPOD BODY PLAN may be roughly described as an elaboration and specialization of a wormlike segmented body plan. No arthropod shows marked segmentation of the organ systems throughout the length of the body such as we saw in nereids. But externally at least, the unspecialized arthropods, like unspecialized annelids, are largely composed of a series of many similar segments bearing similar appendages. In the more specialized types of both groups, we find reduction in the number of segments and increasing differentiation of segments and of segmental appendages, as well as functional coordination of groups of segments into body regions.

Of the notable differences between annelids and arthropods, the most important is the **chitinous jointed exoskeleton** of arthropods. This skeleton serves not only as a protective armor for the soft tissues within, but also as a strong, flexible framework upon which to operate the jointed appendages. Many of the other features that distinguish arthropods can probably be related to the presence of the exoskeleton. The muscles of arthropods are unusual in that all of them are cross-striated. The main body cavity is not a lined coelom but a continuous series of unlined blood sinuses, the hemocoel, which surrounds the organs and permeates the tissues. A heart lies dorsally in one of these blood spaces and pumps blood into a few arteries that empty into the sinuses of an open circulatory system. Blood re-enters the heart through paired openings in the heart walls—an arrangement unique to arthropods and onychophorans.

The chitinous exoskeleton, or **cuticle,** of arthropods is usually much thicker than the cuticle of annelids. Tough outer coverings occur in many groups of animals (for example, the perisarc of hydroids or the cuticle of nematodes or annelids), but in no case are they used so effectively or elaborated into so great a variety of structures as in arthropods. Made of cuticle, in whole or in part, are outer protective shields, biting jaws, piercing beaks, grinding surfaces, lenses, tactile sense organs, sound-producing organs, walking legs, pincers, swimming paddles, mating organs, wings, the linings of the foregut and hindgut, the linings of the air tubes, and innumerable other structures found among the highly diversified insects. This hardened material is to the arthropods what metals and plastics are to industrial nations, and it is partly to the possession of the versatile cuticle that the arthropods owe their success.

Rigid cuticular armor over most of the body minimizes mechanical damage and drying. But the continuous cuticle consists of both hardened areas and thin flexible membranes. In this passalid beetle (from a rotting log in a Peruvian forest) it can be seen that tiny mites are clustered on all the soft cuticular areas along the sides of the head, between segments, and in the larger joints of the legs. (R.B.)

Layers of cuticle are secreted successively by the underlying epidermis. The cuticle is nonliving, but it is the site of much chemical activity and is perforated by many pores and ducts through which secretions may flow out to coat the surface or to modify intermediate layers. It is composed of many different substances, each of which contributes some useful property. Several thin surface layers of cuticle (epicuticle) cover the body uniformly and include in many arthropods a waxy component that makes the cuticle *waterproof*. The bulk of the cuticle beneath is a composite of chitin and proteins, and is both *tough* and *flexible*. The composition of the proteins, and the details of arrangement of the chitin and proteins together, vary over different parts of the body, so that the nature of the cuticle ranges from hard and brittle to soft and expandable. Over most of the surface of an arthropod, the outer portion of the chitin-protein layers is sclerotized (hardened) by the formation of chemical bonds between the proteins and by impregnation with additional materials. In many crustaceans and millipedes, portions of the cuticle become calcified. Such hardened layers are responsible for the *rigidity* of the cuticle, one of the properties that make it so effective as a protective armor and skeleton. Between the hardened areas (sclerites), the cuticle remains flexible and allows movement at the joints in the appendages, or at the junctions between segments of the body. The combination of *protection* and *support* without sacrificing *mobility* is what gives arthropod cuticle so many advantages. In some cases the composition of the proteins allows for great elasticity, more perfect than in any rubber; for example, the elastic cuticular ligaments of insects store and release energy at each beat of the wings. In other cases the specific proteins and the arrangement of the chitin permit a degree of expansion and allow for growth.

Since chitin is not known to occur in vertebrates, or in vertebrate relatives, the chitin in insects appears to be a promising molecular target for pesticide action. And some research is being directed at a better understanding of the action of chemicals on the biosynthesis of chitin.

Areas of hard and soft cuticle are seen as dark sclerites separated by light-colored flexible membranes in this scorpion, an arachnid. The large pincerlike appendages, used for grasping prey, are the darkest and most heavily sclerotized part of the armor. (R.B.)

Soft-bodied caterpillars grow and synthesize cuticle continuously. But no animal with a hard outer covering can grow indefinitely without making some kind of readjustment, and all growing arthropods periodically discard and replace the cuticle. In the course of this **molting,** only the tanned, outer layers (exocuticle) of the old cuticle are shed. The softer inner layers (endocuticle) are dissolved and resorbed, and up to 90% of the materials of the cuticle are recycled in the making of a new and larger cuticle, which forms beneath the old one. At molting, an aquatic arthropod takes large quantities of water into its digestive tract, blood, and tissues; a terrestrial one takes air into its digestive tract and respiratory system. The rapid distension of the animal by extra water or extra air ruptures the old cuticle, which is cast off, a process called *ecdysis.* The new cuticle is light-colored and delicate and often much folded, so that after ecdysis it unfolds and stretches to accommodate the swollen animal. For a few hours or days, until the hardening processes are complete, the newly molted animal is soft and vulnerable and it often hides. It soon resumes its active life, although additional strengthening layers may continue to be added to the cuticle for some time. The extra water or air is gradually expelled, leaving the body some margin of extra space for feeding and for growth during the interval until the next molt. All systems of the arthropod body are affected by the molting cycle, and the complex changes that take place in each are delicately balanced and coordinated by a series of hormones.

Newly molted crab *(below)* is much larger than the exoskeleton *(above)* that it has just shed. *Xantho incisus.* Plymouth, England. (D.P. Wilson)

Molting crab is shedding its pale exoskeleton, which is noticeably smaller than the crab emerging from it. *Macropipus depurator.* Plymouth, England. (D.P. Wilson)

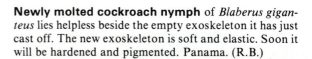

Newly molted cockroach nymph of *Blaberus giganteus* lies helpless beside the empty exoskeleton it has just cast off. The new exoskeleton is soft and elastic. Soon it will be hardened and pigmented. Panama. (R.B.)

Hormonal control of molting has been studied intensively only in certain crustaceans and insects, but the few studies in other groups suggest that all arthropods share some common features. The hormone that regulates molting, *ecdysone,* is a steroid derived from cholesterol and chemically related to many important vertebrate hormones. Although several slightly different chemical variants of ecdysone occur, they are similar enough in all arthropods that ecdysone extracted from a crustacean or insect will induce molting when injected into a spider or a horseshoe-crab. Ecdysone is produced and stored in *molting glands* (Y-organs in crustaceans, and prothoracic glands in insects) and, at molting time, is released into the circulatory system to be distributed to tissues throughout the body. The production, release, and perhaps also the action of ecdysone are controlled by *neurohormones*. These are secreted by *neurosecretory cells* in the brain (the X-organs of crustaceans, located in the optic lobes in the eyestalks, and the pars intercerebralis of insects). The neurosecretory cells convey their hormones along axons to be stored in *neurohemal organs* (sinus glands in the eyestalks of crustaceans, and corpora cardiaca in insects). From these storage sites the neurosecretions are released into the blood. The activity of the neurosecretory cells may be conditioned by external factors such as light and temperature, as well as by internal signals, as from stretch receptors in the body surface. The neurohormones that control molting act in opposite ways in crustaceans and insects. In crustaceans, molting is *suppressed* by neurohormones which are released continuously from the sinus glands except before a molt. Only in the absence of the inhibitory neurohormones is ecdysone released to initiate molting. This can be demonstrated easily by cutting off the eyestalks of crustaceans; such surgery is followed shortly by molting. In insects, molting is *promoted* by neurohormones which are released only before a molt and trigger the release of ecdysone.

In the course of molting, many different neurohormones act together with ecdysone to coordinate specific changes in various metabolic processes: detachment of epidermis and muscles from the old cuticle; dissolution and recycling of materials in the inner cuticular layers; altered metabolism of proteins, carbohydrates, fats, and nucleic acids; altered osmotic balance and calcium metabolism (in crustaceans); cell division; secretion and tanning of new cuticle. Neurohormones also control processes which in arthropods are related to molting, such as differentiation, regeneration, and sexual reproduction. In addition they regulate heart beat and other visceral muscular activity, color changes, movement of pigments during light-and-dark adaptation of the eyes, and a variety of general metabolic processes.

The role of ecdysone in cuticle formation has been neatly demonstrated in a species of cockroach *(Leucophaea maderae)*. From each of the young insects used in the experiment, the middle pair of legs was detached 24 hours after molting, leaving only a short stump (the 2 small basal articles of the appendage). Within each stump, a new leg began to regenerate. After 25 days the tiny regenerating legs, still lacking a chitinous cuticle, were dissected out and removed to glass culture chambers containing a nutrient medium known to support cockroach tissue growth. The 2 legs from each cockroach provided paired experimental and control members; one was treated with a solution of ecdysone in water and the other was treated with water only. After 2 weeks all 14 of the ecdysone-treated regenerates had produced recognizable cuticle. Tests showed that this cuticle contained chitin. Of these 14 regenerates, 8 developed bristles. None of the 14 control regenerates, treated with water only, produced cuticle or bristles.

The cuticle of arthropods serves as an *exoskeleton,* providing an external supporting framework for the tissues within and a surface for the attachment of muscles. Most arthropods, however, also have some *internal skeletal elements.* Infoldings of the chitinous cuticle, the apodemes, penetrate deep into the body as additional surface for muscle attachment; they are pulled out and shed with the outer cuticle at ecdysis. Internal skeletal structures may also be formed by connective tissue, and these are not shed but grow with the rest of the animal.

Body wall of annelid and arthropod contrasted. In annelids the cuticle is thin, and the epidermis is underlain by heavy layers of circular and longitudinal muscles (dotted lines indicate the divisions between segments). In arthropods the cuticle is heavy; the muscles occur in separate bundles.

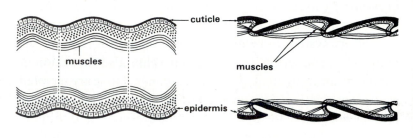

Annelid   Arthropod

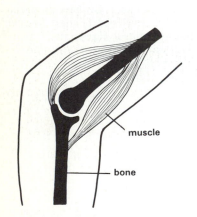

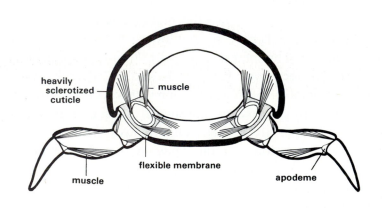

Skeletal and muscular systems of vertebrate and invertebrate contrasted. The bony endoskeleton of vertebrates is surrounded by soft fleshy parts, muscles and skin, and is continuously remodeled and enlarged during growth. The chitinous exoskeleton of arthropods surrounds all the soft parts and requires intermittent growth and periodic molting. We can imagine how it might feel to be an arthropod by mentally putting on an iron suit of armor that adheres closely to our skin, dissolving our bones, and attaching our muscles to the armor instead.

In the chitinous exoskeleton of an arthropod, **hardened plates** are continuous with more **flexible membranes,** which allow expansion of the body, bending between segments, and movement at the many joints of the appendages. The flexible membranes at the joints of the limbs do not act as hinges; articulation is by hard cuticular projections that move on each other. The **exoskeleton** provides a framework for the attachment of muscles. These are fastened to the inner surface of the body wall or to cuticular invaginations, the **apodemes.** (Modified after R.E. Snodgrass)

The muscles are not attached directly to the skeleton, but to the epidermis. Fibrils secreted by the epidermal cells run from the point of muscle attachment through the epidermis to the cuticle and through the cuticle to the epicuticle. At molting, new fibrils are secreted, and the old ones are broken only when the old cuticle is finally shed, so that the animal's ability to move is minimally disrupted by molting.

The **muscular system** of arthropods differs from that of other groups: both skeletal and visceral muscles are *cross-striated*. In vertebrates, the visceral muscles are of the smooth type, slow-acting and able to contract to less than 20% of their resting length, while the skeletal muscles are cross-striated and fast-acting and they contract to only about half their resting length. Compared to the cross-striated fibers of vertebrate muscles, which are relatively uniform in ultrastructure and speed of contraction, arthropod muscle fibers are both varied and versatile. Those of fast muscles tend to be more like the vertebrate type, while those of slow muscles have more widely spaced, less regular striations, and contain different proportions of actin and myosin filaments. Moreover, a single muscle fiber may contain elements that vary in ultrastructure, and the fiber may contract rapidly or slowly in response to stimulation by multiple nerves (see also chapter 21). The muscles that operate the wings of some insects contract only about 1% at each beat (that is, to 99% of resting length), while certain muscles of barnacles can contract to about 20% of their resting length. Insects in particular have evolved diverse and unusual types of striated muscles that serve in flight (see also chapter 24). The small numbers and large sizes of the muscle fibers in certain arthropods make them particularly convenient for experimental work. Such studies have helped us to understand the basic mechanism of contraction, which is probably similar in all muscles. They have also expanded our ideas of the diversity that exists in neuromuscular systems.

To their chitinous exoskeleton arthropods owe their ability to live on dry land, to a large extent independent of the moist refuges to which other land invertebrates must retreat. **Land life** requires, among certain other adjustments, a relatively impermeable outer covering to prevent drying of the moist tissues within and a fairly rigid framework of some kind to support the soft tissues. In vertebrates the covering is furnished by scales or heavy skin, and the supporting framework is a bony endoskeleton. In arthropods the waterproof and rigid cuticle fills *both* requirements and enables the arachnids, myriapods, and insects to occupy the land with little serious competition from the other invertebrate groups, most of which are largely aquatic.

THE NAME ARTHROPODA refers to the most characteristic structures of arthropods, their **chitinous jointed appendages**. The various appendages of arthropods are specialized for some particular function, and much of the biology of arthropods can be learned from studying how the appendages are constructed and what they are equipped to do.

Arthropod segments are grouped into **body regions**. Each such region, or *tagma*, is usually composed of segments with a common specialization. Often the body regions are separated by narrow constrictions that permit free movement between them, and each region may be covered and strengthened by a continuous, rigid capsule of cuticle, as in beetles (photo at right). Or, one region may be marked off from an adjacent one

Attachment of muscles to body wall is an important mechanical feature of arthropod organization. It is easily seen in the transparent aquatic larva of *Chaoborus*, a mosquito-like insect called a midge or gnat. (P.S. Tice)

Thick cuticle, covered by a surface layer of wax, retards water loss in this beetle, a member of the genus *Eleodes*, which is widespread in dry habitats and deserts in the western U.S.A. Beetles of this genus are further sealed against evaporation by sacrifice of the hind wings and cementing of the hardened wing covers (forewings) to the abdomen. A desert species of *Eleodes*, maintained at 40°C, was found to lose only 0.2% of its body weight per hour. And like many other terrestrial arthropods, *Eleodes* supplements its structural adaptations with behavior that minimizes water loss. It hides, during the heat of the day, under stones or wood or leaf debris. (R.B.)

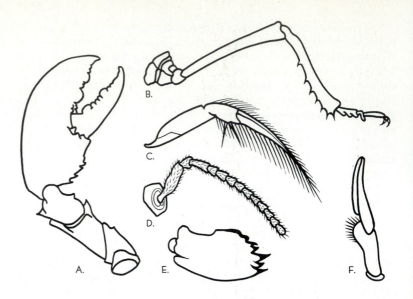

The **joined appendages** of arthropods serve a great variety of functions, even in the same animal. **A.** Pinching leg of a lobster. **B.** Walking leg of a grasshopper. **C.** Swimming leg of a water beetle. **D.** Sensory antenna of a honey bee. **E.** Chewing jaw of a cockroach. **F.** Mating appendage of a male lobster.

Unconstricted body of a harvestman appears to be a single undivided oval, but it consists of 2 body regions. The fused head and thorax, bearing the mouth, mouthparts, sense organs, and 4 pairs of legs, are broadly united to the externally segmented abdomen, which contains most of the viscera. (C. Clarke)

only by the distinctive structure and function of its appendages.

Centipedes and millipedes are divided into 2 regions: a compact head and a long trunk. In the insects and in some crustaceans, the segments are grouped into 3 body regions: head, thorax, and abdomen. In the majority of crustaceans, some or all of the thoracic segments are fused with the head, and there result 2 or 3 body regions. In the arachnids also, 2 or 3 body regions are distinguished. The body regions of the various groups of arthropods are composed of different numbers of segments and are not necessarily homologous. The thorax of an insect does not correspond to the thorax of a crustacean. But within each class, the arrangement of regions, and to some extent the number of segments in each, is uniform and characteristic, except among some of the diverse groups of crustaceans. The grouping of segments into discrete body regions (a process called *tagmatization* or *tagmosis*) is one of the features of the arthropod body plan that seems to have evolved several times independently. Further, in contrast to most annelids, most adult arthropods have a fixed number of segments, and the total number of segments is generally much smaller than in annelids. The same is true of vertebrates, which have a fixed number of segments and a fusion of many of them. We may say that, generally, the more primitive members of a group have a large and indefinite number of repeated but similar parts, while the more specialized members of the group have a smaller and definite number of repeated parts with much division of labor among them, or they have the repeated parts fused into compact masses or organs.

Head of cricket. The compound eyes may be less important than the long sensitive antennas in this nocturnal insect. Between the tips of the jointed maxillary palps is the rounded lobe-like upper lip, or labrum, which covers the mouth and mandibles in this front view of the head. Jerusalem or sand cricket, *Stenopalmatus,* of the U.S. Pacific Coast. (E.S. Ross)

The **head** of an arthropod, or the anteriormost part of the first body region, typically bears paired jointed appendages which are sensory or are used in feeding. There are typically 5 pairs of head appendages in crustaceans, 3 or 4 pairs in myriapods, and 4 pairs in insects. First come the feelers, or *antennas,* of which there are 2 pairs in crustaceans, 1 pair in myriapods and insects. These are followed by the jaws, or *mandibles,* which usually serve for biting and chewing. The next 2 segments may each bear a pair of *maxillas,* accessory mouth parts which aid in feeding, particularly in handling the food and in holding it to the mouth. The mandibles and maxillas are often highly modified and one or more pairs may be absent. Arachnids, horseshoe-crabs, and pycnogonids have no antennas or mandibles, and the first pair of appendages are typically pincer-bearing *cheliceras.* The second appendages, called *pedipalps,* are usually sensory and may also be used in feeding, walking, mating, or other activities.

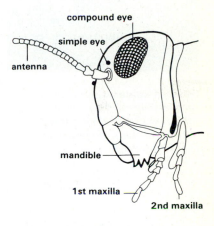

Head of an insect showing characteristic structures. Some insects lack 1 or both pairs of palps. (Modified after R.E. Snodgrass)

Several groupings of arthropods are based on and named for certain head appendages. The crustaceans, myriapods, and insects are sometimes grouped as **mandibulates.** In all insects, and in some myriapods, the second pair of maxillas are modified to form a lower lip, or *labium,* behind the mouth, so these two groups may be classed together as **labiates.** The horseshoe-crabs and arachnids, usually together with the pycnogonids, are designated as **chelicerates.**

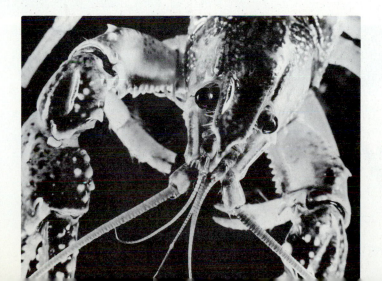

Head of crayfish shows the stalked eyes and jointed antennas, the first pair short and 2-branched, the second pair long and single. Two pairs of antennas are a distinguishing feature of crustaceans. (P.S. Tice)

In annelids, portions of the head (prostomium and sometimes peristomium) and the posterior end (pygidium) are considered not to be segmental. Likewise in arthropods, the portions usually considered non-segmental include the **acron**—the anterior and dorsal part of the head, bearing the eyes—and the **telson**—the posterior end of the abdomen, bearing the anus.

The number of segments, and their arrangement into body regions, can be useful characters for describing and classifying arthropods. In a typical middle-of-the-body segment there is, at least during early development, a one-to-one correspondence of discrete mesodermal and coelomic elements, nerve ganglia, and appendages. Unfortunately, at the head end, where possible homologies between the appendages of different arthropods are especially interesting, this neat correspondence breaks down, and the definition and identification of head segments is especially difficult. After 100 years of detailed morphological and developmental studies, the segmental composition of arthropod heads is still in debate.

Some biologists consider the antennas of myriapods and insects and the first antennas of crustaceans to be segmental appendages; other biologists look on these appendages as part of the non-segmental acron. The presence of an additional head segment without appendages is also disputed.

Because biologists do not completely agree on how to define or count segments, or on whether the segments of different arthropod subphyla correspond in any meaningful way, comparative discussions of segmentation tend to be more confusing than helpful. We will concentrate instead on describing the characteristic appendages in terms of their diverse forms and uses.

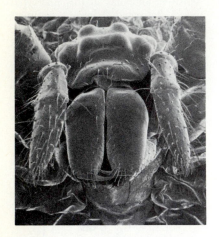

Head of spider with heavy piercing *cheliceras* flanked by bristly sensory *pedipalps.* Female black widow, *Latrodectus mactans.* SEM. (Courtesy, P.B. Armstrong)

Antennas of various arthropods differ not only in structure but also in function. Though most are sensory organs, the second antennas of the spiny lobster *(right)* are heavily armed with spines and used in offense or defense. *Bottom left,* the downwardly directed antennas of the aquatic larva of *Chaoborus* (a tiny mosquitolike but non-biting midge) are used to snatch small prey organisms out of the surrounding water. (P.S. Tice) *Bottom right,* the large second antennas of *Daphnia,* a small crustacean, are used for swimming. The minute first antennas are sensory. (R.B.)

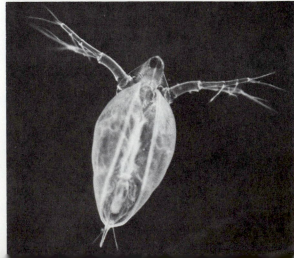

The head end also bears the **eyes,** which share a basic set of functional parts in all arthropods. *Dioptric bodies* transmit and refract light rays and condense them upon light-sensitive cells. The cuticle over the surface of the eye is transparent and usually much thickened to form a lens, and there may be one or more additional refractive structures within the eye. Deeper in the eye is a layer of *light-sensitive cells,* which contain visual pigments similar to our own. From these cells arise nerve fibers that enter the central nervous system. The light-sensitive cells are partially shielded by absorbing or reflecting layers of *screening pigment,* which in some cases can be moved by day or night to alter the amount and direction of the light that impinges on the light-sensitive cells.

Most arthropods have **simple eyes,** which are far from simple but are so called because all of the light-sensitive cells share a single lens. Most crustaceans, horseshoe-crabs, insects, and some centipedes have in addition **compound eyes,** which may be composed of thousands of closely packed units, called **ommatidia,** each with its own dioptric apparatus, bundle of light-sensitive cells, and screen of pigment. Some kinds of simple eyes closely resemble ommatidia, and compound eyes probably evolved from clusters of such simple eyes. Compound eyes of this kind are unique to arthropods, and, except for the "camera" eyes of cephalopod molluscs, are the most highly developed of invertebrate eyes.

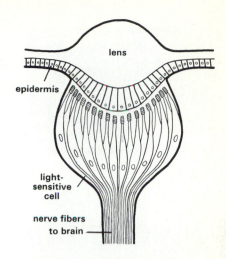

Simple eye of a spider. The light-sensitive cells share a single lens. (After Hentschel)

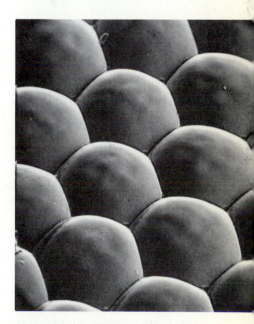

Hexagonal facets in the compound eye of a grasshopper. SEM. (P.P.C. Graziadei)

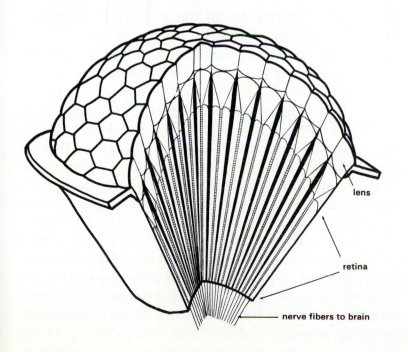

Compound eye of an insect, with a sector cut away, shows the many visual units, each with its own lens. (After R. Hesse)

Vision has been studied mostly in a small number of insects and crustaceans, and even among these, the structures, mechanisms, and capabilities are diverse and impressive. At least some arthropods have *binocular vision,* which depends on the use of both eyes together, as in vertebrates. Some can see light, not only in the *range of wavelengths* visible to us, but also well into the ultraviolet range. Honey bees, for example, are attracted to some flowers primarily by reflected ultraviolet light, and some butterflies recognize each other by patterns on their wings visible only in the ultraviolet range. The spectrum visible to certain butterflies is broader than that known for any other animal. However, the fact that many arthropods can use light over a broad spectral range, and have specialized sets of receptors sensitive to different wavelengths, does not necessarily mean that they perceive colors. *Color vision* is the ability to distinguish wavelength independent of intensity, and this is difficult to demonstrate except in animals that can be trained readily. Probably many arthropods have color vision, but it has been most studied in honey bees, by conditioning the bees to respond to lights of different colors and carefully regulated intensities. A variety of arthropods use the patterns of *polarized light* in the sky to navigate as they walk or fly; some aquatic crustaceans are reported to orient to polarized light underwater.

Two nearly hemispherical eyes give many arthropods a *field of vision* of nearly 360° vertically or horizontally, and small groups of *ommatidia may be specialized* for various visual tasks. For example, polarized light is analyzed by a small group of ommatidia near the top of each compound eye in certain ants; and color and ultraviolet light sensitivities differ over the surface of the eyes in flies. In both crustaceans and insects, small areas of the eye, where a larger number of larger lenses are all pointed in one direction, have more acute vision than the rest of the eye and correspond to the fovea in the center of a human retina; dragonflies and some other insects have more than one fovea, pointed in different directions, as birds do.

Arthropod eyes have *no mechanism to change focus* for near or far vision, and they do not give as sharp an image as a camera eye; perhaps many arthropods see something a little worse than a newspaper photograph as it would look to us under a magnifying glass. With each ommatidium isolated optically from its neighbors by its screen of pigment, only the narrow band of light rays that enters any one ommatidium reaches its light–sensitive cells. These rays correspond to a small portion of the outside scene, and it is thought that the image is thus a **mosaic** composed of as many points of light as there are ommatidia. The more ommatidia, the smaller the angle between them; and the narrower the receptive field of each, the more detailed is the image. The huge eyes of dragonflies have up to about 28,000 ommatidia, but even these must give a coarse pattern compared to that from the more than 6 million cones in a human retina. It has been suggested that for an extremely small eye, the compound design is more acute than the camera type. However, in all animals the qualities of the visual information depend not only on the optics but ultimately on the manner of processing by the nervous system.

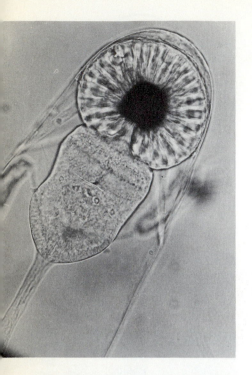

Fused compound eye of a small freshwater crustacean, *Leptodora kindti,* has 300 ommatidia, about 10 times the number in the fused compound eye of the water-flea, *Daphnia,* a smaller and more familiar relative. *Leptodora* is a giant (18 mm) among members of its group (order Cladocera, see chap. 21). Most cladocerans filter-feed on microscopic organisms, but *Leptodora* uses modified legs to seize prey, a process in which vision is important. In this closeup of the anterior part of the elongate, transparent head one can see, between eye and body wall, the pairs of muscles that move the eye about continually. Immediately behind the eye are the optic lobe and the brain, which sends 2 long connectives toward the mouth area. Barely visible are the fine nerves that go to the eye muscles and to the short, stout first antennas of this female. Long bristles on the antennal tips bear chemoreceptors. Males of *Leptodora* have long, slender first antennas that presumably help in finding females. (R.B.)

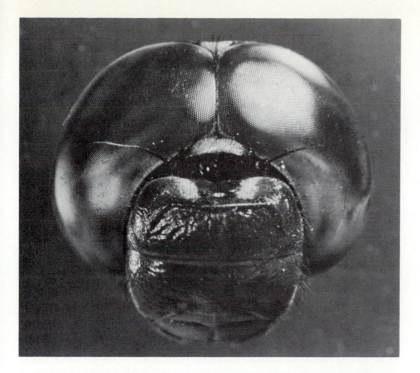

Compound eyes of a dragonfly may cover most of the head, providing a wide field of vision. They can respond to sudden movement 12 m away and play a dominant role in feeding and in finding mates. In some species the number of ommatidia may reach 28,000 or more. Correlated with this high development of the eyes is the minute size and minor role of the antennas, one under each eye. (R.B.)

The numbers of ommatidia in other insects are fewer than in dragonflies, but moths and butterflies have 10,000 to 17,000 and flies have about 4,000. In social insects numbers may vary not only with species but also with caste: 1 to 600 in workers, 200 to 800 in winged females, and 400 to 1,200 in males. When ommatidia are numerous they are crowded and hexagonal in shape. When few, they are separated by areas of cuticle and round in shape.

For example, *interactions between ommatidia* may greatly enhance contrast, and the mechanisms involved have been extensively studied in the eyes of horseshoe-crabs *(Limulus)*, which are perhaps the best understood of any animal eyes (see chapter 22). In any case, arthropods react not so much to details in an image (as we do) as to *motion*. The perception of motion depends not only on the moving object successively stimulating adjacent ommatidia, but also on the neural connections between the ommatidia. The compound eye is admirably adapted for detecting the slightest movement of prey, predator, or mate, and indeed stationary objects may sometimes be seen better when an arthropod is moving than when it is still. Arthropods with simple eyes, as well as those with compound ones, may be observed moving their heads from side to side, "scanning" a scene. This behavior enables a caterpillar, using only a single simple eye, to distinguish at least some details of form. If presented with two upright sticks, the caterpillar will climb the taller one. Held stationary, a simple eye can perceive only light and dark. Fast-flying insects such as bees and flies can distinguish up to 300 separate flashes per second. At the same rate of flashing, human eyes perceive continuous illumination, being unable to separate more than about 50 flashes per second.

In each ommatidium of a compound eye, the dioptric apparatus typically includes a *cuticular lens* and *crystalline cone*. The cone and the light-sensitive *retinula cells* are surrounded by cells containing *screening pigments* which

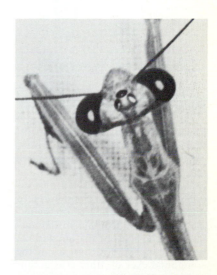

Three ocelli are conspicuous atop the head of this mantis. (R.B.)

absorb or reflect oblique rays of light. Projections (microvilli) from each of the 8 or so retinula cells are directed toward the central axis of the ommatidium and together form the *rhabdom,* which contains the *visual pigment.* Only light absorbed by visual pigment in the rhabdom can serve as a visual stimulus, causing impulses to be generated and carried along the *nerve fibers* from the ommatidium to the optic centers of the central nervous system.

The structure and optics of the compound eyes of arthropods are as diverse as the habits and habitats. Eyes with ommatidia in which the crystalline cone is directly apposed to the rhabdom (called apposition eyes) are characteristic of arthropods active in bright light: insects such as honey bees, flies, and butterflies, and crustaceans such as hermit-crabs and terrestrial or intertidal isopods. The rhabdom is of higher refractive index than the surrounding cytoplasm and acts as a light-guide. Eyes with ommatidia in which the crystalline cone is at some distance from the rhabdom (called clear-zone eyes) are characteristic of arthropods active in dim light; cone and rhabdom are usually connected by a glassy, threadlike or tapering structure, the crystalline tract. Both crystalline tract and rhabdom act as light-guides. In both types of eyes, *in bright light,* each ommatidium is usually isolated from its neighbors by screening pigment. In addition, pigment granules may lie against the rhabdom within the retinula cells, absorbing some of the light that passes along the rhabdom and so partly protecting the visual pigments against continual bleaching. When light-adapted, the eye is generally at its most acute and least sensitive. *In dim light,* light-sensitivity is increased by a rapid accumulation of visual pigment and by movements of pigments and cells. In apposition eyes, the ommatidia remain optically isolated, but pigment granules adjacent to the rhabdom migrate peripherally and are replaced by vesicles of low refractive index; this increases the efficiency of the rhabdom as a light-guide. In some insects with apposition eyes, the receptive field of each ommatidium increases in dim light; this increases the sensitivity of the eye but decreases the acuity of vision. In clear-zone eyes, in dim light, the screening pigment migrates to the ends of the ommatidia leaving the sides of the crystalline tracts and rhabdoms exposed. This improves the efficiency of both crystalline tract and rhabdom as light-guides and also permits any escaping light to be absorbed by adjacent rhabdoms. Thus the animal can see in dimmer light, but the fields of adjacent rhabdoms overlap somewhat, and the resulting image is not as sharp.

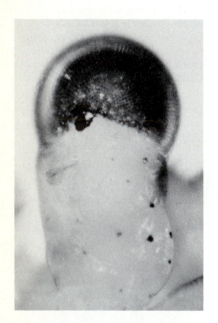

Eye of a shrimp, *Palaemonetes.* In this and some other decapod crustaceans, the ommatidia have *square facets* instead of hexagonal ones and *crystalline pyramids* instead of crystalline cones (see also chapter 21). In these eyes, light is focused on the retina by reflection from mirrorlike surfaces on the sides of the ommatidia, rather than by refraction as in the compound eyes of other crustaceans and insects. (R.B.)

Diagrammatic sections through an **ommatidium,** the
unit of a compound eye. This ommatidium, from the eye
of a honey bee, is of the apposition type, characteristic of
diurnal insects and crustaceans. The *cuticular lens,* visible from the surface of the eye as a hexagonal facet, is not
optically homogeneous but includes 3 layers of different
refractive index. It is secreted by a pair of *inner pigment
cells.* These surround the *crystalline cone,* which is made
up of 4 cells that fit tightly together; each contains a small
nucleus and contributes one quadrant of the cone. From
the cone cells extend 4 slender processes, the *cone extensions,* which descend between the light-sensitive cells and
finally expand to form a 4-parted *basal pigment body.*
Radially-arranged, light-sensitive *retinula cells* bear tightly-
packed microvilli that together form the *rhabdom,* which
contains the visual pigment. The microvilli of the different retinula cells within an ommatidium contain different
visual pigments and have different spectral sensitivities,
enabling the bee to distinguish colors. Each retinula cell
continues basally as a *nerve fiber.* A sheath of many *outer
pigment cells* surrounds and isolates each ommatidium
from its neighbors. (Based on Varela and Porter, on
Gribakin, and others)

One mechanism of light/dark adaptation is represented in this diagrammatic cross-section through an
ommatidium of a honey bee. In *bright light* (left side of
figure), pigment granules of high refractive index closely
surround the rhabdom and absorb some light from it.
Cisterns of the endoplasmic reticulum are small and
dispersed. In *dim light* (right side of figure), large cisterns
of low refractive index surround the rhabdom, increasing
its retention of light. The pigment granules are dispersed
peripherally. This section represents a level, between the
tip of the crystalline cone and the retinular nuclei, where
retinular pigment granules are most abundant. (Based on
Kolb and Autrum, and others)

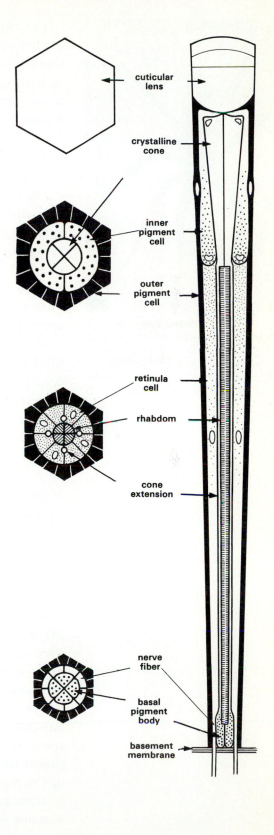

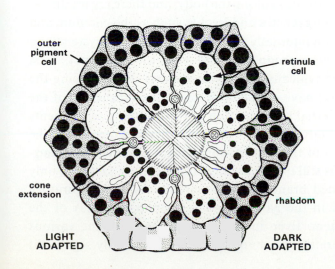

Four pairs of walking legs, typical of arachnids, are seen here in a spider, in which the legs articulate with the muscular part of the cephalothorax (the combined head and thorax). (R.B.)

The **thoracic** or **middle segments** of arthropods most often bear appendages used in walking. In crustaceans the thoracic segments may be numerous, and the appendages serve in feeding and in walking or swimming, as well as having sensory and respiratory components. Arachnids typically have 4 pairs of walking legs. In insects the thorax is composed of 3 segments, each of which bears ventrally a pair of legs. Dorsally there are typically two pairs of wings, borne on the second and third segments.

The **abdominal** or **posterior segments** may or may not have appendages. In many crustaceans the abdomen lacks appendages, but in some, such as lobsters, there is a pair of appendages on every abdominal segment. In arachnids and insects there are practically no abdominal appendages homologous with the appendages of other segments. Familiar exceptions are silk-spinning structures of spiders and reproductive appendages on the most posterior segments of insects.

Internally, the abdomen of arachnids and insects is generally filled with the soft organs of the digestive and reproductive systems, while the middle segments hold most of the muscles associated with the walking legs (and in insects, the wings). By contrast, the abdomen of a lobster is nearly filled with the powerful (and delicious) flexor muscles, and the abdomen of a crab is reduced to almost nothing; in these and many other crustaceans the soft visceral organs are in the thoracic region of the animal. Some crustaceans have viscera in the abdomen.

THE SEGMENTAL ARRANGEMENT OF PARTS that we saw in the annelids, both internally and externally, showed only the beginnings of fusion and specialization at the head end. Some of the tube-dwelling polychetes show a degree of external specialization along the body, and there are modifications among oligochetes and leeches such as the clitellum and the suckers. But internally, except for the digestive system (and sometimes the reproductive and excretory systems), each organ system typically has repeated parts. In discussing the arthropods we have already referred to modifications of the primitive external segmentation in the grouping of segments into distinct body regions, the fusion of head segments, and the specialization of the various appendages. The **internal segmentation** is still more modified and is clearly apparent only in the repeated branches of the circulatory and respiratory systems and in the ganglia and segmental branches of the nervous system. Even in primitive members of the various groups, where these systems have many regularly segmental elements, the segments are not as isolated or as independent as in annelids; and in more specialized arthropods the segmental elements are often reduced in number by elimination or by fusion until segmentation is all but obscured. This reduction of parts seems to have taken place independently in several arthropod groups.

The **nervous system** consists basically of a pair of anterior dorsal ganglia, usually fused into a single "brain" which connects, by a ring of nervous tissue encircling the digestive tract, to the first pair of ganglia of the two ventral nerve cords. In primitive arthropods this system can hardly be distinguished from that of some annelids (though in most annelids the paired ganglia and nerve cords are fused). In more specialized arthropods the paired ganglia of each segment are usually fused, and there are all stages of condensation of the ganglia, reaching a peak in animals such as mites which have all the ganglia fused into one large mass.

The arthropod brain is a composite structure. It includes a number of discrete lobes associated with the various sensory structures, as well as segmental ganglia that fuse with the nonsegmental portion during development. The first ventral ganglion, likewise, almost always represents a number of fused pairs of segmental ganglia.

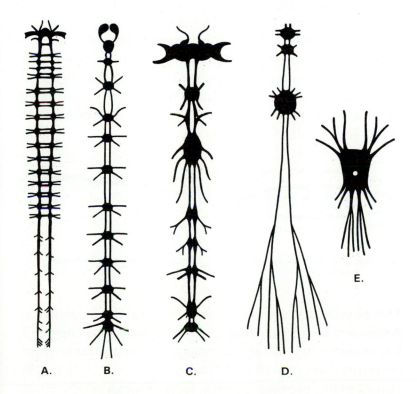

Nervous systems of several arthropods, showing varying degrees of fusion of the ganglia. Within each subphylum, and within the larger classes as well, a spectrum of forms is found, from relatively unspecialized types with many ganglia distributed along the ventral nerve cords to more specialized types with most or all of the ganglia fused into a single mass. **A.** Primitive crustacean. **B.** Caterpillar. **C.** Honey bee. **D.** Water bug. **E.** Mite. (**E.** After W.W. Moss)

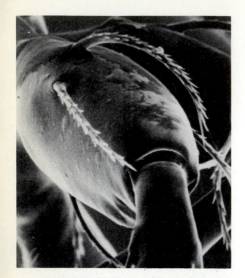

Sensory bristles of arthropods are hollow projections of the body wall, each with a sensory cell at its base. Shown here is a small portion of the leg of a mite, *Oppia coloradensis.* The bristles are about 25 μm long. SEM (T.A. Woolley)

One might suppose that an animal that lives encased in a heavy cuticle would be handicapped in establishing connections between the central nervous system and the external environment. On the contrary, the cellular and cuticular layers of the body wall of arthropods have been modified to form highly specialized **sense organs** of great variety. The *eyes,* sensitive to light, and the *antennas,* sensitive to touch and to chemical stimuli, have already been mentioned. Some arthropods also have *statocysts,* pits containing hard particles and lined with sensory cells, and *auditory organs* which have a flexible membrane stretched across an opening in the hard cuticle. In addition, the surface of the body is covered with a variety of *sensory projections and pits.* The simplest type is a bristle formed by a hollow outgrowth of the cuticle and connected with a sensory cell at its base. Any mechanical stimulus that moves the bristle sets up an impulse in the sensory cell with which it connects. Certain small and slender bristles which are not movable and have thin and permeable walls are thought to be among the receptors of chemical stimuli.

The **coelom** of arthropods is extremely limited. It appears in the embryo as a series of clefts in the mesoderm, and in some groups is represented in the adult by the ducts or cavities of the gonads or by the excretory organs. The main body cavity of the adult is not a coelom but a large blood space, or **hemocoel,** which forms part of the circulatory system.

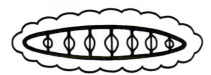 

Closed circulatory system of an annelid.

Open circulatory system of an insect

The reduced circulatory system of insects, as suggested in simplified form in the diagram, is an extreme case among arthropods. Insects, and some other terrestrial arthropods, have a system of air tubes that deliver oxygen directly to the tissues and carry carbon dioxide away; so the respiratory function of the blood is less important. In arthropods with localized respiratory organs, such as aquatic crustaceans with gills, the system of blood vessels is much more extensive, with branching arteries leading from the heart to main regions of the body.

The **circulatory system** is an *open* one, as in molluscs. Arthropod hearts vary from small rounded sacs occupying a few segments to long tubular structures with a linear series of chambers that run most of the length of the body. All have the characteristic openings in their sides through which blood returns to the heart from the sinus in which it lies. The heart is suspended by elastic ligaments from the dorsal body wall and attached underneath, by ligaments, to the horizontal sheet of tissue that partially separates the pericardial sinus from the rest of the hemocoel. When the heart contracts, the elastic ligaments are stretched and the flaps of the heart valves that guard the openings in the heart walls are brought together, so that blood cannot leave the heart except through one or more arteries. No matter how extensively these may or may not branch and rebranch, they do not end in capillary beds that connect with veins. Instead they empty into spaces among the tissues, directly bathing the cells with blood. These spaces drain into a series of large connecting sinuses. The flow of blood lacks the direction imparted by the blood vessels of a closed circulatory system, but the blood is shunted into channels by a series of membranes. Sometimes these provide entirely separate channels for the arterial flow and the return flow to the pericardial sinus. When the heart relaxes, the elastic ligaments shorten, pulling apart the heart walls, opening the valves, and admitting blood into the heart.

The circulating fluid that fills the hemocoel of arthropods is called **blood** or **hemolymph.** The latter term is a reminder that this fluid combines most of the functions served separately in many groups by blood in a closed vascular system and by interstitial lymph that comes into direct contact with the cells of all the tissues. In the serum of arthropod blood are nucleated cells of several types. A dissolved respiratory pigment may also be present: hemocyanin in some crustaceans (malacostracans), horseshoe-crabs, and arachnids—and hemoglobin in some other crustaceans and a few aquatic and parasitic insects. Terrestrial insects and myriapods, in which respiratory exchange is served by a separate system of air tubes, have no special oxygen-carrying pigment in the blood.

The blood of marine arthropods is very much like that of other marine invertebrates. The osmotic pressure and the relative proportions of sodium and chloride ions are close to that of seawater. In freshwater and terrestrial crustaceans, arachnids, myriapods, and the more primitive insects, the osmotic pressure of the blood is lower, but still depends primarily on high and similar proportions of sodium and chloride ions. However, in many insects, especially in the most specialized orders, the osmotic role of the inorganic ions is partially replaced by small organic molecules such as amino acids. In the saturnid moths, for example, amino acids contribute 40% of the osmotic pressure as compared with only 1% or less in most invertebrates and vertebrates.

The structures concerned with **respiratory exchange** demonstrate again the great adaptability of the arthropod body wall. Most aquatic arthropods have *gills,* thin-walled extensions of the body wall through which carbon dioxide and oxygen pass readily. In terrestrial arthropods respiration is usually served by a system of air-filled *tracheal tubes* or *book lungs,* formed by tubular or sheetlike ingrowths of the ectoderm and lined with cuticle. Air enters or leaves through small

Although the circulatory systems of molluscs and arthropods are both described as *open,* they are not the same. In both phyla, blood leaves the heart through arteries and then flows through unlined hemocoelic sinuses that permeate the tissues. But in arthropods, blood finally returns to the pericardial cavity, a hemocoelic sinus surrounding the heart, and enters the heart through valved openings in the heart walls. In molluscs, blood returns through veins that enter the heart; the pericardial cavity is a coelomic cavity and is not part of the circulatory system.

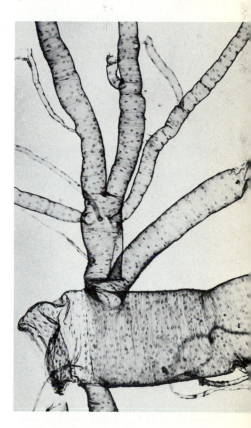

Tracheal tubes of insects are strengthened by a cuticular lining. This preparation shows only the larger tubes. Photomicrograph. (A.C. Lonert)

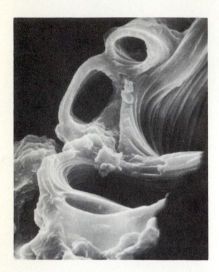

Cut-open tracheal tubes of a mite, *Ixodes ochotonae,* show the spiral cuticular thickenings that strengthen the walls of these fine air passageways. SEM (T.A. Woolley)

Two copepods. The larger one, *Centropages,* is just under 2 mm, a common size for copepods. The smaller one shown here, *Acartia,* is 1.4 mm long. Others may range from 0.3 to 10 mm or more. England. (D.P. Wilson)

Parasitic copepods span a greater size range; dwarf males of some species are only 0.1 mm long, while large females (*Pennella balaenopterae*) parasitic on whales reach 32 cm in length.

openings and, in insects, myriapods, and some arachnids, is piped directly to the tissues, partly or almost completely replacing the respiratory functions of the circulatory system.

Ectoderm and cuticle also line the anterior and posterior regions of the **digestive system**. The cuticle of the foregut may be produced into hard teeth that grind up the food. Large digestive glands connect with the midgut in most crustaceans and chelicerates; myriapods and insects generally lack such glands. In many arthropods, cells of the gut perform **excretory functions,** and the anus thus serves as the exit for excess salts and nitrogenous wastes, as well as for digestive residues. In most terrestrial arthropods, the principal excretory organs are **Malpighian tubules;** these lie in blood spaces and secrete fluid into the hindgut, where controlled reabsorption of some of the water, salts, and other materials takes place, the remainder being eliminated. Excess nitrogen is excreted mostly in the form of insoluble compounds; thus its excretion involves little water loss. However, this means of saving water can be expensive, expecially for an animal with a high protein diet, as considerable amounts of energy and carbon are required to produce the excretory compounds. A blood-sucking tsetse fly must sacrifice about half the dry weight of its food in the process of excreting the excess nitrogen. Aquatic insects, with no water conservation problems, simply excrete the nitrogen as ammonia. In myriapods and insects, the Malpighian tubules are usually derived from the ectodermal hindgut and they excrete mostly uric acid; in arachnids the tubules are derived from the endodermal midgut and the principal nitrogenous excretory product is guanine. Except in winged insects, the excretory functions of the Malpighian tubules and gut are often supplemented by excretory organs of coelomic origin. In crustaceans coelomic organs are the main excretory structures and Malpighian tubules are absent; nitrogen in the form of ammonia also diffuses out freely though the surfaces of the gills.

THE RANGE IN SIZE of aquatic arthropods is from minute crustaceans like most copepods to lobsters that may weigh 20 kg and measure over 60 cm long. Giant spider crabs measure about 45 cm across the body, and their spindly outstretched claws may span 3.6 m. Such large size is practical in the ocean, where the water supports most of the weight of the animal. It is more startling to see big coconut-crabs, up to 32 cm long and with heavy powerful claws, walking over the land and climbing tall palm trees. But in general terrestrial arthropods are limited

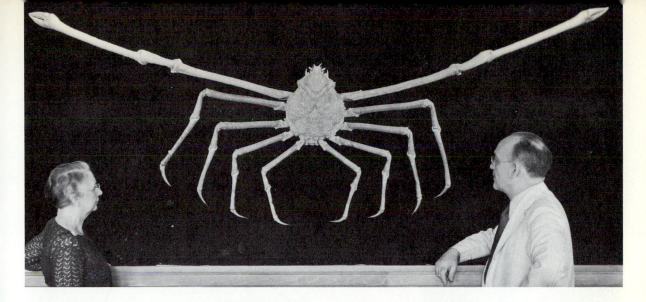

Giant crab, *Macrocheira,* from the waters off Japan. (Buffalo Museum of Science)

to a relatively **small size,** which is not without its compensations. Small size, combined with great development of the muscles, makes for active habits and easy escape from predators.

Almost any small crevice will afford shelter, and small arthropods (like small molluscs) can occupy the innumerable ecological niches on land or sea that are not available to larger animals. In addition, small size requires relatively little growth, and many forms develop from the egg to the sexually mature adult in a few days or weeks. Such a **short life-cycle** permits many generations in a year; and this means that such animals may undergo rapid evolution, which explains, in part, the great numbers of species of terrestrial arthropods, particularly insects and mites.

The eggs of many arthropods (some freshwater crustaceans, arachnids, centipedes and millipedes, and some insects) undergo direct **development,** and the young hatch out looking like miniature versions of the adults. On the other hand, most marine crustaceans hatch from the egg as free-swimming larvas, which must undergo stepwise changes, with each molt, to attain the adult form. In many insects, the larvas (caterpillars, grubs, etc.) are so different from the adults in both form and habit that it is not possible to have a gradual stepwise metamorphosis. The larva surrounds itself with some kind of protective material and becomes transformed into a *pupa,* an externally quiescent stage. Internally, however, active developmental changes produce the sexually mature adult which finally emerges. Winged insects and most arachnids do not molt after the adult stage is reached, but many crustaceans continue to grow and molt throughout life.

Largest terrestrial arthropod is the coconut-crab, *Birgus latro.* Coconut-crabs feed on fruits, carrion, and on crabs. On uninhabited islands in the western Pacific, they are active in the daytime, but where humans make their presence felt, coconut-crabs are nocturnal. The adult has a thick exoskeleton, reduced gills, and a heavily vascularized gill chamber, and drinks fresh-water. These features allow it to range several kilometers inland. It returns to seawater only to release its eggs, already containing developing young, which hatch, feed, and grow before emerging to live on land. Young coconut-crabs reveal their identity as hermit-crabs by taking up a snail-shell house, which they carry about and use as a retreat (see chap. 21). When larger, they abandon the shell. (K. Gillett)

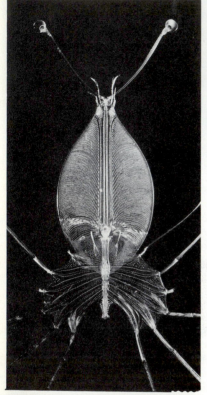

Adult spiny lobster *(above)* produces eggs which hatch into a **free-swimming larva** *(left)* which in turn gradually changes with successive molts into the adult. (Adult, Shedd Aquarium; larva, W. Beebe)

Adult house flies laying **eggs** in rotting organic matter.

Larvas, called maggots, grow rapidly, storing food.

Pupas, in which radical changes result in the development of the winged adult. (A.C. Smith)

In many insects which undergo metamorphosis there is a marked **division of labor among the different phases of the life-history.** In butterflies, for example, the caterpillar larva has chewing jaws and feeds on leaves, consuming large quantities of food in a relatively short time and growing rapidly. Thus, its role in the life-history is *feeding* and *growth*. The pupa undergoes profound changes in structure, and its function may be said to be that of *transformation*. The adult is a winged form which has no chewing jaws and feeds by extracting nectar from flowers with a long sucking tube. Its ability to fly makes it important in *dispersal,* but its chief role is that of *reproduction,* for it does not grow and much of its food goes to produce the gametes. Some adult insects, such as the mayflies, never feed at all but mate soon after emerging from their pupal cases; the females lay eggs and the adults die. Among crustaceans, some of these roles are switched around, with the short-lived larva being the chief agent of dispersal and with growth taking place mostly during the long adult phase.

AMONG DIFFERENT ANIMAL GROUPS we have seen more and more complex *levels of structural organization*. We saw, for example, specialization within single cells, among cells, among tissues, and among organs with the protozoans, sponges, cnidarians, and flatworms, respectively. Repeated segmental organization of organ systems is well represented in annelids, and now in arthropods we see elaborate specialization of different segments and their associated appendages. With such complex specialization, arthropods are often viewed as a major peak of invertebrate evolution.

That we can follow increasingly complex levels of organization through different animal groups does not mean, however, that these groups evolved in sequence, one form from the other. Sponges, for example, may best be considered as highly specialized animals organized to filter tiny food particles from water, and with little or no relationship to any other animal group. Similarly, cnidarians, with their unique nematocysts, are highly specialized, and only remotely related to other animals. The body plans of sponges and cnidarians, however, provide us with living models of increasing organizational complexity among animals, and help us think about the course of evolution of complex animals long ago.

Most biologists agree that arthropods arose from annelid-like ancestors, that is, animals with segmental organization. Similarities in the development and organization of many structures in annelids and arthropods suggest these animals share common ancestry. Biologists disagree, however, as to whether arthropods arose only once from a single annelid-like animal, or two or more times from several remotely related annelid-like stocks. Chelicerates, with the predaceous chelas or fangs of their first pair of jointed appendages, seem fundamentally different from other arthropods, which generally have sensory antennas as their first pair of jointed appendages, and use mandibles for feeding. Moreover, the mandibles of crustaceans develop from the first article of one of the jointed appendages, while those of myriapods and insects develop by a shortening and fusion of a whole jointed appendage. In addition, most crustacean appendages are divided, or biramous, while myriapod and insect appendages are not so divided; instead they are described as uniramous. Many biologists see these differences as so fundamental as to rule out any single common ancestral stock for the different arthropod groups. They find it easier to believe that the Uniramia (myriapods and insects), Chelicerata, and Crustacea may have been derived from separate annelid-like stocks, and that they may be no more closely related to each other than they are to living annelids. Such reasoning has led some to consider uniramians, chelicerates, and crustaceans as separate phyla. Perhaps more detailed information about biochemistry and development will eventually help to resolve these phylogenetic questions.

The term Uniramia was first introduced as the name of a phylum including onychophorans as well as myriapods and insects. The term was chosen to distinguish these animals with unbranched (uniramous) appendages from crustaceans and from chelicerates (and from the extinct trilobites), groups with appendages considered to be originally two-branched (biramous). We use "Uniramia" in a restricted sense in the classification presented below, which represents arthropods as a single phylum. More detailed classifications of the arthropod groups follow each of the next 4 chapters.

CLASSIFICATION: Phylum ARTHROPODA

Subphylum CRUSTACEA

Class Remipedia
Class Cephalocarida
Class Branchiopoda, fairy-shrimps, etc.
Class Malacostraca, lobsters, crabs, etc.
Class Maxillopoda
  Subclass Mystacocarida
  Subclass Tantulocarida
  Subclass Copepoda
  Subclass Branchiura, fish-lice, pentastomids
  Subclass Cirripedia, barnacles, etc.
Class Ostracoda, seed-shrimps

Subphylum CHELICERATA

Class Merostomata, horseshoe-crabs
Class Arachnida, scorpions, spiders, etc.
Class Pycnogonida, sea-spiders

Subphylum UNIRAMIA

Superclass MYRIAPODA

Class Chilopoda, centipedes
Class Diplopoda, millipedes
Class Symphyla
Class Pauropoda

Superclass INSECTA (HEXAPODA)

Class Protura
Class Collembola, springtails
Class Diplura
Class Thysanura, silverfishes, etc.
Class Pterygota, winged insects
  Subclass Exopterygota
  Subclass Endopterygota

Chapter **21**

Crustaceans

THE NAME Crustacea was originally used to designate any animal having a hard but flexible "crust," as contrasted with one having a shell like that of a clam or oyster. Since nearly all arthropods have a hard, flexible exoskeleton, we now use more distinctive criteria for assigning an animal to the subphylum **CRUSTACEA.** Most crustaceans are aquatic arthropods with gills, mandibles, 2 pairs of antennas, and 2 compound eyes, typically stalked. The familiar lobsters and crabs with heavy calcareous armor are giants among crustaceans; the great majority are small animals, less than a few centimeters long, with thin noncalcified exoskeletons. Among the roughly 35,000 species, there are many common freshwater and even terrestrial kinds, but crustaceans are predominantly marine and are so abundant in the oceans that they have been called "the insects of the sea." There is hardly any way of life that is not followed by some member of this varied group.

The earliest crustaceans probably had along the whole length of the segmented body a series of similar appendages, each of which served in locomotion, food collection, respiration, and sensory functions. Various conditions approaching this are found in several classes of living crustaceans that have, in addition to some specialized appendages, a series of similar, multifunctional ones. By comparing such relatively primitive crustaceans to a type such as a lobster, in which the appendages have become quite diversified, and by tracing the development of the corresponding parts of each, we are better able to understand how a generalized, multipurpose appendage can, through gradual changes, become a chewing mandible or a crushing pincer.

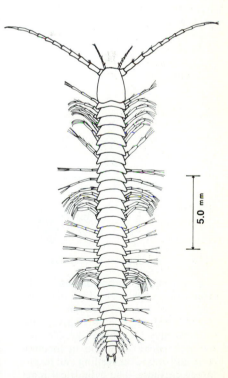

Series of similar appendages occurring along the entire trunk is a primitive character of *Speleonectes* (class Remipedia), a small marine cave crustacean. (J. Yager)

481

482

Lobster at the entrance to its shelter in an aquarium. *Homarus gammarus.* Roscoff, France. (R.B.)

This description of a lobster is based mostly on the American lobster, *Homarus americanus* (class Malacostraca, order Decapoda, infraorder Astacidea), found along the east coast of North America from Labrador to North Carolina. Aside from minor details, lobsters are so much like freshwater astacidean crayfishes that the description of a lobster applies, in general, to both kinds of animals. More differences will be found in applying this account to spiny lobsters and other palinurans (see classification).

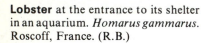

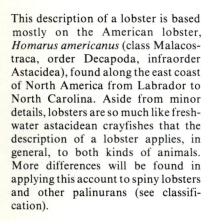

Multipurpose appendages of the trunk of a cephalocarid (in diagrammatic cross-section) give some idea of what appendages of primitive crustaceans may have been like. The flat outer branches function in locomotion and create the feeding and respiratory currents. The cylindrical inner branches are ambulatory. The many projecting bristles are sensory. The stoutly bristled apposed surfaces of the appendage bases (gnathobases) manipulate food. (After H.L. Sanders)

LOBSTERS

THE BODY of a lobster consists of a large **cephalothorax,** which represents the combined head and thorax, and an **abdomen.** A single large shield, the **carapace,** grows out from the last head segment, spreading posteriorly until it extends the length of the cephalothorax and completely covers the dorsal surface and sides. The carapace fuses dorsally with the segments but remains free laterally, leaving the segmentation visible on the ventral surface. The segments of the abdomen are clearly marked, both dorsally and ventrally.

The **cuticle,** secreted by the underlying epidermis, covers every part of the body, forming a jointed exoskeleton which is hardened by sclerotization of proteins and by **calcification,** infiltration with calcium carbonate and calcium phosphate. Calcification of the cuticle of the carapace results in a rigid protective shell over the cephalothorax; but in the abdomen, the calcified cuticle is interrupted by folded, soft cuticle between segments, allowing for flexibility. The cuticle also furnishes some internal support in the form of thin infoldings (apodemes) that increase the area for the attachment of muscles and protect soft organs.

Each of the **appendages** consists, at least in early stages, of a basal part, or **protopod,** which bears at its free distal end an outer branch, or **exopod,** and an inner branch, or **endopod.** The numbers of articles and joints in these three elements may vary, but the plan of this two-branched (biramous) appendage is the most common and widespread pattern among crustacean appendages. In lobsters the two-branched plan is clearest in the appendages of the abdomen. It is modified in the appendages of the head and thorax by the presence of additional extensions of the basal part or by the loss of the outer branch. In embryonic stages of lobsters, however, the appendages arise as simple two-branched structures. Such corresponding structures with similar development in the segments of an animal are said to be **serially homologous.**

The cephalothorax of a lobster is a continuous unit, but the **head** is usually considered to include an anterior nonsegmental portion (acron), bearing a pair of **compound eyes** on the ends of jointed movable stalks, plus 5 appendage-bearing segments. The eyestalks are not considered to be serially homologous with the other appendages, as they arise in a different way. The first and second head appendages are **2 pairs of antennas,** a character that distinguishes crustaceans from other arthropods (which have only one pair or none). The **first antennas** are shorter and more slender than the second pair and end in 2 delicate sensory filaments. The large **second antennas** have only a single filament, about as long as the lobster's body. The third head appendages are a pair of toothed **mandibles,** which meet in the midline over the mouth and crush the food by moving from side to side (instead of up and down as do vertebrate jaws). Finally, the fourth and fifth head appendages are the **first** and **second maxillas,** which pass food forward to the mouth and help to shred soft food. Each second maxilla bears a thin, lobed plate that beats continuously, driving a respiratory water current over the gills.

The **thorax** has a pair of serially homologous appendages on every segment. The first 3 bear the **first, second,** and **third maxillipeds,** which function as feeding appendages. They are sensory and also serve to handle food, shredding it first and then passing it on toward the mouth. The maxillipeds and other feeding appendages (mandibles and maxillas) manipulate the food by means of **gnathobases,** bristly or toothed processes on the basal (or proximal) part of the appendage.

The appendages of the fourth thoracic segment are the **chelipeds,** so called because they bear the large pincers, or **chelas,** used in both offense and defense. The large chelas are not symmetrical in lobsters over 3 cm long. In a small lobster both of them are slender and have sharp teeth; but as the animal grows and molts, they differentiate. One chela becomes heavier than the other, and its teeth fuse into rounded tubercles; it is used for crushing. The other remains more slender, the teeth become still sharper, and it is used especially for seizing and tearing the prey.

Each of the next 4 segments has a pair of **walking legs.** The first 2 pairs of walking legs end in small chelas, with which they grasp pieces of food and pass them forward to the maxillipeds. Together with the third maxillipeds, these legs are also used to loosen and shovel sediment as the lobster modifies a space among boulders into a secure shelter or excavates a burrow as a home. A lobster in its shelter, with the formidable large

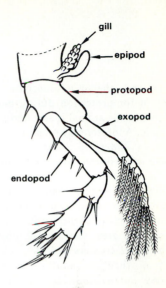

Biramous appendage of the first larval stage of a lobster, the left fourth thoracic appendage. The **protopod** bears a developing gill and epipod (gill separator). The outer branch, or **exopod,** is used in swimming; flattened and fringed with feathery bristles, it beats actively like a little oar. The spiny inner branch, or **endopod,** is used in seizing and handling food. Later the exopod will be lost and the endopod will bear a large chela, which is here just beginning to form at the tip. (After F.H. Herrick)

chelipeds (keé-li-peds)
chelas (keé-las)

In some crustaceans, not all of the thoracic segments are incorporated into the cephalothorax. Those that remain free and distinct are together recognized as a separate body region, called the **pereon;** their appendages are called **pereopods** and are usually locomotory. (Perversely, in lobsters and other decapods, the large chelipeds and legs are called pereopods even though there is no pereon.) The abdomen of crustaceans is sometimes called the **pleon** and its appendages, **pleopods.**

A lobster's shelter is especially important for protection during molting, when the animal is soft and vulnerable. A lobster may prepare for its molt by adding fortifications to its home and will remain within it until the new exoskeleton is hard. When shelters are scarce, competition for them may be keen. One week after artificial cement shelters were placed in a sandy area near a large lobster population, lobsters had moved into 60% of them; a month later and throughout the following year, almost all of the shelters were occupied by 1 or more lobsters.

chelas guarding the entrance, is secure against almost any predator or competitor. All of the walking legs are covered with sensory bristles, and the legs of the last (eighth) thoracic segment bear bristly brushes for cleaning the abdominal appendages. In all the thoracic appendages, except the first and last, the basal piece bears a thin flap (epipod), to which is attached a gill. The flaps separate and protect the gills, and the walking movements of the legs move the gills and stir up the water in the respiratory cavity under the carapace.

The **abdomen** has a pair of appendages on every segment. Those on the first abdominal segment are small and are different in the two sexes. In the male they are modified to form a troughlike structure used for transferring sperms during mating. In the female they are unspecialized. The next 4 segments all bear similar two-branched appendages, the **swimmerets,** which aid in forward locomotion and in the female serve as a place of attachment for the eggs. The sixth and last abdominal appendages are the **uropods,** which resemble modified and enlarged swimmerets. Together with the flattened terminal **telson,** they form a tail-fan, used in backward swimming.

Lobsters furnish a striking example of *specialization among appendages* of different segments and, in the case of the large chelipeds, between the right and left sides of the same segment. While the flattened two-branched swimmerets are not very different from the appendages of a hypothetical crustacean ancestral type, a lobster has at the same time such specialized appendages as the mandibles. In the development of the lobster appendages, we see how a series of originally similar parts can become gradually differentiated into highly specialized and dissimilar structures which, though no longer serving the same functions, are still homologous.

Extensive powers of **regeneration** enable lobsters to replace or repair almost any lost or injured appendage, although the new part may require several molts to reach normal size. If one of the large chelipeds or walking legs of a lobster snaps off, the separation usually occurs at a preformed breakage plane close to the body. At this level, the remaining stump is covered by a strong membrane, perforated only by a narrow passage for nerves and blood vessels. Even this small opening is quickly blocked by a valve, which is closed and held in place by blood pressure within the limb base. Such a break therefore results in little loss of blood and minimal tissue damage. Also, regeneration occurs more readily at or near this level than anywhere else along the limb. Reflex separation of a limb, or **autotomy,**

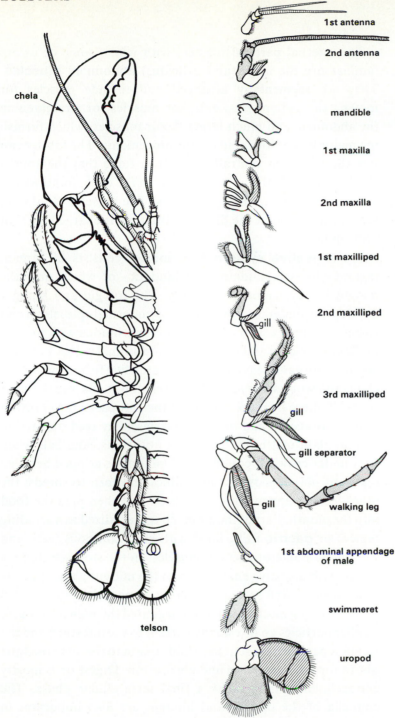

chela

1st antenna

2nd antenna

mandible

1st maxilla

2nd maxilla

1st maxilliped

2nd maxilliped

gill

3rd maxilliped

gill

gill separator

gill

walking leg

1st abdominal appendage
of male

swimmeret

telson

uropod

Appendages of a lobster show a marked division of labor. Some have extra lobes or other processes, and some lack the outer branch; but they develop according to a common basic plan. In the drawing of each appendage the *endopod* (inner branch) is stippled, the *exopod* (outer branch) is shaded with diagonal lines, and the *protopod* (basal part) and its processes are left unshaded. The first antennas develop in a manner different from that of all the other appendages, and their homologies are uncertain.

The first antennas of crustaceans do not develop as biramous appendages. Their development, their mode of innervation from the brain, and their preoral position in the embryo all suggest that they are not serially homologous to the other appendages. (Similar arguments apply to the antennas of myriapods and insects.) In lobster embryos the first antennas remain single until long after the other appendages have become biramous, and at hatching the inner filament is still represented only by a small bud from the base of what finally becomes the outer filament. Thus, whether or not the first antennas are serially homologous with the other appendages, their two-branched condition is not. Two-branched first antennas are unique to lobsters and some other members of the class Malacostraca.

by contraction of muscles in the limb base, always occurs at the preformed breakage plane. A lobster will autotomize one or both of the large chelipeds if they are severely damaged or caught firmly in the grip of a predator, such as a fish or octopus. The lobster may thus escape with its life at the expense of only a minor injury and the temporary deprivation of replaceable parts.

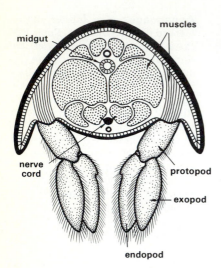

muscles

midgut

nerve cord

protopod

exopod

endopod

Cross-section of the abdomen of a lobster shows mostly muscles, surrounded by blood sinuses. Most prominent are the bulky muscles that flex the tail in the fast backward escape response.

In the various crustacean groups the foregut, midgut, and hindgut do not always form analogous structures. For example, in lobsters and other malacostracans the stomach is ectodermal (foregut), while non-malacostracan crustaceans may have an endodermal (midgut) stomach or no distinct stomach region. Also, ceca of diverse numbers, shapes, and sizes sprout from various parts of the midgut in different groups.

The internal parts of lobsters with which some of us are familiar are the large (and delicious) abdominal **muscles**. They are segmentally arranged and include muscles for moving the swimmerets, extensor muscles for straightening the abdomen, and much larger flexor muscles, which furnish the major power for rapid escape movements. The lobster can flex the abdomen ventrally with such force that the animal shoots backward through the water. In the cephalothorax are muscles for moving the appendages. The muscles of lobsters are striated, even in the digestive tract, as is characteristic of arthropods.

The **digestive system** of a lobster consists of 3 main regions—foregut, midgut, and hindgut—of which only the midgut has an endodermal lining. The foregut and hindgut develop as tubular ingrowths of the ectodermal epithelium and so become lined with a cuticle that is continuous with the rest of the exoskeleton and is shed when the animal molts. Lobsters are predators and scavengers; they catch live fishes and dig for clams and worms; they attack mussels and snails, breaking off the shell, piece by piece, to obtain the soft parts; they prey on other crustaceans and on echinoderms; they feed on carrion and on plant matter; and they are notorious cannibals when kept under crowded conditions. The food is gripped between the third maxillipeds and mandibles and torn to shreds; the other feeding appendages also shred and then pass the food into the mouth. Part of the stomach is specialized as a grinding region, or **gastric mill,** lined with hard chitinous teeth and worked by numerous sets of muscles. In the stomach, food is mixed with digestive enzymes from the midgut gland, pulverized, strained, and sorted. The coarsest particles are returned to the grinding mechanism or regurgitated through the mouth; smaller particles go in a steady current to the intestine; and the smallest particles are sent in a fluid stream to the large **midgut gland** for final digestion and absorption. Digestion is mostly extracellular, but includes a final intracellular phase. The epithelia of the midgut and hindgut are also important in regulating absorption of water and salts.

The anterior portion of the stomach is large and bulbous and serves chiefly for storage. The posterior part is mainly for sorting and straining. The grinding region lies between the two. Lobster relatives (malacostracans) that eat large pieces of food have similar grinding mechanisms in the digestive tract, while other crustaceans that filter feed on microscopic food lack such devices. In either case, the digestion of fine particles and absorption by the midgut glands or ceca decreases the work of the intestine, making it less important than in many other animals. This explains how lobsters and other crustaceans can get along with such a short uncoiled intestine.

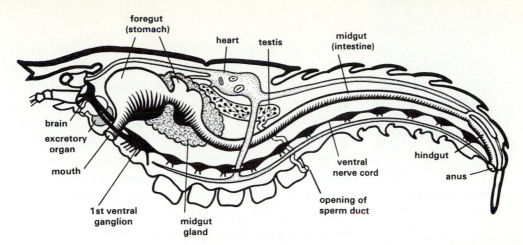

Internal anatomy of a lobster. As in other relatively specialized segmented animals, most of the organ systems consist of a single large organ or pair of organs, instead of small local units repeated in every segment. In a female lobster, the gonad (ovary) is located as in this male, but the oviduct opens at the base of the second walking leg, or sixth thoracic appendage. (Based partly on F.H. Herrick)

The extensive respiratory surface needed to supply the demands of a large and active animal like a lobster is furnished by many **gills,** feathery expansions of the body wall, which are filled with blood channels. The gills are attached to the bases of the legs, to the membranes between the legs and trunk, and to the wall of the thorax. They lie on each side of the body in a cavity enclosed by the curving sides of the carapace. Water enters the cavity under the free edges of the carapace, passes upward and forward over the gills, and is directed out anteriorly in a current maintained by the constant beating of the flattened plates on the second maxillas. Additional respiratory surface is provided by the inside of the carapace.

The important respiratory structures are not always the thoracic gills as they are in lobsters. Gills may be present in other locations, as on the abdominal appendages of isopods; or gills may be absent, respiratory surface being provided by the inner side of the carapace or by other specialized areas. Small crustaceans with thin exoskeletons may have no specialized respiratory structures but effect exchange through the general body surface. In terrestrial and semi-terrestrial crabs and hermit-crabs, the gill chamber is filled with air and the inner carapace wall serves as a lung; terrestrial isopods have evolved a system of air-filled tubes within the abdominal appendages, similar to the tracheal systems of insects and spiders.

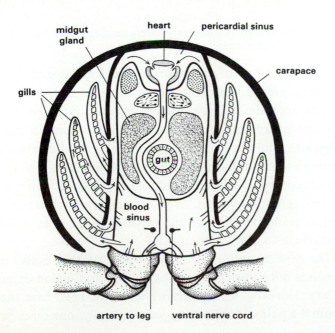

Cross-section of the thorax of a lobster to show the gills beneath the carapace and the path of the blood through some of the main blood channels. The gills are attached close together on or near the leg bases (they do not extend as high on the body wall as they have been drawn here in order to show the circulation). Sandwiched in the carapace is a thin layer of tissue (not shown) which is abundantly supplied with blood and provides additional surface for respiratory exchange. The cuticle is thick on the outer surface of the carapace and thin on the inner (respiratory) surface.

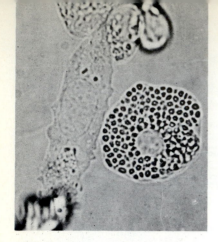

Blood cells from a spiny lobster were kept alive for a considerable time outside the body, at 25°C, using tissue culture methods. The cell on the right was 20 μm in diameter and moved about actively in ameboid fashion. Bermuda. (R.B.)

The heart varies, even within the class Malacostraca, from the compact globular shape seen in lobsters to a tube that runs the length of the body with paired openings to admit blood, and arteries to distribute it, in nearly every segment. Other classes are similarly variable, and many groups lack a heart (for example, cirripeds and many copepods and ostracods). Non-malacostracans usually have few or no arteries. The respiratory pigments are hemocyanin in malacostracans, hemoglobin in non-malacostracans; they are always dissolved in the blood, not in blood cells, although hemoglobin may occur in cells of other tissues.

The **circulatory system** is an open one, as in other arthropods. The muscular heart lies dorsally in a sinus filled with blood (hemolymph). In the sides of the heart are 3 pairs of openings through which blood from the pericardial sinus enters the relaxed heart. When the heart contracts, valves prevent the blood from going out these openings; instead, it is driven into arteries that go to the tissues of the body. The smallest branches of the arteries open, not into capillaries or veins, but into blood sinuses in the tissues. Blood returning from the tissues collects in a large ventral sinus and from there enters the gills, where it gives up carbon dioxide and takes up oxygen. Then it is returned, through a number of channels, to the pericardial sinus. Dissolved in the blood is a copper-containing protein, hemocyanin.

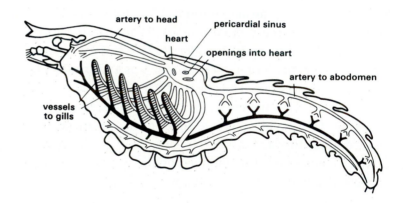

Circulatory system of a lobster, showing the main blood channels. (Modified after C. Gegenbauer)

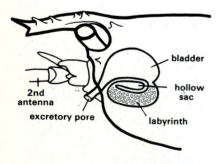

Excretory organ of a lobster. Paired excretory organs occur in different crustaceans either at the bases of the second antennas (antennal glands) or at the bases of the second maxillas (maxillary glands), rarely both.

The single pair of **excretory organs** are sometimes called *antennal glands,* because of their location at the base of the second antennas, or *green glands,* because of their color. Each consists of a hollow sac in which fluid collects, a glandular labyrinth, and a coiled tube leading to a bladder, which opens through a pore at the base of the second antenna. The antennal glands control the volume and ionic composition of the blood and excrete some wastes. Most nitrogenous waste, however, is lost as ammonia through the surfaces of the gills, which also contribute to regulation of salts.

The general pattern of the **nervous system** is like that of annelids but shows more cephalization and fusion of ganglia. A large dorsal ganglion (brain) lies in the head near the eyes. From it a pair of connectives pass ventrally, one on either side of the esophagus, and join a compound ganglion, the first ventral ganglion, from which the double nerve cord extends backward, enlarging into paired ganglia in every segment.

No nervous system is simple. Compared to the 10^{10} to 10^{11} nerve cells in a mammalian nervous system, however, the mere 10^5 to 10^6 nerve cells in a lobster or crayfish do offer a relatively simpler situation in which to study interactions among nerves and between nerves and muscles. Arthropod behavior, although sufficiently complex to be interesting, is somewhat stereotyped, and therefore easier to analyze than the more flexible responses of vertebrates. In addition, the very large nerve and muscle cells of many crustaceans lend themselves to physiological studies; giant fibers in the ventral nerve cord are common, and in giant barnacles *(Balanus nubilus)* and Alaskan king-crabs *(Paralithodes camtschatica)* single muscle fibers may be 2 or 3 mm or more in diameter, compared to a maximum diameter of about 0.1 mm for mammalian muscle fibers. Studies on lobsters, crayfishes, and other crustaceans have indeed helped to clarify aspects of neuromuscular organization that are similar in many animals.

However, crustacean systems have not turned out to be merely simpler versions of mammalian ones. Many features that have been regarded as "normal," from long study of mammalian nerves and muscles, turn out to be unusual compared to those of most animals, including other vertebrates. In a mammalian muscle, each motor nerve cell sends branches to relatively few muscle fibers, and each muscle fiber receives impulses from only a single nerve cell. Inhibition of a muscle fiber occurs through inhibition of its motor cell by another nerve cell, within the central nervous system. The speed of response of the muscle fiber is fixed, and smoothness and flexibility of muscular action depend on the enormous number of nerve and muscle fibers. In a crayfish muscle, a single nerve cell may send branches to all of the fibers in the muscle, and even to more than one muscle. A single muscle fiber may receive branches from up to 4 or 5 nerve cells and respond differently to each. One nerve cell may stimulate a rapid contraction with a single impulse, while many impulses from another nerve cell will produce a slow, facilitated response in the same muscle fiber. A third nerve cell may prevent contraction; thus inhibition takes place peripherally, acting directly on the muscle, as well as in the central nervous system. The nervous organization and the versatile responses of the muscle fibers permit a remarkable degree of flexibility with an extremely small number of nerve and muscle fibers.

The most conspicuous **sensory structures** are the antennas and the compound eyes. There are about 12,000 ommatidia with square corneal lenses in each compound eye of an adult lobster, and they are of the clear-zone type, suited to gather light under dim conditions (see chapter 20). In the eyes of lobsters and other crustaceans with square facets, light is focused on the retina by reflection from mirrorlike surfaces on the sides of the ommatidia, rather than by refraction as in the compound eyes of insects and of crustaceans with hexagonal facets. As lobsters are most active at night, and live at depths where even in the daytime there is not enough light for clear vision, the eyes are probably secondary in importance to the sensory bristles that are distributed all over the surface of the antennas, other appendages, and trunk—from 50,000 to 100,000 bristles occurring on the large chelas and walking legs alone. The various bristles and other sense organs respond to

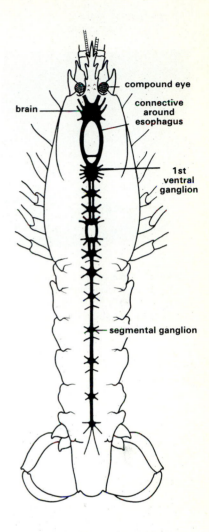

The **nervous system** of a lobster includes a large dorsal ganglion, a ring of tissue around the esophagus, and a ventral, double, ganglionated cord. Lobsters show a moderate amount of condensation and lateral fusion of ganglia; 6 pairs of appendages are innervated from the first ventral ganglion. Crabs have a much more condensed and fused central nervous system, while in fairy-shrimps the paired ganglia of each segment are even more discrete than in some annelids.

Square facets of the compound eye of a lobster, *Homarus americanus*. Each facet is about 60 μm across. SEM. (J.A. Nowell)

Large male lobsters may reach a body length of over 60 cm, and their large pincers may measure almost 40 cm; such an animal, weighing close to 20 kg, is estimated to be over 30 years old. Large females may be as long but have smaller pincers and weigh far less. Differential growth of males and females begins approximately when they become sexually mature.

Various observations and preliminary experiments suggest communication by **sex pheromones** in a number of marine crustaceans. The active substances are usually released from the antennal glands of newly molted females and appear to stimulate the males to search, help them to locate the females, and elicit courtship and mating behavior.

touch, water currents, pressure waves, and diverse chemicals. Long tubular processes abundant on the first antennas detect chemicals in the water; chemoreceptors on the tips of the mouthparts and walking legs respond on contact with food. A lobster is informed about its own position and movements by proprioceptors, located in the soft membranes between the segments of the body and between the articles of the appendages, and by a pair of statocysts in the basal articles of the first antennas. Each statocyst is a liquid-filled sac that opens to the outside by a small pore. On the floor and walls of the sac are rows of fine sensory bristles in contact with a statolith made of numerous tiny sand grains cemented together. Any movement of the lobster bends the bristles (because of the relative inertia of the liquid in the statocyst) or causes the statolith to shift over them, so that the statocyst yields information about the lobster's movements as well as its orientation with respect to gravity. Vibrations in the substrate are also detected.

To study functions of the statocysts in shrimps, one investigator performed a very ingenious experiment. He obtained a shrimp that had just molted and therefore had no sand grains in the statocyst. He put the animal in filtered water and supplied it with iron filings. The shrimp picked up the filings and placed them in the statocyst. Then, when the investigator held a powerful electromagnet above the animal, it turned over on its back.

The **reproductive system** consists of a pair of ovaries or testes, which lie in the dorsal part of the body and from which a pair of ducts lead to the external openings at the bases of the second walking legs (sixth thoracic appendages) in females, fourth walking legs (eighth and last thoracic appendages) in males. Lobsters are unusual among crustaceans in that the female takes the initiative in choosing a mate. About a week before she is ready to molt, she leaves her shelter and moves in with a male, preferably a large one with a stable residence. A lobster in its home shelter is generally aggressive toward an intruder, and this is the male's first reaction; but the female is thought to identify herself as a potential mate by her behavior and by releasing a sex pheromone that inhibits aggression and promotes courtship. Shortly after she has molted and while she is still soft, mating occurs. The male turns the female on her back and deposits the nonflagellated, nonmotile sperms into her sperm receptacle, in the midline just behind the oviduct openings. After mating, the female remains with the male for about a week, and during this time he protects her against predators and drives away other males. When her skeleton has hardened, she leaves and resumes life on her own, but may not spawn her eggs until many months later. Covered with a sticky

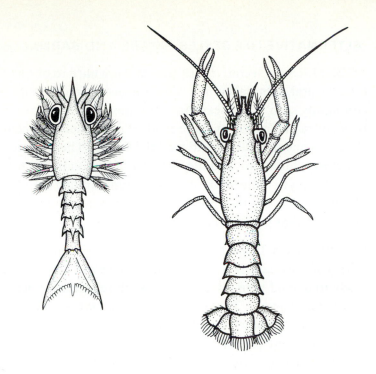

Left, the **first larval stage** of a lobster is about 8 mm long. The appendages are two-branched, similar structures. The swimmerets at this stage are only small buds beneath the cuticle, and the larva swims about at the surface by the rowing action of the flattened, fringed outer branches of the thoracic appendages. *Right,* the **fourth larval stage** of a lobster is about 15 mm long and resembles an adult lobster. Like the first stage, it swims at the surface, feeding on small organisms; but forward swimming is now by means of swimmerets. The statocysts have developed, and locomotion is much improved. The outer branches of the legs are reduced and no longer visible; the inner branches are differentiated, though right and left large pincers are still similar. Mineralization of the skeleton has begun. (After F.H. Herrick)

secretion, the eggs become fastened securely to the female's swimmerets, which keep the embryos well aerated during the 10 to 11 months that she carries them. Hormones coordinate the reproductive and molting cycles, and the female will not molt again until after the eggs hatch.

The young lobster hatches from the egg as a free-swimming **larva** and goes through a series of molts and of intermolt stages, or **instars,** before it comes to resemble the adult. Lobsters continue to molt and grow throughout their lives.

Hatching of the larvas of *Homarus gammarus* occurs shortly after dark. Young of a single female may hatch in batches on successive nights for periods of 2 to 6 weeks. If a female is kept in constant darkness in the laboratory, the nightly hatching rhythm continues, though slightly out of phase. Keeping a female in constant light will delay hatching for 2 or 3 days; and when it occurs, it is erratic. It appears that the onset of darkness synchronizes the hatching rhythm but that there must be a "biological clock" which controls the hatching, in the mother, in the larvas, or in both.

First larval stage of European lobster. *Homarus gammarus.* Length, 1 cm. (D.P. Wilson)

ALTERNATIVE LIFE STYLES: CRABS AND BARNACLES

The great majority of crabs are marine, and unless otherwise designated, all those illustrated in this chapter are marine.

LOBSTERS are secretive creatures and are hard to observe in their natural, subtidal habitat. The crustaceans most easily and commonly seen at the seashore are crabs, skittering over the rocks or racing lightly over the sand, and barnacles, cemented solidly to one spot for the entire adult lifetime. The contrasting structures, life styles, and life histories of crabs and barnacles barely begin to indicate the diversity found among the classes of crustaceans, but they do illustrate two extremes among the many different paths of crustacean evolution.

CRABS, when faced with a human threat, will discreetly retreat if they can. If not, even small ones are astonishingly ready to defend themselves against such relatively enormous creatures as humans. That we should disparage this courageous behavior as "crabby" is perhaps an effort to deny our secret respect for these plucky crustaceans.

In common usage, almost any crustacian is described as some kind of "shrimp" or "crab." However, biologists usually reserve these names for certain decapods. In this book, "crabs" are usually brachyurans ("short tails"). In general, hyphenated names denote animal names that do not conform to this restricted convention, for example, hermit-crabs are not brachyurans (as is the blue crab), but anomurans.

In **structure,** crabs are much like lobsters and comprise a closely related group (class Malacostraca, order Decapoda, infraorder Brachyura). In all decapods the segments and appendages correspond, one to one. But crabs can usually be picked out at a glance by their distinctive shape and their habit of running sideways. The cephalothorax is stout and compact and typically broader than it is long. The small, flat abdomen is visible only ventrally; it turns under and fits into a shallow depression on the underside of the cephalothorax. The antennas are short and the eyes often large.

Dorsal and ventral views of a Dungeness crab, *Cancer magister,* the principal commercial crab of the U.S. West Coast. (R.B.)

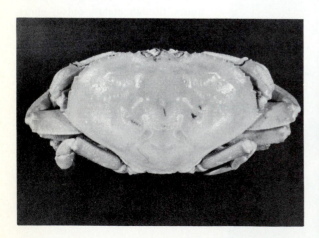
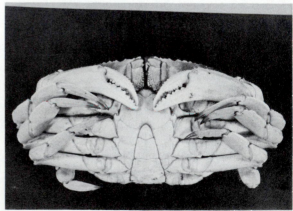

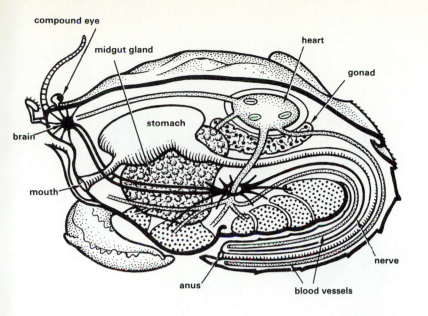

Internal anatomy of a crab is much like that of a lobster or other decapod. The nervous system is highly condensed, with a large compound ganglion in the thorax that represents all the segmental ganglia fused together; it supplies all the appendages from the mandibles through the walking legs. The thorax contains the large muscles of the chelipeds and walking legs. Into the flat abdomen extend only a few nerves and blood vessels and the intestine. (Partly after Pyle and Cronin)

Most crabs live active **life styles,** foraging busily for food while always ready to flee from danger. Their feeding habits tend to be flexible and opportunistic. Of the roughly 4,500 crab species, many are scavengers and omnivores; and even those with primarily predatory or herbivorous tastes, and somewhat specialized feeding equipment, are not above varying their diet. The swift and mostly predatory swimming crabs use their slender but strong chelas, lined with sharp teeth, to seize fishes and squids. Swimmers have relatively light armor, and their last legs are modified as flattened paddles. Slower, heavier species clamber about on rocky subtidal bottoms or burrow in sand or mud. Some have blunt and massive chelas for crushing the armor of other crabs and crustaceans or the heavy shells of clams and snails. Others feed mostly on plants, scraping algal films from rocks or picking off pieces of seaweeds. The many intertidal species have equally diverse habits. Lacking the buoyant support of water, they are usually light in build, fleet, and nimble. Some routinely scale sheer cliff faces during their scavenging forays; others climb 10 m or more into the tops of mangrove trees, where they feed on the vegetation. On sand and mud flats, fiddler crabs emerge from their burrows at low tide and feed by taking shallow scoops of the substrate, extracting the fine particles of organic matter, and rejecting the coarse sediment in the form of pellets left behind in a trail as the crab walks slowly along. In the humid tropics, land crabs manage to live all of their adult lives out of water by foraging at night, mostly on plant material, and hiding during the heat of the day in their burrows or other moist retreats.

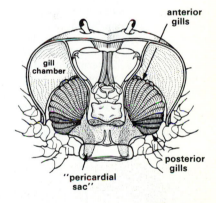

Land crab, *Gecarcinus lateralis,* can absorb salts and water through the gills (shown in dorsal view with top of carapace and midgut gland removed). Up to 87% of the surface of the last 2 pairs of gills represents salt- and water-absorbing tissue. Water is conducted from the substrate, to fine bristles on the underside of the crab, to the convoluted surface of large "pericardial sacs" that extend into the gill chambers, to the posterior gills. The anterior gills and the vascularized walls of the air-filled gill chambers serve more in respiratory exchange. (Modified after D.E. Bliss)

Land crab, *Cardisoma guanhumi,* roams freely on land but has its burrow close to a source of fresh- or seawater. The more terrestrial *Gecarcinus lateralis* need never enter water and can survive only a few hours immersed in freshwater, about a day in seawater. In air, land crabs lose water through the cuticle less rapidly than marine ones but faster than most insects or spiders. Land crabs are able to take up water occurring as dew or as moisture in soil or sand, by capillary forces and by suction; the "bailer" on the second maxilla (which in aquatic crabs pumps water through the gill chamber) pumps air out, creating a partial vacuum that draws water into the gill chamber, to be absorbed by the gills. Land crabs are tropical and subtropical, and feed on plant matter and carrion. Panama. (R.B.)

Freshwater crab, *Pseudothelphusa sonorensi.* Unlike crayfishes, which get rid of excess water by producing copious dilute urine from the excretory organs, freshwater crabs depend on a relatively impermeable cuticle to keep water out. Salt requirements are reduced by keeping blood and scant urine at a salt concentration only half that of marine crabs; salts are absorbed by the gills. Permeable estuarine crabs in low-salinity waters also take up salts through the gills, while increasing their urine output. Most marine crabs have little osmoregulatory ability. Arizona. (M.S. Keasey)

Semi-terrestrial crab, *Ocypode quadrata,* never strays as far from the ocean as do land crabs but lives on sandy beaches in burrows above high-water mark. Members of the genus *Ocypode* are called "ghost crabs" from their pale coloring, nocturnal habits, and the way in which they seem to vanish down their burrows. They are light in build, swift, and among the most active of crabs. Ghost crabs are mostly tropical, but *O. quadrata* has been seen as far north as Rhode Island. 6 cm across. Caribbean. (K.B. Sandved)

Swimming crabs have the last pair of legs modified as flattened paddles. They have been seen far out at sea but typically swim more briefly, spending most of their time on the bottom in coastal waters. Left, *Macropipus depurator,* 5 cm across. England. (D.P. Wilson). Right, *Euphylax dovii,* found from Panama to Chile. (American Museum of Natural History)

Massive chelas are used to crush the thick strong shells of snails and clams. Predatory crabs with crushing chelas probably exerted a major selective pressure in the evolution of mollusc shells with thick walls, structured to resist crushing. As shells became heavier and and stronger, so did crab chelas—an interactive process termed "co-evolution." *Cancer pagurus.* England. (D.P. Wilson)

Burrowing crab, *Corystes cassive-launus,* avoids predatory fishes by burrowing into the sand during the day, with only the tips of the second antennas protruding. The long, stiff antennas have 2 rows of bristles that project inward and interlock when the antennas are closely apposed, forming a tube through which the crab sucks clear respiratory water down to its gill chambers when it lies buried. The flow leaves the gill chambers at the bases of the legs and bubbles up through the sand. This is the reverse of the direction of respiratory flow in other crabs, or in *Corystes* when it is not buried. England. (D.P. Wilson)

Only **reproduction** ties land crabs to the sea. They may travel long distances to the shore, where they release their swimming larvas. These feed and grow in shallow waters until, after several molts, they are ready to emerge onto the land as juvenile crabs. Freshwater crabs, on the other hand, are independent of the sea; they pass through a smaller number of brief larval stages in freshwater or bypass larval life entirely, undergoing direct development into juveniles. These and all other crabs (like most other crustaceans) occur as separate male and female individuals. The fertilized eggs are attached to the female's abdominal appendages, and the embryos are brooded beneath her abdomen until they hatch. The swimming larva, called a **zoea,** has a short rounded cephalothorax and a long, slender, segmented abdomen. The segmentation is at first incompletely developed; only after several zoeal molts are all the segments and appendages present and functional. Crab zoeas feed actively and must swim continuously just to stay in place, although long spines on the carapace help to slow their rate of sinking through the water. They molt to a more crablike stage called a **megalopa,** which can swim and also walk on the bottom. Eventually, having found a suitable site, megalopas molt into juvenile crabs.

Life history of the blue crab, *Callinectes sapidus,* in Chesapeake Bay. **(1) Immature crabs** (in their first spring) and mature males take advantage of rich feeding areas in shallow side-bays and estuaries of low salinity. **(2) Pairing** takes place in shallow waters in summer and fall between a female, ready to molt for the last time, and a mature male. After her molt, they mate. **(3) She migrates,** aided by downstream surface currents, toward the mouth of the Bay, into the high-salinity water needed for development of eggs and larvas. **(4) She extrudes her eggs** after overwintering and carries them attached to her abdomen until they hatch in summer. **(5)** The planktonic **zoea** feeds, grows, and molts 7 or 8 times. **(6)** The **megalopa** settles to the bottom. **(7) Young crabs,** aided by deep upstream currents, gradually move up the Bay again. (Modified after Cronin and Mansueti)

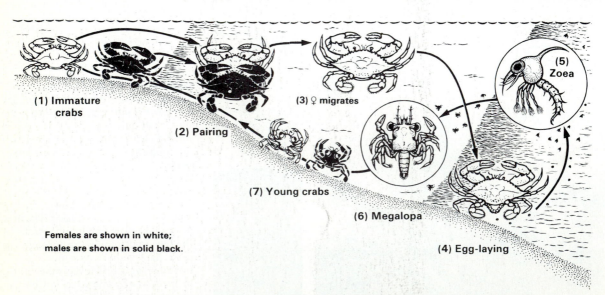

(1) Immature crabs

(2) Pairing

(3) ♀ migrates

(5) Zoea

(7) Young crabs

(6) Megalopa

(4) Egg-laying

Females are shown in white;
males are shown in solid black.

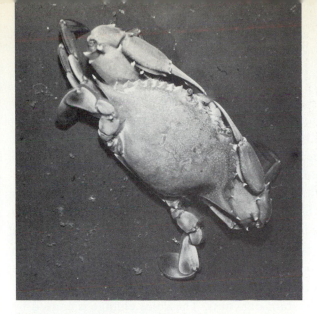

Blue crab, *Callinectes sapidus,* a swimming crab, is the object of the largest crab fishery in the U.S. Ironically, blue crabs are also pests to another fishery, for they prey on young oysters. Although best known from Chesapeake Bay, blue crabs are caught and eaten along coasts from Massachusetts to Texas. Full-grown adults are about 15 cm across and colorfully embellished with olive-green on the carapace and bright-blue or red on the claws. Duke University Marine Laboratory, Beaufort, N.C. (R.B.)

Dungeness crabs mating *(Cancer magister).* In crabs such as these that mate immediately after the female's molt, while she is still soft, pairing occurs several days before mating. The male carries the female and often assists her to molt. During mating, the soft female is underneath; the male on top supports and protects her. Afterwards he may guard her for several days until her skeleton hardens. In those crabs that mate when both male and female are hard-shelled, the female is on top and responsible for support of both partners during mating, and the male need only attend to sperm transfer. A male crab uses the first abdominal appendages to introduce sperms into the female's sperm receptacles. Sperms from a single mating may be stored for months or years and used to fertilize multiple batches of eggs. Coos Bay, Oregon. (R.B.)

Zoea has large compound eyes and swims, dorsal spine foremost, with the first 2 pairs of thoracic appendages (maxillipeds). Several zoeal or equivalent larval stages occur in the development of most decapods; crabs have from 1 to 8 zoeas. Second zoea of the velvet swimming crab, *Liocarcinus puber.* Length, 2 mm. Plymouth, England. (D.P. Wilson)

Megalopa, the last larval stage, bridges the transition between pelagic and benthic habitats. The thoracic appendages are no longer used in swimming but are now developed as feeding maxillipeds, chelipeds, and walking legs. The megalopa swims with its abdominal appendages. The abdomen remains extended during swimming, but can be flexed beneath the thorax when the megalopa walks. The stalks of the compound eyes are better developed. Almost all crabs have one megalopal stage. Megalopa of *Carcinus maenas,* 3 mm long. Plymouth, England. (D.P. Wilson)

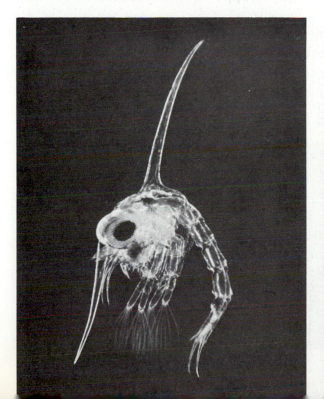

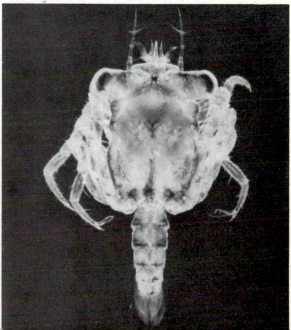

Although some crabs are heavy and slow-moving, others are among the most swift and agile of invertebrates. Their lively and complex **behavior** has attracted much study, especially in accessible species active at low tides, such as the fiddler crabs *(Uca).* One conspicuous feature of the behavior of fiddlers held under constant conditions in the laboratory is that they continue to be active during the twice-daily periods of low tides. This tidal activity rhythm persists for up to several weeks in the apparent absence of any external cues. Fiddlers also continue for weeks to change color daily, darkening during the daytime and turning pale during the night, even though they are kept in constant darkness. The color change is caused by dispersal or concentration of pigments within chromatophores in the tissue beneath the transparent cuticle; this response probably provides camouflage under natural conditions. Both the circadian rhythm of color change and circatidal rhythm of activity are coordinated through rhythmic release of neuro-hormones produced by the eyestalks and brain. The nature of the "clock" that the crabs use to maintain the rhythms is still unknown.

The chelas of crabs are used not only in feeding and in defense against predators but in courtship and other social behaviors such as the encounters that determine social status in many crab populations. In lined shore crabs *(Pachygrapsus)* the large chelas of an adult male are so vital to his social status that he will risk his life rather than autotomize them, whereas a female or a juvenile of either sex will autotomize and regenerate the chelas readily. The relative social status of two crabs—as defined by which one retreats from an encounter—is closely related to size, especially chela size. But behavioral traits, which are sometimes learned, also seem to weigh in the balance.

Experiments with the European hairy crab *(Pilumnus),* suggest that the success of large crabs in winning encounters with smaller ones reinforces aggressive behavior in the dominant individuals. When such socially dominant crabs were transferred into populations that included larger dominants, the transferred individuals continued to behave as aggressively as they had in their own groups and won higher status than would be predicted by size alone. Subordinate crabs transferred into populations of smaller individuals retained their habit of submissive behavior, failing to take advantage of their increased relative size, and they assumed status lower than expected for their size.

Male fiddler crabs have one enormously enlarged chela, either left or right, which often weighs as much as the rest of the animal. It cannot be used in feeding and is too clumsy for effective defense, but is waved conspicuously by the male, as he stands beside his burrow entrance, to advertize his sex and status to other crabs of both sexes. A waving male is approached by receptive females, and avoided or challenged by other males. Combat between males is usually nonviolent and highly ritualized. The waving displays of *Uca* are species specific and may help prevent interbreeding of different species. Females have small, equal-sized chelas. *Uca pugilator.* Carapace width, 25 mm. NW Florida coast. (R.B.)

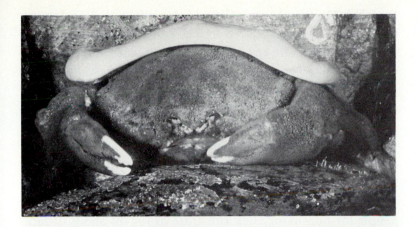

Sponge-carrying crab, *Dromia vulgaris.* The sponge does not attach to the carapace but can be set aside when the crab molts and later taken up again, held in place by the last 2 pairs of legs. Looking like a well-fitted beret, the sponge partly conceals the crab and grows along with it but at a slower rate. After a few molts the crab outgrows its sponge and replaces it with a larger piece, cut to fit and pried loose from its attachment by the chelas. Offered various coverings, the crab selects the sponge *Suberites,* which apparently provides chemical protection also; in an aquarium, *Octopus* will not attack *Dromia* unless *Suberites* has been removed. If nothing better is available, a dromiid will cut itself a hat of paper, using its chelas like scissors. Dromiids also carry encrusting tunicates or zoanthids, and some species adopt clam shells. England. (D.P. Wilson)

Box crab, *Hepatus epheliticus,* on the NW Florida coast is 6 cm across. Presumably receives some protection from the sea anemones on its back and from its chelas. Crabs of the genus *Lybia* carry an anemone in each chela and take advantage of the cnidarian nematocysts for prey capture as well as defense. (R.B.)

Pea crab *(arrow)* in the mantle cavity of a clam. Pea crabs (family Pinnotheridae) also live in tunicates, in the burrows of certain polychete worms and callianassid shrimps, and on sand dollars. These tiny round-bodied crabs filter food from the feeding currents of their hosts or even feed directly on food collected by the host. Though soft-shelled and sedentary when under the hosts' protection, pea crabs are hard-shelled, active swimmers during their 2 free-living stages: first as juveniles seeking hosts and later as mature crabs seeking mates. Bimini, Bahamas. (R.B.)

BARNACLES are marine filter feeders that live all their adult lives attached to rocks, wooden pilings, ships and other floating objects, and the bodies of many other animals. They secrete a skeleton of heavy calcareous plates; and their activities consist mostly of opening or closing the plates and of extending and retracting the appendages during feeding. This sessile life style and monotonous behavioral repertoire provides barnacles with a good living and a secure, widespread niche on most rocky shores. But it contrasts sharply with the mobility and complex behavior of crabs.

The heavy skeletons of barnacles are mistaken by most people for non-living rock substrate or for the shells of molluscs. Early zoologists, too, classified barnacles with molluscs. But the jointed, chitinous appendages, which are visible as soon as a barnacle begins to feed, mark these animals as arthropods. Now recognized as crustaceans (class Maxillopoda, subclass Cirripedia, order Thoracica), barnacles number about 950 living species. By this measure, they are far less diversified than crabs. But barnacles can be extremely abundant. On one stretch of rocky intertidal in England the number of individuals in the densest part of the "barnacle zone" was calculated at over 107,000 per m², and such a zone of almost solidly barnacle-covered rock may extend along temperate coasts for many kilometers.

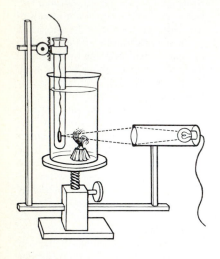

Automatic recording of barnacle activity by a photoelectric system in which movements of the appendages or skeletal plates interrupt a light beam. Records show that barnacle activity was sensitive to motion, gas content, and pH of the water. No consistent daily or tidal rhythms were detected; intertidal barnacles evidently respond directly to wetting or drying during tidal changes. (After Southward and Crisp)

Stalked barnacles, *Lepas anatifera,* on a floating bottle. The stalk is morphologically the anteriormost region of the body. In this species it is flexible and muscular and up to 80 cm long. The rest of the animal is encased in calcareous plates, which open to permit extension of the feeding appendages. England. (D.P. Wilson)

Stalkless barnacles, *Semibalanus balanoides,* are common in W. Europe and on the Atlantic coast of N. America. Plates are in close contact with substrate. In this species the base is membranous, and the barnacle, though sessile, continues to reattach if other growing barnacles push it aside. Submerged specimens reach shell diameters over 25 mm; intertidal ones decrease in size at progressively higher levels on the shore. (D.P. Wilson)

Substrate for attachment, such as the shell of this snail *(Littorina),* may be provided by almost any hard surface in the ocean. Diam. of barnacles, 3 mm. NW Florida coast. (R.B.)

Various kinds of barnacles attach to specific animal hosts: whales, sea turtles, sea snakes, fishes, crustaceans, molluscs, jellyfishes, corals, gorgonians, sponges, and others, often to a specific site on the host.

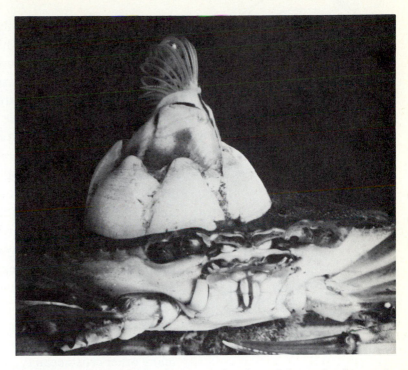

Large barnacle on blue crab suggests that the crab is overdue for a molt. Regular molting keeps crabs and many other crustaceans relatively free of organisms that settle on their exoskeleton. This crab may be inhibited from molting by infection with a cirriped parasite (see rhizocephalans). NW Florida coast. (R.B.)

Stalked barnacles on crab gill take advantage of protection in the gill chambers and have reduced calcareous plates. *Octolasmis mulleri,* shown here on gill of large spider crab *Libinia,* is 1 to 3 mm long (excluding stalk) and is distributed around the world on gills, or on walls of gill chambers, of various crabs. NW Florida coast. (R.B.)

Gray-whale barnacles *Cryptolepas rhachianecti* live embedded in the skin of their host. The ratio of oxygen isotopes in the carbonate of the barnacle's shell is related to the temperature and salinity of the water in which the shell is deposited. By analyzing successive growth zones in the shell, one can reconstruct the successive temperatures of the water through which the whale has migrated, and hence, for example, the number of north-south migrations made by the whale host during the life span of the barnacle. Diameter, 3 cm. (W.E. Ferguson)

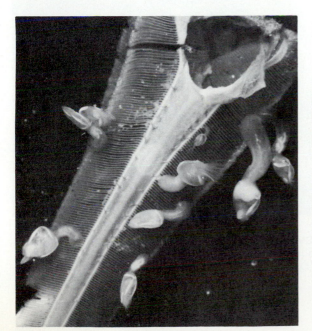

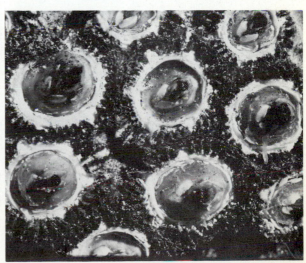

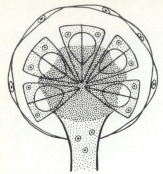

Compound eye of cyprid stage is lost at metamorphosis.

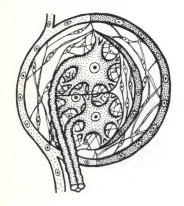

Simple eye of adult barnacle is one of 3 photoreceptors thought to be derived from the 3 parts of the naupliar eye. (Mod. after D.E. Fales)

Barnacles (and especially the parasitic members of the subclass Cirripedia) are so highly modified for their particular life styles that their relationship to other crustaceans was not obvious until their larval stages became known. The adult lacks typical crustacean features such as 2 pairs of antennas. In fact, it has no recognizable antennas at all and is organized quite unlike any other crustacean. But the larva is a **nauplius,** the same larval form that occurs in at least some members of all crustacean classes. A nauplius has a compact body with only 3 pairs of appendages (first antennas, second antennas, and mandibles). There is a median **naupliar eye,** usually with 3 clusters of retinula cells in pigment cups. The naupliar eye persists in the adult of many crustaceans and is thought to give rise to the adult photoreceptors of some barnacles. Barnacle nauplii swim and feed in the plankton, molting repeatedly. The final naupliar molt results in a **cyprid,** a distinctive cirriped larva. Cyprids have a large bivalved carapace and a sensory tail-fork, or **caudal furca,** features shared with adults of many other crustacean types. They swim or crawl by means of 6 pairs of biramous thoracic appendages but do not feed. In addition to the median naupliar eye, there is a pair of lateral compound eyes. The second antennas have been lost, and the first antennas end in an adhesive disk with which the cyprid attaches at metamorphosis.

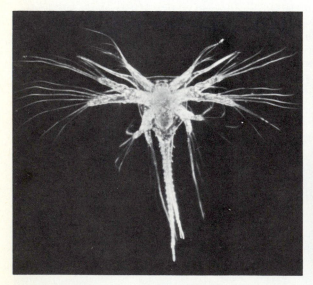

Nauplius of a barnacle has distinctive frontolateral "horns." *Lepas*. Length, 0.7 mm. Plymouth, England. (D.P. Wilson)

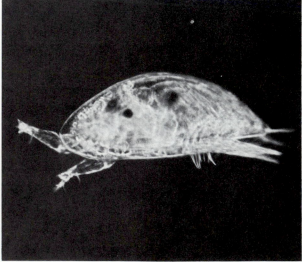

Cyprid of a barnacle has a short abdomen, lost in the adult. Length, 0.5 mm. Plymouth, England. (D.P. Wilson)

Site selection is a life-or-death matter for a young barnacle, and many factors come into it; but the most important is the presence of other barnacles, especially of the same species. Such neighbors not only are the best indicator of a site favorable for survival but also are potential mates. Recognition is by contact with specific proteins (arthropodin) in the cuticle or base. Other factors are water currents, substrate texture (rough or grooved surfaces are preferred), light, and hydrostatic pressure. A cyprid can attach temporarily to test a site, but once it has cemented itself to a spot, there is no second chance. As a barnacle grows, however, it can slowly re-orient the shell to take better advantage of prevailing water currents.

In the course of **metamorphosis** to the adult, the first antennas are buried beneath the basal disk; the thoracic swimming appendages are transformed into feeding appendages, called cirri; and the tissue of the cyprid carapace becomes the adult mantle, which encloses the body and secretes protective calcareous plates. In stalked barnacles, the portion of the body between mouth and basal disk is elongate, whereas in stalkless barnacles, the calcareous plates fit close against the substrate or base. The plates do not generally fuse with each other or with the base; and as a barnacle grows, it adds to base and plates around their margins and on their inner surfaces. The cuticle of the enclosed body and appendages is molted periodically as in other arthropods.

In the primitive cirriped order Ascothoracica, the adults have a large bivalved carapace like that of cyprids. Cyprids also resemble the adults of another class of crustaceans, the Ostracoda, and were named for their resemblance to the common ostracod genus *Cypris*.

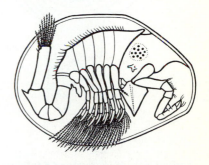

Ascothoracican of the most generalized kind. Except for the notably well-developed abdomen, the segmentation corresponds to that of other cirripeds, and the life history includes nauplius and cyprid larvas. Adults are parasites, attaching with the large first antennas to cnidarians and echinoderms. *Synagoga sandersi*, 4 mm long, from 5,200 m in the SW Atlantic. (Modified after W.A. Newman)

Metamorphosis of acorn barnacle, *Balanus amphitrite*. (After Barnard and Lane)

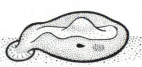

1. Attached cyprid.

2. Shedding of larval exoskeleton.

3. Soft body reorganizing tissues.

4. Young acorn barnacle.

Metamorphosis of stalked barnacle, *Lepas*. (Modified after E. Korschelt)

1. Cyprid explores substrate.

2. Attached cyprid has begun metamorphosis.

3. Vertical body has 3 developing carapace plates.

4. Young stalked barnacle with calcified plates.

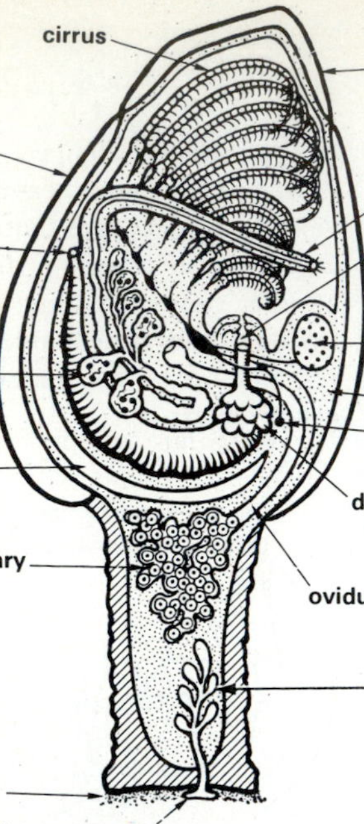

Diagram labels, left figure (Stalked barnacle):
cirrus, tergal plate, carinal plate, penis, mouth, scutal plate, anus, adductor muscle, testis follicle, mantle, eye, mantle space, digestive gland, ovary, oviduct, cement glands, cement, first antennas

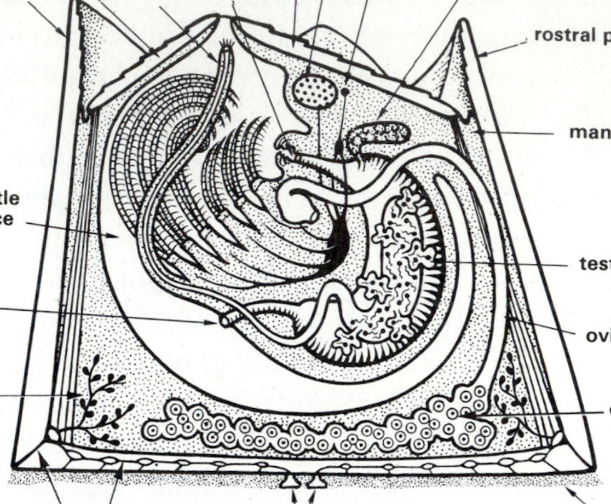

Diagram labels, right figure (Stalkless barnacle):
scutal plate, adductor muscle, carinal plate, penis, mouth, eye, digestive gland, rostral plate, mantle, mantle space, testis follicle, anus, oviduct, ovary, cement, cement duct, first antennas

Stalked barnacle *(left)* and **Stalkless barnacle** *(right),* in lateral view. The **mantle** is homologous to the carapace tissue of other crustaceans; it is unrelated to the mantle of molluscs but is analogous in that it secretes and lines the shell. In the mantle tissue, near the ovaries, lie unicellular **cement glands,** which connect by ducts to the base. Remnants of the first antennas (here exaggerated) are embedded in cement near the center of the base. The base of stalkless barnacles may be calcareous (as here) or membranous. When crowded, stalkless barnacles may grow an elongate basal portion, equivalent to that of stalked barnacles. The two living suborders of stalkless barnacles are thought to have evolved independently from stalked barnacles. (Nervous system shown in solid black).

Stalked barnacle ingesting a brine-shrimp. Barnacles feed mostly on minute planktonic organisms and particles of detritus that they comb from the water with heavily bristled thoracic appendages. But in the laboratory, *Lepas* has been seen to take offered prey larger than itself, and presumably can do so in nature. One forced down a brine-shrimp *(Artemia)* while clutching 7 copepods in the posterior appendages. (After Howard and Scott)

In **feeding,** barnacles extend the thoracic appendages and sweep them through the water or, when in a favorable current, simply hold the appendages extended and filter food from the water as it passes through them. Small food particles are trapped by fine bristles that fringe the appendages, but large particles or zooplankters may be captured as well. The appendages are extended hydrostatically by blood. Although the **circulatory system** consists of a well-defined series of channels and sinuses, there is no heart comparable to those of other crustaceans; blood is circulated at considerable pressures by rhythmic contractions of muscles which compress some of the sinuses. The **respiratory structures** are also unlike those of lobsters, crabs, and many other crustaceans; the gills are convoluted folds in the inner mantle wall. Intertidal barnacles can carry on respiratory exchange in both water and air.

The mode of **reproduction** is likewise unusual among crustaceans, in that most barnacles are hermaphroditic, a feature associated in many animal groups with sessile life styles, sparse populations, or other conditions that complicate finding a mate. Self-fertilization is possible in some hermaphroditic barnacles, but in other species an isolated individual cannot reproduce. Each barnacle has paired testes and ovaries. When acting as a male, the animal protrudes its

Giant barnacle with thoracic appendages extended in feeding position. *Balanus nubilus* and *Concavus (= Balanus) aquila* reach diameters of over 10 cm on the California coast. Their individual muscle fibers are enormous, up to several millimeters in diameter, and are much prized by physiologists for research. Hopkins Marine Station, Monterey Bay, California. (R.B.)

long wormlike penis and inserts it into the opening of any neighbor within reach. A barnacle's reach can be impressive, for example, up to 50 mm for an 8-mm *Chthamalus*. A mass of sperms is deposited near the openings of the oviducts at the bases of the first thoracic appendages. The ovaries are peculiarly situated in the anteriormost part of the animal, the stalk or base. They release eggs into the mantle space, where fertilization and development proceed and from which nauplius larvas are eventually expelled. In some species, tiny males accompany hermaphroditic or female individuals, attached either inside the mantle space or to the outside of the animal. The appendages, digestive tract, and shell of such males are usually reduced or absent.

Mating barnacles do not exchange sperms simultaneously as do some hermaphroditic animals. Instead, one individual acts as a "male," inserting its long penis into the mantle cavity of another barnacle acting as a "female." Later, the role may be reversed. These two barnacles mated several times, alternating roles. *Balanus crenatus*, attached to a whelk shell. England. (H. Angel)

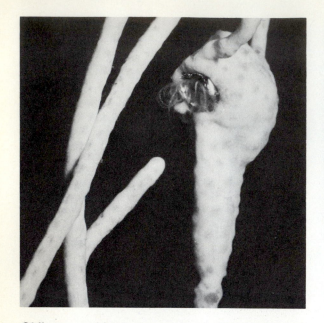

Obligate symbiont on gorgonians is this balanomorph barnacle, *Conopea galeata*. The cyprid settles on an area of gorgonian skeleton laid bare by a predator or mechanical injury, and gorgonian tissue grows around the barnacle. Hermaphroditic individuals are accompanied by small, nonfeeding males that attach to the shell. NW Florida coast. (R.B.)

Human interest in barnacles, other than that of curious biologists, extends to edible types such as these large Chilean barnacles *(Megabalanus psittacus),* which are unloaded at Puerto Montt, southern Chile. They are enjoyed for their delicate flavor, either raw with lemon, baked with cheese, or in chowder. Dental researchers are keenly interested in barnacle cement for its properties of strength and durability and its capacity for sticking to wet surfaces. But most interest in barnacles centers on getting rid of them when they foul shipbottoms, harborworks, and pipes. Ships not treated with antifouling coatings have been known to acquire loads of over 270,000 kg of barnacles, reducing their cruising speed by half. (R.B.)

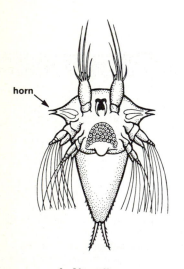

1. Nauplius.

horn

RELATED to the barnacles are perhaps the strangest crustaceans of all, the **rhizocephalans** (subclass Cirripedia, order Rhizocephala). These parasites are so greatly modified that they are unrecognizable as cirripeds, or even as crustaceans, except by their nauplius and cyprid larvas. A female rhizocephalan cyprid somehow locates a crab or other suitable crustacean host and attaches by one first antenna to a thin place on the cuticle. The cyprid then molts, shedding the thoracic appendages, carapace, and other structures. The saclike succeeding stage (the kentrogon) consists of the remaining tissue enclosed in a new cuticle with a sharp, hollow projection at one end, which penetrates the host cuticle. Kentrogon tissue is injected into the host and there proliferates.

Larval stages of rhizocephalan *Sacculina carcini*. (After Y. Delage)

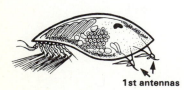

1st antennas

2. Free cyprid.

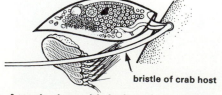

bristle of crab host

3. Attached cyprid sheds and reorganizes.

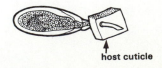

host cuticle

4. Kentrogon.

The injected rhizocephalan tissue produces a rootlike system of nutrient-absorbing tubules (the interna) that branch throughout the host's blood spaces, even to the tips of the legs. After some time, there appears on the host's ventral surface a large bulbous growth (the externa) containing the parasite's eggs. Further development depends on the arrival of a male cyprid, which attaches to the externa and injects tissue that produces sperms. The eggs are fertilized, and the young are released as swimming nauplius larvas that do not feed, but are filled with nutrients derived from the parental host. This store of nutrients supports all the larval stages until the cyprid finds a new host of its own.

The rhizocephalan manipulates the hormonal state of the host to its own advantage, so that host resources go largely into growth of the parasite. In male crabs, rhizocephalan infection causes degeneration of the androgenic glands responsible for growth of the testes and sperm production. Both structure and behavior of males become feminized. Parasitized young female crabs take on the physical appearance of mature ones, but ovaries and eggs fail to develop normally. Finally, adding insult to injury, the parasites inhibit the host crabs from molting, so that when the externas appear, they are in effect protectively brooded beneath the broadened abdomens of both feminized male and barren female crabs.

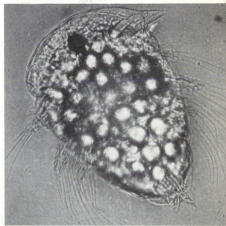

Nauplius of *Sacculina carcini* with many nutrient storage cells. Length, 0.2 mm. Roscoff, France. *Sacculina* infects up to 70% of some crab populations. (R.B.)

Rhizocephalan externas on crab host. Three bulbous, egg-laden sacs of *Loxothylacus texanus* have erupted through the cuticle at the base of the broad abdomen of a blue crab, *Callinectes sapidus.* Gulf of Mexico. (R.B.)

CRUSTACEAN GROUPS

CRUSTACEANS generally have a head, or head region, with a basic set of 5 pairs of appendages (2 pairs of antennas, a pair of mandibles, 2 pairs of maxillas), but even this rule has exceptions. The rest of the body varies greatly; and in number and arrangement of body regions, segments, and appendages, crustaceans are far more diverse than chelicerates or uniramians. Formal descriptions of the crustacean groups and the technical distinctions between them are therefore discouragingly heavy with details of these different arrangements. In the brief introduction to the groups on the following pages, we have attempted only to present a few representatives of the major groups and to convey the most readily discernible differences among them in form and habit.

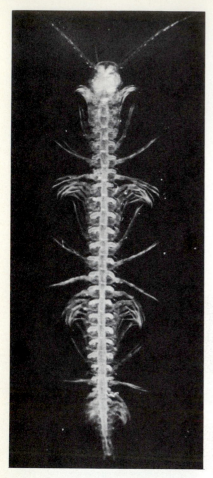

Primitive cave crustacean, *Speleonectes,* class **REMIPEDIA** *(remi,* "oar" + *pedia,* "feet") has a pair of similar biramous swimming appendages on almost every segment of the long trunk, a character that is unique to this newest class of crustaceans, and that has long been presumed to be the condition of the primitive crustacean stocks. The class was first reported in 1980 from several specimens captured in seawater in a submerged cave in the Bahamas. Length, 20 mm. (J. Yager)

Primitive crustacean, class **CEPHALOCARIDA,** has second maxillas that are marked as such by openings of the maxillary glands but are otherwise indistinguishable from similar multipurpose appendages on the thorax—exactly as we might imagine an early stage in the evolution of the 5 differentiated pairs of head appendages that characterize crustaceans. Cephalocarids resemble current concepts of the stocks that gave rise to the diversity of modern crustacean groups. Like all "primitive" animals, they also have specialized features: they lack abdominal appendages and are hermaphroditic. Despite rudimentary compound eyes buried beneath the exoskeleton, cephalocarids show little evidence of a visual sense or a need for one. The few known species have been found on soft marine bottoms, filter feeding on suspended detritus. *Hutchinsoniella macracantha.* Length 3 mm. (Preserved specimen, Scripps collection, courtesy W.A. Newman.) (R.B.)

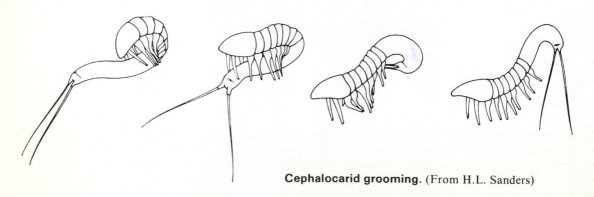

Cephalocarid grooming. (From H.L. Sanders)

The class **BRANCHIOPODA** (*branch* = "gill," + *poda* = "feet") is unique among crustacean groups in having only a few marine members; branchiopods favor shallow fresh or saline lakes and especially small temporary ponds or pools—habitats without large populations of predatory insects or fishes. Most branchiopods are under 2 cm long with delicate bodies bearing a series of similar, leaflike appendages on the trunk. These flattened, lobed appendages serve as paddles for swimming and create the feeding and respiratory currents. Branchiopods feed mainly by filtering small organisms and organic particles from the water; a few are predators or scavengers. The class includes about 800 living species in the orders Anostraca, Notostraca, Conchostraca, and Cladocera.

Fairy-shrimps, order ANOSTRACA, swim continuously and gracefully on their backs. Most species inhabit temporary freshwater ponds or pools. Unlike other branchiopods, anostracans have no carapace. Those shown here are all females with egg sacs. *Eubranchipus,* 25 mm long. Northeastern USA. (R.B.)

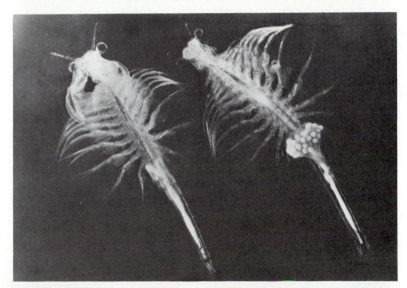

Brine-shrimps, *Artemia,* are anostracans that live in highly saline lakes. As in other anostracans, the male *(left)* is marked by enlarged, clasping second antennas. The female *(right)* carries the eggs in a ventral brood chamber. When released, the eggs are partially developed; there are thin-shelled eggs which proceed to hatch into nauplii, as well as thick-shelled eggs which may remain viable in the dried-up bottom of a pond for up to 5 years, or be carried by winds or birds' feet to new localities, hatching within a few hours of being wetted again. Eggs, nauplii, and adults of *Artemia* are sold as food for aquarium fishes and for planktivorous invertebrates in laboratories. Length, 15 mm. (R.B.)

Tadpole-shrimp, of the order NOTOSTRACA, is protected by a shieldlike carapace that covers most of the body dorsally *(left)*. Turned on its back *(right)*, it shows an elongate trunk bearing up to 70 pairs of similar leaflike appendages. Long extensions on the first pair serve as feelers; the antennas are very small. In addition to feeding on small particles, tadpole-shrimps prey or scavenge on other crustaceans, soft insect larvas, worms, and amphibian eggs or tadpoles. Some species have separate sexes; others are self-fertilizing hermaphrodites. The eggs are carried briefly, then dropped to the bottom. Eggs resting in dry mud for 15 years have been hatched by adding water. The animal shown was from a shallow puddle filled with busy tadpole-shrimps after unusual summer rains in Death Valley, California. *Triops,* 2 cm. (R.B.)

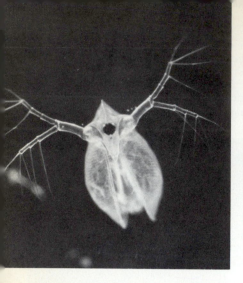

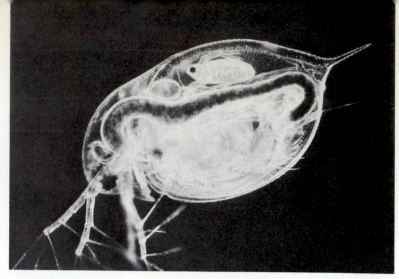

Water-flea, order CLADOCERA, only 2 mm long, swims by rapid jerks of the large 2-branched antennas. The dark spot on the head is a median compound eye. This head-on view of a cladoceran, *Daphnia,* reveals how the folded carapace covers the compact trunk. Apart from a few marine species, these tiny crustaceans (mostly 0.2 to 3 mm long) live in fresh and brackish waters from shallow temporary ponds to lakes. Cladocerans feed on bacteria, protists, and organic particles, and are themselves a major source of food for both young and adult fishes. (R.B.)

Daphnia and other cladocerans have a flexible life history that allows them to respond to their changeable surroundings by taking on strikingly different shapes during different seasons. During much of the year cladocerans replicate by parthenogenesis, females producing female offspring and males being rare or absent. The females molt as often as every 2 days, each time freeing the brooded young and immediately releasing eggs into the dorsal brood chamber. This individual has only one daughter in the brood chamber, but there may be more. Under adverse conditions, usually overcrowding, decreased food, or low temperature, males are produced. If the conditions continue longer, many females produce haploid eggs, which are fertilized after these females mate with the males. Fertilized eggs, usually no more than 1 or 2 of them, pass into the brood chamber and it becomes thick, hard, and dark in color, forming a protective capsule in which the "resting eggs" can survive drying and freezing. The remarkable differences in body shape developed in different members of a single parthenogenic clone are presumed, as in rotifers (chapter 13), to confer special advantages in the uncertain conditions under which these tiny animals flourish. (P.S. Tice)

Hardened brood chamber containing 2 fertilized eggs is seen in this *Daphnia,* taken from a California pond. At the next molt the encapsulated brood chamber (ephippium) will separate from the carapace and sink to the bottom. In the spring these "resting eggs" will give rise to at least part of the small spring population. Because of these resistant eggs, cladocerans are easily carried overland to new habitats, and most species are very widely distributed, some cosmopolitan. (R.B.)

Clam-shrimp, order CONCHOSTRACA, usually does not molt the bivalved, calcified carapace, which is marked by growth rings and encloses the entire body. Clam-shrimps swim, burrow, and climb with their large second antennas. The sexes are separate; some species reproduce by parthenogenesis. Eggs and sometimes young are brooded dorsally within the carapace. Carapace, 8 mm long. California lake. (R.B.)

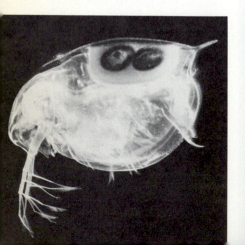

The class **MALACOSTRACA** dominates the subphylum Crustacea; it includes the greatest number of species (over 20,000) and almost all of the large and conspicuous types. Malacostracans share a series of characters already seen in lobsters: head with paired compound eyes and 2-branched first antennas, thorax with 8 segments (female openings on the 6th, male openings on the 8th), and abdomen with 6 (rarely 7) segments plus a telson which, together with uropods, often forms a tail-fan; most of the segments bear appendages. Despite this common plan, members of the class come in many shapes and sizes, habits and habitats.

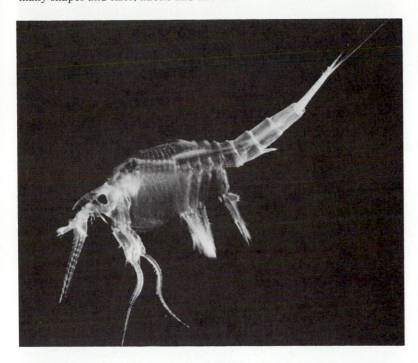

Primitive malacostracans, members of the order LEPTOSTRACA, look and live much like branchiopods. Beneath the large folded carapace, 8 similar pairs of thoracic appendages paddle constantly, creating respiratory and feeding currents. Most leptostracans are found on muddy sea bottoms rich in organic material, and (like cephalocarids and branchiopods) they filter feed on small organisms and organic particles. The abdominal appendages are used in swimming, and the strong first antennas help the animal to burrow busily through seaweeds and sediments. Eggs and young are brooded under the mother's carapace and juveniles are released. Leptostracans differ from more advanced malacostracans in having 2 pairs of excretory organs and a 7-segmented abdomen ending in a caudal furca. *Nebalia pugettensis,* 6 mm long. N. California. (R.B.)

Mantis-shrimp, order STOMATOPODA, takes its common name from large raptorial limbs (second thoracic appendages) held close to the body as the animal waits in ambush for prey. All 700 or so species of stomatopods are pugnacious predators on marine bottoms, mostly in shallow subtropical or tropical waters. Divers who poke fingers into holes in coral risk a nasty cut or even loss of part of a finger to the resident stomatopod. The moveable terminal claw folds back into a groove in the adjacent article, much as the blade folds back in a jackknife. In "spearing" stomatopods, such as *Squilla empusa* (shown here), the strike of the claw is among the quickest animal movements known, completed in 4 to 8 milliseconds. The barbed claw impales soft prey such as worms and fishes. Many Pacific islanders commonly use the claws as fish hooks. Spearing stomatopods dig burrows in mud or sand and tend to be sizeable, up to 38 cm; *S. empusa* is 12 cm long. "Smashing" stomatopods, with a bladelike terminal claw that is enlarged at its base and heavily calcified, use the claw like a club to break open snails, bivalves, and crabs. Such stomatopods are smaller, limited in size by the availability of holes in rock or coral, which they use as homes and defend aggressively. Their brightly colored markings serve in threat displays against rivals or predators and also to impress prospective mates. Prey tissue is shredded by the mandibles and several pairs of maxillipeds. These last are also used by females to hold developing eggs. Behind the maxillipeds are 3 pairs of walking legs, abdominal swimmerets bearing gills, and the large uropods and telson. NW Florida. (R.B.)

Malacostracans of the superorder PERACARIDA ("pouch-shrimps") brood their eggs in a pouch formed by plates on the basal articles of some of the thoracic appendages. All have a cephalothorax, but it may include as few as one thoracic segment fused to the head, the rest of the thorax remaining free as a separate body region (pereon). Most of the 11,000 peracaridan species belong to 2 orders: amphipods and isopods.

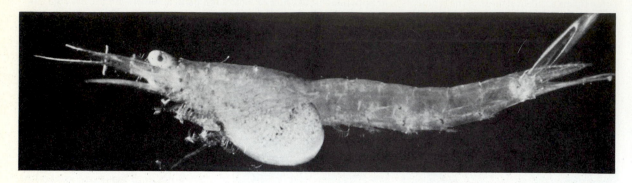

Opossum-shrimp, order MYSIDACEA, is named for the brood pouch, present in all female peracarids but especially conspicuous in this order, which comprises about 450 species. Most mysidaceans swim near marine bottoms, filter feeding and sometimes taking prey or scavenging on carrion. The inner surface of the carapace serves in respiratory exchange; some mysidaceans (lophogastrids) also have gills. Most primitive of the peracarids, lophogastrids resemble leptostracans in having 2 pairs of excretory organs and 7 abdominal segments. Length, 1 cm. Monterey Bay, Calif. (R.B.)

A laterally compressed body and great diversity of appendages characterize the order AMPHIPODA. Thorax and abdomen bear 14 pairs of appendages. On each side are: 1 maxilliped, 2 grasping gnathopods, 2 forwardly directed legs, 3 backwardly directed legs (here typically flexed up against the trunk), 3 swimmerets, and 3 backwardly directed uropods. With this versatile equipment, amphipods cling, walk, scull along on one side, burrow, swim, and jump, besides handling a variety of foods. The legs bear gills and, in females, projections that form the brood pouch. Most amphipods are scavengers and detritus feeders. Among the 5,500 species, marine ones predominate, but many live in freshwater and a few in damp terrestrial habitats. *Gammarus minus* lives in caves, springs, and streams in the eastern and central USA; this 5-mm individual has demolished a leaf. (R.B.)

Beachhopper, *Orchestoidea californiana,* with long second antennas and (as in amphipods generally) unstalked eyes. Beachhoppers execute impressive jumps by forcefully extending the abdomen and pushing with the uropods. They feed on washed-up seaweeds at night and retreat to moist burrows in the sand by day. A beachhopper puts up a good fight when defending its burrow against takeover by another. Oregon. (R.B.)

Caprellids, sometimes called skeleton-shrimps, are amphipods with a long slender thorax and almost no abdomen. They cling to hydroids (as here) and seaweeds, snatching small plankters out of the water or scraping from the substrate whatever organic material they can find. A related group, the cyamids, or whale-lice, are ectoparasites on dolphins and whales. Highly modified for clinging to their hosts, they resemble sea-spiders in form (see end of chapter 22). NW Florida. (R.B.)

Hyperiidean amphipod, *Hyperoche,* found inside a salp (see chapter 29) by a diver at 15 m depth in Monterey Bay, California. Hyperiideans are all marine and pelagic or live attached, like this one, to other pelagic animals. Many hyperiideans have enormous eyes. (R.B.)

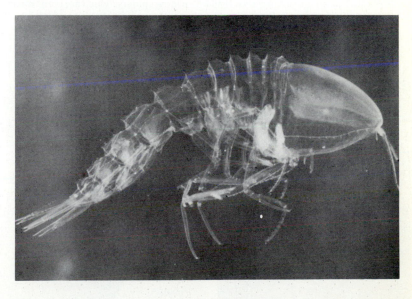

Large pelagic amphipod, *Cystosoma,* up to 15 cm long, is found at moderate depths (100 to 400 m or more). As in most hyperiideans, the cephalothorax is large and bulbous. (B. Robison)

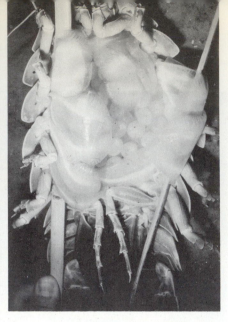

Giant carnivorous isopod, *Bathynomus giganteus,* found deep in the Gulf of Mexico and in the Caribbean, may exceed 35 cm in length and is the largest known member of the pericaridan order ISOPODA. Most of the 4,000 species of isopods are under 2 cm. In contrast to the laterally compressed amphipods, isopods are dorsoventrally flattened. Behind the small cephalothorax (head plus 1 thoracic segment), there are typically 7 pairs of similar walking legs *(iso =* "equal" + *poda =* "legs"), though some of these become variously specialized. The flattened abdominal appendages serve in respiratory exchange and sometimes in swimming. The last segment, or last several segments, are fused with the telson, and in the isopod shown here the uropods are part of a broad tail-fan. *Left,* dorsal view. *Center,* ventral view. *Right,* this female isopod was carrying many large yellow eggs, revealed by opening the brood pouch between the bases of the legs. (R.B.)

Land isopods, called sowbugs, pillbugs, or woodlice, are the mostly highly land-adapted crustaceans. Unlike land crabs, land isopods can reproduce far from any body of water, for the brood pouch in which the eggs develop is kept filled with fluid, and fully terrestrial young are released. As in aquatic isopods, however, the abdominal appendages serve as gills and so must be kept moist. The uropods, dipped into dew, take up water by capillarity, and any rain drop that falls on the animal is channeled at once to the abdominal respiratory surfaces. In addition, many land isopods have a system of invaginated air-filled tubes in the abdominal appendages (visible as white patches in ventral view), a notable evolutionary convergence with the air-filled tracheal systems of myriapods, insects, and arachnids. Common in temperate gardens, as well as in the litter of forest floors, land isopods hide in moist refuges by day and are active by night as scavengers on fallen leaves and other plant material. When abundant they are important in humus formation. They have few defenses, but some produce secretions that discourage spiders; and pillbugs such as *Armadillidium,* shown here, can roll up into a neat sphere when threatened. The main defense is to remain inconspicuous, and most come in drab earthy colors. However, a European *Armadillidium* with large red spots is presumed to gain some protection from its resemblance to black-widow spiders, which are common in the same areas where the isopods occur. Central California. (R.B.)

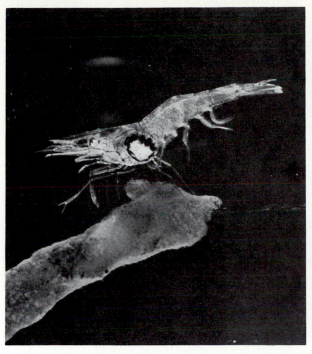

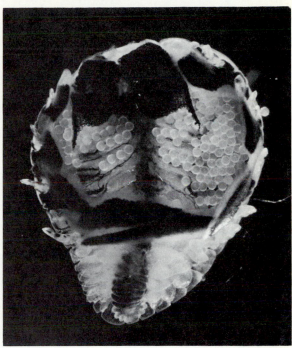

Left, **parasitic isopod on a shrimp.** The isopod, *Probopyrus,* has fastened itself to the inside of the host's carapace and has pierced the thin cuticle of the body wall to suck hemolymph from the shrimp, *Palaemonetes.*
Right, **ventral view of the female isopod** removed from the shrimp reveals a brood of eggs and a tiny dark male near the female's rear end. The first young bopyrid isopod attaching to a shrimp grows and develops into a female; a second arrival attaches to the established female and becomes a male. Gulf of Mexico. (R.B.)

Left, **parasitic isopod on a fish** is sucking blood from the soft tissues beneath the host's gill cover. Many fishes rid themselves of such pests by visiting "cleaners," species of fishes and shrimps that specialize in removing and eating fish parasites. *Right,* **flattened bodies** and sharp claws help these parasitic isopods cling to the host fish. NW Florida coast. (R.B.)

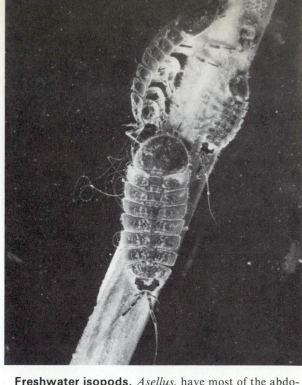

Rocky shore isopod, *Ligia,* is swift and agile. Found on rock faces and ledges in the high intertidal zone and above, this semiterrestrial isopod visits tidepools for water. Oregon. (R.B.)

Freshwater isopods, *Asellus,* have most of the abdomen fused with the telson. The uropods project behind. Freshwater isopods are omnivorous scavengers in caves, springs, and streams. Tennessee. (R.B)

Tanaidacean, *Leptochelia dubia,* order TANAIDACEA, is a peracarid related to isopods. The large, pincerlike second thoracic appendages are used by the males to fight with other males and to open the females' dwelling tubes, built with sticky secretions spun from the thoracic appendages behind the pincers. In some tanaidaceans, including this species, females can change into males. There are about 800 described species, all aquatic and mostly marine. Small (up to 25 mm) and benthic, they feed on plankton or detritus. Some occupy snail shells and may be mistaken for small hermit-crabs. Length, 5 mm. Monterey Bay, California. (R.B.)

Large chelas are prominent in this tanaidacean. (Preserved specimen, Scripps Collection, courtesy W.A. Newman) (R.B.)

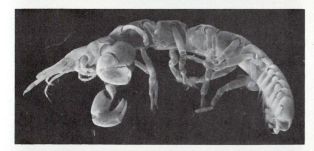

Cumacean, *Colurostylis occidentalis,* order CUMACEA, shows the distinctive hoodlike carapace and slender abdomen typical of this peracaridan group. The carapace encloses a large gill chamber on each side. Most cumaceans burrow in soft bottoms from the intertidal to the deep sea (over 7,000 m), eating microorganisms and organic material found in the substrate. Sand dwellers feed by picking up the grains one at a time and scraping off edible matter. Mud dwellers feed by filtering suspended particles. Cumaceans remain buried by day but are sometimes found swimming freely in the water at night. Of the nearly 900 species, only a few are brackish or freshwater. Length, 10 mm. S. California. (Preserved specimen, Scripps collection, courtesy W.A. Newman.) (R.B.)

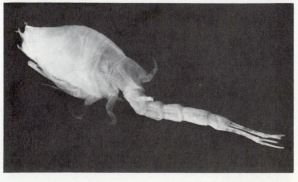

Euphausiid, *Euphausia superba,* order EUPHAUSIACEA, is a large (50 mm or more), bright red, Antarctic species and the dominant herbivore of the Southern Ocean. It occurs in enormous schools, in which all individuals are roughly the same size, stage, and orientation. Luminescence from light organs along the body could help in maintaining the schools, but schooling behavior appears to depend primarily on rheotactic cues from the wake of preceding animals. Aggregations of euphausiids have been reported to stretch for many square kilometers. Known as krill, dense concentrations of euphausiids in the Arctic and Antarctic are devoured by many invertebrates, by fishes (including herrings, mackerels, and other commercial species), by penguins and other seabirds, and by the great baleen whales. Blue whales, largest of the world's animals, live on euphausiids, and a whale's stomach can contain 4,000 kg (more than 4 tons) of a single species of krill, usually *Euphausia superba.* Fisheries of Norway, Japan, and Russia are being developed to provide krill products as food for humans and other animals, as fertilizer, and as bait. Euphausiaceans are all marine; they hatch as swimming nauplius larvas and are also pelagic as adults. Most of the 85 species feed by filtering plankton from the water with bristle-fringed thoracic appendages; some are predaceous. Euphausiaceans belong to the malacostracan superorder EUCARIDA, in which the carapace is fused dorsally to all thoracic segments, and they probably resemble the primitive stocks from which arose the dominant eucarids, the decapods. (Preserved specimen, Scripps collection, courtesy W.A. Newman). (R.B.)

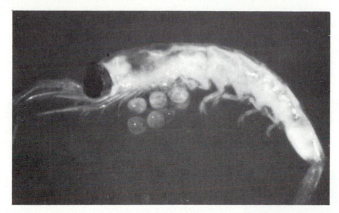

Predaceous euphausiid, with large eyes. This egg-carrying female, 12 mm long, was taken in a night tow off the eastern Florida coast. *Stylocheiron carinatum.* (P.M. Mikkelsen)

The order DECAPODA (*deca* = "ten", *poda* = "feet") with about 9,000 species, is the largest order of crustaceans and includes the shrimps, lobsters, and crabs. As their name suggests, decapods typically have 5 pairs of walking legs. These eucarid malacostracans are, from the human viewpoint, among the most edible, most familiar, and most studied crustaceans.

Peneid shrimp from the Gulf of Mexico is 15 cm long and one of a number of commercially important species of *Penaeus.* Many peneids occur in enormous numbers on shallow sandy ocean bottoms, where they scavenge bits of dead and dying plant and animal material. Peneid and sergestid shrimps, infraorder PENAEIDEA, are the only decapods that freely spawn their eggs, which hatch into nauplii. (R.B.)

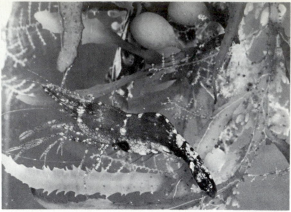

Cleaning shrimp *Stenopus hispidus,* infraorder STENO-PODIDEA, is white with bands of iridescent blue-green, red, orange, and purple, and waves its long white antennas conspicuously, all of which presumably helps fishes to find and recognize it as a cleaner. When a fish presents itself for cleaning, moving in close and taking up a distinctive pose, the shrimp probes among the gills and over the body surface for dead tissue and parasites, which are picked off and eaten. This and related species live in the tropics and subtropics. Members of one stenopid species, *Spongicola venusta,* live in pairs inside glass sponges (chapter 3). Caribbean. (K.B. Sandved)

Shrimp larva shows the transparency and relatively large eyes characteristic of many larval crustaceans. This is a late larval stage of *Processa canaliculata,* a deep-water caridean. England. (D.P. Wilson)

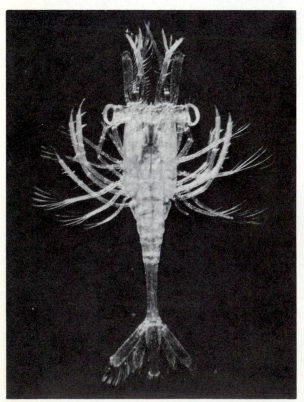

Sargassum shrimp *Leander tenuicornis,* infraorder CARIDEA, shows the 2 pairs of chelas and sharply-bent abdomen typical of carideans, which are the largest group of decapod shrimps (about 1,600 species). This shrimp characteristically lives on floating clumps of the brown alga *Sargassum,* although also found in benthic habitats. The members of the sargassum community feed on the sargassum, on surrounding plankton, and on each other. They also fall prey to passing fishes and seabirds, and like other sargassum species, *L. tenuicornis* is camouflaged by patchy yellowish-brown coloration and spiny outlines. (R.B.)

Snapping shrimp *Alpheus lottini* extends both its chelas from a crevice in its host coral (*Pocillopora*) and reaches for a cushion star (*Culcita*), which feeds on corals. The small snapping shrimps and crabs (*Trapezia*) that live with the coral seem to be amazingly effective in driving off the large coral-eating sea stars *Culcita* and *Acanthaster.* Corals from which the crustaceans have been experimentally removed are eaten much more frequently than are adjacent corals still occupied by their crustacean symbionts. Guam. (P. Glynn)

The larger of the 2 chelas of a snapping shrimp may be almost as long as the body. When forcefully snapped shut, it makes a loud noise and, underwater, produces a pressure wave that can stun a small fish. "Snapping" is thought to function in predation, defense, and intra-specific communication.

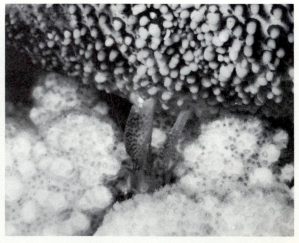

Crayfish, 10 cm long, closely resembles a small lobster; both are in the infraorder ASTACIDEA. Crayfishes live in freshwater lakes, ponds, and streams. They are omnivores, scavenging on dead organic matter and catching small fishes. Lake Pymatuning, Pennsylvania. (R.B.)

Cave crayfish, white and blind like many cave animals, depends on its especially long sensory antennas. *Cambarus.* Body, 8 cm Missouri. (R.B.).

Swamp crayfish, *Cambarus,* beside the chimney that surmounts its burrow. The burrows are commonly up to a meter or more deep and have at the bottom a water-filled cavity; they are built in swamps and meadows, often far from any stream. (C. Clarke)

Eggs of a crayfish, *below left,* develop while attached to the abdominal appendages, as in most decapods.

Young crayfishes, *below right,* hatch at an advanced stage and cling to the mother for some time. After only 2 or 3 molts, they show all adult features. (Both, C. Clarke)

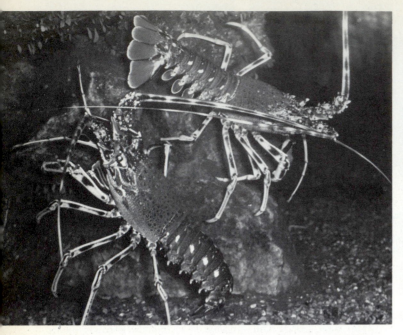

Spiny lobsters, infraorder PALINURA, are relatively gregarious and several can sometimes be found sharing a rock cave. Shown here are European spiny lobsters, *Palinurus vulgaris.* A spiny lobster has no large claws but will defend itself, its food, or its territory against both conspecifics and other species (including unwary human divers) by lashing out with its strong, spiny second antennas. At the same time, it often issues a warning, a rasping sound made by stridulating with the second antennas. Foraging mostly by night, spiny lobsters are

Slipper lobster has the second antennas modified as broad, platelike structures. Between them are the small, conventional first antennas. Like spiny lobsters, slipper lobsters are palinurans, have phyllosoma larvas, and are mostly tropical and subtropical. *Scyllarus.* Misaki, Japan. (R.B.)

omnivorous scavengers and also take live prey. Their larvas, called phyllosomas (see photo in chapter 20) are transparent and leaflike; they live and feed in the plankton. In some places in the southwestern N. Atlantic, thousands of spiny lobsters (*Panulirus argus*) migrate together over a period of a few days in the fall. Closely-linked chains of up to 65 individuals each, marching in single file, may be seen crossing open, shallow areas day and night. Traveling together in lines has been shown to reduce drag as the animals proceed through the water, and may aid in orientation and defense, but the significance of the migrations is not known. (D.P. Wilson)

Burrowing-shrimps, infraorder THALASSINIDEA, feed partly on worms and other burrowing animals, partly on tiny organisms and particles of organic matter sifted from the mud or filtered from the respiratory current of water that they pump through their burrows. These may be quite extensive and elaborate, with multiple branchings, chambers, and openings. Burrowing-shrimps are sometimes abundant enough to play a major role in turning over sediments. *Thalassinia* is a pest in Indopacific rice fields, its burrows admitting saltwater into the fields and the excavated soil of each burrow forming a mound 4 m in diameter. In parts of West Africa, prodigious numbers of *Callianassa* are caught and consumed during periodic swarms. Above left, *Callianassa subterranea.* Below left, *Upogebia deltaura.* Plymouth, England. (D.P. Wilson)

Hermit-crab, infraorder ANOMURA, eating a dead stomatopod. Hermits are scavengers and/or filter feeders. A hermit usually protects its soft slender abdomen in a snail shell. The asymmetrical abdomen curls to fit the right-handed spiral of most shells and has appendages only on the left side. The cephalothorax can be protruded, and the animal walks about carrying its shell, into which it retreats at any sign of danger. Only 2 pairs of legs are well developed for walking. When the cephalothorax is withdrawn into the shell, the heavy chelas block the opening, which they grow to fit exactly. Eventually, however, the animal grows too big for its shell and must either find another empty one or try to steal one from another hermit (hermits only rarely evict snails). If many shells are available, a hermit may be choosy as to fit, weight, and other shell characters. Often, however, there are few shells, and the interactions of hermit species have become a classic ecological subject for studying competition and partitioning of a limited resource. One important lesson from such studies is that hermits do not always share the human bias that bigger, stronger, more perfect shells are better. Given a choice of shells, hermits of some species select ones which are thin, broken, and skimpy. Perhaps these confer advantages such as mobility or camouflage. When no snail shell is available, hermit-crabs substitute other coverings, and some species with straight abdomens regularly carry about hollow pieces of mangrove or bamboo or scaphopod shells. Others inhabit immovable shelters such as calcareous worm or vermetid tubes and holes in rock or coral. Both larval and adult structures indicate that the hermit-crab form and habit evolved independently at least twice: in the mostly temperate Paguroidea such as the *Pagurus* in this photo, and in the tropical and subtropical Coenobitoidea such as *Paguristes, Dardanus,* and *Coenobita* (photos on this and next page) and the coconut-crab *Birgus* (photo, page 477). (R.B.)

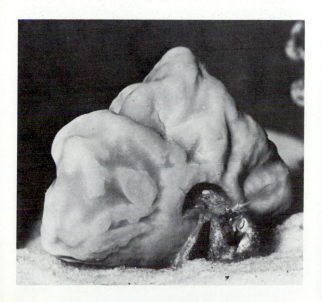

Well-disguised hermit-crab, *Paguristes hummi,* has bright orange sponge *Xestospongia halichondrioides* covering the adopted shell. But it is less heavily and bizarrely laden than some of its fellows. On soft bottoms, the shells of species of *Paguristes* and some other hermits are among the few suitable hard substrates for a great variety of organisms that require them—sponges, hydroids and sea anemones, bryozoans, barnacles, tunicates, various seaweeds, and others. The activities of the hermit are essential, protecting its shell symbionts from predators and from burial by sediment. This hermit, a rather sluggish filter feeder, is less disadvantaged by its heavy load than a more actively scavenging hermit would be, and the water currents created by its filter feeding and walking further benefit many of the symbionts. The effects of the symbionts on their hermit host are complex. Some weaken the shell while others strengthen it. All add to the shell's weight and decrease mobility, but they provide camouflage and, in the case of stinging cnidarians, protection. Hermits with towering growths of branching sponges or lopsided piles of large barnacles must do a continual balancing-act, but a well-balanced mound of sponge can make the shell more stable and harder for a predator to turn over. Height, 4 cm. NW Florida. (R. B.)

Hermit-crab carrying sea anemones is *Dardanus arrosor* of the Mediterranean. This hermit takes an active part in establishing the anemones, *Calliactis parasitica,* on its snail-shell. The hermit vigorously pokes and pulls at the anemones until they detach, then holds them against its shell until they take hold again. One hermit will often steal anemones from another. It has been shown that *Calliactis* protects hermits against predation by octopuses and crabs in the laboratory. The hermit *Pagurus bernhardus* of northern Europe, which also carries *Calliactis,* plays no active role in forming the association. The anemones respond to substances in the surface of the hermit's snail-shell (most commonly *Buccinum*) and climb on by themselves. (R.B.)

Land hermit-crab *Coenobita rugosa,* of East African coasts, is often seen foraging in large numbers. These hermits climb bushes and trees, eating fruits and vegetation, carrion, and any prey slow enough to be caught. *Coenobita* shows many adaptations for terrestrial life, such as chemosensory structures on the first antennas that differ from those of marine hermits and resemble in some respects those of terrestrial insects. The gills are so small that the animal drowns in water, but there is a special, abundantly vascularized abdominal area that effects respiratory exchange in air. Land hermits visit water regularly and will select a suitable mixture of fresh-, brackish-, and saltwater sources from which to fill the shell, a handy water jug. Young stages develop in the sea. Species of *Coenobita* are all tropical or semitropical and are sometimes sold as pets. (R.B.)

Galatheid *Pleuroncodes planipes* was photographed when thousands washed ashore in Monterey Bay, California, their bodies turning the beaches red. Sea gulls flocked to feast on them. Rarely do swarms of *P. planipes* appear this far north, but they are commonly seen off southern California and Mexico, and other species of *Pleuroncodes* replace them off Ecuador and Peru. Called lobster-krill, their great abundance and habit of swarming makes them, like euphausiids, an important food item for albacores and tunas and also for baleen whales when the great mammals migrate from the Arctic and Antarctic to more temperate latitudes during their respective winters. Length, 4.5 cm. (R.B.)

Mole-crab, or sand-crab, *Emerita,* is one of the few sizeable invertebrates of open sandy beaches that manage to live in the zone of breaking waves. The smooth egg-shaped body, with abdomen curled under, offers a minimum of resistance to moving water and sand. The animal burrows backward into the sand, facing incoming waves, and lies at an angle, with plumed antennas protruding and serving as a net for screening plankton out of the backwash of each wave. By moving up and down the shore with the tides, they increase the time available for feeding from receding waves. *Emerita analoga* of the U.S. west coast often selects areas where water continually drains down the beach and feeds also on algal cells. Shown here is the mole-crab of the U.S. east coast, *Emerita talpoida* (30 mm long), which feeds only where there is strong wave action. (R.B.)

Porcelain-crab, *Petrolisthes armatus,* is a filter-feeding anomuran. Crablike anomurans have independently evolved body forms like those of brachyuran crabs, but have (besides the large chelipeds) only 3 pairs of walking legs, the 4th legs being much reduced and specialized. A soft abdomen and long antennas lateral to the eyes also distinguish many anomurans. Carapace width, 2 cm. NW Florida. (R.B.)

Brachyuran crab, infraorder BRACHYURA, has 4 pairs of walking legs. This kelp crab, *Pugettia producta,* lives and feeds mostly on kelps and other brown seaweeds, and often matches perfectly their olive-brown color. If the usual plant food is not available, kelp crabs will switch to eating sessile invertebrates such as barnacles and hydroids. Carapace width, 6 cm. Long Marine Lab., Santa Cruz, California. (R.B.)

Most crustaceans can be divided into two groups, perhaps representing an ancient divergence. The classes Cephalocarida, Branchiopoda, and Malacostraca can be grouped as *thoracopodans,* crustaceans in which currents for filter feeding are typically created by a series of thoracic appendages and food particles are passed forward in a ventral groove to the mouthparts of the head region. Among malacostracans this mode of feeding is seen most clearly in leptostracans, generally viewed as closest to original malacostracan stocks, as well as in filter-feeding mysids and euphausiids. Thoracopodans tend to have a relatively large number of trunk segments.

The classes Maxillopoda and Ostracoda can be characterized as *maxillopodans,* crustaceans in which the maxillas are the important food-gathering appendages. The number of trunk segments is relatively small, and members of the **class MAXILLOPODA** share a basic plan of 10 trunk segments (6 thoracic and 4 abdominal) plus a telson. This class includes the copepods, mystacocarids, tantulocarids, branchiurans, and cirripeds (barnacles and related groups). Of these, copepods best illustrate the maxillopodan feeding mode. Barnacles, on the other hand, feed with their thoracic appendages (although not as thoracopodans do) and best illustrate the irrepressible flexibility of crustacean evolution.

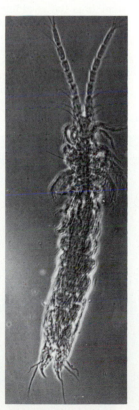

Mystacocarid, *Derocheilocaris typicus,* is an interstitial maxillopodan, somewhat resembling a harpacticoid copepod. Only a few species in one genus have been described in the subclass MYSTACOCARIDA. The largest is only 0.5 mm long. Little is known of their biology. Mystacocarids are interesting for the primitive, almost larval appearance of the adults, which suggests that they might have evolved by progenesis, precocious sexual maturation at a morphologically larval stage. The same has been proposed for copepods, because they also retain, as adults, features usually associated with larval stages, such as the nauplius eye. (Preserved specimen, Scripps collection, courtesy W.A. Newman) (R.B.)

Tantulocarid, *Basipodella atlantica.* Members of the subclass TANTULOCARIDA are ectoparasites on other crustaceans of the deepsea benthos (2,000 to 5,000 m), and they are well-named: *tantulo* = "so small," *carid* = "shrimp." The copepod in the drawing below is only 1 mm long, and the tantulocarid is the minute globular bit hanging on near the end of its host's first antenna. This adult female tantulocarid, 0.15 mm long, is saclike; but young stages have a segmented thorax and abdomen and biramous thoracic appendages, features that are lost at metamorphosis. These animals have been classified with copepods or with cirripeds, and only after recent detailed studies has separate-but-equal rank been proposed for them. (After Boxshall and Lincoln)

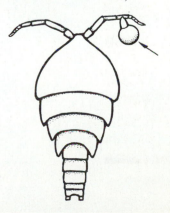

Calanoid copepod, *Calanus helgolandicus,* 3 mm long, belongs to the subclass COPEPODA. It feeds mainly with the second maxillas, filtering small phytoplankton from the water or scooping up larger types one by one. Although the number of species is modest (about 7,500), copepods have been estimated to be the most numerous multicellular animals in the world. They are among the foremost consumers of phytoplankton in both marine and fresh waters, and are in turn fed on by a great variety of invertebrates, fishes, and baleen whales. Not content with their starring role as primary consumers, most filter-feeding copepods such as this one also use the second maxillas to take prey, and many copepods are largely predaceous. As predators, they devour various small or larval invertebrates, as well as fish eggs and larvas. The prey thus include many species which, at later stages, eat copepods. In addition, whole orders of copepods are parasitic on a variety of invertebrate and vertebrate animals, especially fishes. Calanoids, often recognizable by their extremely long first antennas, are free-living and are the dominant marine, pelagic copepods. (D.P. Wilson)

Cyclopoid copepod *Cyclops* was named for the median naupliar eye (dark spot at front of head), found not only in this common freshwater genus but generally present in copepods, which have no compound eyes. Mature females usually carry two egg sacs. The eggs hatch into nauplii, and the larvas undergo many molts before they become adults. *Cyclops* and some other copepods serve as intermediate hosts for tapeworms and nematodes that eventually find their way to human and other vertebrate hosts (see chaps. 9 and 12). Many cyclopoids are themselves parasitic, but the majority are free-living, benthic or pelagic in marine, brackish, and fresh waters. Cyclopoids do not filter feed but use the mandibles, both pairs of maxillas, and a pair of maxillipeds to feed on a variety of animal and plant food, including mosquito larvas. (P.S. Tice)

Harpacticoid copepod, *Tigriopus californicus,* 1 mm long, lives in high tidepools where great fluctuations in salinity and temperature exclude predators such as fishes. It feeds mainly by scraping organic material from the rocks, but can also catch small animals and filter-feed. Harpacticoids are mostly free-living and benthic, and they are more various in shape than pelagic types, sometimes flattened like minute isopods or amphipods. The many interstitial species are elongate and cylindrical, like other interstitial animals. After the nematodes, harpacticoids are the most numerous animals of the seabed. Absence of harpacticoids may be an indicator of high concentration of heavy metals in bottom sediments. Monterey Bay, California. (R.B.)

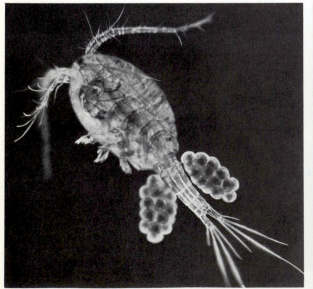

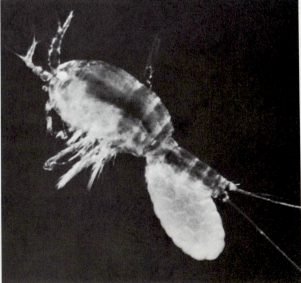

The subclass BRANCHIURA has usually included only the **fish-lice,** ectoparasites of fishes. But recent evidence links fish-lice with another wholly parasitic group, the **pentastomids,** endoparasites of various vertebrates, mostly reptiles. Modifications for parasitic life have made the affinities of pentastomids hard to make out. Most zoologists agree that pentastomids are related to arthropods because of the structure of the periodically molted, chitinous cuticle of the body surface, foregut, and hindgut; the absence of cilia; the way in which gametes are formed; the hemocoel; the cross-striated muscles; the evidence of segmentation (coelomic sacs during development, ganglionated ventral nerve cords, paired appendages); and some other arthropodlike characters.

As long as there seemed to be no clear way to decide which arthropods they should be grouped with, the easy way out has been to treat pentastomids as a separate phylum. But their placement in the Branchiura now seems reasonable. Sperms of fish-lice and pentastomids show many detailed similarities and are unlike any known in other animals. A plausible case also has been made that pentastomid ancestors were ectoparasites on fishes, as fish-lice are today. Ingestion by aquatic fish-eating reptiles would then provide an easy route to endoparasitism in new hosts, and these would finally come to include terrestrial reptiles, the hosts of most modern pentastomids.

Like the other maxillopodan subclasses, the Branchiura is sometimes listed as a class in its own right.

Fish-louse, *Argulus,* 6 mm long, order ARGULIDA, resembles some parasitic copepods but has paired compound eyes and a broad, flat carapace *(dorsal view, left).* It attaches to fishes, and sometimes to tadpoles or newts, by means of hooks or suckers *(ventral view),* and feeds on blood. A few fish-lice cause little damage to a large fish, but when a single fish bears hundreds or thousands, it is quickly weakened, and soon the fish-lice must swim off to find a new host. Mating takes place on a host, which the female leaves to deposit the eggs on a submerged rock or plant, instead of brooding them as most crustaceans do. About 130 species of argulids are found in marine and fresh waters. (Preserved specimens, Scripps collection, courtesy W.A. Newman.) (R.B.)

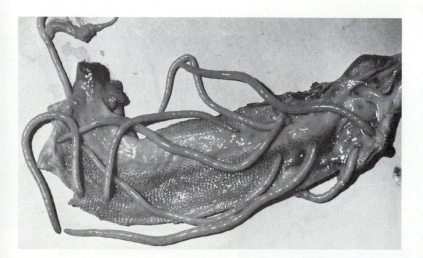

Adult pentastomids, *Kiricephalus pattoni,* attached by a row of hooks on the head to the lining of the lung of a snake. Taiwan. (R.E. Kuntz)

The 90 or so species in the order PENTASTOMIDA feed on blood and tissue fluids in the lungs or air passages of their vertebrate hosts, most of them reptiles, a few birds and mammals. Pentastomids are found mostly in the tropics, but a few extend into cold and temperate areas in warm-blooded hosts. The head attaches to host tissue with hooks, sometimes borne at the ends of 4 stumpy appendages. These, together with the snout that bears the mouth, inspired the unfortunate order name, which means "5-mouthed." The superficially ringed body, up to 16 cm long, contains little besides a straight gut and (as in most parasites) a generous reproductive system. A pair of large glands produce secretions that may protect the parasite against host defenses.

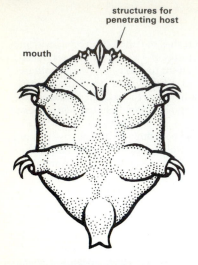

structures for
penetrating host

mouth

Larval pentastomid, *Porocephalus,* with stumpy clawed appendages. Even the larva of the pentastomid genus *Subtriquetra,* which has a brief free-swimming stage, must finally penetrate and migrate through the tissues of the intermediate host. So the differences between a pentastomid larva and a typical, swimming nauplius are not surprising. (Partly after Penn)

Male and female pentastomids mate within the final host, and millions of eggs pass out in the host's nasal secretions, saliva, or feces. When eaten by an intermediate host, the eggs hatch into tiny 4- or 6-legged larvas that bore through the gut wall and enter various organs, where they feed and grow. Eventually, they encyst and wait for the intermediate host to be eaten by some predator, which becomes the final host. The young pentastomids quickly emerge from their cysts and attach to the walls of the nasal passages or lungs, where they mature to adults.

The intermediate hosts are mostly vertebrates (fishes, amphibians, reptiles, small mammals), but cockroaches serve as intermediates for a pentastomid found in lizards. When mammals serve as final hosts (dogs and other carnivores), the intermediate hosts are also mammals, including domestic stock such as cattle, sheep, and goats.

Humans sometimes become infected by ingesting food or water contaminated with eggs. Such infections are common in parts of South America, Africa, and the Middle and Far East, but unless the number of larvas is unusually great, no symptoms are detected. Since people are rarely eaten by a suitable final host, the larvas eventually die within their cysts. Humans also become infected by eating the undercooked liver of an intermediate host. The young pentastomids that emerge cause temporary irritation and swelling of mouth, throat, and nasal passages, but do not survive.

Barnacles are members of the maxillopodan subclass CIRRIPEDIA, order THORACICA. Shown here are several buoy-making barnacles, *Lepas fascicularis.* These stalked, open-ocean barnacles secrete a frothy substance that keeps them afloat. Natural size. (D.P. Wilson)

Primitive cirriped, *Synagoga sandersi,* order ASCOTHORACICA. The adult retains a bivalved carapace, present only in the cyprid larva in other cirriped orders. The right valve of this specimen has been removed. Length, 4 mm. (Preserved specimen, courtesy W.A. Newman.) (R.B.)

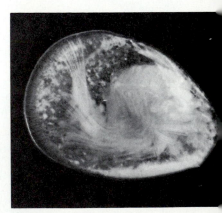

Other cirripeds shown earlier in this chapter were parasitic types in the order Ascothoracica and order Rhizocephala. A fourth group is the order Acrothoracica, which includes small, shell-less cirripeds that burrow in calcareous substrates.

Ostracods, class **OSTRACODA,** are mostly under a few millimeters long and have a clam-like, bivalved carapace, hinged at the top and with the outer walls calcified. The body is short and compact and not obviously segmented, with only 5 to 7 pairs of appendages, plus a caudal furca. With these few tools, and despite their rather uniform appearance, ostracods move and feed in many different ways; over 2,000 living species are known from fresh and marine waters, plus about 10,000 fossil ones. The well-developed antennas are sometimes the only appendages that protrude from the carapace. They may be used for the several kinds of locomotion as well as in sensory and feeding functions. The fifth appendages are also multipurpose and serve for grasping food or for filter feeding in various ostracods, and for walking in others. Ostracods feed as predators, herbivorous grazers or filter feeders, omnivorous scavengers, detritus feeders, or parasites. When the carapace valves are closed by adductor muscles extending between them, the small "seed-shrimp" is completely enclosed. When the valves are open, the animal extends its appendages to scurry busily about, walking, climbing, burrowing, or swimming; a few even clamber through moist terrestrial habitats. Most ostracods release or deposit their eggs, though a few brood in the carapace. Freshwater ostracods (like branchiopods) are commonly parthenogenic and produce eggs resistant to drying. Ostracod nauplii have a distinctive bivalved carapace. (R.B.)

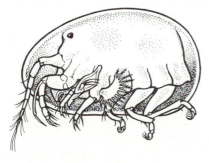

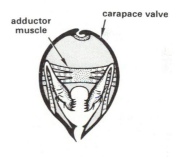

Ostracod in lateral view, with the left carapace valve removed, shows 7 appendages. The dark spot in this podocopid ostracod is a naupliar eye; myodocopids, which are all marine, may have paired compound eyes in addition. (Modified after G.O. Sars)

Transverse view of ostracod, shows adductor muscles running between hinged carapace valves. (Mod. after J.P. Harding)

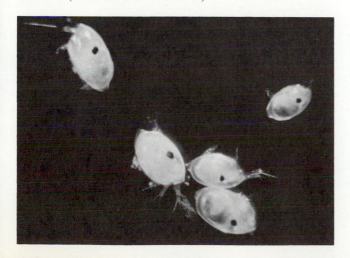

Luminescent ostracods, *Cypridina serrata,* react to light or mechanical stimulation by emitting secretions, a luciferin and a luciferase, from 2 separate glands in the head. The secretions mix and, in the presence of oxygen, produce a bright cloud of bluish luminescence in the seawater, persisting for several seconds. The cloud, 2 to 15 cm long, is many times larger than the 1.6 mm ostracod and is presumed to startle or distract potential predators, allowing the ostracod to escape. Other luminescent ostracods produce blue, yellow, or green light. If kept dry, powdered *Cypridina* will keep indefinitely and will luminesce brilliantly when moistened. (Courtesy F.I. Tsuji)

CLASSIFICATION: Phylum ARTHROPODA, Subphylum CRUSTACEA

Class REMIPEDIA

Class CEPHALOCARIDA

Class BRANCHIOPODA
Order Anostraca, fairy-shrimps
Order Notostraca, tadpole-shrimps
Order Conchostraca, clam-shrimps
Order Cladocera, water-fleas

Class MALACOSTRACA

Subclass PHYLLOCARIDA
Order Leptostraca

Subclass HOPLOCARIDA
Order Stomatopoda, mantis-shrimps

Subclass EUMALACOSTRACA

Superorder Syncarida
Order Anaspidacea
Order Bathynellacea

Superorder Pancarida
Order Thermosbaenacea

Superorder Peracarida
Order Spelaeogriphacea
Order Mictacea
Order Mysidacea, opossum-shrimps
Order Amphipoda, beachhoppers, etc.
Order Isopoda, sowbugs, etc.
Order Tanaidacea
Order Cumacea

Superorder Eucarida
Order Euphausiacea, krill, etc.
Order Decapoda
  Suborder Dendrobranchiata
   Infraorder Penaeidea, shrimps
  Suborder Pleocyemata
   Infraorder Stenopodidea, shrimps
   Infraorder Caridea, shrimps
   Infraorder Astacidea, lobsters
   Infraorder Palinura, spiny lobsters
   Infraorder Thalassinidea, burrowing-
    shrimps
   Infraorder Anomura, hermit-crabs, etc.
   Infraorder Brachyura, crabs

Class MAXILLOPODA

Subclass MYSTACOCARIDA
Subclass TANTULOCARIDA
Subclass COPEPODA
Order Calanoida
Order Harpacticoida
Order Cyclopoida
(6 other orders, mostly parasitic)

Subclass BRANCHIURA
Order Argulida, fish-lice
Order Pentastomida, tongue-worms

Subclass CIRRIPEDIA
Order Ascothoracica
Order Thoracica, barnacles
Order Acrothoracica
Order Rhizocephala

Class OSTRACODA, seed-shrimps

Subclass MYODOCOPA
Subclass PODOCOPA
Subclass PALAEOCOPA

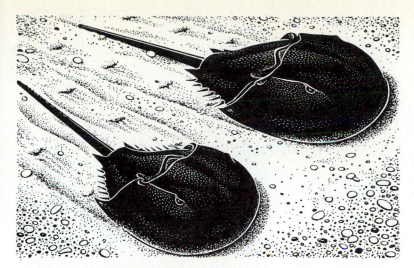

Chapter 22

Chelicerates

Horseshoe-crabs
Arachnids
Pycnogonids

MOST OF US have had to learn at some time that spiders are not insects. And many people on the east coasts of North America and Asia have seen "horseshoe-crabs" on sandy ocean beaches and have been told that these are not crabs. But the fact that spiders and horseshoe-crabs are related to each other still comes as a surprise. They are grouped, along with other spider relatives such as ticks, harvestmen, and scorpions, in the arthropod subphylum **CHELICERATA.** Usually included in this group are odd spindly animals called pycnogonids, or "sea-spiders"; and these, as may no longer come as a surprise, are not spiders.

The name of the subphylum is derived from a characteristic pair of appendages, the cheliceras (ke-lis'-e-ras).

Chelicerata (ke-lis'-e-ra'-ta)

HORSESHOE-CRABS

A FEW SPECIES of horseshoe-crabs are the only living representatives of the class **Merostomata.** These animals are often referred to as "living fossils" because they differ so little from fossil merostomes that lived nearly 400 million years ago. Modern horseshoe-crabs were once included in one genus, *Limulus,* along with some fossil forms from rock layers 200 million years old. Although the oldest of these fossil species of *"Limulus,"* and 3 of the 4 living ones, are now assigned to other genera, the similarity among these animals remains remarkable. How can we explain the uniformity and persistence of this small ancient group? We could suppose that they are somehow resistant to mutation and peculiarly lacking in genetic variability, and therefore unable to change. However, an analysis of various proteins, assumed to indicate genetic variability, showed as much diversity in *Limulus polyphemus* as in species from animal groups that are said to be evolving at more usual, faster rates. The explanation may be that horseshoe-crabs occupy a steadily available type of habitat within which they accept a diversity of food and tolerate large fluctuations in salinity and temperature. They are so well

Fossil limulid, *Mesolimulus walchi,* 87 mm long, from Upper Jurassic limestone (some 140 million years ago) in Solnhofen, Bavaria, West Germany. Carnegie Museum. (R.B.)

529

Horseshoe-crab has conspicuous compound eyes, fixed in the carapace, that look out from each side. The broad front edge of the carapace is used to dig into soft bottoms in search of prey or shelter. Horseshoe-crabs burrow deep in the substrate if left exposed by a receding tide, or during much of the winter in cold regions. *Limulus polyphemus* reaches a length of more than 60 cm, takes 9 years or more to mature, molts many times, and lives 15-20 years. It is abundant along the Atlantic coast of North America and is a major predator on worms and clams. (R. B.)

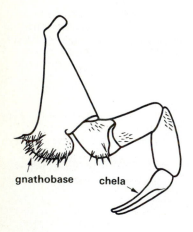

First walking leg of *Limulus*, showing the chewing process.

Note that *chelicera* and *chela* (kee'-la) are distinct. A chela is any arthropodan pincer, such as the large claw of a lobster. Cheliceras often bear chelas, and then are described as *chelate*.

The second pair of appendages are called *pedipalps*. They are walking legs or are specialized, in different orders of chelicerates, for sensory functions, for food capture, for defense, and for mating. In horseshoe-crabs, the pedipalps of the mature male are specialized for clasping the carapace of the female during mating; the pedipalps of immature males and females of all stages resemble the other walking legs.

In the various chelicerate groups other appendages may be similarly specialized but have usually not been dignified with special names.

suited to their environment, without being highly specialized, that almost any gross morphological change would be disadvantageous.

Horseshoe-crabs live in shallow water along sandy or muddy seashores. The body consists of **cephalothorax** and **abdomen,** and the shieldlike carapace that arches over the dorsal surface is hinged at the junction of the two regions. From the posterior end of the abdomen there extends a long tail spine. The animals swim through the water ventral side up by flapping the flattened appendages on the abdomen. But they spend most of the time burrowing in the soft substrate for the worms and molluscs on which they feed.

Like all chelicerates, and unlike other arthropods, horseshoe-crabs do not have antennas. The first pair of appendages, in front of the mouth, are feeding appendages called **cheliceras.** In horseshoe-crabs the cheliceras, and all except the last of the *5 pairs of walking legs,* bear pincers (chelas) that firmly grasp any food organism encountered in the soft bottom. There are no appendages like the mandibles of other arthropods, exclusively specialized for chewing the food, but the walking legs bear **gnathobases,** processes on their basal parts that are specialized for manipulating food, much as in primitive crustaceans. The spiny gnathobases of the first 4 pairs of walking legs coarsely shred the food and pass it to the mouth. The bases of the last pair can be brought together forcefully to break the shells of bivalves and bite or crush tough food. Pieces of food are swallowed together with sand grains and bits of shell and finely ground up in the strong gizzard. Surrounding

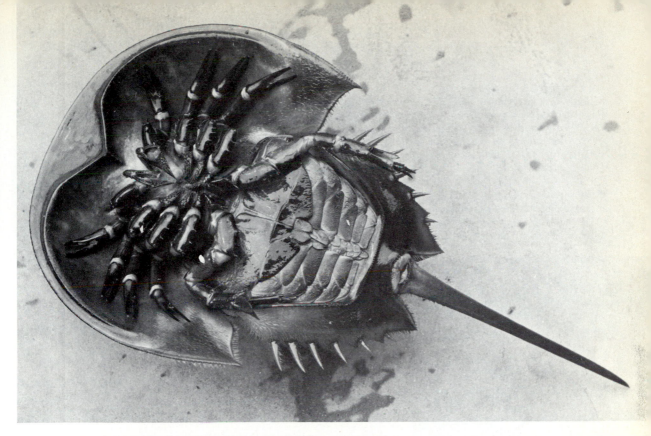

Underside of horseshoe-crab, *Limulus polyphemus*, shows the appendages that lie beneath the protective carapace. The small cheliceras and first 4 pairs of walking legs end in pincers that pick up food from the substrate. In mature males the first walking legs are modified for clasping females. The last pair of walking legs provide the force for digging and bear flat processes that keep them from sinking into the substrate. The flattened abdominal appendages are used in swimming and scuttling along the bottom, and bear the leaflike gills. The anus can be seen at the base of the tail spine. Delaware Bay. (R.B.)

the foregut is a wide collar of nervous tissue that represents a condensation of a large portion of the nervous system. The gut probably serves an excretory role, supplementing that of the paired **excretory organs** (coxal glands) at the bases of the walking legs.

Flattened abdominal appendages, used in both swimming and fast scuttling along the bottom, also bear the **book gills,** groups of thin plates well supplied with blood. The blood circulates throughout the body in an extensive system of

Blood cells of horseshoe-crabs aggregate at the site of a wound and release granules containing coagulable protein, forming a clot that stops the bleeding. This same system responds to minute amounts of bacterial endotoxins and serves to contain invading bacteria. Extracts of horseshoe-crab blood cells are used in diagnosing human bacterial diseases and in checking for the presence of bacterial endotoxins in drugs and intravenous solutions.

Underside of cephalothorax. The small pincers of the cheliceras, in the midline, are directed backwards over the mouth. Behind these, on each side of the mouth, are the spiny gnathobases of the walking legs. This animal appears to have more than its share of light-colored, arrowhead-shaped turbellarian flatworms of the genus *Bdelloura*, commensals often seen on *Limulus* (see chapter 9). Northwest Florida. (R.B.)

An overturned horseshoe-crab on a sandy beach, trying to right itself. Righting is readily accomplished under water, but on a dry beach an overturned animal may quickly die in the hot sun before the tide returns. Delaware Bay. (R.B.)

Because of the long spine, the group of merostomes that includes horseshoe-crabs is called the order **Xiphosura** (*xiphos* = "sword," *ura* = "tail").

Newly laid eggs are often washed out by waves, from the shallow depression in which they were laid. A large female may lay as many as 10,000 eggs, but relatively few hatch.

Mating pair of horseshoe-crabs. The smaller male clings to the larger female and is towed along, sometimes for many days, until they spawn together. Delaware Bay. (R.B.)

arteries and veins as well as in an open system of blood sinuses, and is colored pale blue by the copper-containing protein hemocyanin, which carries oxygen as in many crustaceans.

A long **tail spine** hinged to the posterior end of the abdomen is used in righting, which, because of the animal's habit of swimming on its back, is more than an occasional necessity. An overturned horseshoe-crab arches up on the tip of its tail spine and rolls from side to side while beating the abdominal appendages; eventually it shifts its center of gravity far enough to one side to topple over, right side up. The length of the tail spine of *Limulus* has been found to be exactly optimal for righting; spines of either longer or shorter proportions are less effective in righting, a matter of life or death when a hungry predator is lurking close by. A seagull preparing for a horseshoe-crab dinner first flips the victim, then often bites off and discards the tail spine before beginning to eat, thereby preventing the "dinner-plate" from turning over. A horseshoe-crab without a tail spine cannot right itself and also flips over more often. Normally, the tail spine is

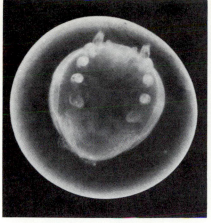

Young embryo of *Limulus* lies on sphere of yolk and is enclosed by heavy egg covering. Cephalothorax shows rudiments of cheliceras and 5 pairs of walking legs. Abdomen shows 2 pairs of leaflike appendages. (R.B.)

"Trilobite" larva swims about actively or burrows in the bottom. This dorsal view shows median and lateral eyes on the cephalothorax and the rudiment of the tail spine at the rear of the abdomen. (R.B.)

Ventral view of "trilobite" larva shows dark branches of digestive glands, a pair of cheliceras in center, and 5 pairs of walking legs. The larva swims ventral side up by flapping the abdominal appendages. (R.B.)

braced against the bottom to resist the lifting, flipping action of waves and obstacles.

During mating, the male clasps the female's carapace with his modified second appendages, and she tows him around with her. When ready to lay her eggs, she digs a depression in the sand near mid or high tide mark. As she deposits the eggs in the depression, the male sheds sperms over them. The eggs hatch into free-swimming young called **"trilobite" larvas** because they look strikingly like trilobites. They are similar to adults but lack the long tail spine. The larval eyes later degenerate, and two pairs of adult eyes form, simple eyes at the front of the carapace and **compound eyes** at the sides. The compound eyes of horseshoe-crabs, although similar in principle and in gross structure to those of crustaceans and insects, differ in detail and are clearly independent inventions. Horseshoe-crabs are the only living chelicerates with compound eyes.

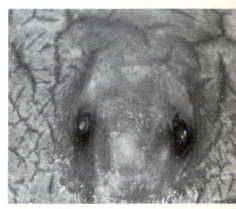

Simple eyes of *Limulus* lie in the midline. (R.B.)

The compound eyes of *Limulus* are large and convenient for neurophysiological studies and may be better known than the eyes of any other animal. It has been found that as each ommatidium responds to a light stimulus, it inhibits the activity of nearby ommatidia. The brighter the light, the greater is the response and the more powerfully are neighbors inhibited. Thus if one ommatidium is illuminated more intensely than another nearby, the former will not only respond more strongly but will also inhibit its neighbor, and thus the difference between the two will be increased. This phenomenon is called **lateral inhibition,** and its effect is to enhance the *contrast* of the image. When an object moves across the visual field of a horseshoe-crab, the neural response to the moving boundary of contrast is greater, so that lateral inhibition is thought to be important in the perception of *motion*.

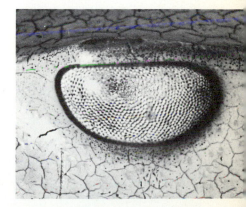

Compound eye of *Limulus*. (R.B.)

534

Typical spider body form is seen in the shamrock spider, *Araneus trifolium,* common in N. American meadows. It is an orb-weaver (family Araneidae). The female reaches 18 mm, the male 6 mm. With over 1,500 species, *Araneus* is the largest genus of spiders. It is found worldwide. (P.S. Tice)

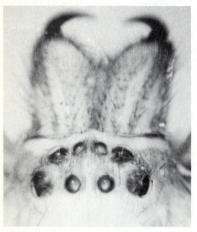

Most spiders have **8 simple eyes,** often arranged in 2 rows, but arrangements and numbers vary. (P.S. Tice)

Fanged cheliceras on underside of spider head. Below them extend, from the bases of the pedipalps, a pair of flat ridged plates that can be brought together to hold and squeeze prey as the spider sucks up prey fluids. (P.S. Tice)

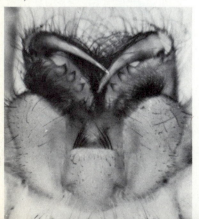

ARACHNIDS

THE CLASS **Arachnida** includes the spiders, scorpions, ticks and other mites, harvestmen, and several smaller groups. No other class of animals is less loved by most people. There is some basis for this dislike, in that scorpions and some spiders can inject a poison which produces painful, though usually not serious, results in humans; some ticks suck human blood and spread disease, and chigger mites are maddeningly irritating parasites in human skin. But few people in large cities have ever had a single unpleasant experience with an arachnid. The sinister reputation of a group like the spiders, which do little harm and vast good (by killing insects we consider pests), is based mostly on a vague fear of animals that run rapidly and live in dark places.

Hardly any description will fit all the orders but, in general, arachnids are terrestrial arthropods that have the body divided into two main regions: a **cephalothorax** bearing 6 pairs of appendages, of which 4 pairs are usually walking legs, and an **abdomen** that has no locomotory appendages, though it may have some other kind. The *4 pairs of walking legs* often serve as a convenient way of distinguishing arachnids from insects, which have only 3 pairs; but many arachnids have specialized the first pair of legs for sensory functions and walk on only 3 pairs. Arachnids differ from most crustaceans, insects, and horseshoe-crabs in having *no compound eyes,* only simple ones. Like horseshoe-crabs, but unlike other arthropods, arachnids have *no antennas* and *no mandibles.* In front of the mouth is a pair of **cheliceras,** which end in pincers or (in spiders) sharp fangs. The next pair of appendages, behind the mouth, are the **pedipalps,** which are leglike and sensory, as in spiders, or are used for seizing prey, as in scorpions. Except for some mites and harvestmen, arachnids are predators. Among the various orders either the cheliceras or the pedipalps are the important weapons of offense, but never both in the same animal.

OF THE more than 70,000 described species of arachnids, about half are **spiders** (order Araneae). In a spider the cephalothorax is covered by a shield, the carapace, on which are set the simple eyes, usually 8 in number. The movable cheliceral fangs are sharp and pointed and are used in capturing and then paralyzing the prey by injecting a poison. Ducts from a pair of poison glands lead through the cheliceral fangs and open near their tips. The pedipalps look like legs but are sensory, and their expanded bases help to hold and compress the prey. In mature males the pedipalps are modified for transferring sperms to females. The 4 pairs of walking legs end in curved claws. The abdomen seldom shows any external evidence of segmentation and has no segmental appendages except the silk spinnerets, of which there are usually 3 pairs.

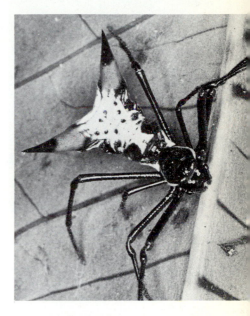

Black and yellow argiope *Argiope aurantia* (family Araneidae) is an orb-weaver common in gardens in the U.S. and Canada. The zig-zag pattern in the center of the inconspicuous web warns birds not to fly through it, saving the spider from frequent web rebuilding. The female measures 25 mm. The male, only 6 mm long, builds his own web, but when mature seeks out and joins a female. (C. Clarke)

Spiny-bellied spider, *Micrathena,* in Panamanian rain forest. There are many species of these spiny araneid orb-weavers in the American tropics, only a few in N. American woods and gardens. (R.B.)

Huntsman spider. The best-known member of this family, the Heteropodidae, is *Heteropoda venatoria,* found worldwide in the tropics and subtropics. It extends into the southern U.S. and may turn up in northern states along with bunches of bananas. (O. Webb)

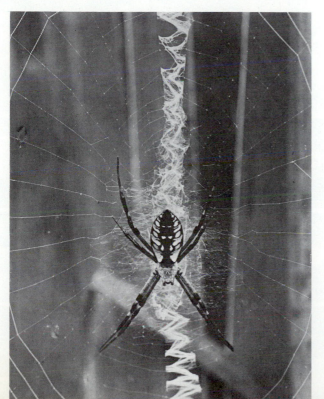

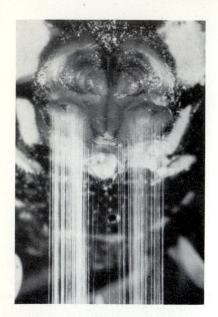

Spinnerets extruding silk.
(K.B. Sandved)

Silk, which issues from the spinnerets as a fluid, is a protein that polymerizes under tension; thus the silk hardens as it is stretched into a thread. The spinnerets are jointed fingerlike organs that have at their tips a battery of minute spinning tubes, sometimes a hundred or more on each spinneret. The spinning tubes connect with several kinds of silk glands that produce at least 6 different kinds of silk for spinning various parts of the web, binding the prey, forming the egg-sac, etc. Some of the tubes produce not silk but a sticky fluid that makes certain threads of the web adhesive. Almost all spiders constantly lay a silk dragline for safety as they run and climb, and many spiders build a silken home into which they can retreat, concealed from predators, when they are resting and especially when they are molting. Silk is also important in reproduction. A male spider spins a special tiny web (made by glands and silk tubes in the anterior part of the abdomen). On this sperm web he deposits a drop of semen that is then transferred to his pedipalps. Once arrived at the web of the female he is courting, he may spin a "mating line" onto which he must coax her. A female spider constructs an elaborate silken sac for her eggs and, in some species, a "nursery web" for the young spiders.

All spiders spin silk but only a few, less than 10%, weave the exquisitely geometrical orb-webs, so eye-catching that we

Cribellate silk of uloborid spiders as seen in scanning electron micrographs. The silk emerges from the cribellum (**A**), a flattened plate on the underside of the abdomen, just in front of the spinnerets. Very fine strands of silk emerge from thousands of microscopic spigots (**B**) that dot the cribellum. Together these strands form a silk sheet that is then pulled through a comb of curved spines, the calamistrum, one on each of the hind legs (arrow points to a calamistrum in **C**). This combing teases the silk into fuzzy strands, that entangle the hairs of the prey (**D**). Silk from a non-cribellate orb-weaving spider traps prey with evenly-spaced sticky droplets (**E**). (B.D. Opell)

The family Uloboridae is the only cribellate family that weaves orb-webs, but these differ from those of the "true" orb-weavers, the Aranaeidae. The uloborids are also the only spiders without poison glands. They wrap the prey in silk but do not bite it until they start to feed on it.

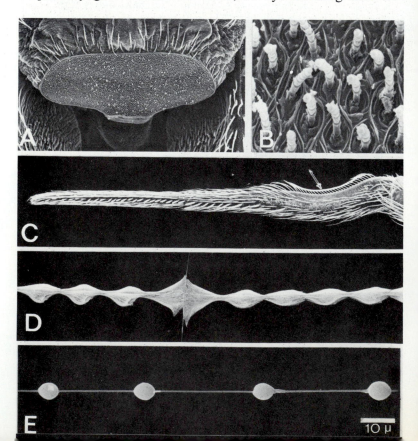

usually think of them as typical. Some spiders build sheetlike webs with a maze of trip-lines strung just above; when an insect blunders into the trip-lines and falls onto the sheetweb, the spider runs quickly along the undersurface, free of trip-lines, and seizes its prey through the sheet. Or, threads stretched between a scaffolding and the ground may bear sticky droplets just above ground level to snare crawling insects. Some spiders spin cobwebs with threads strung irregularly in all directions; still others manage to feed themselves using only single threads strung between twigs. A spider with one of these simpler snares cannot just wait for an insect to entangle itself but sits alertly holding its line and when it feels an insect hit, shakes the line vigorously or throws it into coils that entangle the prey.

The use of silk for trapping prey may have evolved when insects began to jump and fly, and many spiders followed their insect prey upward, finally building silken snares on the vegetation. The role of spiders in keeping insects in check has been known since ancient times; European and Asian folklore has many warnings of bad luck that follows the killing of a spider. One web may snare 500 minute insects in a single day, and an acre of undisturbed British meadow may harbor 2,000,000 spiders. Spiders outrank birds and other insect-eaters in control of insect numbers, but the effectiveness of all these natural predators is diminished wherever they are themselves killed by pesticides.

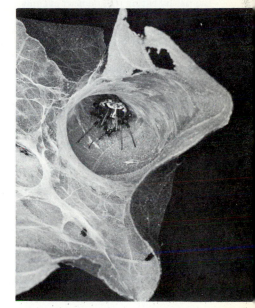

Irregular web on a leaf in a Brazilian rain forest. The web of this cribellate spider is a framework of plain silk strands that support a loose network of the hackled band (fuzzy silk) that entangles small insects. The spider awaits its victims in a tunnel-shaped silk retreat. (R.B.)

Zig-zag bands of silk are conspicuous patterns of loosely woven strands at the hub of orb-webs of orb-weaving and uloborid spiders— but only of species that spin durable webs lasting through the daylight hours. Each conspicuous silk marker is a visual warning to birds of sticky threads in their flight path. Spiders are spared needless destruction of webs. *Above left* is X-shaped marker of *Argiope argentata*. (R.B.) *Below left* is circular marker (E.S. Ross). Vertical marker of *Argiope aurantia* was shown earlier.

There are about 50 spider families; only 12 are cribellate, having a cribellum on the abdomen and a calamistrum on the metatarsus (next to last article) of the 4th pair of legs. In mature males these structures are lost or become vestigial. The cribellate families probably evolved separately from the others.

Ogre-faced spiders (photos on opposite page) are named for the large pair of median posterior eyes that are characteristic of the cribellate family Dinopidae; the other 6 eyes are small. These spiders are nocturnal and can react accurately to the shadows of passing prey even in very dim light. After an unsuccessful night of patient waiting, they recycle the proteins of the net by eating it. The next evening they spin a new web. The family is represented worldwide in the tropics and subtropics, and in the U.S.A. by *Dinopis spinosa*.

The jumping spiders form one of the largest of spider families, the Salticidae, with about 2,800 species. There are some 300 species in the U.S. and Canada, and many in Europe, but most are tropical.

The list of webs and snares could go on and on, but many spiders do not build traps at all. Some specialize in stealing from other spiders' webs, or in eating the web-builders. Others hunt actively for insects or lie in wait to ambush their prey. These spiders often have large eyes and keen sight, in contrast to the web-builders, which depend mostly on sensing the vibrations of a trapped insect to locate it in the web. Ogre-faced spiders hang from a twig or thread, holding at the tips of their front legs a small rectangular net of silk; when an insect crawls within range, the spider drops, quickly spreading the elastic net and throwing it over the prey. Or, the spider may station itself above an ant trail and neatly pick off the ants, one at a time, by reaching down and touching them with its sticky net. Crab spiders, named for their crablike posture and habit of scuttling sideways, specialize in ambush, sitting motionless until an insect approaches close enough to seize. They are often cryptically colored, brown ones resting on bark, green ones on leaves, brightly colored ones on flowers; some can even change color to match their background. The appealing jumping spiders, mostly tiny but often attractively colored and patterned, are active by day and can spot their prey from some distance. They turn ever so cautiously toward the victim and creep slowly forward, then suddenly jump on it. Among other hunters are the strong, fast wolf spiders (including the European "tarantula") and the large hairy mygalomorph spiders often called "tarantulas" in the U.S. Spitting spiders are hunters that can shoot out zig-zags of sticky thread from a distance of 2 cm away, pinning the victim to the ground.

Left, **spider with katydid prey.** *Right,* **remains of katydid.** Panama. (R.B.)

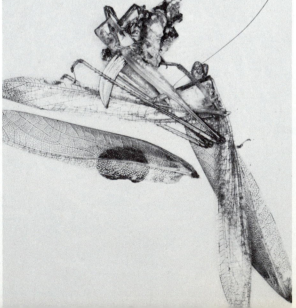

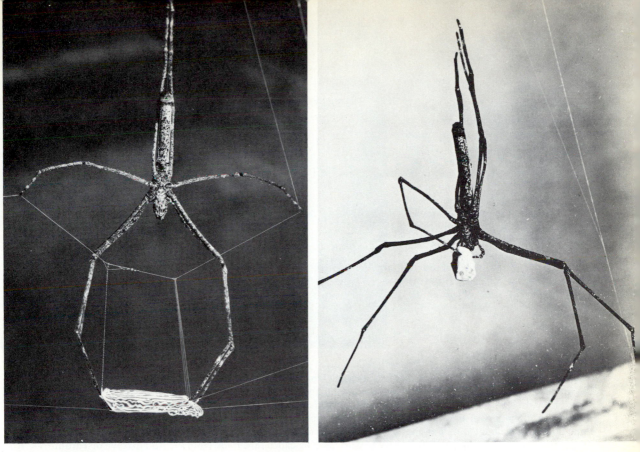

Ogre-faced spider, *Dinopis,* in tropical forest on Barro Colorado Island, Panama. Both views of the spider show the ventral surface. *Left,* the spider supports itself by silk lines attached to the branches of a shrub. It holds outstretched its rectangular net of parallel hackled bands. *Right,* the spider has touched the sticky net to a passing insect and has then thoroughly wrapped its prey in secreted silk before biting it and sucking out its body fluids. (R.B.)

Crab spider walks more readily sidewise than forward or backward. The first 2 pairs of legs are much longer and stouter than the others and are held out at the sides, ready to grasp passing prey. Crab spiders do not spin webs. *Misumenops,* shown here, is a "flower spider" and slowly changes color to match the flower on which it sits. This almost white specimen was found in the hollow of a white calla lily, lying in wait for the bees, flies, and other insects attracted to the floral spike. (R.B.)

Jumping spider (attacking another kind of spider) is typical of salticids in having a short, hairy body and stout legs, of which the 4th pair provide most of the power for leaping. The 2 large median eyes, and 6 smaller ones, afford high visual acuity. Some species, in bright light and secured by a dragline, can jump out from a perch to catch an insect in flight. (E.S. Ross)

A **tarantula** can make a satisfying pet for someone who knows how to handle it gently and who doesn't mind the skin irritation that may be caused by shed or rubbed-off body "hairs." The bite of this Arizona tarantula feels like a sharp pinprick and is not dangerous; the larger tarantulas of tropical America are more poisonous, but the symptoms are usually localized around the bite. Tarantulas stalk their prey, mostly large beetles and other insects, at night. Of about 30 species in the U.S., the majority are in the dry or desert areas of the Southwest. (R.B.)

Tropical tarantula, shown life size, has just emerged from its burrow in Panama. Large tarantulas add to their insect diet an occasional nestling bird, amphibian, lizard, or snake. There are 8 glistening, closely grouped eyes, but these spiders hunt mostly by touch. The large hairy spiders called tarantulas in the U.S. and in tropical America are the hairy mygalomorph spiders of the family Theraphosidae, predominantly tropical. There are none in Europe, where the name tarantula originated. It was applied to an unrelated spider, *Lycosa tarentula,* a wolf spider whose bite became associated in medieval times with the sprees of wild dancing called tarantism from the Italian town of Taranto. (R.B.)

Just after molting a trapdoor spider, *left,* has a white delicate cuticle and is helpless; but in a day or so the cuticle hardens and darkens. The old cuticle, *right,* was first loosened, then split along the sides and shed. Trapdoor spiders are among the primitive spiders that molt as adults. Most spiders do not. (L. Passmore)

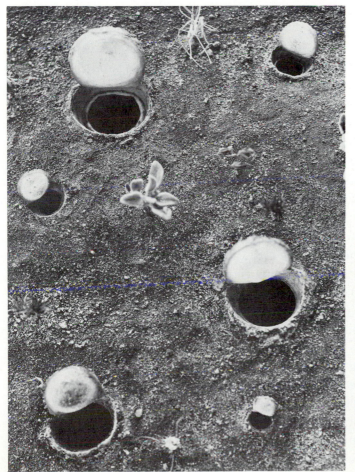

Trapdoor spider *Bothriocyrtum californicum* (family Ctenizidae) lives in a silk-lined burrow and waits, just beneath the trap door that closes the burrow, for passing prey. *Below,* the spider is pouncing on a land isopod. When closed, the silk-hinged trap door matches its surroundings perfectly. *Left,* the doors of the spider-occupants of various sizes have been opened by the photographer, against the pull of the imbedded fangs of the cheliceras on the underside of the door. Measured with a spring scale, one 3-g spider exerted a pull 140 times its own body weight. (L. Passmore)

Submerged spider, *Dolomedes,* glistens on abdomen and legs because of light reflected from the minute air bubbles trapped by body "hairs." (W.H. Amos)

Cellar spider, *Pholcus phalangioides,* 8 mm long, hangs upside down in a loose irregular web of nearly invisible threads that snare mosquitoes, flies, little moths, silverfishes, ants, and other small insects or spiders. In this photo a dead ant, recently sucked dry, hangs just above the tip of the long slender abdomen of the spider. In enlightened households this helpful spider is never disturbed as it hangs motionless, awaiting its prey in dark corners between ceiling and walls of cool rooms or cellars. There are more than 200 species of such long-legged members of the family Pholcidae, sometimes called "daddy-long-legs spiders." Most live outdoors in dry, mild climates, where they shelter under stones. (R.B.)

A number of spiders are able to run on the surface of water or even to submerge completely, crawling along the bottom or on underwater vegetation as if quite at home. The hairs of their bodies trap and hold a layer of air, on which they depend for respiration, while they hunt for aquatic insects, small fishes, and tadpoles. Several species build silk-lined, watertight burrows in crevices among intertidal rocks on seashores; and one kind of freshwater spider lives almost its whole life under water. It builds an air-filled silk "diving bell" to which it can retreat, but often swims freely through the water, glistening with the bubble of air covering its body.

When an insect or other small animal is captured, a spider bites it with the cheliceral fangs, which inject a paralyzing poison. A few hunters and many web-builders use a more indirect method, throwing swathes of silk over the prey so that it is safely immobilized before giving it a bite, especially if the prey is large and formidable enough to bite back. However, butterflies and moths (which can sometimes escape from a spider web by leaving the loose scales of their wings and body sticking to the silk) are seized and bitten at once. Some kinds of prey appear to be extremely distasteful to spiders; the unlucky spider that bites such a catch can sometimes be seen staggering to the edge of its web to wipe its mouth on a handy leaf. Later it may learn to recognize unpalatable species by odor or touch or by their distinctive movements in the web.

A spider has a minute mouth and does not swallow solid food; instead it injects digestive fluid through the wound made by the cheliceras, or crushes the prey with the cheliceras and the bases of the pedipalps, mixing it with digestive fluid. The liquefied tissues of the prey are then sucked up by means of a muscular pharynx and **sucking stomach.** Beyond the sucking stomach the digestive tract gives off several pairs of pouches, which increase the digestive and absorptive surface, and a large digestive gland, which branches extensively and occupies a large part of the spider's abdomen. This gland is the main organ of digestion and is capable of taking up large quantities of food at one time and storing it, until it is gradually used. This enables spiders to go for long periods without taking food, though they must have water quite often.

The circulatory system is open, as in other arthropods. The heart lies dorsally in a large sinus and receives blood, with hemocyanin in solution, through openings in its sides. Blood pressure is used to extend the legs, which have no extensor muscles, and to rupture the old cuticle at molting. The excretory system, as in most other terrestrial arthropods, consists of

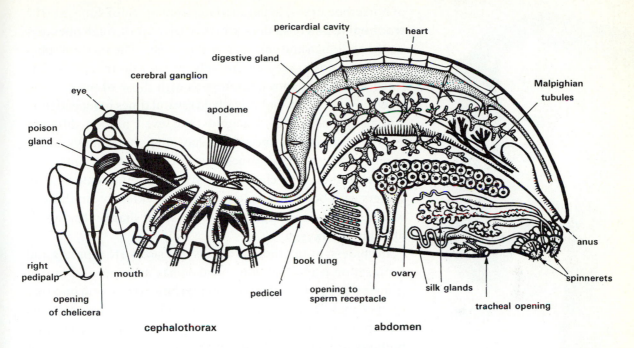

Generalized diagram of a spider. The tubular heart sends an aorta to the cephalothorax, with arteries to the legs, and also sends a caudal artery posteriorly. The dilator muscles of the sucking stomach are attached dorsally to a thickened infolding (apodeme) of the body wall. (Adapted mostly from C. Warburton)

Malpighian tubules, which open into the intestine. In many spiders there are also excretory sacs that open near the bases of the legs. The respiratory organs are of two types. The **book lungs** consist of blood-filled folds of the body wall that hang in an air-filled chamber communicating with the external air through slitlike openings. These folds, or "leaves," which suggested the name of the organ, are held apart by supports so that air circulates freely between them. The blood spaces within the leaves communicate with the blood sinuses of the abdomen. Thus the leaves of the book lung are simply another device for exposing a large amount of respiratory surface to the air. The structure and development of book lungs suggests strongly that they were derived from organs like the book gills

Extending the legs by hydraulic pressure, without extensor muscles, may allow for greater development of the flexor muscles important in seizing prey. The high pressure produced during vigorous activity is not due to the heart, which cannot even approach such levels and actually stops beating at times of maximum activity. The pressure is produced primarily by the contraction of muscles in the cephalothorax.

Measuring the blood pressure in the leg of a spider. A living spider *(Tegenaria)* is held with one leg sealed into a tube. While watching the turgid membrane of one joint of the leg, through a microscope, the experimenter slowly raises the pressure in the tube by admitting compressed air. When the external pressure in the tube just exceeds the internal pressure of the leg, the membrane is seen to collapse. Normal pressures in a quiet spider were approximately 5 cm mercury. When the spider was stimulated, pressures up to 45 cm mercury were recorded. Intermediate pressures extend the legs during normal walking. (Mod. after Parry and Brown)

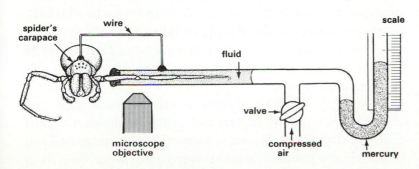

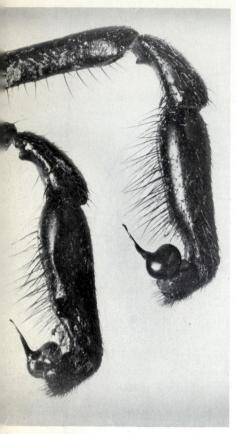

Pedipalps of male spider. The bulbous tips of the pedipalps, charged with sperms, are inserted into the openings of the sperm receptacles on the underside of the female's abdomen. Trapdoor spider, *Bothriocyrtum californicum*. (L. Passmore)

As insurance against unproductive interspecific mating, female spiders of most species have genital openings with species-specific contours. To deposit sperms, the male must have palps of corresponding contours. The female openings occur in a hardened plate, the epigynum, whose sculpturing is often used by spider taxonomists as the most reliable character for distinguishing species.

A pair of closely related species of wolf spiders recognize mates in another way. Females of the 2 species cannot be distinguished by their appearance, genitalia, or sex pheromone. They are courted by males of both species, yet mate only with those of their own species, responding to species-specific acoustic signals that the males transmit through the substrate.

of horseshoe-crabs. In part derived from the book lungs are the **tracheal tubes** of spiders, which receive air through openings on the abdomen and convey it to the tissues. The smallest tubes usually do not branch extensively, as they do in insects. The tracheal tubes are not homologous with those of insects, and tracheal systems probably evolved several times independently among the orders of arachnids and even among different groups of spiders. Some spiders have only book lungs, and a few have only tracheal tubes, but most have book lungs and tracheal tubes. The circulatory system is most extensive in spiders with book lungs alone and is less extensive in those in which a well-developed tracheal system shares the job of distributing oxygen.

In spiders, as in other arachnids, the sexes are separate and fertilization is internal; but the male has no penis and cannot transfer sperms directly from his reproductive system into that of the female. Instead he spins a tiny web on which he deposits a drop of semen. Then he collects the sperm-laden fluid into bulbous organs at the ends of the pedipalps, specialized for mating only in the mature male, and with these he introduces the sperms into the female's sperm receptacles. First, however, he must find her, often by tracking her dragline. A male wolf spider, for example, can discriminate from the chemical and mechanical characters of a female's dragline not only her species and sex but also which way she was traveling. A male orb-weaver must leave his web and wander until he contacts a web that he recognizes as belonging to a mature female of his own species. Any male spider, on finding the usually larger female, is in real danger of being taken as prey and must quickly establish his identity and intentions by appropriate **courtship,** which also acts to dispose the female to mate. The keen-eyed jumping spiders engage in elaborate visual displays, the male energetically dancing and waving his brightly colored or striped legs in front of the female. A male web-builder more often sends distinctive chemical or tactile signals, plucking the strands of the female's web or special strands that he adds to it. As always, there are those ready to take unfair advantage of any situation; one species of spider preys on the females of another, gaining entrance to the web by imitating the male's courting movements.

In contrast to most insects, which lay their eggs and abandon them, many spiders and other arachnids are good mothers. The silken egg-sac is painstakingly constructed and often fiercely guarded, whether hung in or near the web or hauled about by the mother wherever she goes. A mother wolf spider opens the egg-sac after a set time has passed, allows the tiny spiderlings to climb onto her back, and carries them for about a week. Fishing spiders place the egg-sac in a "nursery web" built among leaves or stones, and the spiderlings remain there for some days under their mother's 8 watchful eyes. The young of some web-building or burrow-dwelling spiders remain in the mother's home for extended periods, in some cases up to 2 or 3 years, sharing her food; and babies of some species drink drops of predigested food from their mother's mouth. When the young spiders are ready to disperse, they often climb up the stems of grasses or to the tops of bushes and spin a long thread which is caught by the wind and carries them away, sometimes to great heights and distances, a venture known as "ballooning." When they finally alight they will begin to hunt or to build webs on their own.

Ballooning spider will be carried aloft by the pull of the wind on the emerging strands of silk. Tiny spiders like this jumping spider can balloon as adults; in most groups that balloon only the newly hatched spiderlings can do so. (C. Clarke)

Egg sac of black widow spider, *Latrodectus mactans*. (L. Passmore)

Newly hatched spiderlings of black widow spider. (L. Passmore)

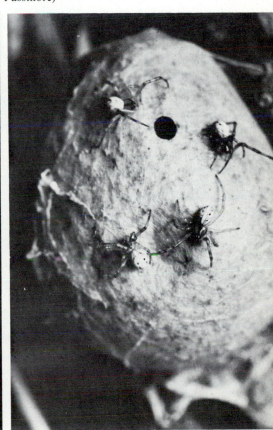

Black widow spiders, *Latrodectus mactans,* mate on the coarse web spun by the large female. In most parts of the USA the globose and shiny black abdomen of the female has on the underside a red hourglass-shaped mark. When sexually mature the small male neither bites nor feeds but searches out the female in her web and transfers the sperms stored in his (visibly enlarged) pedipalpal tips. Afterwards the male usually dies, though occasionally he may be killed and eaten by his "widow," a trait more typical of certain other spiders. All widow spiders are poisonous, but most bites in humans are by the black widow. The injected neurotoxin causes extreme pain, muscular rigidity, and other systemic effects. Medical attention should be sought promptly. Deaths are rare, occurring mostly in children. In the USA about 5 deaths a year are attributed to all spider bites.

Latrodectus mactans occurs in warm regions around the world, but with differing red or yellow markings. In the USA black widows and other species of *Latrodectus* (family Theridiidae) are most common in the South but extend into northern states. (L. Passmore)

In Texas the silk of the black widow spider is favored over all others for use as cross hairs in instruments such as levels or transits used in surveying. The silk strands are of uniform thickness and even in diameter.

Brown spider (family Loxoscelidae) has only 6 eyes. Members of all species of *Loxosceles* are poisonous to mammals but most, such as *L. arizonica* from Arizona (seen here), seldom come in contact with people. Exceptions are the South American brown spider, *L. laeta,* of Chile, Argentina, and Peru, and the brown recluse spider, *L. reclusa,* of North America, including the south and central U.S. These two species inhabit houses and other buildings, and may bite people. The wound ulcerates and heals poorly; it may require skin grafting. Rarely, the bite is fatal in children. The brown recluse spider often hides among clothing in closets and can be recognized by a dark violin-shaped marking on the cephalothorax. The body is about 15 mm long. (M.S. Keasey)

In some kinds of spiders, the young do not promptly disperse but instead remain as a group for some time, with or without their mother. Together they build a web and share the captured prey, but eventually they separate to lead solitary lives. Some **social spiders,** which spend their whole lives in a spider community, probably evolved from such temporary associations. Social spiders' webs and adjacent retreats, built cooperatively, may cover an area of some square meters and house hundreds of spiders of both sexes and all ages. Up to 30 small spiders may be seen together subduing a large active insect which none could have captured alone, and still more may join in the communal feast. Some of the smaller prey are carried into the retreats, injected with digestive juices, and given to the groups of young spiderlings sheltered there. The egg-sacs may be placed in common chambers, and the mothers feed not only their own babies but also those of other community members. The mutual attraction and cooperation of social spiders presents a sharp contrast to the characteristic aggressiveness of solitary spiders toward any other individual.

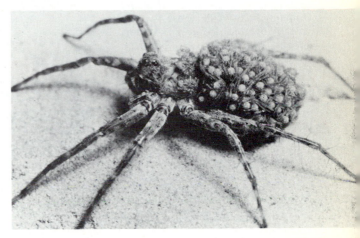

Wolf spider with young on back. California. (E.S. Ross)

Left, **communal web of social spiders** is spread on bushes. Pucalpa, Peru. (K.B. Sandved) *Right,* **social spiders** collected from a communal web that stretched over 20 m² of vegetation lining a Panamanian shore. A grasshopper that flew into the web was promptly immobilized and then wrapped up in silk by many of these small spiders working together. (R.B.)

Scorpion with posterior part of the abdomen held high over the body comes out of its burrow at night to catch insects and spiders. This one has seized a spider in its left pincer. Scorpions that live under stones or bark usually carry the stinger-tipped "tail" curled to one side. *Uroctonus mordax,* shown here, occurs in California and Oregon. Scorpions are found in climates as cold as Alberta, Canada, the southern European Alps, and the southern Andes of S. America, close to the snowline. Some live in moist forests, but most are found in hot, dry climates. In the U.S. those that are most dangerous are of an Arizona species, *Centruroides sculpturatus.* Species of *Centruroides* in Mexico have caused fatalities, especially in children. In scorpion country it is best to avoid walking barefoot in the house at night and to shake out one's shoes in the morning before putting them on. (E.S. Ross)

Some biologists classify scorpions, not as arachnids, but in the class Merostomata, close to the fossil eurypterids, which closely resemble scorpions in body form and segmentation. Scorpions do differ from all other arachnids in many respects. For example, there are 4 pairs of book lungs, in segments 20-13, whereas other arachnids either lack them or have at most 2 pairs, in segments 8 and 9. The lateral eyes, except for their simple lenses, most closely resemble the lateral, compound eyes of horseshoe-crabs. The embryology of scorpions also links them to horseshoe-crabs and divides them from other arachnids. Scorpions, probably aquatic, are the first arachnids in the fossil record. But the first clearly terrestrial arachnids, with well-preserved book lungs, were not scorpion-like; and scorpions may have invaded land later, independently from other arachnids.

SCORPIONS (order Scorpiones) are easily recognized by their large **pincers,** borne on the armlike pedipalps, and by the **stinger** at the tip of the narrowed and jointed posterior portion of the abdomen, which looks like a tail and is often held high and curving forward over the back. The stinger injects a poison that is paralyzing or lethal to small prey but seldom used by most scorpions unless the prey struggles vigorously; scorpions sting primarily in defense. The prey consists of insects and other small arthropods, occasionally small lizards or tiny mice. Prey is seized by the large pedipalpal pincers and torn apart by the small pincers of the cheliceras. As in other arachnids, the food is partially digested by enzymes secreted onto it before it is sucked up. Scorpions appear to detect prey by air movements, to which long fine hairs on the pedipalps are sensitive, or by contact. There is a pair of simple eyes in the middle of the carapace that covers the cephalothorax, and several pairs of smaller eyes at the forward edge. Vision is not thought to be acute in these nocturnal animals, but the eyes may play some secondary role; blinded scorpions, although observed to react to prey, did not attack it. During the day scorpions rest in crevices beneath rocks or logs, under bark, or in burrows. They can survive long periods of starvation, even a year, and remain in hiding during cold seasons. Most species live where they can be active throughout the year, in tropical or subtropical regions, in wet forests as well as in dry deserts.

Scorpion, handled by an expert. A sting is often painful, but only a few species can seriously injure or kill a person, and these are not the most formidable in appearance. They are found in some areas of the southwestern U.S., Mexico, South America, and North Africa, and in these regions antivenins are sometimes available. Tempe, Arizona. Specimen and handling, courtesy H. Stahnke. (R.B.)

Scorpion with young. These newly born young stay with their mother until after their first molt. This Mexican species of *Centruroides* can deliver a very toxic, even dangerous sting. (E.S. Ross)

At **mating,** after a male and female scorpion have completed the preliminary courtship rituals, they embark on their famous "promenade," which sometimes lasts for hours. It may look romantic, but seems to consist mainly of the male hauling the female about while he tests the ground with his *pectens* ("combs"), abdominal appendages unique to scorpions and used in finding a suitable surface on which to attach the *spermatophore*. A scorpion's spermatophore is a large and complex structure with a stalk that must be attached to a firm substrate. At the top of the stalk is a packet of sperms and an automatic opening mechanism that ejects the sperms once the tip is properly placed in the female's genital opening. When the male has finished secreting this device, he attempts to position the female over it, pulling and lifting her until the spermatophore is in place. This must be at least as hard as it sounds, for it usually requires several trials. Spermatophores are standard equipment for several orders of arachnids.

The female does not lay eggs but gives birth to young. In some species, the embryos simply use the abundant yolk in the eggs, which are retained in the mother's oviduct. In other species, the eggs contain little yolk, and the embryos are nourished by food continuously supplied by the mother through a complex arrangement analogous to a mammalian placenta. The newborn young climb onto the mother's back, where they remain under her protection until after their first molt. Then they may disperse, or may stay with the mother and share her food for up to two months before taking up an independent life.

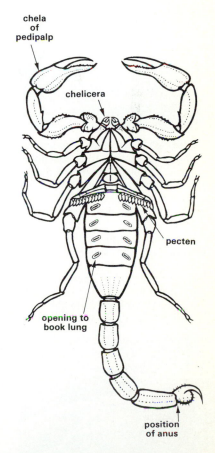

Ventral view of scorpion showing pectens and 4 pairs of openings to the book lungs. (After various sources)

Oribatid mite, *Oppia coloradensis,* only 0.3 mm long. Oribatids are all minute (0.2 to 1.5 mm), heavily sclerotized, and often called "beetle mites" from a superficial resemblance. This one remained alive for 40 minutes in the electron beam and vacuum of a scanning electron microscope. Oribatid mites are all free-living and found around the world in soils and leaf litter and under bark or stones. They occur by the millions in grass pastures, and when swallowed by grazing sheep, some species transmit sheep tapeworms. (T.A. Woolley)

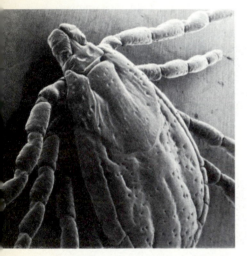

Ixodid tick, *Dermacentor andersoni,* is one of the vectors, from rodents to humans, of tularemia and of Rocky Mountain spotted fever. Both SEMs are of a ♀. *Dorsal view, above,* shows projecting feeding region, and just behind it the hard shield that covers only the anterior portion of the ♀ body. *Ventral view* shows genital area in midline between 2nd legs. Light-colored valves guard anus near rear. The 2 plates bearing respiratory openings are partly obscured by the bases of the 4th legs. (T.A. Woolley)

MITES (order Acari, or Acarina) are among the smallest of all arachnids, the most widely distributed, the most diverse in form and habit, the most numerous in both individuals and species, and the most dangerous to humans. More than 30,000 species have been described, but estimates of the species yet to be described range to 500,000. Most are free-living.

All acarines are fluid feeders, imbibing their food by means of a sucking pharynx. Some suck animal or plant tissue fluids. Others liquefy solid food by pouring onto it enzymes secreted by the salivary glands. The undivided (and usually unsegmented) body has an anterior, movable projection, composed of cheliceras, pedipalps, and other parts, which together form the feeding apparatus.

The most dangerous mites, which include only a small fraction of the group, are found among the parasitic species, especially the **ticks,** relatively large mites (2 to 6 mm) that suck blood from vertebrates, including humans. Tick bites sometimes cause a progressive paralysis of the human victim, beginning with the lower part of the body and spreading upward until respiratory failure results in death. The paralysis is due to a toxin that interferes with conduction in motor nerve fibers. If the tick is discovered and removed before the victim's breathing stops, recovery is prompt and complete. Cases of paralysis are rare, however, and the main danger to humans from ticks and other parasitic mites is that they transmit a great variety of disease organisms. These include viruses (encephalitis, Colorado tick fever), rickettsias (Rocky Mountain spotted fever, scrub typhus, Q fever), bacteria (tularemia), spirochetes (relapsing fever), and sporozoan protozoans (babesiasis). In addition to those that directly affect humans, many tick-borne diseases endanger domestic animals, for example, redwater fever of cattle, caused by sporozoans. Parasitic worms are also spread by mites. Some of these are blood-sucking species that transmit filarial nematodes to rodents. Other mites, which serve as intermediate hosts of tapeworms, remain in soil during the day but climb up on vegetation to feed in the evening and are then accidentally swallowed by grazing sheep, cattle, and horses, the definitive hosts of the tapeworms.

In addition to spreading disease, mites directly cause afflictions that are not life-threatening but can be intensely irritating. Mites in household dust are one of the main causes of dust allergies. The "chiggers" of the southern U.S., and other warm parts of the world, are parasitic immature stages of mites that are free-living predators as adults. Picked up from the ground or low vegetation, the young mites attack the skin, producing a tiny crater or tube filled with lymph and partly digested tissue, on which they feed. The secretion they introduce causes severe itching long after the young mites have dropped off. The skin disease called "scabies" is caused by a mite that completes its entire life cycle burrowed into the skin and is passed from person to person by contact. Species with similar habits are responsible for "mange" on dogs, cats, horses, and other domestic mammals and for a variety of skin disorders of birds, including domestic poultry. Mites that invade the tracheal system of honeybees, producing "Isle of Wight disease," kill many bees in European hives.

Lists of mite parasites are heavily weighted with those that attack humans and domestic animals, only because these are the first to be noticed and studied, but other animals of many phyla suffer their own mite afflictions. For example, many owners of dogs and cats are aware of the ear mites that cause the irritation, scratching, bleeding, and excessive wax secretion that together result in dark waxy accumulations in their pets' ears. On the other hand, few people know about the widespread and much more serious cases of ear mites in moths. Moth ear mites cause deafness in the infested ear, no small matter for the moth, whose ears are tuned to the high frequency sounds emitted by bats in hunting. Good hearing is vital to the moth's chances for escaping these predators. However, if a deafened moth is caught by a bat, its ear mites also perish. This selective pressure seems to have resulted in a system whereby the mites regularly infest only one ear or the other, but not both, leaving the moth with its hearing only slightly impaired. The first ear mite on a moth leaves some kind of track, probably pheromonal, which all subsequent mites follow. Even when the infested ear becomes too crowded, the mites do not enter the other ear but will leave the host moth and take their chances at finding another.

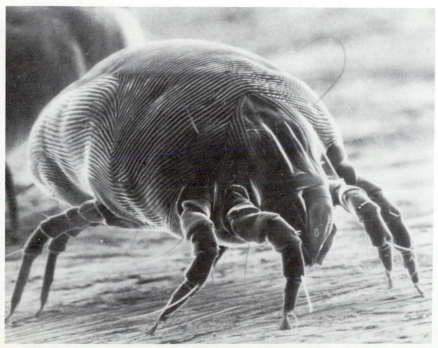

House dust mite, *Dermatophagoides farinae,* is thought to be a major cause of dust allergies. SEM. (W.C. Lane, courtesy G.W. Wharton)

This mite belongs to a group of 40 families (Astigmata) which lack a respiratory system. They are slow moving and small (0.2 to 1.5 mm), and diffusion of oxygen through the integument suffices. Many feed on fungi, decaying vegetable matter, and grains. Where humidity is 70 to 95%, the flour mite, *Acarus siro,* destroys stored grains or flour. Parasites include mites that cause scabies and seborrheic dermatitis in humans. Also in this group are ear mites of cats and dogs, and feather mites of birds.

Hard tick probing human skin. Anyone walking in tick-infested forests should examine the entire body for ticks immediately afterwards. The name "hard tick" refers to members of the family Ixodidae, in which a heavily sclerotized dorsal shield covers almost all of the body in the young and in males. Most feed on mammals; and larva, nymph, and adult may each parasitize a different host, so that ticks are prime vectors of disease organisms. E. Africa. (E.S. Ross)

Soft ticks (family Argasidae) parasitize birds mostly, but also snakes, lizards, and turtles, and some mammals including humans. *Ornithodoros turicata,* which lives in dry caves in the southwestern U.S.A., is a hazard to those who explore caves or camp in cave entrances. It transmits the spirochetes that cause relapsing fever. The ticks shown here had lived without food in the U.S. Public Health Service laboratory for 5 years, and were still infective. (Science Service)

Mating pair of hard ticks. The greatly swollen female may have taken a week to complete her huge blood meal. She is shown here lying on her back with the male in position over her genital opening. He expels a spermatophore and then pushes it in with his cheliceras, as he has no penis. Panama. (R.B.)

Feeding parts of ixodid tick project from a broad base. This anterior region, the gnathosoma, is not a head as it bears neither brain nor eyes (the 2 oval porose areas are not eyes). The base extends forward, in the center, ensheathing 2 elongate cheliceras that can be extended to make a wound in the skin. These overlie the hypostome, a long trough (with serrated edges) that directs blood into the mouth at its base. When the feeding parts are well inserted into the host, salivary enzymes aid in liquefying host tissue. The wide pedipalps, pushed to the sides when the tick feeds, act as counter-anchors. SEM. (P. Armstrong)

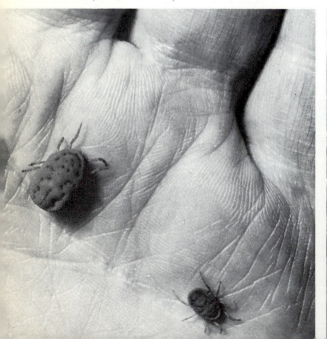

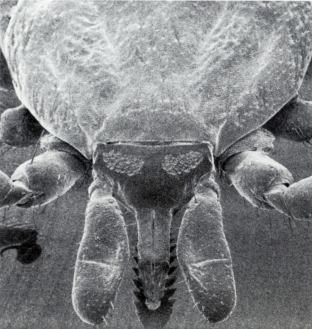

Most arachnids are predatory, and mites parasitic on larger animals are in one sense specialized predators. But many mites are exceptional in feeding exclusively on plant material, piercing the cells and sucking out the contents. Mites also spread viral and fungal infections from plant to plant. Some mites are important pests of crops and gardens, such as the "red spiders" and other spider mites, which do great harm to fruit trees, cotton, and other crops. The drying up of leaves of house plants, especially of ivy in heated houses in winter, is caused by "red spiders," which can be seen by shaking the plant over a large piece of white paper. Some spider-mites spin silken webs for shelter. Other mites, which attack grain stores, cause damage not so much by what they eat as by introducing mold spores and favoring growth of molds by raising the humidity.

Not all the activities of mites are harmful from a human viewpoint, however. In parts of Europe, certain mites are deliberately introduced into cheese, to give it a characteristic taste, fragrance, and appearance; perhaps the result is here also due to molds introduced by the mites. Much more important is the job of pest control done by the many free-living predatory mites. For example, many of the mites found in stored grain are feeding on molds or on grain-eating mites and insects. Enormous mite populations are present in leaf litter and soil, and these play an important role in breaking down plant and animal organic materials in such habitats. We usually think of the insects as dominating small terrestrial niches, but in samples taken from a forest floor, there may be more mite species and more individual mites than insects and all other arthropods put together.

Gnathosoma, in ventral view, of *Ixodes ricinus,* a Eurasian tick that feeds on cattle, sheep, and horses. The long cheliceras each have at their tip a small pincer with teeth on both the rigid and the moveable digit. The pincers cut the skin of the host and tear away at the flesh with little slashing motions that deepen the wound. The ventral surface of the hypostome is covered with recurved spines that serve as a holdfast when cheliceras and hypostome are inserted into host tissue. This is what makes dislodgement of ticks so difficult. It should always be done slowly. SEM. (T.A. Woolley)

Three **free-living mites** from various habitats. (P.S. Tice)

Wood mite from rotten wood has a white body. 0.8 mm long.

Field mite (ventral view) is red in color, 6 mm long, and densely covered with rows of long bristles.

Aquatic mite (ventral view), from freshwater, has long legs fringed with bristles, an adaptation for swimming. Body length 1 mm.

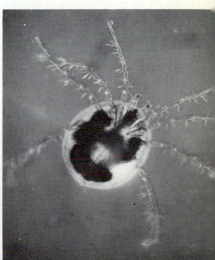

Tarsal claws of *Dermacentor andersoni.* SEM (T.A. Woolley)

Sense organ called Haller's organ is characteristic of ticks. It occurs on the dorsal surface of the tarsus of each first walking leg. If this portion of the forelegs is cut off, the tick ceases to respond to chemicals or to variations in humidity to which it was very sensitive before the operation. In this SEM of the tarsus of *Dermacentor andersoni* the narrow opening in the foreground leads to a capsule containing sensory bristles. Immediately beyond it (the tip of the tarsus is out of the picture at the top) is a shallow depression with erect sensory bristles. It has been suggested that the open pit is a hygroreceptor, and the enclosed one a chemoreceptor. (T.A. Woolley)

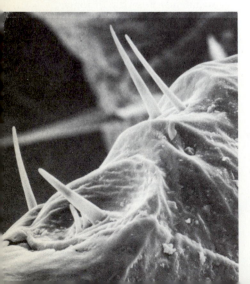

The diverse life styles of mites are reflected in **specializations of structure**. For example, the cheliceras bear pincers in predatory and scavenging forms but have become styletlike, for piercing animal and plant tissues, in some parasitic species. The 8 legs characteristic of arachnids have been reduced in some parasitic mites to 6 or fewer, and even 8-legged mites are peculiar among arachnids in that they hatch from the egg, or are born, with only 6 legs, adding the last, hindmost pair only after the first molt. Among aquatic mites, the legs may be modified for swimming.

Except for the ticks, most mites are very small, less than a millimeter long, and the proper name of the order is apt, coming from a Greek word meaning tiny, too small to be divided. The bodies of mites show no division into segments, except in a few primitive forms. Nor is there division into body regions, except for the presence of a small anterior portion called the gnathosoma, or capitulum, which bears the cheliceras and pedipalps. This is not a head, for it does not include the brain or eyes. The "brain" of mites is not clearly set off from the other ganglia of the central nervous system, which are fused into a single large mass. Like other very small animals, mites have a large surface area in proportion to their body volume, and the smaller mites can carry out respiratory exchange without any additional respiratory structures at all. Larger mites have a simple tracheal respiratory system. The large surface of mites also makes them especially vulnerable to drying. Some are restricted to moist habitats, and others manage to survive in dry conditions by periodically retreating to moist microniches, where they absorb water from the air. Another feature of small size in mites is the reduced circulatory system, which usually lacks a heart and consists only of irregular blood sinuses.

Mites are found in all parts of the earth. In addition to the majority of terrestrial forms, some mites live in freshwaters and in the oceans. They have been found in hot springs and deserts, in deep caves, and on high mountains, from northernmost Greenland to within a few degrees of the South Pole. One way in which these tiny, wingless, slow-moving arthropods have become so widely distributed is by blowing in the wind. Another is by their habit of "hitch-hiking," not only with the hosts on which they feed, but also with a great variety of flying insects to which they cling firmly, sometimes with specially modified appendages, just for the ride.

A **whipscorpion,** *Mastigoproctus giganteus* (order Uropygi), has stout pedipalps for grasping prey and uses its sensitive whiplike first legs as feelers. All live in the tropics or subtropics. This species, found in the southern U.S., reaches 8 cm in length and is the largest of the group. Whipscorpions are shy and nocturnal, spending their days hidden under stones or logs, or in burrows which they dig with their pedipalps. If threatened, they spray a stream of irritating fluid from the rear end, and the spiny pedipalps can draw blood. The defensive secretion of *Mastigoproctus,* mostly acetic acid, suggested the name "Vinegaroon." Whipscorpions (and their relatives, the tailless whipscorpions) mate, as scorpions do, by means of a stalked spermatophore attached to the ground; the female carries the eggs and young at least until after the first molt. Arizona-Sonora Desert Museum, Tucson, Arizona. (R.B.)

A **tailless whipscorpion** (order Amblypygi), like the whipscorpion, has a flattened body and strong spiny pedipalps, and walks on its last 3 pairs of legs, moving either forward or sideways. The sensory "whips" of the first pair of legs are extremely long; they are used to monitor everything within a wide circle around the animal, exploring all objects and surfaces with a delicate tapping motion. The pedipalps serve not only to capture prey, but are used in defense and can deliver a sharp pinch. Amblypygids require a humid environment and are found in the tropics and subtropics. Body lengths range from 8 to 45 mm. Panama. (R.B.)

A **windscorpion** (order Solifugae) is easily identified by its enormous cheliceral jaws. Windscorpions are exceptionally active arachnids, able to run very fast, "like the wind," and are equipped with an extensive tracheal system more like that of an insect than like that of any other arachnid. As they run, they hold both pedipalps and first legs out in front as feelers. The long, abundant bristles that cover the body and appendages are probably the principal source of sensory information, but there is also a well-developed pair of eyes. Some windscorpions can climb trees or walls or even a smooth vertical glass surface, and most can dig in sand or loose soil, excavating burrows in which they rest, molt, or hibernate. Most species are found in the tropics or subtropics, especially in deserts, but some live in more temperate regions. Central California. (R.B.)

The **jaws of a windscorpion** are formidable weapons, and their owners are not shy. Diurnal species, known as "sun spiders," may boldly frequent city streets, and the nocturnal majority are attracted to lights, so encounters with humans are not uncommon in areas where windscorpions are abundant. The bite is painful but the serious consequences occasionally reported are thought to be due to infection and not to any venom. Windscorpions are voracious predators and will eat until their bodies are so distended that they can hardly move. Their diet consists mostly of insects, some specializing in termites and one species preferring bedbugs, but they also devour other solifugids, large spiders, scorpions, earthworms, snails, toads, lizards, small birds, and mice. Large solifugids may measure 5 cm long. (R.B.)

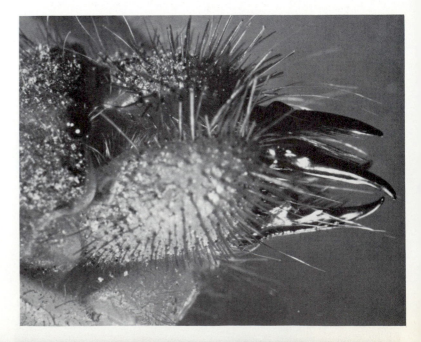

A pseudoscorpion (order Pseudo-scorpiones) has poison glands, which open at the tips of the pedipalpal pincers, but the largest of these small arachnids is just over 1 cm long, and none is dangerous to humans. They prey mostly on other small arthropods. Pseudoscorpions live in the leaf litter of forest floors or beneath the bark of trees; under stones or logs, or in cracks among rocks; in the nests of various birds and mammals, or of ants and bees; or in almost any terrestrial habitat that provides the shelter of small crevices. None is aquatic but some seashore species are regularly submerged by the tides. The animal in this photo, from Monterey Bay, California, is of a species, *Garypus californicus,* usually found in or above the high intertidal zone. A few pseudoscorpions live in nests of silk, spun from the tips of the cheliceras. Most species, however, construct silk nests only for molting, hibernation, or brooding the young. When two pseudoscorpions of the same species meet, they rapidly vibrate the pedipalps or make other specific movements that may identify them or communicate warning. The movements of both prey and other members of the same species are probably detected mostly by long sensory bristles on the pedipalps, for the eyes are small and simple. Some pseudoscorpions have elaborate courtship behavior. In one species the male constructs two converging fences of silk threads, forming between them a pathway that leads the female to his spermatophore. When a female is ready to lay her eggs, she secretes a brood pouch on her ventral surface. Into this pouch she extrudes first the eggs and then a fluid that nourishes the embryos throughout their development. (R.B.)

Riding on insects or other arthropods, especially harvestmen, is a common means of distribution for pseudoscorpions. This one, about 2 mm long, was found clinging firmly to a leg of a fly. When the fly died, probably from the poison of the pseudoscorpion, the minute arachnid promptly abandoned the fly, but returned later to feed on it, at times disappearing completely into the carcass. (R.B.)

Harvestman is so named because most species mature in July and August and are seen at harvest time. The long legs make it easier to climb among leaves. All the legs have several kinds of sense organs, but especially the longer second legs, which may be held high in the air and used like antennas to explore the surroundings. One of the 8 legs is missing, a common condition, and the main strategy for escaping predators. Missing legs are not regenerated. Also defensive are odors, sometimes quite strong, produced by the scent glands of harvestmen. *Leiobunum exilipes,* dark pigmented variety. 8 mm long. California. (R.B.)

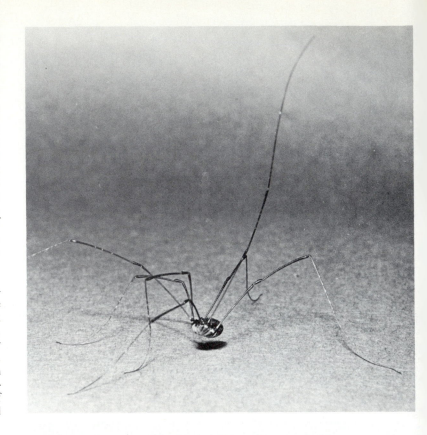

Not all members of the order Opiliones (or Phalangida) are of the "daddy-long-legs" type. Some look like little mites only 2 mm long. Others are flattened and have sturdy legs about as long as the body. However, all have an anterior region (bearing mouth, eyes, chelate cheliceras, pedipalps, and 4 pairs of walking legs) which is broadly joined to a clearly segmented abdomen.

Aggregation of *Leiobunum* was first noticed in September in a California garden. More than 100 were densely crowded, high on a fence, under a mass of ivy that had to be lifted to make the photo. This caused them to spread out, but with legs still touching. When further agitated they swayed rhythmically in unison. Some departed before sunset; and by 9 pm all were gone into nearby vegetation hunting for live insects or fallen fruit. At dawn all had returned to aggregate in their daytime resting place. (R.B.)

The strongly rhythmic activity of harvestmen is probably associated with light and other physical factors, but there is also an internal component. It has been determined, in *Leiobunum longipes,* that brain and intestinal tissues produce 5-hydroxy-tryptamine (a compound found widely in animals and plants). The curve of neurosecretory activity in intact animals, and in cells cultured as long as 80 days, follows the curve of observed behavioral activity from dusk to dawn, both peaking at 2 am.

PYCNOGONIDS

THE SEA-SPIDERS, the class **Pycnogonida,** are peculiar
marine arthropods with tiny bodies and long clumsy legs. They
feed mostly on soft-bodied invertebrates, especially cnidarians, bryozoans, and sponges, but also molluscs, echinoderms,
tunicates, and various worms. Some eat algal material or
detritus. Pycnogonids occur in benthic habitats all over the
world from the intertidal zone to abyssal depths and are almost
always found clinging to their prey or moving slowly along the
bottom. Many can swim slowly, rising up into the water by
beating the legs up and down, and may be carried over long
distances by water currents. One species described as "bathypelagic" is probably not really free-swimming but lives on
some larger bathypelagic organism such as a medusa. Most
pycnogonids are small and inconspicuous. Though intertidal
forms occur commonly in hydroid and bryozoan colonies, and
in mussel beds and eelgrass, they are usually hard to discern.
The larger species are found in the cold waters of polar regions
and of deep seas, and in some of these the body may reach 6 cm
in length and the legs span half a meter.

At first glance it is hard to tell head from tail in a pycnogonid. The **proboscis,** a filtering structure with the mouth at
its tip, may be longer and larger than the rest of the body and
appear more like an abdomen. Where one looks for the
abdomen, there is usually only a tiny nubbin, bearing the
anus. The short tubular **cephalothorax** may be conspicuously
segmented, and each segment bears a pair of prominent lateral
projections, with which the legs articulate. Members of most of
the roughly 600 species of pycnogonids have, like arachnids,
8 walking legs. However, anyone watching one of the

A **large Antarctic pycnogonid,**
Colossendeis scotti, with a small
anemone on the tip of its proboscis
(which here faces the viewer). This
species feeds on cnidarians, mostly sea
anemones and stoloniferans. Leg
span, about 30 cm. McMurdo Sound.
(G. Robilliard)

Heads or tails? It may take a moment or two to decide which end is which in a living pycnogonid, but in a fossil pycnogonid, the question can be even more difficult. Some Devonian fossils, first described as isopods and named *Palaeoisopus problematicus,* were later interpreted as pycnogonids with a relatively large rounded abdomen and a long slender proboscis with several segments *(left)*. Still later, additional specimens revealed that the rounded "abdomen" was a pair of stout pincers surrounding a short proboscis, and that palps, ovigers, and a tubercle with 4 eyes were probably also present *(right)*. (Redrawn from J.W. Hedgpeth, from reconstructions *left,* by F. Broili and *right,* by W.M. Lehmann)

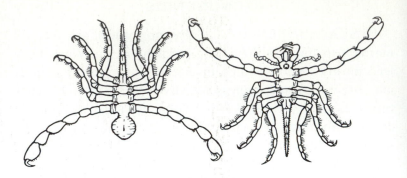

common, temperate, intertidal pycnogonids stumble along is likely to conclude that the legs must be better adapted for something other than walking. These pycnogonids spend most of their lives clinging to the sessile or sedentary invertebrates on which they feed. Larger, longer-legged species walk in somewhat more coordinated fashion. The legs sprout from the sides in a nearly radial arrangement, and pycnogonids can move forward, backward, sideways, or at any angle, shifting direction quickly and smoothly and without turning the body. The length and structure of the legs also allow the animal to sway in all directions, reaching a large area of the surface on which it is feeding, without risking a change in foothold that might cause it to be washed away or attract the attention of a predator. A number of species have been described with individuals having 10 or even 12 legs but otherwise resembling 8-legged forms. Whether these extra-legged animals really represent separate species is debated, but they are certainly more than occasional freaks and demonstrate a plasticity of form unique among arthropods.

A stout, short-legged **pycnogonid,** *Pycnogonum stearnsi,* usually found on intertidal sea anemones *(Anthopleura* and *Metridium)* and hydroids *(Aglaophenia)* along the U.S. Pacific coast. This one, on *Aglaophenia,* is marked as a male by the presence of slender ovigers ventrally; females of *Pycnogonum* lack ovigers. The 4 conspicuous black dots in the midline, behind the prominent proboscis, are the eyes. Body length, 4 mm. (R.B.)

Head-on view of *Pycnogonum stearnsi.* Scanning electron micrograph. (J. Lombardi and J.P. Wourms)

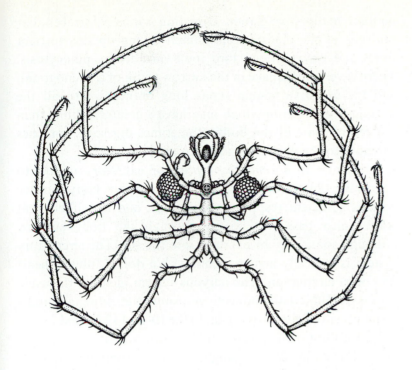

Dorsal view of a **pycnogonid,** *Nymphon rubrum,* one of the species that has 3 pairs of anterior appendages, in addition to the 4 pairs of walking legs. Each oviger of this male specimen holds a large egg mass; specialized combs of spines for grooming are present near the oviger tip. The 4 eyes, on a dorsal tubercle, look in all directions. The legs are more than 3 times as long as the central part of the body, which measures 4.5 mm; in some species the proportions are even more extreme. The pose of the animal in this drawing is not natural; the legs were laid out to display anatomical details. (Combined from drawings by G.O. Sars)

Besides the 8-12 walking legs, there may be 3 other pairs of anterior appendages. The first pair, the **cheliceras** (or chelifores), usually bear pincers that grasp the prey or tear off bits of food. The second appendages are sensory **palps;** in large species feeding on soft bottoms, the long palps move back and forth across the bottom just in front of the proboscis. Behind the palps are the **ovigers,** on which the males carry the egg masses and sometimes the young larvas. In larger forms, the ovigers are used by females and by nonbrooding males in cleaning the other appendages and the proboscis. Brooding males are unable to groom themselves with their egg-laden ovigers and often become heavily encrusted with various organisms. Any of the anterior appendages may be reduced or missing in various species, except that the ovigers are usually present in adult males. In species without cheliceras or palps, the proboscis is often better developed and more mobile, with strong rasping ridges around the mouth and numerous sensory bristles.

In **feeding,** whether or not cheliceras are present to macerate the food initially, the proboscis sucks up the soft prey tissues and reduces them to a fine particulate suspension by means of hard teeth and spines on the proboscis walls. Any remaining large particles are strained out by the spines, and only small particles and fluids pass into the midgut. There are

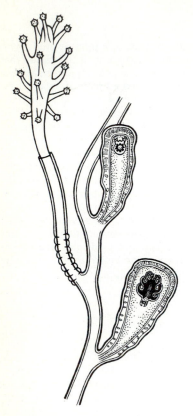

Developing pycnogonids, *Phoxi-chilidium,* in hydroid host, *Syncoryne.* (After G.J. Allman)

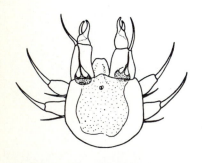

Protonymphon. This pycnogonid larva appears typical but belongs to one of the extra-legged forms, *Penta-nymphon antarcticum,* and will grow up to have 5 pairs of walking legs. (From J.W. Hedgpeth)

no discrete digestive glands. Digestion is mostly intracellular, the cells of the digestive epithelium taking up the nutrient fluids and tiny particles into food vacuoles by pinocytosis. There is very little space in the narrow trunk of a pycnogonid, and the digestive system sends long branches into all the walking legs, each of which may have a greater volume than the central part of the body. Sometimes digestive branches extend into the cheliceras and proboscis as well.

There are *no special respiratory or excretory organs* to aggravate the space shortage. The extensive body surface permits sufficient respiratory exchange, and it is likely that wastes also pass out largely through the body surface, although some may be excreted into the gut. The circulatory system is typically arthropodan, with a dorsal tubular heart and open circulation. The nervous system includes the usual dorsal ganglion with connectives around the esophagus and a chain of paired ventral ganglia. In the 10- and 12-legged forms, there are additional pairs of ganglia corresponding to the extra legs. Above the dorsal ganglion, on a prominent tubercle projecting from the dorsal surface, are 4 simple eyes, but these are lacking in many deep-sea species.

Like the digestive system, the reproductive system branches into the legs. The gametes are shed from pores in the leg bases, and when a male and female pair off and spawn together, the eggs are fertilized as they emerge from the female. The male gathers the eggs onto his ovigers, and males of one species have been seen with as many as 14 egg clusters, each containing all the eggs of one female. In some other species, the female also mates repeatedly, entrusting only a portion of her eggs to each mate. When the eggs are close to hatching, the male may deposit them on a hydroid colony or other home suitable for the larvas, which commonly live on, or within, various cnidarians. In some species the male carries the eggs through hatching, and the young cling to their father until they are well developed. A pycnogonid larva, or **protonymphon,** has only 3 pairs of appendages. The young pycnogonid passes through several molts, adding segments one by one, until it has all its appendages and acquires the adult form. The cheliceras present in the larva either grow with each molt, or are dropped at some stage and are lacking in the adult. The larval palps and ovigers do not simply grow but are lost and then, in those individuals that have them as adults, are replaced during the course of molts. Most adult pycnogonids do not molt and cannot regenerate lost parts.

THE RELATIONSHIPS of pycnogonids to other arthropods
are a matter of controversy. It is generally agreed that pycnogonids are most closely related to horseshoe-crabs and arachnids, especially the latter. Pycnogonids and arachnids resemble each other in the use of the first appendages, which
commonly bear pincers, for feeding; in the lack of antennas
and mandibles; in the possession of 8 walking legs; and in
certain structural features of the legs, eyes, brain, and digestive
tract.

Because pycnogonids have a number of distinctive characters, however, some biologists prefer to place them in a
subphylum by themselves. The ovigers have no clear homo-
logs among the appendages of horseshoe-crabs and arachnids.
The presence of the 10- and 12-legged forms is another peculiar
feature; when this phenomenon is better understood, its significance may be easier to evaluate. The greatly reduced
abdomen, the multiple genital pores on the legs, the mode of
development (with addition of segments in successive molts),
and the protonymphon also set pycnogonids apart from their
chelicerate relatives, the horseshoe-crabs and arachnids.

Some of the features that separate pycnogonids from horseshoe-crabs and
arachnids have close parallels in other arthropod groups and have not been
considered sufficient grounds for setting these groups aside as new subphyla.
A greatly reduced abdomen is seen in members of several groups of
crustaceans, for example, caprellids and certain other amphipods with
sedentary habits, such as the "whale -lice," amphipods parasitic on whales.
Like pycnogonids, these are adapted for clinging to a host and have not only
a similarly reduced abdomen but so greatly resemble pycnogonids in almost
every aspect of body shape that whale-lice and pycnogonids were once
classified together. In several arthropod groups (trilobites, crustaceans,
myriapods, and some primitive wingless insects) the young of some species
hatch without the full complement of segments or appendages, although
among chelicerates, only the trilobite larva of horseshoe-crabs adds several
pairs of appendages after hatching. (Mites hatch without the last pair of legs
but this is a secondary loss; the beginnings of all 4 pairs are present in the
early embryo.)

The pycnogonids illustrate once again the great adaptability of the arthropod body plan that makes it difficult for us
to be sure of relationships.

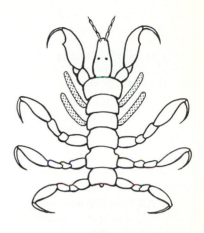

Similar life-styles have led to
similar body forms in this crustacean
(*Cyamus,* above) and pycnogonid
(*Pycnogonum,* below). Both are
adapted to cling to large soft-bodied
hosts. Cyamids are "whale lice,"
amphipods parasitic on whales and
dolphins. Members of the genus
Pycnogonum feed on sea anemones
and other soft-bodied invertebrates.
(*Cyamus* modified after Y.-M. Leung.
Pycnogonum modified after G.O.
Sars)

CLASSIFICATION: Phylum ARTHROPODA, Subphylum CHELICERATA

Arthropods with the body usually divided into cephalothorax and abdomen. The first pair of appendages are feeding *cheliceras,* not sensory antennas; mandibles are absent. Mostly predatory; exceptions are indicated below.

Class MEROSTOMATA. Aquatic, with gills on abdomen. Includes two orders, one extinct: **order Eurypterida,** the sea-scorpions; and one living:

Order Xiphosura, the horseshoe-crabs. With large, horsehoof-shaped carapace and long tail spine. Simple and compound eyes. Cheliceras with pincers; 5 pairs of walking legs; 5 pairs of abdominal appendages bearing book gills. "Trilobite" larva. Feed on worms and molluscs in soft shallow marine bottoms. *Limulus, Tachypleus, Carcinoscorpius.*

Class ARACHNIDA. Almost all terrestrial, usually with book lungs and/or tracheal tubes for air-breathing. Usually with 4 pairs of walking legs and one or more pairs of simple eyes.

Order Scorpiones, the scorpions. Small cheliceras with pincers; armlike pedipalps with large pincers; 4 pairs of walking legs. Posterior part of abdomen narrow and tail-like, with stinger at tip that injects paralyzing poison. Book lungs. Mostly tropical and subtropical. *Vejovis, Buthus, Centruroides, Pandinus, Euscorpius.*

Order Uropygi, the whipscorpions. Small cheliceras; stout pedipalps with small pincers; first legs are long and sensory; other 3 pairs used in walking. Abdomen ends in long slender whiplike projection. Can spray irritating defensive secretions. Book lungs. Tropical and subtropical. *Mastigoproctus* (vinegaroon), *Thelyphonus.*

Order Schizomida. Small forms, similar in most respects to uropygids, differing in various details such as more slender pedipalps, lack of eyes, fewer book lungs. Tropical. *Schizomus, Trithyreus.*

Order Amblypygi, the tailless whipscorpions. Small cheliceras; large spiny pedipalps; first legs are extremely long sensory whips; other 3 pairs used in walking. Book lungs. Tropical and subtropical. *Admetus.*

Order Palpigradi, the microwhipscorpions. Small forms, probably related to uropygids, schizomids, and amblypygids. Slender cheliceras with pincers; first legs are long and sensory; other 3 pairs and pedipalps are used in walking. No eyes. Respiratory structures are simple sacs. Abdomen ends in long slender projection. *Eukoenenia.*

Order Araneae, the spiders. Most conspicuous group of arachnids, due to worldwide distribution, large number of species, and habit of spinning silk webs. Cheliceral fangs inject paralyzing poison; pedipalps sensory and modified in male for sperm transfer; 4 pairs of walking legs. Book lungs and/or tracheal tubes. *Aphonopelma* ("tarantulas"), *Bothriocyrtum* (trapdoor spiders), *Pholcus, Loxosceles* (brown spiders), *Latrodectus* (widow spiders), *Theridion, Lycosa* (wolf spiders), *Dinopis* (ogre-faced spiders), *Uloborus* (orb-weaving cribellate spiders), *Araneus* and *Argiope* (acribellate orb-weavers).

Order Opiliones (or Phalangida), the harvestmen. Cheliceras with pincers; pedipalps may be strong and spiny, or slender and sensory; 4 pairs of walking legs, often extremely long. Eyes usually on a tubercle. Tracheal tubes. Defensive secretions with strong odors. Predatory and scavenging on plant materials. Worldwide. *Leiobunum* ("daddy longlegs").

Order Acari (or Acarina), the mites. Cheliceras with pincers or styletlike; pedipalps mostly small. Walking legs, 4 pairs or fewer. Respiratory system of tracheal tubes, or absent. Some forms produce silk. Predatory, scavenging, parasitic on plants and animals. Some freshwater and marine species. Worldwide. *Oppia* ("beetle" mites), *Ixodes* and *Dermacentor* (ticks), *Eutrombicula* and *Neotrombicula* (chiggers), *Tetranychus* (spider mites), *Dermatophagoides* (dust mites), *Sarcoptes* (scabies mites), *Zercon* (free-living predaceous mites), *Acarus* (storage mites).

Order Ricinulei. Relatively rare forms, found in forest litter and caves in scattered tropical and subtropical regions. With heavy exoskeleton and anterior hood that covers mouth and cheliceras. Small pincers on cheliceras and pedipalps; 4 pairs of walking legs, the third pair modified in the male for sperm transfer. No eyes. Tracheal tubes. *Cryptocellus.*

Order Pseudoscorpiones, the pseudoscorpions. Small forms, looking like tiny tail-less scorpions. Cheliceras with pincers, spin silk from tips; armlike pedipalps with large pincers and poison glands, which open near tips; 4 pairs of walking legs. Tracheal tubes. Worldwide. *Chelifer* (found in houses, libraries), *Garypus.*

Order Solifugae (or Solpugida), the windscorpions or sun-spiders. Very large cheliceral pincers; pedipalps and first legs are sensory; other 3 pairs used in walking. Body relatively flexible and fast-moving, covered with long sensory bristles. Tracheal tubes. Mostly tropical and subtropical. *Eremobates, Galeodes.*

Class PYCNOGONIDA (or Pantopoda), the sea-spiders. Sometimes placed in a separate subphylum. Marine, slender-bodied forms, with reduced abdomen, without special respiratory or excretory structures. Cheliceras usually with pincers, or absent; palps slender and sensory, or absent; ovigers, appendages used to carry the eggs and unique to this class, are present in the male, sometimes also in the female. Usually 4 pairs of walking legs, sometimes 5 or 6. Simple eyes on tubercle. Protonymphon larva. Worldwide. *Nymphon, Pallenopsis, Pycnogonum, Colossendeis, Ammothea, Achelia, Phoxichilidium, Endeis.*

Chapter 23

Myriapods

Centipedes
Millipedes
Symphylans
Pauropods

NOT COUNTING the number of legs (which saves a lot of of trouble), myriapods are much like insects without wings. Both myriapods and insects have a single pair of antennas, a pair of mandibles, and a hardened head capsule. Both have tracheal respiratory systems and various other similarities in internal anatomy, for example, a straight simple gut, usually lacking digestive ceca, but with Malpighian tubules. Both have unbranched (uniramous) appendages, a feature represented in the name of the subphylum **UNIRAMIA,** which myriapods and insects together comprise.

Despite the many similar features—and even these are probably not all homologous—myriapods and insects also display critical differences which have made them a focus of controversy in questions of arthropod evolution. For example, one might expect to find among myriapods some clues to the origins of the compound eyes of insects. Instead, the simple eyes of myriapods, and the units of the compound eyes of one group of centipedes, are quite different from both insect and crustacean ommatidia and therefore probably evolved independently. Likewise, tracheal systems have almost certainly evolved separately in the arachnids and the uniramians as adaptations to terrestrial life, as have Malpighian tubules. Both these sets of organs also vary enough among uniramian groups to suggest several independent origins *within* this subphylum.

Four classes of many-legged uniramians are conveniently grouped in the superclass **MYRIAPODA.** Of these, the principal members are the **centipedes** and **millipedes,** which make up the great majority of species. The minor members, in terms of both number of species and size of individuals, are the **symphylans** and **pauropods.** Myriapods are often extremely abundant and number a respectable 11,000 species, with representatives in both tropical and temperate lands. They are ground-dwellers, taking shelter in moist retreats beneath rocks, logs, and leaf litter, or burrowing in soil. If they have been somewhat neglected by biologists, it is perhaps because they tend to be overshadowed by their flashier arthropod relatives, or because only a few pose any medical or economic threat to humans.

Taken in the strict sense of "ten thousand feet," the name myriapod is admittedly an exaggeration. But in the sense of "countless" or "innumerable feet," it seems apt to anyone who tries to count the legs of a millipede as it walks by.

Though the name refers to the multiple legs, myriapods are also similar in having a body composed of only 2 regions: head and trunk.

Stone centipede, *Lithobius forficatus,* shown in ventral view. It hides under objects, especially stones. If handled roughly, it inflicts a bite that feels like the prick of a needle; the mild pain subsides quickly. The clumps of simple eyes are not image-forming, and prey capture depends mostly on the sensitive antennas. If these are cut off, the animal does not take prey. *Lithobius forficatus* reaches 32 mm. Preserved specimen. (P.S. Tice)

Poison claws show in this ventral view of a 15-cm scolopendromorph centipede from Bermuda. The poison gland of each claw extends into the base of the appendage, and the 2 bases are united ventrally into a single large plate. In this preserved specimen, the poison claws are pulled to the sides, revealing the 2 pairs of maxillas. The lobelike first maxillas obscure the mandibles. The leglike second maxillas are seen arching over the first pair.

Centipedes should be handled carefully, for even some small ones can deliver a "bite" with the poison claws that is painful, although not serious for humans. The bite of a large scolopendromorph centipede can hospitalize an adult and be dangerous to children. There have been rare reports of fatalities. (P.S. Tice)

Myriapods seldom offer any special attraction either, except for various large centipedes, said to be delicious; some Polynesian islanders enjoy them toasted over a fire, while other peoples prefer to eat them raw.

CENTIPEDES

ABOUT 2,800 species comprise the class **Chilopoda,** the centipedes. All are predators, feeding on soft insects, land isopods, various arachnids and other small arthropods, snails, slugs, and worms. Large tropical centipedes sometimes even catch small lizards, snakes, birds, and mice. Sight plays little role in prey capture, for centipedes hunt in the dark and most are blind or have only small clusters of simple eyes. They seem to find their prey mainly by chance encounter and strike rapidly when their antennas have contacted something recognized as edible. The prey is seized by the **poison claws,** the paired appendages of the first trunk segment. These have perforated tips through which a poisonous secretion from a pair of glands is injected into the prey. The biting mandibles are assisted by 2 pairs of maxillas.

Each of the trunk segments, except for the first and the last two, has a pair of walking legs. The number of pairs ranges from 15 to 177 and is always odd, so no centipede is exactly "hundred-legged," as the name implies. Centipedes run rapidly and the smooth operation of so many legs is an impressive feat of coordination. The loss of one or more legs causes no perceptible limp; the gait is at once modified to compensate. *Scutigera* readily autotomizes its long legs and can run rapidly along on only half the normal number. Lost appendages are regenerated and appear at the next molt.

The legs are moved in a regular order, each slightly out of phase with its neighbors, so that waves of movement appear to be passing along the body. As a centipede increases its speed, each stride becomes longer and more quickly executed and fewer feet are in contact with the ground at any one instant. For example, in a 42-legged scolopendromorph centipede running at top speed, only 3 feet at a time are on the ground. The fastest centipedes are the scutigeromorphs, with only 30 very long legs; a house centipede *(Scutigera coleoptrata),* 2.2 cm long, was timed running 42 cm/sec.

Centipedes can make themselves very flat dorso-ventrally, which enables them to escape quickly into narrow crevices and to take shelter readily under stones or bark. Such humid hiding places are essential to survival, for centipedes have little capacity for **water regulation.** Though terrestrial, these animals are extremely vulnerable to drying. A centipede such as *Lithobius* will die in a few hours in an uncovered dish of dry earth. Water is lost rapidly through the permeable cuticle and through the respiratory system. The air-filled tracheal system opens through spiracles that have no closing mechanism. Water is also lost in the excretion of large quantities of ammonia since only a small fraction of the excretory waste is uric acid, the relatively insoluble product excreted by insects to conserve water. Most of a centipede's water conservation measures consist in several kinds of behavioral responses. The dominant one is a direct, positive response to moisture in the soil or other substrate and to air of high humidity. Another is a strong positive response to contact; a centipede will seldom come to rest unless its body is in contact with a surface on at least two sides. Also important is a negative response to light; centipedes avoid light when possible and are active almost exclusively in the dark. Even eyeless centipedes avoid light and so do those whose eyes have been experimentally covered; other parts of the body must be sensitive to light. These behaviors combine to keep centipedes in moist, confined refuges much of the time. Water loss is made up by drinking and by fluids in the food.

Even though moisture is of primary importance to centipedes, and may partly explain why the majority of species live in the humid tropics and subtropics, no centipedes are aquatic and they usually avoid entering water. *Lithobius* can survive many hours completely immersed in water, but its body swells. The burrowing geophilomorph centipedes, on the other hand, do not swell in water, having waterproofing in the cuticle that protects them against waterlogged soils. Some geophilomorphs live in the rocky intertidal and can remain submerged in seawater for up to 40 hours.

That the habit of keeping under cover is a positive reaction to contact as well as a negative response to light can be shown by a simple experiment. A centipede placed in a glass dish will run about ceaselessly; but if some transparent glass tubing is placed in the dish, the animal will soon come to rest in the tubing, which affords a maximum of contact with the surface of the centipede.

House centipede, *Scutigera coleoptrata,* is often found in buildings, frequenting moist areas, usually basements and bathrooms. It feeds on silverfishes, cockroaches, flies, and other small arthropods, and should be welcomed in any household. The last pair of legs, twice as long as the 25-mm body, are used only as feelers, adding to the sensory information provided by the other legs and by the long threadlike antennas. Unlike other centipedes, scutigeromorphs may run in the open, often in full light, and have large compound eyes that probably help in catching fast prey such as flies. Also unique to scutigeromorphs are the spiracles along the back, each leading to a compact mass of short tracheal tubes from which oxygen is taken up and distributed by the blood. Other centipedes have lateral spiracles and longer, more branching tracheal tubes that distribute oxygen throughout the body. (R.B.)

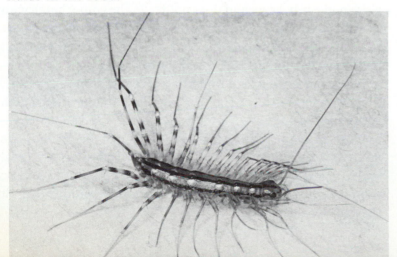

Scolopendromorph centipede, *Scolopendra morsitans,* is yellow with green cross-bars. It has been distributed, by commerce, around the world in warm climates. In the U.S. it occurs in southern states. The largest tropical members of the order Scolopendromorpha may be longer than 25 cm. Aside from their painful and sometimes dangerous bite, scolopendromorphs may pinch with the spiny last pair of legs. In running across the skin they make tiny cuts with the tips of the feet. If alarmed, they secrete an irritating substance that enters the cuts. S.W. Africa. (E.S. Ross)

Marine centipede inhabits rocky California shores. This geophilid centipede, 5 cm long, hides in crevices by day, emerging during the night at low tide to feed on insects and mites in the high intertidal. Most members of the order Geophilomorpha live in soils and feed on insect larvas and on worms. They have up to 177 pairs of legs. *Nyctunguis heathii.* (R.B.)

Reproduction in centipedes involves a sedate tactile courtship of up to several hours; the male and female of *Lithobius* pat each other on the hind end with the antennas. The last 2 segments, which lack walking legs, bear the genital appendages and openings. Fertilization is internal, but centipedes do not copulate. Sperm transfer is indirect and similar to the process in many arachnids. In most centipede species, the male spins a small irregular web on the ground or on the burrow wall and deposits in it a spermatophore. When the spermatophore is in place, the male may signal the female to approach it by tapping her with his antennas or he may lead her toward it. The female introduces the spermatophore into her genital opening, and the sperms are released internally. The male of *Scutigera* gently pushes the female toward the spermatophore, then picks it up in his mouth and places it into the female's genital opening.

Centipedes handle their eggs and young in two different ways. Some centipedes (scolopendromorphs and geophilomorphs) lay all their eggs in one batch, then coil tightly around the eggs, keeping them clean and protected for up to 3 months (see chapter heading). While brooding, the female remains hidden in her burrow and does not feed. The young hatch with the full adult number of segments and appendages but are at first helpless, and the mother continues to brood them until they have molted several times. Other centipedes (lithobiomorphs and scutigeromorphs) lay their eggs one at a time, covering them with soil and leaving them. These young hatch with fewer than the adult number of segments and appendages, adding them at successive molts. Young of both kinds are nourished by stored yolk and cannot feed until after several molts. Some centipedes complete their life history in a year; others take up to several years to mature and continue to molt throughout their lives.

MILLIPEDES

THE NAME millipede means "thousand-legged," and though this is an exaggeration, millipedes can have prodigious numbers of legs. The record count is 752 legs, but most millipedes have about a hundred; some kinds, which are short and wide and resemble land isopods, have only a few dozen legs. Millipedes are typically burrowers, and the large number of legs, pushing at the same time, provide power for bull-dozing through the soil. The head is curled under during burrowing, and the millipede pushes with a calcified shield of cuticle behind the head. The cuticular rings of the body of most millipedes are also calcified for added strength.

Even so, the long body that should normally go with so many legs would probably buckle under the force of burrowing if the body were not shortened by a peculiar feature of millipedes. During development, fusion occurs between pairs of trunk segments so that in the adult there are *2 pairs of legs to each body ring* (except a few anterior and posterior ones). This double-legged feature of all millipedes is recorded in the name of the group they comprise, the class **DIPLOPODA**. In addition, each body ring is very short, compared to a segment of a centipede.

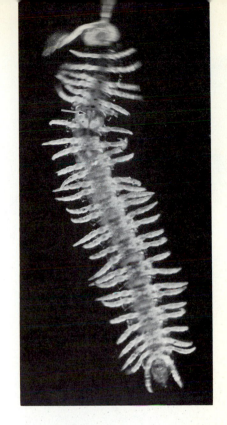

Ventral view of a wriggling millipede shows the double pairs of legs and the anterior position of the genital area *(arrow)*. (P.S. Tice)

The head of a millipede has the same appendages as that of a centipede except that the first maxillas fuse to form a kind of lower lip (gnathochilarium) and there are no second maxillas. A second-maxillary segment appears in the embryo but never develops appendages; it contributes to the pushing shield (collum) behind the head. The head may bear 2 clusters of many simple eyes set close together, as here, or eyes may be absent. A short distance behind the head, on the ventral surface, are the genital openings. In most millipedes the male uses modified legs to transfer sperms into the female's genital opening. The **posterior end** of the animal shows the terminal anus and a number of typical body rings (diplosegments), each with 2 pairs of legs. (P.S. Tice)

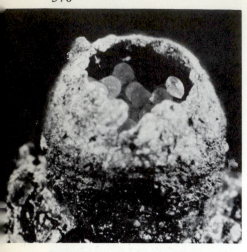

Millipede eggs are often laid in a chamber made of earth or in a cocoon of silk, constructed and often guarded by the mother. Some millipedes also build protective chambers in which to molt. (P.S. Tice)

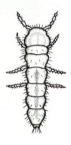

A newly hatched millipede usually has only 3 pairs of legs. Several body rings and leg pairs are added at each successive molt until the adult number is reached. This individual *(Amplocheir sequoia)* belongs to one of various species of luminescent millipedes. (After Davenport, Wootton, and Cushing)

Spiral coiling is not only a defensive behavior but also reduces dessication in resting millipedes. (R.B.)

Giant millipede from East Africa, photographed in the London Zoo. This one was 20 cm long. The largest millipedes reach 28 cm. (R.B.)

Of the 8,000 species of millipedes, the great majority are herbivores, feeding mostly on decaying vegetation and contributing significantly to soil formation. They sometimes eat the living roots of young plants, becoming garden and agricultural pests. Efforts to banish them with insecticides are more likely to eliminate instead the susceptible centipedes, which might have aided in pest-control. Occasionally, under conditions not well understood, a local millipede population suddenly expands and enormous numbers of migrating millipedes are observed.

Millipedes do not bite and are mostly quite safe to handle, since their first line of defense is to curl up into a tight ball or spiral. Many can also secrete toxic or disagreeable substances, including hydrogen cyanide, from pores along the body. Some large tropical species can spray their defensive secretion 30 cm or more; and it is best not to hold any millipede too close to one's eyes.

Rolled-up millipede looks much like a land isopod. When rolled, this pale orange Guatemalan millipede (order Glomeridesmida) was about 1 cm in diameter. Rolling up is also seen in the order Glomerida of N. America and Europe. In the order Sphaerotheriida of S. Africa, Asia, and Australia, a rolled-up millipede may reach golfball size. (R.B.)

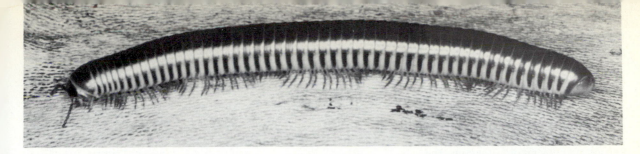

The shapes of millipedes go with distinctive habits. Members of several orders that burrow through homogeneous substrates tend to be cylindrical, like the 8-cm Florida millipede *above*. Millipedes that move into or enlarge existing crevices use the body as a wedge and tend to be flat-backed, often with lateral keels on the dorsal plates, like the Guatemalan millipede (order Polydesmida), *below left* and the mass of pink Costa Rican millipedes (order Platydesmida) *below right*. Some platydesmids are said to be colonial. (R.B.)

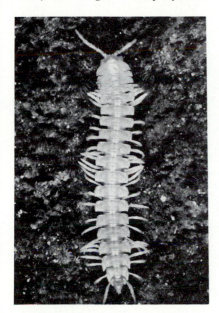

The class **SYMPHYLA** comprises 120 species of small blind myriapods, all less than 1 cm long. Symphylans look like little centipedes but have only 12 pairs of legs. The body is extremely flexible, enabling the animal to twist and turn as it makes its way through tiny cracks and channels in soil or rotting wood, among leaf litter, or under stones. With 3 pairs of mouthparts that resemble those of insects, symphylans feed on soft plant material, both live and decaying. They are sometimes called garden-centipedes and can be troublesome garden pests. Females may reproduce by parthenogenesis, but even in sexual reproduction, male and female need never meet, for the males simply deposit stalked spermatophores. When a female encounters a spermatophore, she takes it into her mouth and stores the sperms in special cheek pouches. When ready to lay she takes an egg from her genital opening (anteriorly located, as in millipedes), attaches it to a moss plant or other substrate, then works it over with her mouth, smearing it with sperms. The young hatch with 6 or 7 pairs of legs and acquire the full number only after several molts. Symphylans may live 4 years or more and continue to molt throughout life.

Symphylan. Length, 2 mm. Central California. (R.B.)

Female symphylan laying an egg. 1. She curls her head under to reach the genital opening and takes the egg in her mouth. **2.** She attaches the egg to a moss plant, then smears it with sperms stored earlier in her cheek pouches. *Scutigerella.* (From L. Juberthie-Jupeau)

572

The class **PAUROPODA** comprises 350 species of tiny blind myriapods found in soil, leaf litter, decaying wood, and other moist situations. The soft, pale body suggests a slow creeper or wriggler, but pauropods run rapidly for their size (less than 2 mm long), holding the body straight and turgid. Little is known of their habits, but they have been reported to feed on fungal filaments, carrion, and organic matter in soil. Pauropods resemble millipedes in their 2 pairs of mouthparts and in the anterior location of the genital openings. As in some other minute arthropods, there is no heart and usually no tracheal system.

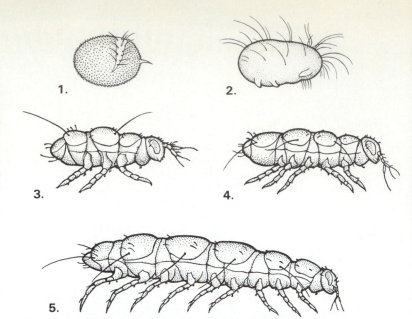

Development of a pauropod, *Pauropus silvaticus.* The egg **(1)** hatches into a quiescent stage **(2)**, showing only the beginnings of appendages. A molt produces the first active stage **(3)** with 3 pairs of legs. Segments and legs are added at successive molts **(4)** until the adult form with 9 pairs of legs is reached **(5)**. What looks like an eye on the side of the head is a sense organ thought to detect auditory or other vibrations. (After O.W. Tiegs)

CLASSIFICATION: Phylum Arthropoda, Subphylum Uniramia, Superclass MYRIAPODA

Class CHILOPODA, centipedes. Predaceous, with 2 pairs of maxillas, plus poison claws (maxillipeds). Genital area located posteriorly; sperm transfer indirect (spermatophores).

Superorder Epimorpha. Eggs brooded by mother; young hatch with full number of segments and legs.

Order Scolopendromorpha. With 21 or 23 pairs of legs. Includes largest forms. *Scolopendra, Cryptops.*

Order Geophilomorpha. With 31 to 177 pairs of legs; no eyes. Many burrowing forms. *Geophilus, Nyctunguis.*

Superorder Anamorpha. Eggs not brooded; young add segments and legs at successive molts.

Order Lithobiomorpha. With 15 pairs of legs. *Lithobius.*

Order Scutigeromorpha. With 15 pairs of very long legs; compound eyes. *Scutigera.*

Class DIPLOPODA, millipedes. Herbivorous, with 1 pair of maxillas (1st maxillas, fused as gnathochilarium). Genital area located anteriorly.

SUBCLASS PENICILLATA (PSELAPHOGNATHA). Bristly, non-calcified cuticle; sperm transfer indirect (spermatophores).

Order Polyxenida.

SUBCLASS CHILOGNATHA. Calcified cuticle; sperm transfer direct.

Superorder Pentazonia. Short-bodied, can roll into a ball.

Orders Glomerida, Sphaerotheriida, Glomeridesmida.

Superorder Helminthomorpha. Medium- to long-bodied, can coil spirally. Includes great majority of millipedes.

Orders Craspedosomatida, Stemmiulida. Can spin silk; varied body forms.

Order Polydesmida. Medium-length, flat-backed body form.

Orders Julida, Spirobolida, Spirostreptida. Long, cylindrical body form.

Superorder Colobognatha. Can coil spirally, varied body forms, wedge-shaped anterior end.

Orders Platydesmida, Polyzoniida.

Class SYMPHYLA, symphylans. Herbivorous, with 2 pairs of maxillas (2nd maxillas fused as labium). Genital area located anteriorly; sperm transfer indirect (spermatophores). *Scutigerella.*

Class PAUROPODA, pauropods. With 1 pair of maxillas (1st maxillas). Genital area located anteriorly; sperm transfer indirect (spermatophores). *Pauropus.*

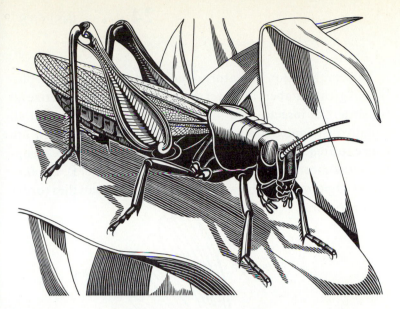

Chapter 24

Insects

THE SUCCESS OF ARTHROPODS as measured by sheer numbers of species is overwhelmingly due to the insects, which outnumber other arthropods by about 8 to 1. Although many factors must contribute to the profusion of insect species, the key elements are probably small size, wings, and metamorphosis. These features allow an insect to complete its whole life history by specializing in the use of a single limited resource or by combining several resources widely separated in space or time. A single plant that would provide only a few bites for a large animal can support hundreds of small insects comprising many species, each of which takes advantage of some different plant part. Individuals of different species specialize in eating the leaves, buds, flowers, seeds, fruits, bark, or roots; others drink the nectar or collect the pollen; and yet others suck the sap or feed on plant tissues in various stages of decay. Wings permit insects to exploit specialized food sources that are rare or widely scattered, or to fly to different localities, as resources change with the seasons, and draw on several resources, any one of which, by itself, would be insufficient. Likewise, insects with larvas and adults that have different eating habits can profit from a combination of resources and habitats. The larvas of the apple blossom weevil, for example, feed on the stamens, pistils, and other parts of the unopened apple flowers, a food source available for only about 3 weeks each year; the adult beetles feed on the leaves of the apple tree during the summer and, after overwintering, on the buds in the spring. Such specific associations between insects and thousands of species of seed plants are the basis for much of insect speciation. In addition, many species of predatory and parasitic insects specialize in exploiting other particular insects.

The largest living insect, if measured by its wingspan, is a noctuid moth, a species of *Erebus,* that is 28 cm from wingtip to wingtip. The longest walkingstick insect is 28 cm. And the most massive insect, a beetle, is 15 cm long. The smallest insect (a "fairyfly," a parasite in insect eggs) is only 0.21 mm long.

If there were only about as many species of insects as there are of crustaceans or chelicerates, the arthropods would still be the largest phylum of animals, but they would not outnumber all other animals by better than 3 to 1. Of the roughly 900,000 described species of living arthropods, about 800,000 are insects. These numbers are not precise, but efforts at more precision would be relatively meaningless in view of the many millions of insect species that entomologists believe remain to be described. Some estimate the total number of living insect species to be 10 million. If the rain forests of the world continue to be destroyed at the present rate, a large proportion of the undescribed insects will never be known. In a recent collection of insects from a forest canopy in Peru, 80% of the insects had never been seen before.

The beetles, with about 350,000 described species, are by far the largest order of insects and include over 3 times as many species as the second largest *phylum* of animals, the molluscs. One single family of beetles (the Curculionidae, mostly weevils) numbers over 60,000 described species, more than any animal phylum except the molluscs.

The combination of features that seems to favor the evolution of great numbers of insect species does not characterize all the orders of insects. And in fact, insect species are not at all equally distributed among the 30 or so orders. Over 85% are concentrated in 4 dominant orders: the beetles (order Coleoptera), the moths and butterflies (order Lepidoptera), the flies (order Diptera), and the wasps, bees, ants, and sawflies (order Hymenoptera).

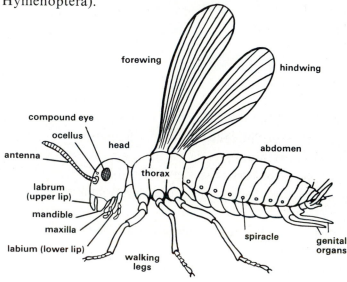

Diagram of a **generalized insect**. Most insects have fewer abdominal segments, owing to loss or fusion at the posterior end. (After R.E. Snodgrass)

Ventral view of grasshopper shows body regions and attachment of legs to thorax. (R.B.)

THE NAME "insect" comes from a Latin word meaning "incised" and refers to the fact that insects generally have a sharp division between the head and thorax and between the thorax and abdomen. The head of an adult insect is all in one piece, but separate segments can be identified in the embryo and at least some of them are indicated in the adult by the paired appendages they bear. The thorax, which has undergone less specialization and fusion than the head, has 3 segments, each bearing a pair of appendages, usually walking legs. The second and third thoracic segments each bear a pair of wings (except in certain primitively wingless or specialized forms). The abdomen is the least specialized region of insects, being composed of relatively similar segments, generally without appendages except for certain structures at the posterior end. The external anatomy of insects varies so much that it is less satisfactory to generalize than to describe a particular insect such as a grasshopper, which is usually considered to be a fairly typical representative of the superclass **INSECTA**.

GRASSHOPPERS

THE **head** of a grasshopper has 2 compound eyes and 3 simple ones (ocelli). The compound eyes are similar in structure to those of a lobster or other crustacean; and while they are not on stalks, they have a broad field of vision because they occupy a relatively large area and curve around the sides of the head. There are 4 pairs of appendages. The first of these are a pair of long, jointed sensory *antennas*. The "upper lip," or *labrum*, in front of the mouth, does not form from paired appendages but is a median outgrowth. The second pair of appendages are the hard, toothed jaws, or *mandibles*. The third pair are accessory feeding appendages, or *maxillas*, each with a jointed sensory palp. The fourth pair form the "lower lip," or *labium*, which has on each side a sensory palp. The labium of the adult appears to be a single plate, but in the embryo it arises as a pair of structures which later fuse in the midline. Thus each half of the labium is serially homologous to one of the maxillas.

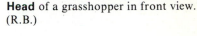

Head of a grasshopper in front view. (R.B.)

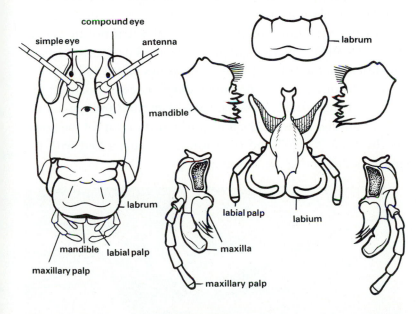

Front view of the head of a grasshopper.

Mouth parts removed from their attachments. (After R.E. Snodgrass)

The **thorax** is partly covered dorsally by a chitinous shield and by the wings; but 3 segments, with a pair of legs on each, are clearly visible at the sides. The *legs* are composed of a characteristic series of articulated parts, or articles. Each leg ends in 2 curved claws between which there is a fleshy pad that aids in clinging to surfaces. The first two pairs of legs are typical walking legs. The third is specialized for jumping; one article, the femur, which contains muscles for jumping, is enlarged out of proportion to the others. The *2 pairs of wings*

With each step of a walking insect, a foreleg and hindleg touch the ground on one side and a middle leg on the other side. Such a tripod of legs appears to be the minimum number required to maintain stability or firm contact in an animal as small as an insect. And striding along consists of moving from one tripod to the next. Hexapody, the use of 6 legs in walking, has probably evolved independently in the various groups of wingless insects and in winged insects, as well as in some other arthropods.

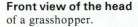

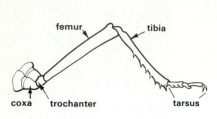

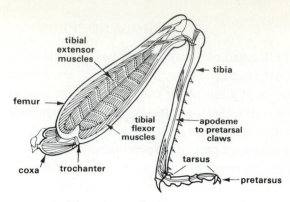

Middle leg of a grasshopper, used in walking, shows in relatively unspecialized form the parts of an insect leg. (After R.E. Snodgrass)

Jumping leg of a grasshopper. Extensor muscles occupy most of the femur and are inserted by a broad apodeme onto the base of the tibia. When these powerful muscles contract, the tibia is suddenly extended, pushing the tarsus against the ground with enough force (from both legs) to launch the insect into the air. When the grasshopper is at rest, the tibia is flexed under the femur. Small muscles in femur and tibia are inserted onto a long, tendonlike apodeme that runs all the way to the claws of the pretarsus, serving to retract them. They are extended by the elasticity of the pretarsus. (After R.E. Snodgrass)

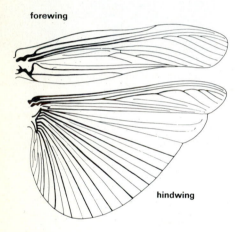

Wings of a grasshopper. Only the larger veins are shown. (After R.E. Snodgrass)

Degree of development of wings varies in grasshoppers. Both sexes may include normal or short-winged forms, or only the female has short wings. Certain groups of grasshoppers are exclusively short-winged or wingless, or in one group the forewings are vestigial and the hind wings vary in length. Florida. (R.B.)

are different from each other, although both function in flying. The forewings are narrow and hardened and serve as a cover for the hindwings when the grasshopper is at rest. The hindwings are quite broad in flight, but when not in use are folded like a fan and fit under the first pair. The wings are made of cuticle and are stiffened by thickenings called *veins*.

The **abdomen** has no appendages except those at the posterior end which are associated with mating and egg-laying. The abdomen contains the bulk of the soft organs of the body, as the head is small and the thorax is nearly filled with the muscles that move the legs and wings.

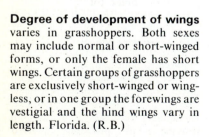

The **digestive tract** consists of three parts: foregut, midgut, and hindgut. The *foregut* starts at the mouth, receives a secretion from the salivary glands, and runs on as a narrow esophagus, which leads to the *crop,* a large thin-walled sac in the thorax. On the inner walls of the crop are transverse ridges armed with rows of spines which probably serve to cut the food into shreds. The crop is mainly a storage sac which enables the grasshopper to eat a large quantity at one time and afterward digest it leisurely. From the crop the food passes into a muscular *gizzard,* lined with chitinous teeth. At the posterior end of the gizzard is a valve, which prevents the food from passing into the midgut before it is thoroughly ground and also prevents food in the midgut from being regurgitated. Digestion begins in the crop, for the food entering that organ is already mixed with salivary secretion, and it also receives some digestive juices which pass anteriorly from the midgut.

Study of the crop content of insects should preceed any expensive investment of time, energy, or toxic chemicals for pest control. For example the meadow grasshopper was long considered to be a general feeder on vegetation and an enemy of agriculture. Careful examination of the crop content revealed a diet consisting mainly of pollen grains and seeds, together with some insect parts. Thus the assessment of the meadow grasshopper, as it impinges on the human food supply, was changed from an insect that is harmful to one that is neutral or perhaps even beneficial.

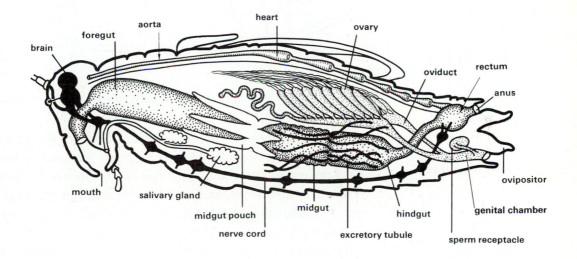

Internal anatomy of a female grasshopper. (Based on R.E. Snodgrass)

The whole foregut is lined with cuticle, and little, if any, absorption of food occurs there. The *midgut,* which lies mainly in the abdomen, has no cuticular lining and serves as the main organ of digestion and absorption. Opening into the anterior end of the grasshopper midgut are 6 pairs of pouches which secrete digestive juices and also aid in absorption, but in most insects the midgut is a simple tube. The junction of the midgut with the *hindgut* is marked by the attachment of long excretory tubules. The hindgut is lined with especially thin cuticle. It receives the waste materials of digestion and the nitrogenous secretions of the excretory tubules.

The insect digestive lining does not secrete mucus, as in most animals, as a protection against abrasion by solid food. The foregut and hindgut are protected by cuticle, but the midgut is not. Instead it is protected by a proteinaceous film, laced with chitinous fibers, called a **peritrophic membrane,** which is continually replaced. The membrane is lacking in most fluid-feeding insects.

Water loss is a serious problem for insects that live in deserts or even in temperate areas that have hot dry seasons. The waxy layer of the cuticle retards evaporation through the body surface but does not completely prevent it. Thus the small size of insects becomes an important factor in water loss. A desert beetle that is 1/2 the length of another beetle of similar shape will have only 1/4 as much body surface exposed to the evaporating power of the surrounding air. But it will have only 1/8 the fluid-filled body volume and so will dry up twice as fast.

The tubules of the **excretory system** are called *Malpighian tubules,* after their discoverer. They lie in the blood sinuses and extract nitrogenous wastes from the blood, mostly in the form of uric acid. Fluid from the Malpighian tubules, containing uric acid and salts, is poured into the hindgut. Here, especially in the rectum, much of the water and some of the salts are reabsorbed, leaving crystals of uric acid which are expelled by way of the anus as dry excretions. Dry wastes are characteristic of small land animals, which thus conserve water.

Oxygen for **respiration** is not distributed by the circulatory system. Instead, air is piped through branching **tracheal tubes,** which ramify to all parts of the body. The tubes open to

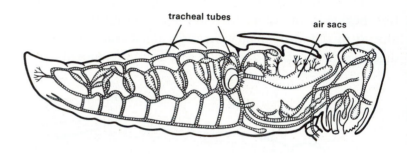

Tracheal system of a grasshopper. Only the main tracheas and air sacs are shown. (After Vinal)

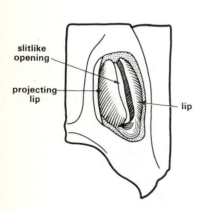

External view of spiracle.

the outside through paired **spiracles** that lie along the sides of the thorax and abdomen. The openings are guarded by hairs that keep out dirt and by a valve that can be opened or closed to regulate the flow of air. From the spiracles, air passes first into a system of major longitudinal and transverse trunks, from which smaller tubes branch freely until they end blindly in the tissues. The terminal groups of fine branches (tracheal capillaries, or tracheoles), less than 1 μm in diameter, are produced within single cells, which continue to surround them. Cuticle lines the entire tracheal system, and spiral thickenings of the cuticle prevent collapse of the delicate tubes. The cuticular lining is shed at each molt, except from the tracheoles. Here and there the tracheal tubes widen into large air sacs. Muscular breathing movements, especially of the abdomen, alternately compress the air sacs and then allow them to expand, thus aiding in changing the air. In grasshoppers the anterior spiracles open at inspiration, and the posterior ones at expiration; coordination of the breathing movements with the opening and closing of the spiracles effectively creates a one-way flow of air in the main longitudinal trunks.

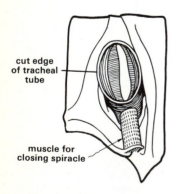

Internal view of spiracle.

Closing mechanism of the second thoracic spiracles of a grasshopper consists of 2 movable hardened lips that can be drawn together by a closing muscle. The spiracle is opened by elasticity of the membranes when the closer muscle is relaxed. (After R.E. Snodgrass)

In the deepest branches oxygen moves, by diffusion alone, first along the tubes, and then into the surrounding blood spaces and tissues. Carbon dioxide leaves by the reverse route (but small amounts may also escape through the thin parts of the body surface). Within the tissues, diffusion of oxygen is almost a million times slower than in air, and the distance to be traveled must be kept very short. Tracheoles surround almost every cell and even make deeply indented channels into large muscle fibers so that no part of these active cells is more than a few micrometers from a tracheole.

There is no mechanism known for making a membrane permeable to oxygen but impermeable to water vapor, and most of the water loss suffered by a grasshopper is through its large respiratory surface. This loss is reduced partly by keeping the spiracles closed as much as possible. In a grasshopper at rest, only the first and last pairs open; when the insect is more active, additional pairs open. The flight muscles, when active, automatically improve their own air supply by pumping on the large thoracic air sac; the 2 pairs of spiracles that supply these muscles remain open continuously, and water loss increases sharply. (During prolonged migratory flight, locusts make up for this loss by increased production of metabolic water.) The spiracles are closed more, and breathing movements are reduced, when a grasshopper finds itself low on water, surrounded by air of low humidity, or newly molted and thus especially vulnerable to dehydration. Even under these conditions, however, high levels of carbon dioxide will cause the spiracles to open.

If intake of food and water is limited and the surrounding air is dry, insects lose water steadily in spite of all their water conservation measures (waxy cuticle, regulation of the spiracles, dry wastes). In such circumstances the best they can do is to minimize the rate of loss. In addition, some insects tolerate substantial losses in blood volume, thus partially protecting the tissues against dehydration. In one species of grasshopper studied, the blood volume decreased 20% over 24 hours. Cockroaches have survived losses of up to 60% of their initial blood volume over a period of 8 days.

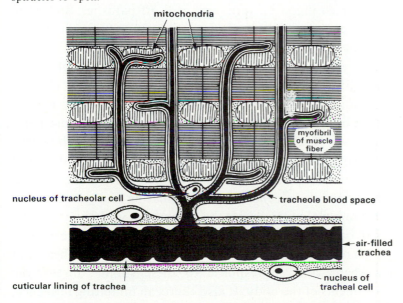

Tracheoles branch among the cells of all the tissues of an insect. The more active tissues are more abundantly supplied with tracheoles, and thus with oxygen, than are tissues that have lower metabolic rates. In the most active tissues, such as the flight muscle of a wasp, represented here, tracheoles occur not only *among* the large muscle cells, but deep *within* each cell. The muscle cell membrane is not pierced; it invaginates to form narrow, branching tunnels in which the tracheoles run, passing between the myofibrils along cytoplasmic channels that are filled with large mitochondria, the site of respiration. Tracheoles bring oxygen to within a few μm of each mitochondrion and occasionally even into invaginations within a mitochondrion. (Based on various sources)

The blood vessels of the **circulatory system** are much less extensive in a grasshopper than in a lobster or a spider. There is only one vessel, the long contractile *dorsal vessel,* composed of the tubular *heart,* which pumps the blood forward, and its anterior extension, the *aorta.* In each segment through which it passes, the heart is dilated into a chamber perforated on each side by a slitlike opening through which blood enters. Blood leaves the heart through a series of excurrent openings and through the aorta, which carries the blood into the head and there ends abruptly. The blood flows out into spaces among the tissues and makes its way back into the thorax, where it bathes the thoracic muscles. From there it enters the abdomen and bathes the various organs, absorbing food from the midgut and giving up wastes to the excretory tubules. Then it returns to the heart. In an open system of this kind, the rate of flow is relatively low. However, this is no disadvantage, even in an active insect such as a grasshopper, since the distribution of oxygen has been taken over largely by the system of air-filled tracheal tubes. There are no respiratory pigments for carrying oxygen. Besides serving to distribute food and collect wastes, the blood acts as a reservoir for food and water; it contains cells which ingest bacteria and wall off parasites; and when under pressure, the blood aids in hatching from the egg, in molting, and in the expansion of the wings.

The course of the blood is really more definite than this brief sketch has intimated. There are partitions which deflect the blood so that it enters one side of each leg and emerges on the other. In the abdomen there are 2 large horizontal partitions which aid in directing its course, and the dorsal one separates the cavity containing the heart from that in which the other viscera lie.

The **nervous system** is a ventral, double, ganglionated cord like that of a lobster. A grasshopper has a fairly generalized system, as compared with that of some insects, which have all the thoracic and abdominal ganglia fused into one mass. The *dorsal ganglion,* or "brain," lies above the esophagus and between the eyes. It is joined to the first ventral ganglion by a pair of connectives which encircle the foregut. The brain has no centers for coordinating muscular activity; after removal of the brain, the animal can walk, jump, or fly. As in other invertebrates, the brain serves as a sensory relay which receives stimuli from the sense organs (on the head and elsewhere) and, in response to these stimuli, directs the movements of the body. It also exerts an inhibiting influence, for a grasshopper without a brain responds to the slightest stimulus by jumping or flying—a very unadaptive kind of behavior. Even in the absence of any external stimulation, the animal displays an incessant activity of the palps and legs. In addition the brain is the seat of certain complex behavior patterns and their modification by learned responses. The *first ventral ganglion* controls the movements of the mouth parts and exerts a general excitatory influence. The *segmental*

The insect nervous system is affected by chemical changes in the surrounding blood, but only after some delay. Free exchange between blood and nervous tissue is evidently prevented by some kind of barrier, which probably lies in the sheath that covers insect nerves, for desheathed nerves are rapidly affected by changes in the surrounding medium. In having such a **"blood-brain barrier,"** insects resemble vertebrates; this characteristic has not been found in any other invertebrates studied.

ganglia are connected and coordinated by nerves which run in the cords, but each is an almost completely independent center in control of the movements of its respective segment (or segments) and appendages. In some insects these movements have been shown to continue in segments which have been severed from the rest of the body. An isolated thorax is capable of walking, and an isolated abdominal segment performs breathing movements. In a grasshopper a complex of co-ordinated movements such as flying, is possible without input from the brain and it is largely independent of sensory feedback from the wings and flight muscles. In a vertebrate, coordinated movements depend on continuous and detailed sensory feedback from the moving parts. In a grasshopper the segmental ganglia have genetically built-in nervous connections that generate the whole coordinated sequence of basic flight movements while receiving only random stimulation. Yet the sequence can be modified by sensory input. For example, stimuli from stretch receptors at the base of each wing normally modify the frequency of wing beats. Or a grasshopper that has lost a wing can use sensory information about its lopsided course to correct its flight movements, so that it flies quite well again (but for this, the brain is necessary).

The **reproductive system** of a male grasshopper includes a pair of *testes,* which discharge sperms into a *sperm duct.* Into the duct, glands secrete a fluid in which sperms are conveyed to the female. The sperm duct opens through an *intromittent organ* near the posterior end of the body. In the female there is a pair of *ovaries,* with *oviducts.* The two oviducts converge into a *genital chamber.* During copulation the male introduces sperms into a sac in the female, the *sperm receptacle,* in which they are stored until the time for egg-laying. The mature eggs pass down the oviduct and, contrary to the usual procedure in animals with internal fertilization, the yolk and shell are put on before the egg is fertilized. A small pore is left, however, through which the sperm enters. As the eggs pass into the genital chamber, they are fertilized by sperms ejected from the sperm receptacle. A set of stout appendages, which together form an *ovipositor* near the posterior end of the abdomen, are used for digging a hole in the ground and depositing the eggs. The abdomen stretches much more than shown in the figure on the right. An abdomen only 2.5 cm long can dig a hole 8 to 10 cm deep. The soft intersegmental membranes, said to owe their elasticity chiefly to their component of chitin, stretch 1000% as the valves of the ovipositor, and the muscles attached to them, pull the abdomen down.

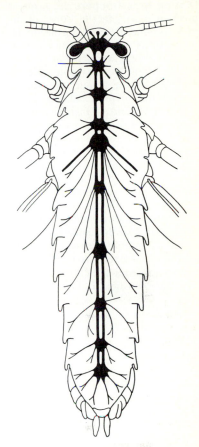

Nervous system of a grasshopper, showing brain and optic lobes; first ventral (subesophageal) ganglion with nerves of mandibles, maxillas, and labium; 3 fused pairs of thoracic ganglia; and 5 fused pairs of abdominal ganglia (the last a fusion of 3 pairs of ganglia). (Modified after R.E. Snodgrass)

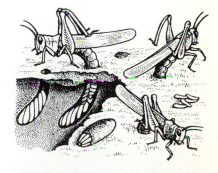

Egg-laying of *Melanoplus,* which deposits its eggs in the ground, where they pass the winter. The egg pod seen lying on the surface has been dug up and one end opened to expose the compact mass of eggs. (Modified after C.V. Riley)

Mating grasshoppers are brought together mostly by the "song" of the male, produced by rubbing the minute projections on each hind femur against the hardened radial vein on the forewing of the same side. If accepted by the female, the smaller male mounts his mate and they oppose the posterior ends of their abdomens. Copulation takes place as described earlier, the intromittent organ of the male inserting the sperms into a receptacle of the female. The sperms are further protected by enclosure in a proteinaceous sac, or spermatophore. Insects that do not produce spermatophores have a long penis which penetrates to or near the female receptacle. These adaptations ensure that the sperms of these terrestrial animals are never exposed to the drying effects of external exposure. (P. Leibman)

The **development** of insects is strongly influenced by the large amount of yolk in the egg. Instead of dividing into 2 cells and then into 4, and so on to form a blastula, the zygote nucleus divides many times without the division of the cytoplasm. The nuclei then move to the periphery, and cell membranes appear between them, thus forming a layer of cells. Subsequently, along a strip on the ventral side, some of the cells pass inward and give rise to mesoderm and endoderm; the remaining outer layer is ectoderm. A large portion of the ectodermal layer never becomes part of the embryo but instead forms membranes which surround it. The rest of the ectoderm forms the epidermis, nervous system, tracheal system, foregut, and hindgut. These last three develop from invaginations of the ectoderm; like the epidermis, they secrete a cuticle, which lines the lumen. The midgut develops from endoderm. The mesoderm becomes divided into segments, and paired coelomic sacs appear within the mesoderm of most of the segments. These coelomic sacs later break down and do not form the adult body cavity, which is not a coelom but a hemocoel.

Embryonic development of an insect. **1**. Fertilization. **2**. The nuclei migrate to the periphery. **3**. Cell membranes appear between the nuclei, resulting in a single-layered embryo which corresponds to the blastula of other animals. **4**. The segments develop roughly in sequence from front to rear. **5**. As development proceeds, the anteriormost segments are incorporated into the head and their appendages differentiate as mouthparts, while the 3 thoracic segments enlarge and their appendages elongate as legs; the abdominal segments form rudimentary appendages, most of which disappear before hatching. (Modified after R.E. Snodgrass)

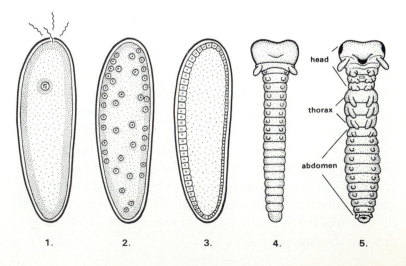

head

thorax

abdomen

1.　　　2.　　　3.　　　4.　　　5.

The young grasshopper, known as a **nymph,** hatches from the egg in a form which resembles the adult except in the relatively large size of the head, as compared with other parts, in the lack of wings and reproductive organs, and in certain other details. It feeds upon vegetation, grows rapidly, and molts at intervals. With each successive molt, of which there are 5 in most grasshoppers, differentiation continues, so that the final molt results in the adult grasshopper.

Newly hatched grasshopper nymph has a pale, delicate cuticle. Length, 3.5 mm. (R.B.)

Grasshoppers grazing on grasses and herbaceous plants are common in meadows everywhere, but large numbers in agricultural fields can cause serious economic losses for farmers. (A. and E. Smith)

Certain kinds of grasshoppers, called **migratory locusts,** are polymorphic: a non-migratory *solitary* type develops when the nymphs are uncrowded, and a migratory *gregarious* type develops under crowded conditions (see also p. 604). The 2 forms differ in shape and color, in physiology, and especially in behavior. The gregarious forms begin to migrate in groups even as nymphs, and swarms of adults fly great distances. Wherever they settle, they devastate crops and vegetation, even stripping the bark of young trees. A single large swarm may contain billions (10^{10} to 10^{11}), and each consumes its body weight daily. Important genera include *Locusta, Schistocerca, Nomadocris, Dociostaurus, Patanga,* and *Chortoicetes.* No great migrations of the N. American *Melanoplus* have arisen for over a century, perhaps due to changes in agriculture.

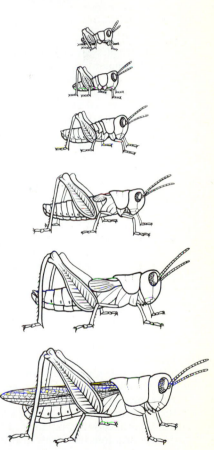

Metamorphosis of a grasshopper. Between any 2 successive stages, or *instars,* the animal molts. The early nymphal instars have a relatively large head and no wings. In the last instar, the *imago* (adult), the head is smaller in proportion to the rest of the body and the wings are fully developed,

VARIATIONS IN INSECT STRUCTURE

TO HAVE LOOKED at insects at all is to have noticed that they vary tremendously—from flattened, crawling cockroaches that feed as scavengers, to swimming beetles that capture animal prey, or flying butterflies that suck nectar from flowers, or the broad spectrum of parasitic insects. The vast array of insects is generally thought to be derived from myriapodlike ancestors by a reduction and fixation of the number of segments, a reduction of appendages on the abdomen, and a specialization of the appendages of the anterior part of the body. Among the variations that are most important in adapting the insects to an incredible variety of terrestrial niches are those of the sense organs and appendages.

Primitively wingless hexapods—proturans, collembolans, diplurans, and thysanurans—are not closely related to each other or to the winged insects. All may have arrived at the 6-legged condition independently, evolving from different myriapodlike ancestors. Unlike the winged insects, members of these 4 wingless classes bear segmental appendages along the abdomen; they deposit spermatophores on a substrate instead of copulating; and they continue to molt after reaching sexual maturity. The tiny proturans differ further by their lack of antennas and by their mode of development after hatching: they add a body segment at each of 3 successive molts. Of the 4 wingless groups, the thysanurans are most like the winged insects, sharing characters such as well-developed compound eyes and mouthparts that stick out freely from the ventral side of the head. Because of differences in these characteristics, and some others, proturans, collembolans, and diplurans are sometimes excluded from the Insecta. The primitively wingless hexapods are conspicuous exceptions to many generalizations that apply to all winged insects.

Variations in insect shape. *Upper left,* dorso-ventrally flattened cockroach. Panama. (R.B.) *Lower left,* laterally compressed katydid. Peru. (E.S. Ross) *Right,* stick insect. Cambodia. (R.B.)

Day-flying wasp, *left,* builds nest, collects food, and works during the daytime; it sleeps at night. Its niche is occupied during the night by a **night-flying wasp,** *right,* which is adapted to night activity; it has larger ocelli *(arrow),* than the ocelli of the day-flying species, and its body has numerous light yellow markings. This species is inactive on its nest during the daytime. Panama. (R.B.)

The **eyes** of insects are of two main types. In addition to a pair of compound eyes (discussed in chapter 20), there are 3 small **simple eyes,** the dorsal ocelli. These simple eyes often complement the image-forming function of the compound eyes by providing special information about changes or differences in light intensity. Interactions between the two kinds of eyes have been investigated in many kinds of insects by covering either the simple or compound eyes with opaque paint and then recording the behavior. In this way it has been found, for example, that grasshoppers cannot orient toward a light source using the simple eyes alone. Orientation is possible with the compound eyes only but is more accurate when both sets are working together. A flying grasshopper reacts to a sudden decrease in light intensity (such as might be caused by the shadow of a predatory bird) with a burst of speed. If either the simple or compound eyes are covered, this defensive response is much reduced. Covering the simple eyes makes many insects walk or fly more slowly and generally react less promptly to visual stimuli received by the compound eyes. However, the same experiments performed at a higher or lower light intensity or on a different kind of insect may produce opposite effects or none. The simple eyes seem to have evolved different kinds of complementary interactions with the compound eyes in the various insect orders. Their role is still not well understood in any insect.

Simple eyes are notably absent in most wingless insects, even those with close relatives having both wings and simple eyes. This suggests that the simple eyes may have some widespread role in flight. If so, fresh questions arise as to how some flying species, including most beetles and some moths and flies, get along without simple eyes.

In phototaxis experiments, bees were found to go toward the brighter of two lights that differed 2-fold in intensity. With the simple eyes covered, the bees made this discrimination only when the lights differed 8-fold in intensity. Such bees also foraged only at higher light intensities, beginning later in the morning and stopping earlier in the evening than normal workers. Experiments suggested that the simple eyes are important for navigation by polarized light at the low intensities of dawn and dusk.

In flying insects such as migratory grasshoppers and dragonflies the ocelli are said to serve as horizon detectors helpful in stabilizing the course of flight. In a walking insect, the desert ant *Cataglyphis,* the ocelli can read compass information from the blue sky. When the ant's compound eyes are covered, and both sun and visual landmarks obscured, the ants use the pattern of polarized light in the sky as a compass cue in finding their way home.

The **antennas** are the chief site of the sense of *touch*. In some insects they are very long, as in cockroaches, crickets, and katydids. In others, such as dragonflies, which depend more on visual information, the eyes occupy nearly the whole head, while the antennas are relatively minute. The touch receptors of the antenna are the fine bristles with which it is clothed. The antennas also bear receptors for the senses of *smell* and *taste,* of *humidity,* and of *temperature.* Occasionally the antennas serve in unexpected ways, as in the aquatic larvas of some midges that use theirs in seizing prey. The males of certain beetles, collembolans, and fleas use their antennas to hold onto the females.

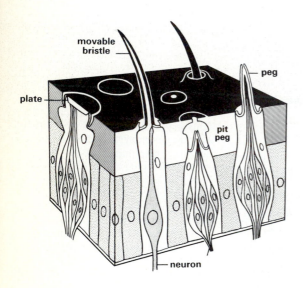

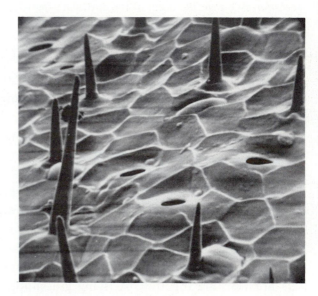

Portion of an insect antenna, a body area dense with external sense organs (sensilla). Each microscopic sensillum usually consists of an epidermal cell that secretes a cuticular component, and one or more neurons whose axons run to a ganglion. The long **bristle** with thick cuticular wall, is freely movable, as the base is set into a thin, flexible membranous socket (produced by a separate epidermal cell). Such articulated bristles are *mechanoreceptors,* responding to touch, vibration, or wind when in flight. (Movable bristles, when located at leg or body joints, are pressed on by body movements and act as *proprioceptors,* supplying information on the position of body parts). Other antennal sensilla shown here are a thin-walled **peg,** innervated by several neurons; it has an opening at its tip and may have pores along its length; it is a *chemoreceptor,* sensitive to taste or smell. The **pit-peg** sunken into a pit below the surface and communicating with the exterior through a surface pore in the cuticle, is an *olfactory organ,* sometimes specialized to detect the levels of CO_2. The flat circular **plate** is an *olfactory receptor;* it has pores on its surface and in the membranous groove around its periphery. (Combined and adapted from N.E. McIndoo and from R.E. Snodgrass)

Surface view of antenna of a grasshopper, seen here magnified 1,100 times, in a scanning electron micrograph (SEM), shows the outlines of the epidermal cells that secrete the cuticle of the antenna and the cuticular components of the external sense organs. Bristles, pegs, openings to pit-pegs, and thin sensory domes that act as olfactory receptors are all conspicuous. (P.P.C. Graziadei)

That odors applied to a sense organ induce nervous impulses along its neuron is electrophysiological evidence of its function. Bees trained to collect sugar-water from distinctively scented cards can no longer distinguish odors when the terminal 8 articles of both antennas are cut off. That the trauma of the operation is not the explanation is shown by the fact that other bees, trained to go to dishes of sugar-water set on colored cards, still return to cards of the correct color, even after the same surgery.

Though most concentrated on the antennas, tactile bristles are abundantly scattered over the entire body of most insects. Taste receptors (contact chemoreceptors) occur not only on the antennas and in the mouth, but also on the palps, the tips of the legs (tarsi), and the ovipositors. A female grasshopper, digging a hole with her ovipositor in moist sand, can detect the salt content; if it is high, she will seek another site for her eggs. Insects show responses to a variety of sugars, salts, acids, bitter substances, and plain water. Other sensory receptors may also occupy rather unusual places. For example, the auditory organs of a grasshopper are on the sides of the abdomen just above the base of the third legs, while those of a katydid are near the upper end of the tibia of the first pair of legs.

Auditory organ of katydid communicates with the exterior through a slitlike opening in the base of the tibia of the foreleg. (K.B. Sandved)

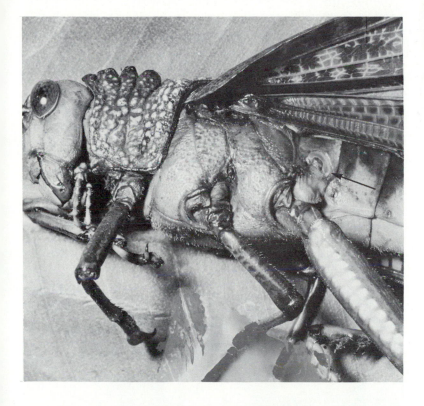

Tympanum (eardrum) of a grasshopper is a tightly stretched membrane covering an opening in the side of the first abdominal segment. When sound waves strike the tympanum *(arrow)* it vibrates, transmitting impulses to a chordotonal sense organ attached to its inner surface. (R.B.)

Beetles of the genus *Melanophila* have the remarkable ability of detecting the infrared radiation emitted by forest fires, which attract these insects from miles around. The infrared receptors are maximally sensitive to a narrow band of infrared wavelengths that falls within the range of the maximum energy radiation from burning wood and maximum transmission by the atmosphere. The receptors are clustered in a pair of pits at the bases of the middle legs, a position in which they are constantly exposed to radiation as the beetle flies toward a burning forest. Having reached the fire, the insects probably avoid flames and hot surfaces by means of conventional temperature receptors on the antennas. With the flames still burning close by, the beetles boldly lay their eggs on the freshly killed trees.

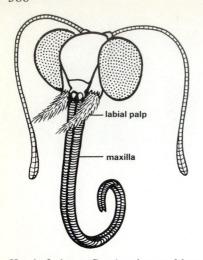

Head of a butterfly, showing **sucking mouthparts.** The mandibles are vestigial. The two maxillas are greatly elongated, each forming a half-tube, so that when they are held together, they form a long sucking proboscis, through which nectar or other liquids are drawn up by muscular pumping.

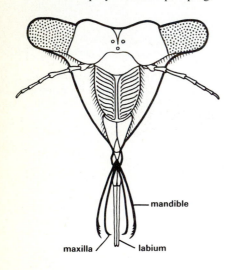

Head of a cicada, showing **piercing and sucking mouthparts.** The long tubular labium is not inserted into the plant and serves only as a sheath for the other mouth parts, being grooved on its anterior surface to form a channel in which lie the mandibles and maxillas, which do the piercing. The mandibles are long, fine stylets with minute teeth at the end. The maxillas are similar but hooked at the tips; each is crescent-shaped in cross-section, and the two are fastened together by interlocking grooves and ridges to form a channel through which the plant juices are drawn up by the sucking action of a muscular pump in the forepart of the head.

The **mouthparts** of grasshoppers, which represent a fairly generalized group of insects, are of the *biting and chewing* type. These are the most primitive kind, and they are present also in beetles and in many other insects that devour plants or prey on animals. The mandibles usually do the biting and may take the form of sharp pincers for seizing prey. But in dragonfly nymphs, and in certain predatory flies that have lost the mandibles, it is the labium that bears sharp mandiblelike pincers. Two other main types of mouthparts are common in insects that feed exclusively on liquids such as nectar, sap, or blood; *sucking* mouthparts, as in butterflies; and *piercing and sucking* mouthparts, as in cicadas. Here again, different appendages have become adapted to similar functions. For example, the tube through which the insect sucks up liquid food is formed by the mandibles in predatory water-beetle larvas, by the maxillas in butterflies and cicadas, by the mandibles and maxillas in antlion larvas, by the maxillas and labium in bees, by the labrum and maxillas in fleas, and by the labrum and any of several other parts in various kinds of biting flies. In mayflies and other insects that do not feed when mature, the mouth parts are much reduced in the adult.

Mandibles of male stag beetles are branched like the antlers of a stag and may be half as long as the body or more. The larger the male, the greater the proportion of the total body length represented by the mandibles. Useless for chewing food, the mandibles in these males serve in sexual competition with other males of the species, as in the struggle seen here. Stag beetles feed on honeydew and exudations from plants, which they gather with a hairy, flexible labium. Most species are tropical and large, the males up to 10 cm long. (K.B. Sandved)

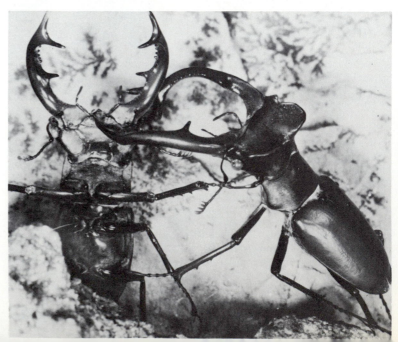

Grooming legs of nymphalid butterflies are tiny, brushlike structures, close under the head. They serve in cleaning the eyes and antennas. Nymphalids have only 4 walking legs. *Colobura dirce.* Panama. (R.B.)

Seizing legs of a praying mantis are long, stout, muscular, and spiny— well equipped to seize prey with lightning speed and to hold on to it firmly. As these legs are the first pair and project from an elongated prothorax (first segment of the thorax), they are well situated to reach far forward when prey comes within striking distance. (R.B.)

The thoracic **legs** of insects are modified in a variety of ways, but are usually composed of the same basic parts (figured and named in the diagram of a grasshopper leg). The forelegs and middle legs of a grasshopper are unspecialized *walking* legs, with terminal pads and claws for clinging to vegetation or other objects; the hindlegs are modified for *jumping.* Water beetles have flattened legs, fringed with bristles, for *swimming.* But the legs may serve other functions besides locomotion. Cicada nymphs and mole crickets use their broad, stout forelegs for *digging* in soil. A praying mantis uses its large forelegs in *grasping* prey and walks on its other two pairs; and in an adult dragonfly, all of the legs are so specialized for grasping prey or erect vegetation that they can no longer be used for walking. Legs are also adapted in various insects for *producing sound* and for *grooming.* In some sedentary forms, such as scale insects, the legs are much reduced or absent.

Grasping legs of a sucking louse (order Anoplura) hold tightly to body hairs of a Chinese water buffalo. (E.S. Ross)

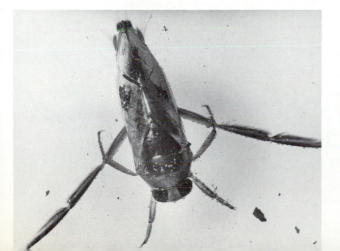

Swimming legs of a backswimmer, a freshwater bug (*Notonecta,* order Hemiptera). The flat, broad surface of the oarlike legs is enhanced by close-set rows of long bristles that fringe tibia and tarsus. The short forelegs seize small aquatic prey, and the middle legs help to hold the prey to the mouth as the bug's sharp beak sucks prey juices. The backswimmer is shown here with convex dorsal surface up, but when the bug is rowing itself about on its back, with the flat ventral surface up, its body shape is like that of a rowboat in the water. (E.S. Ross)

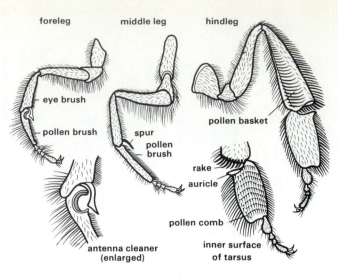

Pollen-collecting legs of a honey bee are the hindlegs, equipped with a pollen basket on the outside of each tibia. (C. Clarke)

The **legs of a honey bee** are modified for collecting pollen. Each pair is different from the others, so that, together, they constitute a complete set of tools for manipulating the pollen upon which the bee feeds. On the **foreleg**, along one edge of the inner surface of the tibia and the large first subarticle of the tarsus, is a fringe of short, stiff bristles that form a *pollen brush* for collecting the pollen grains that become caught among the feathery bristles of the body when the bee visits flowers. The pollen brushes of the forelegs clean pollen from the foreparts of the body, including the compound eyes. The tarsus also bears a semicircular notch that is lined with a comblike row of bristles and is closed by a clasplike projection on the end of the tibia. Comb and clasp together are called the *antenna-cleaner*. The antenna, held firmly in place by the clasp, is cleaned by drawing it through the notch. The **middle leg** is the least specialized of the three. The large first tarsal subarticle is wide and flat and covered with stiff bristles which form a brush for removing pollen from the forelegs and the thorax. The **hindleg** is the most specialized and carries the load of pollen. Rows of *pollen combs* on the inner surface of the large and flattened first tarsal subarticle scrape the pollen from the middle legs and abdomen. The *rake,* a series of stout spines on the lower end of the tibia, removes the pollen from the combs of the opposite leg, and it falls onto the *auricle,* a flattened plate on the upper end of the first tarsal subarticle. The leg is then flexed slightly, so that the auricle is pressed against the surface of the tibia, compressing the pollen and pushing it onto the outer surface of the tibia and into the *pollen basket*. The pollen basket is formed by a concavity of the tibia, which has, along both edges, long bristles that curve outward. The pollen clings together and to the basket hairs because it is moistened with sticky nectar regurgitated from the mouth. When the baskets are loaded, the bee returns to the hive and deposits the pollen in special wax cells. Combined with sugars and other substances, the pollen mixture becomes "beebread," which provides a source of protein for both adults and larvas.

Foot of honey bee has, on the last article of the tarsus, 2 sharp claws and between them a soft, blood-filled adhesive pad. When applied to a rough surface, as at *left,* the claws catch in surface irregularities. When applied to a smooth surface, as at *right,* the claws are drawn back and the pad adheres to the surface. Detailed study of a pad on the tibia of a bug *(Rhodnius)* shows that it is covered with fine tubular hairs lubricated with an oily secretion from glands at the base. The hair tips end obliquely so that oil flows readily and they do not adhere when sliding forward; when the pad is drawn backward, the film of oil breaks down and the firmly applied hairs adhere to the surface by molecular forces of attraction. (Adapted from various sources)

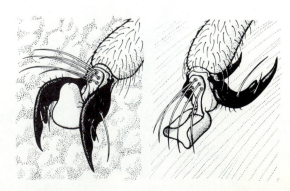

The **wings** of insects are flattened, two-layered expansions of the body wall and at first consist of the same parts: cuticle and epidermis. Later the two opposing layers meet, except along the channels in which lie nerves, air-filled tracheal tubes, and blood spaces. In later stages these channels form the "veins" of the wing. The epidermal cells finally degenerate, and the adult wing is almost completely made of cuticle, though in some species it may have a circulation of blood and nerves connected to sense organs.

Not all insects have wings. The 4 *apterygote* ("wingless") groups (Collembola, Protura, Diplura, Thysanura) have never developed them. In some of the more specialized *pterygote* ("winged") orders, both pairs of wings have been lost secondarily, as in for example, fleas and lice. Besides, in most of the orders which typically have two pairs of wings, there are some secondarily wingless species. In the social insects certain castes lack wings. And in some species the males have wings while the females are wingless (as in certain moths, fireflies, and army ants), or the opposite (as in some parasitic ants and wasps, and some stoneflies).

Above left, **two pairs of wings** are typical of insects. In dragonflies, both pairs are membranous and have a dense pattern of veins. (R.B.)

Above right, **hardened forewings** of beetle are wing-covers, or elytra, that protect the membranous hindwings folded underneath when the beetle is at rest. The forewings move only a little during flight but contribute stability and additional surface for lift. The hindwings beat actively. (P.S. Tice)

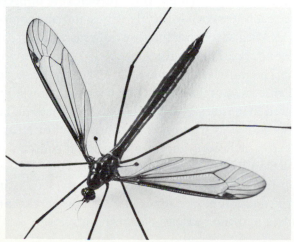

One pair of membranous wings suffices in the flies (order Diptera), some of the fastest of insect fliers. The hindwings are reduced to **halteres,** tiny gyroscopic organs easily seen in this crane fly. The halteres vibrate freely and contain sense organs that check any tendency to roll or to deviate from flight course. If the halteres are cut off, the insect cannot fly. Halteres develop from wing buds on the third thoracic segment, and in *Drosophila,* a vinegar fly, certain mutations cause these wing buds to develop into hindwings with veins instead of into halteres. (R.B.)

In insects such as dragonflies and grasshoppers, the 2 pairs of wings beat independently, though at the same frequency, and must be exactly coordinated so that the turbulence created by one pair does not interfere with the other. In many insects, such as butterflies, bees, and wasps, the forewings and hindwings are mechanically linked together, by tiny interlocking spines or hooks, and function as a single pair. In beetles, the hindwings beat most actively; the forewings, which are thick stiffened wing covers when the beetle is at rest, move only a little during flight but contribute stability and important additional surface for lift, as do the rigid wings of an airplane. In flies, the forewings do the flying and the hindwings are reduced to tiny stalked knobs, the halteres, with important sensory and stabilizing functions.

The powerful **flight muscles** are mostly not attached directly to the wings but to the inside of the exoskeleton of the thorax. As opposing sets of these muscles contract, changes in the shape of the thoracic walls flip the wings up and down. The exoskeleton of the thorax, the hinges of the wings, and the flight muscles themselves are all highly elastic, so that much of the energy expended by the muscles in moving these parts is promptly recovered. In a locust, about 86% of the energy of the upstroke is stored in these elastic components for use in the downstroke. Without this storing and recovery of energy, the muscles could not supply enough power for flight. In order to take advantage of the powerful elastic recoil, the timing of the wingbeats must be matched to the mechanical properties of the thorax and wings. Although many factors determine the optimum frequency of the wingbeats for each insect at any moment, large insects generally beat their wings relatively slowly and smaller insects more rapidly.

Even among large insects, wingbeat frequencies are impressive. We can see each stroke of butterfly wings, moving sometimes as slowly as 4 beats/second; but those of grasshoppers, at 15-20 beats/second, or of dragonflies, at 20-30 beats/second, are only a blur. The muscles are able to contract and relax so often in these insects, not because they contract unusually rapidly, compared to muscles of other animals, but because the absolute distance moved is so short. The control of the rhythm is straightforward; each contraction of the flight muscles is elicited by nerve impulses. This system of *synchronous control* can operate up to a frequency of about 100 beats/sec (as in some moths), the maximum frequency at which the nerves can transmit impulses and the muscles can respond.

Though insects are **ectotherms,** like all other invertebrates, many insects have high body temperatures, during activity, that compare with those of the endothermic vertebrates, the birds and mammals (32° to 40°C). A study of 50 species of butterflies revealed that the mean body temperature during activity was 35°C and the range was from 28 to 41°C.

Behavioral regulation of temperature is important in butterflies. No matter their size or color, practically all butterflies must have a thoracic temperature of 27°C or higher in order to take off on controlled flight. Some noctuid moths generate heat in their flight muscles by shivering and reduce heat loss by a thick "furry" coat of bristles on the thorax. They can raise the temperature of the thoracic muscles to 30°C and can take off even when the external temperature is near 0°C.

Different families are often characterized by various wing positions when they bask in the sun. Monarchs, swallowtails, and fritillaries usually hold the wings wide open, fully exposing the thorax to solar radiation. In very hot weather they bring the wings together to shade the thorax and if this is not enough they shelter in the shade. Sulfur butterflies cannot bask with wings wide open; they tilt sideways, exposing the ventral surface of the hind wings to the sun.

The **indirect flight muscles** are those which are not attached directly to the wings, but to the inside of the thorax, shown here in diagrammatic cross-section. **1. Upstroke:** Contraction of dorsoventral muscles pulls the roof of the thorax down and forces the wings up, pivoting them on projections from the sidewalls. **2. Downstroke:** Contraction of dorsal longitudinal muscles bows up the roof of the thorax and forces the wings down. The movements of the thorax and muscles are exaggerated in this diagram; the muscles may shorten and lengthen by as little as 1%.

The indirect muscles are assisted by the elasticity of the thorax and, in butterflies, grasshoppers, and beetles, by direct flight muscles (attached to the wings) that contribute power during the downstroke. In dragonflies, the downstroke is executed almost completely by direct muscles. Direct muscles also rotate the wings during each stroke, constantly adjusting the angle at which air flows over the surface. Or, in insects that beat their wings at very high frequencies, the wing base is structured so as to adjust the angle of the wing automatically during each stroke.

Even so, many small insects do beat their wings much faster. Bees and wasps often fly at about 200 beats/sec; mosquitoes top 600 beats/sec; and tiny midges reach 1000 beats/sec. These insects employ *asynchronous control;* nerve impulses supplied to the muscles at a relatively low frequency do not stimulate individual contractions but serve only to maintain the muscles in an active state and to modify flight. The rapid rhythm of wingbeats is maintained by the interactions of the muscles with the elastic thorax and the ability of the special type of muscle in these insects to contract when stretched. When one set of flight muscles contracts, the other is automatically stretched and contracts in turn, thus stretching the first, and so on. Asynchronous control and the special type of muscle that it requires seem to have been evolved independently by small flying insects of many different orders.

In some insects, the structure of the wing attachments is such that the wings are stable only when fully raised or fully lowered. Displaced only slightly from either extreme, the wings automatically "click" into the alternate position. This **click mechanism** is thought to increase the velocity of the wingstroke and the aerodynamic efficiency. In insects with asynchronous control, click action is an important factor in producing the rhythm of the wingbeat, but click mechanisms are also found in insects with synchronous control and may have a variety of functions. In flies and beetles, when the wings are folded, the click mechanism produces sound. In cicadas, there is another kind of click mechanism in the sound-producing organs, located in the abdomen and completely separate from the wings; in this system, control may be synchronous or asynchronous.

Fringed wings, as in this thrips (order Thysanoptera), have been independently evolved by various minute insects such as beetles and hymenopterans with wingspreads of a few millimeters or less. Students of aerodynamics do not agree whether such wings produce enough lift to support flight or whether these small insects are "swimming" in air by using drag forces. The tiny "fairyflies" (order Hymenoptera), smallest of all insects, fly in air and swim underwater by means of fringed wings. (R.B.)

The metabolic rate of a flying honey bee, about 320 cal/g/hr, falls in the middle range for most insects (100-500 cal/g/hr). By contrast, for a small mammal running at maximum speed, the rate is at most about 32 cal/g/hr; for a fast-running human, 16 cal/g/hr. However, the metabolic rate of a hovering hummingbird is about 200 cal/g/hr, comparable to the rates of insects. Most flying birds and running mammals increase their metabolic rate only about 7-14 times over resting rates.

Cerci are usually simple, jointed sensory appendages found at the tip of the abdomen. They are prominent in the little embiopteran shown *above,* and can be seen in the photos of many of the paleopteran and orthopteroid orders of insects illustrated later in this chapter; they are found also in the order Mecoptera. Sometimes they are highly modified, as in the earwig shown *below;* a male of *Forficula auricularia.* The embiopteran shown here was found in its silken case under a stone in a California meadow. The earwig was from a California garden. (R.B.)

The intense, rapid activity of flight depends on relatively high temperatures, and small insects cannot fly in cold air. Many larger insects can generate enough heat, by contracting their muscles during preflight "warm-up exercises," to be far more independent of external temperatures. The "furry" appearance of bees and moths comes from insulation that helps retain heat, and in hot weather such insects must cool themselves, by taking measures such as changing the pattern of blood circulation, as mammals do. The metabolic rates of actively flying insects are the highest known in any animal and reach 50-100 times resting rates. The tracheal system supplies oxygen so abundantly that metabolism during flight is completely aerobic; there is no oxygen debt such as that built up in human muscles during vigorous activity. The high respiratory demands of cells of the flight muscles are met by huge mitochondria that may occupy up to 40% of the volume of the cell. For bees making short trips to and from the hive, sugar is the principal "quick energy food"; but insects that fly for many hours, such as migrating locusts or butterflies, depend on fat, which yields more energy for its weight.

The **abdomen** of most adult insects bears appendages only at the posterior end; these are specialized for copulating or egg-laying, or as sensory projections (cerci), or in other ways. In the female grasshopper they are used for digging a hole in the ground in which to lay eggs. In ichneumons (wasplike hymenopterans) that parasitize wood-boring beetles, the long, sharp ovipositor is used to drill into a tree and deposit eggs in the body of a beetle larva detected deep within the wood. When the eggs hatch, the young ichneumon larvas feed on the beetle larva. In the honey bee, the ovipositor is modified as a sting connected with poison glands, as some of us know from painful encounters with these insects. In earwigs, the cerci are large, strong forceps used in defense and prey capture; those of the male are curved and toothed and may play some role at mating time.

Only in some of the primitively wingless insects do the adults bear paired appendages, serially homologous with the thoracic legs, on most of the segments of the abdomen. A series of rudimentary abdominal appendages is present in most insect embryos, but usually disappears in the course of development. Many larval insects bear abdominal appendages (for example, the legs of caterpillars or the gills of aquatic mayfly nymphs), but it is not always clear which of these are serially homologous to the other segmental appendages and which are secondarily acquired.

COMPARED TO the tremendously diverse external characters of insects, their insides appear much alike. Variations in the **digestive tract** are related mostly to what the animals eat. In cockroaches, which feed on solid food, the gizzard is well developed and its lining is armed with hard plates and spines. Insects that suck plant juices have no gizzard, but some have long convoluted guts with special adaptations to dispose of the large quantities of fluid ingested. Honey bees suck up nectar into a honey-stomach, which corresponds to the crop of grasshoppers. The nectar, destined for storage in the nest, is prevented from passing into the midgut by a valve (a structure which corresponds to the gizzard of grasshoppers).

In the great majority of insects, the **respiratory system** consists of air-filled tracheal tubes that open to the outside through spiracles. Open spiracles of aquatic insects, both immature stages and adults, are guarded against the entry of water by fine hairs or valves that open when the insect periodically surfaces to breathe. Aquatic fly larvas often bear all the spiracles at the tip of a long, retractable tube that remains at the surface or a pointed projection that taps into air-filled spaces of submerged plants. Many aquatic bugs and beetles carry air bubbles down with them. Others maintain a *plastron,* a permanent layer of air held next to the body by hairs or other cuticular elaborations; as long as the insects are in water well-aerated enough that oxygen diffuses into the air layer as fast as the insects use it, they need never surface. Most aquatic larvas and nymphs have no open spiracles, and oxygen diffuses through the cuticle into the air-filled tracheal system, which branches abundantly just beneath the body surface. Additional respiratory surface is commonly provided by *tracheal gills,* thin-walled projections containing air tubes. These may beat actively to create respiratory currents or may help to propel the insect. They can occur almost anywhere on the body, usually along the abdomen in mayflies, at the hind end in damselflies, and within the rectum in dragonflies. Insect larvas that live as endoparasites in the fluids or tissues of other animals gain access to oxygen much as aquatic insects do. Some parasites have open spiracles and maintain an air channel to the outside or into the respiratory system of an air-breathing host. Others have a closed respiratory system and obtain oxygen by diffusion through the cuticle. Some of the tiny, wingless proturans and collembolans, and a few endoparasitic insects, have no air tubes; respiratory gases are exchanged between blood and air through the body surface.

Tracheal gills project along both sides of the abdomen of the aquatic larva of the dobsonfly, *Corydalus cornutus.* (P.S. Tice)

Bot fly larvas, *Gasterophilus,* attached to the stomach lining of a horse, live in a medium of low oxygen content. They survive by taking advantage of air bubbles swallowed by the horse. When in contact with an air bubble, the larva opens a special posterior chamber from which lead tracheal tubes that branch extensively among masses of cells containing hemoglobin. Enough oxygen can be bound to the hemoglobin and stored in these cells to supply the larva for about 4 minutes. The few insects that have hemoglobin are either aquatic or parasitic forms. (U.S. Bur. Entomol.)

Mated damselflies in tandem. The male is still holding the female with claspers at the tip of his abdomen and will stay with her until she completes egg-laying. (See p. 622) (E.S. Ross)

Copulating beetles. The male is astride the larger female. Buprestid beetles. Malaya. (R.B.)

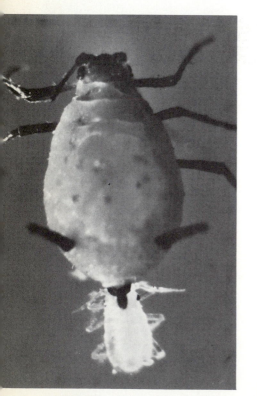

Aphid giving birth to nymph. (L. Brubaker)

In most insects **reproduction** follows the pattern described for grasshoppers. Separate male and female individuals copulate, and the females lay fertilized eggs that eventually hatch into a new generation of males and females. There are several common modifications of this standard life history, and many insects practice one or another of these, but the aphids (order Homoptera) seem to have tried to take advantage of all of them. Although male and female aphids mate in the fall and produce fertilized eggs in the standard way, the nymphs that hatch from these eggs in the spring are *all female*. Each nymph undergoes several molts and grows quickly but never develops wings, devoting her energy to producing all-female offspring by *parthenogenesis,* that is, from unfertilized eggs. The young develop within the mother's ovaries, growing rapidly on nutrients supplied through a placenta-like arrangement, and are born alive *(viviparity)*. An aphid may give birth to about 100 offspring, at the rate of several per day, and each becomes a mother only a week or so after her own birth. Even before birth, the female embryos have developing offspring within their own bodies; and this *precocious reproduction* (progenesis), combined with parthenogenesis and rapid development, results in extraordinary rates of multiplication, as any gardener will attest. The plant on which the aphids sit, drinking sap, soon becomes crowded, and crowding (acting through tactile stimulation by aphids bumping into and crawl-

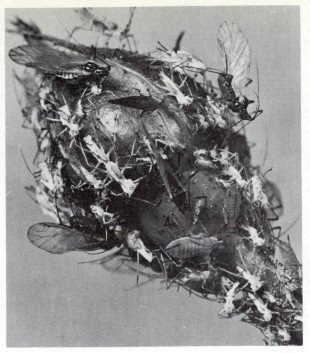

Aphids are prolific, and this rose bud is covered with young molting nymphs and their shed skins, large nymphs, wingless adults, and winged adults, all sucking plant fluids. (R.B.)

Winged and wingless aphids and nymphs suck fluids of rosebush. (E.S. Ross)

ing over one another) causes increasing numbers of individuals, or their offspring, to develop wings. Changes in food quality, due to the advancing season or the impact of the aphids themselves, also contribute to the appearance of winged forms. The winged aphids fly to new plants of suitable species and there each begins a new colony of aphids, giving birth parthenogenically to wingless offspring, which are usually more prolific than winged individuals. Throughout the summer, generations of winged and wingless parthenogenic females spread and multiply. Shortening days and falling temperatures in late summer or fall cause male and female sexual forms to be produced again. In addition to the *polymorphism* of sexual and parthenogenic, winged and wingless forms, there may be, for example, specialized forms that produce the sexual forms. Some aphids expand available resources by completing the life history on two successive plant species, and the forms that live on each plant may be quite distinct; so that as many as 20 forms with different behavioral and reproductive capacities have been identified in the life history of a single aphid species. On the other hand, many aphids in the tropics have a simplified life history, lacking any sexual phase at all and reproducing solely by generation after generation of parthenogenic females.

It is not surprising that, in such a large and diverse group as insects, the events of **embryonic development** are varied, and there are exceptions even to the extremely general account that was given for the typical yolky eggs of grasshoppers. For example, a few insects (primitively wingless collembolans and highly specialized parasitic hymenopterans) lay eggs that have little yolk and that cleave completely into separate cells. Early cleavage in insects has been found to be *indeterminate,* so that if nuclei are experimentally damaged or removed, the remaining ones can still produce a whole normal embryo. Later development becomes less flexible in the sense that even very localized damage results in an incomplete embryo; but the stage at which this occurs is another large variable in the spectrum of insect development. Some of the different patterns seen after hatching are described in the next section.

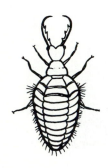

The **antlion** is a larval form that undergoes a striking metamorphosis. The wide-bodied larva *(above)* traps its prey by digging a pit in the sand and lying buried at the bottom. Only the sickle-shaped, piercing and sucking mouthparts protrude, ready to grasp any ant that falls in *(right)*. The adult **antlionfly** is a slender night-flying hunter that eats insects with its biting mandibles.

METAMORPHOSIS

NOT ALL insects undergo a metamorphosis that involves radical changes. **Ametabolous insects,** such as the primitively wingless bristletails (class Thysanura), hatch from the egg looking like miniature adults. They grow larger, change in body proportions, and mature sexually, but such changes are no greater than those undergone by most animals in their development. In **hemimetabolous insects,** such as grasshoppers (order Orthoptera), the young, or *nymphs,* lack wings but otherwise generally resemble the adults in having compound eyes and in using similar mouthparts to eat the same type of food. The developing wing buds are external and visible at an early stage, and though many changes take place at the final metamorphic molt, the most conspicuous external change is the full development of the wings.

The metamorphosis of hemimetabolous insects is sometimes described as *direct* or *incomplete,* that of holometabolous insects as *indirect* or *complete.*

Dragonflies (order Odonata) are like other hemimetabolous insects in that the immature stages have external wingbuds and resemble the adult in general body form. However, dragonfly nymphs show special features related to their aquatic habitat. Their metamorphosis thus involves more changes than that of grasshoppers. Both young and adult dragonflies are carnivores with chewing mouthparts and large compound eyes. But the dragonfly nymph lives in water and feeds on aquatic animals, catching them with its long grasping labium; the winged terrestrial adult preys on flying insects, catching them with its legs. Also, the nymph has a closed tracheal system and extracts oxygen from the water by means of tracheal gills in the rectum, thin-walled ridges containing tracheoles in the rectal wall. At regular intervals, the water in the rectum is expelled and fresh water is drawn in. Occasional rapid expulsion serves as a means of jet-propelled locomotion when speed is called for. The adult has spiracles that permit free exchange of air.

Dragonfly nymphs take from a few months to as much as 4 years to complete their growth and development. Both the size and the proportions of the various parts change during this period, as can be seen in this photo by comparing the wingbuds of the smaller, more immature nymph with those of the larger one. The mature nymph is starting to climb the stem of a water plant and will ascend above the surface of the water, in preparation for the final molt to the adult stage. (L.W. Brownell)

Emergence of an adult dragonfly. The cuticle of the nymph splits open along the back, and the adult struggles out, leaving the empty nymphal cuticle clutching the plant stem. Even the cuticular linings of the old tracheal system are shed and show up here as long white filaments sticking out from the dorsal thoracic surface. As the crumpled wings gradually expand and dry, the exquisite network of veins becomes visible. The newly emerged adult continues to cling to the old nymphal cuticle until the wings have fully hardened and then flies off. Spain. (E.S. Ross)

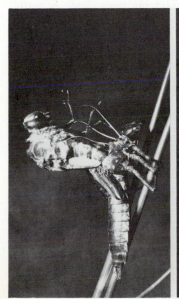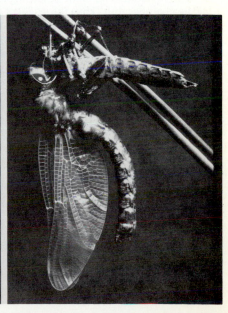

Egg of Monarch butterfly is 0.5 mm long and glued to the underside of a milkweed leaf. In midsummer the developed larva may hatch in 3 or 4 days. The larva of a butterfly chews its way through the egg membranes, but in various other insects breaking through the egg may be aided by swallowing air, by muscle contractions, or by special hatching spines. (Photos by C. Clarke)

The most striking changes of all are those undergone by the **holometabolous insects,** in which the larvas are so different from the adult, not only in habit but also in gross structure, that metamorphosis is a two-step process. The example that immediately comes to mind is the development of a caterpillar into a butterfly. In development of this type the larval stages and the adult stage are separated by an externally quiescent *pupal stage,* intermediate in appearance. In the pupa, larval tissues are modified or break down and become reorganized to various degrees. In butterflies and moths (order Lepidoptera), for example, many larval parts remain largely intact and are modified only to the adult form, whereas in flies (order Diptera) most of the larval tissue breaks down and is replaced.

Larva of a Monarch butterfly has biting mandibles and feeds only on milkweed leaves, almost all poisonous. It becomes distasteful to most birds, which learn to avoid caterpillars with the conspicuous colors and banding of the Monarch. The forms, colors, legs, and sensory projections of various kinds of insect larvas differ in relation to habitat and mode of feeding rather than with the form of the adult. Thus the origin of the word larva, which means "masked."

Monarch **larva changing into a pupa.** The larva at the *near right* has fed, grown, and molted several times. After spinning a pad of silk onto the underside of a branch, it has fastened the curved spines of its last pair of stubby larval legs into the silken pad and hangs head down. Behind the head are 3 pairs of clawed legs that become the legs of the adult. *At right* the larval skin is being shed and the developing pupa inside is partly revealed.

Developing **pupa** at the *left* shows spiracles in the abdominal segments. *At center* is a fully-formed pupa, sometimes called a chrysalis in butterflies. The **newly emerged adult,** *right,* clings to the empty pupal case while its wings expand and dry.

New adult structures such as the wings, legs, and antennas form through the growth of certain cells set aside early in development as *imaginal disks*. They develop gradually during the larval phase, but are invaginated and not visible externally. When the larva changes to a pupa, the adult structures abruptly evert and continue to develop, appearing in finished form at the final molt when the adult emerges. There is some tissue reorganization at each molt of hemimetabolous insects like grasshoppers, and the difference between their gradual metamorphosis and the dramatic metamorphosis of holometabolous insects like butterflies is largely a matter of degree.

According to their mode of development, the orders of winged insects are divided into two groups, believed to have diverged early in insect evolution. In the **Exopterygota,** the young are nymphs with *external wing buds* and mostly they undergo hemimetabolous development. In the **Endopterygota,** the young are larvas (caterpillars, maggots, grubs, etc.) with *internal wing buds* and they undergo holometabolous development.

In both types of development, the **hormonal control of molting** is similar, and metamorphosis involves a significant hormonal change. As described earlier, secretion of a neurohormone from the brain initiates molting by stimulating the prothoracic glands to release the molting hormone, ecdysone. Throughout the larval period, however, the action of ecdysone is modified by the presence of another secretion called **juvenile hormone** and produced by the *corpora allata,* two glands lying close behind the brain. The principal effect of juvenile hormone is to prevent metamorphosis. As long as it is present, the epidermis and other tissues continue to show larval features and cannot differentiate into the pupal or adult form. If the corpora allata are surgically removed from a larva, even at an early stage when it is still very small, metamorphosis occurs precociously, and miniature pupal and adult stages may be formed. Under normal circumstances, however, metamorphosis does not occur until the larva has reached a critical size and perhaps received specific environmental or physiological cues. Then the corpora allata stop secreting juvenile hormone; ecdysone is secreted and acts in the absence of juvenile hormone for the first time; and metamorphosis proceeds. Apparently juvenile hormone also has a role in maintaining the prothoracic glands, for following the final molt to the adult in the absence of juvenile hormone, the prothoracic glands degenerate and no further molting occurs.

A few exopterygotes have independently evolved types of holometabolous development. In whiteflies and scale insects (order Homoptera), the young, after a brief motile period, quickly become sessile and are barely recognizable as insects. A young whitefly sucks plant juices, molts several times, and finally produces a pillbox-shaped structure from which emerges an active 4-winged adult. Female scale insects remain sessile as adults, but the males of some species are holometabolous and emerge from the final molt looking something like little flies.

Thrips (order Thysanoptera) show an intermediate sort of development. The nymphs are unremarkable, but one or more pre-adult instars may burrow underground and spin a cocoon, from which a winged adult thrips eventually emerges.

Exceptional development also occurs among endopterygotes. In strepsipterans, the winged male pupates like other holometabolous endopterygotes, but the wingless grublike female usually matures without passing through a pupal stage.

Once metamorphosis has begun, juvenile hormone assumes other, remarkably different roles. It seems to play a part in maintaining the prothoracic glands, for following the final molt to the adult in the absence of juvenile hormone, the prothoracic glands degenerate and no further molting occurs. Juvenile hormone is mostly absent in the last-instar nymph of hemimetabolous insects and in the pupa of holometabolous insects, but it begins to be secreted again in the adult; and its effects at this stage are to sustain growth of the eggs in the female, secretion by the accessory reproductive glands in both sexes, and various kinds of behavior associated with reproduction. And this list does not include all the actions of juvenile hormone, which can coordinate many diverse and complex activities within the animal, apparently because the tissues respond to it differently at different times. Perhaps there are separate sets of larval, pupal, and adult genes, each of which is activated at the appropriate phase of the life history. Experimental manipulation of hormonal balance sometimes produces larval/pupal or pupal/adult intermediates, which might result from two gene sets acting simultaneously. One interpretation of the action of juvenile hormone is that it determines which set of genes may be activated at any given time. The hormone ecdysone is a steroid, and it is believed to act directly on the genes in the nucleus of the cell to initiate the production of specific products, as do steroid hormones in vertebrates.

Hormonal control of metamorphosis in a moth. Before each molt a **neurohormone** is secreted by the lateral neurosecretory cells of the *brain* and stored in the **corpora cardiaca** (see molting in arthropods, chap. 20). When the hormone is released into the blood and reaches the **prothoracic glands,** these are stimulated to secrete **ecdysone,** the molting hormone. During most of the larval period: (**1**) the action of ecdysone is modified by **juvenile hormone,** secreted by the **corpora allata.** At each larval molt, a new larval cuticle is secreted and, at ecdysis, the larger larva of the next instar emerges. During the final larval stage (**2**), the corpora allata become inactive, ecdysone is secreted in the absence of juvenile hormone, and the cells of the epidermis manufacture pupal cuticle. During the pupal stage (**3**), juvenile hormone is absent. If it is experimentally introduced, the pupa will molt into another pupa. Normally, secretion of ecdysone leads to production of adult cuticle and other adult structures. Both juvenile hormone and ecdysone are thought to act on the chromosomes of the epidermal cells, stimulating the production of RNA which then specifies the type of protein synthesis and the cuticle production. Based mostly on the cecropia silk moth, *Hyalophora cecropia* (shown), and on the tobacco hornworm moth, *Manduca sexta.* (Combined from various sources)

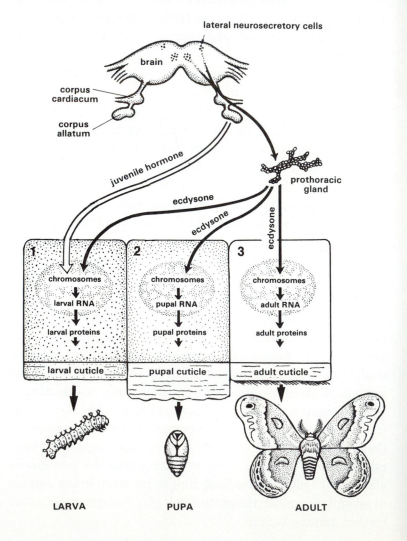

Since normal insect development depends on balanced hormonal control, it should be possible to use **insect hormones as insecticides.** The juvenile hormone produced by insects is unstable and is active in all insects, and therefore unsuitable. However, many similar substances that act like juvenile hormone have been isolated from plants or produced synthetically. Like native juvenile hormone, the synthetic hormones can cause fatal abnormalities if introduced at inappropriate times, not only during metamorphosis, but also during development of the eggs, when applied either to the parents or to the eggs themselves. More subtle effects, such as causing females to mature their eggs out of season, could be equally devastating. The advantages of these substances over native juvenile hormone and conventional insecticides are that each synthetic form has a degree of specificity for certain insect types, such as one family of bugs; they appear to be non-toxic to other animals (non-insects) and plants; they are biodegradable but stable enough for use; and insects cannot readily become resistant to products that closely resemble their own hormones, as they do to conventional insecticides. Major problems are the control of dosage and timing. To be fatal to a pest at metamorphosis, the hormone must reach the insects at a susceptible stage of development, usually a very brief period shortly before metamorphosis. Otherwise it either has no effect, or it may only delay metamorphosis, allowing the larvas to continue to grow and producing giant but otherwise normal adults! Another approach is through antijuvenile hormone substances which have been isolated from plants. These cause early larval stages to metamorphose prematurely and abnormally, sterilize adults, and potentially interfere with any process that requires juvenile hormone.

Hormonal control of the life history has been shifted, in certain fleas, from the hormones of the insects themselves to the sex hormones secreted by the adrenal glands of their vertebrate hosts. Sexual maturity and reproductive behavior are controlled by and dependent upon the rising and falling levels of corticosteroids, and some other factors, in the blood of the host. The European rabbit flea does not mature sexually unless it is sucking blood from a pregnant rabbit, and the fleas do not pair off and copulate unless they transfer to the newborn rabbit young.

Many insects survive seasons of cold or drought or other unfavorable conditions in a state of reduced activity and delayed development called **diapause.** Like so many other facets of insect development, diapause is under hormonal control. It may begin only when conditions are unfavorable, or may be cued by independent environmental indicators such as shortened daylength, which in temperate climates gives reliable advance warning of cold weather to come. Likewise, emergence from diapause sometimes simply follows the return of favorable conditions, but it may require a critical daylength or an intervening period of cold. In some species, diapause occurs regardless of conditions as an obligatory part of the life history. Eggs, larvas, pupas, and adults are all capable of entering diapause, but in any one insect species it can usually occur only at one stage of development. Diapause at the late larval or pupal stage may serve to synchronize the emergence of many adults and improve chances of finding a mate. Diapause in the adult can mean either a deep dormancy that is necessary for its own survival, or only a delay in reproduction that results in more favorable conditions for its young.

PHEROMONES

HORMONES secreted *within* the body act as coordinators, carrying information among the tissues and organs and producing specific responses. Pheromones are chemical coordinators secreted by an organism *to the outside,* carrying information to other members of the same species and eliciting specific responses in them. Sometimes, as in the control of reproduction, systems of hormones and pheromones interact. Mature male locusts, for example, secrete a pheromone that accelerates sexual maturation of immature individuals. The pheromone excites receptors on the antennas; these probably signal the central nervous system, which in turn stimulates the corpora allata to secrete juvenile hormone, which is necessary for sexual maturation. Removal of the corpora allata prevents both sexual maturation and pheromone secretion. The presence of very young locusts in a group, or of individuals without corpora allata, retards sexual maturation in males, probably through another pheromone. Maturation of the group as a whole is thus partially synchronized.

Pheromones acting as sex attractants and stimulants can produce spectacular effects, bringing males to females over distances of several kilometers. A single female pine sawfly held in a cage in a pine tree attracted over 7,000 males in 5 hours. Starving male cockroaches, allowed to choose between food odor and female sex pheromone, rushed gallantly to the latter. Sex attractants are especially common and well studied among butterflies and moths. Females of the silkworm moth secrete a fatty alcohol named bombykol. The large featherlike antennas of the male bear thousands of receptors capable of responding to a single molecule of bombykol and responsive only to this substance or closely similar ones. The males respond by moving upwind until the concentration increases sufficiently to provide a gradient. Then they follow the gradient directly to the source of the bombykol; and if this is a drop on a piece of filter paper, they will crowd around the paper and completely ignore a nearby female in an airtight glass container. Male armyworm moths, seeking their females by scent, sometimes find instead a bolas spider which secretes a mimic of the female moth's sex pheromone. The spider, armed with this irresistible bait, needs no elaborate web and catches the eager male moths using only a little sticky ball suspended by a single silken thread.

Many male insects also produce sex pheromones which both attract the female over short distances and serve as "aphrodisiacs," exciting her to cooperate in mating. In the courtship of

Antennas of male moth are densely branched in this species, *Hyalophora euryalis,* and provide enormous surface for the location of odor-receptive cells, important in detecting the air-borne pheromones emitted by females, sometimes perched at great distances. (E.S. Ross)

grayling butterflies the male must bow before the female so that her antennas touch two scented patches on his open wings before mating can proceed. Both male and female grain beetles (mealworm beetles) produce pheromones which attract and elicit sexual behavior in the oppposite sex, stimulate further pheromone release, and accelerate egg growth in young females. In certain butterflies and bees the male reduces his competition by a secretion that inhibits the response of other males to the female pheromone.

Another kind of pheromone secreted by insects is attractive to both sexes and may result in large aggregations of certain caterpillars or beetles. Often these insects are conspicuously colored and distasteful to predators. The advantage of aggregating, for any individual, lies in the likelihood that the bad taste of some *other* individual in the aggregation will teach a predator its lesson.

Pheromones that serve as alarm substances, or to mark trails, or to influence the development of different types of individuals within a single species are uniquely well developed among the social insects.

Hair-pencils are seen here protruding from the tip of the abdomen of a male butterfly in the rain forest of southern Thailand. Such bundles of bristles, occurring in some butterflies, are sprinkled with aromatic particles, or pick up scented scales from the wings, then scatter them over the females during courtship, so increasing the attractiveness of the male. (R.B.)

SOCIAL INSECTS

THE PHENOMENON of **polymorphism,** as we saw it in the colonial cnidarians, was a division of labor among the structurally differentiated subindividuals of an asexually produced colony. Some insects show another kind of polymorphism in which closely related individuals live together as cooperating members of a **social colony,** and different members of the colony are often structurally specialized for performing different functions. The social insects include wasps, bees, and ants (order Hymenoptera) and termites (order Isoptera). In all of these, the continuity of the society depends on one or more *reproductives,* fertile individuals that continually produce new members for the colony throughout their lives and often do nothing else. They are fed and cared for by their offspring, which remain in the nest as more or less sterile *workers.* The workers also clean, repair, and protect the nest, collect the food, and raise each new brood of sisters and brothers.

All species of **termites** are social. The sterile, wingless castes include not only workers but *soldiers,* individuals with large jaws for biting invaders or pointed "snouts" that squirt sticky defensive secretions; soldiers cannot feed themselves and must be fed by workers. In wood-eating termites, young nymphs become workers first but may later develop into soldiers or into reproductives. These last are either wingless types that remain within the colony or winged forms that fly out in swarms to start new colonies. On landing, they shed their wings and pair off. Each male and female, or "king" and "queen," together build the beginnings of their nest and then mate. The royal pair remain together in the nest throughout their lives and mate repeatedly. In some termite species, males and females share equally in all functions; both sexes serve as workers, soldiers, and reproductives. In various other species, male and female workers may be of different sizes and have different chores, while soldiers may be exclusively of one sex or the other.

There are fundamental differences between colonies of wasps, bees, and ants and those of termites. The former group are holometabolous and their eggs hatch as helpless larvas that must be cared for by members of the colony, as are the pupas until the adults emerge. Therefore the functioning members of the colony are all adults. In contrast, termites are hemimetabolous, and they hatch from the egg as juveniles that look much like miniature adults. After the first molt they may care for themselves and begin to contribute to the work of the colony, which is done almost entirely by immature members (child labor).

Damp-wood termites of the genus *Zootermopsis* live in decaying trunks and logs of conifers. The small-headed juveniles seen here do the work of the colony but can differentiate into any of the more specialized castes. At the top are 2 developing soldiers with larger heads. At the left is a well-developed soldier with dark, strongly chitinized, toothed jaws. There is no special nest, and the working members excavate tunnels in the wood. When the wood is exhausted, the colony moves on, so the queen in such a colony has to be mobile. Though larger than members of the other castes, she is not enormous, as in the tropical species that build large, long-lasting mounds. Friday Harbor, Washington, USA. (R.B.)

Soldier of a dry-wood termite, *Kalotermes,* is well equipped for its aggressive role in the colony. As with damp-wood termites, any of the nymphs that do the work of the colony may develop into a soldier or a winged form. Florida. (R.B.)

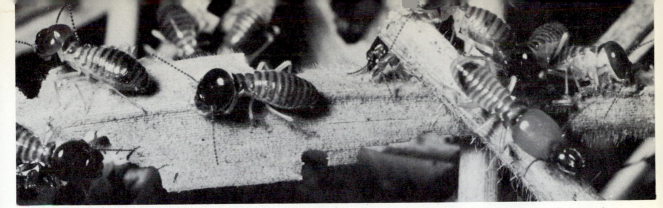

Workers and soldiers of a species of *Macrotermes* are foraging on grass. The small-headed workers chew off pieces of grass that are taken back to the nest and eaten or stored in chambers as a reserve for the dry season, when foraging in the open is not possible. A large-headed soldier, with long mandibles, stands guard at the right. Species of *Macrotermes* that live in open savanna and forage on grasses must do so under the cover of earthen tubes they build over the stems. The grass or other vegetation collected by species of *Macrotermes* is eaten and the fecal pellets used to grow "fungus gardens" in chambers in the nest. The fungi attack the lignin in the pellets and break it down to simpler substances which the termites (and their symbiotic intestinal bacteria) can digest. N.W. Tanzania. (E.S. Ross)

Termite queen dominates this photo of the opened "royal cell," a thick-walled chamber (usually deep within a termite nest) that houses the queen and king. The royal pair have compound eyes and were once winged. They are the primary reproductives of the nest, having founded the colony after a nuptial flight, shedding of wings, and mating. In the chamber they mate repeatedly and she lays eggs at intervals of seconds. The eggs are carried away by the numerous small workers to special brood chambers. The workers also feed and clean the royal pair. The queen lies here with her small, dark head and thorax at the left, both dwarfed by her large white abdomen, enormously swollen by the enlargement of her ovaries. The king, usually by her side, is seen here retreating from the light of the opened chamber. He is larger and more sclerotized than the light-colored workers that surround the queen. No large-headed soldiers are visible. *Macrotermes.* Liberia. (E.S. Ross)

Winged forms of *Reticulitermes hesperus* emerging from an infested house. These sexually capable males and females will take off as a nuptial swarm, settle at some distance, and drop their wings. The males, probably attracted by a pheromone from the females, will follow them and each pair will then excavate a chamber, mate, and start a new colony. Mill Valley, N. California.(E.S. Ross)

Social hymenopterans—wasps, bees, and ants—differ from termites in that *all regularly contributing members of the colony are females.* Males may be present only during limited periods and are entirely peripheral to the colony's activities, except at mating time. Mating of the winged reproductives of both sexes occurs during their "nuptial flight" from the nest, and the males then die. The females never mate again. A new colony is usually started by a fertilized female, either alone, or with one or more others, or together with sterile workers. **Ants** are always social, except for some species that have become parasites and live in the nests of other, usually closely related ant species. Ants closely resemble termites in many ways. Their sterile castes are wingless and include both soldiers and workers, sometimes of several sizes. Only the reproductives have wings, and the queen sheds her wings after mating.

Soldier of army ant colony, *Eciton hamatum,* is 14 mm long and the largest of the several sizes of workers. The large head and long mandibles make this soldier a formidable antagonist, even for humans. Once the mandibles penetrate the skin they are difficult to remove. Soldiers carry no food but run along the moving column of smaller workers, meeting any threat. Peru. (E.S. Ross)

Queen of carpenter ant colony *(Campanotus).* (E.S. Ross)

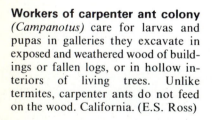

Workers of carpenter ant colony *(Campanotus)* care for larvas and pupas in galleries they excavate in exposed and weathered wood of buildings or fallen logs, or in hollow interiors of living trees. Unlike termites, carpenter ants do not feed on the wood. California. (E.S. Ross)

Nest of army ants, *Eciton,* is a temporary bivouac, consisting of the living bodies of the ants themselves, all clinging to each other. It is usually assembled under an overhanging log, as in this photo. Within the outer layers of ants, containing many long-mandibled soldiers, are living runways leading inward to chambers where young are cared for and the queen resides. Such a bivouac may have as many as 700,000 ants. Barro Colorado Island, Panama. (R.B.)

Column of army ants leaves in early morning when sunlight strikes the outer layer of ants of the bivouac and stimulates them to activity. This moving column of workers, following a pheromone trail, and guarded by large-headed soldiers along its flanks, will attack any insects or other arthropods in their path, biting the prey, pulling it to pieces, and carrying the softer pieces back to the bivouac. After a time the column becomes a two-way stream of ants leaving the bivouac and ants returning with booty from distances of as much as 100 meters. (R.B.)

During a nomadic phase the colony migrates at night to a new bivouac site, carrying the brood of larvas. The routine of massive daytime feeding raids and nighttime migrations is followed for 2 to 3 weeks. Then, the colony settles down to a quieter (statary) phase in which fewer ants go out on feeding raids and the colony remains at the same site for about 3 weeks (see page 613) before another nomadic phase starts. (R.B.)

Yellowjacket wasp nest, made of masticated wood, taken out of its underground cavity and torn open to reveal the tiers of hexagonal cells in which successive broods have been raised. The upper tiers of smaller cells housed workers. The lowest tiers, with larger capped cells, still house developing queens. Late in summer, males are produced. The sexual forms mate. Then all die except the mated queens. These overwinter and in the spring start new colonies. Yellowjackets are aggressive in collecting insects to feed their larvas, and they sting readily at any threat. *Vespula.* (G. Pickwell)

The proverbial "busy bee," despite its reputation for ceaseless labor, actually spends much of its time inside the hive quietly resting or wandering about the combs, doing nothing obviously productive. The same has been observed of ants. Such high levels of unemployment would seem to be inefficient. However, the welfare of the colony may depend in the long run on a large reserve of workers that can take advantage of transient but abundant food sources or respond in great numbers to attack by an enemy.

Temperature regulation is another activity that sometimes demands the attention of many honey bee workers. During most of the year the core temperature inside the hive is near 36°C, just below human body temperature, and varies by little more than half a degree. In hot weather, the bees cool the hive by fanning with their wings and by bringing in water to spread over the combs. In cold winter weather, the bees cluster together tightly to conserve the heat of their bodies. Within the cluster, the temperature remains mostly between 20° to 30°C, even when outside readings are below freezing to -20°C or lower. The construction of many ant and termite nests is such as to provide automatic ventilation and broad control of both temperature and humidity. Workers achieve finer control for the immature stages by moving them from place to place within the nest.

Among **wasps** and **bees,** only a small proportion of the species are social. Most live solitary lives, while others show intermediate behaviors that suggest how full sociality may have evolved, by association of sisters or by association of daughters with their mother. Social wasps and bees are not divided into numerous distinctive castes. All individuals have wings, and workers and queens may differ only slightly in size and other subtle details. In some wasps and bees, typically living in small colonies, the queen and workers can be distinguished only by observing differences in their behavior. Sometimes there is a hierarchy of females, from one that lays most of the eggs and does chores only within the nest, down to lower ranking individuals that seldom or never lay eggs and do most of the foraging.

Domestic **honey bees** live in large colonies with many thousands of members and have a single, distinctive queen that lays all the eggs and does no other work. Division of labor among the workers is not by caste but by age. Newly emerged workers spend most of their working time cleaning the cells in which the immature stages develop. As a worker grows older, she spends more time in feeding the larvas and the queen, then in building combs, in guarding the hive entrance, and finally in foraging for pollen and nectar. Except for overwintering, workers live little more than a month, while queens survive several years. The males, called "drones," do no work and are fed by the workers until they leave the nest to mate.

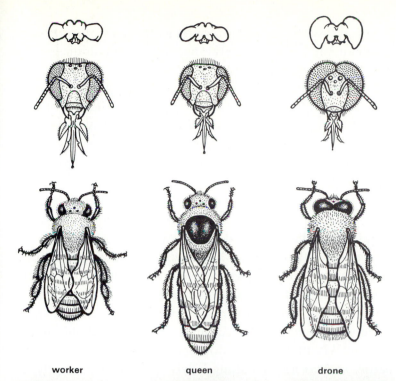

worker   queen   drone

Castes of the honey bee *Apis mellifera* (order Hymenoptera) differ in both structure and behavior. The most obvious external variations are in size and shape of body and wings, in distribution of body bristles, in antennas, eyes, and mouthparts, and in specializations of the legs. The top row shows the brains of the 3 castes. The large eyes of the drone, and the correspondingly large optic areas of the brain, are presumably related to his task of seeking out a queen with which to mate. (Based on various sources. Brains, after C.R. Ribbands)

Honey bee queen surrounded by workers, touching her with their antennas. The queen has inserted her abdomen into a cell of the comb, apparently laying an egg. (E.S. Ross)

Social insects use **pheromones** as a means of integrating behavior and regulating development in their colonies, which are sometimes vast. A single colony of African driver ants may include more than 22 million individuals, together weighing over 20 kg. A foraging worker ant that finds a source of food will pick up a load and head for her nest, laying behind her *trail-marking pheromones* that guide other workers to the food. Trail pheromones are secreted by a variety of glands in different ant species and may be emitted from the sting apparatus, anus, or last pair of legs. Marking pheromones are also produced by termites, and by some wasps and bees, which release pheromone at food sources and at the nest entrance. *Alarm pheromones* are widely used by social insects to warn colony members of danger, such as a predator entering the nest. The effect may be to summon other colony members and excite them to attack, or, on the contrary, to cause them to disperse. When a honey bee stings, she releases a pheromone that quickly draws other workers to the site, primed to do battle, so that anyone stung is wise to retreat promptly. In some termite species, the king and queen produce *pheromones of caste recognition and control.* Workers that lick and groom the royal pair ingest small amounts of these pheromones and pass them on to other workers in the course of mutual feeding that goes on constantly among all members of the colony. The pheromones inhibit the development of workers into reproductives. When a king or queen grows old or dies, the amount of inhibitory pheromone circulating in the colony decreases, and a new royal individual of the proper sex develops. If several of one sex develop at the same time, they recognize each other by means of pheromones and may fight until only one remains, although in some species multiple reproductives are tolerated. In very large colonies, the amount of inhibitory pheromone may drop below the necessary level and extra royal couples may develop in outlying parts of the colony. In honey bee colonies, the queen likewise produces pheromones that announce her presence and inhibit the rearing of new queens. When the old queen dies, or just before she leaves the nest to lead a swarm, the amount of inhibitory "queen substance" drops. In this case, the absence of pheromone does not automatically allow a new queen to develop, as in termites. Instead, it causes the colony to set about rearing a replacement. Several ordinary young female larvas are fed on "royal jelly," a substance that the workers secrete. This special diet causes the larvas, which would otherwise become workers, to develop into queens.

If honey bee workers are denied contact with the queen (from whom they would normally receive inhibitory "queen substance"), their ovaries develop and they lay unfertilized male-producing eggs.

Pheromones from the queen are probably the most important chemical source of colony integration in an army ant nest, but odors from the developing brood and from workers are also strong determinants of ant behavior. Here workers of several sizes attend the queen in a colony maintained in the laboratory. The night before, the queen had been plucked from a migrating column (not without multiple bites to the bold hand of T. Schneirla). As she was in the migratory phase, her abdomen is contracted. In the forest, a queen entering the statary (non-migratory) phase undergoes enlargement of her abdomen for about a week. She may produce as many as 100,000 eggs. By the end of the third week, as a new generation of workers emerges from the pupal stage, the queen's eggs hatch into larvas. This sets off a new nomadic phase. *Eciton.* Panama. (R.B.)

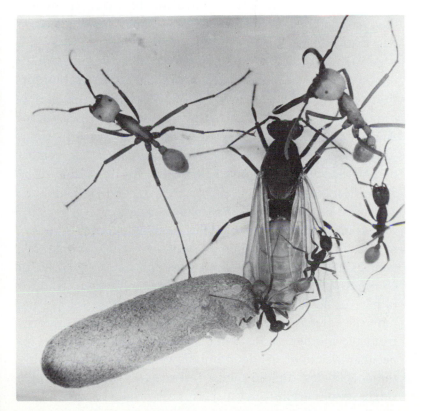

Male army ant, just emerged from pupal case, is attended by workers of various sizes. The largest of these are soldiers that have long mandibles and are capable of warding off threats to colony members. The reproductives mate, and the queens found new colonies. This colony of *Eciton* was maintained in the laboratory by T. Schneirla at the Smithsonian Tropical Research Institute on Barro Colorado Island, Panama. (R.B.)

The **evolution of sociality** in insects has long been a puzzle, especially the evolution of sterile workers and soldiers. How can natural selection favor individuals that leave no offspring? Moreover, complex societies including sterile castes are thought to have evolved independently at least 11 times among the hymenopterans—wasps, ants, and bees—and only once in any other group of insects—the termites. What is it about the biology of hymenopterans that predisposes them to evolve in this way?

The self-sacrificing labor and sterility of whole castes of social insects is best understood as an extreme example of **altruism,** the term applied to any behavior by which an animal, at some cost to itself, benefits another individual. We will measure cost and benefit, not by survival, but in terms of *reproductive success,* the number of offspring surviving to reproduce the parent's genes. However, genes are shared not only with offspring but also with other family members— siblings, cousins, nieces, and nephews. Therefore we would expect an animal to behave so as to benefit such kin, at some cost to its own offspring, whenever increased reproduction of shared genes by the kin more than makes up for the cost. We would not generally expect altruistic behavior to evolve toward unrelated individuals. Observations bear out these expectations.

As the degree of relatedness between kin becomes more remote, and the probability of shared genes decreases, **kinship theory** predicts that altruism will evolve only if the cost is proportionately smaller or the benefit greater. Thus, for an animal to be altruistic to its sibling, which shares ½ its genes (on the average), the cost to its own reproductive success must be less than ½ the sibling's benefit. But for an animal to be altruistic to its first cousin, which shares only ⅛ of its genes, the cost must be less than ⅛ the cousin's benefit.

This requirement for a high degree of relatedness is the key to sociality in hymenopterans. Among all hymenopterans, females develop from fertilized eggs and receive two sets of chromosomes, one set from the mother and one set from the father, as do most animals. Males, however, develop from unfertilized eggs and receive only one set of chromosomes, from the mother. Males have no fathers and no sons. This peculiar system of sex determination is called **haplodiploidy** (haploid males, diploid females). Because males have only a single set of chromosomes, sperms are produced without the usual meiotic divisions or random assortment of chromosomes into different sperms. All sperms of a male hymenopteran

Ironically, the selfless devotion of parents to their offspring does not qualify as altruism, by our definition. A parent caring for its young, even at the sacrifice of its own life, contributes directly to its own reproductive success. Parental care evolves only if it results in more offspring surviving to reproduce the parent's genes, including the genes for parental care.

Thus the genetic basis for the evolution of altruism and of parental care is exactly the same. If cost and benefit are measured in terms of the welfare or survival of the individual, instead of its reproductive success, then parents (especially mothers) are the most conspicuous altruists of all. However, individual survival is important in evolution only to the extent that the individual contributes to the perpetuation of its genes either through its own offspring or through its relatives.

carry his entire chromosome set and are identical. Hence the offspring (all daughters) of any one male and female are *unusually closely related*. They are more closely related to each other than they are to their own daughters and more closely related to their sisters' offspring than to their own brothers!

Asymmetrical degrees of relationship are found in wasps, ants, and bees. This diagram shows the inheritance of 4 traits on different chromosomes (symbolized by different shapes). The scheme given is only one of many possible outcomes, determined by chance during the meiotic divisions that randomly assort the chromosomes among the eggs of the diploid female. The degree of relatedness (values shown beside the arrows) is the probability that any gene of an individual is present also in a related individual. In a diploid species, the degree of relatedness between parents and offspring and among offspring of both sexes is completely symmetrical and is generally ½, that is, half the genes of any individual, on the average, are present also in that individual's parents, siblings, and offspring. Values of relatedness in haplodiploid species are quite different. They may be determined directly from this diagram by counting the number of identical "genes" (same shape and shading) that any individual shares with another and dividing by the total number. Because the total number for males is half that for females, relatedness is not symmetrical between individuals of opposite sex. Males are related by ½ to both their brothers and their sisters. Females, on the other hand, are related by ¾ to their sisters and only by ¼ to their brothers. It is the possiblity of taking advantage of this asymmetry, by favoring the production of closely related sisters (and their offspring), that is the basis of the altruism of the female worker caste. (Based on W.D. Hamilton and on Trivers and Hare)

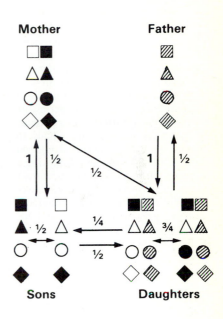

By analyzing the unusual and asymmetrical kin relationships that result from haplodiploidy, evolutionary biologists are able to account for, and even to predict, many of the major consistent features of social hymenopterans. Most conspicuous are the large castes of exclusively female workers (and in ants, soldiers also), most of which never reproduce but spend their lives caring for their siblings. These altruistic females are provided with a "communal stomach," in which they carry food that is freely shared, but more freely with their sisters than with their brothers. Worker wasps and bees give their lives when they sting, for the sting remains embedded and their bodies are torn. The jaws of biting ants likewise often remain suicidally clamped, in altruistic defense of other colony members. Males, on the other hand, which bear no unusually close family ties, are wholly immune to this prevailing spirit of self-sacrifice. They contribute nothing to the colony but are especially aggressive in demanding food, saving all their energies for the moment when they will fly out to mate.

Termites, with their parallel social organization but conventional diploid males and females, have workers of both sexes. They are thus the "exception that proves the rule." Their

Cooperative activities in social hymenopterans primarily involve females, but males sometimes render services to their nestmates without lowering their own individual fitness. Males of the social bumble bee *Bombus griseocollis* straddle developing pupas, maintaining the temperature of the pupas at 4° to 6°C above that of ones not incubated. This increase in temperature is not as great as that produced by workers or queens, but it is important in warming pupas.

independent social evolution shows that haplodiploidy is not necessary to sociality but is probably connected with the unique features that have repeatedly evolved in the social hymenopterans.

BEHAVIOR

THE SHORT LIFE HISTORIES of insects, the small size of insect brains, and the limited number of cells in such brains, all suggest that the behavior of insects is mostly instinctive and stereotyped. But as experimental studies of learning have been few, it is difficult to know how much our lack of evidence for learning in insects reflects their inability to learn and how much it results from our inability to teach them. To demonstrate learning in an animal, its behavior must be understood well enough for the experimenter to devise a situation in which the animal can sense the necessary features and be motivated to learn.

Cockroaches have learned simple mazes when offered as a reward only the modest privilege of being allowed to enter a small dark cup smelling of cockroach. Escape from light and access to a cup of cereal have been used to motivate grain beetles. When grain beetle larvas (mealworms) were trained to turn left in a T-maze, the adult beetles learned a left turn faster, and a right turn slower, than individuals with no training as larvas. Some effect of the larval training seems to have persisted through metamorphosis, despite evidence that during the insect pupal phase, most of the larval nerve cells are replaced.

Social bees and ants have been the subjects of most learning experiments. Ants have learned to find their way quickly through complicated mazes with many turnings. If the direction of a correct turning was repeatedly reversed by the experimenter, the ants mastered the reversal slowly at first and then more rapidly each time; they seem to have learned to learn.

Honey bees can be taught readily, and learning plays a major role in their foraging for food. The bees remember not only the scent of flowers from which they have collected nectar and pollen, but also their location. When a honey bee worker that has been collecting nectar and pollen returns to the hive, it carries (in the nectar and on its body) the odor of the kind of flower it was visiting, for example, clover. Workers in the hive that have collected from clover recognize the scent and are stimulated to leave the hive and fly to the various places where

The small brains of insects limit memory storage, and both bees and butterflies are known to show a preference for extracting nectar from flowers with which they have had a previous learning experience. In experiments on the cabbage butterfly, *Pieris rapae*, the time required by individuals to find the source of nectar decreased with successive attempts, following a learning curve. Learning to extract nectar from a second kind of flower interfered with the ability to feed from the first species of flower and increased the time and energy needed for obtaining the same amount of food.

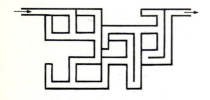

Maze-training of ants (*Formica incerta*) has shown how circumstances may affect behavior. For example, this maze pattern was always mastered faster by ants on their way home from a food box to the nest than by ants on their way from the nest to a food box. When 8 ants were compared to 8 rats in this maze pattern (built on a larger scale for the rats), the ants as a group took 3 to 4 times as many trials to master it. But differences in individual ability also emerged; the most successful ants took only about 2 times as many trials as the least successful rats. Comparisons between animals as different as ants and rats must be made cautiously, however, as both the specific behaviors and the underlying neural mechanisms appear to be fundamentally different. (After T.C. Schneirla)

they have learned to find clover. Other workers, lacking experience with clover, do not immediately associate the odor with food and are recruited to leave the hive in search of clover only if they come in contact with a returning bee that performs a "round dance," consisting of repeated circling, or a "waggle dance" in which the straight path followed in the middle part of the dance is accompanied by waggling of the abdomen and by sounds. The inexperienced potential recruits surround the dancing bee, touch it with their antennas, and thus sense not only the odor of the clover but also other odors from the vicinity of the collecting site.

Besides stimulating recruits to forage and providing odor cues, the dancing bee may offer information about the direction and distance of its collecting site. The direction (angle between the sun and the site) is related to the orientation of the waggle dance. The distance between the hive and the site is related to the speed of the dance, to the duration of the sounds and the number of waggles made by the dancer during the straight middle part of the dance, and to several other of its features. The array of potentially informative features in the waggle dance (that can be measured or counted by a human observer) has been referred to as the dance "language." However, it is not known which of the described features, if any, can be used by the bees, or by what means they extract information from the dance in the darkness of the hive. The "language" does appear to be used successfully by a limited number of bees, always together with odor and visual cues.

Perhaps bee "language" is most used when a new food source, with a scent not associated with food by workers in the hive, is first discovered and "announced" by a dancing scout. In any case, most bees, most of the time, appear to search for a dancer's source by odor cues. When the bees have once located a patch of the flowers they were seeking, they quickly learn its location and thereafter will fly directly to it over and over again, depending on their memory of visual landmarks, odors, direction, and distance. (The relationship between the sun's position and the time of day, necessary to determine direction, is also said to be learned.) Finally, the foraging bees learn to recognize different flowers visually; they learn the hours at which they open and close; and they learn techniques for collecting from each type. Most of the foraging that supplies the colony is by experienced workers going repeatedly to their remembered sources.

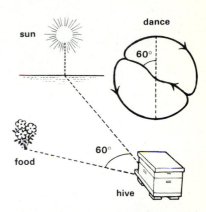

The direction of a food source is incorporated into the "waggle dance" of a foraging honey bee on its return to the hive. When the bee dances on a vertical comb within the dark hive, the angle between the straight part of the dance and a vertical line is approximately equal to the angle between the food source and the sun. (Based on K. von Frisch)

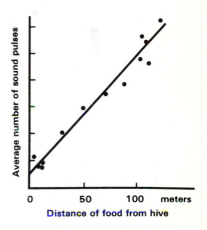

The **distance** of a food source from the hive is related to the duration of the "waggle dance," to the number of waggles of the bee's abdomen, and to various several other features of the dance. Graphed here is the relationship between the distance of the food source and the average number of sound pulses produced by a bee during the straight part of the dance. (Based on A.M. Wenner)

The analysis of the features used by searching honey bees has occupied researchers for decades. Much careful work was first necessary to establish that information on direction and distance of a food source is contained in honey bee waggle dances, that dances are necessary to recruit bees to search for a newly discovered food source, and that numbers of bees do find the same source visited by the dancer. All this does not show, however, that the bee recruits *use* the information. Flies, moths, and various other solitary insects display behaviors, after foraging flights, that are similar in certain respects to honey bee dances; they even contain directional and distance information. There is no indication that these animals are communicating with other members of their species; honey bee dances may have had similar origins.

Experiments in which bees visit artificial food sources, and then return to the hive to dance, have yielded results consistent with the "language" hypothesis. But because odor cues are necessary (and often sufficient) for bees to locate a food source, and because they could account for the same experimental results, it is extremely difficult to determine whether direction and distance information are ever used along with odor cues. In various experiments, when two or more food sources were offered and made as similar in odor as possible, more bees went to the dancers' source than to any other. However, the same results were found when the dances were not properly oriented, as when the sun was directly overhead, or when the dancers were made to perform on a horizontal platform under diffuse unpolarized light.

In other experiments, a light was placed at different angles within the hive which would cause the bees, if they used the dancers' information, to misinterpret it in a predictable way. (The dancers were insensitive to the experimenter's light because their simple eyes were painted over.) More bees came to the food source indicated by the misinterpreted dances than came to several adjacent ones or to the isolated source really visited by the dancers, which was in another direction altogether and so could not have provided odor cues. Thus it appears that at least some bees did use the dance information. However, when the dancers' real source was placed near the others, the total number of bees arriving increased severalfold, and the majority went toward the real source, presumably by odor, rather than following the dance information.

The importance of most bees' preference for odor cues, when experimentally opposed to dance information, is difficult to evaluate, because under ordinary circumstances, the two would generally coincide. Moreover, the relative importance of odor and "language" may change from minute to minute with the weather, may vary in different strains of bees, and may depend strongly on the environment and immediate history of the hive or on the smallest details of methods used in different experiments. Recent findings indicate that many apparent conflicts in the data, and real conflicts in the interpretations of different investigators, reflect small but crucial differences in circumstances or experimental design.

As with any scientific question, we can never be absolutely certain whether future experiments will merely refine and reconcile present interpretations of bee foraging or will cause us to revise them altogether. Readers who go to the scientific literature to follow in more detail the sequence of experiments and discussions on bee foraging will find logical challenges more exciting than those in any mystery story. They will also learn that scientists, like other detectives, may favor some lines of investigation over others and thus tend to find only those clues for which they are looking.

THE ORDERS OF INSECTS

THE SUPERCLASS INSECTA is divided into 5 classes. Four of these contain a relatively small number of primitively wingless insects, while all the remaining insects, winged or not, are included in the great class **Pterygota,** the winged insects. The orders of pterygote insects fall neatly into two subclasses. The **Exopterygota,** with young stages that have external wing buds, change into adults gradually during successive molts. The **Endopterygota,** with distinct larval stages in which the wing buds develop internally, grow but change little in form until they go through a pupal stage and abruptly metamorphose into the adult form.

Despite these differences in development, the major orders of the two pterygote subclasses form two strikingly parallel series in terms of their life histories and ecology. For example, the mayflies in the Exopterygota and the dobsonflies in the Endopterygota are each considered to be among the most primitive members of their respective subclasses. Both have long-lived, feeding juveniles found in freshwater, breathing with tracheal gills, and short-lived adults that emerge, mate, lay eggs at the water's edge, and die. Among herbivorous insects, the exopterygote grasshoppers and the endopterygote caterpillar larvas of moths and butterflies have no rivals at chewing up vegetation. Termites among the exopterygotes, and bees and ants among the endopterygotes, have evolved amazingly similar insect societies, which perform great feats of construction. Even the sucking lice of the exopterygotes have their wingless ectoparasitic counterparts among endopterygotes in the fleas.

Bugs (hemipterans) comprise the most diverse group of exopterygotes. Many are herbivores and important agricultural pests, but others are predators feeding on a wide variety of prey, including agricultural pests. Some bugs, such as water boatmen and backswimmers, are completely aquatic their whole lives, but the adults can fly from pond to pond. Others such as water-striders, skim along the surface of water. Similarly, beetles make up the most diverse group of endopterygotes with a wide variety of herbivores, such as weevils, which are important pests, and predators, such as ladybird beetles, which are valuable predators on insects. Many beetles are also completely aquatic their whole lives, flying from pond to pond as adults, while others, such as whirligigs, swim on the surface.

The photos and brief accounts on the following pages are presented as an introduction to the orders of insects.

Flies (order Diptera) appear to be the one large group of endopterygotes without a corresponding order among the exopterygotes. This may be because a major feature of most flies is their maggots—legless, wormlike larvas—that burrow deep into decaying organic material to feed. Exopterygotes, which do not have larval stages much different from the adult, are largely excluded from this life style.

Collembolans are usually only a few mm long. In moist soils and ground litter, they may reach a density of 100,000 per m² and can be the most important consumers in the community. They are often called "springtails," because most have a forked structure, the *furcula*, with which they propel themselves into the air. *Left*, a springtail at rest with the furcula folded forward and held under tension by a clasplike structure. *Right*, the furcula is suddenly released and springs backward, launching the collembolan. (P.S. Tice)

Marine collembolans are abundant on seashores, ranging down into the upper intertidal zone. Seen here on the water surface is *Anurida maritima*. Woods Hole, Massachusetts. (K.B. Sandved)

Primitively wingless insects, or apterygotes ("without wings") make up 4 orders or classes of minute to small insects: PROTURA, COLLEMBOLA, DIPLURA, and THYSANURA. Most favor damp protected situations, as in soil or ground litter, beneath, bark, or decaying logs. **Proturans** (not shown) are unique among hexapods in having no antennas, and also lack eyes. Proturans hold the sensory forelegs elevated like antennas, and walk slowly on 4 legs. Representatives of the other 3 groups are shown on this page. (See also p. 584)

Lepismatid, the firebrat, *Thermobia domestica*, 13 mm long, has brown and pale scales and tail filaments longer than the abdomen. It frequents warm, dry places. Its two common silverfish relatives are a uniform silvery or slate gray and are seen in cooler, moister parts of houses. All feed on coatings of paper, and on other starchy materials. These thysanurans have small compound eyes and have exposed mouthparts as in winged insects. (R.B.)

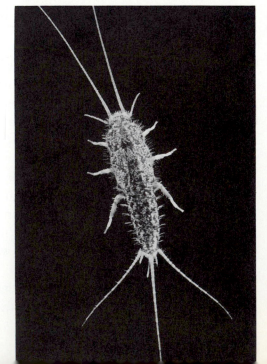

Dipluran, with the 2 tail appendages for which the group is named. This one is a campodean, the most frequently encountered dipluran. Length, 6 mm. Calif. (R.B.)

Machilid, or bristletail, may jump when disturbed. Its large compound eyes mark it as a member of the Archeognatha, primitive thysanurans with mouthparts more like those of the other apterygotes and sometimes made a separate group. Machilids can range into dry woodlands and grassy areas. California. (E.S. Ross)

Beginning on this page with paleopteran orders and continuing through the orthopteroid and hemipteroid orders, the insects presented belong to the subclass **EXOPTERYGOTA**. The exopterygote orders have a hemimetabolous life history, with young nymphs that resemble adults.

Mayflies, order EPHEMEROPTERA, include more than 2,000 species with delicate winged adults that do not feed and are as ephemeral as the order name suggests. All the food-gathering is done by the long-lived freshwater nymphs. When ready to metamorphose, the last aquatic instar rises to the surface of the water, its outer cuticle splits, and a winged but sexually immature subadult emerges. Within a few minutes (in species that live only a few hours) or within a day (in species that live a few days), the dull subadult molts, shedding the delicate cuticle covering both body and wings. There emerges a shiny and more colorful, sexually mature adult that will mate. Mayflies are the *only* insects having a winged form that molts. In some temperate zone species, large dense swarms of males take flight, rising up and floating down in unison in a "nuptial dance." They are joined by the females, and soon pairs depart from the swarm and mate in the air. The female deposits her egg masses in water, and both sexes die.

Subadult mayfly will soon molt and the adult will emerge. **Mayflies** are readily distinguished from other insects with net-veined wings by the disparity between the large forewings and small hindwings, and by the 2 long many-jointed cerci. In some species, the synchronous emergence, mating, and death of millions of mayflies may occur in a single evening, and so many are attracted to lighted windowscreens as to create something of a nuisance. Near the Great Lakes, dead mayflies have been reported to accumulate in deep piles on roads, causing traffic problems. The millions of adults that fall into the water when they die, as well as the aquatic nymphs, form an important part of the diet of freshwater fishes. Many of the artificial "flies" used by fishermen are modelled after mayflies. The sharp decline that follows pollution of lakes and streams is sometimes due, not to direct effects on fishes, but to the serious impact on mayflies and other aquatic insects on which fishes feed. *Hexagenia limbata.* (E.S. Ross)

Mayfly nymph. The nymphs live in lakes, ponds, and streams, and feed with large biting mandibles on plants, organic debris, and sometimes live animals. Nymphs of various species swim, crawl on the bottom, burrow in mudbanks, or cling to the underside of stones. They have leaflike or plumed tracheal gills along the sides, and 2 or 3 tail filaments. A few mayfly species complete their life history in a single summer, but most live for 2 or 3 years, molting dozens of times before metamorphosing. (R.B.)

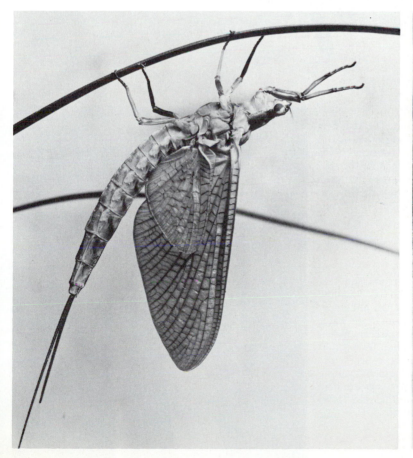

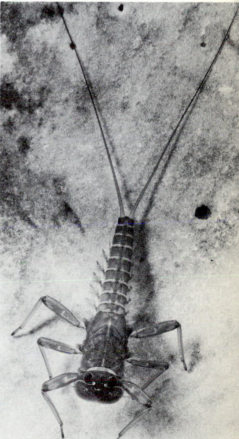

Dragonflies and damselflies, order ODONATA, comprise about 5,000 species of predaceous insects with enormous eyes, narrow bodies (2 to 13 cm long), and 2 pairs of densely net-veined elongate wings of nearly equal size. These attractive insects obligingly consume great quantities of midges, mosquitoes, and other flies. The long, forwardly directed legs are well adapted for seizing prey and for perching, but not for walking. When the insect is in flight, they are held close to the body until a victim is sighted; then the spiny legs spread slightly, and form a basket for catching the prey, which they pass on to the chewing mouthparts.

Odonates are seen most often near the streams, marshes, lakes, and ponds in which their aquatic, predaceous nymphs live and feed. However, many dragonflies are powerful fliers and roam far afield, returning to water only to mate and lay eggs. Mating usually takes place in flight, and pairs flying "in tandem" are a common sight—the male using claspers at the tip of his abdomen to grasp the female by head or thorax, and the female curving her long abdomen down and forward to reach the male copulatory organs near the front of his abdomen (a location unique among insects). A male patrolling his home airspace will usually succeed in driving off any intruding male. The frequency of such conflicts is high and seems to be unrelated to the abundance of food. It is likely that territorial behavior is primarily related to sexual competition.

Odonates cannot flex their wings down over the abdomen. Among living orders of winged insects, only the mayflies (order Ephemeroptera) share this primitive wing trait, but it is seen also in fossils of several of the extinct insect orders of the Paleozoic period. Folding the wings tightly over the back enables insects such as cockroaches, beetles, and wasps to hide in holes or crevices.

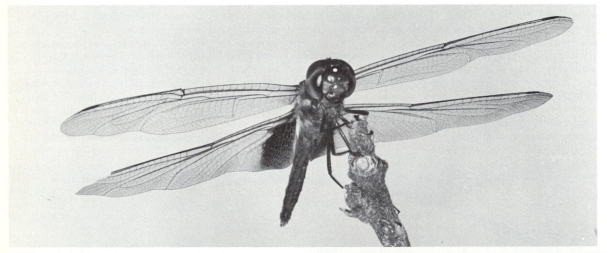

Dragonfly, *above,* suborder Anisoptera, holds its wings straight out at the sides, even when at rest. The abdomen projects stiffly behind the capacious thorax that houses the powerful flight muscles. Dragonflies are more robust than damselflies and have larger eyes, so close together that they almost touch. Active hunters, they fly out in search of prey and catch it on the wing. (R.B.) (See metamorphosis of dragonfly, page 599)

Damselfly, suborder Zygoptera. This one had just emerged from its nymphal cuticle, which has 3 leaflike tracheal gills projecting from the tip of the abdomen. Before the final molt, the nymph crawls out of the water and takes in air through newly opened spiracles. When the adult emerges, it leaves the gills behind. Damselflies differ from dragonflies in having smaller eyes set farther apart and more slender bodies, and in quietly waiting for prey to fly or crawl by. At rest, the wings are held together above the body. Western Australia. (E.S Ross)

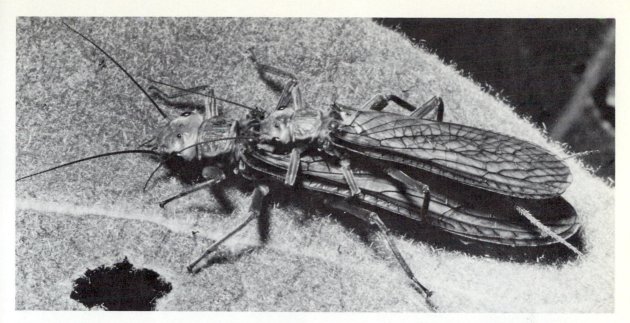

Stoneflies mating, with the male astride the female. The fertilized eggs are dropped into a stream or lake, where they hatch into aquatic nymphs, which are abundant on the underside of stones in stream beds and have suggested the common name of the group. The adult that emerges from the last molt is a weak flier and rests or crawls on stones, bark, or foliage, feeding mostly on algas or lichens. Many have vestigial mandibles and appear not to feed during the few weeks of adult life. The formal name of the order PLECOPTERA (*pleco* = "pleated") refers to the fanwise folding of the rear part of the hindwings, when the 2 pairs of membranous wings are laid over the back of the resting insect. There are more than 1,500 species of plecopterans around the world. The many-veined wings, similar pairs of simple walking legs, many-jointed antennas, and long many-jointed cerci (all conspicuous in this photo) suggest that plecopterans are the least specialized of the orthopteroid orders. Klamath River, California. (E.S. Ross)

Rock crawler, order GRYLLOBLATTODEA. The more than a dozen species included in this small order are wingless, elongate, brown or gray, and 15-30 mm long. Like most wingless insects, they lack ocelli. Compound eyes are absent or reduced. The legs are all alike. Rock crawlers have toothed chewing mandibles and are omnivores. They are found under rocks, in moss, in ice caves, and on snow—always at high altitudes. So far they have been described from western Canada, northwestern USA, Japan, and Siberia, USSR. As their name indicates, grylloblattids combine features of both cockroaches and orthopterans such as crickets. They have many primitive characters. *Grylloblatta*. Mt. Baker, Washington. USA. (E.S. Ross)

Stonefly nymph. (E.S. Ross)

Katydids, crickets, and grasshoppers are the 3 main types among the almost 20,000 described species of the order ORTHOPTERA ("straight wings"). Most orthopterans have 2 pairs of wings, thickened forewings and large membranous hindwings. When the insect is at rest, the pleated hindwings are folded like fans and the forewings serve as wingcovers. Compound eyes are prominent; simple eyes may be absent. Orthopterans are best represented in the tropics but are widely distributed around the world in all but the coldest climates. Except for members of one family that frequent margins of ponds and streams and are good swimmers, orthopterans are terrestrial, both as adults and as nymphs, and live in ground vegetation or in trees. Though there are omnivorous and predatory types, the vast majority are plant-eaters of moderate size and, not suprisingly, they have a considerable effect on vegetation. Migratory grasshoppers, called locusts, are among the most destructive of all agricultural pests.

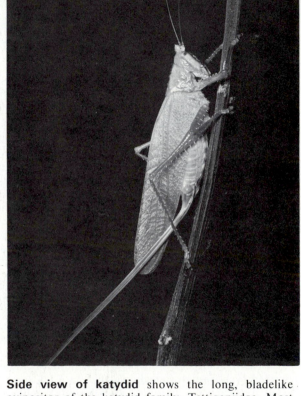

Katydid, in face view, shows two characters shared by crickets but not by grasshoppers: the great length of the delicate antennas and the location of the auditory organs on the tibias of the forelegs. Panama. (R.B.)

Side view of katydid shows the long, bladelike ovipositor of the katydid family, Tettigoniidae. Most have stridulatory organs and many are noted for their singing; each species has a characteristic song. Most are plant feeders; a few prey on other insects. (R.B.)

Field cricket is common in meadows, pastures, and roadsides, and many enter houses. Most field crickets chirp, and they sing day and night. The long ovipositor is cylindrical in crickets, family Gryllidae. (R.B.)

Tree crickets are whitish or pale green. All are excellent songsters, and their high-pitched trills are among the most conspicuous insect songs of summer nights. Family Gryllidae. (C. Clarke)

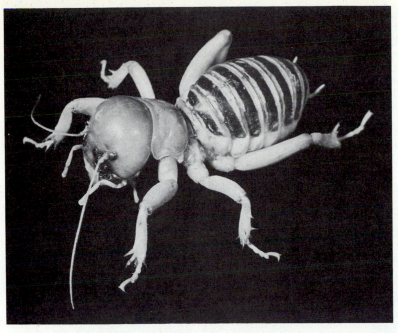

Wingless orthopteran, *Stenopelmatus,* digs in the ground, feeding on roots and tubers. Also known as the Jerusalem-cricket, "Child of the Earth," and "Old Bald-Headed Man," it wanders about at night and may enter houses; it is often encountered by gardeners turning over soil. Its large size (up to 5 cm long), spiny digging legs, giant mandibles, and oversized head give it a threatening appearance, but it is harmless. Western N. America. (R.B.)

Earwigs, order DERMAPTERA, include about 1,200 species whose most distinctive feature is a forceps formed by the 2 unjointed cerci at the tip of the abdomen. The forceps serves in defense and in prey capture and plays a role in mating. Earwigs have chewing mouthparts, and most are omnivorous. Typically there are 2 pairs of wings, inconspicuous when not in use; and the most familiar earwigs in Europe and N. America rarely fly. The short forewings are leathery. Tucked beneath them, with only the lower margin showing, are the delicate hindwings, large and semicircular with radiating veins. They are folded like a paper fan, then folded twice crosswise and stowed under the forewings. Earwigs are largely nocturnal and in the daytime hide in plant litter, among leaves, or under bark, stones, or logs. Most species are tropical or subtropical, but a few species are common in Europe and N. America.

European earwig, *Forficula auricularia,* is black or brown and 12 to 15 mm long. This species prefers flowers, fruits, and other plant parts. A familiar plant pest in Europe, it was carried in ships' cargo to N. America and has found a good life there, especially in California and the southern USA. It has also been introduced by commerce into S. America, Australia, New Zealand, and S. Africa. The male, *above,* has a large, heavily sclerotized forceps, strongly curved and toothed. The female, *below,* has a smaller forceps, straighter and without teeth. She lays a batch of eggs in the soil and rests over or near them, licking the eggs to reduce growth of fungi and later guarding the young nymphs against intruders. (E.S. Ross)

Phasmids include walkingsticks and leaf insects, moderate-sized to large, slow-moving herbivorous insects. Their striking resemblance to the leaves or twigs of the shrubs or trees on which they live is enhanced by behavioral posturing. The order PHASMIDA includes about 2,500 species. Most of them are tropical. All those in N. America are sticklike, and all the U.S. species wingless except for one short-winged species in S. Florida. Another species found only in the South and Southwest reaches a length of 178 mm, the longest insect in the U.S. Aside from their camouflage they defend themselves from the grasp of enemies by readily shedding limbs and by spraying from the thorax a foul-smelling substance. Walkingsticks are not usually numerous enough to do much damage to cultivated plants, but when they occasionally do occur in great numbers in woodland they can defoliate large areas of trees.

Stemlike walkingstick with elongate shape, cryptic coloring, and quiet behavior, is well concealed on a tree trunk among vines. Panama. (R.B.)

Twiglike walkingstick *Pterinoxylus spinosus* is very difficult to see when it is at rest among twigs. (R.B.)

Webspinners, order EMBIOPTERA, or EMBIIDINA, are unique among insects in producing silk from the enlarged tarsi of the forelegs. About 300 species are described; most are small (4 to 7 mm), elongate, and black or brownish. They usually spin silken tunnels among plant debris, in soil or log crevices, or under stones. Most species live in aggregations and weave a maze of tunnels in which they shelter and mate and in which the females tend eggs and young nymphs. This communal home protects against predators and in at least one species permits some control over humidity and temperature. Nymphs and adult females, all wingless, emerge at night to chew on plant materials such as mosses, lichens, bark, or leaf litter. When male nymphs mature and develop wings, they fly away and join another aggregation. Short-lived, they do not feed but use the mandibles to grasp the female's head during mating. Most embiopterans are tropical, but some extend into warm temperate areas in the USA and elsewhere.

Male webspinner with 2 similar pairs of membranous wings. The wings are flexible and can be doubled over the back when the animal moves backward in its tunnel. *Antipaluria.* Venezuela. (E.S. Ross)

Female webspinner in its silken tunnel. Adult females are wingless and not much different externally from the nymphs. Even the first-instar nymphs spin silk. *Pararhagadochir trachelia.* NW Argentina. (E.S. Ross)

Giant tropical cockroach, *Blaberus giganteus,* 7.5 cm long, in a Panama rain forest. It may enter houses and fly about at night. (R.B.)

Cockroaches, order BLATTODEA, have a flat, oval body and a shieldlike pronotum (the chitinous plate over the prothorax) that conceals the head from above; only the long antennas protrude. Thickened forewings protect the membranous hindwings, which are folded in pleats when at rest. In the various species, wings may be well developed, reduced, or absent; females often have shorter wings. The approximately 3,500 species of cockroaches are primarily tropical and live on the forest floor, though some are arboreal. The flattened body, pronotum, and often reduced wings help these robust insects to hide in narrow crevices of wood or rock. In human dwellings, they hide in molding gaps, under loose wallpaper, and in crevices of drawers and cupboards, emerging at night to feed. Their droppings, and unpleasant odors they sometimes leave behind, make them unwelcome house guests. But only about 1.5% of cockroach species take advantage of human food and shelter. Cockroaches, especially the large *Periplaneta americana,* are hardy laboratory animals, well suited for studies on physiology, on behavior and nervous system, or on exchange of pheromones.

Diurnal cockroach is a handsome insect and displays its colorful wings in broad daylight. Peru. (E.S. Ross)

Nocturnal cockroaches retreat into crevices in wood when exposed to light. Like many cockroaches, these are wingless and blackish in color, the nymph not yet darkened. Panama. (R.B.)

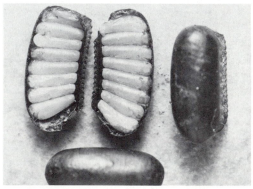

Oriental cockroach, *Blatta orientalis,* originally from Asia, is about 25 mm long. *Left,* a female with a hardened egg case. *Right,* egg case cut open to expose the double row of eggs. (L. Passmore)

German cockroach, *Blattela germanica,* multiplies prodigiously and is the most familiar cockroach in northern cities of the USA and Europe. It is only 12 mm long and pale brown with a pair of dark stripes on the pronotum. This individual was feeding on a crumb in a kitchen at night. (E.S. Ross)

Praying mantids, order MANTODEA, are usually recognized by the long prothorax and the large grasping forelegs. The coxa of the foreleg, small in most insects, is elongated in mantids. And the femur and tibia are spined; the tibia can be folded back against the femur so that the opposing spines hold prey securely. The forewings are narrow and hard, the underlying hindwings membranous, broader, and pleated. The nymphs are similar to the adults.

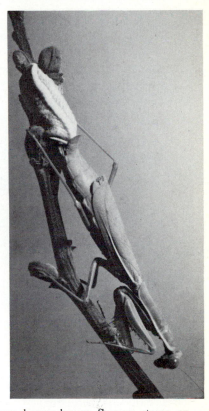

Above, **mating mantids.** Many observers have reported seeing the female begin to eat the male during mating; and he, though headless, continues to copulate. The behavior is dramatic but probably occurs with abnormal frequency in captive animals that are disturbed and inadequately fed. One set of observers who took pains to feed their mantids, and to prevent the insects from observing the observers, seldom saw cannibalism, but instead recorded a courtship behavior that appeared to inhibit aggression. *Right,* **female mantid with egg mass** encased in a hardened envelope. *Stagmomantis californica.* Body 6 cm long. California. (E.S. Ross)

Mantids are mostly tropical and enormously varied to resemble green leaves, brown leaves, flowers, stems, or vines. Panama. (R.B.)

Termites are small to moderate-sized, soft-bodied insects, usually whitish in color. They always live in colonies, either tunneling in wood, or building nests underground or aboveground. They become structurally differentiated into various castes that play different roles in the life of the colony. Usually the castes are of both sexes. The order name ISOPTERA refers to the equal size and shape of the 2 pairs of membranous, many-veined wings in the winged forms. The name termite comes from a Greek word meaning "a worm that bores into wood." This perception is not too different from that in the warmer or sometimes temperate parts of the USA, where termites are seen mostly when opening up logs in a forest or when they tunnel into the dry wood of buildings, striking terror into home owners. In the tropics termites are much more widespread and more destructive. (See also the account and photos of termites in the earlier section on social insects.)

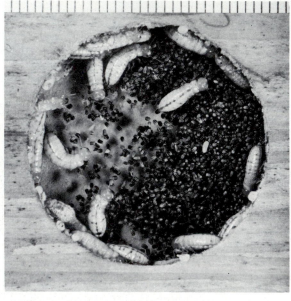

Nasute soldiers, instead of ones with pinching mandibles, defend the colony of *Nasutitermes ephratae* and related termites. When a tree nest was poked with a pencil, making a hole about 1 cm in diameter, a group of nasutes rushed out to do battle. The head of a nasute is prolonged into a pointed nozzle that shoots out a liquid secretion. On contact with air, the secretion becomes viscous and sticky enough to immobilize small enemies. Panama. (R.B.)

Experimental maintenance of small colonies in a hole in a piece of plywood between two sheets of glass makes possible study of physiology, development, and social interactions. The termites feed on the wood, which is digested with the aid of flagellates in the gut. The egg in this colony was laid by a reproductive developed from a worker in the absence of inhibiting pheromones from a queen. The scale is in mm. *Kalotermes.* (R.B.)

Tree nest of termite colony is built around branches of a tree. The nest is made of chewed and partly digested wood that dries hard. Nests of the same species may differ greatly in different ecological areas. *Nasutitermes ephratae;* its nasutes seen in the above photo. Peru. (R.B.)

Small nest in tropical rain forest in Zaire has umbrella-like cap that sheds rain. In dry areas *Cubitermes* makes a simple cap. (R.B.)

Termite mound of *Macrotermes* (fungus growers) houses more than a million termites. *Macrotermes bellicosus* may build mounds 30 m in diameter and 6 m high. (E.S. Ross)

Zorapteran without wings, and without compound eyes or ocelli, is the commonest of the 2 forms regularly found in the 22 species so far described for the small order ZORAPTERA. The winged forms have 2 pairs of membranous wings, compound eyes, and 3 ocelli. Both forms occur in both sexes, and all are minute insects less than 3 mm long. They are gregarious and are found in leaf litter, under bark, in rotting logs, in piles of sawdust. Chewing mouthparts, cerci, and other external characters ally zorapterans with other orthopteroids, but internally they share characters with the hempteroid insects and could be classified with either group. Panama. (E.S. Ross)

Psocids are small or minute soft-bodied insects, most of them under 6 mm long. They comprise the order PSOCOPTERA (so-cop'-te-ra), with close to 2,000 species. As in some other insect groups, there are winged forms that have ocelli and wingless forms that do not. The winged forms can easily be mistaken for tiny flies, but psocids have 2 pairs of wings and are more reluctant to take flight. The convex and prominent compound eyes stick out from the head, and the antennas are relatively long. Most psocids are scavengers, and a majority live outdoors: on bark or leaves of trees and shrubs, under bark or stones, among fungal or algal growths, or in lichens. The psocids most familiar to people are those that enter warehouses and feed on stored cereals, or the wingless species that live in houses and feed on the fungi in old books or the paste in book bindings. These last are called book lice, and this common name is sometimes applied to the whole order. Many psocids favor damp shelters, and winged species that enter houses can often be seen in bathrooms. (R.B.)

Biting lice are small (1.5 to 10 mm), wingless, flattened, external parasites on birds or mammals. They feed with pincerlike mandibles on bits of feather or hair, dead skin, and skin secretions. There are close to 3,000 species in the order MALLOPHAGA. Biting lice mostly infest birds and are sometimes called "bird lice." Some add blood to their diet by biting into living skin and puncturing the living quills of pinfeathers. The common chicken body louse, especially when there are as many as 35,000 on a chicken living in dirty or overcrowded conditions, causes debilitation and susceptibility to disease. Biting lice that attach to the hairs of domestic mammals usually cause only annoying skin irritations, expecially in neglected animals or young puppies. Mallophagans do not attack humans. However, the dog biting louse harbors larval tapeworms and can be the source of tapeworm infection if a louse is accidentally swallowed, especially by children who fondle dogs while eating.

The eggs are attached to feathers or hairs, and the entire life history takes place on the host. Biting lice do not survive long away from their hosts, and most species live on only one or a few closely related hosts—often in a restricted area, like the head or back, that is not easily reached in preening or grooming. Transmission from host to host is usually by direct contact.

The orders Mallophaga and Anoplura (sucking lice) are sometimes combined in the single order Phthiraptera (thir-ap'-te-ra), thought to be evolved from the same scavenging ancestral stock as the Psocoptera.

Biting louse is distinguished from a sucking (anopluran) louse by a head that is broader than the thorax and by strongly sclerotized, pincerlike mandibles. *Cygnus columbianus.* Mounted specimen. Calif. (E.S. Ross)

Sucking lice, a homogenous group with about 400 species, comprise the order ANOPLURA. They resemble biting lice, to which they are closely related, in having a wingless, flattened body with a large segmented abdomen and clawed legs that cling to hair; in both groups the eyes are reduced or absent. Anoplurans feed exclusively on the blood of placental mammals, using piercing and sucking mouthparts that can be withdrawn into the head.

Common human louse, *Pediculus humanus,* is the more medically important of 2 species of anoplurans that suck human blood. The common louse favors cool climates that require wool clothing and people who live in unhygienic, crowded conditions. The male is 2-3 mm long, the female 3-4 mm. This species occurs in 2 varieties that interbreed freely in the laboratory but, on the human body, differ in size, proportions, and distribution. The head louse is smaller and lays fewer eggs (often called "nits"), cementing them to shafts of hair. Washing and combing of hair can reduce or eliminate them. Body lice are a bigger problem because they live mostly in clothing and can survive up to 10 days away from the host. Their larger and more numerous eggs are attached to or scattered in the clothing, which must be heat-sterilized or fumigated. Extreme infestation may involve tens of thousands of body lice. Weeping of the skin, redness, swelling, and low fever are probably caused by allergic reaction to both bites and louse feces. The most serious louse-borne human diseases have been typhus, trench fever, and relapsing fever. The rickettsias or spirochetes are not introduced by the bite of the louse but by the human victim who inadvertently rubs infected louse feces or a crushed louse into an itching bite. During great epidemics that accompanied famines or wars, millions of people died. Epidemic typhus was checked during World War II by use of DDT; since then lice have developed resistance to DDT. (U.S. Army Med. Mus.)

Crab louse eggs, cemented to a shaft of human hair. A female lays 10-30 eggs, and the entire egg-to-egg cycle is completed on the host, within 22 to 30 days at usual temperatures. Prepared slide. (R.B.)

Human crab louse, *Phthirus pubis,* is a blood sucker. It lives mostly among coarse hairs of pubic or anal areas, but may spread to the armpits. In rare cases it may occupy the body from ankles to beard, eyelashes, and eyebrows, but does not invade the fine hair of the head. The broad, short body and the clawed appendages give a crablike appearance to this grayish white louse only 1.5 to 2 mm long. The bites can cause intense irritation and itching. Crab lice move from one host to another by direct contact or by way of clothing or blankets. Neither crab lice nor head lice have been shown to be vectors of disease. (SEM, courtesy P.B. Armstrong)

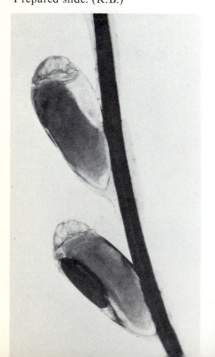

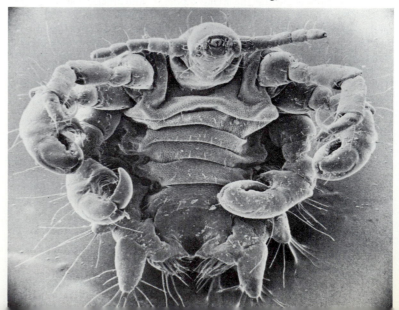

Homopterans include about 35,000 species, many of them familiar garden insects such as leafhoppers, cicadas, spittlebugs, aphids, whiteflies, scale insects, and mealybugs. All have piercing beaks and suck plant juices. In many species, members of either or both sexes may be wingless, but there are typically 2 pairs of wings held rooflike over the abdomen. The order name HOMOPTERA refers to the homogeneous texture of the forewings, which are either entirely membranous like the hindwings or entirely thickened. (This is in contrast to the partly membranous, partly thickened forewings of the bugs, or hemipterans, which have similar piercing beaks. The 2 groups are closely related, and homopterans are often placed with bugs in the order Hemiptera.)

Left, **cicada laying eggs** in a slit in a tree twig, made with her sawlike ovipositor, is *Magicicada septendecim,* the 17-year periodical cicada of the northern USA. The egg-laying damages the twigs, and in years when these cicadas appear in great swarms, whole forests and orchards turn brown. The eggs hatch in about 6 weeks into tiny (1 mm) nymphs that fall to the ground, dig in, and begin to suck the fluids of fine roots. (J.C. Tobias) *Right,* in late May, 17 years later, the **mature nymphs** (25 mm) emerge, climb the nearest tree or shrub, and undergo the final molt. The adults mate, the females lay eggs, and within 6 weeks all are gone. (C. Clarke)

Scale insects and mealybugs are homopterans in which the females are wingless and the tiny males usually have 1 pair of wings. When numerous, they do great damage as they suck the fluids of plants. Female scale insects *(below left)* become immobile with the mouthparts permanently inserted in the host plant and cover themselves with a waxy secretion. These female tortoise scales *(Saissetia)* are legless and have a hard waxy dome beneath which they shelter their eggs and nymphs. The nymphs have been released from a dislodged scale and will crawl away. (R.B.) Female mealybugs *(below right),* covered with a powdery secretion, are exceptions in the group (superfamily Coccoidea), in that they remain mobile and distinctly segmented. Citrus mealybug, southern USA. (C. Clarke)

Left, **spittlebug nymphs** at the base of an artichoke. The frothy excretion protects them from predators, parasites, and drying. Related to treehoppers (p. 787) and leafhoppers, they mature into small spindle-shaped adults. (R.B.)

Hemipterans are the only animals that entomologists call bugs, though most people use the word indiscriminately for almost any small arthropod. The order name HEMIPTERA ("half-wings") refers to the heterogeneous texture of the forewings, which typically are thickened at the base and membranous at the tips. Both pairs of wings are usually held flat across the abdomen, but the wings are variable and may be absent. Hemipterans number about 20,000 species and are commonly flattened in form. When disturbed, members of most species can give off a strong defensive scent, often unpleasant to humans. Many species are amphibious or aquatic. The most consistent feature of hemipterans is the piercing beak, with which they suck plant or animal juices. In this feature they resemble homopterans, which, however, are strictly plant feeders. When both of these closely related groups are combined into the order Hemiptera, bugs comprise the suborder Heteroptera.

Bug is recognized by the X-pattern of its half-thickened, half-membranous forewings, folded across the back. Cactus-joint bug on cactus. (E.S. Ross)

Giant water bugs live in ponds, lakes, and streams, and reach 12 cm in the tropics. They often float at the surface with the tip of the abdomen above the water surface, supplying air to the tracheal system and attracting small fishes and other unwary animals, which take shelter under such a floating "leaf" and are seized and devoured. The female bug attaches the eggs to the back of the male, who carries them until they hatch. *Abedus.* (L. Brubaker)

Marine water strider, *Halobates sericeus,* holds prey in its forelegs and uses the long middle legs like synchronous oars to move on the surface film. There are many freshwater water striders, but only *Halobates* manages to survive on the open ocean, wingless and often hundreds of kilometers from shore. (L. Cheng)

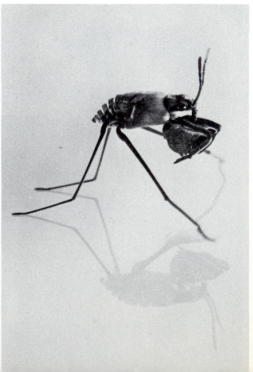

Bed bug, *Cimex lectularius,* drinking human blood. As the bug takes its fill, the expansible abdomen will swell. The slender mouthparts are inserted painlessly under cover of night; only later, when the bug is safely hidden away, does the site become irritating. The hosts of most species are bats and swallows, suggesting that bed bugs have been with us since we lived in caves. (E.S. Ross)

Mexican cone-nose bug, *Triatoma phillosoma pallidipennis,* sucking blood from the hand of the photographer. Species of *Triatoma* (family Reduviidae) transmit Chagas's disease, a form of trypanosomiasis (see p. 23), not through the bite, but when the victim rubs the bug's fecal material into the eyes or mucous surfaces. (E.S. Ross)

Thrips, order THYSANOPTERA ("fringed wings") include 5,000 or so species of mostly herbivorous and minute insects, usually no more than 1 or 2 mm long. A thrips has 2 pairs of narrow straplike wings, each edged with a broad fringe of flexible bristles and similar to those of some minute species in other orders. Many flower thrips can be found easily by tearing open a daisy, dandelion, or clover and watching for dark, shining specks. The wings may glint in the light, but to best see the fringes one should use a microscope (see photomicrograph of a mounted specimen, p. 593). In some species one or both sexes (but usually the males) lack wings.

The metamorphosis is unusual in having features of both main types. The first 2 instars are active nymphlike insects that feed; since they have no external wing buds, they are sometimes called larvas. The next 2 or 3 instars are non-feeding pupalike forms with external wing buds. They are quiescent, and some species pupate within a silken cocoon, but others will move about if disturbed.

The asymmetrical piercing and sucking mouthparts are unique; the strongest of the stylets is the lone left mandible, narrow enough to tear open a single leaf cell or pollen grain. Thrips are important pests of flower gardens, orchards, and field crops, damaging the plants and spreading viruses and bacteria that cause plant diseases. Occasionally, locally abundant thrips irritate human skin, but only a few species can penetrate the skin and suck blood. Herbivorous thrips can be helpful in weed control, and predatory species reduce certain crop pests.

At right are several last-instar nymphs and a dark, slender adult. Amazon Basin, Peru. (E.S. Ross)

The insects on this page and succeeding ones of this chapter represent the great majority of insect species, which belong to the subclass **ENDOPTERYGOTA**. The endopterygote orders have a holometabolous life history with distinct larval, pupal, and adult stages.

Among the most primitive endopterygotes are members of the order NEUROPTERA ("nerve-wings"), named for the network of many longitudinal and cross-veins in the 2 similar pairs of membranous wings. Neuropterans number about 4,700 species and probably represent 2 or 3 natural groups which are sometimes designated as separate orders: (1) alderflies and dobsonflies; (2) snakeflies; and (3) antlionflies, various lacewings, and related families.

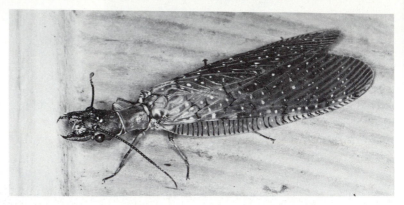

Dobsonflies are nocturnal, and these were attracted to a lighted windowscreen at the Smithsonian Tropical Research Laboratory on Barro Colorado Island, Panama. Similar ones occur in N. America. Despite the prominent biting mandibles in this female *(above)*, and the enormously elongated ones in the male *(left)*, adult dobsonflies do not feed and live only a few days. They are usually seen close to the streams from which the aquatic larvas emerged. The larva (see photo, p.595), called a dobson or hellgrammite, is found under rocks in swift water and uses its strong mandibles to catch nymphs of mayflies and stoneflies or other small freshwater animals. In the water the dobson breathes through tracheal gills on the abdomen. When out of the water, it exchanges air through open spiracles and can survive under stones when streams dry up. After almost 3 years, the larvas leave the water and pupate on or near the shore. In less than a month the adults emerge and mate, and the females lay eggs on objects overhanging the water. (R.B.)

Snakefly has a mobile, elongate head and first thoracic segment that enable it to strike at prey in a snakelike manner. The female shown here has mandibles widespread as she munches on an aphid. The long ovipostor inserts the eggs into slits in bark. The larvas move about under loose bark, preying on small insects. The pupas strongly resemble the adults and are the most primitive type among endopterygotes. After a resting period, they leave their protective chamber and move about actively, seeking a suitable spot in which to settle down until the emergence of the adult. In N. America snakeflies occur only west of the Rocky Mts., and the larvas are found especially under the bark of conifers and, in California, eucalyptus trees. (E.S. Ross)

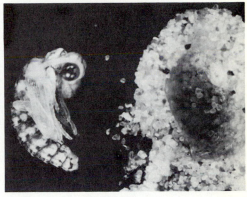

Antlionfly, *Myrmeleon,* raised in captivity. Its larva, an antlion, was collected in the Indiana sand dunes and fed an abundant diet of ants until it pupated. The slender body of the adult measures 4 cm. Antlionflies are nocturnal and shortlived as adults. Hiding in the daytime in low vegetation, they fly out at night and feed on insects with their small biting mandibles. Most species are tropical but some are common in dry places in Europe and the USA. (R.B.)

Antlion, *top,* larva of the antlionfly, builds a pit in sand and lies buried at the bottom with its jaws protruding and prepared to seize any ant or other small arthropod that slips down the slope in an avalanche of loose sand grains. The toothed, sickle-shaped jaws (mandibles and maxillas) pierce the prey, inject digestive fluids, and take in the partly predigested food. When full-grown, the larva spins a silken cocoon. **Pupa,** *directly above,* has been removed from its sand-covered cocoon, about 1 cm diam. (R.B.)

Brown lacewing, *Hemerobius,* eating an aphid *(below left).* Although some lacewings give off a disagreeable odor when handled, these attractive insects should be welcomed in gardens for their help in controlling aphids. The adults and especially the larvas (called aphidlions) feed heavily on these pests. Lacewings are abundant in low vegetation and on the leaves of shrubs and trees throughout summer. (E.S. Ross) The **eggs** of brown lacewings are laid directly on a substrate; those of green lacewings *(below right)* are stalked. (K.B. Sandved)

Beetles make up over 40% of the known species of insects. With close to 350,000 described species, the order COLEOPTERA ("sheath wings") is by far the largest order. The "sheaths" referred to in the order name are the hardened forewings that cover the membranous hindwings. Most beetles are herbivorous, but there are whole families of predatory beetles, many scavengers, some commensals, and even a few parasitic forms. Subsocial behavior occurs in at least 9 families of beetles. Passalid beetles (see photo p. 459) form colonies consisting of a mated pair and their progeny. The adults excavate galleries in damp, rotten logs, and the chewed wood, mixed with digestive secretion, becomes food for the larvas. The colony is kept together by almost continuous stridulatory sounds. In addition to the terrestrial forms, there are many kinds of freshwater beetles. Only a few are marine, and these live in the intertidal.

Resting beetle has hardened forewings meeting in the midline. They are form-fitted to the abdomen so that they completely cover the membranous hindwings that lie folded underneath. (R.B.)

Beetle about to fly has spread the hardened forewings (also called wing-covers or elytra) and has just begun to unfold the membranous hindwings. In flight the forewings serve as stabilizers and provide lift. Panama. (R.B.)

Japanese beetle, *Popillia japonica,* is among the worst of plant pests. Its many relatives in the family Scarabaeidae include the similar but less destructive June beetle, common at lights in early summer. Japanese beetles are attractive bronzy-green insects, native to China and Japan, but scarce there. Accidentally introduced into the U.S. in 1916, they were suddenly freed of their native enemies and spread rapidly. The adult feeds on flowers, fruits, and leaves, defoliating shrubs and trees. Females lay eggs underground. The egg hatches into a whitish grub that feeds on roots, especially of grasses. When present in numbers, the grubs destroy large areas of turf in pastures, lawns, and golf courses. Many beetles have grub-type larvas, but some have elongate ones, hard and brown, also with 3 pairs of legs. Those called wireworms are among the worst of agricultural pests. (U.S Bur. Entom.)

Predatory beetle, *Cicindela oregona,* is called a tiger beetle because of its extreme alertness, aggressive feeding, and the striped pattern of many members of its family, the Cicindelidae. Cicindelids are the most agile of beetles; they fly well and are handsome, with metallic green and bronze coloring and stripes or spots of yellow. The predaceous grublike larva may rise from an opening of its vertical burrow to seize a passing insect. (E.S. Ross)

Spotted cucumber beetle, *Diabrotica,* yellowish green with black spots, eats leaves and flowers, especially of cucumbers, melons, and squashes. In the northern and eastern U.S., they spread bacterial wilt disease by making wounds through which the bacteria enter the plants. (E.S. Ross)

Saw-toothed grain beetle, *Oryzaephilus surinamensis,* is one of many tiny beetles of several families that live in stored flour, cereals, nuts, etc. Worldwide distribution. (R.B.)

Masses of ladybird beetles, *Hippodamia convergens,* hibernate in the hills and canyons of California. In spring they disperse to the valleys and feed on aphids, as do their larvas. They are welcomed by gardeners and fruit-growers. Their red wing-covers warn predators that they are distasteful. (E.S. Ross)

Dung beetle forms a ball of dung, rolls it some distance, and buries it. Balls made early in the year are fed upon by the beetle. Later other balls are made and buried, and an egg is laid on each. The larva feeds on the dung and pupates. A similar scarabid beetle was sacred to the ancient Egyptians and Romans. (E.S. Ross)

Fireflies are beetles (family Lampyridae). Males are attracted to females by exchange of a species-specific code of light signals from thousands of photocytes in the light-colored segments of the abdomen, where many tracheoles supply abundant oxygen for oxidizing luciferin + ATP in the presence of Mg ions and the enzyme luciferase (see p. 371). (C. Clarke)

Darkling beetle is a member of the large family Tenebrionidae. The common name is based on their nocturnal habits, but typically they are a uniform black. This stink beetle, *Eleodes,* raises its abdomen when threatened and emits a disagreeable odor that repels predators. (See p. 463). (E.S. Ross)

Weevil is a member of the family Curculionidae, the largest family in the animal kingdom, with 60,000 species of snouted beetles. The head is prolonged into a long snout that has at its tip a pair of tiny chewing mandibles. About midway it bears the clubbed and elbowed antennas. As almost all weevils are plant-eaters, it is not surprising that thousands of species are agricultural pests as adults or as larvas or both. Familiar weevils are those that feed on cotton bolls, sweet potatoes, strawberries, white pine, and many kinds of stored grains and beans. The plum curculio and its larvas damage plums, cherries, pears, peaches, and apples. Brazil. (R.B.)

Male strepsipteran has just alighted on the abdomen of a wild bee and is poised over a minute female whose abdomen is embedded in that of the bee. He will mate with her, and the fertilized eggs will hatch within her body into 6-legged, active little larvas that crawl out onto the surface of the bee. As the bee makes its rounds, some of the larvas are rubbed off onto flowers and will attach to other, uninfested bees. When the bee returns to its nest, the strepsipteran larvas burrow into the fat bee larvas, molting into legless grubs that absorb nourishment, grow, and finally pupate, along with the developing host. Female strepsipterans never leave their embedded pupal case. The males emerge and fly off to mate. Shown here is *Stylops pacifica,* on the bee *Andrena,* a frequent host. Insects that are stunted or have atrophied gonads because of parasitic strepsipterans are said to be "stylopsized." (E.S. Ross)

Twisted-winged insects, thought to have evolved from beetlelike ancestors, comprise the order STREPSIPTERA (strepsi = "twisted"). Most strepsipterans are parasitic species in which the tiny female is without wings, legs, eyes, or antennas. She lives with her abdomen inserted into the abdomen of her insect host, often a wasp or bee; only her vestigial head and thorax protrude between the host's segments. The winged male leaves the host and devotes the whole of his brief life to searching out a female with which to mate. Strepsipterans occur on all continents and include over 300 species.

Scorpionflies, order MECOPTERA *(meco* = "long," *ptera* = "wings")* now include fewer than 400 species, only a remnant of the varied fossil forms of late Paleozoic times when mecopterans were one of the 3 largest orders of insects. Scorpionflies reach 25 mm or more, and most have 2 similar pairs of long narrow wings, usually with conspicuous bands or spots. The most distinctive feature of this order is a prolongation of the head capsule and chewing mouthparts into a tapering beak with the mandibles toothed near the tip. Scorpionflies may be scavengers, predators, or herbivores. They are found most often along the banks of shaded streams and in damp woods, where in some areas they may be the most abundant of the sizeable insects. Most species lay their eggs in soil. The caterpillarlike larvas have compound eyes. These eyes, and the cerci of the adults, are unusual and primitive features among endopterygotes.

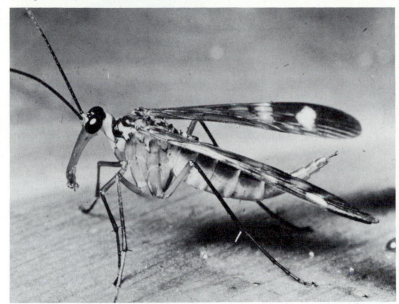

Winged scorpionfly, *Neopanorpa tuberosa,* is typical of the most widespread group. In the males the tip of the abdomen is bulbous and recurved, suggesting the sting of a scorpion, hence the menacing common name applied to all members of this harmless order. In the female, seen here, the last segments of the abdomen are telescoped, and the 2 cerci can be seen at the tip. Khao-Yai National Park, Thailand. (E.S. Ross)

Wingless scorpionfly, *Apterobittacus apterus,* is a California species with no wings. It belongs to a long-legged group with prehensile tarsi. They are called hangflies because they spend much of their time suspended by their forelegs while using the other legs to catch small flies, aphids, and caterpillars or sometimes spiders. (E.S. Ross)

Flies differ from most other insects in having only one pair of wings, the 2 membranous forewings referred to in the order name DIPTERA. The hindwings are represented only as small stalked knobs, the halteres, which serve as organs of equilibrium (see photo 591). Flies are the fourth largest order of insects, with about 85,000 species. The mouthparts are adapted to taking liquid food, either by lapping or sponging it up or by sucking it up, often after piercing the surface to suck blood, liquefied tissues, or body fluids. Fly larvas are legless and wormlike, either terrestrial or aquatic. The abundance of flies, the annoyance some cause, and the serious diseases they transmit (malaria, trypanosomiasis, dengue, viral encephalitis, filariasis, etc.) make it hard to appreciate the helpful roles played by both adults and larvas. Many pollinate useful plants. The terrestrial larvas called maggots are important scavengers that recycle decaying plants and animals. And many flies prey on or parasitize harmful insects.

Hover fly, or flower fly, is one of many syrphid flies common in gardens and meadows. Feeding on nectar and pollen, these dipterans pollinate many plants. (R. B.)

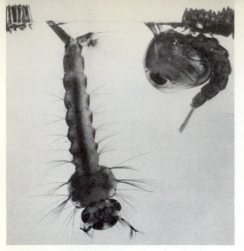

Metamorphosis of a mosquito. *Left,* 2 rafts of **eggs** float at the water surface. An elongate **larva** hangs head down, with a breathing tube at the posterior and emerging through the surface film. Two pairs of small tracheal gills on the last abdominal segment project into the water. At the fourth molt, the larva becomes a **pupa** with a large head and thorax, not distinctly separated. Two small trumpet-shaped breathing tubes project from the thorax to the surface. The abdomen is slender and flexible and at its tip is a pair of leaflike appendages with which the pupa swims about, but it does not feed. *Right,* after 2 or 3 days the cuticle splits down the back and the winged **adult** works itself out. It rests on the floating pupal cuticle until its wings dry. This mosquito, *Culex tarsalis,* transmits western equine and St. Louis viral encephalitis. (E.S. Ross)

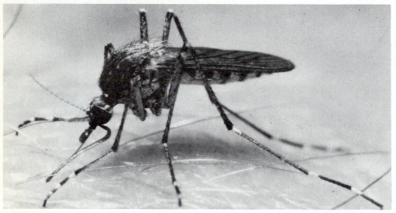

Female mosquito sucks blood with a proboscis composed of 6 long styletlike parts: a sharply pointed labrum that forms an inverted gutter up which blood is sucked, 2 thin piercing mandibles, 2 maxillas with saw-toothed tips, and a hypopharynx that encloses a salivary channel. As these are inserted through the skin, the troughlike labium that encloses the proboscis is pushed up and folded into a loop, as seen here. Male mosquitoes suck nectar and ripe fruit juices, and so do females when no blood is easily available. This Australian mosquito belongs to the genus *Aedes;* a related species, *Aedes aegypti,* transmits yellow and dengue fevers. (E.S. Ross)

Pinecone willow galls are plant growths induced when gall midges, *Rhabdophaga strobiloides,* lay their eggs in the twig tips of the heart-leaved willow *(Salix cordata).* The larva remains in the gall through summer and winter, pupating in spring. Shortly after, the adult emerges and lays eggs in the new willow buds. (L.W. Brownell)

Robber fly has been called the hawk of the insect world because of its sharp vision and the suddenness with which it leaves its perch to attack any suitable insect that flies by. Here it feeds on a leafhopper (homopteran), an easy prey. But this large fly can take on even bumble bees, wasps, dragonflies, tiger-beetles, and grasshoppers. The pointed proboscis stabs the prey and injects saliva that liquefies prey tissues, which are then sucked up. (E.S. Ross)

Right, **head of a drosophilid fly** has large red eyes, short antennas with sensory bristles, and sponging mouthparts that are suspended from a forward extension of the head, the rostrum. At the distal end of the rostrum are 2 tiny maxillary palps and the labium, which has at its tip 2 large soft lobes. Transverse grooves on the lower surfaces of the lobes serve as food channels. Though often called fruit flies, drosophilids do not feed on fruit, but on yeasts that grow in damaged or fermenting fruit. No other animal has been as important as *Drosophila melanogaster* in genetics, the study of heredity, chromosomes, and genes. (R.B.)

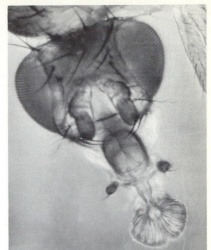

Mating tsetse flies, *Glossina,* the smaller male on top. The larvas develop in the mother's uterus and are deposited on the ground in a shady area. The larva burrows in and pupates; the adult emerges in about a month. Tsetse flies transmit trypanosomiasis (p. 23-24) in tropical and subtropical Africa; the disease occurs only near patches of damp, shady bush. Tanzania. (E.S. Ross)

Mediterranean fruit fly (medfly), *Ceratitis capitata,* rests with outstretched wings. When its eggs are laid in fruits, the feeding larvas can devastate orchards. (S. Whiteley)

Luminescent larva, *Arachnocampa luminosa,* known as the New Zealand glowworm to visitors who see the spectacular luminescence in Waitomo Caves. Their light attracts insect prey that are trapped by sticky droplets secreted along hanging silk threads. The larvas live in caves or windless forest niches, where their hanging snares will not get tangled. The transparent larva *(arrow),* 4 cm long, is barely visible in its trap. The adult gnats do not feed. (R.B.)

Giant tropical fly, a pantophthalmid. Some members of this tropical American family reach lengths of 5 to 8 cm. Fortunately, they do not bite, as do the related tabanids, or horse flies. Panama. (R.B.)

Fleas, order SIPHONAPTERA, are tiny, laterally flattened, and wingless ectoparasites that suck blood intermittently from their warm-blooded vertebrate hosts. Almost 95% of flea species feed on mammals and roam freely among the hairs; most of the rest feed on birds, and a few attack both. Despite their feeding habits, fleas are not closely related to sucking lice. Fleas have holometabolous development and probably come from the same ancestral stock as dipterans. The whitish wormlike larva has biting mandibles; it scavenges on organic matter, especially the feces of adult fleas, and preys on minute insects. When fully developed, the larva spins a cocoon and pupates. The adult sucks blood with a proboscis that has 3 piercing stylets. Fleas are not very specific in their choice of hosts, and people who share their homes with furred pets must be resigned to sharing a few little siphonapterans. Fleas can survive off their hosts for weeks or even months without feeding, and they usually leave the host to lay eggs in the debris of the host lair or nest. Fleas are medically important as carriers of plague and endemic typhus because many rodents serve as reservoirs of these diseases, and their fleas can readily transfer to humans. The usual vector is not the human flea but the oriental rat flea, *Xenopsylla cheopis.* Fleas serve as intermediate hosts for certain tapeworms and occasionally pass them on to humans (see p. 255).

Human flea, *Pulex irritans,* also parasitizes pigs and other domestic animals. The legs are long and stout, and the basal article (coxa), small in most insects, is enormously enlarged. The hindlegs are adapted for jumping to escape danger or to mount a new host. *Pulex irritans* is a champion among fleas, observed to jump 33 cm horizontally and 20 cm vertically (equivalent to a high jump of about 85 m for a person of average height). Found in all parts of the world, it flourishes in climates with cool humid summers and mild wet winters, and so shares with many humans a special fondness for northern California. Preserved specimen. (R.B.)

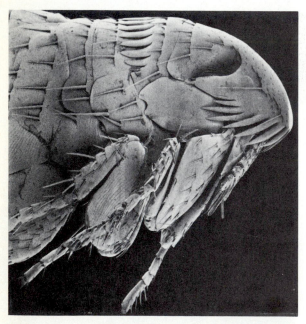

Anterior end of **cat flea,** *Ctenocephalides felis.* The body cover of backwardly directed bristles ease forward movement in hair but prevent slipping backward, and usually frustrate removal by the desperate scratching of the host. Cat fleas and dog fleas *(C. canis)* are the fleas most likely to make life miserable for humans. Getting rid of an infested dog or cat, without clearing the premises of the bereft and hungry fleas, may result in sudden infestation of humans where none occurred before. SEM. (Courtesy P. Armstrong)

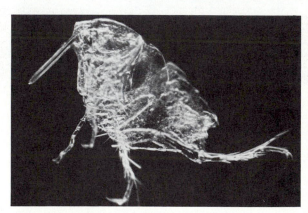

Chigoe flea, *Tunga penetrans,* a reddish-brown pest of subtropical and tropical America, occurs as far north as the southern USA and has been spread to the West Indies, Africa, and elsewhere. The males and young females are about 1 mm long and have habits like other fleas. They favor the feet of humans and pigs, but will attach to dogs, cats, or rats that come along as the fleas lie in wait in shaded sandy soil or the earthen floors of houses. It is the mature female chigoe that causes real trouble. When she attaches tightly to the skin, commonly under a toenail, the irritation causes swelling of host tissue until the flea, now with an egg-filled abdomen the size of a small pea, becomes completely surrounded, with only the tip of the abdomen reaching the surface as the eggs are expelled. Though the female is finally evicted by the host, the infected wound may lead to severe complications. Wearing shoes is an effective precaution. (R.B.)

Caddisflies are small, mothlike insects thought to be close to the ancestral stock of moths and butterflies. The 4 wings are covered with hairlike bristles that suggested the order name TRICHOPTERA. There are about 5,000 species, ranging in size from 1.5 to 40 mm long. Caddisflies are found near almost any body of freshwater, especially between May and September. Most live less than a month and probably do not feed. These insects would be of little ecological importance except that their aquatic, feeding larvas, which overwinter, form a significant part of the food supply for many fishes and other aquatic animals.

Female caddisflies usually lay their eggs in the water, but some deposit the eggs on nearby vegetation, out of reach of aquatic predators. The aquatic larvas, popularly called caddisworms, usually secrete a silken, tubular or trumpet-shaped case and cover it with sand grains, minute pebbles, or bits of plant debris. The cases of many species or families are distinctive. This habit probably gave rise to the name "cadiseworm," in 15th or 16th century England, when cotton and silk were often referred to as "cadise," and "cadise men" were itinerant vendors who advertized their trade by attaching to their coats many bits of cotton or silk ribbons, braids, and yarns.

Most caddis larvas drag their cases along as they walk about, feeding mostly on decaying organic matter. Predaceous forms either build no case and roam freely, or construct a nonportable retreat of pebbles, near which they spin silken snares for prey. When ready to pupate, the larva seals its case or secretes a special pupal one. Some 2 weeks later, the fully developed pupa uses its mandibles to cut its way out of the case, swims to the surface, and crawls out of the water onto a stone or stick. The winged adult then emerges.

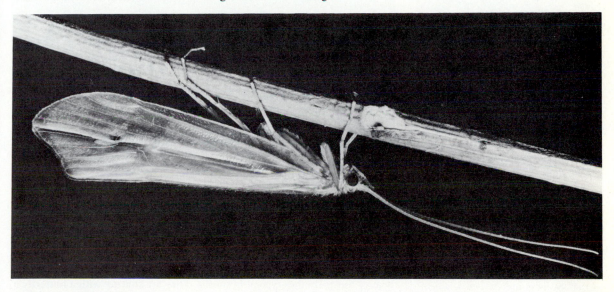

Caddisfly adult resembles a small moth. It holds the wings rooflike over the body when at rest, has long threadlike antennas and large compound eyes, and feeds at dusk or during the night. The larval mouthparts are of the chewing type, but the adult mandibles are much reduced, and those adults that feed must lap up liquids. *Psychoglypha.* California. (E.S. Ross)

Caddis larva with case made of plant debris. When it moves, the larva protrudes its head and 3 pairs of thoracic legs from the anterior opening of its case (at right). At the posterior end, a pair of hooked appendages anchor the larva to its case. Most case-building larvas have tracheal gills along the abdomen, but in some species these can be removed without seriously reducing respiratory exchange, which takes place mostly through the body surface. Lake Pymatuning, Pennsylvania. (R.B.)

Colored pebbles supplied by a student were incorporated by this caddis larva into its case. This made it possible to identify each individual and to follow its movements and behavior. Univ. of Pittsburgh, Lake Pymatuning Laboratory of Ecology, Pennsylvania. (R.B.)

Moths and butterflies, with more than 100,000 species, make up the second largest order of insects, the LEPIDOPTERA ("scale wings"). Overlapping rows of scales cover the wings, and body and legs are covered by similar scales or by long hairlike scales or by hairlike bristles. Lepidopterans are noted for the beautiful colors and patterns of their wings, but the wings themselves are membranous and colorless; the color is almost always in the scales, which are easily detached and enable lepidopterans to escape from spider webs. The adults have large compound eyes with many facets, and usually also 2 ocelli; their mouthparts typically form a tubular proboscis through which liquid food, mostly flower nectar, is sucked up. In contrast, the larvas have only simple eyes, commonly 6 on each side; and they feed with chewing mandibles on leaves or other plant parts. From modified salivary glands that open on the labium, the larvas spin silk used in the construction of larval shelters or cocoons in which to pupate. Most butterflies have no silken cocoon but pupate in a membranous covering (chrysalis; see metamorphosis of monarch butterfly, p. 600).

Wing scales cover much of a butterfly wing in neat overlapping rows like the tiles on a roof. In patterned areas (as in this closeup), the scales vary in shape, size, color, and arrangement. *Parnassius smintheus.* Colorado. (K.B. Sandved)

Cecropia moth, *Hyalophora (= Platysamia) cecropia.* Below, a pupa, exposed by opening its heavy silken cocoon; the branched antennas and developing wings are pressed close against the body. (L. Keinigsberg). Right, a newly-emerged adult rests near its now-empty cocoon while its wings dry. Wingspread 16 cm. (R.B.)

Distinctions between moths and butterflies are not absolute. Butterflies are mostly dayfliers, hold the wings together and vertical when at rest, and have clubbed antennas. Moths—the great majority of the order—are mostly nightfliers, hold the wings spread down when at rest, and have filamentous or feathery antennas. The 2 dayflying moths in this mid-day photo follow only the last of these rules, but they do display the heavy, "furry" bodies typical of moths. They are sharing with a butterfly a patch of wet soil, all sipping water and minerals from the wet mud. Peru. (E.S. Ross).

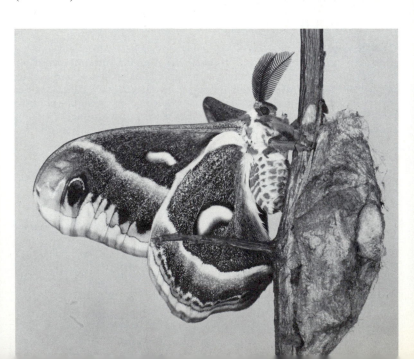

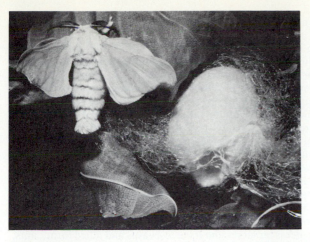

Larva of a lepidopteran has 3 thoracic segments, each bearing a pair of clawed legs; these become the legs of the adult. On the 10 abdominal segments are 5 pairs of "prolegs," including (on the last segment) a pair of claspers. There are 9 pairs of spiracles. Corn-ear-worm. (P.S. Tice)

Silkworm moth, *Bombyx mori,* is the only completely domesticated insect; it is no longer known in the wild. The silk obtained by unwinding the cocoon of this moth still brings a high price. In addition, the caterpillars (silkworms) can be infected with a virus that has been altered to carry a human interferon gene. As silkworms can be grown in great numbers, this system shows promise for producing human interferon to be used in medical treatment. (W.E. Ferguson)

Butterfly laying eggs. She bends the abdomen sharply to deposit the eggs on the underside of a leaf, where they are protected from the harmful effects of direct sun. California. Checkerspot butterfly, *Euphydryas chalcedona.* (E.S. Ross)

Aggregation of monarch butterflies, *Danaus plexippus,* overwintering in a pine tree along the Pacific coast in California, where the nights are frostless and the winter climate cool and moist. The butterflies drink at streams and sip nectar in nearby fields and gardens. Their gonads do not mature until spring. Monarchs belong to a tropical family but have managed to move northward, over millions of years, following the northward spread of various milkweed species on which the butterflies lay their eggs and the larvas must feed. Unlike northern butterflies, monarchs have no developmental stage capable of surviving periods of below-freezing temperatures. As fall approaches they migrate long distances to the south. Those that feed on various flowers in grasslands, pastures, and gardens west of the Rocky Mountains, in the USA and southern Canada, overwinter in coastal California. Those from east of the Rockies migrate even farther to a forest area at high altitude in central Mexico. (E.S. Ross)

Sawflies, wasps, ants, and bees are all included in the 100,000 or so species of the order HYMENOPTERA. The order name (derived from Hymen, the Greek god of marriage) was suggested by the coupling of the 2 wings on each side. The hindwings are smaller and each has along its front edge a row of minute hooks that fasten it to the forewing so that the 2 beat together. Chewing mandibles are universally present, but except in predaceous hymenopterans the mandibles are more often used for cutting the insect's way out of the pupal case or for nest-building rather than for feeding. In the more highly evolved forms (such as the bees) labium and maxillas form a long "tongue" that takes up liquid food. The ovipositor is well developed and in the more evolved forms is modified into a sting. Only in certain families do the females sting, and of these the most familiar are the honey bees, yellow-jackets, and hornets. Larvas of hymenopterans are mostly grublike (with thoracic legs only) or maggotlike (with no legs). The pupas may be in cocoons or in special cells; or in parasitic forms they may develop in the host. The fertilized eggs of hymenopterans usually develop into females, the unfertilized eggs into males. Unlike the termites, with workers of both sexes, all hymenopteran workers are females.

The hymenopterans are among the most economically important and varied of the insect orders, and many show complex behaviors, including the social organization of certain wasps and bees, and of all ants.

The order is divided into 2 suborders. The smaller and more primitive one includes the sawflies and their allies (named for the sawlike ovipositor in the commonest group). In this suborder the larvas are caterpillar-like, with both thoracic legs and abdominal prolegs. The other suborder is far larger and includes wasps and allied forms, ants, and bees.

Solitary wasp, *Ammophila*, is one of the few animals listed as tool-users. This thread-waisted (sphecid) wasp is using a pebble to tamp down the soil over a vertical burrow that she has provisioned with lepidopteran larvas, on each of which she has laid an egg. When the wasp eggs hatch, the young will feed on the prey provided by the mother and prepared by the most elaborate method known for insect provisioning. The wasp stings the prey in many sites and then repeatedly compresses the neck. The toxin introduced by the stinger paralyzes the prey without killing it, so that the tissues stay fresh for many weeks. Utah. (E.S. Ross)

Braconid, *Apanteles*, belongs to a large family of wasplike parasites that are especially beneficial to the human economy as they parasitize many serious insect pests of plants. The braconid larvas that developed inside this caterpillar of a noctuid moth in California have gnawed through the skin of the back to pupate externally. One adult has just emerged from its silken cocoon. (See also p. 457) (E.S. Ross)

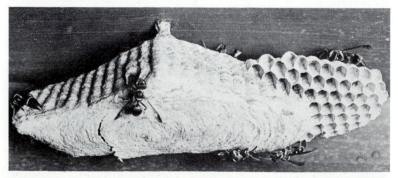

Above right, **social wasp workers** on a nest constructed of masticated wood and suspended from the underside of a palm leaf in a Panama rain forest. The nest of this species consists of a single comb of hexagonal cells. Those at the right are open at the bottom and shelter larvas. The cells at the left were closed at the bottom when that brood of larvas began to pupate. The workers feed mostly on nectar and fruit, but they go out to collect insects and other small animals that they feed, in partly masticated portions, to the growing larvas. Workers maintain the nest and care for the brood, the egg-laying queen, and other female and male reproductives. Most "paper wasps" (vespids) are tropical or subtropical, but similar small vespid colonies, as well as the large paper nests of yellow-jackets and hornets, are familiar in temperate climates. (R.B.)

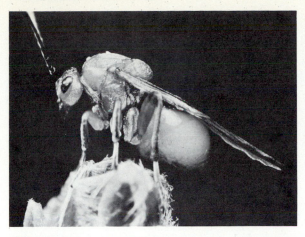

Oak gall wasp *Andricus californicus* is a cynipid wasp (a member of a family of mostly parasitic, parasitoid, and gall-inducing wasps). Seen here on the tip of an oak bud, the wasp is only 5 mm long and reddish brown with transparent wings. This species lays its eggs in the twigs of white oaks, from Washington to Mexico, inducing a large tumorous plant growth, or gall, known as an *oak apple*. Seen here in summer, an oak apple induced by *Andricus* is large (up to 10 cm in diameter), succulent, glossy, and red. In spring it was smaller, kidney-shaped, and green. By fall it will be dry, brown, and papery. The developing larvas occupy separate cells at the center. In fall the adults emerge and lay eggs that will overwinter and develop the next spring. California. (E.S. Ross)

Mud-dauber wasp *Sceliphron caementarium* is a solitary, thread-waisted sphecid wasp. The female builds a nest of kneaded mud attached to sites protected from rain, such as the underside of an overhanging rock (as here) or to an eave, a wall, or a ceiling of a building. The nest cells, each about 25 mm deep, are arranged in a row and then covered over with a layer of mud. The **opened nest of a mud-dauber wasp** shows, from left to right, a pupa, a mature larva beginning to pupate, and 2 other cells each provisioned with spiders. A young wasp larva is seen clinging to a leg of the larger crab spider. The spiders have been stung and are paralyzed. California. (L. Passmore)

Digger wasp, *Eucerceris,* is a solitary, burrowing sphecid wasp. It has stung and paralyzed a small weevil and will store it in its burrow, with other weevils, as food for its larvas. Many species of digger wasps take only certain prey, for example, weevils or buprestids. California. (E.S. Ross)

Ants are all social. The more than 7,600 species comprise one large family, the Formicidae. The workers are wingless; the reproductives are usually winged when they first emerge, but the queens shed their wings after mating. In primitive ants, such as the ponerines of the American tropics, there may be only a few dozen members in a colony and little size differentiation between the castes. In the complexly organized colony of a driver ant species (African counterpart of the army ants of tropical America) the workers have been estimated at 22 million. Large ant colonies may have well differentiated workers of several sizes, large-headed soldiers, a huge queen, and a seasonally produced brood of male and female reproductives that leave the nest to mate. The queens found new colonies. Ants evolved from wasps but are distinguished by a hump on the narrow waist of the abdomen and by the elbowed antennas, which usually have a long basal article. Panama. (R.B.)

Fire ant queen with her progeny of workers and brood of larvas and pupas. Ants do not care for their young stages in separate cells, as do wasps and bees, but tend them in common nest chambers. Ants in many species move the brood from one chamber to another with changes in temperature and humidity. This red species, *Solenopsis invicta,* introduced into the southern USA with lumber shipments from Brazil, has made millions of hectares unusable for farming. Chicks, piglets, and calves succumb to the potent venom of the sting, and humans suffer severe pain and medical problems, occasionally with fatal result. Control has leaned mostly on some highly toxic chemicals. (K.G. Ross)

Predatory ant, *Daceton,* on a forest floor in Peru, has seized a young grasshopper nymph in its long mandibles. Many ants are carnivorous and are ecologically important both as scavengers and as predators, especially on other insects, including many pests. In England, a study of a wood ant nest, monitored by electronic counters, showed that in just 1 hour the ants brought back to their nest 102,000 insects. Trees ringed with grease, to prevent ants from preying on insects in the foliage, grew 30% less. (E.S. Ross)

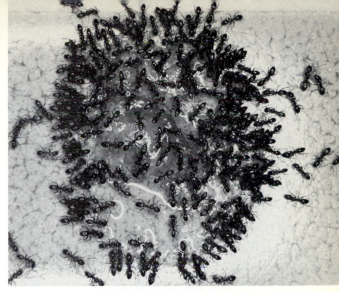

Slave-making ant, member of a raiding party of *Polyergus.* The raiders have just killed the workers in another ant nest *(Formica)* and are stealing the pupas, perhaps eating a few but carrying most back to the *Polyergus* nest to emerge there as slave workers that do all the chores. The slavemakers have saber-shaped mandibles suited for fighting, and they are entirely dependent on their slaves for nest repair, brood and queen care, and collection of food. As the slaves cannot reproduce, they must be regularly replaced by raids. When a fertilized female of *Polyergus* founds a new colony, she enters an alien nest, usually of *Formica,* and kills the queen. The *Formica* workers then adopt the conquering queen, and she proceeds to lay eggs in the nest. California. (E.S. Ross)

Argentine ants, only 1 to 2 mm long, are feeding on a drop of honey in a kitchen, having entered after a rain that disturbed their nest in the ground outdoors. Established around the world, they were introduced to warm parts of the USA by coffee ships from Brazil docking at New Orleans. *Iridomyrmex humilis.* (R.B.)

Leaf-cutting ants, *Atta,* returning to their underground nest, carry pieces of leaves cut from trees or shrubs. The tiny workers seen hitch-hiking on leaf pieces protect the larger leaf-carrying workers from parasitic phorid flies by snapping their mandibles and fencing with their hindlegs. Back in the nest, the tiny workers chew up the leaf pieces and add them as fertilizer to fungus gardens in the nest. The fungal mycelia produce round bodies on which the ants feed. Panama. (R.B.)

Ants attending aphids, a mutualism. Aphids exude a sweet fluid, honeydew, as they suck plant fluids. Ants gather it from leaves or suck it directly from the rear end of aphids, which appear to welcome the ants' attentions and secrete more honeydew when stroked by the ants' antennas. Ants defend their aphids from predators, and some carry aphids away to safety, build shelters over them, or take them underground in winter and replace them on their food plants in spring. Calif. (E.S. Ross)

Bumble bee, *Bombus,* is social. Bumble bees have a long tongue and can take nectar from deep flowers inaccessible to honey bees, so are important pollinators. Body bulk and "furry" covering permits bumble bees to live even north of the Arctic Circle. Bees (superfamily Apoidea) eat nectar and pollen, feeding a similar diet to their young. Of the 20,000 species of bees, most are solitary, each female caring for her brood, usually in a burrow in the ground. (R.B.)

CLASSIFICATION: Phylum Arthropoda, Subphylum Uniramia, Superclass INSECTA (HEXAPODA)

WINGLESS INSECTS (APTERYGOTA)

Class PROTURA. Proturans.

Class COLLEMBOLA. Springtails or garden fleas.

Class DIPLURA. Diplurans.

Class THYSANURA. Bristletails, silverfishes, firebrats.

WINGED INSECTS

Class PTERYGOTA

Subclass EXOPTERYGOTA. Hemimetabolous life history.

PALEOPTERAN ORDERS

Order Ephemeroptera. Mayflies.

Order Odonata. Dragonflies and damselflies.

ORTHOPTEROID ORDERS

Order Plecoptera. Stoneflies.

Order Grylloblattodea. Grylloblattodids.

Order Orthoptera. Grasshoppers, crickets, katydids.

Order Dermaptera. Earwigs.

Order Phasmida. Stick-insects.

Order Embioptera (Embiidina). Web-spinners or foot-spinners.

Order Blattodea. Cockroaches.

Order Mantodea. Mantids.

Order Isoptera. Termites.

HEMIPTEROID ORDERS

Order Zoraptera. Zorapterans.

Order Psocoptera. Booklice.

Order Mallophaga. Biting lice.

Order Anoplura. Sucking lice.

Order Homoptera. Cicadas, aphids, leaf-hoppers, tree-hoppers, spittlebugs, scale insects, mealybugs.

Order Hemiptera (Heteroptera). Bugs.

Order Thysanoptera. Thrips.

Subclass ENDOPTERYGOTA. Holometabolous life history.

NEUROPTEROID ORDERS

Order Neuroptera. Dobsonflies, snakeflies, lacewings, antlions.

Order Coleoptera. Beetles.

Order Strepsiptera. Twisted-winged insects.

MECOPTEROID ORDERS

Order Mecoptera. Scorpionflies.

Order Diptera. Flies.

Order Siphonaptera. Fleas.

Order Trichoptera. Caddis flies.

Order Lepidoptera. Butterflies and moths.

HYMENOPTEROID ORDER

Order Hymenoptera. Sawflies, braconids, ichneumons, chalcids, wasps, ants, bees, etc.

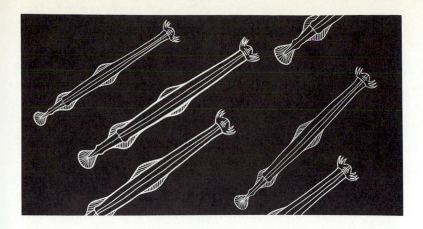

Chapter 25

Chetognaths

Planktonic predators

C RYSTAL-CLEAR and colorless, the streamlined little torpedo-shaped "arrow-worms" are nearly invisible as they dart, by the billions, through the surface waters of all seas. In deeper waters, down to 1,000 m or more, they may range in color from delicate pink to red, as do so many animals from the depths. Although the number of species in the phylum **CHAETOGNATHA** is small (estimates range from about 50 to 80), chetognaths *(keet'-og-naths)* are among the most abundant members of the plankton and are important predators, feeding especially on copepods and other small or larval crustaceans, on young fishes (including commercial species), and on each other.

The transparency of chetognaths helps to conceal them from prey and predator alike as they float motionless in the water, stabilized by horizontal fins (1 or 2 lateral pairs and a tail fin). There are 2 clusters of simple pigment-cup eyes on the dorsal side of the head, but they are not image-forming; chetognaths detect and locate their prey by means of bunches of sensory cilia distributed along the body. When the ciliary receptors are stimulated by vibrations in the water, such as are made by a copepod swimming close by, the chetognath darts forward and strikes, seizing the prey with large grasping spines on the head that give the phylum its name (*chaetognatha* = "bristle-jawed").

The turgid cylindrical body is divided into head, trunk, and tail by transverse septa. There are no special respiratory, circulatory, or excretory systems. The nervous system contains 2 principal ganglia, a dorsal ganglion in the head and a large ventral ganglion in the trunk, which communicate by means of long connectives around the digestive tract. The straight gut extends the length of the trunk, and the anus opens just in front of the trunk-tail septum. The gut is suspended by dorsal and ventral mesenteries that divide the trunk coelom into halves, each containing an ovary. The tail coelom is likewise divided by a longitudinal septum, and each half contains a testis. All

Chetognath catching a copepod. Rows of small teeth on the head assist the grasping spines in prey capture. The victim will be swallowed whole. Western Samoa. (K.J. Marschall)

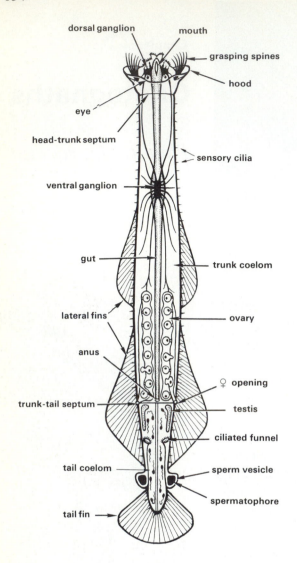

Chetognath in ventral view. During swimming the spines are folded tightly against the head; and the hood, an epidermal fold, is pulled over them for streamlining. Chetognaths commonly measure 2 to 3 cm long; a few are over 10 cm.

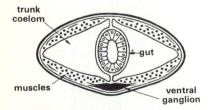

Cross-section through trunk of a chetognath shows the ventral ganglion, the dorsal and ventral mesenteries suspending the gut in the trunk coelom, and muscles of the body wall. (Mod. after Delage and Hérouard)

chetognaths are hermaphrodites, but it is not known to what extent self- and cross-fertilization occur under natural conditions.

Like many fragile planktonic animals, chetognaths are difficult to observe under natural conditions or to keep alive in the laboratory. Much of their biology and behavior must therefore be pieced together like a puzzle with indirect and fragmentary clues. Plankton samples are collected at a great number of locations, depths, and times of the year. The chetognaths present are identified and measured, and their reproductive state determined. The other members of the planktonic community are noted. And water conditions such as temperature, salinity, and nutrient content are also recorded. From such data, the diets, life histories, and distributions of the animals may be cautiously inferred. For example, the size and maturity of individuals of many species are found to increase with depth, implying that the animals either sink or actively swim downward as they grow. Often superimposed on this pattern are daily cycles of vertical migration. Chetognaths are sensitive to temperature and salinity; different species are thus good indicators of water and fishing conditions and are used by marine hydrographers to trace and identify water masses.

Individuals of *Sagitta elegans* are distributed in relatively shallow depths from the Arctic through the north Atlantic and north Pacific oceans. Those from temperate latitudes seem to grow rapidly, become mature at relatively small sizes, and produce 5 or more generations in a single year. Those in Arctic waters grow slowly, become much larger, and require 2 years for a single generation. Since such large animals can produce many more eggs, their slow growth is not necessarily a disadvantage, but may rather be a positive adaptation to the more seasonal food supply or other conditions at high latitudes. Similar patterns have been recognized in other species of chetognaths and of several other phyla.

The reproductive system of chetognaths is unusual in that the gonads have no direct connection to the outside. The testes release cells that develop into mature sperms as they circulate in the tail coelom. Ciliated funnels that open into the coelom collect the sperms, which pass through ducts to the sperm vesicles to be packaged into spermatophores. At spawning the spermatophores are exposed by rupture of the epidermis. The

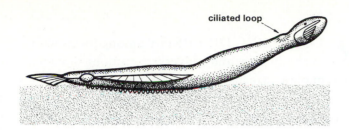

ciliated loop

sperms (whether from the same or another individual) are introduced through the female openings and fertilize the eggs within the ovary, reaching them through stalks that attach the eggs to the ovary wall. The fertilized eggs pass through the ovary wall, apparently squeezing through small spaces between cells, and into a duct from which they are spawned to the outside in clusters held together by a gelatinous secretion. The eggs of pelagic chetognaths may be carried on the parent's body or may float free. They develop directly into young chetognaths that gradually attain adult morphology as they grow.

It is difficult to relate chetognaths to any other animals. They are usually placed close to the echinoderms, to be treated in a later chapter, because of similarities in early development (the fertilized egg undergoes radial cleavage; the mouth opens at the end opposite to the blastopore; and the coelom arises as paired cavities cut off from the developing gut). The later development and adult structure of chetognaths, however, show a baffling mixture of characters that conceals their affinities. In life style, chetognaths show many parallels with ctenophores. Transparent and pelagic throughout their lives (except for a small benthic minority in both phyla), these hermaphroditic hunters show us two different paths to becoming planktonic predators.

Benthic chetognath, *Spadella cephaloptera,* is an oddity in this almost wholly planktonic phylum. About 1 cm long, the animal is found in tide pools from Britain to the Mediterranean, attached to seaweeds or to the bottom by adhesive cells along the ventral tail region. It deftly snatches passing copepods without budging from its attachment, but can move to new sites. The ciliated loop, on the dorsal side of head and trunk, is characteristic of chetognaths and usually considered sensory. (After D.A. Parry)

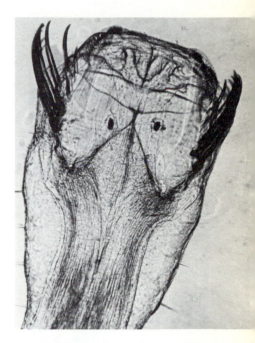

Head of a chetognath, showing large grasping spines on each side and rows of small teeth. Two dorsal clusters of simple eyes show as dark spots. Bands of longitudinal muscle run down the center. Bimini. (R.B.)

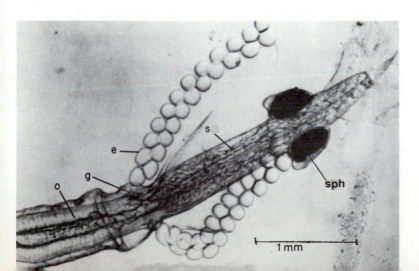

Eggs being extruded by an individual of *Sagitta hispida* (only the rear half of the animal is shown). **sph,** spermatophore; **o,** ovary; **g,** ♀ female genital opening; **e,** egg cluster; **s,** sperms. (M.R. Reeve)

Chapter 26

Lophophorates

Phoronids
Brachiopods
and Bryozoans

and
Kamptozoans

lophophore = "crest or tuft bearer"

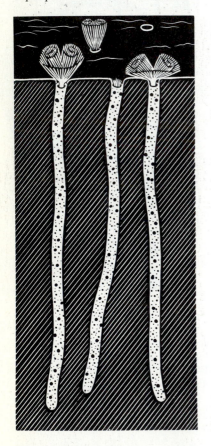

Phoronids in sand-encrusted tubes, within burrows in soft sediment. Only the tentacles extend into the overlying water.
Phoronis is a surname of *Io* and *Isis* in Greek and Egyptian mythology, reflecting the importance of the classics in the education of the 19th century naturalists who described and named many animals.

RECOGNIZING DIVERSITY among forms with a common body plan, and convergence among those with quite different plans, are major themes in the study of invertebrates. But, like naturalists of the early 19th century, readers may still be startled to learn that some tentaculate worms, clamlike animals, and hydroidlike forms are neither polychetes, molluscs, nor cnidarians, but instead they are probably related to each other. All are sessile or sedentary as adults, and they feed by means of a system of delicate, ciliated tentacles, that collect phytoplankters, as well as other small organisms and particles of organic detritus, suspended in the water. The tentacular system is called a *lophophore,* and the 3 groups are collectively referred to as the **lophophorates.**

Cilia on the tentacles draw water into the lophophore, where food particles are caught and transferred to the mouth. The greater the number of tentacles, the more water they can process—and the more food they can catch. The three forms of lophophorates have increased the number of tentacles by means of different arrangements. In the wormlike **phoronids** the tentacles are borne on an elaborate spiralled ridge curling over nearly the entire dorsal surface of the animal. In the clamlike **brachiopods** the tentacles are borne on long coiling "arms" that are protected within dorsal and ventral shells. And in the hydroidlike and coral-like **bryozoans,** small zooids replicate to produce colonies of interconnected individuals, each with a ring of tentacles around the mouth.

All lophophorates are marine except for about 40 freshwater species of bryozoans. With about 4,000 marine species, bryozoans are by far the largest and most ecologically important group. Brachiopods were major components of the faunas of ancient seas, but most became extinct nearly 225 million years ago and relatively few survive today. Phoronids, though they are of minor importance in numbers of species and individuals, and apparently always have been, provide the clearest introduction to most of the features characteristic of lophophorates.

PHORONIDS

ALTHOUGH wormlike in form, members of the phylum **PHORONIDA** do not have the familiar anterior-posterior axis of annelids and most other worms. Rather, the end capped

656

with the lophophore bears also both mouth and anus. Nearly all of the body is an extension and elongation of the ventral portion, carrying a long U-shaped gut. Most phoronids are small, only a few centimeters long, but individuals of one Californian species can reach nearly 25 cm in dorso-ventral "length" while being only 2 mm or less in diameter. The animals live in burrows in shell, rock, wood, or in soft sediment. Chitin, secreted by the body wall and usually encrusted with sand grains and debris, lines the burrows, and the animal moves freely up and down, extending the lophophore from the open end to feed. The tentacles of the lophophore are filled with coelomic fluid that serves as a hydrostatic skeleton holding them upright.

In the smaller phoronids, the lophophore is a circular or a horseshoe-shaped ring of tentacles around the mouth on the upper end of the worm. In larger species, the ring coils around on itself, to the left and right, resembling a scroll in cross-section. The more the ring coils, the greater the number of tentacles that can be fitted into the lophophore.

The coils of the lophophore in the largest species of phoronids, *Phoronopsis californica*, form spirals that extend the tentacles in the inner coils above those on the outer coils. The upper limit of phoronid size is probably set by the spatial limitations of how many tentacles can be added by coiling and spiraling the lophophore.

Sand-encrusted tube with the upper end of the phoronid protruded, showing lateral whorls of lophophore. *Phoronis architecta*. Panacea, Florida. (R.B.)

Aggregated clump of phoronids, *Phoronis vancouverensis,* with expanded lophophores. Monterey Bay, California. (R.B.)

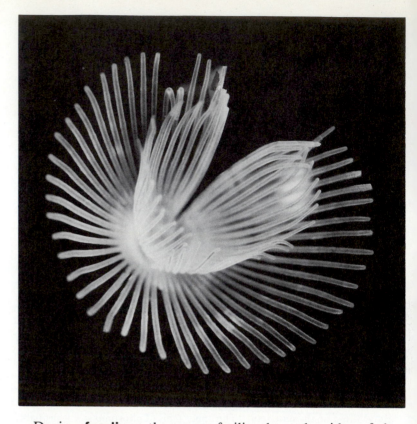

Tentacles of phoronid lophophore form a horseshoe-shaped ring. Note 2 tentacles in *upper left* bent into the lophophore. *Phoronis vancouverensis.* Monterey Bay, California. (R.B.)

The tentacles of the lophophore, as well as the body wall in general, probably absorb dissolved organic material directly from the seawater, supplementing nutrition.

Many kinds of burrowing or tube-building, sedentary animals have a U-shaped design that places the mouth and anus near the surface while most of the body is kept safely in the tube. Some, such as the echiuran *Urechis caupo,* live in a U-shaped burrow; others, such as some polychetes, have the body folded on itself in a straight tube; and a few, such as sipunculans, phoronids, bryozoans, and pterobranchs (chapter 28), have a U-shaped gut in a worm-shaped body.

During **feeding,** the rows of cilia along the sides of the tentacles draw water down into the lophophore and expel it between the tentacles. Food particles that are detected as they pass by the tentacles stimulate a local reversal in ciliary beat, knocking the particles back into the lophophore. Cilia on the inner surface of the tentacles beat dowward, concentrating the food particles. A ciliated groove on the floor of the lophophore, overhung by a ridge, the *epistome,* transports the food particles around the coils to the mouth. This mechanism of feeding, using localized ciliary reversal, is characteristic of all lophophorates, as well as the larvas of echinoderms and enteropneusts, and also the adults of pterobranchs (see chapters 27 and 28).

The **digestive tract** of phoronids is a simple tube that leads from the mouth down the length of the animal, makes a U-turn, and then returns to terminate with the anus located behind the mouth, just outside the lophophore.

The phoronid **coelomic system** is divided into three parts: an anterior coelom, the *protocoel,* supports the epistome; a middle coelom, the *mesocoel,* surrounds the esophagus and supports the lophophore, extending into each tentacle; and paired but irregularly divided coelomic sacs, the *metacoels,* form the main body coelom in which the gut and most other organs are suspended.

Phoronids have an **excretory system** with a pair of ciliated funnels, the *nephridia,* that open into the metacoel and empty through two pores, one on each side of the anus. In addition to removing metabolic wastes from the body, these funnels serve as ducts that carry sperms and fertilized eggs out of the body.

The **circulatory system** is well developed, with two blood vessels around the base of the lophophore, one sending blood into the tentacles, the other receiving the return flow. Respiratory exchange occurs through the surface of the tentacles, which serve both as gills and as feeding structures. The vessel receiving oxygenated blood from the tentacles proceeds down

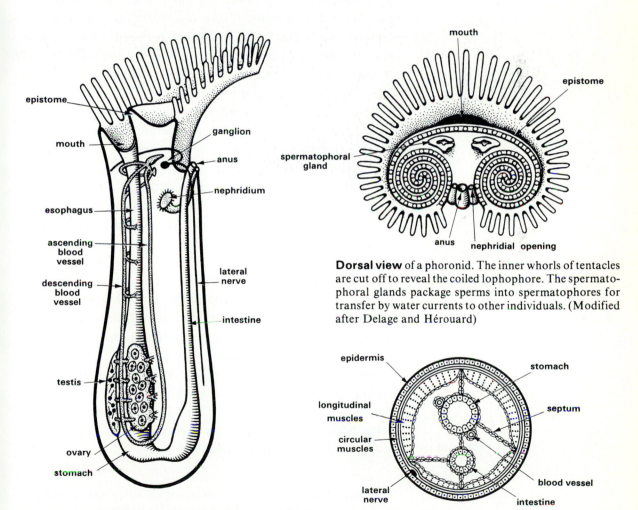

Dorsal view of a phoronid. The inner whorls of tentacles are cut off to reveal the coiled lophophore. The spermatophoral glands package sperms into spermatophores for transfer by water currents to other individuals. (Modified after Delage and Hérouard)

Diagrammatic longitudinal aspect of a phoronid showing the feeding, digestive, circulatory, excretory, nervous, and gonadal systems. The coelomic system (not shown) includes the protocoel in the epistome, the mesocoel around the esophagus and extending into the tentacles, and the metacoel that forms the main body coelom. (Combined from various sources.)

Diagrammatic cross section of the body of a phoronid, showing the muscle layers of the body wall. Septa divide the metacoel and suspend the gut and main blood vessels. (Mostly after K. Heider)

the length of the body into capillary beds around the stomach and gonads. Blood is collected from the capillary beds by a contractile ascending vessel that pumps it into the first lophophore vessel and on to the tentacles. Phoronid blood contains both colorless blood cells and red blood cells with hemoglobin.

The **nervous system** of phoronids is not well developed, as might be expected for an animal that moves little during its adult life. Movement is sluggish and is usually restricted to traveling up and down the tube, although when removed from their tubes, some phoronids can burrow into soft sediments and secrete a new tube. There is a nerve ring around the esophagus and a small dorsal ganglion between the mouth and anus. A single (left) or pair of nerves extend down the lateral sides of the animal and branch to the trunk muscles, coordinating rapid retraction into the tube.

The species with the smallest individuals, *Phoronis ovalis* and *P. psammophilia,* regularly undergo **asexual replication.** Small clones can develop by repeated transverse fission, or by budding, but the individuals do not remain in physical contact. Under stressful environmental conditions, phoronids may cast off the lophophore and associated structures; a new lophophore is regenerated by the remaining body. In one species even the cast-off lophophore regenerates a complete body.

The **gonads** are located in the main body coelom, where the gametes accumulate, and there are both hermaphroditic species and those with separate sexes. Fertilization is internal, and the zygotes are either released into the sea or protected during early development among the coils of the lophophore. In most species a free-swimming larva, the **actinotroch,** develops and feeds on plankton while drifting in the sea, using cilia on the *larval tentacles* to feed in the same manner as adults. Swimming is mainly by the action of a ring of cilia around the anus, the *telotroch.*

Early cleavage pattern and mesoderm formation in phoronids is similar to that in echinoderms and hemichordates, while the formation of the mouth and coelom is more like that in molluscs and annelids. This mixture of embryological features confounds attempts to align phoronids and other lophophorates with other phyla of animals (see chapter 30).

The anterior part of the actinotroch larva contains a protocoel, the larval tentacles contain extensions of a mesocoel, and the posterior part of the body has a large metacoel; these develop into the adult coelomic sacs. A pair of closed excretory tubules, the larval *protonephridia,* develop after metamorphosis into the adult nephridia.

The actinotroch has a large sac, the *internal sac* (or metasomal sac), in its ventral surface. This sac everts at **metamorphosis,** growing ventrally and carrying the gut with it, to form the main body of the sedentary adult. The wall of the internal sac, now the main part of the body wall, contains gland cells that secrete the tube. In most species larval structures, including the tentacles, are transformed into adult structures, but in a few species the larval tentacles are shed, and new juvenile tentacles develop from buds in the larva.

The dramatic metamorphosis seen in phoronids, with ventral hypergrowth distorting the anterior-posterior axis, is similar to that in sipunculans, suggesting a convergence of both development and body form to exploit a sedentary, suspension-feeding lifestyle by these two groups.

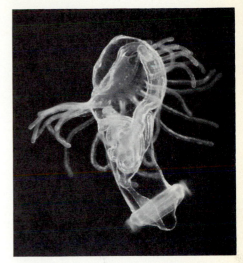

Actinotroch larva of a phoronid with long feeding tentacles. The ciliated ring at the posterior end *(lower right)* is used for swimming. *Phoronis.* England. (D.P. Wilson)

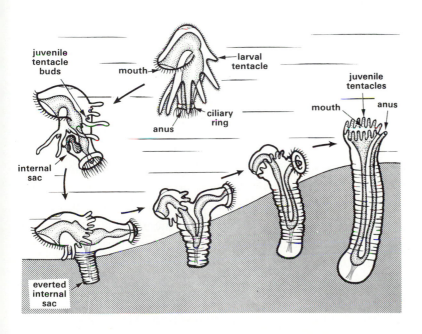

Settlement and metamorphosis of a phoronid larva. Rapid eversion of the internal sac changes the growth axis of the animal. And as the animal grows ventrally, it burrows into the substrate. (Mostly after A. Meeck)

There are only about a dozen species of phoronids, divided between two genera: *Phoronis* and *Phoronopsis.* Most of the species are widely distributed on tropical and temperate shores. Small and inconspicuous, they are rarely noticed but are not uncommon.

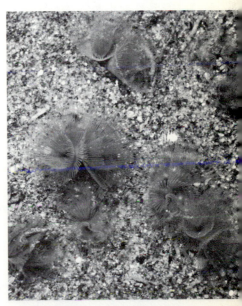

Patch of green phoronids, *Phoronopsis viridis,* in a shallow estuarine mud flat at low tide. Diameter of individual lophophores, about 5 mm. Elkhorn Slough, California. (R.B.)

Articulate brachiopods attached to a rocky outcrop with the pedicle emerging from the ventral valve.

The name brachiopod means "arm-foot," reflecting the early confusion about these animals with molluscs bearing names such as gastropods, pelecypods, and cephalopods.

One common name for brachiopods is "lamp shells" because some resemble ancient oil lamps.

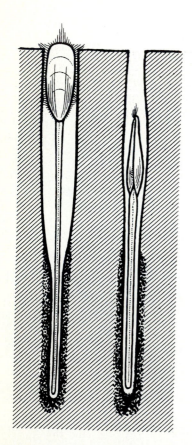

Inarticulate brachiopods extended and retracted in burrows in soft substrate. (After P. Francois)

BRACHIOPODS

MEMBERS of the phylum **BRACHIOPODA** resemble bivalve molluscs in the possession of a *bivalved calcareous shell* that is secreted by a mantle and encloses nearly all of the body. However, the valves are left and right in bivalves, dorsal and ventral in brachiopods. Moreover, rather than containing a muscular foot and ctenidial gills, the enclosed mantle cavity of brachiopods contains the coiled arms, or **brachia,** of an elaborate lophophore that collects suspended food particles. The body, while not elongated in the dorsoventral axis, is organized more like that of phoronids than like that of any other phylum, suggesting a common ancestry.

There are two groups of brachiopods. In the **Articulata** the shell valves are composed mainly of calcium carbonate with very little organic material, and they are joined by a hinge with interlocking teeth. In the **Inarticulata** the valves are unhinged and are composed mainly of calcium phosphate with a high content of chitin and protein. Most members of both groups have a stalk, or *pedicle,* emerging posteriorly through a groove or hole in the ventral valve. In articulates the pedicle is solid and relatively inflexible. It is usually firmly attached to a hard surface such as a rock, and the animals are limited in their ability to move or even change position. Almost all the body muscles are attached to the valves, and they both close and open the valves; muscles between the valves and pedicle twist the body in different directions, displacing accumulated sediment from the valves. The generally larger lophophore of articulates is supported by stiff connective tissue and by a calcareous skeleton, the *brachidium,* which is attached to the dorsal valve and is often complex in its growth and shape.

Many inarticulate brachiopods live in mud or fine sand. The pedicle of these is thick, fluid-filled, and muscular, and is used both for burrowing and for extending or retracting the animal in the burrow. The muscles that attach the two valves together are effective only in closing them, and the valves are opened

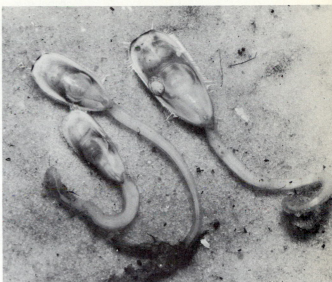

Articulate brachiopod attached to a cluster of barnacle shells. Note tuft of lophophore between the valves. *Magellania venosa*. Strait of Magellan. (R.B.)

Inarticulate brachiopods on sand after being dug from their burrows. One has a gastropod slippershell on its shell. *Glottidia pyramidata*. Panacea, Florida. (R.B.)

indirectly by hydrostatic pressure when the lophophore is pulled posteriorly, pressing on the body cavity. The lophophore of inarticulates is held in position mainly by muscles and connective tissue, and there is no brachidium.

The **digestive system** of brachiopods is similar to that of phoronids, but there are glands branching from the stomach that digest and absorb most food. An anus opens posteriorly or on the right side in inarticulates. Articulates have a blind intestine and no anus; undigested materials are periodically expelled out of the mouth into the mantle cavity, carried to the edge of the mantle by cilia, and then ejected when the valves snap shut.

Brachiopods, like bivalve molluscs, expel material from the mantle cavity by rapidly snapping the valves shut. There is evidence that some fossil brachiopods used this behavior, as scallops do today, to swim briefly above the bottom, perhaps escaping potential predators.

The complex **coelomic system** in brachiopods is comparable to that of phoronids, with one or two pairs of ciliated nephridial funnels that open into the metacoel and serve both to excrete metabolic wastes and to expel the gametes. The metacoel is subdivided by septa and extends into the mantle tissues to form an intricate system of *mantle canals,* ciliated tubules that end near the outer edge of the mantle.

When full-grown some articulate brachiopods detach and their weight keeps them securely in place, even on muddy bottoms; others cement the valve directly to rocks. Many species of inarticulates never develop a pedicle but cement the ventral valve to rocks when they settle.

The direct opening and closing muscles of articulates take much less space than the indirect system of inarticulates which depends on a large fluid-filled coelomic cavity, and inarticulates have much less room within the valves for the development of the lophophore than do articulates.

The tentacles of the lophophore in brachiopods can be arranged within the mantle cavity to form incurrent and excurrent chambers that efficiently partition and filter water, as in bivalve molluscs.

"Mantle" is a general term referring to a fold of the body wall that envelops much of the body and that usually secretes a calcareous shell, as in both molluscs and barnacles, as well as in brachiopods.

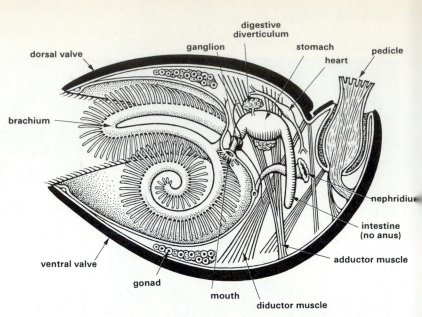

dorsal valve

digestive
diverticulum

ganglion

stomach

heart

pedicle

brachium

nephridium

intestine
(no anus)

ventral valve

adductor muscle

gonad

mouth

diductor muscle

Diagrammatic longitudinal aspect of an articulate brachiopod showing major organ systems. Contraction of the adductor muscles closes the valves, while contraction of the diductor muscles raises the dorsal valve. Coelomic cavities are not shown. (Combined from various sources)

A small brachial fold about the mouth is said to be an epistome and in some inarticulates it contains coelomic spaces that may be a *protocoel*. The lophophore and region around the esophagus contain a *mesocoel* with an intricate system of tubes in both the brachia and tentacles of the lophophore. The main body coelom, the *metacoel*, encloses most of the digestive tract and other organs, and extends into the mantle and, in the inarticulates, into the pedicle.

There is a **circulatory system** in most brachiopods, with a contractile *heart* suspended in the metacoel above the stomach. Blood vessels supply the tentacles of the lophophore and follow the coelomic tubules into the mantle, but the capillary beds are generally open.

Some of the cells in the coelomic fluid contain the respiratory pigment hemerythrin, similar to that in sipunculans. Its presence suggests that the coelomic vessels, rather than the blood vessels, are the main transport system for respiratory gases. Hemoglobin has not been found in brachiopods. The respiratory rate of brachiopods is one of the lowest known for any animal, and they can survive long periods without oxygen or food, making them "metabolically minimal" organisms.

The **nervous system** includes a small ganglion around the esophagus, with nerves extending into the brachia and mantle folds. There are no well-developed sense organs but, as in bivalve molluscs, the edge of the mantle has sensory lobes that probably detect touch and chemicals in the water. Brachiopods depend mainly on their shells for protection from predators, and the shells of some fossil species were elaborately armed with spines. Distasteful chemicals in the tissues probably provide important protection for modern species, as does their generally cryptic habit. Predators of brachiopods include sea stars and sea urchins.

Brachiopod larvas have pigmented eyespots and the pelagic juveniles of inarticulates have statocysts, fitting their more active existence; the adults of one species of particularly motile inarticulate brachiopod are reported to have statocysts.

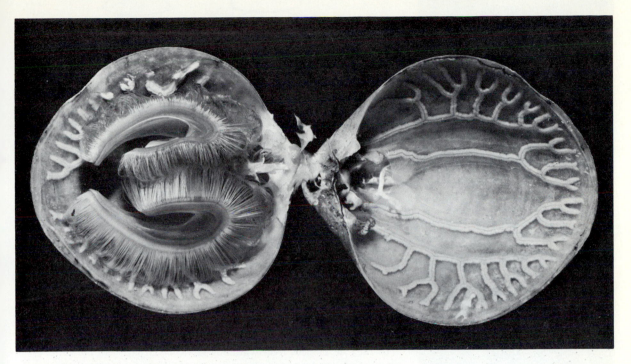

Opened articulate brachiopod with the lophophore in the dorsal valve (left) and the mantle canals (containing eggs) of the coelom showing in the ventral valve (right). *Magellania venosa.* Strait of Magellan. (R.B.)

All brachiopods are solitary, and there is no form of asexual replication. In **sexual reproduction** gametes are released from the multiple gonads into the metacoels and discharged through the nephridia. Most species have separate sexes, and articulate brachiopods produce yolky nonfeeding embryos and larvas that develop in the plankton before settling and metamorphosing into sessile juveniles.

Although there is no internal sac in brachiopods, the inner surfaces of the mantle lobes of articulate larvas may be homologues of the internal sac. As with the internal sac of phoronids (and some bryozoans), these surfaces are rapidly exposed at metamorphosis to form most of the outermost epithelium of the juvenile, and they secrete the external shell.

More than 30,000 species of brachiopods flourished in Paleozoic seas. Inarticulates appear to have preceded articulates slightly and are usually considered closer to the ancestral forms; but species of both groups have been found in the Cambrian, near the base of the fossil record of animals. It is not known why most species of these highly successful animals became extinct at the end of the Paleozoic era, but the hiatus they left appears to have been largely filled by bivalve molluscs, which, although present throughout most of the Paleozoic, did not really thrive until after most brachiopods had disappeared.

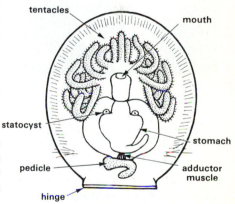

Inarticulate "larva" is a pelagic juvenile that swims and feeds in the plankton. The organs can be seen through the transparent shell, and the pedicle remains coiled in the mantle cavity until the juvenile settles and begins benthic life. *Lingula.* (After N. Yatsu)

The swimming inarticulate juveniles of *Pelagodiscus atlanticus* were described long before the adults were found at abyssal depths.

Larva *(above)* and **juvenile** one day after metamorphosis *(below)* of the articulate brachiopod *Terebratalia transversa.* SEMs. (S.A. Stricker and C.G. Reed)

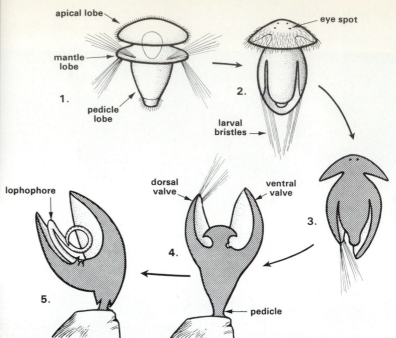

Settlement and metamorphosis of an articulate brachiopod larva. **1.** The fully developed, free-swimming larva has 3 parts, an *apical lobe,* a *mantle lobe,* and a *pedicle lobe.* **2.** Dorsal and ventral folds of the mantle lobe grow posteriorly to enclose most of the pedicle lobe. **3.** Section of fig. 2. **4.** At metamorphosis the pedicle lobe attaches to a hard surface and develops into the pedicle; the folds of the mantle lobe are thrown anteriorly, enveloping the apical lobe as they begin to secrete the enclosing shell valves. **5.** The apical lobe develops into the lophophore and most of the enclosed body. (**1** and **2** after A. Kowalevski; **3, 4,** and **5** after Delage and Hérouard)

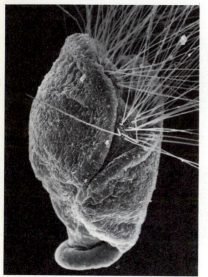

CLASSIFICATION: Phylum BRACHIOPODA

Of the more than 10 orders of brachiopods in the Paleozoic, only 5 are living today. These include fewer than 300 species, mostly in shallow temperate and polar seas.

Class INARTICULATA. Lack a hinge between the valves, which open hydrostatically; lack an internal skeleton in the lophophore; anus present. Two orders.

Order Lingulida have elongate valves and a long pedicle. Burrow in soft substrates. The genus *Lingula* appears nearly unchanged since the Ordovician and at 350 million years old is the oldest living genus of any animal known. *Glottidia, Lingula.*

Order Acrotretida live on top of surfaces, have more rounded shell with a small pedicle or none at all, and cement the ventral valve to the substrate. *Crania, Discinisca, Pelagodiscus.*

Class ARTICULATA. Valves lock together with a hinge and are opened directly by muscle contraction; lophophore usually with internal skeleton; no anus. Three orders.

Order Strophomenida lack a pedicle and cement the ventral valve to the substrate. *Lacazella.*

Order Rhynchonellida have spiral brachia with a simple skeleton. *Hemithyris.*

Order Terebratulida have much-looped brachia supported by a complex skeleton. This order appeared last in the fossil record (in the Silurian) and has most of the living species. *Argyrotheca, Gryphus, Laqueus, Magellania, Terebratalia, Terebratella, Terebratulina.*

Freshwater bryozoans with tube-like bodies. The horseshoe-shaped lophophore is extended in some, retracted in others.

BRYOZOANS

ALTHOUGH superficially resembling hydroids or corals, members of the phylum **BRYOZOA** are more similar to minute colonial phoronids, with bodies less than a millimeter long. Each body, or *zooid,* has a circular or horseshoe-shaped lophophore and is largely covered with a thin chitinous cuticle that usually encloses a gelatinous, leathery, or calcified exoskeleton. Bryozoans increase in size not by enlarging an individual lophophore or by growth along a dorsoventral axis, but by extensive replication of the zooids. The zooids, in most cases, remain discrete but maintain intimate communication with each other throughout the encrusting or upright branching colonies that grow on aquatic plants, rocks, and other surfaces.

In addition to growing mainly by replication, bryozoans differ from phoronids in an eversible lophophore that can be completely withdrawn into the body, displacing the coelomic fluid of the metacoel. Contraction of the *retractor muscles* rapidly withdraws the lophophore, while contraction of the muscles encircling or attached to the body wall exerts pressure on the coelomic fluid and pops the lophophore back out to feed in the overlying waters. The chitinous cuticle and enclosed exoskeleton adhere tightly to the epidermis (unlike the tube of phoronids) so that the colonies look like sheets or branches of tiny boxes. A cuticular flap, the *operculum,* often closes over the withdrawn lophophore, and a rich assortment of spines and armaments usually adorns the exoskeleton.

When first observed by 19th century naturalists, the gut and lophophore of bryozoans appeared to be a discrete body within a second body, the protective body wall. The gut and lophophore complex was named the **polypide** because it resembles a hydroid polyp, and the body wall was named the **cystid** ("sac"). Although the "polypide" often degenerates and is regenerated from the walls of the "cystid," these two portions of the body are similar to comparable portions of the bodies of other coelomate animals. The term **zooecium** is sometimes used instead of cystid. Or more narrowly, it may refer to the exoskeleton that persists in dead and fossilized bryozoans.

Bryozoa means "moss animals" and refers to the mosslike appearance of many of the colonies. Another name for the group is **Polyzoa,** meaning "many animals." In addition, **Ectoprocta** is the name often used to distinguish bryozoans, in which the anus is located *outside* the ring of tentacles, from kamptozoans (entoprocts), in which the anus is *within* the ring of tentacles.

Eversion of the feeding tentacles by bryozoans is comparable to that by sipunculans and holothuroids, but is much more rapid.

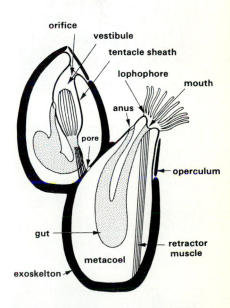

Diagram of **extended and retracted lophophore** of a bryozoan showing prominent retractor muscle. Mechanism for everting lophophore is not shown. (Modified after Delage and Hérouard)

Zooids in species of the genus *Monobryozoon* do not remain together, but separate soon after budding, so that clones rather than colonies are formed. These minute interstitial marine animals cling with small stolons to sand grains and shell fragments; more than one feeding zooid would overwhelm their chosen particle of substrate.

Colony **growth** is by budding. Thin portions of the body wall grow out as small vesicles or tubes, and the whole of a new zooid, including gut, gonads, lophophore, and the exoskeleton if present, develops from the body wall of the bud. Budding patterns reflect both the genetics of the individual animal and factors such as current flow and substrate. These factors determine the enormously varied shapes of the colonies, which occur as thin sheets, delicate convoluted folds, massive coral-like heads or gelatinous growths, and bushes or tangles of upright or dangling tufts.

Although there are no pelagic bryozoans, one Antarctic species grows on pieces of ice and forms large floating gelatinous colonies over deep water hundreds of miles from the shore.

1. *Bugula*

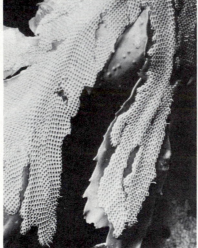

2. *Membranipora*

3. *Phidolopora*

4. *Schizoporella*

Bryozoan colonies have a large variety of shapes: **1**. upright branching forms attached only at the base; **2**. thin encrusting sheets; **3**. lacy, delicate folds; **4**. massive, calcified coral-like forms; **5**. massive gelatinous growths. (R.B.)

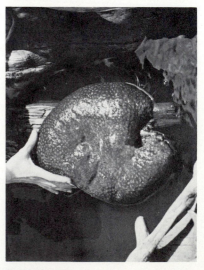

5. *Pectinatella*

Most bryozoan colonies adhere tightly to the substrate throughout their lives and are completely sessile, but colonies of a few species can move slowly over the substrate. Free-living marine colonies of *Cupuladria* move through and upon sandy surfaces by coordinated rowing movements of long vibracula (see below), and gelatinous freshwater colonies of *Cristatella* creep over aquatic plants on a muscular sole at rates of 2 to 3 cm per day.

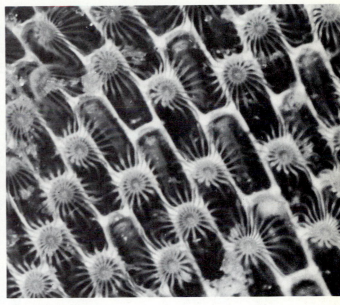

1. *Bugula*   2. *Membranipora*

Bryozoan lophophores are similar in structure and function, despite the variety in colony appearance.

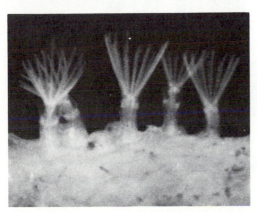

3. *Alcyonidium*   4. *Pectinatella.*

In **feeding,** bryozoans collect food particles from the overlying water in much the same manner as do phoronids, and the ingested material passes through a U-shaped gut. In addition, the upper third of whole tentacles often flick particles back into the lophophore toward the mouth. Tentacle flicking during feeding is also seen occasionally in phoronids and brachiopods, and it is commonly used for feeding by crinoids and pterobranchs.

In sheetlike colonies, such as in the genus *Membranipora,* incoming water that passes through the tentacles flows down over the surface of the colony and then rises in "chimneys" above groups of nonfeeding zooids. Such "chimneys" are regularly spaced, so that there are incurrent and excurrent areas on the colonies.

Unlike phoronids and brachiopods, bryozoans lack well-defined circulatory and excretory systems, probably a consequence of their minute body size. Nevertheless, mesenteric strands (funicular cords) extend through the metacoel between the gut and body wall, and between zooids in the colony, and these sometimes enclose small channels that probably serve as rudimentary blood vessels. Moreover, in some species there is a ciliated coelomic duct from which the eggs emerge.

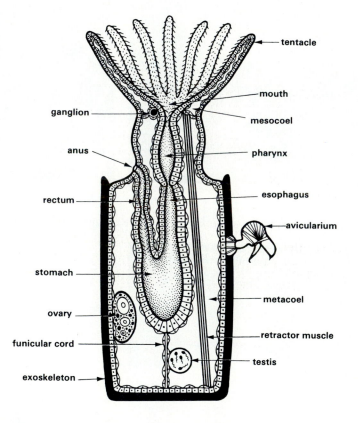

Diagrammatic section of a bryozoan zooid with the lophophore extended and attached avicularium (see below). (Combined from various sources)

There are three classes of bryozoans. Zooids are organized most like a phoronid body in the exclusively freshwater **Phylactolaemata,** and these have a gelatinous exoskeleton of various thicknesses. The zooids of the exclusively marine **Stenolaemata** are enclosed by cylindrical calcified exoskeletons. In the largest class, the mainly marine **Gymnolaemata,** the squat, boxlike or cylindrical zooids lack circular or annular muscles of the body wall, and in most species are partially or almost completely encased in a calcified exoskeleton.

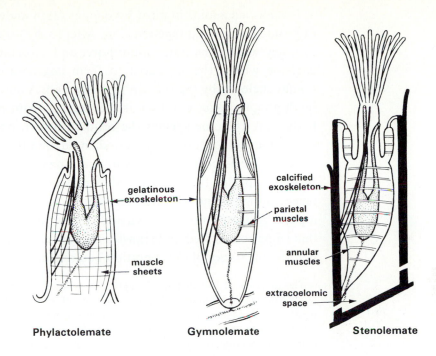

Lophophore eversion differs in the three classes of bryozoans. In *phylactolemates* (left) circular sheets of body wall muscles contract and exert pressure on the coelomic fluid, popping the lophophore out. Because the metacoels are freely open among the zooids, the mechanism of independent lophophore eversion is not completely understood. In *stenolemates* (right) the zooids are enclosed by cylindrical calcified exoskeletons; the metacoel is separated from the outer body wall by an extensive extracoelomic space (a pseudocoel), and contraction of annular muscles around the metacoel is largely responsible for increasing the hydrostatic pressure and everting the lophophore. *Gymnolemates* (center) lack circular or annular muscles of the body wall, and contraction of parietal muscles that cross the metacoel (shown here on a tubular zooid of a noncalcified species) deforms the body wall, increasing the hydrostatic pressure. (Mostly from P.D. Taylor)

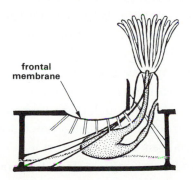
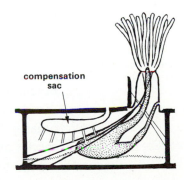

Most gymnolemates are boxlike and have calcified exoskeletons that cannot be deformed to evert the lophophore. In some species *(left)* only the side and basal surfaces of the body wall are calcified and the upper or *frontal* surface is membranous. When the lophophore is retracted the frontal membrane bulges outward from the displaced coelomic fluid, and when the parietal muscles pull the frontal membrane inward the lophophore is everted. The thin frontal membrane is vulnerable to attack by a variety of predators, and much of the diversity seen in gymnolemate bryozoans reflects the growth of protective spines and calcified plates that grow under or over the frontal membrane. In some of the most common species *(right)*, the frontal wall is nearly completely calcified and only a small pore opens into an invaginated frontal membrane, the *compensation sac*. Seawater is expelled from the compensation sac when the lophophore is retracted, and drawn into it when muscles enlarge it and evert the lophophore. (From P.D. Taylor)

Established zooids in most bryozoans often undergo cycles of disintegration and regeneration. After living for a few days to a few weeks, the gut, lophophore, and associated tissues coalesce within the metacoel to form degenerating *brown bodies* enclosed by the remaining body wall. A new functioning lophophore-gut complex is regenerated from cells of the body wall. In some species the residual brown body is held within the metacoel of the renewed zooid, and the number of cycles can be determined simply by counting the brown bodies. In other species the brown bodies are enclosed within the regenerating gut and are defecated. Within an actively functioning and growing colony, brown body formation is staggered so that some zooids feed and transfer nutrients to others that are in different stages of degeneration and regeneration.

The function of brown body formation is unknown, but it appears to rejuvenate the zooid and may serve to dispose of accumulated metabolic wastes. Similar cycles of zooid disintegration and regeneration are known for colonial hydroids and tunicates (chapter 29), and the senescent cells of many animals commonly disintegrate, leaving accumulations of lipofuscin-rich "brown spots" in the tissues.

In freshwater phylactolemates the gut and lophophore of the zooid periodically disintegrate, as in other bryozoans, but they do not regenerate in the same place in the colony. Instead, new zooids are continually formed from the general body wall of the colony, replacing those that disintegrate.

Not only do zooids within a colony transfer nutrients to each other and coordinate their budding patterns to form well-defined colony shapes, but their activities are coordinated in many species by a common nerve network that rapidly transfers impulses throughout the colony. Disturbance of one lophophore can result in rapid retraction of the lophophore by most or all the zooids in a colony. Each zooid has a small ganglion between the mouth and anus, as in phoronids and brachiopods, and nerves from this ganglion communicate with those of adjacent zooids. Although there are no discrete sense organs, bristle-like zooids (vibracula, see below) in some species may be specialized tactile receptors.

Like hydroids and some other colonial organisms, many species of bryozoans display **polymorphism.** Most of the zooids of a colony, *autozooids,* have a well-developed lophophore and feed; other types of zooids are various kinds of *heterozooids* that feed little or not at all but have functions that presumably benefit the colony by providing protection from predators or fouling organisms, by attaching the colony to the substrate or adding structural support, by holding nutrient stores, or by producing gametes or protecting embryos.

The retractor muscles of bryozoans are among the most rapidly contracting muscles known, shortening more than 20 times their length per second, nearly twice as fast as the fastest vertebrate muscle.

Among the various kinds of heterozooids, *avicularia* (singular, avicularium) appear to protect colonies from predators and settling larvas of other organisms. Some have an enlarged operculum that serves as a pinching "jaw" of a stalked birdhead-like structure, while in sessile avicularia the operculum forms a long rod that can clamp trespassers to the colony surface. In *vibricula* the body wall and other parts of the zooid all but disappear, and the elongate, bristlelike operculum sweeps the colony surface and may also be sensory. In some branching colonies, the body wall of a zooid forms long and tubular *kenozooids*, which are stolons, stalks, and attachment units for the colony. Closed zooids, called *kleistozooids*, apparently provide a few species with convenient nutrient storage compartments. Dwarf zooids, *nanozooids*, with only a single tentacle, are found in a few stenolemates, but their function remains unknown. Among many stenolemates and a few gymnolemates, some zooids lose the ability to feed and become little more than enlarged gonads and embryo brood-chambers, termed *gonozooids*. In most gymnolemates, however, the autozooids retain the ability to produce gametes; and specialized chambers, *ovicells*, which are usually formed in part by heterozooids, brood the embryos.

In addition to avicularia, many bryozoan colonies have well-developed protective spines. In some species the spines form only on zooids growing near the edge of a colony after the presence of nudibranch predators is detected, or when the edge of another colony is met; the spines provide protection from slow-moving predators such as nudibranchs, and they are an effective barrier against overgrowth by other colonies.

Several species of bryozoans have intimate symbiotic associations with tiny hydroids of the genus *Zanclea*. The bryozoans provide surfaces for the hydroids to attach and grow, and the hydroids provide protection with their nematocysts.

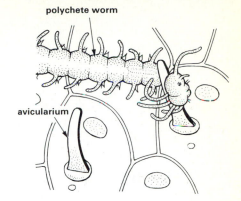

polychete worm

avicularium

Sessile avicularium of the bryozoan *Reptadeonella costulata* clamps a small polychete to the surface of the colony, trapping it. (After J.E. Winston)

Another way to discourage animals and plants from settling on and overgrowing a colony ("fouling") is practiced by at least one bryozoan: *Cupuladria doma* grows on sandy bottoms as small, unattached, domed colonies that periodically molt their chitinous exoskeletons from the upper and lower surfaces, ridding themselves of accumulated foulers.

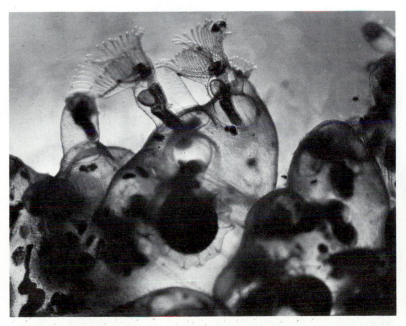

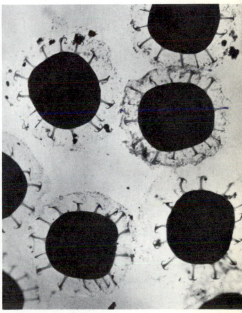

Statoblasts, resistant cysts that are functionally similar to gemmules in freshwater sponges, are asexually produced by phylactolemate bryozoans and accumulate in the metacoel *(left)*. These can survive long periods of freezing and drying, enabling a colony to survive many years in seasonally variable lakes and ponds. Some kinds float and are carried down streams, or are blown or carried from pond to pond, spreading individuals of a colony over very large areas. *Right,* close-up of statoblasts. Lake Pymatuning, Pennsylvania. *Pectinatella.* (R.B.)

Modes of **sexual reproduction** are highly variable and poorly known in bryozoans. In many stenolemates nonfeeding heterozooids produce either sperms or eggs in different colonies; however, colonies may change from one sex to the other. In colonies of most other bryozoans, both sperms and eggs are produced with the same autozooids. Mechanisms of spawning and fertilization are largely unknown, but in some cases, sperms are released through pores in the tips of the tentacles and caught by the tentacles of other colonies. Eggs are fertilized as they are released.

Stenolemates are among the few animals known that normally seem to develop by *polyembryony*. One or a few embryos within a gonozooid grow and then divide repeatedly to form hundreds of genetically identical embryos, each of which is eventually released as a swimming, non-feeding coronate larva.

In a few marine species a swimming, feeding **cyphonautes larva** develops and drifts in the plankton for weeks or longer. Embryos are more often held within specialized chambers where they develop and grow, sometimes being nourished by the parent through a placenta. Upon release, the yolky **coronate larva** usually settles and metamorphoses within a few hours.

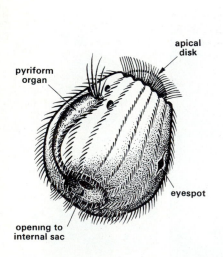

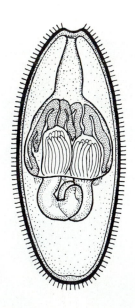

A **coronate larva** of *Bugula* is similar to most non-feeding larvas of bryozoans. The yolky larvas swim only a few hours before settling. (Adapted from W.F. Lynch)

Phylactolemate "larva" of *Plumatella* is a small colony with a few zooids enclosed in a ciliated fold of tissue used for swimming and lost after settlement. As in most freshwater animals, the embryos are retained by the parent, and there is no distinct larval stage comparable to other bryozoan larvas. (After P. Brien)

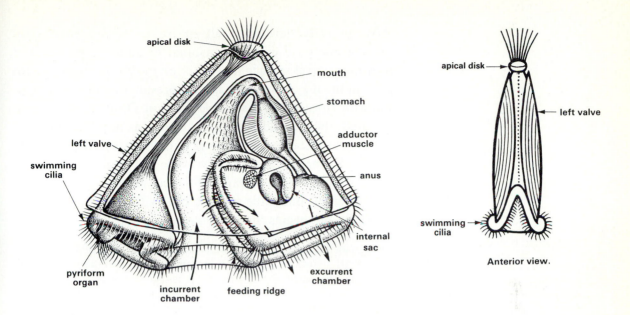

Cyphonautes larva.

(labels in figure: apical disk, mouth, stomach, adductor muscle, anus, internal sac, left valve, swimming cilia, pyriform organ, incurrent chamber, feeding ridge, excurrent chamber, Lateral view., apical disk, left valve, swimming cilia, Anterior view.)

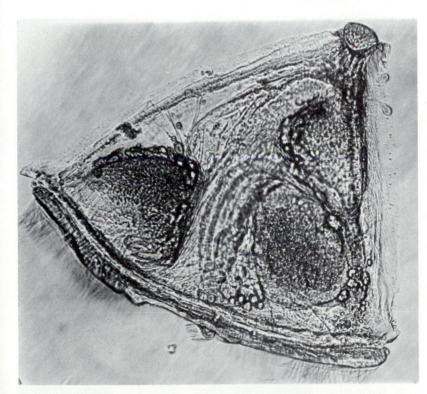

Cyphonautes larvas are found in relatively few bryozoans, and these are not all closely related *(Membranipora, Electra, Conopeum, Farrella, some species of Alcyonidium).* Individuals of these bryozoans produce large numbers of small eggs, and the larvas feed in the plankton for days to weeks or more. The adults typically live on seaweeds, short-lived substrates that can disappear completely during storms or unfavorable conditions. Cyphonautes larvas are capable of wide dispersal to distant places where they may encounter healthy young stands of seaweeds. In contrast, coronate larvas do not feed, swim only a few hours or less, and tend to settle near their parents, which typically grow on more permanent substrates such as rocks.

Cyphonautes larva of a bryozoan collected from the plankton. These triangular, flattened animals with lateral chitinous valves feed on phytoplankton collected by cilia on ridges within a large vestibule, and they can live in the plankton for long periods before settling. Bimini. (R.B.)

When fully developed all bryozoan larvas possess an *internal sac* (also called metasomal or adhesive sac) tucked into the ventral surface. An anterior *pyriform organ,* is also present in all gymnolemate larvas, presumably functioning to sense the substrate for a suitable location. Once such a location is found, the internal sac rapidly everts, as in phoronids, and secretes a substance that adheres to the substrate and, in some species, forms a cuticle that covers the metamorphosing larva. During metamorphosis most of the larval tissues are destroyed, and a new body, the **ancestrula,** with one or two zooids, develops and begins the new colony.

Bryozoan metamorphosis is complex, involving the disintegration of most larval tissues and extensive movement and rearrangement of the remaining cells that form the ancestrula. It has been followed carefully only a few times, and varies considerably among species in such details as which cells remain and which disintegrate. In many stenolemates and in *Bowerbankia* the internal sac disintegrates after attachment, and the epithelium at the other end of the larva forms the epithelium of the ancestrula. In some other species the epithelium of the internal sac folds over the larval body, reminiscent of the mantle fold of articulate brachiopods, and forms part or all of the body wall of the ancestrula. In all cases, the ciliated larval epithelium folds inward and disintegrates, and the cavity that remains becomes the metacoel of the ancestrula. Such a mode of coelom formation appears to be unique among animals.

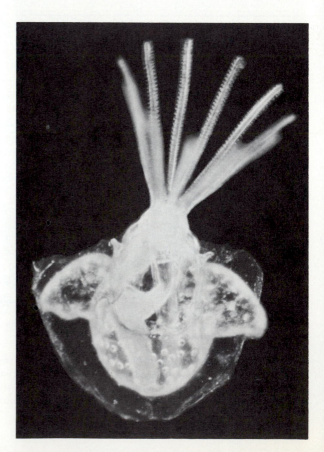

Ancestrula of a gymnolemate bryozoan soon after metamorphosis with a functional lophophore and two lateral buds. (D.P. Wilson)

Like brachiopods, bryozoans have a rich fossil history, and of the 15,000 to 20,000 described species, only about 4,000 are alive today. Stenolemate bryozoans of 4 different orders flourished in the Paleozoic seas; species of only one of these orders survive today. Uncalcified gymnolemates were also present in the Paleozoic, but calcified gymnolemates did not appear until the Mesozoic, after most stenolemates had disappeared. Phylactolemates lack a hard exoskeleton that can be fossilized and, like most freshwater animals, have a poor fossil record.

CLASSIFICATION: Phylum BRYOZOA (ECTOPROCTA)

Class STENOLAEMATA. Exclusively marine bryozoans with narrow cylindrical zooids (stenolaemata = "narrow throat") and a circular lophophore; all with calcified body walls. Lophophore eversion mainly by contraction of dilator muscles of the orifice and annular muscles around the metacoel; large extracoelomic spaces (pseudocoels) between coelomic lining and epidermis. Limited polymorphism. One living order.

   Order Tubuliporata (= Cyclostomata)
Crisia, Lichenopora, Tubulipora.

Class GYMNOLAEMATA. Mainly marine bryozoans with squat or cylindrical zooids and a circular lophophore (gymnolaemata = "naked throat"). Lophophore eversion depends on deformation of the body wall by muscle contraction. Two orders.

Order Ctenostomata: zooid walls membranous or gelatinous, not calcified; limited polymorphism. *Alcyonidium, Bowerbankia, Flustrellidra, Monobryozoon, Victorella, Zoobotryon.*

Order Cheilostomata: zooid walls calcified and orifice closed with an operculum; polymorphism common; includes most living species. *Bugula, Conopeum, Cupuladria, Electra, Hippothoa, Membranipora, Phidolopora, Reptadeonella, Schizoporella.*

Class PHYLACTOLAEMATA. Exclusively freshwater bryozoans with cylindrical zooids possessing an epistome (phylactolaemata = "guarded throat") and a horseshoe-shaped lophophore similar to that in phoronids; all with membranous or gelatinous body wall containing muscles, and with metacoels widely open between zooids. No polymorphism. *Cristatella, Fredericella, Pectinatella, Plumatella.*

THE THREE PHYLA of lophophorates, although different in general appearance, possess many common features. In all lophophorates, the lophophore functions in a similar manner in feeding. At least some members of each phylum have a 3-part coelomic system with a protocoel, mesocoel, and metacoel. And all have a similar organization of the digestive, circulatory, and nervous systems. Early development also has similar components; and although the larvas differ, metamorphosis involves eversion of an internal sac (or in articulate brachiopods a comparable structure), and the epithelium of the internal sac often forms the juvenile epidermis that secretes a tube or cuticular exoskeleton. These features link the three lophophorate phyla together, and some workers have argued that they should be joined as one phylum, the Lophophorata, with the Phoronida, Brachiopoda, and Bryozoa each being a subphylum or class.

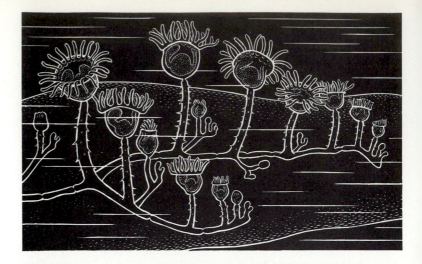

Kamptozoan colony of *Pedicellina* with upright stalked zooids growing out of a branching stolon.

Kamptozoa means "flexible animals" and refers to the bending and nodding motion characteristic of some members of the group. Another name for the group is **Calyssozoa**, meaning "cup animals."

KAMPTOZOANS

AS IF it were not difficult enough to distinguish sessile polychetes from phoronids or hydroids from bryozoans, there is one more group of small polyplike animals that confuse matters even more. Indeed, when first noticed in the early 19th century, the members of the phylum **KAMPTOZOA** or **ENTOPROCTA** were considered to be an unusual type of bryozoan. Although marked differences were noted between the two groups over a century ago, even today some biologists believe they should be united as one phylum. Other biologists see the similarities as the result of convergence, and the differences marked enough to put the kamptozoans far from bryozoans. Kamptozoans do not even seem to be related to lophophorates, and we include them in this chapter only because they share a similar appearance and life-style with bryozoans, and because the two groups are often collected together and studied by the same people.

Kamptozoans have a cuplike body, the *calyx,* perched on top of a muscular *stalk* that sometimes has a "foot disk." Within the calyx there is a U-shaped gut, similar to that in bryozoans, but surrounded by cells and tissue fluid ("hemo-lymph") in a *pseudocoel,* and there is no trace of any coelomic cavities or of blood vessels. Also within the calyx are a pair of protonephridial tubules that usually open through a single pore posterior to the mouth. The circle of tentacles forms a *vestibule* into which both the mouth and anus open. Connective tissue and muscles fill the tentacles, which are unable to retract into the body, but rather fold down over the vestibule. Lateral cilia on the tentacles draw water *up* between the tentacles, as in tentaculate polychetes, and opposite to the flow in all lophophorates.

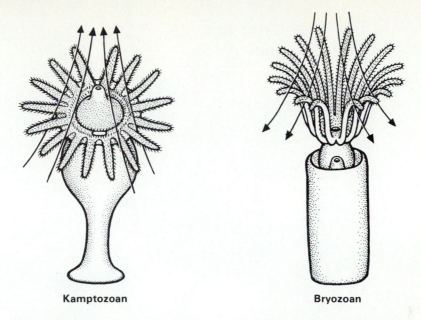

Kamptozoan   Bryozoan

Kamptozoa and **Bryozoa** also have been named **Entoprocta** and **Ectoprocta,** respectively, because in the former the anus opens within the ring of tentacles while in the latter it opens outside the ring. The direction of the feeding currents also differs between the two groups, and the position of the anus allows for efficient removal of feces in each group.

Most people working with bryozoans never accepted the term Ectoprocta, perhaps from a disinclination to be called ectoproctologists. The few people working with kamptozoans have not had such strong feelings. Most English- and French-speaking workers currently use the term Entoprocta, but German- and Russian-speaking workers generally call the group Kamptozoa. We also use Kamptozoa because without Ectoprocta, Entoprocta loses some of its discriminating meaning, because Kamptozoa nicely describes the "nodding head" behavior of some of the animals, and because it is a more euphonious word.

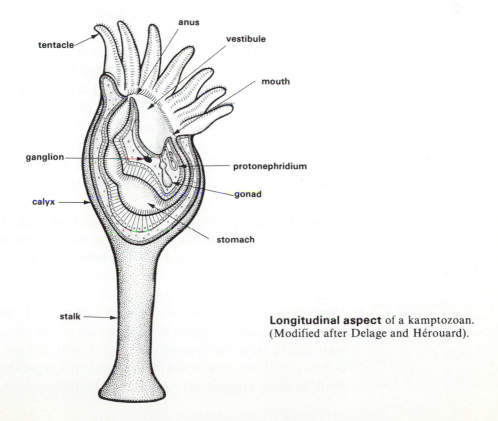

Longitudinal aspect of a kamptozoan.
(Modified after Delage and Hérouard).

Colony of pedicellinid kamptozoans with calyxes raised up on long slender stalks. Beaufort, N. Carolina. (R.B.)

Kamptozoans of most species are tiny, only a few millimeters long, and they live on the bodies or tubes of other animals such as polychetes, sipunculans, bryozoans, ascidians, and sponges. They can move about on the "foot disk," and position themselves in currents generated by the host, in a sense becoming "current commensals." Larger kamptozoans permanently attach to rocks or plant material, and the only movement is waving and bending up and down mediated by contraction of muscles in the stalk. As in most sessile and sedentary animals, the nervous system of kamptozoans is rudimentary and consists of a small central ganglion and radiating nerves.

Kamptozoans replicate extensively by **budding**. Sometimes buds even form on buds, resulting in chains of two or three buds of different ages. As in bryozoans (and some pterobranchs and ascidians) all the tissues in the new bud, including those of the gut, are derived from the body wall. Most kamptozoans have tiny solitary bodies, and budding occurs from specific regions near the mouth on the calyx. The buds detach from the "parent" and crawl away to form large clones, usually on the body of the host animal. Larger kamptozoans develop attachment stolons, and budding occurs from

Budding can be so intense that in some species fully-formed adults are precociously budded from larvas while they are still swimming, and clones can form even before the "parent" larva settles.

the stalks or stolons. Although the bodies remain connected by the stolons, forming colonies, there is little evidence of communication or nutrient flow between zooids, so characteristic of bryozoan colonies, nor is there any polymorphism.

As might be expected for sessile animals that grow by budding, kamptozoans have considerable regenerative ability, and in colonial forms calyxes are occasionally cast off and regenerated. Moreover, thick-walled cysts serve as resting bodies, which can survive adverse conditions.

 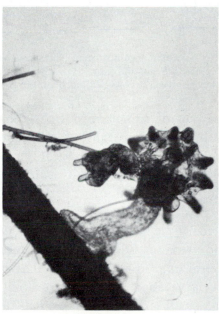 

Budding by a **loxosomatid kamptozoan.** The bud growing from the side of the "parent" *(left),* attaches to an algal filament *(center),* and eventually breaks free *(right).* Western Samoa. (K.J. Marschall)

Most or all species of kamptozoans are hermaphrodites, but usually eggs and sperms are produced at different times within an animal. Sperms are released freely into the water, and fertilization occurs internally, apparently from sperms caught in the tentacles. Embryos develop within the vestibule, often in special brood chambers, and swimming larvas are released. As in some molluscan trochophores, the flattened ventral surface of kamptozoan larvas is heavily ciliated, and they are able to move over surfaces, perhaps sensing suitability for settling with their large preoral organs. Settlement and metamorphosis does not involve extensive degeneration of larval tissues as in bryozoans. After settling, the gut and associated organs rotate during metamorphosis until the ventral surface with the mouth, anus, and tentacles face into the overlying water.

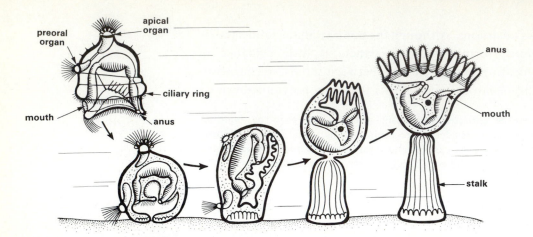

preoral organ

apical organ

ciliary ring

mouth

anus

anus

mouth

stalk

Settlement and metamorphosis of a pedicellinid kamptozoan. The larva settles on the ciliary ring, and its sense organs disintegrate. The gut and associated organs rotate until they face into the overlying water. Tentacles develop and the young animal begins to feed. (Modified after C. Cori)

Features of kamptozoan embryology and early development are similar to those found in molluscs and annelids. Moreover, the larva undergoes only a modest metamorphosis. One view of kamptozoans is that they evolved from a trochophore-type ancestor, perhaps from a larva that swam and crawled close to surfaces, continued to feed with the ciliated girdle after settling, and precociously developed gonads without fully completing metamorphosis (into the ancestral annelid or mollusc?). The lack of a coelom may reflect larval ancestry or be the result of a loss, perhaps related to the small size of kamptozoans, rather than a primitive character. At any rate, kamptozoans are an enigma among invertebrates, and although they are of little if any ecological or economic importance, they will continue to challenge those biologists intrigued by more unusual creatures.

CLASSIFICATION: Phylum KAMPTOZOA (ENTOPROCTA)

There are only about 100 species of kamptozoans in about a dozen genera usually divided among 3 families. Higher levels of classification have not been necessary.

Family Loxosomatidae. Marine, solitary individuals forming clones through budding; nearly all commensal living on the bodies of other animals. *Loxosoma, Loxosomella.*

Family Pedicellinidae. Marine colonies with the zooids connected to attached, branching stolons by a muscular stalk. *Barentsia, Myosoma, Pedicellina.*

Family Urnatellidae. Freshwater colonies with beaded stolons fastened to a small attachment disk. Sometimes combined with the Pedicellinidae. *Urnatella.*

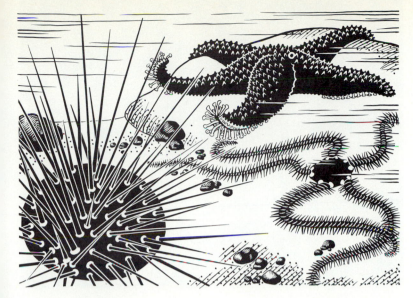

Chapter 27

Echinoderms

Asteroids
Ophiuroids
Echinoids
Holothuroids
Crinoids
Concentricycloids

A PERCEPTIVE seventeenth century Frenchman wrote "It is well to comprehend clearly that there are some things which are absolutely incomprehensible." But his admonition has not deterred zoologists, from his time to ours, in trying to trace the origins, and to understand the functional anatomy of the "spiny-skinned" animals that comprise the phylum **ECHINODERMATA**.

The structure of adult echinoderms is utterly different from that of animals belonging to other phyla. It is only in the embryos and larvas that we see clues of possible distant relationships between the echinoderms and, ironically, the phylum to which humans and other vertebrates belong. Of this, more will be said in the next chapter.

Echinoderms comprise one of the major phyla, with about 6,500 living species. All are marine, and the adults, with a few exceptions, inhabit the sea floor. Some 20 classes flourished in ancient Paleozoic seas, many with bizarre shapes and features, but only 6 classes now survive: **asteroids** (sea stars), **ophiuroids** (brittle stars and basket stars), **echinoids** (sea urchins, heart urchins, and sand dollars), **holothuroids** (sea cucumbers), **crinoids** (feather stars and sea lilies), and **concentricycloids** (sea daisies).

ASTEROIDS

SEA STARS (also called starfishes) are the echinoderms most familiar to everyone, even to people who have never been near the seashore. Many are brightly, even gaudily colored in hues of yellow, orange, red, blue, and purple. Most of the more than 1,800 species are basically similar in structure, and they vary mainly in their skeletal organization and habits.

The Greek root *aster* (="star") appears in the names of many sea stars, and the class name **Asteroidea** means "starlike."

The account given here is based mostly on the common sea stars of rocky temperate seashores such as *Asterias* of the north Atlantic and *Pisaster* of the northeast Pacific.

Sea stars in tidepool. *Asterias vulgaris* is a familiar sight on rocky shores of both sides of the N. Atlantic. These carnivorous echinoderms mostly prey on molluscs, especially bivalves and snails, but also act as scavengers on whatever dead animals they may encounter. Mount Desert Island, Maine. (R.B.)

Exposed intertidal rocks covered with sea stars clinging tightly to the rock surface with their many tube feet. Not until the tide rises again will they move off to feed, most often on mussels. In the subtidal zone, they roam actively about in search of snails and other prey. *Pisaster ochraceus,* the ochre star of the Pacific coast of N. America, is sometimes colored to fit its name; more often it is purple, brown, or orange. Oregon. (R.B.)

Subtidal gravel bottom at about 10 m, densely covered with antarctic sea stars, *Odontaster validus,* bright red on top and yellow beneath. Large populations of these stars prey or scavenge on other benthic animals and play a large part in shaping the bottom community. Those in the central cluster were eating a sea urchin. McMurdo Sound, Antarctica. (P.K. Dayton)

5 arms, or rays, is the most common number for sea stars. *Protoreaster,* rosy beige with black calcareous protuberances, is common on sandy shores throughout the Indo-Pacific. Nha Trang, Vietnam. (R.B.)

Many-armed star. In the typical species of 5-armed stars, unusual individuals are found with from 4 to 7 arms. In addition, there are species of stars that regularly have 6 arms and others with many arms, up to nearly 50. This species, *Solaster endeca,* has from 7 to 13 arms. Mount Desert Island, Maine. (R.B.)

Pentagonal sea star with large marginal plates and without discrete, elongate arms is *Ceramaster placenta.* This small, rather stiff sea star was dredged from 150 m near the Bay of Biscay. (D.P. Wilson)

Pillow star, *Culcita novaeguineae,* is a leathery sea star with an inflated shape, up to the size of a soccer ball. These tropical sea stars feed on corals. A symbiotic fish is often found in the coelom, but how it enters or leaves is not known. Palau. (N. Haven)

Flexible and limber as it turns over or walks about, a sea star can quickly become rigid if picked up or disturbed. This changeable consistency depends not on muscular contraction, but on a type of collagen peculiar to echinoderms. By regulating the amount of calcium in the collagen fibers that bind together the skeletal ossicles, the animal is able to control the stiffness of the fibers. This is the cobalt-blue *Linckia laevigata.* New Caledonia. (R.B.)

Vertebrate bone is composed mainly of calcium phosphate (apatite), while each ossicle of an echinoderm skeleton is a single crystal of magnesium-rich calcium carbonate (calcite).

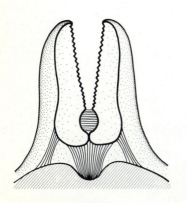

A pedicellaria. (After L. Cuénot)

The body consists of a central **disk** from which radiate **5 arms.** The **mouth** is in the center of the disk on the lower (oral) surface. There is no head, and the animal can move with any one of the arms in the lead. Rare individuals of the 5-armed species have 4, 6, or 7 arms, and some species normally have up to several tens of arms, but such numbers are unusual. The 5-sided, or **pentamerous radial symmetry** is characteristic of sea stars and all other living classes of echinoderms. Because of their radial symmetry, echinoderms were once grouped with cnidarians and ctenophores, but we now believe that the *radial symmetry of echinoderms is secondarily derived,* both during embryological development and evolutionary history. Larval echinoderms are bilaterally symmetrical, while the oldest known echinoderm fossils are neither radial nor clearly bilateral.

The bodies of sea stars appear to be quite stiff, but they can bend and twist into all sorts of shapes to fit the contours of a rocky bottom. A meshwork of small **calcareous plates,** or ossicles, from which project numerous movable and immovable **calcareous spines,** is embedded in the tissues of the body wall. The plates are joined by tough connective tissue, which unites them into a flexible skeletal frame, and by small muscles, which in contracting can change the animal's shape. A series of closely fitted plates, the **ambulacral ossicles,** extend in two rows along the underside of each arm. They constitute the most prominent part of the skeletal system, and protect the nerves and vessels of the arm. Such an internal skeleton, like our own bony one, is an *endoskeleton.* The ossicles of echinoderms are like bone in being permeated with spaces containing body fluid and active, living cells, and in this they differ from the exoskeletons of arthropods and the shells of molluscs, which contain no living cells and usually lie outside the soft tissues of the body. Sea stars and other echinoderms often appear hard and rough, but the outer surface is a living, ciliated epidermis.

In many sea stars, some of the movable spines are modified as tiny 2- or 3-valved pincers, the **pedicellarias,** which occur scattered over the body surface or in clumps around the bases of larger spines. When a small animal creeps over the surface of a sea star, or the larva of a barnacle or other invertebrate attempts to settle there, it is caught and crushed by the toothed jaws of the pedicellarias. This helps to keep the surface free from foreign growths. In sea stars without pedicellarias, the water currents created by the cilia of the epidermis must serve to clear the surface of intruders and debris. Some sea stars have

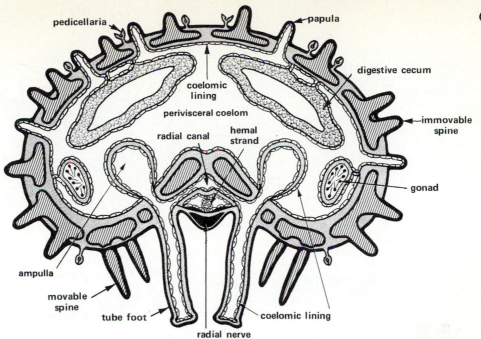

Cross-section of arm of sea star. The only parts of the hemal and perihemal systems shown are the main radial hemal strand and the surrounding perihemal canal.

so many pedicellarias on their bodies that they can catch prey with them. One species has been seen catching unwary fishes that imprudently rested on the sea star's surface and were seized by hundreds of pedicellarias, passed to the mouth, and eaten.

The inner surface of the body wall is lined with ciliated epithelial cells that enclose the **perivisceral coelom** (somato-coel). This coelom provides a spacious fluid-filled cavity into which all the main organs, including the gonads and digestive organs, protrude. As in other coelomates, the organs are enveloped by coelomic lining, suspended by thin mesenteries, and bathed by coelomic fluid. This coelom, in contrast with that of annelids, is not a hydrostatic skeleton but serves major circulatory functions.

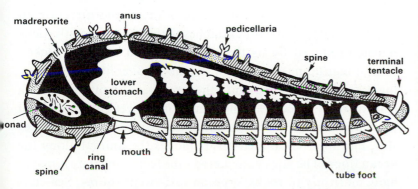

Surface of arm shows large white pedicellarias; spines covered with tiny pedicellarias; and many fingerlike gills (papulas). *Pycnopodia helianthoides*. Friday Harbor, Wash. (R.B.)

Longitudinal aspect of a sea star, through the disk and one arm. Perivisceral coelom is in solid black. Nervous system, hemal system, muscles, and coelomic linings are not shown.

Water vascular system of sea star.
The 5 pairs of Tiedemann's bodies on
the ring canal are not shown. Many
sea stars have in addition one or more
large Polian vesicles on the ring canal.
The function of the madreporite is
obscure. It can be plugged or removed
with little immediate effect on the sea
star's ability to move. Moreover, in
some echinoderms, the madreporite
or its equivalent does not open to the
outside but to the perivisceral coelom.
It is believed that the madreporite
may serve to equalize pressure between the water vascular system and
the surrounding seawater, as during
tidal changes, or function as part of a
pressure-sensitive mechanism.

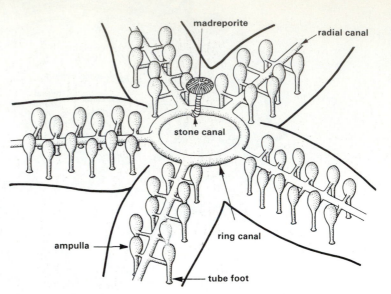

Burrowing beneath the sand by
pushing the grains aside with spines
and non-suckered tube feet is a gray
sand star, *Luidia clathrata*. NW
Florida. (R.B.)

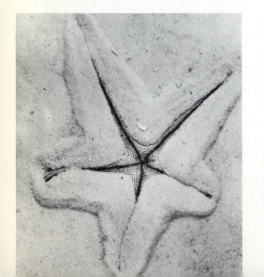

Locomotion of sea stars is by means of a kind of hydraulic-
pressure mechanism, known as the **water vascular system,**
unique to and characteristic of all echinoderms. The system
consists of a separate coelom (tentacular coelom, or hydrocoel) in the form of interconnecting fluid-filled tubes or canals.
A **ring canal** encircles the mouth and gives off **5 radial
canals,** one running along the underside of each arm and
ending at the tip in a terminal tentacle, thought to be sensory.
From the radial canals extend thin-walled cylindrical tentacles, called **tube feet,** or podia (singular, podium). Each
tube foot connects to a rounded muscular sac, the **ampulla,**
which protrudes inward, between the ambulacral ossicles, into
the perivisceral coelom. The tube feet with their ampullas
make up closed units, each separated from the fluid-filled
radial canal by a small valve. When the ampulla contracts, the
fluid it contains is forced into the tube foot, extending it. Small
muscles direct the extended tube foot in one direction or
another, and the tube foot attaches to the substrate with sticky
mucus produced by its tip, which is flared out as a sucker.
Next, longitudinal muscles of the tube foot contract, shortening it, forcing fluid back into the ampulla, and pulling the sea
star forward. Of course, one tube foot is a weak structure, but
there are hundreds or even thousands of them in a large sea
star, and their combined effort moves the animal along the
ocean floor. The sticky, suckered tube feet allow the animal to
hang on firmly to rocks, even on surf-swept shores. Sea stars
living mainly on soft sand or mud have little use for flattened,
suckered tube feet and instead have pointed tube feet which
can be used for walking on "tippy toes" and in burrowing.

Although the ampullas protrude between the ambulacral ossicles into the coelom, the radial canal runs along the outside surface of the ossicles in a V-shaped groove, called the **ambulacral groove**. This arrangement of parts is termed *open*, because of the apparently exposed and vulnerable position of the radial canal. However, when the animal is disturbed, rows of movable spines along the edges of the groove can be brought together to protect the radial canal and retracted tube feet, as well as other structures.

In many sea stars, ampullalike sacs, the Polian vesicles, protrude off the ring canal into the main body coelom; these are thought to hold reserve fluid for the water vascular system. Other small structures on the ring canal, Tiedemann's bodies, may produce hormones or cells. The ring canal is connected to the outside through the **madreporite,** or sieve plate, located on the upper (aboral) surface of the animal and connected to the ring canal by a tube, the **stone canal,** so named because its wall is stiffened by calcareous rings. The madreporite is perforated by many small holes and slits through which water can flow in either direction between the water vascular system and the surrounding sea.

The most condensed part of the **nervous system** is a **nerve ring** that encircles the mouth beneath the ring canal and connects with **5 radial nerves** that extend into the arms beneath the radial canals in the ambulacral grooves. The primarily sensory radial nerves, together with motor nervous elements just above them, receive information from the sensitive tube feet and direct their movements. If the radial nerve in one arm is cut, the arm will move in a direction different from that of the other arms, sometimes even tearing itself off from the rest of the animal. The movements of the body wall and its spines and pedicellarias are coordinated by a sensory **nerve net** running just beneath the epidermis of the animal and by motor components deeper in the wall. **Sense organs** are poorly developed. The terminal tentacle and several other tentacles at the arm tips are thought to be sensitive to food and other chemical stimuli. At the base of the terminal tentacle is a cluster of bright-red simple eyes. And sensory cells occurring all over the body surface as part of the nerve net are also sensitive to light, touch, and other stimuli. Although the nervous system of sea stars appears relatively simple, and there is no well-defined central nervous system, experiments have shown that sea stars can learn to associate various stimuli, such as light, with the presence of food.

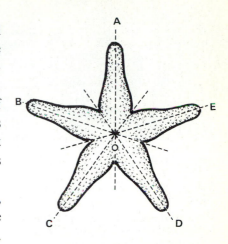

The **radial symmetry** of a sea star is modified by the madreporite, which is usually a conspicuous feature of the upper surface. The madreporite marks a point through which the animal can be divided into two symmetrical parts and by which the different arms can be recognized. Conventionally, the arm opposite the madreporite is labelled *A* while the remaining arms, counterclockwise from above, are referred to as *B, C, D,* and *E.*

Cluster of simple eyes shows as a dark spot near the tip of the arm of *Asterias rubens.* The eyes have a role in phototaxis in at least some stars; they also may detect light intensities or monitor photoperiod. Helgoland, West Germany. (R.B.)

Overturned sea star attaches 2 arms, which curl and begin to move under the body *(left)*. The 2 adjacent arms are drawn up and over *(center),* and finally the last arm releases its hold and swings over *(right)*. Most fairly flexible stars use this "somersault" method of righting. *Nardoa.* New Caledonia. (R.B.)

"Inverted tulip" method of righting is demonstrated by this specimen of *Luidia.* The overturned star pushes itself up on the tips of its arms until it topples over on one side. Then the arms of that side attach and begin to move off, and the rest of the star is pulled right-side-up after them. Plymouth, England. (R.B.)

2-armed sea star, with 3 tiny regenerating arms, can still right itself when turned over. Echinoderms have no statocysts and presumably determine their orientation by the pull of gravity on various structures of the body. *Asterias forbesi.* Beaufort, N.C. (R.B.)

Agonistic behavior is often displayed between two members of the same species, as with these two individuals of *Patiria miniata.* Upon contact, they engage in "arm wrestling," and the animal which places its arm farthest over the other "wins." The "loser" retreats. Monterey Bay, Calif. (D. Wobber)

Sea stars move rather slowly over the ocean floor in search of prey, **feeding** mainly on slow-moving, sedentary, or sessile molluscs such as snails, limpets, clams, mussels, and oysters. Anyone who has ever tried, barehanded, to open a live clam will wonder how a sea star can perform such a feat. In fact, the force of many tube feet acting together can pull the shell valves slightly apart. The delicate lower part of the sea star's **stomach** then turns inside out through the mouth and slips through the opening between the valves, a slit so slim as to defy belief—0.1 mm suffices. In addition, the edges of most bivalve shells are slightly irregular, leaving very small openings between the shell valves. A sea star is able to manipulate the prey with its tube feet until one of these minute openings is next to the mouth and can insert the stomach through such an opening to digest the clam. If the bivalve is attached, as with a mussel, the many tube feet tear the prey from the substrate, and the sea star's stomach is inserted through the opening where the byssus threads emerge. Digestive enzymes are carried into the prey along ciliary tracts on the everted stomach lining, and they begin digesting the bivalve's soft tissues, including the powerful adductor muscles, which relax and allow the shell to gape open for further entry of the sea star's everted stomach.

Rows of tube feet emerge from the open ambulacral grooves that radiate from the central mouth of a sea star. Although the tips of the tube feet look like suckers, they attach mostly by means of an adhesive secreted by the flattened sole of the foot, then release by secretion of another substance which destroys the adhesive. Mount Desert Island, Maine. (R.B.)

Sea stars feeding, with the lower portion of the stomach everted through the mouth.

A sea star *(Asterias vulgaris)* that was humped over a clam in the intertidal has been picked up to show the everted stomach. Maine. (R.B.)

A bat star *(Patiria miniata)* against the glass of an aquarium with the stomach everted over a chiton. California. (R.B.)

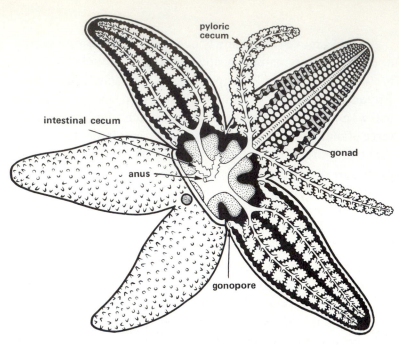

Digestive system of a sea star with body wall cut away. In the arm on the upper right, the two pyloric ceca have been spread apart to show the rows of ampullas and the gonads. The upper part of the stomach is unshaded; the lower part, stippled. In many sea stars saclike intestinal ceca extend from the short intestine; their function, as with many other echinoderm structures, is unknown.

Crown-of-thorns sea stars, *Acanthaster planci,* are famous because of their formidable aspect and because at times they appear in great numbers on coral reefs. They feed mainly on living corals by everting the stomach over the coral polyps and digesting them. At peak densities, the stars have killed substantial portions of the reef corals in localized areas of the tropical Pacific.

The partly digested material is taken into the upper stomach and then into the 5 pairs of branched digestive diverticula, or **pyloric ceca** (singular, *cecum*), where final digestion and uptake occur. The ceca also act as the major organs for storage of nutrients. Very little indigestible material is taken in by this method; the animals are mainly fluid feeders, making particular use of ciliary currents. There is practically no intestine and the anus is very small, or in some species, absent altogether. Many sea stars engulf prey whole and later eject the indigestible remains through the mouth.

Control measures have been taken, such as killing the stars by injecting formalin. With or without such action, most of the stars disappear, and new corals settle and replace those lost. Great Barrier Reef, Australia. (T.W. Brown, Australian Deep Sea Diving Service)

Many species of sea stars feed on animals other than molluscs, such as sponges, sea anemones, worms, barnacles, other echinoderms including sea stars, various dead invertebrates or fishes, algal films, and even mud rich in organic material. In some species the diet is quite specialized. In addition, the epidermis of sea stars and other echinoderms can absorb small organic molecules, such as simple sugars and amino acids, from very dilute solution in seawater. Significant quantities of these molecules may leak from partly digested prey, or from small organisms crushed by the pedicellarias. This mode of nutrition may be important for the cells of the outer surface, which are distant from the digestive system.

Running parallel to the water vascular system in all echinoderms are strands, filled with loosely constructed tissues and with fluids, comprising the so-called **hemal system.** Hemal strands encircle the mouth between the ring canal and nerve ring, sending branches to the digestive system and gonads and into the arms between the radial canals and radial nerves. A mass of spongy brown hemal tissue, the **axial organ,** surrounds the stone canal and, just below the madreporite, connects to a small pulsating "dorsal sac," or "heart." To complicate matters further, much of the hemal system lies in the **perihemal canals,** which together make up a third and separate coelomic system (the perihemal coelom). Despite its name, there is little evidence that the hemal system is a circulatory system, and its functions are unknown. Perhaps it transports special cells, hormones, or complex large molecules to specific organs. In any case, the principal circulation of nutrients and oxygen takes place within the spacious perivisceral coelom, which is lined with a ciliated epithelium that keeps the coelomic fluid in constant circulation.

There are no specialized **respiratory** or **excretory systems** in sea stars. Metabolic wastes are mainly in the form of ammonia, carbon dioxide, and other products which are readily soluble in water. Excretion of these wastes, and uptake of oxygen from the surrounding water, occurs by diffusion through the thin walls of the tube feet. In addition, most sea stars have many small fingerlike projections of the body wall and perivisceral coelom protruding between the calcareous plates of the surface. These projections, called "skin-gills," or **papulas,** have very thin walls, ciliated inside and out, through which respiratory and excretory exchanges occur. Solid waste materials such as cellular debris are engulfed by ameboid cells that circulate in the coelomic fluid. When loaded with wastes, these cells leave the body through the walls of the papulas, thereby "committing suicide."

The function of the hemal system and especially the axial organ has puzzled zoologists for over a century. Surgical removal of the axial organ has little visible effect on the animal. It is readily regenerated. Currently it is thought that the axial organ is a site for the production of new cells and of hormones, and it may also function in the animal's defense against disease.

The reader may note that sea stars have many structures with unknown functions, while some major functions have not been assigned to well-defined structures. With this in mind, the American zoologist Libbie H. Hyman, in her treatise on the echinoderms, wrote, "I ... salute the echinoderms as a noble group especially designed to puzzle the zoologist."

Lacking specialized excretory organs, sea stars and other echinoderms also lack the means to regulate the ionic content of their coelomic fluids, in which most of the major ions are present in the same concentrations as in the surrounding seawater. Most echinoderms are restricted to marine waters of full salinity, but to the extent that they can tolerate corresponding dilution of their body fluids, certain species do manage to live in brackish waters. There are sea stars in the Baltic Sea where the salt content is only about 25% of full salinity. Intertidal species can tolerate occasional rains, but echinoderms are largely excluded from areas of continually fluctuating salinity, such as estuaries, and are absent from freshwaters.

Regenerating sea star with its 8 arms all of different lengths and stages of regeneration. This species readily drops and regenerates the arms. *Luidia ciliaris.* Roscoff, France. (J. Vasserot)

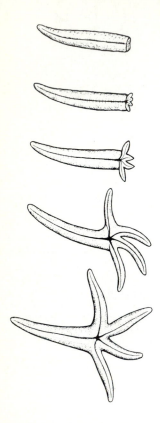

Regeneration from a single arm of *Linckia.* Such replacement is seen in only a few species of sea stars, all in the family Linckiidae. These stars regularly reproduce asexually by rupturing an arm a few cm from the disk. The main body regenerates a new arm, and the cut surface of the cast arm grows out a new disk and a cluster of tiny arms. Such regenerating individuals, called "comets," are slow to achieve normal proportions, and populations of *Linckia* contain few symmetrical members, most being in some stage of regeneration at all times. (After E. Korschelt)

Sea stars have a remarkable facility for **regeneration** of lost parts, including one or more whole arms, and this is adaptive for slow-moving animals that live on surf-swept shores. When an arm is broken off, as by a rolling boulder, the remaining stub seals off the open wound and begins to regenerate. In most species the central disk must be intact for the arms to regenerate, and if the disk is cut in half, both halves die. In some species, however, **asexual reproduction** regularly occurs by division of the disk and complete regeneration of each half.

Sexual reproduction begins with the production of eggs and sperms in 2 or more ovaries or testes located in each arm. Most sea stars, like other echinoderms, have separate female and male individuals. The gonads open to the outside through small ducts leading to pores near the bases of the arms, and the gametes are shed into the seawater, where fertilization occurs.

Eggs and sperms cannot survive long in seawater; unless fertilization is completed promptly, they perish. So it is important that both males and females spawn their gametes at the same time. To do this the individuals of a population must develop their gametes at about the same time, and this timing has been shown in some species to be regulated by seasonal changes in photoperiod. Spawning itself is further synchronized by a variety of factors such as rapid changes in temperature or light. In addition, in some species the first individuals to be stimulated to spawn by physical factors release substances with their gametes which stimulate other individuals to spawn, and this chain reaction assures maximum fertilization.

Spawning of eggs and sperms is partly under the control of a hormone, a polypeptide produced in the radial nerves. This hormone is released and finds its way to the gonads where it stimulates cells in the gonads to produce a second substance, l-methyladenine. This in turn stimulates the eggs to complete meiosis and the gonad wall to contract and expel the eggs or sperms. Injecting a sea star with low concentrations either of l-methyladenine or of radial nerve extracts leads to spawning within 30 to 40 minutes. It is not known what normally regulates the release of the radial nerve hormone in the animal, or how the hormone reaches the gonads.

16-cell stage.

The first sperm that penetrates the egg membrane starts a physicochemical change such that a large amount of fluid collects beneath the egg membrane, elevating it and serving at least in part to prevent other sperms from entering. The egg and sperm nuclei fuse, and **fertilization** is complete. The **zygote** divides into 2 equal cells, and these divide synchronously into 4, 8, 16 cells, and so on. The cleavage planes are alternately vertical and horizontal so that the cells are stacked in rows one upon another, a pattern called **radial cleavage.** As cleavage continues there results a ball of cells (morula) and then a ciliated blastula with a fluid-filled **blastocoel.** The blastula at first tumbles about in the enclosing egg membrane but soon hatches out and swims free in the water. An infolding of the cells at one pole transforms the blastula into a gastrula with ectoderm and endoderm. This crowds out most of the blastocoel and produces the embryonic gut cavity, or **archenteron,** open to the outside through the **blastopore.** From the endoderm near the top of the archenteron arise a pair of outpocketings, which pinch off as small thin-walled sacs on each side. These eventually develop into the **coelom** and its derivatives, including the water vascular system. The remainder of the archenteron differentiates to form an esophagus, a round stomach, and intestine. The blastopore serves as the larval anus, and where the esophagus bends ventrally, and meets an ectodermal ingrowth (stomodeum), the larval mouth breaks through.

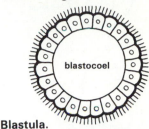

Blastula.

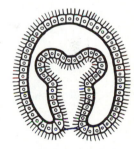

Early gastrula.

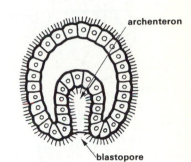

Outpocketing of coelomic sacs from endoderm.

This pattern of development, which differs from that of the protostomes (molluscs, annelids, etc.), characterizes a group of phyla called the **deuterostomes,** which includes the echinoderms and probably the lophophorates, as well as the hemichordates and chordates (treated in the next 2 chapters). The name refers to the formation of the mouth from a second opening (*deutero* = "second," *stome* = "mouth"), *not* from the blastopore. Other defining elements of the pattern include radial cleavage (vs. spiral cleavage) and the formation of the coelom as an *enterocoel* from outpocketings of the archenteron (not as a schizocoel from splitting of a mesodermal mass).

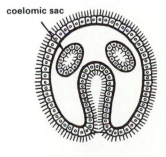

Coelomic sacs pinch off.

Early development of a sea star.
(Mostly after Delage and Hérouard)

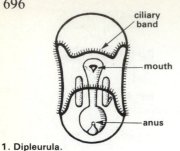

1. Dipleurula.

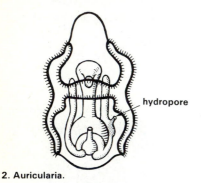

2. Auricularia.

hydropore

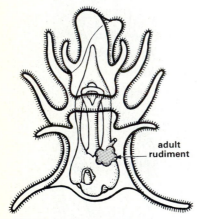

adult
rudiment

3. Bipinnaria.

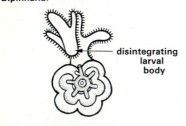

disintegrating
larval
body

4. Metamorphosis.

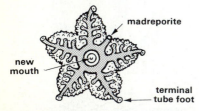

madreporite

new
mouth

terminal
tube foot

5. Juvenile sea star.

Later development of a sea star.
(Mostly after S. Hörstadius)

While the gut and coelom are forming within, the larva elongates and a band of cilia develops around the mouth. This bilaterally symmetrical stage, with a single ciliary band, is called the **dipleurula** (a name originally invented for a hypothetical ancestor of echinoderms). It is typical of all classes of echinoderms except crinoids.

The next stage in sea stars, the **auricularia,** begins with the extension on each side of lobes that carry with them the ciliary band. When the band eventually separates into two bands and the lobes become long arms, the larva is a **bipinnaria**. The cilia propel the larva about and produce feeding currents, trapping phytoplankton and directing them to the mouth.

In many sea stars, the bipinnaria develops 3 small projections on its anterior end. This stage, called a **brachiolaria,** attaches by the projections to some solid object on the ocean floor during **metamorphosis.** By this time, development of the coelomic cavities has already prepared the way for the bilateral larva to transform into the radial adult.

The two original coelomic sacs elongate, extending down both sides of the body, and the anterior and posterior parts may fuse. The right and left posterior portions (somatocoels) enlarge to enclose most of the digestive system, later forming the main body coelom of the adult. Of the two anterior portions (axohydrocoels), the right one degenerates or forms only the tiny "dorsal sac" in the adult. The left one enlarges and sends a tubular outgrowth to the dorsal surface, where it opens by a pore (hydropore) to the outside; this tube is the forerunner of the stone canal. The anterior part (axocoel) forms the perihemal coelom. The posterior part (hydrocoel) forms the whole water vascular system; a tubular outgrowth that will become the ring canal develops 5 lobes, the beginnings of the future radial canals. In larvas ready to complete metamorphosis, the compact rudiment of the ring and radial canals can be seen lying on the left side of the larval stomach.

The final stages of metamorphosis are completed rather rapidly. A new mouth breaks through on the larval left side through the middle of the ring canal, while a new anus opens on the larval right side, thus producing an adult axis at right angles to the larval axis. The 5 radial canals grow out and develop tube feet, the body takes on the adult shape, and the tiny young sea star crawls away to a new life.

Some species of sea stars and other echinoderms, including all crinoids so far studied, do not have feeding larvas. Rather, the eggs are provided with abundant yolk that nourishes the developing embryos and larvas. Yolkiness modifies embryonic development, but the basic pattern of development, including the remarkable metamorphosis, is similar. Some of these yolky embryos are brooded around the mouth or among the spines of the parent. Others are released as larvas, called **vitellarias,** that swim free in the ocean currents and are carried to new sites. These yolky embryos and larvas probably evolved independently many times from ancestral free-swimming, feeding dipleurula-type larvas. Provision of embryos and larvas with abundant yolk limits the number of eggs that can be produced but increases the chance of survival of any individual offspring. If individuals that produce fewer but better endowed eggs have an increased number of surviving offspring, the genes involved in producing such eggs will be perpetuated and become widespread in the species.

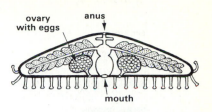

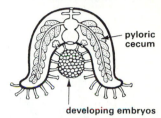

Juvenile sea stars of *Leptasterias* crawling away after being released from the brood pouch. Size about 2 mm. Friday Harbor, Washington. (R.B.)

Development of a sea star, *Asterias rubens.* England. (D.P. Wilson)

Brooding sea star, *Leptasterias*, is a small form which broods its embryos under the mouth. It abstains from feeding and lives on nutrients stored in the pyloric ceca. (After F.-S. Chia)

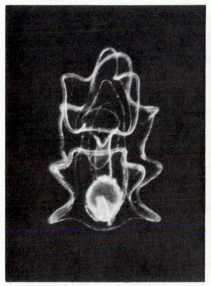

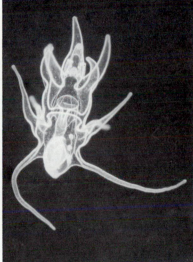

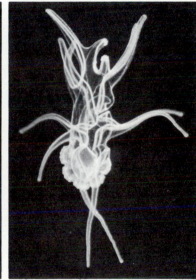

1. Young bipinnaria, ventral view.

2. Early brachiolaria.

3. Brachiolaria, with well-developed adult rudiment; view from right side.

4. Metamorphosing brachiolaria with much of the larval tissue resorbed; view from right side.

5. Late metamorphosis.

6. Young sea star. About 2 weeks after after metamorphosis; 1 mm across; tube feet project beyond spines.

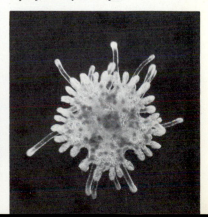

OPHIUROIDS

The Greek root *ophi* ("snake") appears in the scientific names of many brittle stars, and the name of the class, **Ophiuroidea,** means "snake tail-like."

A brittle star. (R.B.)

The row of ambulacral ossicles in the arm form a sturdy endoskeletal axis that looks something like the vertebras of a backbone, and the ossicles are often called "vertebras," which, for an invertebrate, is a classic contradiction in terms. The ossicles are so shaped and articulated that movement is possible (in most ophiuroids) only in the horizontal plane.

BRITTLE STARS inhabit ocean floors the world over. They are the most abundant of echinoderm groups, both in species, which number about 2,000, and in individuals, which in some areas form aggregations of millions that thickly carpet the ocean bottom. Pulled from their hiding places in cracks, under rocks, or just beneath the surface of the sand or mud, they resemble small sea stars except that the rounded or pentagonal **disk** is sharply marked off from the long sinuous arms. The disk is often soft or even flabby, and its surface is smooth and leathery or covered with small scales or spines. At the base of each arm there is a pair of larger disk plates, the radial shields, which in some species cover much of the disk or form a striking pattern. The small **madreporite** is located on the lower surface next to the 5-angled **mouth,** which is framed by 5 wedge-shaped calcareous **jaws,** each bearing a row of teeth. The 5 long, sinuous **arms** (rarely 6 or 7) may be smooth but more often bear vertical rows of spines on each side. The arms are covered above and below with scalelike plates that give them a jointed appearance. These plates, together with the snaky movements of the arms, have earned for these flexible echinoderms the alternative name "serpent stars."

A single series of large cylindrical **ambulacral ossicles** (each formed by the fusion of originally paired ambulacral ossicles) form a central endoskeletal axis within each arm. These ossicles articulate with one another by knobs and grooves at each end, and they are joined by muscles. The arms are the main means of **locomotion** in brittle stars, and the arrangement of muscles and ossicles is such that the arms can hook the substrate with their lateral spines and pull the animal forward. Usually only one or two arms pull the animal along, while the others trail behind. If the arms are seized, they often break off; hence the common name, "brittle star." A new arm regenerates readily.

The pointed tentaclelike tube feet are sensory, and they aid in locomotion by sticking to hard substrates and pushing aside sand grains during burrowing, but their main function is in **feeding.** Brittle stars feed mainly on small organic particles, collected either from water currents or from the surfaces of rocky or muddy substrates. When suspension-feeding, brittle stars raise their arms up into water currents and catch suspended food particles on the waving tube feet. In some species, mucus produced by the tube feet streams out in the water and aids in catching particles. Other species snare planktonic animals with their spines. When feeding from

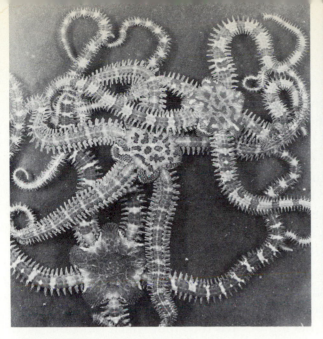

Brittle stars, *Ophiopholis aculeata,* display striking polymorphism in their variety of patterns and colors. This is said to confuse predators such as fishes, making it difficult for them to form a consistent search image. This species is abundant in the Northern Hemisphere around the world. Mount Desert Island, Maine. (R.B.)

Underside of brittle star shows the central mouth surrounded by 5 pairs of buccal tube feet. The 5 narrow arms are bordered by spines; and in each, the ambulacral groove is closed over by a row of plates. In addition to their role in feeding, the tube feet serve in respiratory exchange. Mount Desert Island, Maine. (R.B.)

surfaces, brittle stars sweep the substrate with their flexible arms, and the sticky tube feet pick up food particles to be passed to the mouth. The particles are passed from tube foot to tube foot, as in a bucket brigade, down the arm to the mouth. Specialized tube feet around the mouth (buccal podia), together with the jaws, compact the particles into small pellets before ingestion. Food is digested in the saclike stomach. There is no anus, and undigested remains are expelled through the mouth.

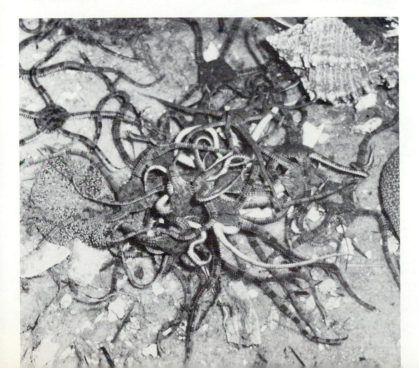

Brittle stars feeding on a dead fish. The carcass is hidden by the mass of brittle stars that have been attracted by scent. Many brittle stars are scavengers or detritus feeders. *Ophioderma.* N.W. Florida. (R.B.)

In one unusual brittle star, *Ophiocanops fugiens,* found in the southern Philippines, the arm coeloms are quite large, and the gonads and outpocketings from the stomach extend into them, as in sea stars. This has been interpreted by some zoologists as evidence for the common ancestry of brittle stars and sea stars.

Nerve cells in each radial nerve are concentrated in the arm segments to form a series of ganglia. Each ganglion communicates only with immediately adjacent ganglia through large nerve fibers, and giant interneurons running through the nerve ring connect the ganglia of adjacent arms. There are no ganglia in the nerve ring. Coordination of activities, even relatively complicated ones such as righting, are achieved by a series of reflexes proceeding from one ganglion to the next. This system of coordination, apparently unique to echinoderms, does not depend on a centralized "brain" but rather might best be described as being more like a "republic of reflexes."

The **perivisceral coelom** is small in brittle stars and is mostly in the disk, surrounding the saclike gut and the gonads, with small tubular coelomic extensions running out in each arm over the ambulacral ossicles. The **water vascular system,** and the associated **nervous** and **hemal systems,** are similar to those in sea stars except that the radial canal, radial nerve, and hemal vessel in each arm are covered below by a series of small plates, so that the ambulacral groove is *closed*. All that remains of the groove is a small tube below the radial nerve, the epineural canal. The tube feet project between the covering plates. There is little space in the slender, compact arms for large ampullas, and instead there is only a small swelling at the top of each tube foot which functions as an ampulla. In some brittle stars, additional small ampullalike bulbs project off the radial canal between the ambulacral ossicles.

Situated between the arms at their junction with the disk are slitlike openings into the **bursas,** sacs which protrude into the coelom of the disk. These sacs, unique to brittle stars, are richly ciliated inside, and water currents are circulated through them for respiratory and excretory exchange. Also, the gonads open into the bursas and release their gametes into the protection of

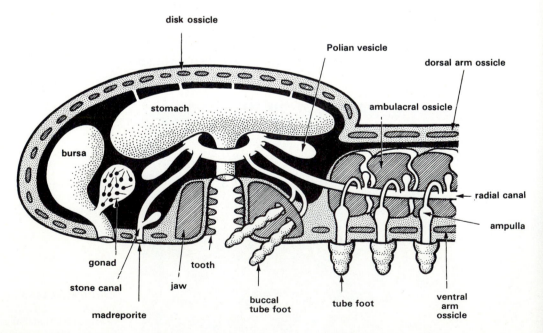

Longitudinal section of the disk and part of one arm of a brittle star. The perivisceral coelom and its narrow extension into the arm are in solid black. Nervous system, hemal system, muscles, and coelomic linings are not shown. (The plane of the diagram is slightly oblique to the central axis of the arm, so that it passes through 2 sets of teeth and a bursa; compare with photo of underside of disk on previous page.)

these sacs. In many species the eggs are fertilized in the bursas, and the embryos are brooded there during development, receiving shelter and sometimes also nourishment from the parent. Other brittle stars shed their eggs into the sea, where fertilization and development occur, as in sea stars with feeding larvas. Free-swimming, feeding larvas of brittle stars differ from those of sea stars, however, in having calcareous rods that support the "arms," elongate lobes bearing the ciliary bands. These exquisite larvas with their 3 or 4 pairs of long arms are called **pluteus larvas** and are remarkably similar to the free-swimming, feeding pluteus larvas of sea urchins.

We have seen that while superficially similar in appearance, sea stars and brittle stars are different in structure and habit. They were probably derived from a common ancestral stock and evolved along divergent lines. Sea stars have retained the open ambulacral groove; and with their locomotory tube feet and broad powerful arms search the ocean floor for relatively large animal prey. Brittle stars, in contrast, having closed the ambulacral groove and narrowed the arms, use the tube feet for feeding on small organic particles.

Sea stars and brittle stars are sometimes placed together in the class **Stelleroidea**; Asteroidea and Ophiuroidea are then considered subclasses.

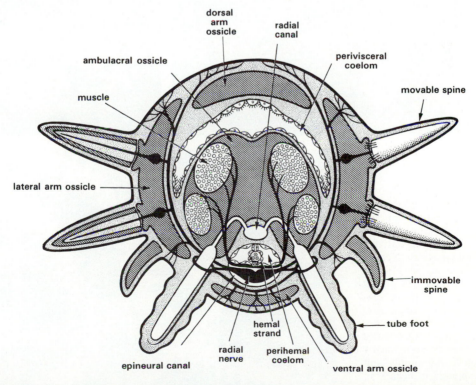

Cross-section of arm of brittle star, passing *between* ambulacral ossicles and cutting through the 4 muscle bands that run from one ossicle to the next. The nervous system is shown in solid black. On the left side of the diagram the spines are sectioned to show the nerves that run in them. Coelomic linings are represented in the main arm coelom and in the perihemal coelom, but not in the water vascular system.

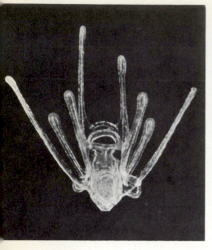

Development of a brittle star. *Left,* ventral view of the fully-developed larva, called an ophiopluteus, with the ciliated bands on long larval arms supported by calcareous rods. *Center,* metamorphosing ophiopluteus. *Right,* newly metamorphosed brittle star with the arms not yet elongated. Black brittle star, *Ophiocomina nigra.* Plymouth, England. (D.P. Wilson)

Left, **long-armed ophiopluteus** and *right,* **newly metamorphosed juvenile** of *Ophiothrix fragilis,* the common brittle star of European Atlantic coasts. Plymouth, England. (D.P. Wilson)

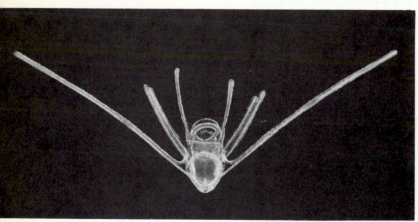
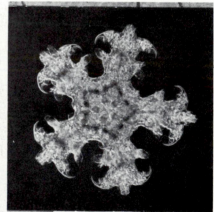

Adult and recently released juvenile of the tiny viviparous brittle star *Axiognathus squamatus.* The adults are hermaphrodites and probably are able to self-fertilize. Up to 7 embryos develop and grow within the bursas of the parent until they reach a relatively enormous size and are ready to crawl away as well-formed juveniles. Both juveniles and adults can travel long distances attached to drifting seaweeds and other debris. Small size, viviparity, hermaphroditism, and a propensity to "raft" are characteristics that make this species both locally abundant in many tide pools and extremely widespread; it occurs in shallow waters in all seas, from subpolar to tropical shores. Adult disk diameter, 3 mm. Monterey Bay, California. (S.S. Rumrill)

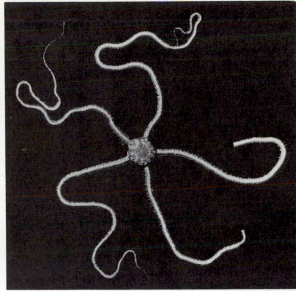

Long-spined brittle star, *Ophiothrix angulata,* is a survivor. It will remain alive in hot, poorly oxygenated seawater long after other animals have died, and is easily maintained in the laboratory. In the dark, it luminesces dimly when stimulated. Though abundant, it is most easily found by collecting the various sponges in which it hides. NW Florida. (R.B.)

Long-armed brittle star, *Hemipholis elongata,* has a relatively tiny, rounded central disk. Great aggregations of these brittle stars live buried in sandy mud, their long arms (15 to 20 cm) extending up into the water. They probably feed on detritus. Both this species and *Ophiothrix angulata* occur from the Carolinas to Brazil. NW Florida. (R.B.)

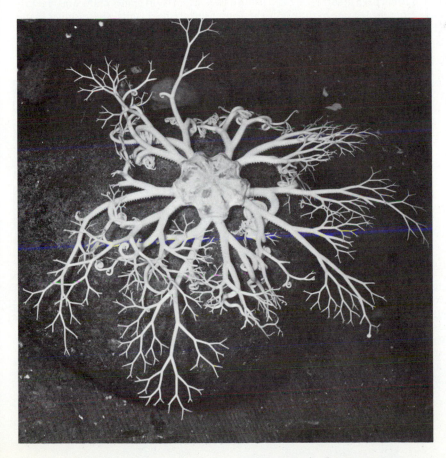

Basket star, *Gorgonocephalus eucnemis,* is a spectacular suspension feeder with an arm-span of over half a meter. The branched arms form an intricate basketwork when extended into the water. Basket stars hide under rocks during the day and collect food at night by climbing up on rocky perches and spreading their arms perpendicular to the currents. The arm branchlets possess little hooks with which they catch large zooplankters or small benthic animals. Puget Sound, Washington. (R.B.)

Sea urchins, several tropical species assembled in an aquarium. *Echinothrix diadema* (upper left) has both thick and thin spines, strikingly banded and provided with fine barbs. The slate pencil urchin, *Heterocentrotus mammillatus* (upper right), uses its heavy spines to brace itself in holes in the coral reef. *Echinometra mathaei* (lower right) also shelters in holes, which it defends against competing urchins of the same species. *Echinostrephus aciculatum* (left center) makes cylindrical burrows up to 10 cm deep in soft coral rock, using the teeth and short spines of the oral region (seen here, as this urchin has bored in from the far side of the rock); the aboral surface bears long, sharp spines that discourage protential predators. Blunt-spined cidaroid, *Eucidaris metularia* (right center), feeds on living corals; unlike the spines of other urchins, cidaroid spines are not covered with an outer epithelial layer and they become encrusted with foreign organisms. Early Mesozoic cidaroids gave rise to all other modern echinoids. Oahu, Hawaii. (R.B.)

Arbacia punctulata, the subject of many classic embryological studies, continues to be in demand by developmental biologists as a reliable source of gametes for experimental investigations. This is the common purple sea urchin of the eastern U.S., distributed along shores of the Atlantic and Gulf of Mexico. NW Florida. (R.B.)

ECHINOIDS

SEA URCHINS and other echinoids appear very different from sea stars, yet they share the same fundamental structure. Sea urchins look like animated burrs with their cover of sharp, movable spines that provide protection and aid the tube feet in locomotion and feeding. Instead of small calcareous ossicles embedded in the body wall, urchins have closely fitted calcareous plates arranged to form a rounded skeleton, or **test,** enclosing the soft parts. The mouth is in an opening on the lower surface of the test, while the anus opens through a hole on top. The plates of the test radiate upward from mouth to anus in 10 series. The 5 series of **ambulacral plates** bear minute pores through which connections to the tube feet project, and these plates correspond to the ambulacral ossicles of the 5 arms of sea stars. Between the series of perforated ambulacral plates are 5 series of large, 5-sided solid interambulacral plates.

At the top of the test, around the anus, is a ring of 10 small **apical plates,** each with a pore. Through 5 of these apical plates, the terminal tube foot of each ambulacral series extends. Through the alternating 5 plates, the gonads open; one such plate is also perforated as the **madreporite.** All the plates of the test bear calcareous **spines,** mounted on ball-and-socket joints and provided with muscles so that they can move in all directions. As in sea stars and brittle stars, the test plates and their spines constitute an *endoskeleton,* permeated with body fluids and living cells, and covered with a thin ciliated epithelium.

The class name **Echinoidea** means "hedgehoglike." There are about 900 species of echinoids.

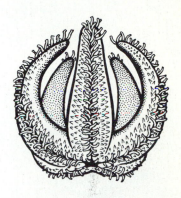

If we imagine the arms of a sea star, with the upper surface skinned off, bending upward to meet at their tips, and if we fill the angles between them with hard plates, we can see how the globular sea urchin is similar to a 5-armed sea star.

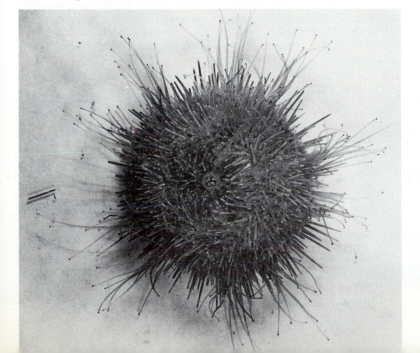

Oral view of a sea urchin shows the mouth with its 5 teeth, in the center of the undersurface. The slender spines and long tube feet are used in locomotion, and the tube feet also attach the animal to the substrate and catch drifting pieces of seaweed. Two rows of tube feet arise from each of the 5 closed ambulacra, separated by interambulacral areas. *Strongylocentrotus droebachiensis.* Maine. (R.B.)

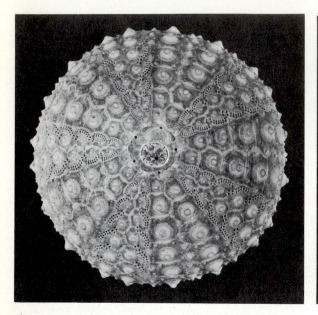

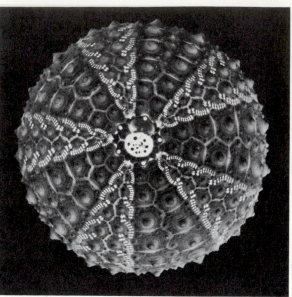

Sea urchin test, in aboral view, with spines removed. The 5 series of ambulacral plates, each with holes for emerging tube feet, alternate with 5 series of solid interambulacral plates. All bear rounded bosses on which the spines pivot. Growth of the calcareous test occurs in the same way as in the closely-fitted bony plates of a human skull. As the animal grows, each separate plate grows along its outer edges and increases in thickness. In sea urchins, new plates are also added from time to time at the top of each series. *At right,* in a test illuminated from the inside, light shines through the holes where the tube feet emerge. In the 10 apical plates, the 5 gonopores glow most prominently, one in the large madreporite plate. A smaller hole shows in each of the 5 smaller alternating plates, through which the terminal tube foot of each ambulacrum projects. Many tiny ossicles surround the anus. *Strongylocentrotus franciscanus.* Monterey Bay, California. (R.B.)

X-radiograph of 2 test plates shows the growth zones that formed as the plates enlarged. The lines form as the urchin grows faster or slower, depositing bands of skeletal material that differ slightly in structure and density. In the laboratory, prominent bands can be induced by alternating periods of feeding and fasting the animals, so that they alternately grow rapidly or slowly. Under natural conditions, many lines are formed each year, and they are not the reliable annual markers that tree rings are. (R.W. Buddemeier)

Skeletal meshwork of high-magnesium calcium carbonate in a sea urchin test. The spaces within the meshwork are occupied by body fluids and living tissue, as in vertebrate bone. Although details of the structure vary among different kinds of ossicles and among taxonomic groups, any bit of echinoderm skeleton (including fossil remains) is at once recognizable by this characteristic structure. SEM of the growing edge of a test plate. (J.S. and V.B. Pearse)

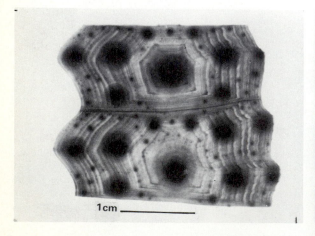

1cm

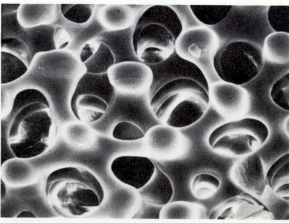

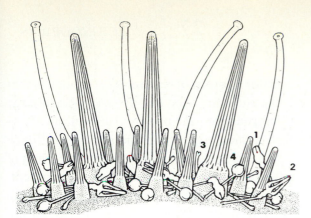

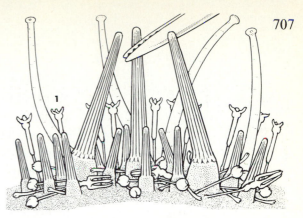

Undisturbed sea urchin. Pedicellarias lie against the surface. Spines and tube feet are straight and extended.

Mechanical disturbance. Large pedicellarias extend and bite at forceps. Spines converge and press against it.

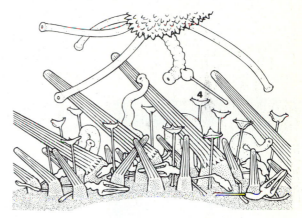

Swimming plankters *(Artemia).* Two kinds of large pedicellarias, and also spines, catch and crush nauplii. Tiny pedicellarias seize ones that fall to the surface.

Predator, a sea star *(Marthasterias).* Tube feet of urchin contract, spines bend away. Toxic, toothed pedicellarias open wide and bite at star's tube feet. The star retreats.

Behavior of different pedicellarias and other surface appendages of a sea urchin, in response to different stimuli. Types of pedicellarias: **1,** ophiocephalus; **2,** tridentate; **3,** triphyllous; **4,** globiferous. *Psammechinus miliaris,* a small portion of the aboral surface. (Based on M. Jensen)

The **pedicellarias** of sea urchins are more complex than those of sea stars. They are on tall moveable stalks, and most have 3 jaws (rather than the 2 jaws more common in sea stars). In some urchins large pedicellarias, acting mainly for defense, contain poison sacs, and the tips of the jaws are grooved or hollow like a hypodermic needle. Other types seize and crush swimming larvas and other plankters, which are not only prevented from settling and growing on the urchin but also provide an extra source of nutrition. Small, flexible pedicellarias probe among the spines, keeping the surface clean and free of foreign growths.

Toxic pedicellarias of the sea urchin *Strongylocentrotus franciscanus* open wide on contact with certain predatory sea stars. Even an isolated tube foot of the sea star, presented to the urchin with a forceps, elicits this reaction. California. (D. Wobber)

Although formidable and capable of inflicting painful wounds, the spines of most sea urchins are not poisonous. Moreover, most pedicellarias are too small to inject poison through the human skin. However, a few kinds of sea urchins such as the tropical leather urchin, *Asthenosoma,* do have poisonous spines, while others, such as *Toxopneustes,* with short spines, bear large pedicellarias which cover the test like delicate 3-petaled flowers. These sea urchins have been known to seriously injure or kill unsuspecting divers who picked them up.

The **water vascular system** of sea urchins is similar to that of sea stars, and the suckered tube feet are attached to large ampullas which protrude into the perivisceral coelom. However, the tube feet, projecting beyond the spines, are much longer and more slender, and they are provided with a ring of calcareous ossicles around the sucker for muscle attachment. The radial canal, along with the radial nerve and radial hemal strands, runs along the *inside* of the ambulacral ossicles, rather than along the outside as in sea stars. This is a *closed* ambulacral system, but of a type quite different from that in brittle stars where the radial canal and associated structures lie between the ambulacral ossicles and outer covering plates.

Sea urchins **feed** mainly on algal material, on encrusting animals such as bryozoans, and on dead animal matter. The spines and tube feet are used to snare large pieces of food. The food is "tested" and manipulated by specialized tube feet (buccal podia) around the mouth, and then chopped into small pieces by 5 sharp pointed teeth, held by a complex set of jaws, with some 35 ossicles and powerful muscles. This chewing mechanism, which prepares the food for easy ingestion, is called **Aristotle's lantern.** The **digestive tract,** lying coiled in the spacious perivisceral coelom, is a long, thin-walled tube which can be divided into pharynx, esophagus, small and large intestine, and rectum. The long digestive tract seems to be associated with the mainly vegetable diet, which requires more prolonged digestion than does animal food. A small tube, the "siphon," branches off from the esophagus, parallels the small intestine, and rejoins the large intestine, allowing excess water ingested with the food to bypass the small intestine. Portions of the digestive tract are enmeshed with hemal strands which may aid in the transport of large molecules to other parts of the body.

Aristotle's lantern of a sea urchin, with its many ossicles and muscles. An arrow marks one tooth. The esophagus shows as a dark tube emerging from the top. *Strongylocentrotus franciscanus.* Central California coast. (R.B.)

The tube feet, with their large ampullas extending into the perivisceral coelom, are the main organs of **respiratory exchange,** but there are also thin-walled projections that serve as "gills" on the soft membrane surrounding the mouth; these structures provide the additional surface for respiratory exchange required by the powerful jaw muscles.

The role of the tube feet and their ampullas in respiration can be elegantly demonstrated with luminescent bacteria which require oxygen to produce light. If a sea urchin is cut in half crosswise, and the upper half turned over and filled with water containing luminescent bacteria, then covered with a piece of glass, light produced by the bacteria will soon be restricted to surfaces of the ampullas.

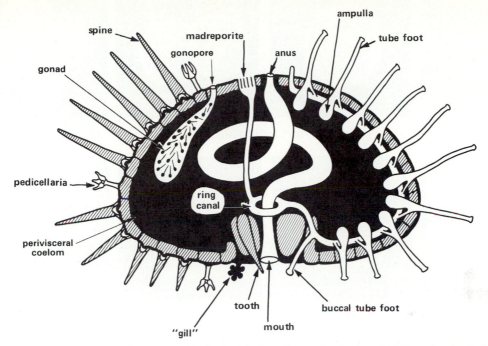

Section of a sea urchin, cutting through an interambulacral row of spines on the left and an ambulacrum on the right. The perivisceral coelom is in solid black. Such a section cuts through only one of the 5 teeth of the Aristotle's lantern. Nervous system, hemal system, muscles, and coelomic linings are not shown.

The most conspicuous organs inside the body of a sea urchin are those of the **reproductive system.** When packed with ripe eggs or sperms, the gonads nearly fill the spacious body coelom. The gametes are spawned into the sea at the slightest disturbance. **Development** of the embryos is essentially like that of sea stars and brittle stars, and the pluteus larvas have a calcareous skeleton as in brittle stars.

As in other echinoderms, before metamorphosis the left anterior coelom grows out to form the ring canal of the water vascular system. Then, in sea urchins, the ectoderm on the left side thickens and sinks inward over the ring canal to form a **vestibule.** The first tube feet protrude into the vestibule, and at metamorphosis the vestibule opens up exposing the new oral surface of the juvenile sea urchin.

Spawning of gametes. This male sea urchin *(Arbacia punctulata)* released sperms after a low-voltage electric current was applied. A 0.55 molar solution of potassium chloride injected into the coelom also causes most species of sea urchins to spawn. (R.B.)

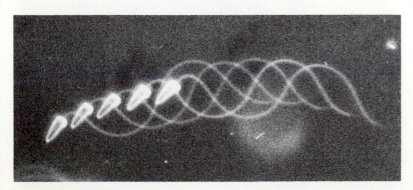

Sperm of a sea urchin, a multiple-flash dark-field photomicrograph showing the head and flagellum in 5 successive positions as the sperm swims toward the left. This technique was used to analyze the biophysics of flagellar propulsion. *Lytechinus pictus.* Scale line, 10 μm. (C.J. Brokaw)

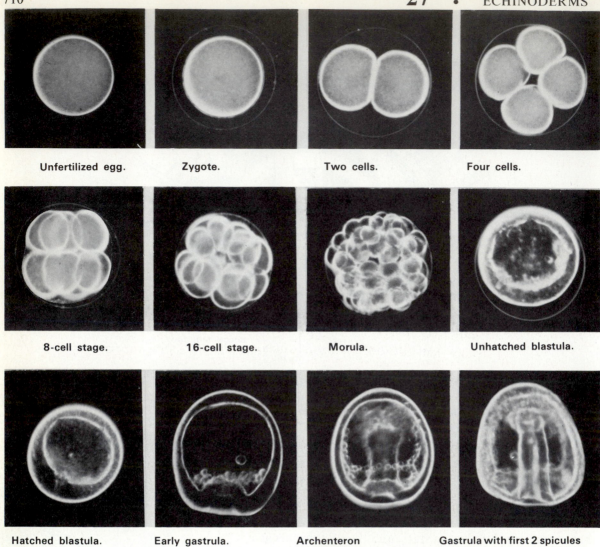

Unfertilized egg.   Zygote.   Two cells.   Four cells.

8-cell stage.   16-cell stage.   Morula.   Unhatched blastula.

Hatched blastula.   Early gastrula.   Archenteron developing.   Gastrula with first 2 spicules forming.

Early development of a sea urchin, *Echinus esculentus.* Plymouth, England.

Gastrula of a sea urchin, dried and broken open, reveals mesenchyme cells scattered around the blastocoel. Attached between the tip of the archenteron and the roof of the blastocoel, they help with the process of invagination. Long diam. of gastrula, about 235 μm. *Strongylocentrotus droebachiensis.* SEM. (C.B. Calloway and R.M. Woollacott)

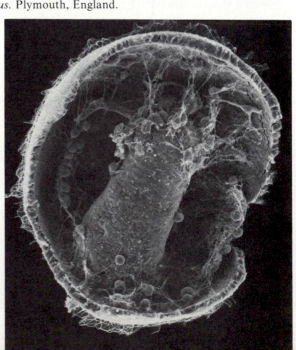

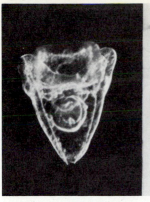 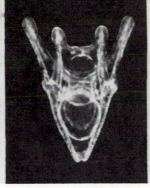 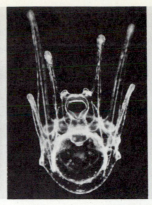 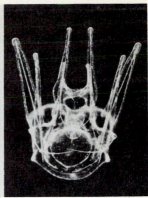

Early echinopluteus.

Fully developed echinopluteus.

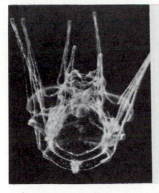 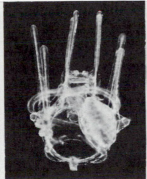 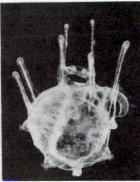 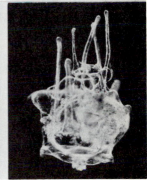

Development of adult rudiment on left side. The larva is now "competent" (ready to metamorphose).

 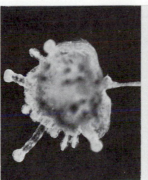  

Metamorphosis proceeds. Larval spines project above, tube feet below. **Young sea urchin.**

Later development of a sea urchin, *Echinus esculentus.* Plymouth, England. (D.P. Wilson)

Because they are so numerous and easily available, the eggs and sperms of sea urchins have provided excellent material for many studies in developmental biology. They can be obtained easily by injecting animals with a solution of potassium chloride or by passing a 6-volt current through the body. The muscles of the gonads contract, expelling the gametes into laboratory dishes. This makes it possible to study in detail the events of fertilization and the important early minutes of development when the embryo begins to manufacture its own ribonucleic acids and enzymes and becomes independent of the molecular machinery provided in the egg. The embryos develop quickly into free-swimming, feeding pluteus larvas which can be raised in the laboratory through metamorphosis to the adult stage, so that the genetics of development can also be studied.

Sea urchins are usually found on hard rocky surfaces, often in crevices and holes which they themselves produce by the action of their spines and teeth. Other echinoids, the **sand dollars** and **heart urchins,** are specialized for living in soft sands and muds. Heart urchins burrow below the surface and pick up organic particles with their tube feet, while partly buried sand dollars produce mucus that catches organic particles settling on the upper surface from the water. Some sand dollars stand up in the sand on the "anterior" edge, oriented so that they lie perpendicular to the water currents that move along the bottom. Sometimes hundreds or thousands of sand dollars can be found packed together standing edgewise on sandy bottoms. The symmetry of sand dollars and heart urchins has taken on bilateral aspects. The location of the anus is shifted from its usual apical position to a site near one edge of the lower surface, establishing anterior and posterior ends in these animals. In heart urchins the mouth also is shifted to the edge of the lower surface opposite the anus. The bilateral symmetry can be seen inside these echinoids in the loss of 1 to 3 of the 5 gonads, so that there are only 1 or 2 gonads on each side of the animal.

Heart urchin, *Echinocardium cordatum,* exposed in its burrow in sand. Specialized tube feet construct the vertical mucus-lined sand tube leading to the surface (top of photo). This channel serves for circulation of water, and long respiratory tube feet extend up into it. Other specialized tube feet around the mouth help to gather fine food particles from the sand. England. (D.P. Wilson)

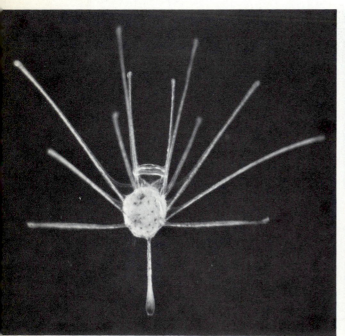

Pluteus of a heart urchin is recognizable by its "aboral spike" projecting below. *Echinocardium.* Plymouth, England. (D.P. Wilson)

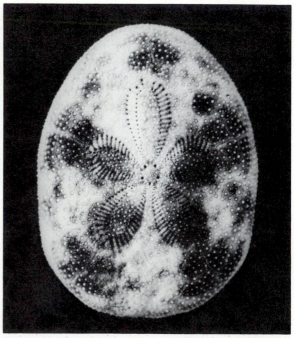

Cake urchin test, aboral view showing the central gonopores and the 5 petaloid ambulacra, with holes for the respiratory tube feet. The many small circles are raised bosses on which the short spines are mounted. The tube feet of the oral surface are specialized for feeding. *Clypeaster reticulatus,* about 3 cm long. (A. Reed)

Bed of sand dollars, *Dendraster excentricus,* was estimated to have a mean density of 400 to 500 individuals per m², ranging up to 1,200 per m² at the dense seaward edge, shown here. Laboratory findings indicate that close spacing of individuals enhances their ability to feed on suspended material. Current velocity determines optimal density, which is adjusted locally and seasonally by active movements of the sand dollars in a mutually advantageous fashion. The higher summer densities, corresponding to times of less wave surge, may also favor fertilization during the summer spawning season. Zuma Beach, S. California. (J.G.Morin)

Five-slotted sand dollar (or keyhole urchin), *Mellita quinquiesperforata,* burrows obliquely into the sand, leaving the posterior edge exposed. Several sand dollar species develop slots, or lunules, in various numbers and patterns; they probably facilitate feeding. *Left,* live animal burrowing in sand. *Right,* test, cleaned of its carpet of short spines. NW Florida. (R.B.)

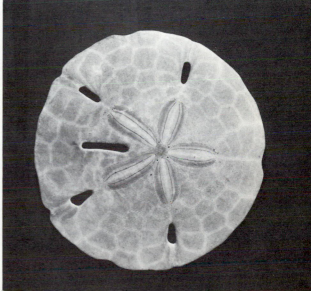

Diadema setosum is one of several species of this handsome genus, conspicuous in the tropics around the world. The long black spines are sharp, brittle, and barbed. Difficult to remove and painful, they are often said to be poisonous, but are probably no more toxic than any other introduced foreign protein. Small shrimps and fishes often shelter among the spines. Although the urchins have no recognizable eyes, they will orient the spines toward an approaching hand or foot. They move toward dark objects and areas, and aggregate in clumps during the day, dispersing at night to forage. Populations of *Diadema* spawn in synchrony, at certain phases of the moon, and Pacific islanders know when to collect them so that the gonads will be at their peak of flavor. Test diameter about 10 cm. Australia. (O. Webb)

Colobocentrotus clings tightly to surf-swept rocks. At the other extreme from slender-spined forms such as *Diadema,* this subtropical urchin has the aboral spines modified as a mosaic of little flattened shields, plus a girdle of heavier paddle-shaped ones. The flattened under surface has a reduced number of spines but an unusually large number of tube feet. All these modifications presumably help the animal to withstand pounding surf. Hawaii. (A. Reed)

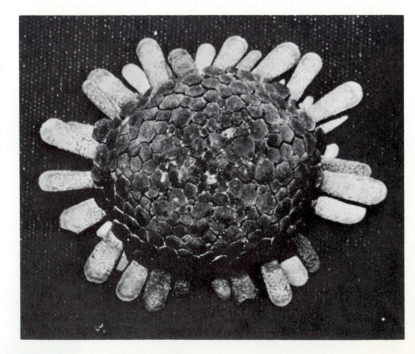

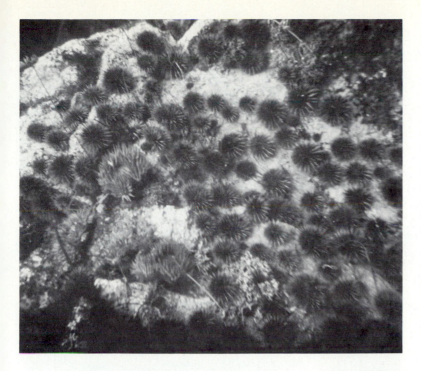

Enormous populations of sea urchins sometimes develop, grazing so voraciously on bottom-growing seaweeds that large areas are stripped nearly barren of upright plants. Such masses of sea urchins can sometimes be attributed to human activity. For example, in the northwest Atlantic heavy fishing of lobsters, a major predator of sea urchins, has been linked to the appearance of large populations of sea urchins and the consequent severe overgrazing of plants. In the northeast Pacific, 19th century fur-hunters nearly exterminated sea otters, another major sea urchin predator; this probably contributed to the development of huge populations of sea urchins and to a decline in the kelp forests on which sea urchins feed. *Strongylocentrotus purpuratus.* 10 m. Papalote Bay, Baja California. (J.S. Pearse)

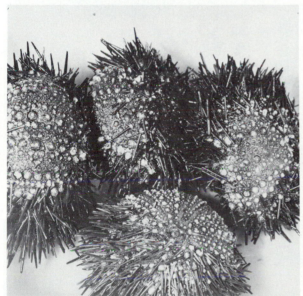

Mass mortalities of sea urchins have been recorded a number of times from various places around the world. Far from providing a happy balance to urchin population explosions, these seem to be unrelated events. They may be local events affecting only a minority of the population, as with the dying sea urchins *(Strongylocentrotus)* seen in this photo from Monterey Bay, California; or they may be widespread, as with the death of all but a few percent of the urchins in affected populations of *Diadema antillarum* in the western tropical Atlantic and Caribbean

Predation is another factor that may severely reduce populations of sea urchins, and in some places humans are significant predators. Sea urchin gonads are enjoyed by people of the Pacific islands, the Mediterranean, and various other parts of the world. This Chilean at Puerto Montt favors his sea urchin on-the-half-shell. Urchins are exported from Chile, the U.S., Australia, and other coasts where they are still relatively abundant to Japan, where they have been overcollected. (R.B.)

in 1983 and 1984. The pattern of spread and species-specificity of this episode suggested a water-borne pathogen. Some instances have been associated with stress, such as unusually high temperatures, which might make the urchins more vulnerable to pathogens. It is not clear whether the various bacteria, fungi, and protozoans found in sick urchins are causing the disease or merely represent secondary infections. (R.B.)

Suspension-feeding sea cucumber attaches itself to the bottom with tube feet that extend along the body in 5 rows. This species commonly hides its pale body in rock crevices or in burrows vacated by other animals, holding its darker colored, delicately branched, dendritic tentacles extended up into the water currents. Fine suspended food particles adhere to the sticky tentacles, which from time to time are inserted into the mouth and wiped off. The tentacles, which are modified buccal tube feet, may be withdrawn when the animal is not feeding or is disturbed. *Cucumaria saxicola.* Plymouth, England. (D.P. Wilson)

Deposit-feeding sea cucumber exploits the organic content of the soft subtidal substrates on which it lives, using its peltate tentacles to scoop sand and mud into its mouth. The small amount of organic material is digested and absorbed, and the great bulk of inorganic sediment is passed through the gut and eliminated through the anus. Where such cucumbers are abundant, much of the surface sediment is processed through their guts several times each year. In this type of holothuroid, there is a pale, differentiated undersurface with 3 rows of many tube feet, contrasting with a darkly pigmented uppersurface on which the remaining 2 rows of tube feet are modified as papillas. *Holothuria.* Plymouth, England. (R.B.)

Detritus-feeding sea cucumber has a long smooth wormlike body, without tube feet. It pulls itself slowly along the bottom with its pinnate tentacles, which also sweep the surface, collecting the fine organic particulate material on which this tropical holothuroid feeds. Faster locomotion involves peristaltic contractions of the body wall, so thin that it reveals the viscera within. *Opheodesoma spectabilis.* Hawaii. (A. Reed)

HOLOTHUROIDS

SEA CUCUMBERS may be thought of as elongated, flabby sea urchins, which have no spines or large skeletal plates and which lie on one side. The mouth is at one end of the cylindrical body, and the anus is at the other. As in most echinoderms, there are 5 rows of tube feet. The 3 rows of suckered tube feet along the lower surface serve for attachment and locomotion; and 2 rows of small pointed tube feet along the upper surface provide for some respiratory exchange. This specialization of the tube feet, and other modifications associated with lying on one side, result in a kind of bilateral symmetry derived from an earlier radial symmetry. It differs from the bilateral symmetry of sand dollars and heart urchins in that it does not involve a shift of the anus or other changes seen in those echinoids.

The arrangement of the tube feet varies among the several families of sea cucumbers. Some have 5 discrete rows as described, but the 3 rows on the lower surface may be condensed to form a well-defined "sole" which firmly attaches the animal to the substrate. Others have tube feet scattered all over the body, while still others lack tube feet except for those around the mouth.

The major skeleton of sea cucumbers is a **calcareous ring** of plates, usually 10, which encircle the pharynx, serving for the attachment of muscles and supporting the ring canal and nerve ring. The calcareous ring may correspond, at least in part, to the Aristotle's lantern of sea urchins. Embedded in the fleshy body wall of sea cucumbers are numerous microscopic calcareous spicules, often with bizarre shapes: tiny perforated plates, tables, buttons, and anchors. Presumably they add strength to the body wall, but the significance of the different shapes is obscure. Five strong bands of **longitudinal muscles** enable the animal to twist and turn, or even to swim short distances by undulating movements. In burrowing, the sea cucumber alternately contracts its longitudinal and circular muscles.

Among the sea cucumbers are found the only pelagic echinoderms. Some are deep-sea forms that swim slowly over the bottom, gently undulating back and forth, with the mouth directed downward. In the icy waters of the Antarctic, they have been encountered at shallow depths by divers. Other pelagic holothuroids appear very like medusas, with a membrane, supported by tentaclelike papillas, extending down around the mouth and with the soft internal organs compacted into a jellylike "bell."

Sea cucumbers, with tentacles around the mouth and cylindrical bodies, look somewhat like large polyps lying on one side. The class name **Holothuroidea** is from Greek roots meaning "polyplike." There are about 1,100 species of holothuroids.

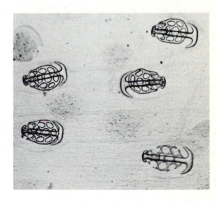

Calcareous spicules in the skin of *Leptosynapta inhaerens*. Photomicrograph. (D.P. Wilson)

Pelagic deep-sea holothuroid, *Benthodytes,* swimming just above the bottom. Photographed from the submersible *Alvin* at about 2,000 m near the Bahamas. (D.L. Pawson)

People of the tropical Pacific have long used pieces of sea cucumber to poison fishes in rocky pools. More recently, some of these toxins have been purified and found to break up vertebrate red blood cells as well as to inhibit the growth of cancer cells. On the other hand, some of these same species of sea cucumbers are boiled, dried, and sold as *trepang* or *bêche-de-mer,* to be eaten as a high-protein delicacy all over the Pacific but especially by the Chinese. Cut into thin strips it is often used in soups, or may be served as a main course.

In Western Samoa two of the authors were served raw sea cucumber viscera that had been relieved of astringency and bitterness by soaking all day in warm seawater.

The spacious rectum and respiratory trees of some sea cucumbers, which are regularly flushed with seawater, make a fine home for other animals, including small crabs and fishes. A fish commonly occupies the main stem of one of the respiratory trees of some tropical species. From time to time, it pokes its head out of the host's anus, and at night it ventures forth in search of food.

The **water vascular system** in sea cucumbers is similar to that in sea urchins, with the ring canal encircling the calcareous ring and the 5 radial canals running along the *inside* of the body wall. Although there are no ambulacral ossicles, the arrangement is similar to the closed ambulacral system of sea urchins, and branches to the tube feet extend through the body wall. Large ampullas protrude into the spacious perivisceral coelom. A stone canal arises from the ring canal, but in most species this does not lead to a madreporite on the external surface; rather it divides and ends internally as numerous **madreporic bodies,** each perforated with many tiny pores that provide communication between the water vascular system and the perivisceral coelom.

The buccal tube feet are greatly enlarged and elaborated as **feeding tentacles.** Many sea cucumbers feed by slowly creeping along the bottom and using the tentacles to pick up organic particles that adhere to sticky mucous secretions. Other species are more sedentary, and their tentacles are branched to form a meshwork when fully expanded into the overlying water. The tentacles coil around and snare large pieces of floating food or catch small particles on sticky secretions. The food-laden tentacles are thrust deep into the mouth, one by one, to be wiped clean by the pharynx.

The digestive tract includes a bulbous stomach followed by a long thin-walled intestine looping within the perivisceral coelom and ending in a rectum, or "cloaca," from which usually arise a pair of **respiratory trees,** thin-walled branching tubes that lie in the perivisceral coelom. Water is drawn through the anus into the rectum, and then into the trees for respiratory exchange with the coelomic fluid in much the same way as in the rectal respiratory systems found in echiurans and dragonfly naiads.

Although circulation of nutrients and dissolved gases undoubtedly occurs within the large perivisceral coelom, as in other echinoderms, many sea cucumbers also have a notably well-developed **hemal system.** Two main vessels running along the digestive tract are interconnected by a multitude of capillaries within the intestinal wall and by the **rete mirabile,** a "wondrous network" of hemal branches between the coils of the intestine and in close association with the left respiratory tree. One of the main vessels contracts rhythmically, like a heart, and pumps hemal fluid throughout the system to most parts of the body.

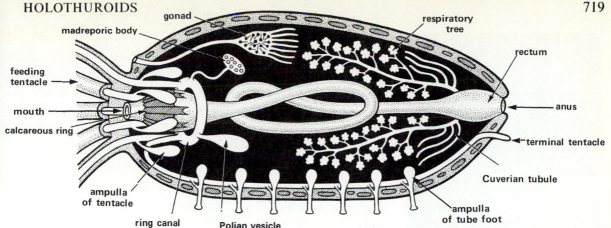

feeding tentacle

mouth

calcareous ring

madreporic body

gonad

respiratory tree

rectum

anus

terminal tentacle

Cuverian tubule

ampulla of tube foot

ampulla of tentacle

ring canal

Polian vesicle

The soft-bodied sea cucumbers lack protective spines or skeletal plates, and one may wonder how they are protected from predators. Many burrow or hide in deep rocky crevices, but others creep around in the open, seemingly oblivious of potential predators. Their defense must be mainly chemical rather than structural, for many sea cucumbers have powerful toxins in the body wall. Moreover, in many common tropical species some of the tubules of the respiratory trees are modified to form toxic and extremely sticky slender threads called **Cuverian tubules.** When disturbed, the animal points its anus at the source of irritation, contracts the body wall, and forces the Cuverian tubules through a preformed rupture line in the rectum and out the anus. The broken tubules lengthen, writhe about in all directions, and thoroughly entangle the offending animal. The lost tubules are readily regenerated.

Longitudinal aspect of a sea cucumber. The perivisceral coelom is shown in solid black. The calcareous ring (marked by diagonal lines) encircles the pharynx. As the radial conals leave the ring canal, they pass forward between the calcareous ring and the pharynx, but are here shown exposed, for clarity. Nervous system, hemal system, muscles, and coelomic linings are not shown.

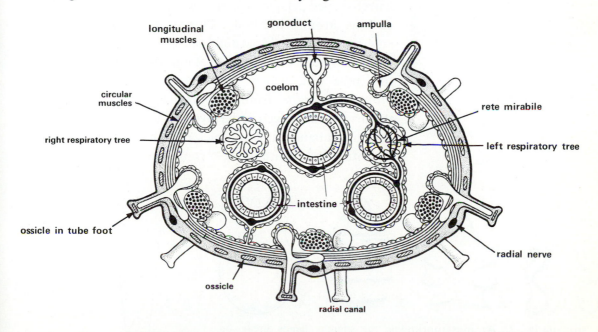

longitudinal muscles

gonoduct

ampulla

circular muscles

coelom

rete mirabile

right respiratory tree

left respiratory tree

intestine

ossicle in tube foot

radial nerve

ossicle

radial canal

Cross-section of a sea cucumber. The nervous system and the hemal system are drawn in solid black.

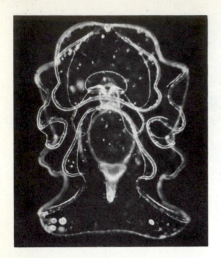

Auricularia. *Labidoplax digitata.* Plymouth, England. (D.P. Wilson)

Pentacula. Bimini. (R.B.)

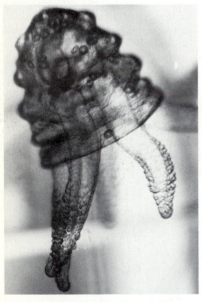

Development of a sea cucumber shows changes from bilateral auricularia to biradial pentacula.

When disturbed or stressed, as in warm or stagnant water, many sea cucumbers have the startling and unattractive habit of **eviscerating.** The body wall contracts and ruptures (at one end or the other), and the gut, respiratory trees, and gonad are all spewed out. This drastic action may deter, or distract, or provide a meal for the predator while the sea cucumber escapes. Or, voiding most of the viscera may lower respiratory demands during periods of stress. Some sea cucumbers eviscerate seasonally, and this may be a means of ridding the body of waste products accumulated in internal organs, a habit reminiscent of brown body formation in bryozoans. An eviscerated sea cucumber, under favorable conditions, will **regenerate** all lost internal organs, and some sea cucumbers are able to **reproduce asexually** by transverse fission of the body followed by regeneration of both halves.

Sea cucumbers **reproduce sexually,** producing gametes in a single, much-branched gonad suspended in the perivisceral coelom. Sexes are usually separate, as in other echinoderms, and the eggs and sperms are spawned into the seawater. Some retain the eggs and brood their young.

Early development, like that of other echinoderms, usually leads from a dipleurula larva to a feeding **auricularia,** similar to the auricularia of sea stars, but with a more extensively looped and sinuous ciliated band. Unlike other echinoderms, sea cucumbers undergo metamorphosis in mid-water to a little barrel-shaped pelagic juvenile, called a **doliolaria,** with ciliated rings around the body. During metamorphosis the left anterior coelom grows around the larval gut to form the water vascular system and the ventral mouth shifts anteriorly. The first tube feet, which will become the feeding tentacles, protrude anteriorly from around the mouth so that the axis of the larva is unchanged in the juvenile. The animal settles, sheds the ciliated bands, and becomes a benthic juvenile, now called a **pentacula** because of the 5 conspicuous tentacles.

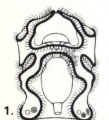    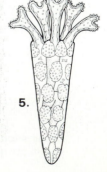

1. Auricularia. 2. At metamorphosis the sinuous bands break up. **3. Doliolaria. 4. Pentacula. 5. 3-month-old pentacula** shows bifurcating tentacles and the many ossicles in the body wall. (**1-4,** *Synapta digitata,* after R. Semon; **5.** *Cucumaria elongata,* modified after Chia and Buchanan)

CRINOIDS

FEATHER STARS gracefully extend slender branching arms in all directions from a small central body, and at first glance do not suggest their relationship with globular sea urchins or chunky sea cucumbers, but they share all the basic echinoderm features. The central body of a feather star is encased in calcareous plates, embedded within the surface tissues and arranged in pentamerous cycles. The arms are supported by a series of internal ambulacral ossicles, and each arm is lined along both edges with **pinnules,** slender side branches supported by smaller ossicles and bearing tufts of tentaclelike tube feet.

Feather stars differ from sea stars in their **orientation;** a feather star is like a sea star turned over, with the mouth directed upward into the water. The animal holds onto the rocky bottom with a ring of rootlike **cirri,** and each cirrus is supported by a series of ossicles. The 5 arms divide repeatedly until there are as many as 200 radiating out from the body in some species. The alternately arranged pinnules give each arm a delicate featherlike appearance, hence the common name.

The number of arms in different species of feather stars is correlated with depth and temperature; those species in shallow tropical seas have the most arms, while species in deeper or more temperate waters have the fewest. Coloration also is related to depth and temperature; and the brilliantly colored shallow-water feather stars of the tropics, with patterns of browns, yellows, and reds, are among the most beautiful of marine creatures.

The class name **Crinoidea** is derived from a Greek root meaning "lilylike." Crinoids that live permanently attached by a long slender stalk number about 75 species and are often called "sea lilies." The stalkless crinoids, called feather stars, are shaped like little ferns, and their common name in Japanese means "sea ferns." There are about 600 species of feather stars.

Feather star, *Antedon,* the common shallow-water crinoid genus of the NE Atlantic and Mediterranean. Its 5 arms branch close to their bases into 10, each of these with many alternating pinnules. The cirri hang free, as this animal moves over the bottom on its arms. Banyuls, France. (R.B.)

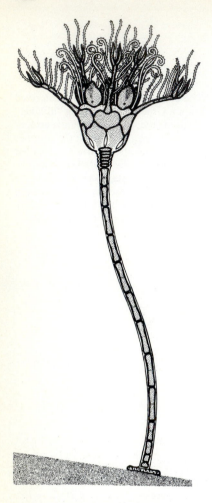

Stalked stage of a developing feather star looks like one of the permanently-stalked crinoids called sea lilies and illustrates the ties between the two groups in this class. Called a "pentacrinoid," this stage feeds with tube feet that spring directly from its 5 arms. The buccal tube feet of the future adult are among the first to develop and can already be picked out as those with curled tips. The young crinoid will develop pinnules and cirri and finally break free, leaving the stalk behind. This 4-month-old individual of *Florometra serratissima* was about 8 mm high. (Mod. after Mladenov and Chia)

Feather stars **feed** on tiny organisms and fine particles in the water, and the outspread arms and pinnules form an elegant food-collecting device. Food particles are caught by the tube feet of the pinnules and passed to tiny ciliated food grooves that run between the tube feet and join the ambulacral groove running along the surface of each arm. The ambulacral grooves join on the upper surface of the body and channel the food into the mouth. The buccal tube feet are thought to "taste" the food before it is ingested. The digestive tract coils once or more within the body before leading to the anus, which opens at the tip of a small projection next to the mouth.

Some feather stars move about constantly, pulling themselves along by their arms. If alarmed, they may even leave the bottom and swim short distances, gracefully waving their arms up and down. Others appear to stay in the same place for months at a time. And still others hide in the crevices of coral reefs during the day, emerging each night and climbing or swimming up to perch on the corals and feed.

A **nerve ring** around the mouth sends radial nerves into the arms along the radial canals, as in other echinoderms; but the **aboral nervous system** is much more extensive than in other classes. The major nervous concentration, centered in the body opposite the mouth, sends nerves through channels in the ossicles of the arms, pinnules, and cirri, coordinating the movements and posture of these structures.

A ring canal encircles the esophagus, as in the **water vascular system** of other echinoderms, and sends radial canals out along the upper (oral) surface of each arm, branching as the arm branches. Small water vascular branches extend into each pinnule and these branch again to form the small groups of tube feet borne by the pinnules. The fingerlike buccal tube feet arise directly from the ring canal and surround the mouth. There is no madreporite, but numerous short **stone canals** lead off the ring canal and open into the perivisceral coelom, which in turn communicates to the ouside through tiny pores that open through the body surface around the mouth.

Feather stars are unusual among echinoderms in that the perivisceral coelom is nearly obliterated by strands of connective tissue that break up the coelomic space into small cavities and channels. Coelomic canals run through each arm, *between* the radial canal (on the oral surface) and the arm ossicles (which are aboral), an arrangement quite different from that in any other echinoderm. The hemal strands and surrounding perihemal coelomic canals are well-developed, with a large

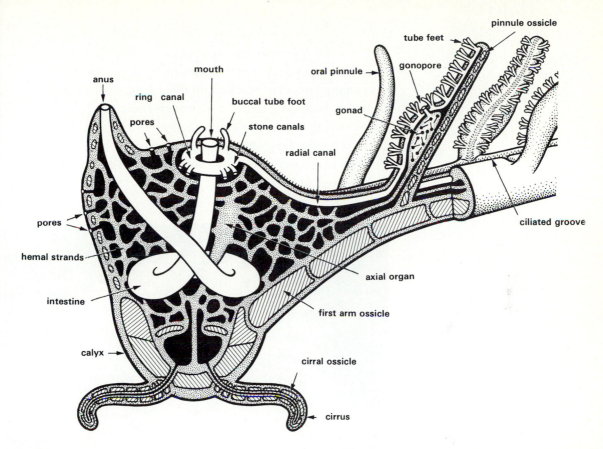

Diagram of a crinoid (feather star) showing the central body and one arm. The perivisceral coelom (solid black) is divided, by strands of hemal and connective tissue, into many communicating spaces. The largest strand, seen descending through the coil of the intestine, contains the well-developed axial organ. The nervous system is not shown,

axial organ, many connections to the digestive tract, and prominent branches in each arm and its pinnules.

Feather stars have many separate **gonads,** located within the individual pinnules of the arms. The eggs or sperms are shed through pores in each pinnule. Fertilization and development may take place free in the seawater, or the young may be brooded on the female's pinnules until released as larvas or juveniles. All feather stars so far studied produce large yolky eggs, which develop into nonfeeding **vitellaria** larvas, encircled with bands of cilia and similar to the doliolarias of sea cucumbers. There is no larval mouth or anus, but after gastrulation, the gut forms as a closed sac, which buds off coelomic pouches. The left anterior coelomic sac develops into the water vascular system, as in other echinoderms, and a vestibule forms over the developing water vascular system. At metamorphosis, the larva settles and attaches by a stalk, the vestibule opens up, and the first tube feet emerge. The juvenile feather star eventually breaks away from the attachment stalk and takes up an independent life.

but its main concentration is within the calyx, the cup-shaped base of the body. Nervous tissue surrounds the large coelomic spaces in the calyx and sends branches through channels in the ossicles of the arms and cirri. The first pinnules (oral pinnules) are probably sensory and protective, and they bear neither tube feet nor gonads. The pinnules and cirri are shown disproportionately large. Not shown are the coelomic linings, and the muscles and ligaments between the ossicles.

SEA LILIES are crinoids similar to feather stars except that they remain attached throughout their lives. The attachment stalk elongates and is strengthened by a columnar series of disc-shaped ossicles, stacked up like coins, with a central core of tissue, including nerves. Although there are relatively few species living today, and almost all are restricted to deep water, they are particularly interesting to zoologists because they are the only survivors among all the kinds of stalked echinoderms that thrived in ancient seas.

Sea lily, a stalked crinoid photographed from a submersible at 200 m off the north coast of Jamaica. The many arms, about 25 cm long, are held perpendicular to the current (which here was coming toward the photographer). The stalk, about 1 m long, bears a tangle of cirri. *Cenocrinus asterias.* (D.L. Meyer, courtesy of D.B. Macurda, Jr.)

CONCENTRICYCLOIDS

SEA DAISIES are the most recently discovered class of animals. When their discovery was announced in 1986, they became the first species of a new living class of echinoderms to be described in 165 years. How could they have escaped detection for so long? Most important, they are members of an unusual fauna—animals that live on sunken wood in the deep sea. For hundreds of millions of years, wood and other debris from land has been washing into the ocean, where much of it eventually becomes waterlogged and sinks to the bottom. There in the darkness and cold, thousands of meters deep, the wood is slowly decomposed by bacteria and shipworms. It becomes the habitat of a specialized fauna, which only recently has been recognized, for sunken wood is rarely collected from the deep sea.

Sea daisies also have eluded even those few scientists who have examined sunken wood from the deep sea because they are small, less than 1 cm in diameter, and they superficially resemble the upper surface of a brittle-star disk. Brittle stars are well known for shedding the upper surface of the disk when disturbed, and it would not be unusual for them to do so after being pulled to the surface from the deep sea.

One of the most distinctive features of sea daisies is the presence of *2 concentric water vascular rings* that encircle the disk. The inner ring, corresponding to the ring canal encircling the esophagus of other classes, is connected by a short tube to the exterior through a hydropore on the upper surface. Four Polian vesicles establish the other interradial positions of the inner ring. The outer ring, unique to this class, bears tube feet and ampullas; in other echinoderms the tube feet are borne on radial canals. Short canals connect the two rings *interradially.* The outer ring, together with an outer nerve ring, is located on the external surface of a series of supporting *ring ossicles,* and (in an arrangement that is similar to the open ambulacral groove of sea stars) the ampullas are connected to the tube feet by short canals between the supporting ossicles.

Another unusual feature of sea daisies is the complete absence of any internal digestive system. A thin membrane, or *velum,* stretches across the lower surface of the animal. This velum probably represents a simple stomach that can be applied directly to the surface of decomposing wood to digest and absorb nutrients.

Within the body cavity of sea daisies are 5 pairs of conspicuous brood pouches, each containing embryos at various stages of development, up to juvenile stages similar in

Concentricycloidea, the class name, refers to the concentric rings of the water vascular system.

There is only one species of sea daisies to date: *Xyloplax medusiformis.* The Greek roots of the generic name refer to the habitat and flattened shape of the animal (*xylo* = "wood," *plax* = "plate"). The Latin roots of the species name refer to the animal's superficial resemblance to a cnidarian medusa.

Several other small echinoderms also have been found on waterlogged wood from the deep sea. These include the peculiar star-shaped ophiuroid *Astrophiura,* with greatly reduced arms, and the recently discovered (1974) disk-shaped sea star *Caymanostella,* first found on wood from the Cayman Trench in the Caribbean Sea. *Xyloplax* was discovered off New Zealand during a search for additional specimens of *Caymanostella.*

It is not known how sea daisies move among pieces of wood scattered over the sea floor, or how they find and colonize new pieces of wood. Their discoverers suggested that perhaps the animals use a parachute-like mode of locomotion in the bottom-water currents.

form to the adults. All of the original 9 specimens were brooding embryos; there are apparently no free-swimming larval forms.

Being so recently discovered, sea daisies have not yet been examined and described in detail, so their exact relationship to other classes is unresolved. They appear to be most similar to sea stars in general body organization. However, the outer ring canal with its interradial connections to the inner ring canal, the series of ring ossicles, and the lack of radial canals and associated structures make concentricycloids unlike any other known echinoderms, living or fossil. It is exciting to realize that a new class of animals as well known as echinoderms can be discovered even today.

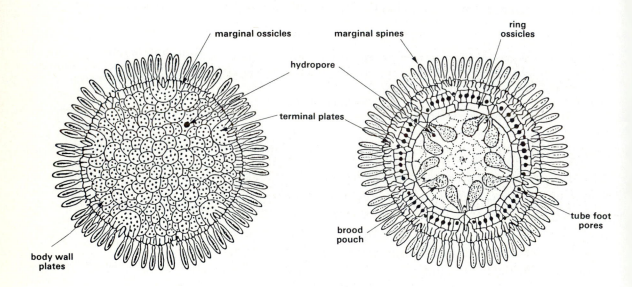

Skeletal arrangement of a sea daisy, showing the upper surface *(left)* covered with small flat plates and fringed with marginal spines, and the lower surface *(right)* with concentric rings of ossicles including the ring ossicles, supporting ossicles for the tube feet, and marginal plates. The 5 pairs of brood pouches are also shown in the view of the lower surface. (From Baker, Rowe, and Clark)

Water vascular system of a sea daisy, showing the concentric inner and outer ring canals connected by short interradial canals. The dotted line indicates the plane of section of the diagram on the next page. (Modified after Baker, Rowe, and Clark)

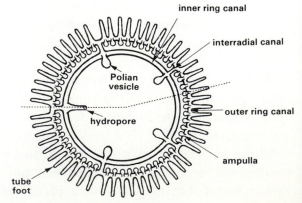

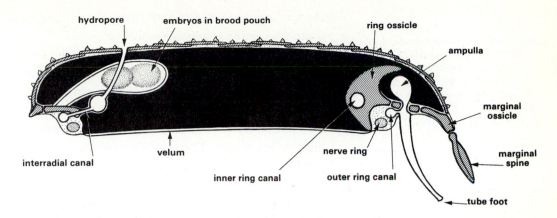

Section through the body of a sea daisy, running through the interradius and hydropore on one side *(left)* and through a ring ossicle, tube foot, and marginal spine on the other side *(right)*. The perivisceral coelom is in solid black and one brood pouch with embryos is shown within. Internal nervous system, hemal system, muscles, and coelomic linings are not shown. (Based on Baker, Rowe, and Clark)

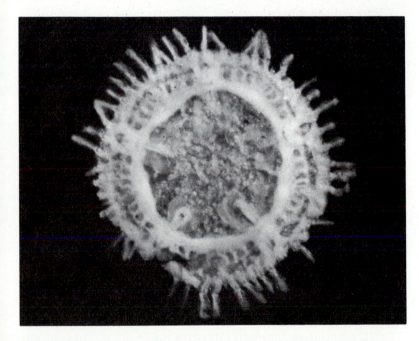

Sea daisy, view of lower surface showing the ring of marginal spines and the 5 pairs of brood pouches seen through the thin velum. Preserved specimen, 3 mm in diameter. National Museum of New Zealand, Wellington. (A.N. Baker)

EXAMINING the 6 living classes of echinoderms as we have done is like looking at the tips of a few branches of a bushy tree. In Paleozoic seas, 400 to 600 million years ago, more than 15 other classes of echinoderms flourished. Members of most of the extinct classes were sedentary or attached to the sea floor, and they probably fed on particles in the water. For this reason, the suspension-feeding sea lilies and feather stars are often viewed as more archaic than the mobile sea stars, brittle stars, and sea urchins. Actually, species in all classes of echinoderms lived in the early Paleozoic seas, and it is unclear which should be considered the more primitive or ancestral. Among

Sedentary or sessile echinoderms such as sea lilies, feather stars, and many fossil forms sometimes are placed into the subphylum **Pelmatozoa,** while the remaining more motile groups are placed in the subphylum **Eleutherozoa.** However, many fossil groups do not easily fall into one or the other category, and the division does not seem particularly useful.

Homalozoans with a single projecting ambulacrum and with conspicuous pores on the upper surface, sometimes compared to the gill pores in hemichordates and chordates. Class Stylophora. Above, *Phyllocystis,* 1 cm in diameter. Below, *Cothurnocystis,* 4 cm in diameter. Ordovician. France. (R.B.)

the earliest known echinoderms, in any case, were animals quite unlike any living echinoderm.

Most of the earliest echinoderms were neither radial nor pentamerous, having 1, 2, or 3 ambulacral tracts (or none) along their bodies. The **helicoplacoids,** for example, are known from fossils collected in the eastern Sierras of California, and date from the early Cambrian period, nearly 600 million years ago. These small animals were spindle-shaped and had 3 ambulacral tracts that wound around the body. Perhaps they burrowed in soft muds. The **eocrinoids,** also known from Cambrian fossils, and the **cystoids,** from the Ordovician period (500 million years ago), were flat to globular, stalked or stalkless animals with 2, 3, 5, or more simple body projections (brachioles) which were probably used for suspension feeding.

Perhaps the most enigmatic fossils are those assigned to the several classes of **homalozoans.** These had flattened bodies of peculiar asymmetrical shapes, heavily encased with calcareous plates; and they had either no recognizable ambulacrum or a single ambulacrum which, in some, was borne on a long projection. These animals were so unlike present-day echinoderms that it is difficult to imagine how they lived, or even to be sure that they really were echinoderms. Some have been interpreted as vertebrate-like animals with echinoderm-like skeletons; called "calcichordates," they have been considered as ancestors to chordates. It has also been suggested that they were modified bilateral animals, lying on their right side, with the upwardly directed left side bearing tentacles, homologous to tube feet, along the ambulacrum. This idea fits neatly with our knowledge of the metamorphosis of living echinoderm larvas, in which the water vascular system develops from the left side of the bilateral larva.

Echinoderm fossils have not provided many clues about either echinoderm origins or relationships among the various classes, but they do offer us glimpses of the extraordinary history that must somehow connect the members of this bizarre phylum. Still abundant and diverse in all seas, modern echinoderms play important ecological roles as predators, grazers, scavengers, and suspension feeders. Their unique body organization, enigmatic physiology, and peculiar development show them to be almost as mystifying as any Paleozoic fossil, and they will continue to challenge imaginative zoologists for many more generations.

Often the classes of echinoderms are divided into 4 subphyla, based mainly on general body form. The generally star-shaped forms in the subphylum **ASTEROZOA** include the extinct class **Somasteroidea,** and 3 living classes: Asteroidea, Ophiuroidea, and (probably) Concentricycloidea. More globular forms in the subphylum **ECHINOZOA** include the extinct classes **Helicoplacoidea, Camptostromatoidea, Edrioasteroidea, Ophiocisti-oidea,** and **Cyclocystoidea,** and 2 living classes: Echinoidea and Holothur-oidea. The subphylum **CRINOZOA** also includes globular-shaped animals but with upwardly directed feeding arms or simple projections (brachioles); most classes are extinct, **Lepidocystoidea, Eocrinoidea, Cystoidea, Parablastoidea, Paracrinoidea, Edrioblastoidea,** and **Blastoidea;** but one is living, Crinoidea. Members of the wholly extinct subphylum **HOMALOZOA** had flattened asymmetrical bodies with the ambulacral system borne on a single armlike projection, or on the side of the body, or absent; 4 classes: **Ctenocystoidea, Homostelea, Homoiostelea,** and **Stylophora.**

Helicoplacoid. Lower Cambrian. White Mts., Calif. *Helicoplacus.* Length 30 mm. (R.B.)

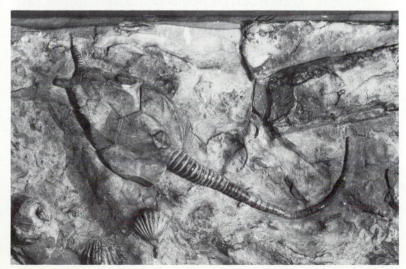

Cystoid, with 2 brachioles and a long stalk. Ordovician. *Pleurocystis.* (R.B.)

CLASSIFICATION: Phylum ECHINODERMATA

More than 20 classes of echinoderms have been recognized, of which only 6 have living representatives. Most of the classes appear to be quite distinct from one another, and we have no clear evidence of evolutionary relationships among them on which to base natural groupings.

Class ASTEROIDEA, the sea stars (or starfishes). Arms short and thick with spacious coelomic cavities containing gonads and branches of saclike digestive system. Open ambulacral grooves. Skeleton with double series of ambulacral ossicles down each arm; many body-wall ossicles, short spines, pedicellarias. Mostly active predators on sedentary animals; some scavengers or detritus feeders. Feeding larvas include auricularia, bipinnaria, and brachiolaria stages. *Luidia, Platasterias, Astropecten, Odontaster, Oreaster, Culcita, Acanthaster, Patiria, Henricia, Solaster, Linckia, Asterias, Pisaster, Leptasterias, Pycnopodia.*

Class OPHIUROIDEA, the brittle stars (or serpent stars) and the basket stars. Coelomic cavities, gonads, and saclike digestive system mostly confined to central disk. Arms long and slender. Ambulacral grooves closed by shieldlike plates. Skeleton with single series of ambulacral ossicles down the center of each arm; arms often covered with plates and spines. Mostly suspension feeders or detritus feeders; some predators on small animals, or scavengers. Pluteus-type feeding larvas. Brittle stars: *Ophioderma, Ophiopholis, Ophiothrix, Ophiura, Amphiura, Axiognathus.* Basket stars: *Gorgonocephalus, Astrophyton.*

Class CONCENTRICYCLOIDEA, the sea daisies. Diskshaped with tube feet around the margin supported by a circle of ring ossicles. Open ambulacral groove. Spacious coelom without internal digestive system, but with 5 pairs of sacs in which embryos develop to juvenile stages. Only one species, described in 1986 from 9 specimens found on waterlogged wood collected from the deep sea off New Zealand. *Xyloplax.*

Class CRINOIDEA, the feather stars and sea lilies. Branching arms with side-branches (pinnules) containing gonads and bearing tube feet. Much reduced and divided coelom containing coiled digestive tract. Globular body partially encased in skeletal plates and sometimes with small ossicles in body wall; single series of ossicles down each arm and pinnule; ossicles also support the attachment stalk (in sea lilies) or cirri (in feather stars). No known feeding larvas. Sea lilies, with permanently attached stalk: *Endoxocrinus, Bathycrinus, Cenocrinus, Rhizocrinus.* Feather stars, stalkless, sometimes with cirri for temporary attachment: *Antedon, Florometra, Tropiometra, Comanthus.*

Class ECHINOIDEA, the sea urchins, heart urchins, and sand dollars. Spacious coelom containing 2-5 gonads, and coiled digestive tract. Closed ambulacral system. Body encased by skeletal test of closely fitted plates covered with spines and pedicellarias. Mostly mobile grazers (sea urchins) or burrowing detritus feeders (heart urchins and sand dollars). Pluteus-type feeding larvas. Sea urchins: *Cidaris, Diadema, Arbacia, Lytechinus, Toxopneustes, Tripneustes, Echinus, Strongylocentrotus, Echinometra, Colobocentrotus, Heterocentrotus* (slate-pencil urchin). Sand dollars: *Echinarachnius, Clypeaster, Dendraster, Mellita.* Heart urchins: *Echinocardium, Lovenia, Spatangus.*

Class HOLOTHUROIDEA, the sea cucumbers. Soft to leathery, elongate body with microscopic or larger, scalelike ossicles in the body wall. Spacious coelom containing a single gonad and coiled digestive tract. Closed ambulacral system; buccal tube feet greatly enlarged as feeding tentacles. Sedentary suspension feeders, or creeping or burrowing detritus feeders. Feeding larva is an auricularia, followed by the doliolaria, a postmetamorphic pelagic juvenile. *Cucumaria, Thyone, Psolus, Holothuria, Stichopus, Pelagothuria, Benthodytes, Opheodesoma, Synapta.*

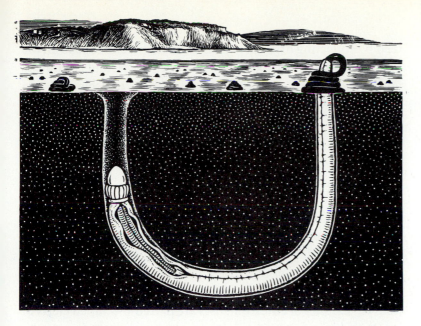

Chapter 28

Hemichordates

Enteropneusts and Pterobranchs

Clues to chordate roots

THE PHYLUM **HEMICHORDATA** consists of about 100 species of soft-bodied marine animals in two widely divergent classes: tiny pterobranchs (mostly colonial) that live sessile or sedentary lives on the sea bottom, and larger wormlike, burrowing enteropneusts ("acorn worms"). Neither is ever seen by most people, and their interest for us lies chiefly in an unexpected assortment of features that they share with lophophorates, with echinoderms, and with invertebrate and vertebrate chordates, presenting tantalizing clues to the still elusive connections between these phyla. That the invertebrates most closely related to our own phylum may be such unprepossessing forms as the hemichordates is intriguing.

ENTEROPNEUSTS

THE CLASS Enteropneusta is the larger of the two hemichordate classes, both in number of species (about 70) and size of individuals (up to 2 m or longer). The soft, wormlike body, covered with cilia and coated with protective mucus, is bilaterally symmetrical. It consists of 3 parts, each with a separate coelomic component. The first part is a rounded or elongate **proboscis** with which the animal burrows in soft marine sediments, mostly in shallow waters. The proboscis connects by a narrow stalk to the short thick **collar,** which serves as an anchor during burrowing, and the rest of the animal extends behind as a long **trunk.**

Some burrowing enteropneusts **feed** by swallowing large quantities of sediment; they digest the organic matter and pass the remains through the gut, depositing piles of castings near the posterior opening of the burrow. Enteropneusts may also

Hemichordates were so named at a time when they were included in the phylum Chordata along with other invertebrate chordates (urochordates and cephalochordates—see chapter 29) and with vertebrates. Now most biologists agree that the structure seen in hemichordates and thought to be a short notochord (the buccal diverticulum) had been misinterpreted and is probably unrelated to the notochord of chordates. Although hemichordates are now placed in a separate phylum, the name calls our attention to the characters common to both phyla and the relationship between them.

Permanent U-shaped burrows like the one shown in the chapter heading are typically built by enteropneusts with a long proboscis (e.g., *Balanoglossus*). Types with a short proboscis (e.g., *Ptychodera)* tend to burrow continuously through the substrate, parallel to the surface, or are found sheltering beneath stones or seaweeds.

Acorn worm is a common name for enteropneusts. The shape of the proboscis in many species, and the way it fits into the collar, does suggest an acorn. When we dug this individual out of its intertidal burrow, its orange secretion, which smelled like iodoform, clung to our hands. This enteropneust, and those in the photo on the opposite page *(Ptychodera flava),* show genital wings, lateral extensions of the trunk that contain the gonads. Brittany, France. (R.B.)

feed by collecting organic particles that adhere to mucus on the proboscis. Backward-beating cilia sweep the strands of mucus and food particles over the surface of the proboscis and into the mouth, which opens ventrally just under the anterior edge of the collar. The mass of mucus-bound food moves through the pharynx into the esophagus and long intestine; undigested material is eliminated through the anus at the posterior end of the body.

The epidermal cilia that move strands of food-laden mucus into the mouth are assisted by a current of water which enters the mouth and flows out through **pharyngeal slits** (also called gill slits). Such slits in the wall of the pharynx are *a character shared only with chordates* and constitute the strongest link between hemichordates and chordates. The pharyngeal slits of enteropneusts are paired U-shaped clefts in the dorsolateral wall of the pharynx, numbering from a few to several hundred. The rim of each pharyngeal slit is lined with beating cilia that together create the water current.

The tissue surrounding each pharyngeal slit is stiffened and supported by **skeletal supports,** thickenings of the basement membrane of the pharyngeal epithelium, so that the slit is held open. The opening leads into a **pharyngeal pouch** ("branchial sac"), a pharyngeal outpocketing that in turn connects to the outside through an excurrent opening. In hemichordates, the pharyngeal slits serve mostly to get rid of the water that enters the mouth; they do not play a significant role in collecting food, although some fine particles are captured from the current leaving the slits and are conducted from the dorsal part of the pharynx, where the slits occur, to the ventral, nutritive part.

An abundant blood supply in the tissue surrounding the pharyngeal slits facilitates **respiratory exchange.** The colorless blood flows in 2 main vessels, dorsal and ventral, which run the length of the animal, and in an extensive system of well-defined but unlined blood sinuses. Thus the **circulatory**

Diagrammatic cross-section of the pharyngeal region of the trunk. Nuclei are shown only in the endoderm lining the pharynx and the pharyngeal pouches. The gonads are shown here as they occur in genera such as *Saccoglossus.* In many other hemichordates, such as species of *Balanoglossus* and *Ptychodera,* the gonads occur in genital wings, lateral extensions of the trunk that commonly fold over the top of the animal, concealing the excurrent openings. (Modified after C. Dawydoff)

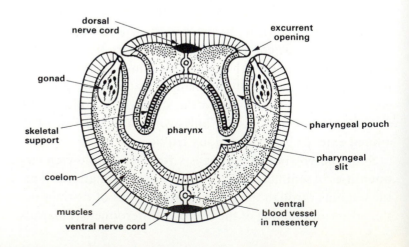

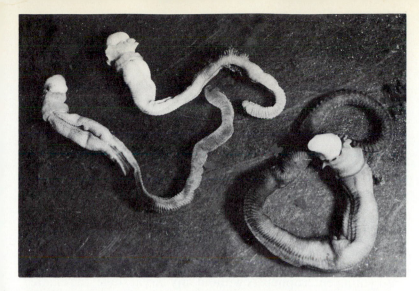

system has both open and closed components. The contractions of the main vessels and of a fluid-filled heart-vesicle in the proboscis help to circulate the blood. In the proboscis, all the blood flows through a large central sinus into a structure called the glomerulus, which is thought to be an **excretory organ**. The excreted wastes are presumably released into the small coelomic cavity of the proboscis, which opens to the outside through one or two pores, recalling the anterior coelom and open pore of echinoderms; the coelomic cavities of the collar also open to the outside. Similar systems are found in invertebrate chordates.

Another chordate-like character is the *dorsal tubular nerve cord* in the collar of some enteropneusts, although the **nervous system** as a whole is primitive and shows little centralization. It includes median dorsal and ventral nerve cords that lie in the epidermis and extend the length of the trunk, connected by a nerve ring anteriorly. There are no specialized sense organs, but a concentration of sensory cells at the base of the proboscis probably tests the water and particles approaching the mouth, which may close to reject large particles before they reach it. The surface of the animal is sensitive to light, especially at the ends.

Enteropneusts have few defenses besides remaining out of the sight and reach of predators. In Hawaii the gastropod *Conus lividus* preys mainly on *Ptychodera flava,* shown here. This species and some species of *Balanoglossus* and *Glossobalanus* luminesce when disturbed, emitting light from photocytes in the epidermis and discharging a luminous secretion. The flashes of luminescence followed by the prompt withdrawal of the enteropneust into its burrow, may startle and confuse a predator. Some enteropneusts accumulate bromine and incorporate it into strong-smelling secretions, which may be defensive. A number of enteropneusts also accumulate quantities of iodine, a habit that is found scattered in several invertebrate groups but is interesting in hemichordates because it is also a chordate character. Hawaii. (R.B.)

Diagram of an enteropneust, anterior portion in longitudinal aspect, shown as if sectioned in the median sagittal plane but with the dorsal and ventral mesenteries that carry the main blood vessels cut away. The water current enters the mouth, flows out of the U-shaped **pharyngeal slits** into the **pharyngeal pouches,** and exits through the excurrent openings. Note the coelomic compartments of the proboscis (protocoel), collar (mesocoel), and trunk (metacoel); these are not simple fluid-filled cavities but are largely occluded by muscle and connective tissue derived from the coelomic lining. The proboscis skeleton, derived from the epidermal basement membrane and coelomic lining, helps support the proboscis and stalk. (After various sources)

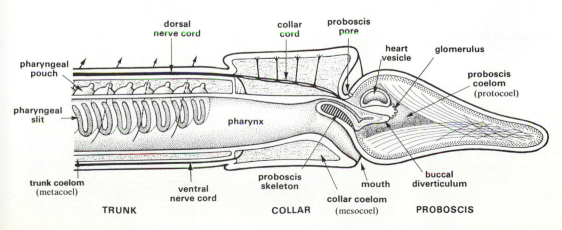

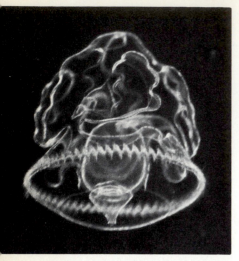

A **tornaria larva** may feed and grow in the plankton for many months and during this time travel long distances. Frequently taken in plankton tows, many kinds have yet to be connected with adults of described species, for example *Planctosphaera pelagica,* which reaches up to 28 mm in diameter. This photo is of a tornaria of *Glossobalanus.* England. (D.P. Wilson)

There may also be special receptors sensitive to substances released during **spawning,** for spawning by one animal appears to stimulate its neighbors to spawn, the eggs or sperms emerging from multiple gonads distributed along the anterior trunk. Females of a species of *Saccoglossus* were observed to spawn first, followed by the males. Fertilization is external, and early **development** resembles that of echinoderms (cleavage is radial, the blastopore becomes the anus, and the coelom develops from outpocketings of the embryonic gut). Large eggs with greater food reserves undergo direct development into young enteropneusts without any larval stage; the ciliated young may swim briefly, but do not feed. Small eggs with lesser food reserves develop into feeding larvas, called **tornarias** because they rotate continually as they swim. Sinuous ciliary bands that loop over the surface of the larva capture phytoplankters and direct them into the mouth, much as in the bipinnaria larva of a sea star. In fact, the feeding larvas of enteropneusts were taken for echinoderm larvas until they were finally reared through metamorphosis and found to develop into young enteropneusts.

Metamorphosis of an enteropneust. When its growth is complete, a tornaria may still delay metamorphosis until it encounters the turbulence and surface contact of a benthic habitat. Even then, it keeps its options open by retaining the main band of swimming cilia until after it has safely burrowed into a suitable soft substrate. (Modified after T.H. Morgan)

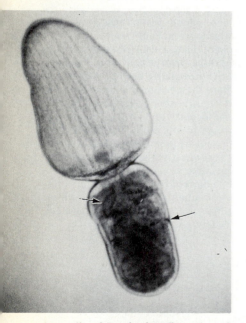

Juvenile of *Ptychodera flava* about 2 mm long, metamorphosed from a tornaria collected in the Hawaiian plankton. Arrows point to the first pair of pharyngeal slits and to a line of pigmented cells marking the site of the main band of swimming cilia, which has been lost. (M.G. Hadfield)

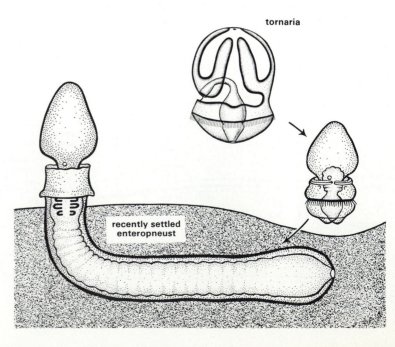

tornaria

recently settled enteropneust

PTEROBRANCHS

THE MEMBERS of the class Pterobranchia do not at first glance look anything like enteropneusts, but rather like bryozoans. Pterobranchs are tiny, usually less than a few millimeters long, and are sessile or sedentary. Most live as colonies, produced by budding from a sexually produced individual, within a protective covering of secreted tubes. As in many sessile tube-dwellers, the body is saclike and the digestive tract U-shaped, so that the anus discharges over the edge of the tube. From the lower end of the body projects a long stalk, a permanent connection with the rest of the colony. Pterobranchs show the same 3 body regions seen in enteropneusts: proboscis, collar, and trunk. The flattened, shield-like proboscis secretes the tube and is used to move within it, like a little creeping foot. The collar is extended dorsally into one or more pairs of tentacle-fringed arms, each containing an extension of the collar coelom. Held high like little wings (*ptero* = *"wing"*), the arms gather tiny organisms and organic particles from water currents, and cilia convey the food down the arms and around the collar to the ventral mouth. In species of *Rhabdopleura* the single pair of tentaculate arms strikingly resembles the horseshoe-shaped lophophore of some bryozoans. Species of *Cephalodiscus* typically have 5 pairs of arms, and this number, taken together with all the other similarities of hemichordates to echinoderms, seems more than a simple coincidence.

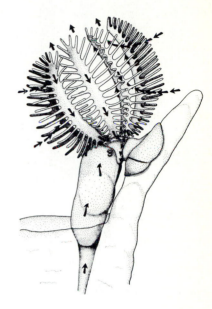

Cephalodiscus feeding, attached to a spine of the colony skeleton. As cilia drive water past the arms, particles caught by individual tentacles are flicked into a central groove of the arm and moved by cilia down to the mouth. Feeding resembles that of lophophorates. (From S.M. Lester)

Pterobranch of the genus *Rhabdopleura*. Compare with drawing. Bermuda. (J. and V. Pearse)

Rhabdopleura, with 2 tentaculate arms. Cilia on the tentacles and in ciliated tracts down the center of each arm direct food particles to the mouth. Note proboscis and collar pores. (Modified after various sources)

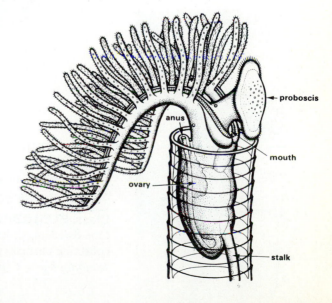

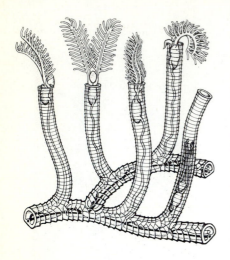

The internal structure of pterobranchs corresponds for the most part to that of enteropneusts except that the pharyngeal slits are absent or represented only by a single pair, and the collar cord is a solid mass. The openings to 1 or 2 large gonads lie near the anus. Surprisingly, male and female members may be found in the same colony, and some species have hermaphroditic members, with eggs in one gonad and sperms in the other. The embryos are protected within the parental tubes and develop into planula-like larvas. These elongate to a wormlike stage that looks very like a young enteropneust, and only later does the digestive tract make a U-turn and the body take on the rounded adult shape.

Colony of pterobranchs, *Rhabdopleura.* The animals are sensitive to any disturbance, muscles in the stalk pulling the body quickly into the safety of the tube. When undisturbed, the animals creep up the tubes again and perch on the rims to feed. In *Rhabdopleura* the colony members are budded from a stolon; in *Cephalodiscus,* clusters of stalked members arise from a basal disk, and the stalk is so extensible that the animals can creep right out of the colonial skeleton onto adjacent objects. In the single collection of *Atubaria,* solitary individuals were found living free, without tubes, clinging to hydroids.

Tube of pterobranch *Rhabdopleura* is secreted in successive rings. Bermuda. (J.S. and V.B. Pearse)

Rhabdopleura **feeding** with its arms and tentacles, each containing a fluid-filled coelomic space, held high and turgid. Arms about 1 mm long. Bermuda. (J.S. and V.B. Pearse)

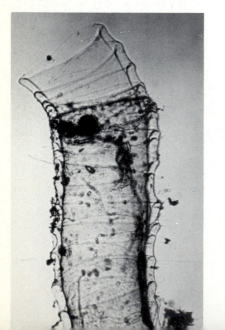

CLASSIFICATION: Phylum HEMICHORDATA

Class ENTEROPNEUSTA, large wormlike burrowers with many pharyngeal slits. Feed by collecting particles in mucus on proboscis, some by ingesting sediment. Tornaria larva or direct development. *Saccoglossus, Balanoglossus, Glossobalanus, Ptychodera.*

Class PTEROBRANCHIA, tiny saclike body, usually colonial tube-dwellers, with single pair of pharyngeal slits or none. Feed by collecting particles on tentaculate arms. Planula-like larva. *Rhabdopleura, Cephalodiscus, Atubaria.*

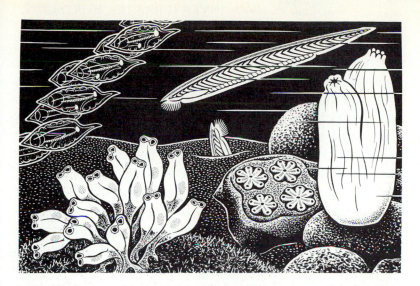

Chapter 29

Invertebrate Chordates

Tunicates and Lancelets

Chordates Without Backbones

TRACING the roots and branches of one's family tree can be embarrassing enough, but anyone sufficiently curious to extend the search phylumwide should be prepared to acknowledge some even more unlikely relatives. The phylum **CHORDATA,** in which humans are classified, has some 45,000 species. The vast majority of them are vertebrates (subphylum Vertebrata)—fishes, amphibians, reptiles, birds, and mammals—all with a skeletal axis of vertebras, separate but articulated bony or cartilaginous pieces arranged in a row along the back. The remaining chordate species (about 3%) have no such vertebral column and are referred to as *invertebrate chordates.* Unlike the actively grazing or predatory vertebrates, which are widespread on land and in freshwaters as well as in the seas, the invertebrate chordates lead relatively passive lives as filter-feeders and are all marine. The tunicates, subphylum UROCHORDATA, mostly have saclike bodies and their behavior is not much more complex than that of sponges; the smallest individuals are less than a millimeter long and not many measure more than a few centimeters. The lancelets, subphylum CEPHALOCHORDATA, look vaguely like small slender worms and live mostly sedentary lives in coarse sediments. Why such creatures as these should be included in the same phylum as brainy humans and massive whales takes a bit of explaining.

TUNICATES

THE BENTHIC TUNICATES, or ascidians (class Ascidiacea), make up the great majority of the 1,300 members of the subphylum **UROCHORDATA** and are the ones that seashore visitors are most likely to encounter. There are also 2 small classes of pelagic species, but whereas these colorless pelagic tunicates are nearly transparent in the water and hard to see,

Solitary ascidian, *Halocynthia papillosa.* The incurrent siphon at the top and the excurrent siphon at the right are both wide open here as the tunicate filters food particles out of the steady current that flows through. *H. papillosa,* 6 cm high, is coral red to orange in color and lives in the sublittoral on sandy bottoms. Marine Institute, Split, Yugoslavia. (R.B.)

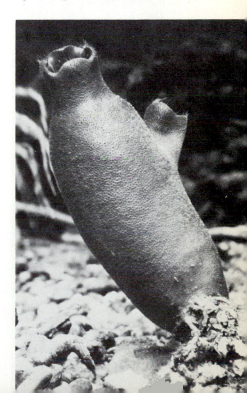

737

the ascidians are hard to miss. Their bright reds, yellows, oranges, purples, and greens, along with pastel and buff hues, white, and black, often combine with striking growth patterns and add richness to rocky shores around the world. On tropical coasts, ascidians are usually limited to subtidal habitats; and even in temperate regions, where they can be seen in the low intertidal zone, they prefer shaded overhangs to surfaces in full sun. Most live attached to rock as adults, but a few dwell on mud, and there are even tiny wormlike interstitial forms.

The **tunic** that surrounds, supports, and protects the saclike body of a tunicate has a gelatinous texture but may be quite tough. Though it is mostly a secretion of the epidermis, it is not simply an inert covering but rather like an external connective tissue, for it contains branches of the circulatory system and several kinds of cells. The tunic can consist of over 99% seawater, but the translucent ground substance of proteins and carbohydrates is strengthened with organic fibers. A common and surprising component is cellulose, which we associate mostly with plant tissues though it has been found also in the connective tissue of humans and other mammals, especially older individuals.

In some species, cells of the tunic contain high concentrations of toxic vanadium compounds and sulfuric acid; iron and other metal compounds may also be accumulated. Tunic cells may secrete calcareous spicules or ornament the surface of the tunic with elaborate spiny projections. Large quantities of organically-bound iodine may also occur in the tunic. We may infer from experiments that some or all of these chemical and structural specializations serve as defenses against predators and against animals that might settle on and overgrow the tunic. In one study, fishes and crabs that would not feed on tunicates readily devoured those in which the acid-containing cells had been burst by freezing and thawing.

Except for a few deep-sea forms, ascidians and other tunicates live by **filter-feeding,** using as part of the filtering apparatus the **pharyngeal slits** which are the most conspicuous chordate character of the adults. Each of the original pharyngeal slits becomes subdivided into many small openings, resulting in a sieve-like pharynx (branchial sac). As in enteropneusts, cilia lining the openings create a steady current of water that enters the mouth and flows out through the pharyngeal slits. The slits open into a cavity, the **atrium,** which may be compared to the pharyngeal pouches of enteropneusts. The atrium almost completely surrounds the pharynx and opens to the outside dorsally. In many ascidians the mouth and atrium open through siphons, tubular extensions of the body surface. A large solitary tunicate may filter several liters of seawater per hour.

Defensive spines, flexible but armed with circlets of recurved bristles, sprout from the tunic of *Halocynthia hilgendorfi igaboja.* Covering the surface like a dense crop of thornbushes, the spines effectively protect this ascidian from the large snail *(Fusitriton oregonensis)* that is a major predator on other ascidians along the Pacific coast of N. America. Removal of the spines renders this tunicate vulnerable to snail predation, despite the fair amount of vanadium in the tissues. Spine length, to 6 mm. SEM. (C.M. Young)

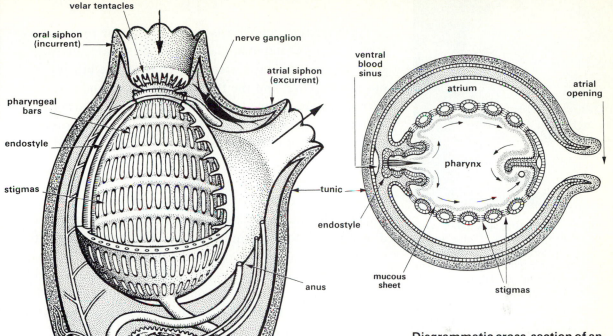

Diagram of a solitary ascidian, viewed from the left side, with part of the pharyngeal wall cut away. Large particles are excluded from the pharynx by velar tentacles (near the base of the incurrent siphon) that coarsely strain the water current. Just below the level of the velar tentacles is the opening of the neural gland, which lies against the nerve ganglion and arises during development as part of the nervous system; it is thought to be related to the vertebrate pituitary gland. (Combined from various sources)

Diagrammatic cross-section of an ascidian at the level of the atrial opening. Arrows show the direction in which cilia move the mucous sheets. Nuclei are shown only in cells of endodermal origin, lining the pharynx. Whereas in hemichordates the pharyngeal pouches develop as many paired endodermal outpocketings, in tunicates the atrium develops as a single pair of ectodermal invaginations. The 2 original atrial sacs and openings fuse into one. The original pharyngeal slits become subdivided during development into many *stigmas,* all of which in any one cross-section are derived from a single pair of pharyngeal slits. (Combined from various sources)

During feeding, a ventral ciliated groove in the pharynx, the **endostyle,** continuously secretes sheets of porous mucoid material to each side. These sheets are held against the pharyngeal walls by the pressure of the water current and are continuously moved dorsally by cilia. The sheets trap food particles but allow the water to pass through. As they meet in the dorsal midline, the food-laden mucous sheets are twisted into a ropy strand and moved along into the esophagus and stomach, to be digested. Digestion is extracellular. The gut is U-shaped and opens through the anus into the atrium, from which wastes are carried away in the outgoing current. At regular intervals, and when disturbed, ascidians contract the body abruptly, forcefully expelling water and any accumulated debris from both mouth and atrial openings. It is this habit that has earned them the name "sea squirts."

An endostyle and a mode of filter-feeding like those of invertebrate chordates are seen also in the larvas of certain primitive fishes and are part of the evidence for ancestral ties between invertebrate chordates and vertebrates. In these larval fishes, the endostyle later develops into the thyroid gland, which secretes the iodine-containing hormone thyroxin, central in the control of metabolism. In tunicates, the endostyle concentrates iodine and binds it into organic compounds identical to precursors of thyroxin, but the function of these compounds in tunicates is not known.

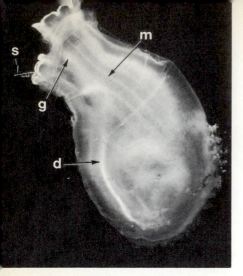

Tunicate releasing sperms through the excurrent siphon. *Ciona intestinalis* spawns at dawn or, in the laboratory, at any time it is exposed to light after a sufficient period of darkness. Strong contraction of the body-wall muscles (**m**), as here, causes more forceful exit of the sperms (**s**) but is not essential to spawning. The sperm duct (**d**) contracts and expels sperms when illuminated, even if removed from the body, isolated from nervous and hormonal control. The duct lacks muscles and recognizable photoreceptors; its epithelial cells respond to light by producing networks of contractile microfilaments in the cytoplasm; contraction of the microfilaments narrows the duct lumen and expels the sperms. Adult tunicates have no eyes, but various organs, including the main nerve ganglion (**g**), are sensitive to light. (R.M. Woollacott)

This squirting behavior is controlled by the **nervous system,** which is centered in a large ganglion situated between the siphons. The nervous system also controls the beating of the pharyngeal cilia; and the flow of the feeding and respiratory current may be stopped or its rate may be altered according to the amount of food or sediment or possibly deleterious substances detected in the water by sensory cells located on the siphons and in the atrial walls.

The nervous system sends no branches to the **heart,** which lies in a small cavity within a pericardial sac. The heart has been much studied because of its unusual habit of reversing direction every few minutes or so. Peristaltic contractions pump blood out of one end of the tubular heart for a while, then the other. Heart reversal is seen in all tunicates, and in a few other animals, but its significance is not clear. The heart pumps blood through an extensive system of unlined vessels and sinuses that abundantly supply the pharynx and other viscera, as well as the tunic. The principal body cavity is a **hemocoel;** no major coelom ever develops in tunicates. The blood has no oxygen-carrying pigment, but contains a variety of cells which participate in distributing nutrients, producing tunic, and storing excretory products. **Excretion** requires no special organs in tunicates; the major excretory product is ammonia, which diffuses into the seawater, but uric acid and other purines (components of nucleic acids) are stored in the form of solid deposits in blood cells and in various vesicles that occur among the viscera or attached to the body wall.

Molgulid tunicates have a prominent, ductless "renal sac," supposedly an excretory organ that accumulates and stores excretory wastes in the form of solid concretions of calcium salts and urate. Recent studies show, however, that the concretions are not permanently stored but are metabolized by fungus-like protists that live in the sac, possibly as mutualistic symbionts enhancing the host's reproductive success.

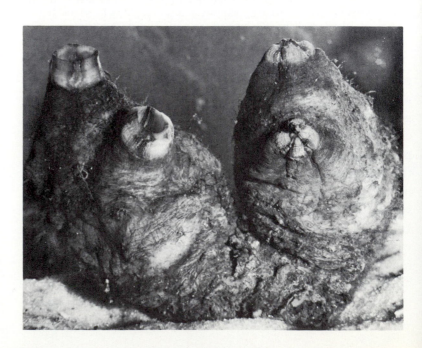

Solitary tunicates, *Styela plicata,* 7 cm tall, tend to grow closely clustered. One has its siphons open; in the other they are closed. Both squirted vigorously on being handled. N.W. Florida. (R.B.)

The blood vessels are often the principal connection between the individuals of **colonial ascidians,** as well as one
source of the cells that form new individuals. In tunicates there
is not such a great difference between solitary and colonial
types as we saw in the hemichordates between solitary
enteropneusts and colonial pterobranchs. Each member of a
tunicate colony is much like a solitary individual, though
usually much smaller. The colony members may be quite
discrete, connected only by long stolons or a basal mat of
tunic; or, they may be partially or completely embedded in a
thick layer of common tunic. The greatest degree of colonial
organization occurs among types in which the members,
embedded in a common tunic, are arranged in groups that
discharge their atrial siphons into a central chamber with a
single opening to the outside. Connected together by blood
vessels that run throughout the tunic, the individuals in such a
colony may live only a few days, during which time they grow,
bud, mature their eggs and sperms, release their young, and
finally die, all in nearly perfect synchrony with each other.
Meanwhile, the buds they have produced repeat the cycle, so
that the colony may continue to exist indefinitely. The tissues
from which buds arise, and the ways in which they develop, are
extraordinarily diverse and seem to break all the rules followed
by the sexually produced embryos of most animals, including
tunicates. A bud can arise, for example, from a small number
of apparently undistinguished cells in the lining of the atrial
wall, develop into all the organs of the body, and produce an
individual seemingly identical to one that develops from an
embryo.

Colonial tunicate, *Botryllus schlosseri,* is the commonest ascidian on
European shores. The starlike clusters of individuals, only 1.75 mm long, are
embedded in a common tunic. Adjacent colonies may be yellow, green,
purple, violet, or blue, with the individuals marked in red or yellow. Each
member has an incurrent opening at its outer tip and shares the common,
larger excurrent opening at the center of the cluster. France. (R.B.)

Solitary ascidian.

Buds from stolons may become
disconnected ...

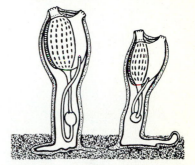

... or may remain permanently
connected.

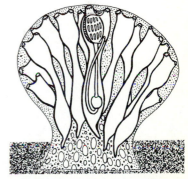

A common tunic may surround
colony members (one shown opened)
with independent feeding currents ...

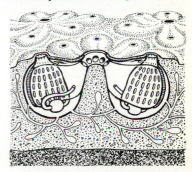

... or systems of colony members may
share an excurrent opening.

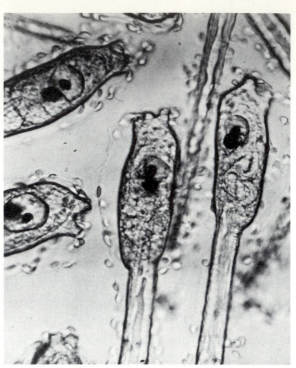

Colonial tunicate, *Botryllus schlosseri*, growing in still water may form pendants, a growth form similar to that of quiet-water sponges. Individual members of the tunicate colony shown here are about 1.75 mm in diameter. England. (D.P. Wilson)

Tadpoles of ascidian, *Ascidia nigra*, show clearly the prominent eyespot and adhesive projections by means of which the larva attaches to a substrate and begins metamorphosis. Bermuda. (K.B. Sandved)

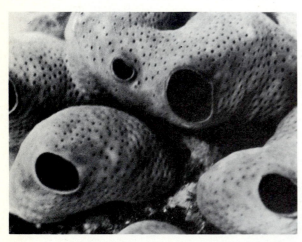

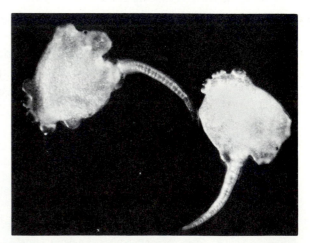

Didemnid colony members share a large excurrent opening. The excurrent chambers are colored green by the presence of photosynthetic prokaryotic symbionts, *Prochloron*. These resemble in certain respects the chloroplasts of all eukaryotic green plants, and they have the same kinds of chlorophylls *(a* and *b)* as do green plants. Didemnid colonies are highly plastic and, if observed carefully over a period of days, they may be found to move, or a colony with multiple openings may divide. This enables the animals to command more territory, and they spread quickly over sizeable areas. Great Barrier Reef, Australia. (J.S. Pearse)

Tadpole-like larva of an ascidian, *Ecteinascidia turbinata,* has conspicuous notochordal cells in the tail. A detailed study of this species showed that the anterior neural tube becomes divided into right and left halves. The pigmented eye (dark spot between the mouth and atrial openings) and the statocyst lie in the right side, which disintegrates at metamorphosis, along with the posterior nerve cord. The left side develops into the adult ganglion and neural gland. This left/right asymmetry is tantalizingly suggestive of the brain lateralization characteristic of many vertebrates and especially well developed in humans. Bermuda. (K.B. Sandved)

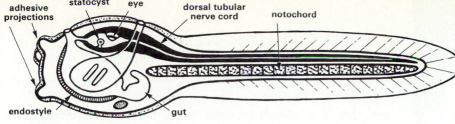

1. **Tadpole-like larva** of an ascidian swims for only a few minutes to a few days until it finds a suitable attachment site.

Sexual reproduction in tunicates provides possibilities for developmental study that are as intriguing as those presented by budding. Tunicates are hermaphroditic, and self-fertilization occurs regularly in some species. In others self-fertilization is inhibited by genetically-controlled barriers, or because the eggs of an individual (or colony) usually mature before the sperms. Many solitary ascidians spawn the eggs through the atrial opening, while most colonial species hold them in the atrial cavity during fertilization and development. The ascidian embryo develops into a **tadpole-like larva,** which does not feed but swims actively until it locates a suitable site for permanent settlement. This step will determine the success of its entire life, if any, and the larva is equipped with a variety of sensory guides—a light-sensitive eye, a statocyst, and chemical and tactile receptors. The larva typically attaches by 3 adhesive projections at the anterior end and rapidly undergoes a radical metamorphosis into the sessile adult form.

The vertebrate-like appearance of the tunicate tadpole is reinforced by its vertebrate-like development and by its possession of several more chordate characters, besides the pharyngeal slits and endostyle seen in the saclike adult. The larva has a well-developed **dorsal tubular nerve cord** which is enlarged anteriorly into a rounded hollow vesicle and which extends into the muscular post-anal **tail,** another chordate character. Lying beneath the nerve cord and extending the length of the tail is a prominent **notochord,** hallmark of the chordates and source of the phylum name. The notochord is central to the tadpole's ability to swim. It consists of a row of about 40 large, turgid, vacuolated cells, surrounded by a sheath. On both sides of it are muscles that, without the notochord, would shorten the tail; but with the notochord to maintain the tail at constant length, alternating muscular contractions cause the tail to undulate and propel the tadpole through the water. All chordates produce a notochord in the course of their development but in adult vertebrates its role is taken over by the vertebral column of cartilages or of bony vertebras.

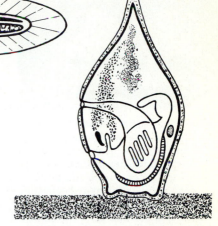

2. Larva attaches at its anterior end and begins **metamorphosis.** Tail, notochord, and most of the nervous system are resorbed.

3. While the ganglion, heart, digestive system, and atrium continue to develop, and pharyngeal slits divide into many smaller openings, ...

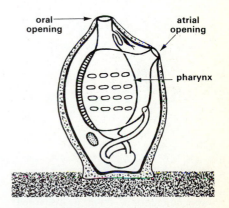

4. ... the siphons and internal organs become rotated about 90° so that the siphons are directed away from the substrate. (Combined from various sources)

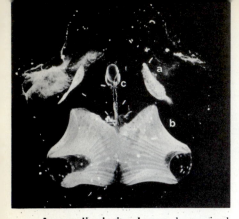

Appendicularian house has paired incurrent openings covered by filters **(a)** that exclude plankters such as large dinoflagellates and diatoms but admit smaller organisms and particles, which are caught on paired collecting surfaces **(b)** within the house, while the stream of water exits. The appendicularian **(c)** sits in the center of the house and drives this water stream by undulations of its tail. Meanwhile, it periodically sucks the sorted, concentrated food from the collectors into its pharynx, where the food is trapped on mucus secreted by the endostyle and excess water exits through 2 pharyngeal slits; this water stream is driven by pharyngeal cilia. When the house becomes clogged with uneaten organisms and fecal pellets, which may happen several times a day, it is discarded and a new one is secreted. *Stegosoma magnum.* (J.M. King)

THE PELAGIC TUNICATES are distributed in marine plankton worldwide and often occur in extraordinary densities, so that their filtering activities strongly affect the plankton community. Members of the class Appendicularia, or Larvacea, retain a tadpole-shaped body throughout their lives. Appendicularians secrete around themselves elaborate mucoid "houses" that not only act as an external filtering device but also protect the small, delicate animal and help to keep it from sinking. Members of the class Thaliacea (salps, doliolids, and pyrosomes) are more like floating ascidians, but with the incurrent and excurrent openings at opposite ends, so that the feeding and respiratory current flows straight through the barrel-shaped body and may serve in swimming as well.

Swarms of thaliaceans containing thousands of animals per m³, extending for hundreds of kilometers and many meters deep, are common; and appendicularians have been reported to occur at local densities of over 25,000 animals/m³. Pelagic tunicates, especially salps, can grow and reproduce at rates that outstrip most of their herbivorous planktonic competitors, which are often conspicuously absent from salp swarms. Not only do they feed at great rates, but they filter out organisms and particles of a greater size range than do other filter-feeders, down to 1 μm for salps and 0.1 μm for appendicularians. (Many kinds of minute plankters that had escaped nets were first described from the stomachs of salps or the filters of appendicularian houses.) Bacteria and other minute plankters thus enter the food chain, becoming available to a great variety of small fishes and invertebrates that prey on pelagic tunicates or on discarded appendicularian houses, their filters clogged with food. Such houses are a large component of "marine snow," the descriptive term given to the diversity of organic aggregates that support significant communities of microorganisms as they sink slowly through oceanic waters like snowflakes through air. Pelagic tunicates also significantly affect nutrient cycles in the oceans by rapidly recycling nitrogen and by packaging large quantities of dispersed organic matter into compact fecal pellets that sink quickly into deeper waters and become available to organisms there.

Appendicularian has a notochord and dorsal tubular nerve cord that extend down its long tail throughout its life. Appendicularians are almost all hermaphroditic; sperms are spawned through tiny ducts but eggs are released by rupture of the body wall, followed by death of the animal. England. (D.P. Wilson)

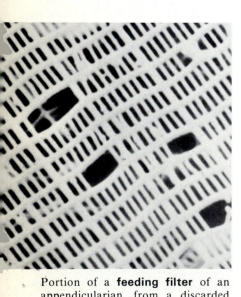

Portion of a **feeding filter** of an appendicularian, from a discarded house collected in "marine snow" at 1,650 m depth off southern California. Average mesh opening, 0.06 x 0.38 μm. SEM. (M. Gowing)

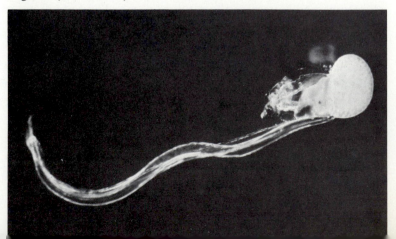

Salps (class Thaliacea) use muscle bands that nearly encircle the barrel-shaped body to drive the feeding and respiratory current, which also serves to jet-propel the animal through the water. The pharyngeal cilia are not important in pumping water, and there are only 2 large pharyngeal slits leading to the posterior atrium. The life history of salps includes both budding and gamete-producing phases. The solitary asexual individual *(above left)* produces many chains of buds from its budding stolon, visible as a white spiral. Each chain consists of tens or hundreds of hermaphroditic individuals *(above right)*. The chains are released to feed and grow; they gradually break up and disperse, as they mature their gametes. The fertilized egg, brooded and nourished by the parent, develops into a solitary budding individual. *Pegea.* (L.P. Madin, Woods Hole Oceanographic Institution)

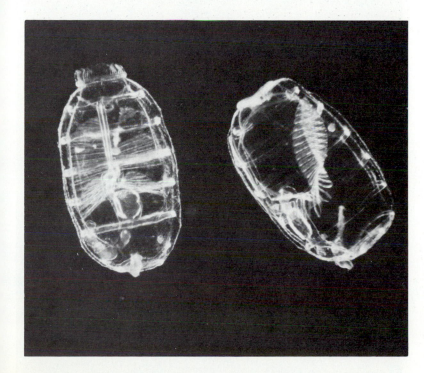

Doliolids (class Thaliacea) look much like salps. The feeding and respiratory current of doliolids is ciliary as in other tunicates, and there are several to many pairs of pharyngeal slits. The well-developed muscle bands that completely encircle the body serve in jet-propelled swimming. The life history of doliolids, like that of salps, includes a solitary budding stage. The buds come to lie on a long tail-like appendage and are of several kinds (feeding, locomotory, and sexual), reminding us of those of siphonophores. Locomotory individuals break away, bearing buds that develop into hermaphroditic gamete-producing individuals and are released to swim freely. Two are seen in this photo. The fertilized egg passes through a tadpole stage and finally gives rise to an asexual individual. *Doliolum nationalis.* England. (D.P. Wilson)

Giant pyrosome colony, *Pyrosoma
spinosum* (class Thaliacea), off New
Zealand, was about 10 m long and
over a meter in diameter. Though it
might appear to be swallowing this
diver, he is merely peering into the
open end of the long tubular colony,
closed at the other end. The colony
consists of a huge number of small
individuals, each about 17 mm long
and much like an ascidian except that
the atrium is posterior, as in other
thaliaceans. The colony members lie
embedded in the walls with their
mouths on the outside of the colony
and their atriums opening to the
inside. Cilia lining the many pharyngeal slits of each individual drive the
feeding and respiratory current, and
the combined currents directed into
the central cavity propel the whole
colony through the water.

The individuals respond to mechanical, chemical, and light stimulation by
stopping the ciliary beat and luminescing brilliantly (*pyrosoma* = "fire-
body"). The light serves as a signal to
surrounding individuals, which stop
beating their cilia and then luminesce
in turn, so that if some solid object or
an unsuitable patch of water is encountered, the colony stops moving forward and a wave of light sweeps along
it. A potential predator may be frighten-
ed away. It has been suggested that
some of the reports of a torpedo
attack in 1964 in the Gulf of Tonkin,
which increased American involvement in Vietnam, might have arisen
from glimpses of luminescent pyrosomes, which are common in that
area. (R.V. Grace)

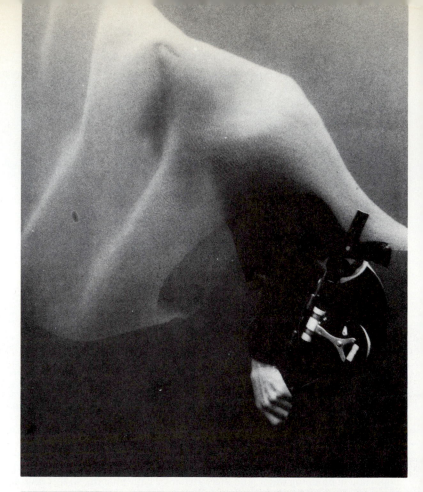

LANCELETS

THE laterally compressed, semitransparent little animals that comprise the subphylum **CEPHALOCHORDATA** have a notochord that extends the full length of the slender body, almost to the tip of the front end, hence the subphylum name. The common name, lancelet, is derived from the body shape, especially the lanceolate tail.

In cephalochordates the diagnostic chordate features are all found in both larva and adult: a tubular nerve cord, dorsal to the notochord; pharyngeal slits; and an endostyle that secretes an iodine compound similar to the thyroxin of vertebrates.

There are about 25 named species of lancelets, all marine and nearly all in shallow tropical and warm-temperate waters. They are bottom-living and relatively sedentary filter-feeders on small organisms such as diatoms and on organic particles. **Feeding** resembles that in tunicates. The feeding and respiratory current of a lancelet is maintained by the beating of the lateral cilia on the bars of the slotted **pharynx,** which is more rigid than in tunicates, having skeletal supports in the tissue between the slits, much as in acorn worms. The water enters the mouth, passes through the pharyngeal slits into the **atrium,** and exits through the ventral atrial opening near the rear of the body.

Anterior end of lancelet has a delicate hood which overhangs the mouth and is rimmed with stiffened tentacles that form a sieve, keeping out particles that are too large. In the cavity of the oral hood the water slows, and any particles that drop from the ingoing stream are caught by mucus on the "wheel organ," a series of ciliated loops. The particles then continue backward along with the main stream. Any unsuitable materials that still remain in the current are screened again by the velum, a circle of smaller tentacles that surround the mouth opening. *Branchiostoma floridae.* NW Florida. (R.B.)

The pharyngeal slits of acorn worms and lancelets develop in almost exactly the same odd way. From the top of each original pharyngeal opening, a "tongue-bar" of tissue grows downward. In acorn worms, the tongue-bar does not quite reach the bottom of the opening, and a U-shaped pharyngeal slit is formed. In lancelets, the tongue-bar reaches and fuses with the bottom of the original opening, dividing it into 2 narrower, parallel slits. The pharyngeal slits of tunicates also become subdivided, but in a different way.

The feeding current is filtered through mucoid sheets that are secreted by a ventral ciliated endostylar groove and are driven up the pharyngeal walls. In the dorsal groove of the pharynx, a strand of food-laden mucus moves steadily backward to the esophagus, and then into the midgut. Digestive enzymes from a midgut cecum become thoroughly mixed into the food strand. Only the smallest particles, already partly digested, enter the cecum and are there phagocytized by the epithelial cells. Thus digestion is both extra- and intracellular. The digestive tract is not muscular; as in other filter-feeders, the food is moved along entirely by the beating of tracts of strong cilia. Undigested remains are expelled from the anus, which is separate from and behind the atrial opening.

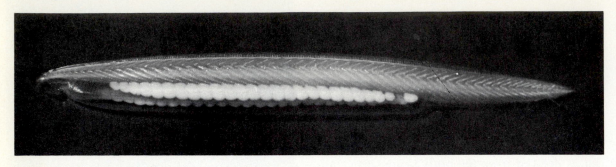

Lancelet of the genus *Branchiostoma* has a row of gonads on each side of the body. In the genus *Asymmetron* only the right side of the body bears gonads. The generic name *Amphioxus* ("sharp at both ends") has been discarded. In many books and reports cephalochordates are still referred to as "amphioxus." Photo is of *Branchiostoma floridae*. 5 cm. Gulf of Mexico. (R.B.)

Diagram of a lancelet, viewed from the left side. Cone-shaped muscle blocks fit into each other, making a chevron pattern, as in fishes.

Clogging of the mouth region elicits a sudden strong contraction of the transverse muscles in the floor of the atrium, ejecting a stream of water from the mouth and cleansing the food channel. A similar reflex, elicited by stimulation of the lips of the atrial opening, produces a posteriorly directed stream. In addition the steady passage of water through a lancelet is interrupted every few minutes, the atrial floor is contracted, and then the water flow is resumed. All these events, which remind us of the squirting of tunicates, are under nervous control.

The **nervous system** has some surprising similarities to that of vertebrates—and many differences. The slight enlargement of the anterior end of the dorsal nerve cord can hardly be considered a "brain." But as the nerve cord continues posteriorly it gives off a series of segmental nerves, both dorsally and ventrally. The ventral nerves, mostly motor, go to the massive blocks of striated longitudinal **muscles** that occupy much of the dorsal and lateral parts of the body. The turgid **notochord** enables coordinated contractions of these segmentally arranged muscles on either side to work antagonistically; and this produces the undulatory, eel-like movements with which cephalochordates swim about, burrow, or move along in loose sand almost as fast as some fishes do in water. The segmentally arranged muscles and nerves, and the lined **coelomic cavities,** are all vertebrate-like features.

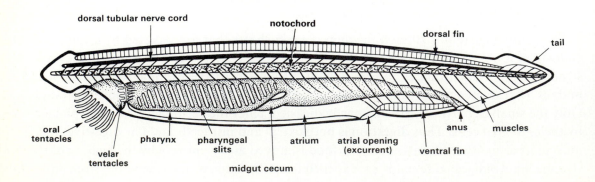

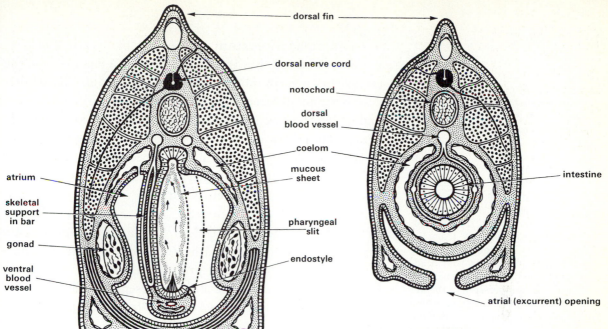

Diagrammatic cross-section of a lancelet at the level of the anterior pharynx, in front of the midgut cecum *(above left)* and at the level of the atrial opening *(above right).* Nuclei are shown only in the columnar endodermal cells lining the pharynx and intestine. Low mesodermal cells line the coelomic cavities. The pharyngeal section is shown as if cut obliquely, exactly parallel to the pharyngeal slits and bars, for comparison with the diagrams of an enteropneust hemichordate and an ascidian tunicate. Because the segmental structures of lancelets do not occur in opposing pairs, but alternate on the two sides of the animal, the section shows a number of asymmetries: a tongue-bar on one side of the pharynx and a pharyneal slit on the other, as well as asymmetries in muscles and nerves. Note, in the posterior section, the relatively large coelom around the gut, such as might be seen in a section through a fish. The atrium of lancelets is an ectodermal inpocketing, as in tunicates, but opens ventrally. (Combined from various sources)

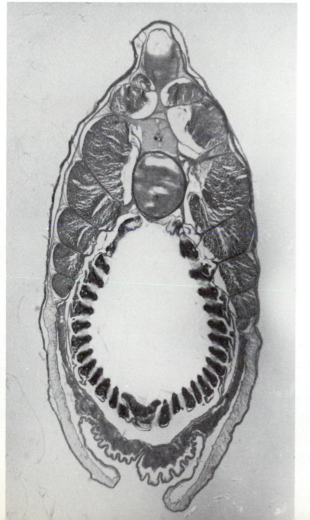

Histological cross-section of a lancelet. As the pharyngeal slits are oriented at an acute angle to the long axis of the animal, a cross-section through the pharynx always cuts through many successive slits and bars, resulting in a misleading resemblance to tunicates. (R.B.)

Lancelets are of no economic importance on most shores, but *Branchiostoma belcheri* is collected by the ton for human consumption at Amoy, on the southeast China coast. The chief predators of lancelets are fishes, which they mostly escape by their habit of remaining buried as they feed during daylight hours. After sunset lancelets may rise from the sand and swim about. If disturbed, lancelets emerge suddenly from the sand and swim about in zig-zag fashion, then plunge in again head-first. On some exposed beaches at a very low tide, compression of the wet sand by walking or by digging with a shovel will cause lancelets to pop out of the sand, then immediately dive back in.

The qualities of different sands affect lancelet behavior. When in coarse sand lancelets can maintain a ciliary respiratory and feeding current while completely buried. In fine sand, however, they must protrude the anterior end. Sand grains sieved in the laboratory have sharper edges and are otherwise less attractive to lancelets than natural sands with rounded edges. In addition, sands are more attractive to lancelets when living microorganisms coat the grains and when there is relatively little decaying organic matter present.

Whether or not the endostyle plays any hormonal role in lancelet metamorphosis remains to be determined. However, a few endostyles from lancelets, if implanted into an axolotl, will induce metamorphosis just as vertebrate thyroxin does in axolotls or in frog tadpoles. (Axolotls are salamander-like amphibians that normally become sexually mature in the larval form, without metamorphosis.)

The **circulatory system** has the same basic pattern of large vessels as in a fish. But there is no heart in a lancelet, and the larger vessels have muscular walls that pulsate. The smaller vessels are not differentiated into arteries, veins, or capillaries, and unlined tissue channels form part of the closed circulatory circuit. The colorless blood has no cells but contains proteins.

The **excretory organs,** in some species up to 90 pairs of them, are segmentally arranged and lie dorsally in the pharyngeal region. These excretory organs are protonephridia, bearing clusters of solenocytes, each a tubular cell that encloses a single beating flagellum. Similar protonephridia are found in various groups (gnathostomulids, gastrotrichs, some polychetes, priapulans) and probably evolved several times. Certain cells in vertebrate kidneys resemble solenocytes of lancelets and may have been derived from similar cells in a common ancestor.

Unlike tunicates, lancelets have separate sexes. The **gonads** are arranged segmentally along the sides of the body in the atrial region. When the gonads are mature, they burst through the atrial wall, and the gametes are shed through the atriopore. Spawning usually occurs at or after sunset and fertilization takes place in the seawater.

The **developing young** go through a free-swimming stage propelled by a covering of long cilia. Even when ciliary locomotion is replaced by muscular movement, the cilia continue to beat, and food particles that settle on the body surface are wafted toward the mouth. The mouth and anus open on the left side of the midline, the first pharyngeal slits on the right. These are only a few of the asymmetries that develop in what was originally a bilaterally symmetrical larva. Although they inevitably remind us of the asymmetrical development of echinoderms and hemichordates, they might also be explained in terms of possible functions in the feeding and swimming of the larva. Another interesting feature of cephalochordate larvas is their ability to attach temporarily to firm substrate by 3 adhesive papillas at the anterior end, strongly reminiscent of those of tunicates. The larvas are more pelagic in their habits than are the adults, but eventually they move to the sea bottom and metamorphose; the mouth moves to a ventral position, and the atrium grows to surround the pharynx, which develops slits on both sides. If carried into deep water, the larvas may delay metamorphosis for periods of up to several months or more, grow to large size, and begin to develop gonads. The fate of such waifs is not known.

LOOKING BACK over the evidences for ancestral ties between invertebrates and vertebrates, we recall that the wormlike burrowing hemichordates, described in the preceding chapter, share certain characteristics with chordates: pharyngeal slits and a dorsal tubular nerve cord. Tunicates and lancelets, the invertebrate chordates treated in this chapter, have in addition several other chordate features: a notochord; a muscular tail region behind the anus; and an endostyle, a ventral pharyngeal groove that is the forerunner of the thyroid gland of vertebrates. Lancelets show all 5 of these chordate characters in the adult stage. Most adult tunicates have only pharyngeal slits and an endostyle to attest to their chordate affinity; but the other chordate characters are present in the swimming tadpole-like larva. A summary of tunicate and cephalochordate affinities must stress that these animals are *invertebrates*. Nevertheless, they are indisputably the invertebrate groups that provide the best suggestions of what the early chordate stock that gave rise to the vertebrates might have been like.

We humans and many other vertebrates develop only rudiments of some of these chordate characters and, like tunicates, lose them at an early stage, but we have our dorsal tubular nerve cord, outsized at the anterior end, to confirm us as accredited chordates. How we developed from the same primitive stock as gave rise to tunicates and lancelets is a whole other story.

CLASSIFICATION: Phylum CHORDATA

Subphylum UROCHORDATA (TUNICATA), tunicates. Invertebrate chordates with a covering tunic. All marine and almost all filter-feeders, using pharyngeal slits and secretory endostyle. Circulatory system with heart and unlined vessels and sinuses; hemocoelic body cavity. Almost all hermaphroditic. Worldwide, mostly in relatively shallow waters.

Class ASCIDIACEA. Sessile and saclike as adults, with the atrium opening dorsally. Tadpole-like larva with notochord in tail and with dorsal nerve cord. Includes great majority of tunicate species. The orders and suborders are based on the location of gonads and structure of pharynx; both solitary and colonial species may occur in any one group.

> **Order Enterogona,** gonads in visceral mass.
> > **Suborder Aplousobranchia.** *Clavelina, Didemnum, Aplidium, Polyclinum.*
> > **Suborder Phlebobranchia.** *Ascidia, Ciona, Perophora.*
>
> **Order Pleurogona,** gonads in body wall.
> > **Suborder Stolidobranchia.** *Styela, Botryllus, Pyura, Halocynthia, Molgula.*

Deep-sea ascidians from all the above groups may be highly modified. Some (e.g., *Culeolus*) lack pharyngeal cilia but appear to filter-feed with a wide-meshed pharynx; the body is supported off the muddy bottom on a long stalk, and perhaps the animals take advantage of bottom currents. Others are carnivorous, having a greatly expanded, lobed oral siphon that traps prey, but still have a sievelike pharynx (e.g., *Octanemus*). Still other abyssal carnivorous tunicates, without a sievelike pharynx, have been separated into another class:

Class SORBERACEA. Deep-sea tunicates without a sievelike pharynx but with a small number of slits that open from narrow pharynx into small atrial cavity. Do not filter-feed but capture motile prey (e.g., copepods, isopods) with grasping lobes of a retractable oral siphon. Rudimentary endostyle. *Hexacrobylus, Gasterascidia, Sorbera.*

Class THALIACEA. Gelatinous pelagic tunicates with the atrium opening posteriorly. Complex life histories of asexual and sexual forms, mostly with direct development.

> **Order Salpida,** salps. *Salpa, Thalia.*
>
> **Order Doliolida,** doliolids. *Doliolum.*
>
> **Order Pyrosomida,** pyrosomes. *Pyrosoma.*

Class APPENDICULARIA (LARVACEA). Pelagic tunicates that are tadpole-like throughout life and collect food by means of mucoid "houses." *Oikopleura, Stegosoma.*

Subphylum CEPHALOCHORDATA (ACRANIA), lancelets. Invertebrate chordates with notochord and dorsal nerve cord throughout life. All marine filter-feeders, using pharyngeal slits and secretory endostyle. Closed circulatory system with contractile and unlined vessels. Coelomic body cavities. Separate sexes. Found in sandy sediments, mostly in warm shallow waters. *Branchiostoma, Asymmetron.*

Subphylum VERTEBRATA (CRANIATA), fishes, amphibians, reptiles, birds, and mammals. Vertebrate chordates with notochord during development, usually transitory. Various modes of feeding, mostly large food. Closed circulatory system with heart and lined vessels. Coelomic body cavities. Usually separate sexes. Worldwide in freshwater, marine, and terrestrial habitats.

Cambrian scene *(above),* a reconstruction of a marine bottom as it probably appeared about 540 million years ago. Swimming and crawling everywhere are a variety of different kinds of worms and arthropods, particularly trilobites (now extinct). Tubular sponges and branching gorgonians are also conspicuous. Reconstruction based on fossils from the Burgess Shale, British Columbia, by the Carnegie Museum, Pittsburgh, Pennsylvania. (R.B.)

Permian scene *(right),* a reconstruction of a swamp as it appeared about 250 million years ago., With a wingspread of about 70 cm, the dragonflies of the time were the largest insects that ever lived (these went extinct at the end of the Triassic period). The forests were dominated by ferns, horsetails, and gymnosperms. Reconstruction based on fossils from Kansas, by the Field Museum of Natural History, Chicago.

Chapter 30

Animal Relationships

THE KNOWN PHYLA of living animals, introduced in the previous chapters, number well over 30, each with a characteristic body plan. Species of these phyla are found in a wide variety of habitats and display great diversity of form. Yet each retains some or all of the distinctive features of its respective phylum, and almost all are unquestionable members of one phylum or another. There are few convincing intermediate species that might serve as bridges between the phyla. How then did the different phyla originate, and how are particular phyla related to each other, if at all?

Most phyla of animals have been on Earth for at least the past 500 million years. Thus, clues to their origins must involve events that occurred long ago, and the exact history of many of these evolutionary events will never be known. However, the clues are numerous, if diverse and fragmented, and they can be used to group many of the phyla together into "superphyla" that relate the phyla within them and indicate possible modes of origin. Such groupings can serve as working hypotheses, subject to minor or even major change as new information emerges.

As in any historical reconstruction, all the evidence about how the different phyla of animals are related to each other is indirect; one cannot directly observe or experiment with past events. And the many types of evidence are derived from different disciplines of science, including astrophysics, geology, and chemistry, as well as the various subdisciplines of biology. In most cases the evidence is *comparative* in nature, based mainly on similarities and differences in the structure, development, or molecular biology of individuals of one phylum as compared with those of other phyla.

A major problem in comparative analysis is how to distinguish between similarities arising from common origin *(homologies)* and those derived from common function *(analo-*

There are 10 phyla that contain many thousands of species each: Porifera, Cnidaria, Platyhelminthes, Nematoda, Annelida, Mollusca, Arthropoda, Bryozoa, Echinodermata, and Chordata. These are the phyla, shown in the figure heading of chapter 1, that comprise the bulk of animal species. Most of the remaining phyla contain only a few hundred species, or less.

gies). For example, many animals have well-developed eyes. Did the eye originate only once in animals and all the different kinds present today result from extensive **evolutionary divergence** of the original eye? If this is the case, one kind of eye such as the simple eyespots in some flatworms and medusas may represent the original primitive form, while more complex single-unit eyes, such as those of octopuses, spiders, and fishes, would be derived, advanced forms. The compound, multiple-unit eyes of arthropods could be even more advanced. Alternatively, because vision has such obvious adaptive value, the eye could have originated independently in different lineages of animals, and the similarity among different eyes could be the result of extensive **evolutionary convergence.** There may be only a limited number of ways to achieve a functional eye, and eyes may have been "invented" many times in different groups of animals. Thus the fact that markedly diverse types of animals have similar eyes is not necessarily good evidence that the animals are related.

In the case of complex single-unit eyes it is generally agreed that there has been considerable convergence among different phyla. Vertebrates, cephalopods, arachnids, pelagic annelids, and cubomedusans all have well-organized eyes with a single lens and retina, but details in the organization of their eyes (as well as in those of the more primitive eyes in the less complex members of each group) provide persuasive evidence that complex single-unit eyes originated independently at least 5 times. Complex compound eyes, found only in arthropods, are similar but not indistinguishable in the different subphyla of arthropods. If the several subphyla of arthropods represent different lines of *parallel evolution,* as may be the case, compound eyes almost certainly originated independently several times.

The most commonly used evidence for establishing relationships among groups has been the **comparative morphology** of adult forms. Relatively large features such as those in the skeletal, nervous, circulatory, and digestive systems are usually compared. Together these make up the characteristic body plan of a phylum, and those that are shared among several phyla are considered as evidence of relationships. Arthropods, annelids, and molluscs, for example, are often grouped together, and assumed to have a common ancestry, because the principal cords of the central nervous system are ventral and the main vessels of the circulatory system are dorsal, while the reverse is true in chordates and some other (presumably related) phyla. Moreover, the body plan of both arthropods and annelids is basically segmented while that of molluscs is not, suggesting that annelids and arthropods are more closely related to each other than either is to molluscs.

Comparative morphology was developed by the French zoologist Georges Cuvier at the beginning of the 19th century. He recognized that all anatomical parts of a body are functionally related to each other, and that by studying the anatomy of one animal in detail a person can predict the anatomy of another generally similar animal from only a few parts. This methodology has been especially useful for reconstructing the morphology of extinct animals from a few fossil bones, and Cuvier is often considered the founder of paleontology.

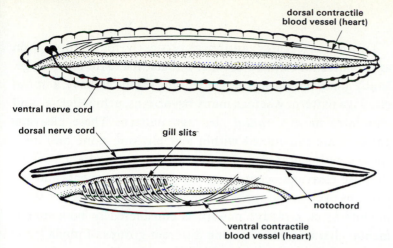

dorsal contractile
blood vessel (heart)

ventral nerve cord

dorsal nerve cord   gill slits

notochord

ventral contractile
blood vessel (heart)

Annelid-arthropod body plan includes a dorsal contractile blood vessel (heart) and a segmented ventral nerve cord.

Chordate body plan includes a ventral contractile blood vessel (heart) and a hollow, non-segmented dorsal nerve cord, as well as a notochord, pharyngeal gill slits, and a post-anal tail.

Studies of comparative morphology are not limited to large structures. Microscopic ones, such as cellular organization, also can be revealing. For example, the photosensitive membranes of the eyes or eyespots of most chordates, lophophorates, and cnidarians are formed by foldings of the cell membrane covering the cilia of the retinal cells. In contrast, the membranes of the eyes and eyespots of most molluscs, annelids, and arthropods are formed by repeated folding of the cell membrane itself and do not involve cilia. Such a contrast in details of fine-structural organization provides further evidence that annelids, molluscs, and arthropods are related to each other, but not as closely related to chordates, lophophorates, and cnidarians. However, with this and other examples of cellular or subcellular comparisons, problems of convergence and divergence may be especially troublesome. It may take very little evolutionary change, for example, for the membrane foldings in the cilium to spread over the rest of the cell, and even for the cilium to be lost.

The comparative morphology of large molecules, or **comparative biochemistry,** also appeared to have considerable promise for revealing animal relationships. Structural molecules such as chitin and connective tissue polysaccharides, oxygen-carrying blood pigments such as hemoglobin and hemocyanin, and certain energy-storing compounds in muscles ("phosphagens") were all thought to contain clues about common ancestry. However, more thorough comparative analyses have shown that such molecules evolved several times in different groups of organisms, and the resulting convergence makes phylogenetic analysis difficult or impossible.

Evidence from comparative morphology of adult structure is usually supplemented with evidence from **comparative embryology.** Except in vegetative propagation (from buds or fragments), multicellular organisms develop from one cell, an egg. The types of stages that are passed through during subsequent development often are characteristic of many or most of the embryos of a particular phylum or group of phyla.

Spiral cleavage   **Radial cleavage**

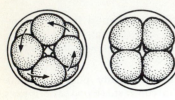

top view

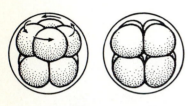

side view

Spiral cleavage in an annelid with the upper tier of cells twisted 45° out of line with the lower tier *(left)*. **Radial cleavage** of an echinoderm with the upper tier of cells directly over the lower tier *(right)*.

Distinctive features can appear very early, for example, even at the time of the third cleavage, which leads to the 8-cell stage. Many platyhelminths, annelids, and molluscs have a *spiral* cleavage pattern, whereas many bryozoans, echinoderms, and chordates have a *radial* cleavage pattern. These cleavage patterns are not mixed within any phylum; none has some species with spiral cleavage and others with radial cleavage. Because cleavage occurs very early in development, and the patterns of cleavage appear to be so fixed and distinctive at the phylum level, cleavage pattern is considered to be a fundamental characteristic dividing different groups of phyla.

In addition to platyhelminths, annelids, and molluscs, at least some gnathostomulids, nemerteans, kamptozoans, sipunculans, echiurans, and arthropods have spiral cleavage; these phyla are grouped together as the **spiralians.** Radial cleavage is typically found in cnidarians, phoronids, brachiopods, and hemichordates, as well as in bryozoans, echinoderms, and chordates; these phyla may be grouped together as the **radialians.**

In many species in some phyla the cleavage pattern during development is neither spiral nor radial. In most such cases, development is believed to be derived from an earlier pattern that did conform to the development characteristic of the phylum concerned. In particular, when the egg is supplied with abundant yolk, cleavage is modified and is either restricted to a small portion of the surface of the egg, as in cephalopods and birds, or does not occur until many nuclei have formed, as in most arthropods. Because cephalopods are clearly molluscs and arthropods are clearly related to annelids, as indicated by comparative morphology, these groups are included within the spiralians even though they do not undergo spiral cleavage. Similarly, on the basis of comparative morphology, birds (and mammals) clearly belong to the radialians, even though they do not display radial cleavage.

All the species of some phyla display cleavage patterns that are neither spiral nor radial. These include nematodes and other micrometazoan phyla with highly determinate but asymmetrical cleavage patterns and ctenophores with a biradial cleavage pattern. These phyla may not belong to either the spiralians or the radialians.

Several other characteristic features are found in the developmental patterns of the spiralian or radialian phyla, and these features also serve to separate the two groups. The fate of the cells and tissues in an embryo is fixed much earlier in development in spiralian phyla than in radialian phyla. Thus, if the cells of 2- or 4-cell embryos of spiralians are separated from each other, each will develop into only a half or a quarter of an embryo and larva; in radialian embryos, each of the cells would develop into a whole, though smaller larva. By the 64-cell stage of the spiralian embryo, when it is only an early blastula, there is one identifiable cell, the *4d cell,* that will give rise to nearly all the mesoderm in the animal. In contrast, in radialian embryos, the mesoderm is not segregated in the embryo until after the gastrula is formed and mesodermal precursor cells bud off from the embryonic gut.

In many species of animals gastrulation by invagination, cell overgrowth, or cell migration leaves an opening, the blastopore, into the embryonic gut. This opening sometimes remains open during development and eventually becomes the mouth. Phyla with species having such a mode of development are referred to as **protostomes** ("mouth first"). Most protostomes are spiralians. In contrast, in most radialian phyla the mouth never develops from the blastopore but rather develops secondarily; in echinoderms and hemichordates, the blastopore often becomes the anus. These radialian phyla are referred to as **deuterostomes** ("mouth second").

There are some exceptions to the correspondence between cleavage pattern and blastopore fate. Cnidarians, for example, undergo radial cleavage, while platyhelminths undergo spiral cleavage, yet the mouth often forms from the blastopore in both groups. However, in both groups the opening into the gut serves as both the mouth and anus, and the problem about whether they are protostomes or deuterostomes is moot.

Phoronids also have radial cleavage and the mouth sometimes forms from the blastopore. In this case, as in some spiralians, the blastopore becomes an elongated slit and the anterior part becomes the mouth while the posterior part becomes the anus. Because of these features, phoronids are often placed within the protostomes and their relationship to other phyla has been uncertain. They have been viewed as an intermediate between the spiralian/protostome and radialian/deuterostome phyla and as a model for how deuterostomous development could be derived from protostomous development. However, most of their other morphological and embryological features align them with the deuterostome phyla, and the phylogenetic significance, if any, of the fate of the blastopore in phoronids remains elusive.

In the other lophophorate phyla, the brachiopods and bryozoans, cleavage is radial, and either the blastopore closes and both the mouth and anus form secondarily (albeit near the site of the blastopore), or, more typically, no blastopore ever forms during gastrulation, which proceeds by delamination (splitting of layers).

In spiralian/protostome phyla, the mesoderm fills the space (blastocoel) between the embryonic gut and the outer epidermis mainly by proliferation of descendants from the *4d* cell. Coelomic spaces then develop in most spiralians from splits that form within mesodermal masses. Such a mode of coelom formation, typified by annelids, is called *schizocoely* ("split coelom"). All coelomate spiralians are schizocoelous animals. In contrast, in most radialian/deuterostome phyla, the mesoderm is formed by a budding off of the wall of the embryonic

The *4d* cell of spiralians is so named because it is one of the 4 cells in the fourth quartet of micromeres (smaller cells) to divide from the 4 macromeres (large cells). In spiralians the other 3 cells of the fourth quartet plus the macromeres will form all the endoderm, while all the remaining cells develop into ectodermal structures.

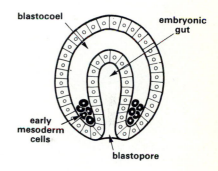

Mesoderm formation. In **spiralians,** as in a mollusc gastrula *(above),* mesoderm cells separate early in development and form discrete cell masses. In **radialians,** as in an echinoderm gastrula *(below),* most of the mesoderm forms by budding off the wall of the embryonic gut.

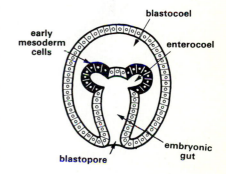

gut. Typically, as in echinoderms, the buds are hollow, and space within each bud contributes to the coelom. This mode of coelom formation is called *enterocoely* ("intestine coelom"). Enterocoelous animals are all radialians.

Although cnidarians do not have a coelom, the gastrovascular cavity of scyphozoans and anthozoans is divided into multiple spaces by septa. These fluid-filled spaces act as a hydrostatic skeleton that supports the body and hollow tentacles of anthozoans; the spaces in the tentacles are at least analogous to the tentacular coeloms of lophophorates and echinoderms. The enterocoels of radialians, including the tentacular coeloms, form during development by a pinching off of embryonic gut spaces and therefore also might be homologous to the interseptal spaces of cnidarians.

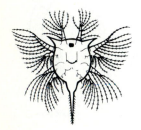

Larval barnacle is free-swimming and planktonic. (After Bernard and Lane)

Adult barnacle, enclosed in calcareous plates, is sessile and benthic. (After E.B. Newhall)

The embryos of many species of benthic marine animals hatch as larvas that are markedly different from the bottom-living adults. These larvas are usually pelagic and can be carried in ocean currents considerable distances from the site of the parents before they settle down and metamorphose into juveniles. Their larval form is well suited for survival and growth in the plankton and may be very different from the benthic design of the juveniles and adults. While resulting from selection for survival in the plankton, larval design also reflects a long genetic heritage. Therefore, **comparative larval morphology** provides additional important information that often can be used to sort out relationships between groups of animals. As when comparing adult morphologies, the difficulty of distinguishing homologies from converging analogies in larval resemblances must always be kept in mind.

The larvas of many species of annelids, molluscs, and other spiralian phyla are very similar in appearance. Called *trochophores,* they are top-shaped animals with an apical tuft of sensory cilia and two bands of cilia around the equator that serve for both swimming and feeding. In contrast, in many radialian species (echinoderms and hemichordates), the larvas are shaped more like kidney beans, with a band of feeding cilia looped around the mouth; although the feeding cilia may also serve for swimming in early stages, separate tufts or rings of cilia (such as the telotroch in hemichordates) often develop for swimming. These larvas have been given a variety of names that are specific for different stages in different classes, but they may be grouped together as *dipleurulas,* the name often given to the earliest stage of echinoderm larvas.

Arthropods are conspicuous among spiralians because none has a larva like a trochophore. Arthropods are characterized as being covered with an exoskeleton devoid of cilia, and this feature is part of the larval design as well. Without cilia, a trochophore would be completely nonfunctional. All arthropod eggs are provided with substantial amounts of yolk, and development carries the embryos to relatively late stages before they hatch as larvas or juveniles equipped with appendages that can serve both for feeding and, if aquatic, for swimming. Similarly, chordates do not have larvas that are like dipleurulas, but development of the relatively yolky embryos carries them to later stages before they hatch; typically, early chordate larvas do not feed, and they use their muscular tail mainly for swimming.

The larvas of some other phyla are quite distinctive and, depending on the viewpoint of the author, are said to be "trochophore-like" or "dipleurula-like." In particular, the larvas of phoronids (actinotrochs) and the other lophophorate phyla are sometimes said to be "trochophore-like" by people who believe they should be aligned with spiralians. However, phoronid larvas, with their long ciliated feeding tentacles and posterior ring of swimming cilia, are a distinct larval type with little similarity to either trochophores or dipleurulas.

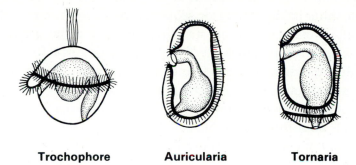

Trochophore　　　**Auricularia**　　　**Tornaria**

Early larval stages are very different in spiralians and radialians. **Trochophores** are typical of polychetes, molluscs, and some other spiralians. **Auricularias,** typical of sea stars and sea cucumbers, and **tornarias,** typical of acorn worms, are dipleurula-type larvas of radialians.

DEVELOPMENTAL CHARACTERISTICS OF THE TWO MAJOR GROUPS OF ANIMAL PHYLA.

	Echinoderm-Hemichordate-Chordate	Mollusc-Annelid-Arthropod
Cleavage	Radial	Spiral
Development	Indeterminate	Determinate
Blastopore fate	Never mouth (sometimes anus)	Mouth
Mesoderm	From endoderm	From *4d* cell
Coelom	Enterocoelous	Schizocoelous
Larvas	Dipleurula type	Trochophore type

Also usually included in the mollusc-annelid-arthropod lineage on the grounds that they possess most of the same embryonic characteristics are: platyhelminths, gnathostomulids, nemerteans, kamptozoans, echiurans, sipunculans, pogonophores, and onychophorans. Also usually included in the echinoderm-hemichordate-chordate lineage are chetognaths.

The lophophorate phyla (phoronids, brachiopods, and bryozoans) have a mixture of characters, and their placement remains uncertain and controversial; on balance they appear to be more closely aligned to the echinoderm-hemichordate-chordate lineage.

Sponges, placozoans, mesozoans, cnidarians, ctenophores, nematodes (and most other micrometazoan and parasitic phyla), and priapulans cannot at present be aligned with one lineage or the other using either adult morphological characteristics or embryological features.

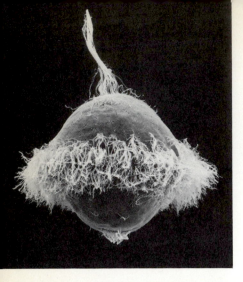

Echiuran larva is a trochophore indistinguishable from an early, presegmented, annelid trochophore larva. One way to account for echiurans is to view them as annelids that grow to sexual maturity while suppressing the development of segmentation. Similarly, sipunculans and perhaps molluscs might be viewed as descendants of annelids that lost segmentation through neoteny. *Urechis caupo.* SEM. (C.B. Calloway and R.M. Woollacott)

Copepod, with a single "nauplius eye" and a metanauplius-like body, is an example of a group that might have arisen through progenesis, that is, by being sexually precocious.

During the development of an organism, its various features appear at different times. As long as they are present when needed for continued survival, there seems to be considerable flexibility as to when they are scheduled within the developmental program. And the schedules for the appearance of various juvenile or adult structures are not necessarily closely linked.

Thus, structures that appear only in adults in some species—in particular, those associated with sexual maturity—may appear earlier, in the larval or even embryonic stages, in other species. Or, larval or juvenile features may be retained in the development of adult stages in some groups but not in others. Such shifts in the timing or sequence of the appearance of larval, juvenile, and adult structures are said to be *heterochronous.* This process has probably been a powerful force in molding the characteristics found in some groups. The recognition that animals may develop mature sex organs and be functionally capable of reproducing while still in the larval or juvenile body form has added new dimensions to the challenging study of comparative morphology.

The term *neoteny* refers to the suppression of the development of adult features while sexual development proceeds, and *progenesis* refers to precocious sexual maturation in a morphologically juvenile stage; both processes are forms of *pedomorphosis* and result in sexually mature animals with some features possessed only by the juvenile stages of their ancestors.

Studies of comparative morphology and comparative embryology strongly indicate that most phyla of animals can be placed within either one or the other of two large groups or "superphyla." (1) The spiralians, or protostomes, include the molluscs, annelids, and arthropods, among others. (2) The radialians, or deuterostomes, include the echinoderms, hemichordates, and chordates (and possibly others). How the phyla *within* each branch of this dichotomy are related to each other, especially in terms of which phyla were ancestral and which descendant, is more difficult to establish. An even more unyielding problem is how the two superphyla are related to each other. Did one superphylum arise from the other—or did each evolve independently from different animal or even protistan ancestors?

The **fossil record** might be expected to reveal the time of origin of particular phyla and which arose from which. After all, fossils are the most direct kind of evidence about life in the past, and the order in which different groups appeared can be determined by *stratigraphy,* the sequence of the occurrence of fossils in the rocks. With modern radiographic dating techniques, the age of the oldest fossils of each group can be determined. Moreover, the order of appearance of fossils of different groups should provide compelling evidence about which phyla came on the scene early and therefore could be ancestral to those that made their entrance later.

Fossils were known to the ancients, and the serious study of fossils for nearly 200 years has provided strong evidence for the time of appearance of many of the phyla. However, that evidence tells us only that all the major phyla appeared at about the same time—500 to 600 million years ago. Complex animals such as arthropods, molluscs, echinoderms, and chordates date from nearly the same time as sponges, cnidarians, annelids, and brachiopods. Many of the fossil members of these groups are similar to living counterparts today, and there is little indication, even within a phylum, of a steady progression of more and more complex forms. Moreover, few, if any, new phyla made their first appearance in the fossil record after about 500 million years ago.

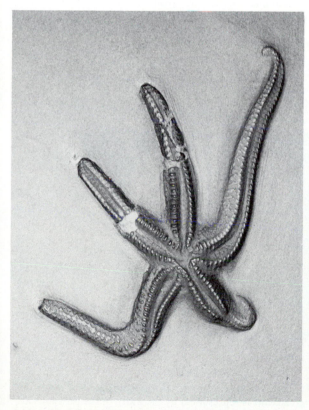

Asteroid fossil from Devonian rocks more than 350 million years old appears remarkably similar to modern forms. *Aspidosoma,* oral view. Bundenbach, Germany. (R.B)

Stalked crinoid fossil from Carboniferous rocks showing arms, calyx, and stem. Most species of stalked crinoids, which flourished in shallow Paleozoic seas, did not survive past the Permian extinction, but some still thrive in deeper water today. (R.B.)

Thus, while providing information about how long the known phyla have been on Earth, evidence from the fossil fauna provides very little information about which group is derived from which. Indeed, our current understanding of the fossil record tells us that either there was a rapid evolutionary divergence of phyla or that they evolved independently from different protistan ancestors at a unique time in the Earth's history when conditions first became suitable for multicellular animal life. And once established, species of the different phyla diversified, perhaps filling all available habitats so thoroughly that no place remained for new phyla to develop.

Few phyla have disappeared completely through extinction, but several times in the history of the Earth mass extinctions have eliminated most species of at least some phyla. The largest mass extinction, the Permian extinction that closed the Paleozoic Era, occurred over a period of many millions of years about 225 million years ago. It coincided with a time when all continental land masses were joined together, reducing shorelines and perhaps adversely influencing the climate. The second largest mass extinction, the Cretaceous extinction that closed the Mesozoic Era, occurred about 70 million years ago. Geochemical evidence indicates that it coincided with a collision of the Earth and a large extraterrestrial body. The resulting fire and smoke that darkened the skies, and the subsequent cooling of the Earth, may have led to large scale extinction over a relatively short period of time. It was followed by a rapid diversification in some groups, especially in molluscs, arthropods, and chordates, perhaps as a consequence of the extinction of previously dominant groups.

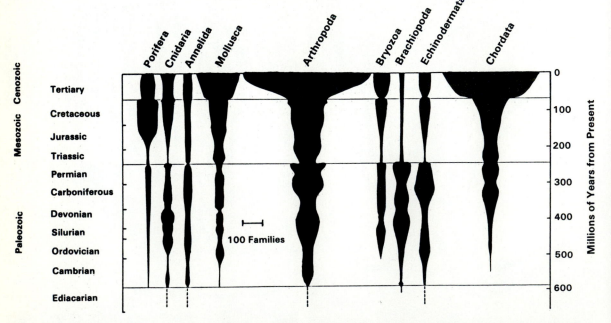

Fossil record: First appearance and changes in number of families through time of the major animal phyla as revealed by the fossil record. The number of families at any given time is represented by the width of the polygons at successive levels. The number of species in different phyla vary; some arthropod and chordate families (of insects and fishes) have large numbers of species while all the brachiopod families have very few species. Geologic time, in millions of years before the present, indicated on right (only fossil data are used; data on living species are not included.) Assignment of the Ediacarian fossils to phyla is uncertain and is indicated by dotted lines. (Based on data of Sepkoski and Hulver)

TABLE OF GEOLOGIC TIME

Eons (Billions of years ago)	Eras (Millions of years ago)	Periods (Duration in millions of years)	Principal evolutionary events among animals.
PHANEROZOIC	CENOZOIC	Quaternary 2	Modern species; insects and gastropods with numerous species on land and in the sea.
	64	Tertiary 62	Recovery of most groups after Cretaceous extinction; most modern genera and families established.
	MESOZOIC	Cretaceous 71	Continuous development, then gradual decline of many groups until abrupt extinction of remaining ammonoids and many species of other groups, possibly from results of impact of a giant meteorite.
		Jurassic 57	Ammonoids abundant; most modern orders well established and flourishing; birds and mammals appear.
	225	Triassic 33	Slow and erratic recovery of groups surviving the Permian extinction.
	PALEOZOIC	Permian 55	Decline of most groups of animals with extinction of trilobites, possibly from major cooling period; dragonflies, beetles, and bugs present.
		Carboniferous 65	Crinoids and blastoids peak and begin slow decline; winged insects (mayflies, grasshoppers, and cockroaches) and reptiles appear.
		Devonian 50	Continued development of many forms on land and in the sea; brachiopods and eurypterids peak; arachnids, wingless insects, and amphibians present on land.
		Silurian 35	Extensive tabulate coral reefs; graptolites and trilobites begin to decline; millipedes appear on land.
		Ordovician 70	Most major classes of animals present. Trilobites, echinoderms, graptolites, and nautiloids near their peak.
		Cambrian 70	Animals with hard skeletons first appear. Most animal phyla present; archeocyathids, trilobites, and brachiopods numerous.
0.65	650	Ediacarian 80	First animal fossils, not clearly recognizable as members of modern phyla but resembling cnidarians, worms, and arthropods.
PALEOPHYTIC 2.0		1,350	No animal fossils but sea floor covered with algal mats (stromatolites); oxygen atmosphere began to develop; earliest nucleated cells with meiotic sex.
PROTEROPHYTIC 2.6		600	Development of bacterium-like forms, some filamentous and photosynthetic; first evidence of free oxygen.
ARCHEAN 3.75		1,150	Earliest known sedimentary rocks, some with microfossils of bacterium-like cells; probable time when life arose.
HADEAN 4.5		750	Formation of the earth; differentiation of earth's rocks; radiometric clocks set.

Ediacarian fossils provide the earliest evidence of metazoan life on Earth. All are impressions of soft-bodied organisms that lived in shallow seas over 600 million years ago, about 50 million years preceding the Cambrian. Best known from sandstones in the Ediacara Range of South Australia, these fossils have been found also in other localities of the world. Many bear little or no similarity to modern animals, but some may have been large cnidarians, such as scyphozoans, chondrophores, and anthozoans. Others, such as those shown below, resemble more complex phyla. Whether they were, in fact, early members of any phyla still living today and possible ancestral forms, or were members of phyla long since extinct, is a question of considerable current debate. At any rate, they shed little light on the question of which phyla were ancestral to other phyla, or if, indeed, animals have a common ancestry.

Spriggina floundersi is segmented like a polychete but has a prominent "head shield" unlike any modern polychete. 2 cm long.

Dickinsonia costata appears to have been a large animal with segments like an annelid, but it lacked a well-defined head and appendages. 8.5 cm long.

Tribrachidium heraldicum is a disk-shaped fossil with 3 curving ridges, similar in appearance to the edrioasteroid echinoderms that were abundant in the Paleozoic (now extinct) but it lacks a calcite skeleton. 2.5 cm in diameter.

Parvancorina minchami may have been an early arthropod with a carapace over the dorsal surface. 2 cm in diameter. (All specimens on this page are from South Australia. M.F. Glaessner)

Insects have flourished in the Cenozoic to the present time, and their fossils are often virtually identical to modern species. *Left*, a carbon film left by a **bee** fossilized in Tertiary rocks from Florissant, Colorado. *Right*, carbon films left by two **termites** in Tertiary amber (fossilized tree resin) of northern Europe. (R.B.)

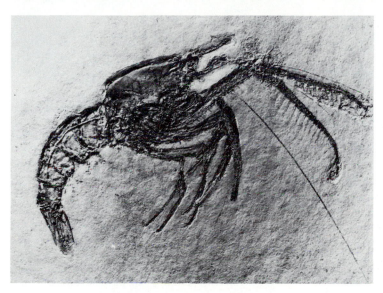

Crustaceans similar to modern forms were common in Mesozoic seas. *Left*, a **shrimp**, *Penaeus speciosus*, assigned to a contemporary genus. *Right*, a **crab**, *Epropinquus*, showing details of the appendages. Both in Jurassic rocks from Solnhofen, Germany. (R.B.)

Trilobite fossils from the mid-Cambrian period; *Neolenus*, from Burgess shale, British Columbia. (G.E. Resser).

Eurypterid fossil with large paddle-like swimming appendages; Silurian period. Trilobites and eurypterids did not survive into the Mesozoic, but eurypterids may have given rise to scorpions and other arachnids. (R.B.)

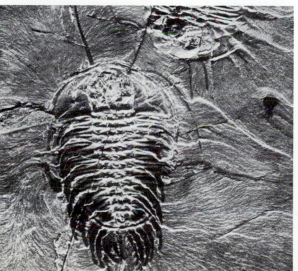

Edrioasteroids belong to an extinct class of sessile echinoderms that were abundant in the Paleozoic. Many were attached to the bodies of other animals, such as to the trilobite shown here, reminiscent of modern barnacles attached to crabs today. (R.B.)

Brachiopods were abundant throughout the Paleozoic but most were extinct by the end of the Permian. Some were so numerous that they formed beds of limestone called "lamp-shell coquina." *Dalmanella* in Ordovician rocks. (M. Fenton)

Graptolites are characteristic fossils of the early Paleozoic. They probably were colonial forms, related to modern pterobranchs, with numerous feeding zooids held in tiny cups. Many had gas floats and may have been pelagic like modern siphonophores. *Diplograptus,* about 10 cm across, from the Ordovician. (R.B.)

Ammonites were shelled cephalopods that thrived in both the Paleozoic and Mesozoic seas, but none survived the Cretaceous extinction. Some coiled species reached almost 2 m in diameter while some of those with straight cone-shaped shells grew to nearly 6 m long—the largest shelled invertebrates that ever lived. The coiled fossil shell of *Coeloceras* (about 6 cm in diameter) is from Jurassic rocks of Whitby, England. (R.B.)

The seemingly "sudden" appearance of most phyla of animals about 570 million years ago, near the beginning of the Cambrian Period, remains a major riddle in our planet's history. Earlier sedimentary rocks are not uncommon, and the ones examined so far contain fossils of unicellular and even prokaryote life, some in rocks over 3 billion years old. There can be little question that a large variety of protists was present long before the Cambrian, and that these could be ancestral to animals (see Chapter 2). But except for the Ediacarian fossils, the earlier rocks are largely devoid of fossils of animals, indicating that animals were indeed absent, rather than that they simply did not leave fossils. There is evidence of extensive glaciation before the Cambrian, and climatic conditions may not have been favorable for multicellular life. Moreover, oxygen, which has been accumulating in the atmosphere as a product of photosynthesis for close to 2 billion years, reached about 20% of present levels in the Cambrian; before that time oxygen levels may have been too low to support multicellular life.

Oxygen levels in the atmosphere stabilized about 100 million years ago, and since then oxygen production by photosynthesis has been largely offset by its consumption in the respiration of organisms, especially animals.

The nearly simultaneous appearance of many or most phyla in the fossil record suggests that they may have evolved independently from different protistan ancestors, and that the animal kingdom is polyphyletic. The idea is countered by comparative evidence from living forms. Details of cellular organization, such as the formation of asters during mitosis and the presence of gap junctions between cells, are very similar among animals, suggesting common ancestry. Moreover, the flagellated sperms of most animals are remarkably similar, and animal development characteristically includes the formation of a blastula followed by gastrulation. These features are not seen in plants, fungi, or even among any modern protists. On the other hand, details of cellular organization and early development are about the only features that sponges share with other animals, and it appears likely that at least sponges originated from protists independently from other animals. Thus, animals probably are at least diphyletic. Whether other phyla originated independently from different protistan ancestors, and if so, how many times, remains even more uncertain at the present time.

Fortunately, techniques are being developed that make it possible to compare the genetic information itself in different species and thereby estimate more directly the degree of relatedness between groups. On the assumption that "one gene produces one enzyme," initial work focused on comparing the enzymes of different species. All enzymes are proteins, and the sequence of amino acids in a protein is determined by the sequence of nucleotides in the desoxyribonucleic acid (DNA) of the corresponding gene. In addition, techniques have been developed to compare the nucleic acids of different species. Nucleotide sequences of both DNA and RNA (ribonucleic acid) can be compared, the RNA being central in translating instructions of the DNA in genes into the production of protein.

Proteins may be compared by using electrophoretic and immunological techniques. Different DNAs also have been compared by determining how tightly single strands anneal to each other; the more closely they anneal, the more similar they are, and the more closely they are presumed to be related. As techniques of molecular biology have developed, it has become possible to establish the sequences of amino acids in proteins or of nucleotides in nucleic acids. These sequences, involving hundreds to thousands of amino acids or nucleotides, can then be compared by computer analysis and the degree of similarity calculated. Comparison of ribosomal RNA (rRNA) from different organisms has proved to be particularly promising because rRNA is abundant in all organisms, is large and therefore not likely to show convergence of sequences, changes slowly and independently of both morphology and environment, and has highly conservative sections that can be used for matching sequences from different species.

When species with well-established fossil records are compared, a time frame can be provided. Surprisingly, the degree of difference in the enzymes or nucleic acids among species is often proportional to the length of time since the species diverged. This finding has provided the basis for believing that these molecules provide a kind of "molecular clock" that can reveal the approximate age of different groups. **Comparative molecular biology,** perhaps better referred to as comparative genetics, thus provides great promise for revealing both the degree of relatedness between phyla (or determining whether they *are* related) and, if the rate of the molecular clock can be determined, the absolute age of the phyla. At this time, the analyses are only beginning, the underlying assumptions are still not all recognized and resolved, and the results are too tentative to be used with any confidence. However, it is reasonable to expect that while many relationships established by more classic methods of comparative morphology and embryology will be supported, others will be challenged and hotly disputed. The future of phylogenetic studies appears to be both exciting and potentially revolutionary to our current understanding of how organisms are related to each other.

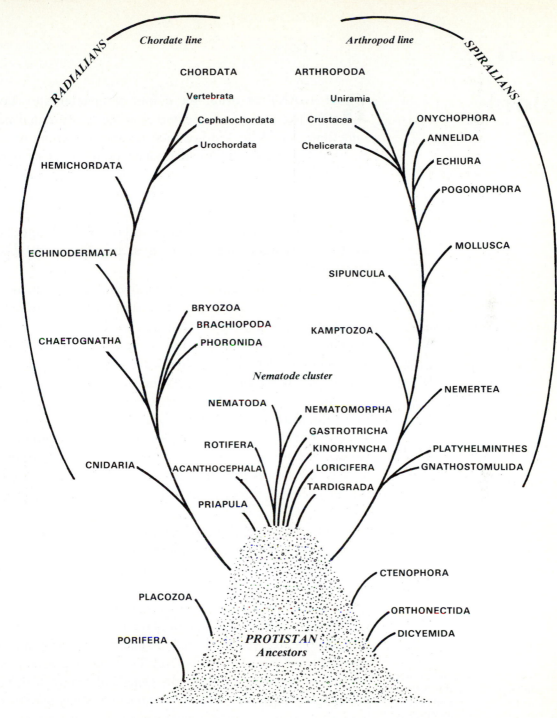

Animal relationships. This diagram is intended to suggest one view of how different phyla might be related to each other, reconstructed from the various lines of current evidence. Many such "phylogenetic trees" are drawn with a single trunk indicating a single, monophyletic origin of all animals. The "tree" shown here is designed more like a "flower patch" to emphasize the current lack of strong evidence for a single origin and the possibility of multiple animal origins.

The two major groups, radialians and spiralians, differ with respect to general morphology and development, as discussed in this chapter. Nematodes and other aschelminths make up a third group of uncertain affinity with each other as well as with other animals. Sponges, placozoans, ctenophores, and mesozoans (orthonectids and dicyemids) are also of uncertain affinity with each other and with other animal groups. The protistan ancestors of animals are unknown, but they are usually thought to have been flagellate-like. How many separate origins of animals occurred remains unresolved.

Chapter 31

Colors of Invertebrates

The plant bug *Psylla mali,* normally found on apple trees, is green. But when found on cherry trees it is brownish-red. The explanation is that the bug takes its color from the symbiotic bacteria it harbors. These bacteria, when cultured in the laboratory, form a bright green pigment if cultured on a medium containing apple juice, but appear bright red if on a medium containing cherry extract. Red color in the water flea *Daphnia* results from infection with a red bacterium. Scarlet red color in the copepod *Calanus* is caused by fluke or tapeworm parasitization.

Surface coloration does not always tell the whole story. In the Namib desert of Africa, 2 tenebrionid beetles *(Onymacius)* live side by side. One has a white dorsal surface, the other a black one, and their activity rhythms are very different. Investigation reveals that the intensity of the UV radiation reaching the inside of the abdomen is about the same, because the externally white beetle has a layer of dark pigment under the exoskeleton.

BRILLIANT or drab, the colors of invertebrates have much to tell us. Even the external surface of an animal may sometimes reveal the kind of food it eats, something of its metabolic processes, the efficiency of its respiration, its general state of health, whether it is approaching a molt, its approximate age, differences between the adult sexes, and their reproductive state. Particular colors may indicate whether an animal contains green or yellowish-brown symbionts, or whether it is parasitized by a virus, bacterium, fluke, or tapeworm.

Freshwater invertebrates are generally dull grays or browns, but no one explanation seems to account for this. Among the few exceptions are some of the small crustaceans, some bright red mites, and a few leeches. In marine waters bright color is seen in many temperate and cold water forms but it is greatly excelled by the variety of pastels and the brilliant patternings of tropical marine invertebrates. On land the snails of the forest floor are mostly monotonously colored, but some tree snails of warm climates have shells that are highly colored and banded or striped. Aquatic insects are subdued in color; their terrestrial relatives in temperate climates are often colorful. But the peak of insect color and pattern is reached in the tropics.

Depth of color may provide clues to the intensity of light, the range of temperature, the humidity to which an animal has been exposed, or whether it frequents green foliage, patchy bark, or brown forest litter.

Light induces the deposition of pigment in many animals, as we well know from sun-tanning and freckles. And invertebrate integuments are usually strongly pigmented on the exposed, dorsal surface, less so on the ventral. The kinds of pigment and the sites of deposition are usually genetically controlled, but the quantity may vary with light intensity. Many species of flatworms are blacker or browner when living near the surface than at greater depths. Cave planarians tend to be white and eyeless; those in the open have eyes and are usually darkly pigmented. Sea anemones that are deep red in shallow waters may be increasingly paler with increase in depth. In pupal stages of some butterflies pigmentation varies with light intensity, but the light acts through the eyes, rather than through the integument, as can be demonstrated by cutting the ventral nerve cord in front of or behind particular ganglia.

Temperature has important effects on pigmentation. Insects and crustaceans bred at low temperatures may have much more melanin (and other pigments) and they absorb more solar radiation. At higher temperatures they have less pigment, may be pale in color, and reflect heat.

Humidity also influences the synthesis of melanin. Butterflies are darker in the wet season, lighter in the dry season. Pupas that usually grow in the dry season, if made to develop in high humidity will produce darker adults.

Crowding can affect melanogenesis, as in migratory grasshoppers. The gregarious phase results in increased activity and rapid movement, and the grasshoppers migrate. The increased metabolic rate stimulates melanin formation and the insects are blackish. In the solitary, less active, non-migratory phase the grasshoppers make little melanin and appear green.

Coloring and behavior may be closely correlated. Dense coloring that absorbs solar energy may be accompanied by thermoregulatory or hygroregulatory behavior that exposes the pigment to the sun at appropriate times of the day or conserves water by hiding in crevices or under rocks. Cryptic coloring is enhanced by quiescence and slow movements, and conspicuous coloring by display behavior that advertizes gender and readiness to mate or that provides ample warning to predators that the potential prey is distasteful, emits an irritating secretion, has a potent sting, or has an exceptionally aggressive response to threat. Some distasteful species of insects have concealing color and behavior in years when they are few in numbers, but are conspicuously colored in years of abundance, when there are enough members to teach the population of predators that such colored insects should be avoided. Concealing color and behavior also aid predators in avoiding detection as they approach prey or lie in wait.

The spectrum of light visible to humans is not necessarily identical with that of other animals and this must always be kept in mind when considering the visual effects of invertebrate colors and patterns. It has been noted that flowers pollinated by birds are predominantly red, as compared with those pollinated by insects, many of which can see into the ultraviolet range. Also, it is possible that an insect wing could have two signals—one in the ultraviolet effective for insect eyes, and one conspicuous (or inconspicuous) to vertebrates.

There are many **approaches to the analysis of animal coloration.** Physicists, organic chemists, and biochemists have provided the foundation on which the various biological

A species of *Daphnia* found in an Alpine lake in Italy has a coal-black head, back, and antennas when the temperature drops to 6°C. After less than a week in a warm laboratory, the little water fleas lose the black pigment, as it is broken down faster than it is synthesized.

The blue pigment of *Velella* and *Porpita*, two surface-floating cnidarians, changes color with rising temperature or lower salinity; these color changes are reversible.

Migrating grasshoppers bask in the sun each morning, giving maximum exposure to their dark integument. This speeds up the attainment of a body temperature adequate for take-off.

Not only some insects but also some fishes are known to see into the UV range. And people who have had their crystalline lens removed can see some of the longer wavelengths of UV as a bluish hue.

In poriferans colorful carotenoid pigments are mostly confined to amebocytes in the mesohyl, while blackish melanin is found only in the pore cells of certain sponges. In cnidarians colored pigments may occur in both cell layers, but the green symbionts of hydras occur only in the endoderm. In stony corals the skeletal cups are always white, the living tissues colored. But in alcyonarian corals it is the skeletal parts that are often colored, the polyps white.

In experiments using Mediterranean shallow-water squids, a red squid becomes yellow and then violet if the animal is stabbed. A gray-brown *Octopus,* on sighting an enemy or being pushed, turns dark brown or brick red. Later the skin turns white, violet rings appear, and from these warts rise. At 12°C the natural color changes are suppressed.

The wavelength of light is expressed in nanometers (nm).
1 nm = 10^{-6}mm

disciplines have made their contributions. Cytologists have determined the intracellular or extracellular sites in which pigment occurs in either the body tissues or fluids, or complexed with calcareous skeletons or chitinous coverings. Geneticists have identified the genes responsible for the color variants of certain species. Physiologists elucidate the roles of colored compounds in screening out potential damage by excess ultraviolet radiation, in respiration, in other metabolic processes, in vitamin synthesis, in photosensitivity and color vision, etc. Those concerned primarily with animal behavior design experiments that test how color affects the visibility of an animal to other species or to other members of its own species, and how significant these reactions are in survival and reproduction or in such special relationships as cleaner shrimps and their host fishes. Endocrinologists and neurophysiologists have demonstrated the nervous controls in cephalopods and the neurosecretory cells and hormones that mediate the color changes of many crustaceans and some insects.

Chromatic changes that take place in fractions of a second in cephalopods, or many weeks in some insects, both involve **chromatophores,** pigment-containing cells in which the pigment can disperse or concentrate, so changing the tint of the animal. Well-developed chromatophore systems are common in cephalopods and crustaceans. Chromatophores occur only sporadically among annelids, echinoderms, and insects.

In the sea urchin *Diadema* the chromatophores appear to act as independent effectors, directly responsive to light. The urchins are pale at night, and turn quite black in the strong sun of daylight hours. When a microbeam of wavelengths close to 425 to 500 nm is trained on a single chromatocyte, the pigment granules disperse. As this action-spectrum is close to that which elicits spine-movements, this raises the question of whether the chromatocytes could mediate the general dermal "light sense." If so, could a chromatophore be independent as an effector but have connections with a superficial nerve-plexus as a *receptor* of light?

As described in chapter 15, the chromatophores of cephalopods are complex structures with a nucleus, an elastic sac filled with pigment, and 6 to 20 radiating uninucleate smooth muscle fibers attached to the sac. When the fibers contract, they flatten and distend the sac. Other animal groups have unicellular chromatophores or groups of these closely arranged. In crustaceans the chromatophores are highly branched cells, and the dispersing pigment spreads into the branches. In some insects epidermal cells may act as chromatophores. In one species of grasshopper each epidermal cell contains blue and brown granules. When the nucleus lies in the center, with brown granules in the top half and blue granules in the lower half, the animal looks dark. If the nucleus moves to the bottom of the cell and the brown and blue granules exchange places, the animal looks light.

There are two kinds of chromatic change. In **morphological color changes,** which occur over weeks, usually in response to the color of the animal's background, as in crustaceans and some insects, there is an increase or a decrease not only in the quantity of pigment deposited but also in the number of chromatophores. In **physiological color changes** the pigment in the chromatophores simply disperses or concentrates in response to various stimuli: light, temperature, humidity, daily hormonal rhythms, psychological challenges, or touch.

Crustaceans, and especially shrimps, are able to adapt their coloring for concealment on either natural or artificial backgrounds in relation to the quality and quantity of light entering the eyes, probably the only receptors for chromatophoric responses in these animals.

Physiological color changes in insects occur in relatively few species. As in crustaceans, the substances that act on the chromatophores are blood-borne. They are produced by neurosecretory cells in the brain and stored in the corpora cardiaca. Cutting the nerves to the corpora cardiaca results in permanent loss of secretory material from these organs and an increase in the brain. Removal of the brain eliminates all color changes, and brainless animals remain pale.

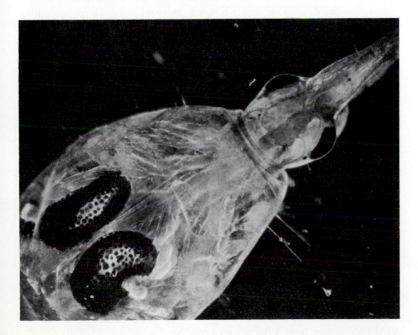

Color changes in *Chaoborus,* **the** aquatic larva of a midge, a mosquito-like dipteran, are controlled by a blood-borne substance originating in the head. If the eyes are covered, the animal behaves as it would in darkness: the melanophores lying on the conspicuous tracheal air sacs, a pair near each end of the body, show a dispersion of the pigment. In the morning the dark melanin pigment in the melanophores concentrates and by mid-day the air sacs pale, probably a thermoregulatory response to light. By late afternoon the pigment disperses and the air sacs darken. *Above,* side view of whole larva. *Left,* dorsal view of head end showing pigment of melanophores partially concentrated. (P.S. Tice)

Coloration may be so distinctive as to serve as an important clue to the identification of a species. More often it varies so with the many factors already mentioned as to be of little help in this respect. Chromatic differences can be as great between genera and species as between higher taxa or even phyla. Seldom is color useful in sorting out evolutionary relationships. A single carotenoid pigment, astaxanthin, imparts orange-red coloring to animals as different as sea-fans, nudibranchs, lobsters, grasshoppers, or fishes.

Neither astaxanthin, nor other yellow, orange, or red carotenoid pigments that are so widespread in animals can be synthesized *de novo* by them. They must be obtained by feeding on green plants or on certain protists and fungi, or by devouring herbivores, or by preying on carnivores that have fed on herbivores or on carotenoid-rich detritus feeders that ingest the particulate remains of plants and animals, or in parasitism. By such pathways various kinds of colored molecules are passed along in food chains and shared by many animals of different phyla.

At some point the question must be raised as to *why some molecules impart color when present in sufficient concentration, and others do not.* The briefest kind of answer is that the difference lies in the arrangement of the valence electrons that link the chains or rings of atoms in large and complex molecules.

In a "colored" molecule the "color-carrying" group of atoms is *unsaturated,* having double or triple-bonded pairs of atoms with unfulfilled valences. In such areas of the molecule enough of the pairs are slowed in their perpetual vibratory motions as to set up a *chemical resonance* in the whole molecule. And it is the resonance frequency of the molecule that determines its color. If resonance frequency matches that of the shortest wavelengths of the visible spectrum (the violet and a portion of the blue region) only these wavelengths will be interfered with and so absorbed and changed into heat. The rest of the visible spectrum will be reflected or transmitted to the eye, and the colored compound will appear to be yellow, the complementary color of the absorbed wavelengths. If the resonance frequency matches the blue and green wavelengths, the compound will appear red or orange. Absorption in the red or orange region gives blue or green color. White compounds are ones that reflect all the visible wavelengths equally and completely, absorbing only wavelengths in the invisible regions of the spectrum. Black compounds, on the contrary, absorb all the visible wavelengths equally and completely.

The chemical instability in colored molecules increases their *reactivity.* And thus color and biochemical activity are linked, so that colored compounds often assume important physiological roles as biocatalysts, of which chlorophyll and hemoglobin are the best known examples.

Color is made visible in animals in two very different modes. Most common by far are the **pigmentary colors** imparted by "colored" chemicals or **chromes.** Biochemists prefer the term *biochromes* for the naturally occurring pigmented compounds in organisms, or *zoochromes* for those found in animals. Zoochromes may be of external or internal origin. For example, though animals cannot synthesize carotenoids *de novo* but must get them from plants, they then produce a number of carotenoid derivatives, such as vitamin A and varied oxycarotenoids and carotenoproteins that are not found in plants.

The second category of animal coloration is of relatively infrequent occurrence and is generated by physical means. Such **structural colors,** also called *schemochromes,* are seen in iridescent pearls or lustrous mollusc shells, in the misty blues of some aquatic cnidarians or crustaceans, and in the brilliant coloring of many iridescent butterflies or beetles. Most structural colors are not visible unless seen against a dark background, as when looking down into deep water or when the iridescent wing scales of butterflies are underlain by light-absorbing pigment of dark color. And structural colors are far outnumbered by **combination colors** that intermingle structural features and pigment. Common examples of this are the wing scales of butterflies, which generate violet or blue color as a result of surface sculpturing but also contain red or yellow chromes in the walls of the scales, so that the wings appear to be magenta or emerald green respectively.

schema = "form"
chrome = "color"

STRUCTURAL COLORS

OF THE FOUR KINDS of structural colors usually recognized, **whiteness** attracts the least attention but is most widespread. Structural whiteness is generated when solid colorless tissues or calcareous crystals of shells or coral skeletons are laid down in layers so thick that they reflect all the wavelengths of visible light, absorbing only the invisible portions of the spectrum. White butterfly wings may be mat white, pearly white, or silvery white, varying with the degree of light-scattering produced by differences in the submicroscopic sculpturing of the wing scales. Whiteness may also be produced by white pterins, pigments closely related to pterins that produce yellow or red coloration. So the boundaries between structural and pigmentary coloration are not always easy to define.

Whiteness is generated when colloidally or other dispersed particles—solid, fluid, or gaseous—exceed a critical diameter of 700 nm.

Inorganic whites in invertebrates are most familiar in the calcareous skeletons of foraminiferans, various spicules of sponges, cups of stony corals, shells of molluscs, tests of some sea urchins, etc.

Blue-scattering is generated by the presence of light-scattering, randomly dispersed colloidal particles or other microbodies with a diameter not exceeding the magnitude of the wavelength of blue light: 400 to 500 nm at the most.

Blue-scattering is widespread in nature, as in the blue of the sky or of deep waters. And it is relatively common in vertebrates, as in the blue irises of human eyes, or the blue feathers or skin of birds. But examples in invertebrates are fewer. Misty blues are seen in colorless medusas, salps, or crustaceans viewed against a background of dark water, and in such special sites as the unpigmented blue spots of the two-spotted *Octopus bimaculatus,* and in various species of damselflies and dragonflies as blue patches on the wings or as brilliant metallic blue abdomens. (Strong blues in invertebrates are seldom caused by blue-scattering. Most are due to the presence of red carotenoid pigments conjugated with protein, as in the deep blues of medusas or of the chondrophores *Velella* and *Porpita* or in lobsters or insects.)

Blue-scattering is only an optical effect in some invertebrates and may have no special significance. In others, investigation may reveal an adaptive function. In certain damselflies in Australia the brilliant blue abdomen changes to a dull black color during the night. Such a color change could enable the damselflies to warm up quickly in the morning sun after a cold night in their mountain environment. They change back to a less heat-absorbing bright blue as the temperature of their environment rises. Laboratory investigation of males showed that they could be heated to 43°C in sunlight without ill effect if in the bright blue phase but that those in the black phase, if heated up too quickly to change color, died within 2 minutes. Examination of the cells of the abdominal surface under a phase contrast microscope revealed that each cell had a peripheral sleeve of dark pigment surrounding a narrow axial region of clear cellular material. This suggests one kind of mechanism by which reversible blue-scattering could serve in adapting to rapid changes of environmental temperature.

The iridescence of pearls arises from the alternation of ultrathin layers of calcium carbonate and films of moisture. That is why the owners of valuable pearls are advised to wear them regularly. The moisture of the wearer's skin keeps the pearls hydrated. Gradual dessication or high heat destroys the luster of pearls.

Interference colors, familiar as iridescence in the comb rows of ctenophores, in pearls or in the nacreous linings of mollusc shells, or in the iridescent bodies, wings, or wing-covers of insects, are among the commonest of the structural colors of invertebrates. Such colors arise from interference between the light waves that enter and those that are reflected back from multiple, ultrathin underlying layers. To generate iridescence the successive surfaces must alternate with thin films of other materials, as air, or water, or proteinaceous materials with a different refractive index.

Interference colors vary with the thickness of the solid layers and with that of the air spaces that alternate with the solid layers. Beetle wing covers may vary in color from violets, blues, and greens, to yellows and reds or, gold or silver with changes in these dimensions. In the chrysomelid tortoise beetles the metallic golden colors are lost at death as the solid laminations shrink due to dehydration. Even in life, if the beetle is disturbed it undergoes some dehydration and the golden color gives way to green, then blue, then violet. If left undisturbed, the cuticular layers are rehydrated, the color changes are reversed, and the golden metallic color restored.

That the beautiful iridescent blue color of the wings of the morpho butterfly is structural can be demonstrated in several ways. One method is to flood the wing with carbon disulfide, displacing the submicroscopic air spaces between the organic layers of the wing. Now the wing, viewed from directly above, appears brown, the color of the underlying melanin pigment. When the carbon disulfide is evaporated by blowing across the surface of the wing the brilliant blue color returns to view. The wing of a morpho butterfly viewed directly from above changes as the angle of vision is increased by tilting the wing. The viewer sees the blue change to blue-green, then blue-violet, and finally to reddish purple as the angle approaches 90° from the perpendicular. Such changes with angle of viewing are also seen in other iridescent insect surfaces or in the iridescent surfaces of some mollusc shells.

Tortoise-shell beetle is a leaf-beetle (chrysomelid) that has an iridescent golden color when undisturbed, but the color can change when the beetle is threatened, and it disappears at death. Tasmania. (R.B.)

PIGMENTARY COLORS

THE MOST SURPRISING THING about pigmentary colors is that there are so few kinds (chemical classes) of them. So the seemingly endless variety of animal colors and color patterns are due to the presence of relatively few pigments as compared with the multitudes of proteins that result from the permutations of about 25 kinds of amino acids and provide unique proteins to each of the millions of living organisms. Proteins in the diet must be digested and broken down, then resynthesized in the construction of new tissues. But biochromes can be passed on from one living thing to another and reused either intact or chemically altered by oxidation or by conjugation of the pigment with a distinctive protein. Thus the limited number of colored atomic groupings get around freely in the organic world, supplying similar or varied pigments, or precursors of chromes.

The main classes of zoochromes are usually listed (with some chemically related ones grouped together) as the carotenoids, the quinones, the indole pigments (indigoids, melanins, and ommochromes), the tetrapyrroles (porphyrins and bilins), and the flavins, purines, and pterins.

The presentation of invertebrate colors in the text of this chapter, and in the pages of color photos that follow, places emphasis almost entirely on integumental zoochromes, those that are externally visible in surface tissues, shells, or chitinous coverings. Little is said of the important physiological roles played by colored compounds that serve as enzymes, as vitamins, as oxygen carriers, or as visual pigments. No mention is made of colored substances that can be seen but are so sparse and diffusely distributed that sufficient quantities cannot easily be obtained for study.

In the plumose anemone, *Metridium senile,* white specimens have little or no melanin and very little carotenoid. In red animals there is no melanin but much carotenoid. In yellow-orange individuals there is scant melanin or none at all, but an abundant supply of carotenoid. In brown and red-brown anemones melanin predominates in the browner specimens, carotenoid in the redder ones.

Carotene is so called because it colors the cultivated carrot, from which it was first isolated.

Though soluble in fats, carotenoids are not soluble in water, and solubility, or lack of it, is one of the criteria used to separate various classes of pigments. Once extracted by such solvents as acetone, ether, or carbon disulfide, carotenoids can be separated out of a mixture by chromatography, a method so effective that it can separate α- from β-carotene, which differ only in the position of one double bond in one of the ring structures at the end of the long molecule. Carotenoids in solution are identified with a spectroscopic peak of absorption of certain wavebands.

The green eggs of the lobster *Homarus* are colored by ovoverdin, astaxanthin conjugated with protein. The green color persists until shortly before the eggs hatch, when the protein portion of the molecule is detached, liberating the orange-red free astaxanthin. (The change from green or blue carotenoproteins in the live lobster to orange-red in the boiled lobster, when there has been denaturation of the protein, is illustrated and described in the color pages on carotenoid pigments.)

Among the various classes of zoochromes that color integuments, by far the two most widespread are carotenoids and melanins, which are often found mixed in different proportions in the surface tissues, providing a range of hues and tints in the color variants of a single species. Though carotenoids and melanins may serve both as screens against damaging amounts of UV radiation, or for heat absorption, photoperception, concealment, or warning colors to predators, they are very different in chemical structure and in their involvement in metabolic functions.

Carotenoids are soluble in fats and can be seen, under the microscope, dissolved in fat globules in various tissues, especially the integument. They often color fat stores yellow, and they lend yellow color to some insect bloods and to such insect products as silk or beeswax. Thus carotenoids are readily extracted for separation of the various kinds and for determination of their chemical structure.

Free carotenoids are linear hydrocarbons in which the carbon atoms are arranged in a long chain, usually with a ring structure at each end. The carbon atoms in the chain are linked to each other by alternating single and double bonds, thus creating an instability that manifests itself as color, as mentioned earlier. The greater the number of alternating bonds in the chain, the deeper the color. The instability also makes carotenoids very reactive chemically, and though most color the integument (epidermis and chitinous exoskeleton) they are also involved in important physiological roles in growth, vision, and mucus secretion. Yet some animals get along with little or no carotenoid. Animals may eliminate carotenoids promptly, store them in fat or in various organs, metabolize them, or modify them. The carotenes, consisting only of atoms of hydrogen and carbon, may be deposited in the integument, stored, or converted into other carotenoids. More often, integumental carotenoids are partially oxidized to form xanthophylls, named for the yellow color of the simplest one (a more informative collective name would be oxycarotenoids). Both free carotenes and xanthophylls provide yellow, orange, and red colorings. But when conjugated with proteins they may also appear colorless or impart purple, violet, blue, green, brown, or black pigmentation. The linkage with protein does more than change the visible color. It stabilizes the molecule, making it more resistant to heat, to bleaching by light, to ionizing radiation, or to mechanical damage. Boiling or chemical agents denature the protein, breaking the linkage with the chrome and revealing a whole range of colors.

Though most carotenoids are integumental, some probably serve as enzymes, and others, derived from β-carotene and known as vitamin A, are visual pigments called retinals because they are essential in the retina for transducing light stimuli to electrical stimuli that initiate impulses in the optic pathways to the brain.

Carotenoids are concerned with taste and with olfactory perception, as well as with photoperception. There are high concentrations of vitamin A and other carotenoids in the antennas of certain male amphipods; these are absent from the antennas of the females. Vitamin A has not been found in protozoans, poriferans, or cnidarians and its biosynthetic pathway may have been of later origin.

Quinones, like carotenoids, have their original syntheses in plants, bacteria, and fungi (with some suspected exceptions). Few quinones are known to occur in significant quantities in animals. Those that do are limited to echinoderms and to certain homopterous insects. Naphthoquinones are red, purple, and brownish chromes that color the tests, spines, endoderm, ectoderm, and perivisceral fluids of sea urchins and sand dollars. Spinochromes are found only in the spines and tests, while echinochromes occur also in soft tissues.

A red napthoquinone colors the connective tissues and ovaries of the feather star *Antedon bifida,* and an echinochromelike compound conjugated with protein is known to occur in the purple body-wall of a species of sea cucumber. And there may be physiological roles for these quinones.

Anthraquinones occur in yellow, red, and purple feather stars and in certain insects. Acidic anthraquinones derived from homopterans have been widely used to dye fabrics since ancient times. Red cochineal is extracted from certain scale insects, red kermesic acid from the female kermes insect, and lac dye (which often accompanies shellac and other lacquers) is derived from several species of scale insects.

Indole pigments include 3 main classes of chromes (indigoids, melanins, ommochromes) that are chemically related in at least some respects and that are all synthesized from amino acids by the animals in which they occur. They are also the first ones presented here to add nitrogen to the hydrogen, carbon, and oxygen of carotenoids and quinones. Nitrogen is bonded to carbon in heterocyclic ring formations.

Indigoids are of minor occurrence in animals, but they have attracted attention since ancient times as the precursors of certain red and purple compounds used to dye fabrics. The

It is surprising that such pervasive chromes as the carotenoids should be only of exogenous (external) origin as far as we know. Most of the other chromes, some quite restricted in distribution, are mostly of endogenous origin, that is, synthesized by the animal from materials obtained in the diet. Animals that synthesize chrome *de novo* may also obtain additional amounts from the diet.

Some animals passively accept chromes from the diet as they occur in the food. Others are very selective in their absorption of colored compounds. The parasitic crustacean, *Sacculina carcina,* attached at various sites on the body of its host crab, *Carcinus maenas,* robs the crab only of β-carotene, even though its internal branching tubules have ready access to various other carotenoids, especially from the rich stores in the crab's digestive gland.

Echinochrome-like pigments are found even in the gastrula stage of the embryos of *Lytechinus variegatus,* and even before gastrulation in *Strongylocentrotus purpuratus.* Such pigments must be derived from a colorless precursor in the cytoplasm and be traceable to the plant food of the mother.

Sea urchins presumably derive their naphthoquinones from their diet of kelp fragments. And marine mammals, such as the sea otter, *Enhydra lutris,* which feed heavily on sea urchins, may have pink or purple skeltal parts, colored by napththoquinones.

colorless secretions of glands in the mantle of certain marine snails, when exposed to air and light and properly extracted and dissolved, provide dyes that were widely used in ancient times in the Mediterranean world and are highly valued by Amerindians of Central America. Marine snails of such genera as *Murex, Mitra, Purpura, and Nucella* have been the sources of the colorless gland secretion and are still often referred to as "purples." No physiological role has yet been identified for the colorless compounds that are the precursors of the dyes; these indigoids may be excreted waste products.

Melanins rival carotenoids in their widespread distribution in animals. Their name is derived from the Greek word for "black," but they also impart brown, reddish-brown, or even yellow colors. They are so insoluble in water and almost all organic solvents that their chemical structure is not yet well known. They are usually complexed with proteins, as melanoproteins, and add mechanical toughness to fibrous tissues or to shells, in which they may be complexed with conchiolin. Melanins have been mentioned earlier in connection with light, temperature, humidity, and chromatic changes. They are among the few pigments that occur almost exclusively in integuments or shells. They are chemically unreactive and have no known metabolic functions. But they are protective against damaging rays of the sun, they provide concealing coloration for animals seen against dark backgrounds, and a deposit of melanin often serves as a reversible curtain that shields or exposes the light-receptor cells of the retina.

When an insect such as a cicada nymph emerges into the light after a long interval underground the enzyme tyrosinase, in the presence of oxygen, accelerates the tyrosine-to-melanin process and the overall body cuticle darkens. Though there is little evidence of involvement in basic metabolic processes, melanophores increase with sexual development, with certain diseased conditions, and in other circumstances. When kept on dark backgrounds, animals elaborate more melanin, and more melanocytes. This implies a biochemical alteration in metabolism of tyrosine in relation to the amount of incident illumination entering the eyes. It has been shown that populations of moths and other lepidopterans become more melanistic in adaptation to changes introduced into the environment by industrial darkening of various backgrounds.

Ommochromes are yellow, reddish-brown, brown, and sometimes black compounds chemically related to melanins in some respects and formerly confused with them. They were

---

Human albinos have no tyrosinase, a recessive genetic defect, and they cannot synthesize melanin. Consequently their eyes appear pink, from hemoglobin in the fine capillaries. In the absence of a dark background of melanin, the hemoglobin masks the effect of blue-scattering from fibrous tissues in the iris.

The British peppered moth, *Biston betularia* is whitish, with irregular black spots and stripes, and it is inconspicuous on lichen-covered bark of trees in the countryside. But it also occurs as a black (melanistic) variant, genetically determined by a single dominant gene. The black moth is conspicuous on gray lichens, and is easily picked off by birds. But in industrial cities where air pollution had killed the lichens and tree trunks were blackened by soot, the black moth was at an advantage, and it outnumbered the light type. When pollution-control measures were enacted, and the trunks of trees became less sooty, the light-colored moths began to reappear in larger numbers in industrial areas.

first identified as retinal screening pigments between the ommatidia of arthropod eyes and so were named for the Greek word *omma*, meaning "eye." They are now also known to be the pigment in the rapidly changing chromatophores of cephalopods (but not of the ink sac, which stores melanin).

Porphyrins are essential to nearly all living cells, and porphyrin structure is said to be one of the first "inventions" of life on earth. Porphyrins are large, organic molecules consisting of a ring of 4 interconnected pyrroles. They are almost universally distributed through the several kingdoms of living things, and though the green chlorophylls of photosynthesis and the vivid reds of hemoglobins are the most familiar of the biochromes, it is the red cytochrome C that is essential for respiration in every living cell except those few that are anaerobic. Cytochrome C is present mostly in the mitochondria, and in such small quantities that it rarely imparts color to tissues. An exception is the reddish-brown color of insect wing muscles, in which large and exceptionally numerous mitochondria in the muscle fibers support a very high rate of respiration during flight.

Hemoglobins occur in all vertebrates except a few fishes, but are distributed only sporadically among invertebrates, particularly in molluscs and annelids that burrow in muddy bottoms or that live in oxygen-poor environments such as stagnant water. Hemoglobins do not occur in the tunicates and lancelets, the invertebrate chordates; these sedentary or sessile animals continually pump oxygen-laden water through the body as they feed.

Free porphyrins (not combined with protein) occur in the body, either through the breakdown of hemoglobin or cytochrome or perhaps as by-products of the synthesis of the protoporphyrin of the heme portion of the molecule. They are eliminated in the feces, used in synthesis of new hemoglobin, or deposited in the integument in animals not normally exposed to light, as burrowing earthworms, or tube-dwelling polychetes. Free porphyrins are photodynamic, a property that makes them useful as receptors of light in some animals but harmful in others. In the presence of light and oxygen, they become strongly activated, and they sensitize other molecules, destroying proteins and nucleic acids, disrupting membranes, and killing cells. Surface tissues containing free porphyrins that are regularly exposed to light probably always have screening pigment, as has been shown in the black land slug, *Arion ater*, or in the sea star *Asterias rubens*. Some molluscs render

The uroporphyrins of planarians are highly concentrated in the rhabdites and may enhance the defensive effect of the rhabdites.

Animals that have free porphyrin, a red pigment, in the integument are sensitive even to visible light. The vulnerability of earthworms and tube-dwelling polychetes was mentioned in the earthworm section in chapter 16. The European black slug, *Arion ater*, which has free porphyrin in the skin, is lured out from its dark hiding places into the light of day by high humidity after rains. But it has a heavy screen of dark melanin pigment in the skin, apparently in proportion to the amount of free porphyrin. The black form has the most, brown variants have less, reddish ones still less, and pale grey slugs none.

The protoporphyrin present in the integument of the sea star *Asterias rubens* varies directly with the amount of carotenoprotein present, being low in the pale yellowish specimens and higher in the darker brown or violet sea stars.

Traces of porphyrins, some of them combined with metals, color oyster pearls. Green pearls contain greater proportions of metalloporphyrins than do pink pearls.

The red bilichrome, haliotisrubin, is a complex calcium salt found in the shells of 12 species of abalones. Its precursor is the red bilin, phyco-erythrin, synthesized by the red algas in the abalone diet. The violet bili-chrome, haliotisviolin, recoverable from 9 of the species of *Haliotis* studied, may also be derived from phycoerythrin in the diet.

free porphyrins harmless by depositing them in the shell, where they may impart pink color to the calcium carbonate.

In **bilins,** formed secondarily from porphyrins, the porphyrin ring of 4 pyrroles is opened out as a linear structure. Invertebrate bilins are not common. The polychete *Neanthes (= Nereis) diversicolor* has hemoglobin, and its breakdown produces biliverdin that lends a green color in addition to the orange and brown of the carotenoids, particularly at the time of sexual maturity, when hemoglobin diminishes in the blood. Bilins also give green color to the earthworm *Allolobophora chlorotica* and probably to leeches. Some of the rhizocephalans parasitic in crustaceans have large quantities of biliverdin, and their rootlike tubules that permeate the host are bright green. Bilins in insects may color the integument and the blood green, as in many grasshoppers, caterpillars, and others. The beautiful permanent blue color of the skeleton of blue corals is a biliverdin that forms a complex with the calcareous skeleton. A red chrome in the shell of the red abalone, *Haliotis rufescens,* and a violet one in the shell of the black abalone, *Haliotis cracherodii,* are both thought to be bilichromes.

Flavins, present in small concentrations as compounds that are pale yellow when dissolved in water, are not conspicuous. But they do occur in all plant and animal cells that have been examined for their presence. Like the carotenoids, they are synthesized originally only by plant cells, and animals must obtain them, directly or indirectly, from their diet. The flavin found most often in tissues is riboflavin, free or conjugated with protein. Riboflavin is an important component of the vitamin B complex.

Purines are colorless, and their place among biochromes can be questioned. But some of them, especially guanine and uric acid, provide an opaque whiteness, sometimes with a silvery sheen or iridescent interference colors, to the color patterns of invertebrates. White or iridescent cells, called guanophores or iridocytes, owe their reflectivity to platelike microcrystalline deposits of guanine. The crystals in such cells do not migrate, and the cells are not really chromatophores. Xanthine, uric acid, and other purines accompany the chemically related pterins in the color patterns of some moths and butterflies. Uric acid, a waste product of the chrysalid stage, is deposited as part of the pattern of the wings of the adult.

Pterins are related in chemical structure to both flavins and purines, but are more colorful. In butterfly wings they may impart whiteness, yellow, orange, and red. Hymenopterans of the Northern Hemisphere often display large yellow spots of a pterin compound, while those of tropical climates or the Southern Hemisphere generally lack such spots. Red and yellow eye-colors of the vinegar fly, *Drosophila melanogaster,* are pterins. The green integuments and bloods of some insects, referred to earlier under bilins, are often a combination of a blue bilichrome and a yellow pterin, xanthopterin. Aside from imparting color and pattern, pterins have many physiological roles.

INTEGUMENTARY ZOOCHROMES serve a variety of functions, as we have seen. They provide protection against damaging radiation. They help to harden exoskeletal materials against mechanical abrasion. They provide chemical defenses. Integumentary chromes are important in regulating heat absorption or heat loss. And they communicate visual messages that affect the behavior of other species. Such visual effects are unrelated to whether the animal itself can perceive colors, or even whether it has eyes. Vividly colored sponges, cnidarians, or echinoderms need not perceive their own colorations. If these are effective as concealment or advertizement they tend to be preserved and perfected by natural selection.

One of the main advantages of color vision is the ability to discriminate between hues (wavelengths) of the same light intensity. This is important to bees and to flies that must learn to recognize the patterns of flowers with a rich supply of nectar.

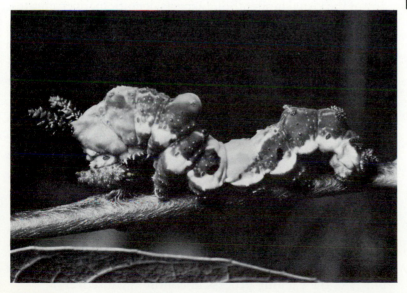

Strong color contrast can be perceived even in black-and-white photos. *Above,* the nymphalid butterfly *Eriboea narcaea* of China (K.B. Sandved) *At left,* a bizarre caterpillar, the larva of an attractive butterfly called Lorquin's Admiral *(Limenitis lorquini),* is olive brown with yellowish markings on the thorax and a conspicuous white patch on the back. To the human eye it strikingly resembles a bird dropping. And it is hard to believe that the distorted shape of this caterpillar, and its strongly contrasting color pattern, have no adaptive value in its relations with visual predators, whether or not they have color vision. (E.S. Ross)

OF THE BIOCHROMES presented here, those most wide-spread in animals are probably, in decreasing order: certain flavins, porphyrin derivatives, carotenoids, pterins, melanins, quinones, and ommochromes. Most phyla of animals have a wide variety, but the spiralian phyla have the greatest variety of biochromes, including some peculiar to them.

The evolution of chromes appears to be very labile, and perhaps by now most of the phyla have explored the main groups of chromes and their possibilities.

At the lowest taxonomic levels (species and genera), differences in chromes are mainly in the integument and concern visible aspects of color, such as providing camouflage, or conspicuous coloring. In contrast, the chromes of oxygen-carriers or those with important biochemical properties are common throughout larger taxons. And metabolic enzymes such as cytochrome C are found throughout all the phyla.

A number of the main groups of biochromes furnish a range of colors so wide that one group could meet all the integumentary needs of any one taxon. Yet most species use many classes of biochromes. Perhaps this spreads the load on any one biosynthetic pathway and makes the best use of available materials and degradation products such as bilins or purines, which would otherwise be wasted.

A number of biochromes are extremely stable, and of all the organic materials found preserved in fossils, biochromes are the most common. Melanin was found in the ink-sacs of fossil Jurassic squids, and quinones in Jurassic crinoids. Recognizable porphyrins date from the Precambrian.

THIS BRIEF CHAPTER cannot deal in more detail with any of the visual or metabolic aspects of color in invertebrates. The color photos and accompanying legends on the pages that follow are intended only to stimulate interest in this fascinating and significant aspect of biology. The subject of color illustrates how necessary it is to study any biological subject at many levels—chemistry, morphology, physiology, behavior, genetics, and evolution—to even begin to understand it. It also illustrates the importance of comparative biology, especially how limited our notions of the zoological world would be without the perspective gained from the study of invertebrates. Anyone who begins to ask even the simplest and seemingly most obvious questions about the sources and functions of color, or virtually any other aspect of biology, quickly learns how little has really been so far discovered. We invite the reader of this book to join other enthusiasts of the invertebrates in the exciting process of continuing discovery.

Predator is camouflaged when prey have good vision. as in the wasps, bees, and flower flies that come to suck nectar from the flower on which this hemipteran, the ambush bug *Phymatia,* lies in wait. Arizona. (E.S. Ross)

Prey is camouflaged when the predators, in this case birds, have good vision. The grasshopper *Melanoplus* is inconspicuous among the grasses on which it feeds. California. (E.S. Ross)

Genetically determined color matches the stalked scyphomedusa *Haliclystus auricula* to the eelgrass on which it lives. It does not move about to feed but finds an ample supply of caprellid amphipods on the eelgrass. Friday Harbor, Wash. (K.B. Sandved)

Acquired colors lent by encrusting sponge, bryozoan, and algal growths disguise the keyhole limpet, *Diodora aspera,* as it moves across substrate covered by lavender, yellow, and orange sponges pigmented by carotenoids. California. (G. Miller)

Ingested food gives color to these unpigmented termites (and to many other white or transparent animals). Jaws of workers and cuticle of queen are darkened by sclerotization. Panama. (E.S. Ross)

Green leaflike katydid is well camouflaged as it rests or feeds in green foliage in a Brazilian rain forest. (E.S. Ross)

Spotted-leaf katydid has a truly remarkable resemblance to the blemished leaves among which it rests and feeds in a rain forest in Costa Rica. (R.B.)

Dead-leaf katydid in a Peruvian forest. Tambopata Reservation. (E.S. Ross)

Lichen katydid, *Dysonia punctifrons*, resembles the gray lichens. Rain forest, Tingo Maria, Peru. (E.S. Ross)

Stone grasshopper, *Phrynatettix tschivavensis*, has minimum visibility as it rests on stony ground in Skeleton Canyon, Arizona. (E.S. Ross)

Orb-weaver spider on bark will escape the eyes of most birds. *Herennis*. Gopeng, Malaya. (E.S. Ross)

Green spiny caterpillar has low visibility among the green mosses that carpet its habitat. Ecuadorian rain forest. (R.B.)

Greenish isopod is almost invisible as it feeds on kelp. Some species of *Idotea* gradually change their color from green to red or brown after they settle on an alga of another color. Oregon. (R.B.)

Thornlike treehopper is well disguised when clinging tightly to plants. *Enchophyllum cruentatium*. Brazil. (K.B. Sandved)

Brown treehopper, *Sphongophorus*, with the covering of the prothorax prolonged backward, escapes the eyes of most bird or lizard predators. Mexico. (K.B. Sandved)

Dead-leaf butterfly, *Kallima inachus,* is famous for its imitation of a dead leaf when the butterfly snaps its wings shut as it alights on a stem. The hind wings bear slender tails that just touch the stem; the front wings point outward. The dark line, with fine lines running from it, suggest the midrib and veins of a leaf. The sudden change from conspicuously colored flying butterfly to a brownish, immobile leaf confuses predators and butterfly collectors alike. *Left,* a dried specimen with wings open. Taiwan. (R.B.) *Right,* Live butterfly with wings closed. Sumatra. (E.S. Ross)

Camouflaged polyclad flatworm, *Hoploplana californica,* closely matches the mottled coloring of the encrusting bryozoan *(Celleporaria brunnea)* on which it lives and feeds. The worm is presumably concealed from predators. *Right,* **flatworm lifted** to show the underside, which lacks concealing colors. Monterey Bay, California. (R.B.)

Pink brittle star on pink sea fan is well camouflaged as it clings to its convenient location for gathering plankton from the passing water currents. Roscoff, France. (R.B.)

Flower mantid nymph is inconspicuous among the flowers on which it rests as it awaits prey. *Hymenopus coronatus.* Gombak Valley, Malaya. (E.S. Ross)

Nudibranch symbiotic with a coral shares the same colors, and the cerata on its back resemble the tentacles of the coral. The nudibranch, *at left,* is almost invisible as it moves among the polyps of *Tubastrea aurea.* (L. Harris)

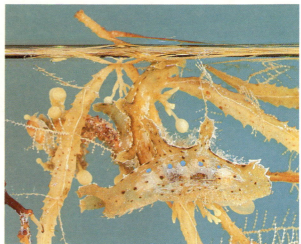

Camouflaged members of a community of invertebrates, and one small fish, are all colored yellow and brown by carotenoid pigments, have disruptive markings, and strongly resemble the floating sargassum seaweed to which they cling. Some feed on the alga; most of them take food from the surrounding water. Many are splotched with white, enhancing the resemblance to *Sargassum* when it has whitish patches of encrusting bryozoans, as at the top of the photo *(at left)* of the snail, *Litiopa melanostoma.* (K.B. Sandved) In addition to the animals shown here, the sargassum community has sea anemones, flatworms, polychetes, pycnogonids, crustaceans, small brittle stars, and others. *Above left,* 2 small crabs, *Planipes minutus. Above right,* a nudibranch, *Scyllaea pelagica,* flattened from side to side, with flat leaflike respiratory lobes that grasp the *Sargassum;* it feeds on the many white, delicate branching colonies of hydroids that grow on the *Sargassum,* and it will take passing cnidarians and ctenophores. *Below left,* 2 caridean shrimps. *Below right,* the little fish *Histrio histrio* with front fins modified for grasping *Sargassum.* The body is colored by carotenoid pigments, but the blue eye is structurally colored by blue-scattering (as may be also the pale blue spots on the nudibranch). Bimini. (R.B.)

Disruptive patterns, shown on this page, break up the contour of an animal and distract potential predators with patches of color that have no recognizable shape. **Land planarian** on forest floor. Brazil. (E.S. Ross)

Polyclad flatworm *Pseudoceros bedfordi* is inconspicuous on patchy intertidal substrate. This one was captured as it swam freely through the water. Heron Island, Great Barrier Reef, Australia. (R.B.)

Porcelain crab *Porcellana sayana.* N. Carolina. (K.B. Sandved)

Painted shrimps, *Hymenocera picta,* work in male-female pairs as they pick away at sea stars, yet are rarely eaten by fishes. Tropical Indo-Pacific. (K.B. Sandved)

Mosaic coloring of shrimp *Saron marmoratus* fuses to a cryptic blur when seen at a distance on coral reefs. (G. Miller)

Grasshopper nymph was inconspicuous among shadows of dense foliage. Tikal, Guatemala. (R.B.)

Camouflage in gastropods may be in shell or integument. The flamingo tongue, *Cyphoma gibbosum,* of Caribbean waters, has a pale shell, conspicuous on the purple or orange gorgonians on which it creeps about. *Left,* the patterned mantle is partly withdrawn, exposing the almost white shell. *Right,* the mantle is fully expanded, completely covering the shell. The pattern of the mantle, seen here against purple sea whips, is not conspicuous when the animal is moving about on latticed sea fans in strong sunlight. (Left, G. Miller; right, R.B.)

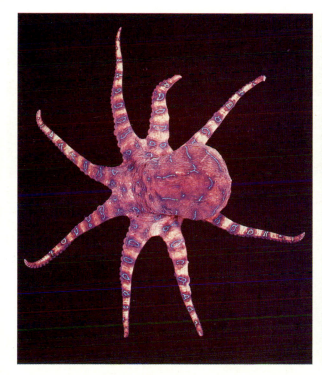

Warning pattern of blue-ringed octopus, *Hapalochlaena maculosa,* of Australian waters off New South Wales, provides strong notice that this is a highly venomous octopus. Despite their prominence, the blue rings and lines are pale, suggesting that they may not be pigmented but consist of clear skin that reflects blue-scattering of underlying tissue, as is known for the blue spots of a California octopus. (K. Gillett)

Color on the undersurface of a chiton is unusual; so is the carnivorous habit of *Placiphorella velata.* It awaits prey with the colored flap, a forward extension of the mantle, upraised. When a small crustacean comes close, the flap is suddenly lowered on the prey. The rosy pigment is excellent camouflage for a chiton resting on a pink encrusting algal mat and surrounded by the waving red algal blades among which it lives. California. (R.B.)

Bright color and strong pattern in nemerteans are usually associated with distastefulness and are warning colors. Some of the brightly colored species are known to come out to feed in the daytime. Drab nemerteans emerge only at night. *(Left) Tubulanus polymorphus,* of the U.S. Pacific Coast. (R.B.) *(Right)* conspicuous nemerteans draped in daylight on an orange-colored gorgonian colony. Panama, Pacific shore. (K.B. Sandved)

Conspicuous patterns in nudibranchs warn of distastefulness. (G. Miller)

Nudibranch *Hypselodoris zebra* lives in grassy shallows, feeding on sponges, so taking up toxic substances. Bermuda. (K.B. Sandved)

Distinctive marking of saddleback caterpillar, in a Panama rain forest, warns predators of stinging bristles. Saddlebacks feed on forest trees in U.S.A. (R.B.)

Bold patterns of black and orange or of black with yellow or red are highly visible to vertebrates with color vision and are complementary colors to the greens of most habitats. Thus they are the predominant warning colors for insects that are distasteful, emit repellant or acrid secretions, sting, or have sharp bristles. Aside from color pattern, aggregation of larvas, nymphs, or adults increases the effectiveness of the warning color to predators. On this page are various examples of noxious insects with warning colors.

Stink bug, *Murgantia histrionica*, destroys cabbage crops. California. (E.S. Ross)

Shield bug, *Graphosoma italicum*. Portugal. (E.S. Ross)

Wasp with bold warning pattern is a wingless female (often called a velvet ant). Her color pattern gives warning of an especially potent sting. *Traumatomutilla*. Bolivia. (K.B. Sandved)

Warningly colored caterpillars, *Schizura concinna*, also have a warning display. When disturbed they throw both front and rear ends up in the air, and a gland on the undersurface of the prothorax can squirt an acid irritating to the eyes of bird, lizard, and other predators, and to human eyes as well. Calif. (E.S. Ross)

Milkweed bug nymphs aggregated on a milkweed pod. Pennsylvania. (R.B.)

Adult milkweed bug, *Oncopeltus fasciatus*. California. (E.S. Ross)

Monarch butterflies, *Danaus plexippus,* are poisonous to birds, having in their tissues cardiac glycosides received from their milkweed-feeding larvas. The queen butterfly, *Danaus gilippus,* which also contains milkweed toxin and has a similar warning color pattern, is a Mullerian mimic of the monarch. (Similar in color pattern but nonpoisonous, the viceroy butterfly, *Limenitis archippus,* is a Batesian mimic of the monarch.) Pacific Grove, California. (R.B.)

Batesian mimicry is exemplified by an *edible* butterfly, *Papilio zagreus* (lower figure). It mimics *Lycorea cleobaea,* a poisonous butterfly (upper figure), which belongs to a different family. Both model and mimic live in the upper Amazon of Peru. (E.S. Ross)

Mullerian mimicry is exemplified by a poisonous West African species of butterfly, *Acraea encedon.* Though it belongs to a different family, it resembles (in both dorsal and ventral aspects) West African members of the genus *Danaus,* the same genus that includes the monarch butterfly. (E.S. Ross)

Orange-and-black mimics occur in many insect orders. *Left,* a poisonous lycid beetle, *Lycus arizonensis,* one of many in its family that are distasteful to birds. The beetle is the poisonous model, and its mimic, *right,* is a dayflying ctenuchid moth. (E.S. Ross)

Wasp model, *Vespula,* a stinging yellow-jacket wasp. It has Mullerian mimics among other stinging wasps and various kinds of Batesian mimics among harmless and palatable insects. (E.S. Ross)

Bee mimic of *Vespula* is this harmless anthidine solitary bee. California. (E.S. Ross)

Flower fly mimic, *Chrysotoxum,* is one of the many harmless syrphid flies that mimic yellow jackets. Oregon. E.S. Ross)

Beetle mimic of *Vespula* is a harmless species, *Strophiona laeta.* Mill Valley, California. (E.S. Ross)

Moth that mimics a **polybiine wasp.** Peru. (E.S. Ross)

Moth that mimics a **wasp** has both structural and pigmentary colors on wings and body. Mexico. (J. Dayton)

Startle display of mantis defending itself on its egg case is effective against predators. Zaire. (E.S. Ross)

Display of poisonous grasshopper suddenly exposes brightly colored hindwings. Natal, Africa. (E.S. Ross)

Large simulated eyes on the thorax of a caterpillar (the small head is tucked under) are enough to make a bird or lizard hesitate. The adult is the pale swallowtail butterfly, *Papilio eurymedon.* California. (E.S. Ross)

Scent glands of this disturbed caterpillar have just been everted from behind the head, and they emit a repellant odor. Anise swallowtail butterfly, *Papilio zelicaon.* California. (E.S. Ross)

Startle display is most effective, as in this large moth, when the forewings are cryptically colored and these are suddenly spread, exposing the large eyelike spots on the hind wings. Panama. (R.B.)

Startle display of *Fulgora lanternaria,* cryptically colored large homopteran. Sudden 30 cm wing spread and 2 eyelike spots on wings are enough to make a predator drop the prey. Panama. (R.B.)

Red-backed cleaner shrimp, *Hippolysmata grabhami,* has a strongly contrasting color pattern that provides easy recognition by fishes seeking removal of external parasites and dead skin. Caribbean. (K.B. Sandved)

Banded coral shrimp, *Stenopus hispidus,* has a distinctive color pattern recognized by the little coral fishes that seek its cleaning services. Caribbean. (K.B. Sandved)

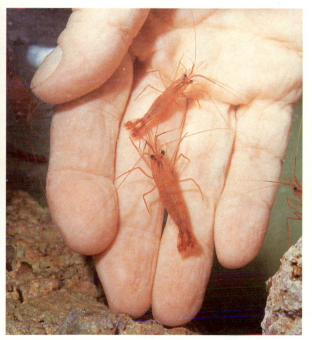

Red cleaner shrimps, *Hippolysmata wurdemanni,* are distinctively lined and they rock to and fro when their client fishes approach. Here they are picking at a hand, presumably for bits of dead skin. N.W. Florida. (R.B.)

Conspicuous color of the large claw of the fiddler crab, *Uca,* augments threatening gestures, made with the claw towards other males of the species. Claw-waving is used in courting females. Australia. (J.S. Pearse)

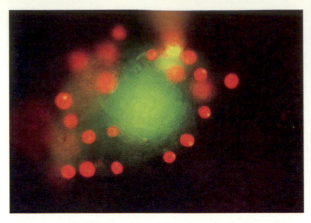

Chlorophyll fluoresces red in the photosynthetic symbionts of a polycystine radiolarian under ultraviolet radiation in a fluorescence microscope. The symbionts are outside the central capsule, as is the rule in this group of radiolarians. (M. Silver)

Green freshwater sponge, colored by symbiotic zoochlorellas, grows faster than do members of the same species on the underside of rocks. (R.B.)

Green hydra, *Chlorohydra viridissima* has zoochlorellas in the endoderm. The living algal cells probably have a mutualistic relationship with their host, but when they die they are digested. (R.B.)

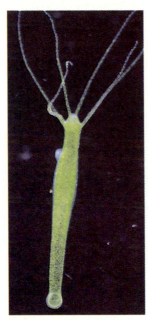

Sedentary rhizostome medusa, *Cassiopeia,* lies upsidedown on bottom, exposing zooxanthellas in oral surfaces to sunlight. Bimini, Bahamas. (R.B.)

Green epidermal pigment screens the yellow-brown zooxanthellas in the endoderm of the sea anemone *Anthopleura xanthogrammica.* Monterey Bay, Calif. (R.B.)

Green and white sea anemones of the same species, *Anthopleura elegantissima,* live side by side in a laboratory experiment. Hopkins Marine Station, Pacific Grove, California. (V.B. Pearse)

Tentacles fully expanded in the sun expose zooxan-thellas of *Anemonia sulcata.* The green color is produced by a green epidermal pigment. Roscoff, France. (R.B.)

Green symbionts *(Tetraselmis convolutae)* live intercel-lularly in the tiny acoel flatworm *Convoluta roscoffensis* (see chapter 9). Roscoff, France. (R.B.)

Greenish sacoglossan feeds on algal filaments and appropriates chloroplasts, maintaining them as endo-symbionts in the digestive gland that ramifies through the frilled mantle lobes. *Tridachia crispata.* (G. Miller)

Tridachia crispata with the mantle lobes opened out, reveals a pattern of blue spots over the dorsal surface. The paleness of the spots suggests that the blue color may be structural. Puerto Rico. (K.B. Sandved)

Green coloring due to chloroplasts is also seen in *Placobranchus ocellatus,* a sacoglossan opisthobranch mollusc. It is shown with the mantle folds open, exposing the chloroplasts in the large digestive gland to the sun. When the light is too bright, or in darkness, the mantle folds can be closed to protect the more delicate tissues. Hawaii. (R.B.)

Structural whiteness is common in nearly unpigmented tissues like those of the tentacular crowns of these tube-worms. (G. Miller)

White commensal nemertean, *Malacobdella grossa,* lives in the mantle cavity of bivalves. Like many other animals that live in dark places, it has no surface pigment. And the structural whiteness of the internal organs shows through the translucent white body wall. Oregon. (R.B.)

Coloration of the cuttle, *left,* is mostly pigmentary, but it is mottled by spots of structural whiteness. When the cephalopod is startled, *right,* its brown chromatophores contract, revealing the structural whiteness of the body tissues. The white spots still show as each is an aggregation of leucophores, cells filled with motile granules of guanine, an excretory product. Aquarium de Noumea, New Caledonia. (R.B.)

Matte whiteness of wing scales of a moth is due to scattering and reflection by the minute sculptural features of each scale. This tiger moth has been disturbed and is displaying its defensive posture. *Estigmene acraea.* California. (E.S. Ross)

Calcareous structures appear white unless the calcium is chemically bound to pigments (examples in later pages). Shown here, beginning in right foreground and proceeding clockwise, are a cowrie shell, a sea urchin test, a coral skeleton, and the partially calcified organic egg case of *Argonauta,* a cephalopod. (R.B.)

Blue-scattering in transparent marine animals is common when they are seen against a dark background. The short wavelengths are reflected from colloidal particles in colorless tissue. *Left,* a chain of salps (J.M. King). *Right,* an amphipod crustacean, *Phronima,* which lives in certain salps. Carotenoid pigment colors eye. (K.B. Sandved)

Blue-scattering from colloidal layers in surface tissues underlain by a black melanin-rich layer is thought to account for the blue color of the oceanic nudibranch *Glaucus atlanticus.* Two are nibbling on the tentacles and two are underneath the float feeding on the polyp of *Porpita,* a cnidarian. The blue coloring of *Porpita* is pigmentary. Florida. (W.M. Stephens)

Brilliant blue of damselfly results from blue-scattering by clear granules in epidermal cells and helps this insect tolerate high temperature in mid-day. In some species of damselflies, the areas of dark pigment expand when the temperature drops, enveloping the abdomen in a sleeve of dull black. The black insect rapidly absorbs the warmth of early morning sun, and gets off to a quick start on cold mornings. When the temperature rises the damselfly turns blue again. Tasmania. (R.B.)

Strong blue color of isopods *Porcellio scaber* accompanies a virus infection. Others not infected may show a slight tinge of blue, due to blue-scattering from the thick, layered exoskeleton. (R.B.)

Interference colors reflected from the nacreous lining of shell of red abalone, *Haliotis rufescens,* are generated by thin layers of calcium carbonate (aragonite) alternating with films of moisture. The iridescent colors change with changes in the angle of viewing. California. (R.B.)

Layered platelets, of submicrocopic dimensions, in the epidermal cells of *Sapphirina,* generate interference colors. This tiny, marine copepod shines like a bit of copper or gold with a violet, blue, or red sheen according to the angle of viewing. Gulf Stream. (K.B. Sandved)

Longitudinal vanes on the wing scales of *Morpho menelaus,* each made up of ultrathin layers, reflect interference colors. Two layers of scales lie shingle-like on the wings, the upper transparent, the under darkly pigmented. Tilting the wings produces a sequence of blue-green, blue-violet, and reddish-purple. Peru. (K.B. Sandved)

Wing scales of a butterfly may have interference structural color, as in the small blue patch here, or pigmented coloring, as in the pink and orange scales. Bolivia. (K.B. Sandved)

Iridescence of snail shell is structural, but the shell is also pigmented. *Calliostoma.* California. (R.B.)

Eyeshine in the multiple eyes that line the mantle edges of scallops is generated by interference colors reflected from a layered tapetum behind the retina. *Aequipecten irradians,* the bay scallop of the Atlantic Coast of the USA, has blue eyes, probably a structural color. (R.B.)

Brilliant interference colors of this tiger beetle, *Cicindela splendida,* are produced by layers of cuticle that alternate with films of air. The thinnest layers generate blue wavelengths from head and prothorax, thicker layers result in green wing-covers and the thickest layers produce the gold of the eyes. Texas, Mexico. (K.B. Sandved)

Blue-green iridescence of this bee owes its high luster to the hard smooth surface, the ultrathin layering of the cuticle, and the underlying dark layer of melanin pigment. This long-tongued bee, *Euglossa exaerete,* has the tongue tip inserted into a flower and is sucking nectar. Ecuador. (K.B. Sandved)

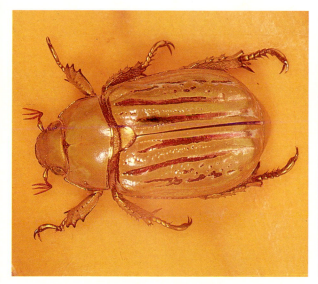

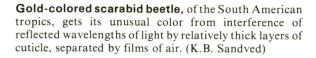

Gold-colored scarabid beetle, of the South American tropics, gets its unusual color from interference of reflected wavelengths of light by relatively thick layers of cuticle, separated by films of air. (K.B. Sandved)

Glittering cuticular scales, each composed of multiple thin layers that generate mostly blue-green interference colors, give a bejeweled appearance to the diamond beetle *Hypomeces.* Philippines. (K.B. Sandved)

Pink or red blooms on the ocean surface, like this one near Shimoda, Japan, owe their color to carotenoids contained in countless millions of dinoflagellates *(Gymnodinium, Noctiluca,* or others). Or they may consist of pink foraminiferans such as *Globigerina,* or of ciliates. Or pink streaks may be caused by synchronous release of millions of packets of gametes of reef corals. (R.B.)

Red water in ponds or ditches may be colored by euglenids or other phytoflagellates. These protists contain carotenoids that mask the green chloroplasts during times of high light intensity. Thailand. (R.B.)

Only one carotenoid, astaxanthin, imparts colors to *Coryphella (= Flabellinopsis) iodinea* . The nudibranch obtains the pigment when feeding on an orange-colored hydroid, *Eudendrium ramosum,* which feeds on plankton containing carotenoid pigment. The pink egg mass, visible inside the nudibranch, is tinted by free astaxanthin. The deep red color of the 2 rhinophores is free astaxanthin combined with 2 esters of astaxanthin. The orange color of the cerata is 80% free astaxanthin, the rest esterified. The blue-violet of the body skin is astaxanthin conjugated with protein. California. (R.B.)

Triophaxanthin, a xanthophyll carotenoid named for the nudibranch *Triopha catalinae,* accounts for 53% of the carotenoid pigment. Analysis of the 3 bryozoan species on which this nudibranch feeds revealed many carotenoid fractions, of which 7 matched those found in the nudibranch. California. (R.B.)

Carotenoids color hydroid colony, *Eudendrium.* Feeding polyps are pink; reproductive polyps each bear 5 orange gonophores that release planulas. (Note 2 caprellids that feed on passing plankton). Florida. (R.B.)

Planula of *Eudendrium.* (R.B.)

Carotenoids in sponges are predominantly carotenes rather than xanthophylls, the reverse of what is found in most other kinds of invertebrates. Here 4 species are competing for space on a rock substrate. New Zealand. (W. Doak)

Bright red encrusting sponge is *Plocamia karykina*, an intertidal sponge fed on by the red nudibranch *Rostanga pulchra*. Both the red nudibranch, and the red egg-ribbon it lays on the sponge, are all but invisible on the sponge substrate. California. (R.B.)

Subtidal sponge, *Dasychalina cyathina*, has pale coloring. (G. Miller)

Finger sponge *Axinella* is deeply colored by the same orange-red pigment, astaxanthin, that gives red carotenoid coloring to lobsters and many other crustaceans. N.W. Florida coast. (R.B.)

Finger sponge is overhung by black coral branches covered with white polyps. New Zealand. (W. Doak)

Green color in sponge *Halichondria panicea* is a combination of a yellow carotenoid and a blue pigment, probably a carotenoprotein. This encrusting sponge also occurs as yellow or orange colonies. California. (R.B.)

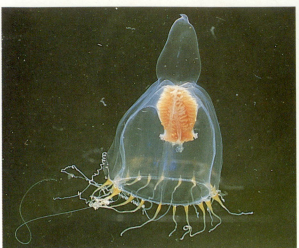

Copepod of strikingly blue color is a common feature of surface plankton in tropical or subtropical waters. The pale blue carotenoprotein pigment is in the exoskeleton; a darker blue pigment colors the epidermis. The egg masses are pigmented by orange carotenoid. Bermuda. (K.B. Sandved)

Carotenoids pigment the gonads of many invertebrates, especially the ovaries. In this little anthomedusa, *Leuckartiara,* the orange gonads are on the sides of the manubrium. The misty blue color of the bell, seen against a dark background, is due to blue-scattering and is structural color. California. (R.B.)

Blue carotenoprotein of *Velella* is red astaxanthin conjugated with protein. The dark blue of the mantle tissue and the lighter blue of the tentacles probably involve different proteins. The mantle tissues screen excess light but allow photosynthesis in the zooxanthellas of the underlying tissue. California. (R.B.)

Blue of *Porpita,* a surface-floating chondrophore of warm marine waters, is a carotenoprotein. As pointed out elsewhere, the blue color of the nudibranch *Glaucus* here feeding on *Porpita,* is a structural color. Florida. (W.M. Stephens)

Carotenoids are the dominant pigments in cnidarians. Five carotenoids color the polyps of the hydrocoral colony *Allopora californica.* The color of the living tissues differs with varying proportions of the main carotenoid, free astaxanthin. California. (G. Miller)

Purple skeleton of hydrocoral *Allopora californica* owes its permanent color to astaxanthin conjugated with protein and bonded to the calcium carbonate of the skeleton. (R.B.)

Blue carotenoprotein imparts strong color to the bells of many scyphomedusas and is the likely source of the deep blue pigment in the dome-shaped bell of the coronate medusa *Periphylla.* Under sea ice, McMurdo Sound, Antarctica. (G.A. Robilliard)

Color variant of *Metridium* takes its color mostly from carotenoids. Other variants range from white to yellow orange, pink, red, or brown. Friday Harbor, Wash. (R.B.)

Orange carotenoid color of this sea fan, a gorgonian, is stored in the minute calcareous spicules scattered in the coenenchyme that surrounds the horny, flexible axial skeleton. When the sea fan is dried it retains the orange color. Aquarium de Noumea, New Caledonia. (R.B.)

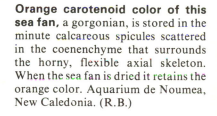

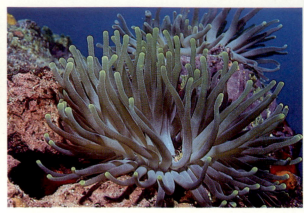

Red or yellow coloring in a sea anemone, *Tealia (= Urticina) piscivora,* is due mostly to free or esterified carotenoids. The anemone has a reddish column and lives subtidally on the U.S. Pacific Coast. (G. Miller)

Green coloring in a sea anemone, *Condylactis gigantea,* is due to red or yellow carotenoids conjugated with protein. It also has yellow-brown zooxanthellas. Disk and tentacles 30 cm in diam. Caribbean. (G. Miller)

Zoanthids. (G. Miller)

Cup corals. *Balanophyllia.* (G. Miller)

Corallimorphs. *Corynactis californica* is most often red, crimson, or pink, but may also be purple, brown, yellow, buff, or nearly white. *Corynactis* reproduces asexually by longitudinal fission, and the clone that results has individuals all of the same color. Some clones cover a square meter or more of substrate. (R.B.)

Pelagic anomuran, *Pleuroncodes planipes,* turns beaches red when great swarms wash up on California shores in some years. Its planktonic diet supplies a rich source of carotenoids. *Pleuroncodes* has more than 80% esterified astaxanthin, about 12% free astaxanthin, and only about 4.5% β-carotene. (R.B.)

European lobster, *Homarus gammarus,* shows great variety in the carotenoproteins of the exoskeleton. The carotenoid component is astaxanthin. The basic coloring of the dorsal surface is blue, the ventral surface mottled white, the antennas bright red. Brought in from the wild, the carapace is blue-black. But in aquariums, where their diet is often inadequate, most appear bright blue. Helgoland Aquarium. West Germany. (R.B.)

Live American lobster, *Homarus americanus,* is normally dark greenish-brown. But genetic variants (mottled, all red, all blue, or black lobsters) occur in the wild, or even "split color" ones—green on one side, red on the other. Inadequate diet in the laboratory produces blue lobsters, carotenoid-deficient diet, white ones. (R.B.)

Boiled American lobster has the red color familiar from restaurant fare. The high heat has denatured the protein, breaking the bond of the carotenoprotein complex and unmasking the red astaxanthin of the exoskeleton. (R.B.)

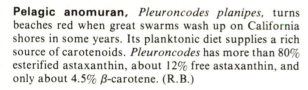

Pigment types in sea stars are limited mostly to yellow, orange, and red carotenoids, but carotenoproteins impart blues, purples, greens, browns, and grays. Both kinds of pigments contribute to the beautiful patterns seen in *Solaster stimpsoni,* which preys mostly on sea cucumbers. Friday Harbor, Washington. (R.B.)

Sexual differences in color for bottom-dwellers without good vision has been reported only for the blood star *Henricia leviuscula* of the U.S. Pacific Coast, in which the males are darker. Its relative, *H. sanguinolenta,* of U.S. and European Atlantic shores is shown here. (R.B.)

Carotenoids move up in food chains. Deep orange-red *Mediaster aequalis* has a varied appetite and is a major predator on orange sea pens. It is eaten in turn by *Solaster dawsoni* (red or orange), a sea star notorious as a predator on other sea stars. California. (R.B.)

Even the tube feet are vividly pigmented in the sea star *Echinaster spinulosus,* seen here in oral view. Gulf of Mexico. N.W. Florida. (R.B.)

Proportion of astaxanthin to other carotenoids in the webbed sea star, *Patiria miniata,* increases with each color phase in the order: yellow, orange, red, brown, or purple. Phases may be solidly colored or mottled, and all live in the same areas of the U.S. Pacific Coast. (R.B.)

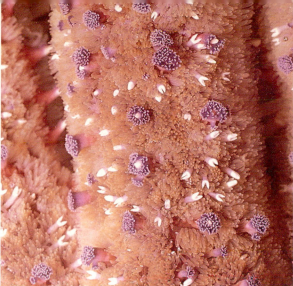

Vivid blue cobalt color of *Linckia laevigata* makes it instantly recognizable in shallow tropical reef lagoons, not only to humans but also to fishes, who avoid it. The strong blue probably warns of distastefulness. And *Linckia* is one of the few sea stars in these lagoons that lies fully exposed in the daytime. New Caledonia. (R.B.)

Pale carotenoid coloring is seen in closeup of surface of *Pycnopodia helianthoides,* a sea star with a very soft surface. The skin gills are pale orange-pink; the clusters of minute pedicellarias are lavender. The large scattered pedicellarias have conspicuous calcareous jaws of structural whiteness. (R.B.)

Carotenoids in brittle stars, many of them carotenoproteins, impart varied colors and color patterns. Presumably these confuse predators, making it difficult for them to learn a fixed prey image. Unlike sea stars, brittle stars have pigment confined to special cells. Blackish forms have an overlying layer of melanocytes (cells with dark melanin pigment). *Ophiothrix* and *Ophioderma.* Puerto Rico. (K.B. Sandved)

Carotenoids in sea cucumbers are usually most concentrated in the gonads. But some red species examined have a high concentration in the skin. Those colored yellow, brown, or black may have melanins.

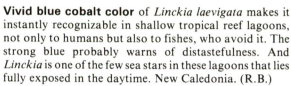

Cucumaria miniata. California. (G. Miller)

Isostichopus. Brazil. (R.B.)

Spinochromes and **echinochromes** are naphthoquinone pigments that confer red, purple, violet, green, and brown colors on the test, spines, skin, and internal tissues of sea urchins. *Left,* large red urchin, *Strongylocentrotus franciscanus.* (S.K. Webster) *Right,* the smaller purple urchins, nestling in holes they have made in rock, are *Strongylocentrotus purpuratus.* (E.S. Ross) Both are found on the West Coast of the USA.

Spinochromes of various species of urchins, though inseparable by various chemical tests, can be separated by different melting temperatures. Spinochrome-F, from the slate-pencil urchin, *Heterocentrotus mammilatus,* melts at 229°C. Hawaii. (R.B.)

Quinone pigments vary in amounts in different tissues of the same species, and in the ovaries they vary with the season. Oral view of *Heterocentrotus mammilatus* shows tube feet and soft tissues darker than the spines. Hawaii. (R.B.)

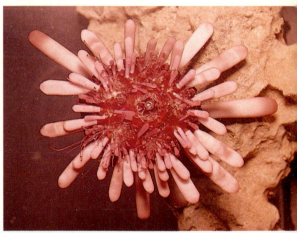

Tests of sea urchins are colored by echinochromes and spinochromes. (R.B.)

Red echinochrome in the test and violet spinochrome in the spines have been demonstrated in some species of sea urchins. *Diadema palmeri.* New Zealand. (W. Doak)

Color intensity varies with depth in many invertebrates, even within the same species. *Lytechinus variegatus* from deep water *(left)* off Beaufort, N. Carolina, has a pale rose-pink test and white spines. The shallow-water specimen *(right)* has a deep rose-pink test and pink spines. (R.B.)

Anthraquinones in feather stars have been recovered from red, purple, and yellow species. This one was resting and feeding on the bottom, under the sea ice, at McMurdo Sound, Antarctica. (P.K. Dayton)

Feather star *Antedon mediterranea* has deep red pigment in the integument and connective tissue but the calcareous skeleton is not pigmented. The red pigment may be an anthraquinone. This species has a yellow variant. Marine Station, Messina, Sicily. (R.B.)

Black pigment in land planarians is thought to be melanin. It is limited to or densest near the dorsal surface. These triclads (unlike most flatworms) have the pigment mostly in the mesenchyme, not in the epidermis. Rain forest near Singapore. (E.S. Ross)

Melanins in sea urchins have been most studied in *Diadema*. There are melanin deposits in test and spines. At night melanin in chromatophores concentrates and the animals pale. In strong sunlight the pigment disperses and the urchins are densely black. (R.B.)

Melanin-pigmented egg capsules of the European cuttle, *Sepia officinalis,* are covered, as they emerge, by the cuttle's ink, a suspension of dark brown melanoprotein. (Sepia ink has long been used by artists.) Netherlands. (R.B.)

Dark pigment in pulmonate snail shells is thought to be melanin, but the pigment is closely associated with the conchiolin of the shell and difficult to extract. The lighter brown shades are probably melanin conjugated with protein. The pigment of the eyes is melanin. *Helix aspersa.* California. (R.B.)

Melanin in insect cuticle adds to the darkening of wings and body by the quinone-tanning that hardens cuticle. Rapid generation of melanin in the cuticle may be a response to light, as when the nymph of a cicada emerges from the ground. Panama. (R.B.)

Melanin is rare in cnidarians, but in the scyphomedusa *Pelagia colorata* the magenta and brown pigments that contribute to the purplish markings on the bell are probably related to melanin. Seen against a dark background, the translucent subumbrellar parts and oral lobes show blue-scattering. Monterey Bay, Calif. (R.B.

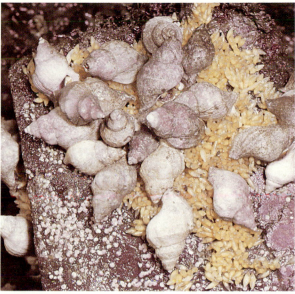

Rapidly changing pigment cells of cephalopods owe their color to ommochromes, pigments related to melanins and like them often conjugated with proteins. Broad-finned squid. New Zealand. (W. Doak)

Indigoid pigments are related to melanins and ommochromes. The yellowish indigoid precursors of the red and purple dyes once used by the Romans and others to dye fabrics were extracted from various genera of snails still called "purples." Among the purples is *Nucella lamellosa,* shown here as an aggregation of snails laying their yellow egg capsules. Oregon. (R.B.)

Ommochromes in insect eyes separate the ommatidia optically, and in the grasshopper nymph *(left),* they impart tan and brown coloring. In the grasshopper *(right)* the eyes are colored by red pterin pigment. Costa Rica. (R.B.)

Chromatic change is almost instantaneous in cephalopods. *Left,* a squid resting quietly close to the bottom. *Right,* the bright light used for the first photo has excited the squid and the skin blanches. *Loligo opalescens.* Calif. (R.B.)

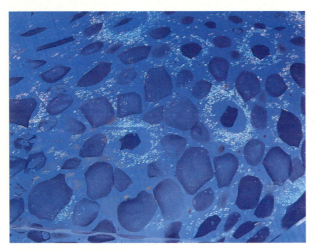
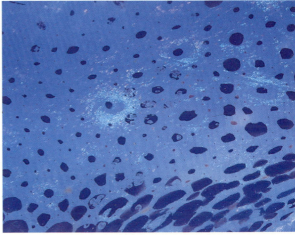

Expanded chromatophores in skin of *Loligo opalescens,* are microscopic bags of ommochrome. Each has radiating strands of muscle under nervous control. With muscles contracted, the spherical cells have been pulled out into flat plates, their pigment fully exposed.

Contracted chromatophores in threatened squid. The muscles that distend the pigment sacs are suddenly relaxed, and the chromatophores return to a spherical shape, diminished area, and minimum exposure of pigment. The iridescent circles are guanine crystals.

Left, **Chromatophores** in newly hatched juvenile squid, *Loligo pealei.* Ommochromes are also present in cephalopod eyes. 3 mm long. Beaufort, N. Carolina. (R.B.)

Right, **Nervous control of chromatophores** in an annelid. In the leech *Placobdella parasitica,* pigment in the red chromatophores is usually dispersed, but stimulation of the nerve cord causes concentration. In green chromatophores, pigment expands in direct response to light or through nervous control. (R.B.)

Free porphyrin is harmful to living tissues exposed to light; its occurrence in them is rare. The common European sea star *Asterias rubens* has protoporphyrin in the integument but not enough to provide coloration. The colors of this sea star, shades of brown, red, or violet, are due to carotenoproteins, and these apparently help to screen light from the protoporphyrin, which is present in low concentration in pale specimens but occurs in increasing amounts in those with deeper colors. Roscoff, France. (R.B.)

Hemoglobin is an **iron-porphyrin hemoprotein** distributed widely but sporadically in invertebrates, especially in annelids and their allies such as the vestimentiferans, which are giant pogonophores. *Riftia pachyptila* has hemoglobin in the blood circulating through the tentacles, permitting these worms to live in the nearly anoxic waters of the Galapagos Rift. (A. Giddings)

Chlorocruorin, the green iron-porphyrin respiratory pigment of many annelids, colors the blood green. *Below left,* tentacular crown of a large sabellid worm, *Sabella spallanzani.* (R.B.) *Below right,* double crown of the Christmas-tree worm, the serpulid, *Spirobranchus giganteus;* the color of the blood is masked by red pigment. (K.B. Sandved)

Green porphyrin pigment in the skin of *Bonellia viridis,* an echiuran, is a dihydroporphyrin (a unique pigment) with a structure that implies an origin other than the chlorophyll in the bottom detritus it gathers in feeding. The skin is toxic, and *Bonellia* is avoided by such predators as *Sepia* or most crabs. Mediterranean coast of France at Banyuls. (R.B.)

Bilichromic compounds, less sensitive to light than porphyrins, color the soft parts of animals that live at the surface. In *Physalia* different bilichromes color the lavender-blue of the float, its often pink crest, the blue tentacles, and green or purplish blues in the feeding and reproductive polyps. Bimini, Bahamas. (R.B.)

Red bilichrome, haliotisrubin, colors the shell of the red abalone, *Haliotis rufescens,* when it feeds on red algas, which synthesize the related red bilichrome, phyco-erythrin. The white stripes are laid down during periods when the abalone feeds on brown kelps. Calif. (R.B.)

Bright blue skeletal color of the blue coral, *Heliopora coerulea,* of Australian and Indo-Pacific equatorial waters, is a biliverdin bonded to the calcium carbonate skeleton. The living colony appears brown from the zooxanthellas in the polyps. (R.B.)

Pink coloring in many marine shells is due to the deposition of free porphyrins in the shell, thus rendering these harmful waste-products inert. Two shells that owe their attractiveness to the coloring of the mouth and flaring lip of the shell are the pink-lipped murex, *Hexaplex erythrostoma,* and the larger queen conch, *Strombus gigas.* (R.B.)

Brownish-red porphyrin that patterns the surface organic layer of *Trochus,* an Indo-Pacific snail, is a uroporphyrin, thus safely disposed of in the shell. When the surface layer is polished away, the ultrathin layering of the shell makes it shine with iridescent interference colors. Its thickness, and lustrous quality once made this shell the basis of a large button-making industry. (R.B.)

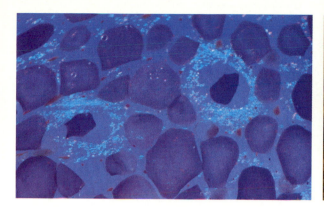

Crystals of guanine, a purine compound, shine with a greenish iridescence in the skin of the squid *Loligo opalescens.* The crystals are contained within iridophores, non-expanding cells that overlie the dark-pigmented chromatophores. When more concentrated than they are here, guanine leucophores produce white patches in the skin of cephalopods and white or silvery colors in other invertebrates. (R.B.)

Siphonal margins of giant clam have iridescent interference colors generated by iridophores containing colorless purine crystals. *Tridacna.* Marshall Islands. (K.B. Sandved)

Purine pigment in *Botryllus schlosseri* provides the prominent silvery bands of the clusters of zooids that compose this encrusting ascidian colony. The purine occurs as granules in the blood cells, and it is under genetic control. Three alleles at one gene locus determine which of 2 types of band will occur, or whether the purine will be dispersed over the dorsal surface. Roscoff, France. (R.B.)

Dispersed purine pigment is masked by 3 more colorful pigments when they are present. An orange pigment dissolved in blood cells is known to be inherited independently of the pattern of silvery bands. The blue and red pigments are presumed to be also under genetic control. The 4 pigments in various combinations produce a varied spectrum of genetically determined colorings in clones that may grow side by side. Roscoff. (R.B.)

Pterin pigments occur most often in insects, and the conspicuous reds and yellows of hemipterans, many of them presumably warning colors, are due to the red pterin erythropterin and the yellow pterin xanthopterin, along with the white pterin leucopterin. These last 2 pigments provide the prominent yellow and white markings of the wasp *Vespula* and its mimics, shown earlier. *Dyadercus andreae.* Florida. (K.B. Sandved)

Nudibranch colorings are usually said to be the most beautiful among invertebrates. The varied and brilliant colors are predominantly carotenoids and carotenoproteins. *Phidiana (= Hermissenda) crassicornis* is variable in color but always has orange areas on the back and a bright blue line along the sides. The white tips of the cerata contain nematocysts from hydroid prey. California. (G. Miller)

Butterfly and moth colorings rival the nudibranchs for variety and brilliance, as do the colors of tropical beetles. Panels of mounted specimens show both pigmentary and structural colors. Note the differences in color of the wings of morpho butterflies viewed from above at different angles. The panels were offered for sale in a forest near Rio de Janeiro, Brazil, a practice contributing to the depletion of lepidopteran populations. (R.B.)

Actiniochrome is a name given to the pigment that occurs in the pink, red, or violet tips of the tentacles of some sea anemones. It is one of the various pigments usually lumped together under "some miscellaneous biochromes," with incompletely known chemical affinities, and possible physiological roles that remain to be explored. *Condylactis gigantea.* Florida. (R.B.)

Red calcareous skeletons of organpipe coral, *Tubipora musica,* at left (R.B.), and **living precious coral,** *Corallium rubrum,* at right (J. Vasserot). The red color is attributed to iron compounds of unknown composition, probably bound to the calcium.

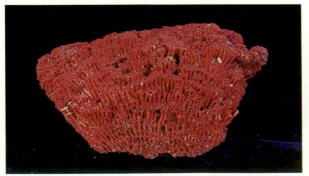

LITERATURE

THE LIST OF BOOKS and other publications below represents a sampling of the wide field of information on invertebrates. It should be looked upon less as a reading list than as a means of entry to the current literature. As each of the sources listed contains references to other works, and these will lead to still others, a reader can rather quickly gain access to the knowledge available in any desired area.

Of course, a given work can cite only others that were published earlier. However, by looking in the *Citation Index* for a paper that one judges to be of interest, publications of more recent dates can also be tracked down. Another method is to search for entries by author or subject in *Biological Abstracts,* which supplies not only references but also brief summaries of biological articles recently published in scientific journals. Yet another useful reference source is the *Zoological Record;* compiled each year, it lists recent publications on each animal group. All of these works can be found in the reference section of most science libraries.

Another approach is simply to browse through some of the dozens of scientific journals that contain articles on invertebrates (for example, *American Zoologist, Biological Bulletin, Marine Biology, Quarterly Review of Biology).* Some specialize in particular groups (for example, *Journal of Protozoology, Veliger, Journal of Crustacean Biology, Annual Reviews of Entomology)* or areas of interest (for example, *International Journal of Invertebrate Reproduction, Journal of Parasitology).* The individual research papers found in such journals constitute the foundation of what we know about invertebrates—a primary body of knowledge that continually grows and changes.

SYSTEMATIC AND GENERAL REFERENCES ON INVERTEBRATES

Barnes, R.S.K. 1984. *A Synoptic Classification of Living Organisms.* Sinauer Associates.

Beklemishev, W.N. 1969. *Principles of Comparative Anatomy of Invertebrates.* Vols. 1 & 2. Transl. from Russian by J.M. MacLennan. University of Chicago Press.

Grassé, P.-P., ed. 1949-1982. *Traité de Zoologie.* Vols. 1-17. Masson et Cie., Paris.

Hyman, L.H. 1940-1967. *The Invertebrates.* Vols. 1-6. McGraw-Hill Book Co.

Kaestner, A. 1967-1970. *Invertebrate Zoology.* Vols. 1-3. Transl. from German by H.W. Levi and L.R. Levi. John Wiley & Sons.

Margulis, L. and K.V. Schwartz. 1982. *Five Kingdoms: An Illustrated Guide to the Phyla of Life on Earth.* W.H. Freeman & Co.

Parker, S.B., ed. 1982. *Synopsis and Classification of Living Organisms.* Vols. 1 & 2. McGraw-Hill Book Co.

SPECIAL ASPECTS OF INVERTEBRATE BIOLOGY

Barrington, E.J.W. 1979. *Invertebrate Structure and Function.* 2nd ed. John Wiley & Sons.

Bereiter-Hahn, J., A.G. Matoltsy, and K.S. Richards, eds. 1984. *Biology of the Integument. Vol. 1, Invertebrates.* Springer-Verlag.

Blackwelder, R.E. 1967. *Taxonomy.* John Wiley & Sons.

Brehelin, M., ed. 1986. *Immunity in Invertebrates.* Springer-Verlag.

Brusca, G.J. 1975. *General Patterns of Invertebrate Development.* Mad River Press.

Bullock, T.H. and G.A. Horridge. 1965. *Structure and Function in the Nervous Systems of Invertebrates.* Vols. 1 & 2. W.H. Freeman & Co.

SPECIAL ASPECTS OF INVERTEBRATE BIOLOGY *continued*

Calow, P. 1981. *Invertebrate Biology: A Functional Approach.* Croom Helm Ltd.

Chia, F.-S. and M.E. Rice, eds. 1978. *Settlement and Metamorphosis of Marine Invertebrate Larvae.* Elsevier.

Cohen, W.D., ed. 1985. *Blood Cells of Marine Invertebrates: Experimental Systems in Cell Biology and Comparative Physiology.* Alan R. Liss.

Corning, W.C., J.A. Dyal, and A.O.D. Willows, eds. 1973-1975. *Invertebrate Learning.* Vols. 1-3. Plenum Press.

Florkin, M. and B.T. Scheer, eds. 1967-1974. *Chemical Zoology.* Vols. 1-8. Academic Press.

Giese, A.C. and J.S. Pearse, eds. 1974-1979. *Reproduction of Marine Invertebrates.* Vols. 1-5. Academic Press. Vols. 6-9 to follow; Blackwell Scientific Publications and Boxwood Press.

Gilbert, L.I. and E. Frieden, eds. 1981. *Metamorphosis: A Problem in Developmental Biology.* 2nd ed. Plenum Press.

Halstead, B.W. 1978. *Poisonous and Venomous Marine Animals of the World.* 2nd ed. Darwin Press.

Hammen, C.S. 1980. *Marine Invertebrates: Comparative Physiology.* University Press of New England.

Jackson, J.B.C., L.W. Buss, and R.E. Cook, eds. 1985. *Population Biology and Evolution of Clonal Organisms.* Yale University Press.

Jagersten, G. 1972. *Evolution of the Metazoan Life Cycle.* Academic Press.

Mayr, E. 1982. *The Growth of Biological Thought: Diversity, Evolution, and Inheritance.* Harvard University Press.

Mill, P.J. 1972. *Respiration in Invertebrates.* Macmillan.

Noble, E.R. and G.A. Noble. 1982. *Parasitology: The Biology of Animal Parasites.* 5th ed. Lea & Febiger.

Olsen, O.W. 1974. *Animal Parasites: Their Life Cycles and Ecology.* 3rd ed. University Park Press.

Rose, S.M. 1970. *Regeneration: Key to Understanding Normal and Abnormal Growth and Development.* Appleton-Century-Croft.

Sawyer, R.H. and R.M. Showman, eds. 1985. *The Cellular and Molecular Biology of Invertebrate Development.* University of South Carolina Press.

Schmidt, G.D. and L.S. Roberts. 1985. *Foundations of Parasitology.* 3rd ed. C.V. Mosby Co.

Shelton, G.A.B., ed. 1982. *Electrical Conduction and Behaviour in "Simple" Invertebrates.* Clarendon Press.

Smith, D.C. and Y. Tiffon, eds. 1980. *Nutrition in the Lower Metazoa.* Pergamon Press.

Stancyk, S.E., ed. 1979. *Reproductive Ecology of Marine Invertebrates.* University of South Carolina Press.

Strathmann, M.F. 1987. *Reproduction and Development of Pacific Coast Marine Invertebrates: The Friday Harbor Laboratories Handbook of Methods and Data.* University of Washington Press.

Wainwright, S.A., W.D. Biggs, J.D. Currey, and J.M. Gosline. 1982. *Mechanical Design in Organisms.* 2nd ed. Princeton University Press.

Wilt, F.H. and N.K. Wessells, eds. 1967. *Methods in Developmental Biology.* Crowell.

NATURAL HISTORY AND IDENTIFICATION OF INVERTEBRATES

Buchsbaum, R. and L.J. Milne. 1962. *The Lower Animals: Living Invertebrates of the World.* Doubleday & Co.

Gosner, K.L. 1971. *Guide to Identification of Marine and Estuarine Invertebrates: Cape Hatteras to the Bay of Fundy.* John Wiley & Sons.

Grzimek, B. 1974-1975. *Grzimek's Animal Life Encyclopedia.* Vols. 1-3. Van Nostrand & Reinhold Co.

Kozloff, E.N. 1983. *Seashore Life of the Northern Pacific Coast.* University of Washington Press.

MacGinitie, G.E. and N. MacGinitie. 1968. *Natural History of Marine Animals.* 2nd ed. McGraw-Hill Book Co.

Meinkoth, N.A. 1981. *Audubon Society Field Guide to North American Seashore Creatures.* Alfred A. Knopf.

Milne, L. and M. Milne. 1972. *Invertebrates of North America.* Doubleday & Co.

Morris, R.H., D.P. Abbott, and E.C. Haderlie. 1980. *Intertidal Invertebrates of California.* Stanford University Press.

Pennak, R.W. 1978. *Freshwater Invertebrates of the United States.* 2nd ed. John Wiley & Sons.

Ricketts, E.F., J. Calvin, and J.W. Hedgpeth. 1985. *Between Pacific Tides.* 5th ed. Revised by D.W. Phillips. Stanford University Press.

Riedl, R. 1963. *Fauna und Flora der Adria.* Verlag Paul Parey, Hamburg.

Smith, R.I., and J.T. Carlton, eds. 1975. *Light's Manual: Intertidal Invertebrates of the Central California Coast.* 3rd ed. University of California Press.

Sterrer, W., ed. 1986. *Marine Fauna and Flora of Bermuda.* John Wiley & Sons.

LABORATORY CULTURE AND DISSECTION MANUALS

Beck, D.E. and L.F. Braithwaite. 1968. *Invertebrate Zoology Laboratory Workbook*. 3rd ed. Burgess Publ. Co.

Berg, C.J., Jr., ed. 1983. *Culture of Marine Invertebrates: Selected Readings*. Hutchinson Ross Publ. Co.

Boolootian, R.A. and D. Heyneman. 1977. *Illustrated Laboratory Text in Zoology*. Holt, Rinehart and Winston.

Dales, R.P., ed. 1981. *Practical Invertebrate Zoology: a Laboratory Manual for the Study of the Major Groups of Invertebrates, excluding Protochordates*. 2nd ed. Blackwell Scientific Publ.

Sherman, I.W. and V.G. Sherman. 1976. *The Invertebrates: Function and Form, a Laboratory Guide*. 2nd ed. Macmillan Co.

Smith, W.L. and M.H. Chanley, eds. 1972. *Culture of Marine Invertebrate Animals*. Plenum Press.

BOOKS AND MONOGRAPHS ON VARIOUS INVERTEBRATE GROUPS
PROTOZOANS, Chapter 2

Chen, T.-T., ed. 1967-1972. *Research in Protozoology*. Vols. 1-4. Pergamon Press.

Corliss, J.O. 1984. The kingdom Protista and its 45 phyla. *Biosystems* **17**:87-126.

Grell, K.G. 1973. *Protozoology*. Springer-Verlag.

Kreier, J.P., ed. 1977-1978. *Parasitic Protozoa*. Vols. 1-4. Academic Press.

Lee, J.J., S.H. Hutner, and E.C. Bovee, eds. 1985. *An Illustrated Guide to the Protozoa*. Society of Protozoologists.

Levandowsky, M. and S.H. Hutner, eds. 1979-1981. *Biochemistry and Physiology of Protozoa*. 2nd ed. Vols. 1-4. Academic Press.

Levine, N.D. (and 15 other members of the Committee on Systematics and Evolution of the Society of Protozoologists). 1980. A newly revised classification of the Protozoa. *Journal of Protozoology* **27**(1):37-58.

Saier, M.H. and G.R. Jacobson. 1984. *The Molecular Basis of Sex and Differentiation: a Comparative Study of Evolution, Mechanism, and Control in Microorganisms*. Springer-Verlag.

Sleigh, M.A. 1973. *The Biology of Protozoa*. Edward Arnold.

Sleigh, M.A., ed. 1974. *Cilia and Flagella*. Academic Press.

SPONGES, Chapter 3

Bergquist, P.R. 1978. *Sponges*. Hutchinson & Co.

Fry, W.G., ed. 1970. *The Biology of the Porifera*. Zoological Society of London Symposium no. 25. Academic Press.

Harrison, F.W. and R.R. Cowden, eds. 1976. *Aspects of Sponge Biology*. Academic Press.

Levi, C. and N. Boury-Esnault, eds. 1979. *Biologie des Spongiaires/Sponge Biology*. Colloques Internationaux du Centre National de la Recherche Scientifique, no. 291. Editions du CNRS, Paris.

Reiswig, H.M. and G.O. Mackie. 1983. Studies on hexactinellid sponges. III. The taxonomic status of Hexactinellida within the Porifera. *Philosophical Transactions of the Royal Society of London* **B301**(1107):419-428. (See also 2 other papers on hexactinellids in this issue).

Simpson, T.L. 1984. *The Cell Biology of Sponges*. Springer-Verlag.

PLACOZOANS AND MESOZOANS, Chapter 4

Grell, K.G. and G. Benwitz. 1971. Die Ultrastruktur von *Trichoplax adhaerens* F.E. Schulze. *Cytobiologie* **4**(2):216-240. (English abstract)

Grell, K.G. 1972. Eibildung und Furchung von *Trichoplax adhaerens* F.E. Schulze (Placozoa). *Zeitschrift für Morphologie der Tiere* **73**:297-314.

Hochberg, F.G. 1982. The "kidneys" of cephalopods: a unique habitat for parasites. *Malacologia* **23**(1): 121-134.

Hochberg, F.G. 1983. The parasites of cephalopods: a review. Memoirs of the National Museum Victoria, no. **44**:109-145.

Kozloff, E.N. 1969. Morphology of the orthonectid *Rhopalura ophiocomae. Journal of Parasitology* **55**:171-195.

Lapan, E.A. and H. Morowitz. 1972. The Mesozoa. *Scientific American* **227**(6):94-101. [dicyemids only]

McConnaughey, B.H. 1963. The Mesozoa. Chapter 11 *in* E.C. Dougherty, ed.: *The Lower Metazoa*. University of California Press.

CNIDARIANS, Chapters 5 & 6

Barnes, D.J., ed. 1983. *Perspectives on Coral Reefs.* Brian Clouston Publisher, Manuka, Australia.

Burnett, A.L., ed. 1973. *Biology of Hydra.* Academic Press.

Friese, U.E. 1972. *Sea Anemones.* T.F.H. Publications, Hong Kong.

Great Barrier Reef. 1984. Reader's Digest, Sydney.

Jones, O.A. and R. Endean, eds. 1973-1977. *Biology and Geology of Coral Reefs.* Vols. 1-4. Academic Press.

Lenhoff, H.M., ed. 1983. *Hydra: Research Methods.* Plenum Press.

Lenhoff, S.G. and H.M. Lenhoff. 1986. *Hydra and the Birth of Experimental Biology—1744.* A translation from the French of A. Trembley's *Mémoires, pour servir à l'histoire d'un genre de polypes d'eau douce, à bras en forme des cornes.* Boxwood Press.

Muscatine, L. and H.M. Lenhoff, eds. 1974. *Coelenterate Biology: Reviews and New Perspectives.* Academic Press.

Mackie, G.O., ed. 1976. *Coelenterate Ecology and Behavior.* 3rd International Symposium on Coelenterate Biology. Plenum Press.

Proceedings of the Fifth International Coral Reef Congress, Tahiti. 1975. Antenne Museum-EPHE, Moorea, French Polynesia. (Look for publications of other coral reef congresses)

Tardent, P. and R. Tardent, eds. 1980. *Developmental and Cellular Biology of Coelenterates.* Proc. 4th International Coelenterate Conference, Interlaken, Switzerland. Elsevier/North-Holland Biomedical Press. (Look for publications of other coelenterate symposiums)

Van Praët, M. 1985. Nutrition of Sea Anemones. *In* J.H.S. Blaxter, F.S. Russell, and M. Yonge, eds: *Advances in Marine Biology* **22**:65-99.

Wood, E.M. 1983. *Corals of the World: Biology and Field Guide.* T.F.H. Publications, Hong Kong.

CTENOPHORES, Chapter 7

Harbison, G.R. 1985. On the classification and evolution of the Ctenophora. Chapter 6 *in* S. Conway Morris, J.D. George, R. Gibson, and H.M. Platt, eds.: *The Origins and Relationships of Lower Invertebrates.* Systematics Association, special vol. **28**. Clarendon Press.

Reeve, M.R. and M.A. Walter. 1978. Nutritional Ecology of Ctenophores. *In* F.S. Russell and M. Yonge, eds.: *Advances in Marine Biology* **15**:249-287.

FLATWORMS, Chapters 8 & 9

Arme, C. and P.W. Pappas, eds. 1984. *Biology of the Eucestoda.* Vols. 1 & 2. Academic Press.

Brøndsted, H.V. 1969. *Planarian Regeneration.* Pergamon Press.

Chandebois, R. 1976. *Histogenesis and Morphogenesis in Planarian Regeneration.* Monographs in Developmental Biology, vol. **11**. S. Karger, Basel.

Erasmus, D.A. 1972. *The Biology of Trematodes.* Crane, Russak.

Prudhoe, S. 1985. *A Monograph on Polyclad Turbellaria.* British Museum (Natural History) and Oxford University Press.

Schockaert, E.R. and I.R. Ball, eds. 1981. *The Biology of the Turbellaria.* Proceedings of the 3rd International Symposium. *Hydrobiologia* **84**. Reprinted by Dr. W. Junk Publ., The Hague. (Look for other turbellarian symposiums)

Smyth, J.D. and D.W. Halton. 1983. *The Physiology of Trematodes.* 2nd ed. Cambridge University Press.

GNATHOSTOMULIDS, Chapter 10

Sterrer, W., M. Mainitz, and R.M. Rieger. 1985. Gnathostomulids: enigmatic as ever. Chapter 12 *in* S. Conway Morris, J.D. George, R. Gibson, and H.M. Platt, eds.: *The Origins and Relationships of Lower Invertebrates.* Systematics Association, special vol. **28.** Clarendon Press.

NEMERTEANS, Chapter 11

Gibson, R. 1973. *Nemerteans.* Hutchinson University Library.

Roe, P. and J.L. Norenburg, eds. 1985. Symposium on the Comparative Biology of Nemertines. *American Zoologist,* **25**(1):1-151.

NEMATODES, NEMATOMORPHS, ACANTHOCEPHALANS, Chapter 12.

Croll, N.A. and B.E. Matthews. 1977. *Biology of Nematodes*. John Wiley & Sons.

Crompton, D.W.T. and B.B. Nickol, eds. 1985. *Biology of the Acanthocephala*. Cambridge University Press.

Levine, N.D. 1980. *Nematode Parasites of Domestic Animals and of Man*. 2nd ed. Burgess Publ. Co.

Lorenzen, S. 1985. Phylogenetic aspects of pseudo-coelomate evolution. Chapter 14 *in* S. Conway Morris, J.D. George, R. Gibson, and H.M. Platt, eds.: *The Origins and Relationships of Lower Invertebrates*. Systematics Association, special vol. **28**. Clarendon Press.

Maggenti, A. 1981. *General Nematology*. Springer-Verlag.

Poinar, G.O., Jr. 1983. *The Natural History of Nematodes*. Prentice-Hall.

ROTIFERS, GASTROTRICHS, KINORHYNCHS, LORICIFERANS, TARDIGRADES, Chapter 13

Donner, J. 1956. *Rotifers*. Transl. from German and adapted by H.G.S. Wright, 1966. Cambridge University Press.

Conway Morris, S., J.D. George, R. Gibson, and H.M. Platt, eds. 1985. *The Origins and Relationships of Lower Invertebrates*. Systematics Association, special vol. **28**. Clarendon Press. Chapter 14: S. Lorenzen. Phylogenetic aspects of pseudocoelomate evolution. Chapter 15: P. Clement. The relationships of rotifers. Chapter 16: P.J.S. Boaden. Why is a gastrotrich?

Hulings, N.C. and J.S. Gray. 1971. *A Manual for the Study of Meiofauna*. Smithsonian Contributions to Zoology **78**.

Kristensen, R.M. 1983. Loricifera, a new phylum with Aschelminthes characters from the meiobenthos. *Zeitschrift für Zoologische Systematik und Evolutionsforschung* **21**:163-180.

Morgan, C.I. and P.E. King. 1976. *British Tardigrades*. Academic Press.

Pejler, B., R. Starkweather, and T. Nogrady, eds. 1983. Biology of Rotifers. Proceedings of the 3rd International Rotifer Symposium. *Hydrobiologia* **104**. Reprinted by Dr. W. Junk Publ., The Hague. (Look for other rotifer symposiums)

MOLLUSCS, Chapters 14 & 15

Abbott, R.T. 1972. *Kingdom of the Seashell*. Crown Publishers.

Abbott, R.T. and S.P. Dance. 1982. *Compendium of Seashells*. E.P. Dutton Inc.

Fretter, V. and J. Peake. 1975-1979. *Pulmonates*. Vols. 1, 2a, 2b. Academic Press.

Morton, J.E. 1979. *Molluscs*. 5th ed. Hutchinson University Library.

Nixon, M. and J.B. Messenger, eds. 1977. The Biology of Cephalopods. *Zoological Society of London Symposium* **38**. Academic Press.

Purchon, R.D. 1977. *The Biology of the Mollusca*. 2nd ed. Pergamon Press.

Rehder, H.A. 1981. *Audubon Society Field Guide to North American Seashells*. Alfred A. Knopf.

Runham, N.W. and P.J. Hunter. 1970. *Terrestrial Slugs*. Hutchinson University Library.

Solem, G.A. 1974. *The Shell Makers: Introducing Molluscs*. John Wiley & Sons.

Wells, M.J. 1978. *Octopus: Physiology and Behaviour of an Advanced Invertebrate*. Chapman and Hall.

Wilbur, K.M., ed. 1983-1985. *The Mollusca*. Vols. 1-10. Academic Press.

Yonge, C.M. and T.E. Thompson. 1976. *Living Marine Molluscs*. Collins.

ANNELIDS, Chapters 16 & 17

Anderson, D.T. 1973. *Embryology and Phylogeny in Annelids and Arthropods*. Pergamon Press.

Dales, R.P. 1967. *Annelids*. Hutchinson University Library.

Edwards, C.A., and J.R. Lofty. 1977. *Biology of Earthworms*. 2nd ed. Chapman and Hall.

Fischer, A. and H.-D. Pfannenstiel, eds. 1984. *Polychaete Reproduction*. Gustav Fischer Verlag.

Hutchings, P.A., ed. 1984. *Proc. 1st International Polychaete Conference,* Linnean Society of New South Wales. Sydney.

Lee, K.E. 1985. *Earthworms: Their Ecology and Relationships with Soils and Land Use*. Academic Press.

Mann, K.H. 1962. *Leeches (Hirudinea): Their Structure, Physiology, Ecology, and Embryology*. Pergamon Press.

Mill, P.J., ed. 1978. *Physiology of Annelids*. Academic Press.

Satchell, J.E., ed. 1983. *Earthworm Ecology*. Chapman and Hall.

ECHIURANS, SIPUNCULANS, POGONOPHORES, PRIAPULANS, Chapter 18

Conway Morris, S., J.D. George, R. Gibson, and H.M. Platt, eds. 1985. *The Origins and Relationships of Lower Invertebrates*. Systematics Association, special vol. **28**. Clarendon Press. Chapter 17: J. van der Land and A. Nørrevang. Affinities and intraphyletic relationships of the Priapulida. Chapter 18: M.E. Rice. Sipuncula: developmental evidence for phylogenetic inference. Chapter 21: M.L. Jones. Vestimentiferan pogonophores: their biology and affinities.

McLean, N. 1984. Amoebocytes in the lining of the body cavity and mesenteries of *Priapulus caudatus* (Priapulida). *Acta Zoologica* (Stockholm) **65**(2): 75-78.

Rice, M.E. and M. Todorovic, eds. 1975, 1976. *Biology of the Sipuncula and Echiura,* vols. 1 & 2. Inst. for Biol. Research, Belgrade, and Smithsonian Institution.

Stephen, A.C. and S.J. Edmonds. 1972. *The Phyla Sipuncula and Echiura*. British Museum (Natural History).

Southward, E.C. 1971. Recent researches on the Pogonophora. *In* H. Barnes, ed.: *Oceanography and Marine Biology Annual Reviews,* **9**:193-220. George Allen and Unwin.

Southward, E.C. 1975. Fine structure and phylogeny of the Pogonophora. Pp. 235-251 *in* E.J.W. Barrington and R.P.S. Jeffries, eds.: *Protochordates*. Zoological Society of London Symposium **36**. Academic Press.

Southward, A.J., E.C. Southward, P.R. Dando, R.L. Barrett, and R. Ling. 1986. Chemoautotrophic function of bacterial symbionts in small Pogonophora. *Journal of the Marine Biological Association of the United Kingdom* **66**:415-437.

ONYCHOPHORANS, Chapter 19

Ghiselin, M.T. 1985. A Movable Feaster. *Natural History* **94**(9):54-61.

(Onychophorans are included in works by Anderson, Gupta, and Manton listed in general arthropod section below)

ARTHROPODS (general), Chapter 20

Anderson, D.T. 1973. *Embryology and Phylogeny in Annelids and Arthropods*. Pergamon Press.

Clarke, K.U. 1973. *The Biology of the Arthropoda*. Edward Arnold.

Edney, E.B. 1977. *Water Balance in Land Arthropods*. Springer-Verlag.

Gupta, A.P., ed. 1979. *Arthropod Phylogeny*. Van Nostrand Reinhold Co.

Gupta, A.P., ed. 1983. *Neurohemal Organs of Arthropods: Their Development, Evolution, Structures, and Functions*. Charles C Thomas, Publ.

Herreid, C.F. and C.R. Fourtner, eds. 1981. *Locomotion and Energetics in Arthropods*. Plenum Press.

Levi-Setti, R. 1975. *Trilobites*. University of Chicago Press.

Manton, S.M. 1977. *The Arthropoda: Habits, Functional Morphology, and Evolution*. Clarendon Press.

Neville, A.C. 1975. *Biology of the Arthropod Cuticle*. Springer-Verlag.

CRUSTACEANS, Chapter 21

Bliss, D.E., ed. 1982-1985. *The Biology of Crustacea*. Vols. 1-10. Academic Press.

Cobb, J.S. and B.F. Phillips, eds. 1980. *The Biology and Management of Lobsters*. Vols. 1 & 2. Academic Press.

Fitzpatrick, J.F. Jr. 1983. *How to Know the Freshwater Crustacea*. Wm. C. Brown Co.

Jegla, T.C., ed. 1985. Symposium on Advances in Crustacean Endocrinology. *American Zoologist* **25**(1):153-284.

Mauchline, J. 1980. The Biology of Mysids and Euphausiids. *In* J.H.S. Blaxter, F.S. Russell, and M. Yonge, eds.:*Advances in Marine Biology* **18**:1-677.

McLaughlin, P.A. 1980. *Comparative Morphology of Recent Crustacea*. W.H. Freeman & Co.

Warner, G.F. 1977. *The Biology of Crabs*. Elek Science.

CHELICERATES, Chapter 22

Barth, F.G., ed. 1985. *Neurobiology of Arachnids.* Springer-Verlag.

Bonaventura, J., C. Bonaventura, and S. Tesh, eds. 1982. *Physiology and Biology of Horseshoe Crabs.* Alan R. Liss.

Foelix, R.F. 1982. *Biology of Spiders.* Harvard University Press.

Fry, W.G., ed. 1978. Sea Spiders (Pycnogonida). *Zoological Journal of the Linnean Society* **63** (1 & 2). Academic Press.

King, P.E. 1973. *Pycnogonids.* Hutchinson.

Merrett, P., ed. 1978. *Arachnology.* Zoological Society of London Symposium **42.** Academic Press.

Milne, L. and M. Milne. 1980. *Audubon Society Field Guide to North American Insects and Spiders.* Alfred A. Knopf.

Obenchain, F.D. and R. Galun, eds. 1982. *Physiology of Ticks.* Pergamon Press.

Savory, T. 1977. *Arachnida.* 2nd ed. Academic Press.

Snow, K.R. 1970. *The Arachnids: An Introduction.* Columbia University Press.

Weygoldt, P. 1969. *The Biology of Pseudoscorpions.* Harvard University Press.

MYRIAPODS, Chapter 23

Blower, J.G., ed. 1974. *Myriapoda.* Zoological Society of London Symposium **32.** Academic Press.

Camatini, M., ed. 1980. *Myriapod Biology.* Academic Press.

Lewis, J.G.E. 1981. *The Biology of Centipedes.* Cambridge University Press.

INSECTS, Chapter 24

Bell, W.J. and R.T. Cardé, eds. 1984. *Chemical Ecology of Insects.* Sinauer Associates.

Blum, M.S., ed. 1985. *Fundamentals of Insect Physiology.* John Wiley & Sons.

Borror, D.J., D.M. DeLong, and C.A. Triplehorn. 1976. *An Introduction to the Study of Insects.* 4th ed. Holt, Rinehart & Winston.

Brian, M.V. 1983. *Social Insects: Ecology and Behavioral Biology.* Chapman & Hall.

Chapman, R.F. 1982. *The Insects: Structure and Function.* 3rd ed. Hodder & Stoughton.

Daly, H.V., J.T. Doyen, and P. Ehrlich. 1978. *Introduction to Insect Biology and Diversity.* McGraw-Hill Book Co.

Downer, R.G.H. and H. Laufer, eds. 1983. *Endocrinology of Insects.* Alan R. Liss.

Eisner, T., and E.O. Wilson, eds. 1977. *The Insects:* readings from *Scientific American.* W.H. Freeman & Co.

Evans, H.E. 1984. *Insect Biology: A Textbook of Entomology.* Addison-Wesley Publ. Co.

Hermann, H.R., ed. 1979-1982. *Social Insects.* Vols. 1-4. Academic Press.

Horridge, G.A., ed. 1974. *The Compound Eye and Vision of Insects.* Clarendon Press.

Insects of Australia. 1970. (Suppl. 1974). Dept. of Entomology, Commonwealth Scientific and Industrial Research Organization, Canberra. Melbourne University Press, Carlton, Australia.

Lanham, U. 1964. *The Insects.* Columbia University Press.

Lawrence, P.A., ed. 1976. *Insect Development.* Royal Entomological Society of London Symposium no. **8.** Blackwell Scientific Publications. (Look for other R.E.S. Symposiums)

Linsenmaier, W. 1972. *Insects of the World.* McGraw-Hill Book Co.

Milne, L., and M. Milne. 1980. *Audubon Society Field Guide to North American Insects and Spiders.* Alfred A. Knopf.

Price, P.W. 1984. *Insect Ecology.* 2nd ed. John Wiley & Sons.

Raabe, M. 1982. *Insect Neurohormones.* Transl. from French by N. Marshall. Plenum Press.

Richards, O.W. and R.G. Davies. 1977. *Imm's General Textbook of Entomology.* 10th ed. Vols. 1 & 2. Chapman & Hall.

Rockstein, M., ed. 1973-1974. *The Physiology of Insecta.* 2nd ed. Vols. 1-6. Academic Press.

Rockstein, M., ed. 1978. *Biochemistry of Insects.* Academic Press.

Saunders, D.S. 1982. *Insect Clocks.* 2nd ed. Pergamon Press.

Wenner, A.M. 1971. *The Bee Language Controversy.* Educational Programs Improvement Corp.

Wigglesworth, V.B. 1964. *The Life of Insects.* Weidenfeld & Nicolson.

Wilson, E.O. 1971. *The Insect Societies.* Harvard University Press.

CHETOGNATHS, Chapter 25

Alvariño, A. 1965. Chaetognaths. Pp. 115-194 *in* H. Barnes, ed.: *Oceanography and Marine Biology, Annual Review,* **3**. George Allen & Unwin.

Ghiradelli, E. 1968. Some aspects of the biology of chaetognaths. *Advances in Marine Biology* **6**: 271-375.

LOPHOPHORATES and KAMPTOZOANS, Chapter 26

Emig, C.C. 1982. The Biology of Phoronida. *Advances in Marine Biology* **19**:1-89.

Larwood, G.P. and M.A. Abbott, eds. 1979. *Advances in Marine Bryozoology.* Academic Press.

Larwood, G.P. and C. Nielsen, eds. 1981. *Recent and Fossil Bryozoa.* Olsen and Olsen, Denmark.

McCammon, H.M. and W.A. Reynolds, eds. 1977. Symposium on the Biology of Lophophorates. *American Zoologist* **17**(1):1-150.

Nielsen, C. 1971. Entoproct life-cycles and the entoproct/ectoproct relationship. *Ophelia* **9**:209-341.

Richardson, J.R. 1986. Brachiopods. *Scientific American* **255**:100-106.

Rudwick, M.J.S. 1970. *Living and Fossil Brachiopods.* Hutchinson University Library.

Ryland, J.S. 1970. *Bryozoans.* Hutchinson University Library.

Ryland, J.S. 1976. Physiology and ecology of marine bryozoans. *Advances in Marine Biology* **14**:285-443.

Woollacott, R.M. and R.L. Zimmer, eds. 1977. *Biology of Bryozoans.* Academic Press.

ECHINODERMS, Chapter 27

Baker, A.N., F.W.E. Rowe, and H.E.S. Clark. 1986. A new class of Echinodermata from New Zealand. *Nature* **321**:862-864.

Binyon, J. 1972. *Physiology of Echinoderms.* Pergamon Press.

Czihak, G., ed. 1975. *The Sea Urchin Embryo: Biochemistry and Morphogenesis.* Springer-Verlag.

Giudice, G. 1973. *Developmental Biology of the Sea Urchin Embryo.* Academic Press.

Jangoux, M. and J.M. Lawrence, eds. 1982. *Echinoderm Nutrition.* A.A. Balkema, Rotterdam.

Jangoux, M. and J.M. Lawrence, eds. 1983, 1986. *Echinoderm studies,* Vols. 1 & 2. A.A. Balkema, Rotterdam.

Keegan, B.F. and B.D.S. O'Connor, eds. 1985. *Echinodermata.* Proc. *5th International Echinoderm Conference,* Galway, Ireland. A.A. Balkema, Rotterdam. (Look for publications of other echinoderm conferences)

Nichols, D. 1969. *Echinoderms.* Hutchinson University Library.

HEMICHORDATES and INVERTEBRATE CHORDATES, Chapters 28 & 29

Alldredge, A. 1976. Appendicularians. *Scientific American* **235**:94-102.

Barrington, E.J.W. 1965. *The Biology of Hemichordata and Protochordata.* W.H. Freeman & Co.

Barrington, E.J.W. and R.P.S. Jeffries, eds. 1975. Protochordates. *Zoological Society of London Symposium* **36**. Academic Press.

Berrill, N.J. 1961. Salpa. *Scientific American* **204**: 150-160.

Goodbody, I. 1974. The Physiology of Ascidians. *In* F.S. Russell and M. Yonge, eds.: *Advances in Marine Biology* **12**:1-149.

Lambert, G. and C.C. Lambert, eds. 1982. Symposium on the Developmental Biology of Ascidians. *American Zoologist* **22**(4):749-849.

ANIMAL RELATIONSHIPS, Chapter 30

Barbieri, M. 1985. *The Semantic Theory of Evolution.* Harwood Academic Publications.

Boardman, R.S., A.H. Cheetham, and A.J. Rowell. 1987. *Fossil Invertebrates.* Blackwell Scientific Publ.

Clarkson, E.N.K. 1979. *Invertebrate Palaeontology and Evolution.* George Allen & Unwin.

Fairbridge, R.W., and D. Jablonski. 1979. *The Encyclopedia of Paleontology.* Dowden, Hutchinson & Ross.

Glaessner, M.F. 1984. *The Dawn of Animal Life.* Cambridge University Press.

Gould, S.J. 1977. *Ontogeny and Phylogeny.* Harvard University Press.

Gutfreund, H., ed. 1981. *Biochemical Evolution.* Cambridge University Press.

Hanson, E.D. 1977. *The Origin and Early Evolution of Animals.* Wesleyan University Press.

Moore, R.D., ed. 1953-1983. *Treatise on Invertebrate Paleontology.* Parts A-W, other volumes to follow. Geological Society of America and University of Kansas Press.

Raff, R.A. and T.C. Kaufman. 1983. *Embryos, Genes, and Evolution.* Macmillan Publishing Co.

Whittington, H.B. 1985. *The Burgess Shale.* Geological Survey of Canada and Yale University Press.

COLORS OF ANIMALS, Chapter 31

Bagnara, J.T. and M.E. Hadley. 1973. *Chromatophores and Color Change: the Comparative Physiology of Animal Pigmentation.* Prentice-Hall.

Burtt, E.H. Jr., ed. 1979. *The Behavioral Significance of Color.* Garland STPM Press, New York.

Fogden, M. and P. Fogden. 1974. *Animals and their Colors: camouflage, warning coloration, courtship and territorial display, mimicry.* Crown Publishers.

Fox, D.L. 1976. *Animal Biochromes and Structural Colors.* 2nd ed. University of California Press, Berkeley.

Fox, D.L. 1979. *Biochromy: Natural Coloration of Living Things.* University of California Press.

Fox, H.M. and G. Vevers. 1960. *The Nature of Animal Colors.* Macmillan.

Kennedy, G.Y. 1979. Pigments of Marine Invertebrates. *In* F.S. Russell and C.M. Yonge, eds.: *Advances in Marine Biology* **16**:309-381.

Needham, A.E. 1974. *The Significance of Zoochromes.* Springer-Verlag.

Simon, H. 1971. *The Splendor of Iridescence: Structural Colors in the Animal World.* Dodd, Mead, & Co.

Vevers, G. 1982. *The Colours of Animals.* Institute of Biology series. Edward Arnold.

Wickler, W. 1968. *Mimicry in Plants and Animals.* Weidenfeld and Nicolson.